Phase Transition Dynamics

Tian Ma • Shouhong Wang

Phase Transition Dynamics

Springer

Tian Ma
Department of Mathematics
Sichuan University
Chengdu, People's Republic of China

Shouhong Wang
Department of Mathematics
Indiana University
Bloomington, IN, USA

ISBN 978-1-4939-4412-5 ISBN 978-1-4614-8963-4 (eBook)
DOI 10.1007/978-1-4614-8963-4
Springer New York Heidelberg Dordrecht London

Mathematics Subject Classification: 35B, 35Q, 37L, 76E, 82B, 82C, 86A10, 92B

Printed on acid-free paper

Springer is part of Springer Science+Business Media (www.springer.com)

For Wenyuan Chen, Ciprian Foias,
Louis Nirenberg, Roger Temam

For Li and Ping

Preface

Most problems in the natural sciences are described by either dissipative or conservative systems. Phase transition dynamics for both types of systems is of central importance in the nonlinear sciences. In this book, the term "phase transition" is to be understood in a broad sense, including both classical phase transitions and general transitions as found in nature. This book is an introduction to a comprehensive and unified dynamic transition theory for dissipative systems and to applications of the theory to a wide range of problems in the nonlinear sciences.

The main objectives of this book are to derive a general principle of dynamic transitions for dissipative systems, to establish a systematic dynamic transition theory, and to explore the physical implications of applications of the theory to a wide range of problems in the nonlinear sciences.

As a general principle, dynamic transitions of all dissipative systems are classified into three categories: continuous, catastrophic, and random. In comparison with the classical classification scheme for equilibrium phase transitions, whereby phase transitions are identified by the lowest-order derivative of the free energy that is discontinuous at the transition, the dynamic classification scheme is suitable for both equilibrium and nonequilibrium phase transitions. Once the type of a dynamic transition is determined for a given equilibrium system, the order of a transition in the classical sense immediately becomes transparent.

The philosophical basis of dynamic transition theory is to search for a system's full set of transition states, giving a complete characterization of stability and transition. The set of transition states is represented by a local attractor near or away from the basic state. Following this approach, dynamic transition theory is developed to identify the transition states and to classify them both dynamically and physically.

The theory is strongly motivated by phase transition problems in the nonlinear sciences. Namely, the mathematical theory is developed with close links to physics, and in return, the theory is applied to physical problems, leading to physical predictions and new insights into both theoretical and experimental studies for the underlying physical problems.

The study of equilibrium phase transitions presented in this book involves a combination of modeling, mathematical analysis, and physical predictions. We adopt the general idea of the Ginzburg–Landau phenomenological approach, and introduce a unified time-dependent Ginzburg–Landau model. Mathematically, this model is a quasigradient flow. Physically, the model is proposed using the Le Châtelier principle. A few typical and important equilibrium phase transition problems are addressed in this book, including the physical vapor transport (PVT) system, ferromagnetism, binary systems (Cahn–Hilliard equation), and the superconductivity and superfluidity of liquid helium (helium-3, helium-4, and their mixture).

The main focus for nonequilibrium transitions is on typical problems in classical and geophysical fluid dynamics, in climate dynamics, in chemical reactions, and in biology, including in particular Rayleigh–Bénard convection, the Taylor problem, Taylor–Couette–Poiseuille (TCP) flow, and the rotating convection problem in classical fluid dynamics; the El Niño–Southern Oscillation (ENSO) and wind-driven and thermohaline circulation in geophysical fluid dynamics and climate dynamics; the Belousov–Zhabotinsky chemical reactions; and chemotaxis and population models in biology.

Our intent has been to address the interests and backgrounds not only of applied mathematicians but also of sophisticated students and researchers working on nonlinear problems including those in physics, meteorology, oceanography, biology, chemistry, and the social sciences. The audience for this book includes second-year or more-advanced graduate students and researchers in mathematics and physics as well as in other related fields. Chapter 1 is designed so that most readers who have completed it will be able to jump directly to the related applications in Chaps. 3–6, after going quickly through the main ideas in this chapter as well as in the general introduction.

The mathematical background for physicists and other researchers in other fields is a good graduate-level mathematical physics course. For example, any student who is capable of studying graduate-level quantum mechanics or statistical physics should be able to follow Chap. 1 and the applications in Chaps. 3–6. Most graduate students in applied mathematics nowadays have to learn some basic physics related to the applied problems they are working on, and they should have no additional difficulty in reading the related topics of their interest in this book.

Sichuan, China — Tian Ma
Bloomington, IN — Shouhong Wang

Acknowledgments

We have greatly benefited from illuminating discussions with Jerry Bona, Mickael Chekroun, Henk Dijkstra, Ciprian Foias, Susan Friedlander, Michael Ghil, Chun-Hsiung Hsia, Paul Newton, Hans Kaper, Louis Nirenberg, Benoit Perthame, Jie Shen, Roger Temam, Xiaoming Wang, and Kevin Zumbrun. We owe a special debt of gratitude to all of them. We are grateful to Honghu Liu and Taylan Sengul for their critical reading, several times, of an earlier draft of the book and for their many constructive comments. We extend our wholehearted thanks to the anonymous referees; their careful reading and thoughtful remarks have led to a more effective book.

Also, we are grateful to Wen Masters and Reza Malek-Madani, of the Office of Naval Research, for their constant support and encouragement. We express our sincerest thanks to Achi Dosanjh, of Springer-Verlag, for her great effort and support on this book project. The research presented in this book was supported in part by grants from the ONR, the NSF, and the Chinese NSF.

The authors owe a debt beyond our ability to express it to Li and Ping for their kindness, love, patience, and dedications to our families. This book could never have been written without their devoted support from both Li and Ping.

Contents

Introduction

Transitions are to be found throughout the natural world. The laws of nature are usually represented by differential equations, which can be regarded as dynamical systems—both finite- and infinite-dimensional. There are two types of dynamical systems: dissipative and conservative. The term "dissipative structure" was coined by Ilya Prigogine. A conservative system is often described by a Hamiltonian structure, and a dissipative structure is closely associated with the concept of absorbing sets and global attractors; see among many others (Lorenz 1963a; Prigogine and Lefever 1968; Foiaş and Temam 1979; Ladyzhenskaya 1982; Temam 1997).

In thermodynamics, a phase transition is understood as the transformation of a thermodynamic system from one phase or state of matter to another. However, many problems in nature involve the study of transitions of one state to new states and the stability/robustness of the new states, and so in this book, the term "phase transition" will be understood in a broad sense, to include both the classical phase transitions in physics and the general transitions found in nature. The study of dynamic transitions also involves the dynamic behavior of transitions and transition states.

This main objectives of this book are threefold. The first is to derive, in Theorem 2.1.3, the following general principle on dynamic transitions for dissipative systems:

Principle 1 *Dynamic transitions of all dissipative systems can be classified into three categories: continuous, catastrophic, and random.*

The second objective is to develop a systematic dynamic transition theory to study the types and structure of dynamic transitions for dissipative systems. The third is to apply dynamic transition theory to explore the physical implications of dynamic transitions in a wide range of scientific problems. The applications include most of the typical types of dissipative systems, including (1) gradient-type systems for equilibrium phase transitions in statistical physics, (2) Navier–Stokes-type equations for incompressible fluid flows, (3) Boussinesq-type equations for geophysical fluid dynamics and climate dynamics, and (4) systems of reaction–diffusion equations modeling chemical reactions, chemotaxis, and population dynamics in biology.

The set of states of a dissipative system can be a complicated object in the phase space, and may include steady states, periodic states, and transients. Hence the starting point of our study is the following key philosophy:

Philosophy 2 *The key philosophy of dynamic transition theory is to search for the full set of transition states, giving a complete characterization of stability and transition. The set of transition states is a local attractor, representing the physical reality after the transition.*

It is natural to understand this philosophy from the physical point of view. For example, according to John von Neumann (1960), the motion of the atmosphere can be divided into three categories depending on the time scale of a prediction about its behavior, namely the motions corresponding to the short-, medium-, and long-term behavior of the atmosphere. Climate, which corresponds to the long-term behavior, should be represented by an attractor containing not only climate equilibria, but also the transients.

The general principle for dynamic transitions for dissipative systems given by Principle 1, along with Philosophy 2, provides a global viewpoint for the study of dynamic transitions. To illustrate the basic motivation and ideas behind this point of view, we consider the following simple example.

Example 3 For $x = (x_1, x_2)^t \in \mathbb{R}^2$, the system

$$\dot{x}_1 = \lambda x_1 - x_1^3, \quad \dot{x}_2 = \lambda x_2 - x_2^3, \tag{1}$$

undergoes a dynamic transition from a basic stable state $x = 0$ to a local attractor $\Sigma_\lambda = S^1$, as λ crosses $\lambda_0 = 0$. This local transition attractor is as shown in Fig. 1.

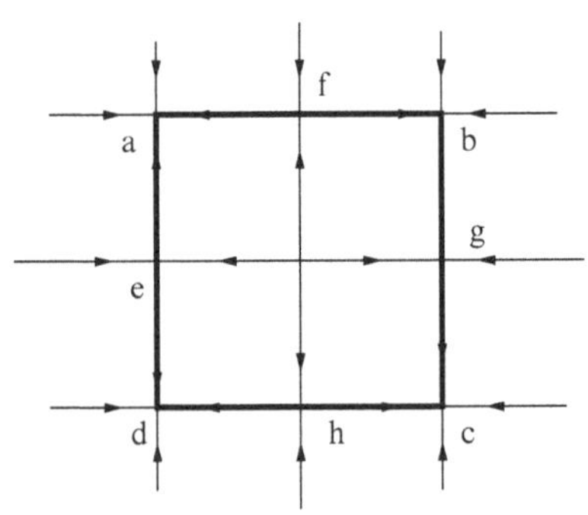

Fig. 1 The states after the transition are given by a bifurcated attractor containing four nodes (the points a, b, c, and d), four saddles (the points e, f, g, h), and orbits connecting these eight points

It contains exactly four nodes (the points a, b, c, and d), four saddles (the points e, f, g, h), and orbits connecting these eight points. The connecting orbits represent transient states.

From the physical transition point of view, as λ crosses 0, the new transition states are represented by the whole local attractor Σ_λ, rather than by any of the steady states or any of the connecting orbits.

To demonstrate this point of view further, we consider another example:

Example 4 For $x = (x_1, x_2)^t \in \mathbb{R}^2$, consider the system

$$\begin{aligned} \dot{x}_1 &= \lambda x_1 + x_1^2 + x_1 x_2 - 10x_1^3, \\ \dot{x}_2 &= \lambda x_2 - 2x_1 x_2 - x_2^2 - 10x_2^3. \end{aligned} \tag{2}$$

In this example, the lowest-order nonlinear terms are quadratic, given by $F(x) = \left(x_1^2 + x_1x_2, -2x_1x_2 - x_2^2\right)^t$. As we shall prove in Chap. 2, the index of the vector field $F(\cdot)$ at $x = 0$ is -2, leading to a dynamic transition as shown in Fig. 2. For $\lambda > 0$ and near 0, there are seven steady-state solutions: the origin $O = (0,0)^t$ and six nonzero points A, B, C, d, e, and f such that the three fixed points A, B, and C are close to the origin O and converge to O as $\lambda \to 0$, and the other three fixed points are away from 0 and do not converge to the origin O as $\lambda \to 0$. The three line segments OA, OB, and OC divide a neighborhood of O into three regions, and for any initial value in one of the regions, the solution converges to one of the three fixed points d, e, and f, away from O. Hence as λ crosses 0, the transition is catastrophic, also called in this book a jump (Type II) transition. The three points A, B, and C near the origin are the three bifurcation points, which do not represent the transition state in any way.

We note that the six nonzero fixed points are connected by heteroclinic orbits, and the region enclosed by these orbits is the global attractor of the system.

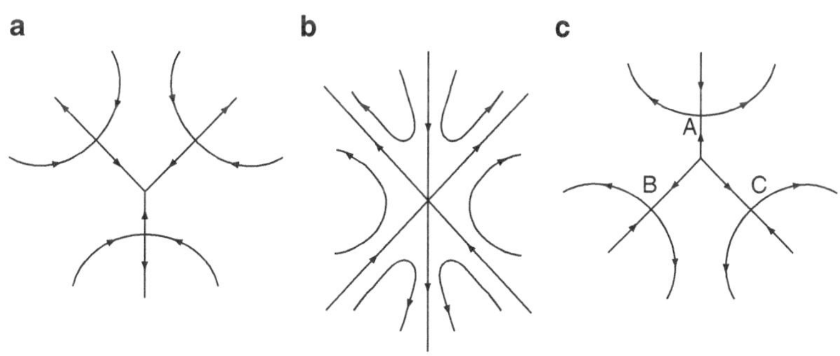

Fig. 2 Transition structure from a singular point with index ind$(F,0) = -2$: (**a**) $\lambda < 0$; (**b**) $\lambda = 0$; (**c**) $\lambda > 0$

The above two simple examples clearly demonstrate the necessity of the key philosophy, which will become more transparent in the wide range of applications of the theory in Chaps. 3–6.

Dissipative systems are governed by differential equations—both ordinary and partial—which can be written in the following unified abstract form:

$$\frac{du}{dt} = L_\lambda u + G(u,\lambda), \qquad u(0) = u_0, \tag{3}$$

where $u : [0,\infty) \to X$ is the unknown function, $\lambda \in \mathbb{R}^1$ is the system control parameter, and X is a Banach space.

We shall always consider u the deviation of the unknown function from some equilibrium state $\overline{u}$. Hence, $L_\lambda : X \to X$ is a linear operator, and $G : X \times \mathbb{R}^1 \to X$ is a nonlinear operator. For example, for the classical incompressible Navier–Stokes equations, u represents the velocity function, and for the time-dependent Ginzburg–Landau equations of superconductivity, we have $u = (\psi, A)$, where ψ is a complex-valued wave function, and A is the magnetic potential.

To address the dynamic transition of a given dissipative system, the first step is to study the linear eigenvalue problem for system (3), which is closely related to the principle of exchange of stability (PES), leading to precise information on linear unstable modes. The precise statement of the PES is as follows. Let $\{\beta_j(\lambda) \in \mathbb{C} \mid j \in \mathbb{N}\}$ be the eigenvalues (counting multiplicity) of L_λ, and assume that

$$\text{Re}\,\beta_i(\lambda) \begin{cases} < 0 & \text{if } \lambda < \lambda_0, \\ = 0 & \text{if } \lambda = \lambda_0, \\ > 0 & \text{if } \lambda > \lambda_0, \end{cases} \qquad \forall 1 \leq i \leq m, \tag{4}$$

$$\text{Re}\,\beta_j(\lambda_0) < 0 \qquad \forall j \geq m+1. \tag{5}$$

Much of the linear theory on stability and transitions involves establishing the PES. There is a vast literature devoted to linear theory, including, among many others, (Chandrasekhar 1981; Drazin and Reid 1981) for classical fluid dynamics and (Pedlosky 1987) for geophysical fluid dynamics.

With the linear theory and the PES at our disposal, we can immediately show that the system always undergoes a dynamical transition at the critical threshold and that Principle 1 holds; see Theorem 2.1.3 for details.

The details of the transition behavior are then dictated by the nonlinear interactions of the system. The dynamic transition theory presented in Chap. 2 provides a systematic approach to the study of nonlinear interactions, leading to detailed information on the types of transitions and the structure of transition states and their physical implications.

The general principle of dynamic transitions for dissipative systems, Principle 1, classifies all dynamic transitions of a dissipative system into three categories, continuous, catastrophic, and random, which are also called Type I, Type II, and Type III in this book. To say that a transition is continuous amounts to saying that the control parameter crosses the critical threshold and the transition states stay in a close neighborhood of the basic state. The transition at $\lambda = 0$ in Example 3 is a continuous transition.

In fact, continuous transitions are essentially characterized by the attractor bifurcation theorem, Theorem 2.2.2, first proved in Ma and Wang (2005c,b). The attractor bifurcation theorem states that when the PES holds and the basic state is asymptotically stable at the critical parameter value λ_0, the system undergoes a continuous dynamic transition, which is described by a bifurcated attractor.

A key assumption in the attractor bifurcation theorem is the asymptotic stability of the basic solution at the critical parameter value λ_0. One important idea of the proof is a refined upper semicontinuity of attractors, which also plays an important role in analyzing the other two types of transitions.

There are many physical systems that can undergo a continuous transition. For example, consider the classical Bénard convection. As the Rayleigh number crosses the critical Rayleigh number, the system undergoes a continuous transition to an attractor, homeomorphic to an $(m-1)$-dimensional sphere S^{m-1}, that consists of steady states and transients. Here m is the number of unstable modes of the linearized eigenvalue problem at the critical Rayleigh number, dictated by the spatial geometry, which also defines the pattern formation mechanism of the problem. As we shall see in Sect. 4.1, steady states occupy only a zero-measure subset of the sphere S^{m-1}, and the transients occupy most the sphere.

When there is no longer asymptotic stability of the basic state at the critical parameter, the system undergoes either catastrophic or random transitions, as dictated by the nonlinear interactions. The dynamic transition theory presented in this book gives a systematic approach to distinguishing these transitions.

Intuitively speaking, catastrophic transition corresponds to the case in which the system undergoes a more drastic change as the control parameter crosses the critical threshold. The transition given in Example 4 and the transitions in many examples given in Chaps. 3–6 are typical catastrophic transitions.

A random transition corresponds to the case in which a neighborhood (fluctuations) of the basic state can be divided into two regions such that fluctuations in one of them lead to continuous transitions, and those in the other lead to catastrophic transitions.

Given the above observations, it is then crucial to determine the asymptotic behavior of a system at the critical threshold. For this purpose and for determining the structure of the local attractor representing the transition states, the most natural approach is to project the underlying system to the space generated by the most unstable modes, preserving the dynamic transition properties. This is achieved with center manifold reduction.

An important part of dynamic transition theory deals with the analysis of reduced systems with close links to physics and with the philosophy of searching for a complete set of transition states. First, the transition type is completely dictated by the flow structure of the reduced system at the critical threshold, leading to precise information on the phase transition diagram. Also, for a continuous (Type I) transition, the reduced system provides a complete description of the set of transition solutions. It is worth mentioning that the far-field solutions in the catastrophic and random cases will have to be derived from the original partial differential equations, although the transition types and phase diagrams are completely determined by the reduced system.

In a nutshell, this book establishes a general principle for dynamic transitions for dissipative systems. Namely, all transitions of a dissipative system can be dynamically classified into one of three types: continuous, catastrophic, and random. A systematic dynamic transition theory is then developed to identify the types and structure of transitions, and is further applied to a wide range of problems in physics, fluid dynamics, climate dynamics, chemistry, and biology.

It is believed that the methods and ideas presented in this book can be readily applied to many scientific problems in related fields as well, as evidenced by the wide range of applications presented in this book.

The main outline of this book is as follows. The philosophy and general principles of dynamic transition theory with applications to a few typical problems are addressed in Chap. 1. Mathematical aspects of dynamic transition theory are addressed in Chap. 2.

Chapter 3 consists of applications of the general theory to a few typical equilibrium phase transition problems, including the PVT system, ferromagnetism, binary systems, superconductivity, and superfluidity. As mentioned earlier, the study of these equilibrium phase transition problems involves a combination of modeling, mathematical analysis, and physical predictions.

Chapter 4 focuses on transition problems in classical fluid dynamics. The transition of the classical Bénard convection problem is studied in Sect. 4.1, leading to a continuous (Type I) transition at the first critical Rayleigh number. For the Taylor problem, the TCP flow problem, and the rotating convection problem, the associated linear operators are not symmetric, leading to the existence of all three types of transitions as well as the existence of periodic solutions. This chapter gives a detailed account of these transitions and their physical implications. In the TCP flow case, for example, we show that the Taylor vortices do not appear immediately after the first transition, and that they appear only after the system has undergone a transition in its structure in the physical space as we increase the Taylor number to another critical number. This result is derived using a combination of the dynamic transition theory presented in this book and the geometric theory of incompressible flows developed recently in Ma and Wang (2005d).

Chapter 5 addresses dynamic transitions in geophysical fluid dynamics and climate dynamics. One important aspect of the study is the introduction of turbulent frictional terms in the model, leading to a transition to circulations over the tropics, consistent with the Walker circulation. Another important component of this chapter is the introduction of a new mechanism of ENSO that is consistent with observations. Dynamic transitions corresponding to typical meridional circulations and the thermohaline circulations are examined as well.

Chapter 6 can be considered an introduction to transition problems in chemistry and biology, focusing on Belousov–Zhabotinsky chemical reaction equations, the chemotactic model, and the population model.

Application-oriented readers may go directly to the study of the scientific problems presented in Chaps. 3–6 after reading through Chap. 1.

Chapter 1
Introduction to Dynamic Transitions

The study of phase transitions is an active field with a long history. The interested reader is referred to (Prigogine and Lefever 1968; Glansdorff and Prigogine 1971; Nicolis and Prigogine 1977; Onuki 2002; Reichl 1998; Stanley 1971; Chandrasekhar 1981; Drazin and Reid 1981; Cross and Hohenberg 1993; Pismen 2006) for some classical treatments and various historical accounts.

This book aims to provide a comprehensive, unified, and balanced account of dynamic transition theory and its applications to partial differential equation models for equilibrium phase transitions in statistical physics and to dynamic transitions in classical and geophysical fluid dynamics, climate dynamics, chemistry, and biology. This theory establishes a dynamic transition principle, Principle 1, following the philosophy of searching for a complete set of transition states.

We present in this chapter a brief introduction to this dynamic transition theory together with an introduction to modeling and to fundamental issues of dynamic transitions motivated by problems in the nonlinear sciences. The application-oriented reader may go directly to the study of the scientific problems presented in Chaps. 3–6 after quickly reading this chapter.

1.1 First Principles and Dynamic Models

1.1.1 Physical Laws and Mathematical Models

Modeling is a crucial step in studying a nonlinear problem. A good model should be derived on the basis of first principles and the key characteristics of the underlying physical problem. The first principles are often given by physical laws:

Model = Mathematical representation of physical laws.

T. Ma and S. Wang, *Phase Transition Dynamics*, DOI 10.1007/978-1-4614-8963-4_1,

Models for a large number of nonlinear problems are presented in the form of differential equations, ordinary or partial. For example, the basic model of a fluid motion with constant density is based on two first principles: the conservation of momentum (Newton's second law) and the conservation of mass. Let $x = (x_1, x_2, x_3)$ be the spatial coordinate system, t the time variable, and ∇, div, and Δ the gradient, divergence, and Laplacian operators. Let $u = u(x,t) = (u_1(x,t), u_2(x,t), u_3(x,t))$ be the velocity field, $p = p(x,t)$ the pressure function, ρ the density function, and f the external forcing applied to the fluid. Then Newton's second law takes the following form:

$$\rho\left[\frac{\partial u}{\partial t} + u\cdot\nabla u\right] = -\nabla p + \mu\Delta u + f, \tag{1.1.1}$$

where the left-hand side is mass times the acceleration, and the right-hand side is the force, including the pressure force $-\nabla p$, the viscous frictional force $\mu\Delta u$, and the external force f. Here μ is the viscosity coefficient.

The conservation of mass takes the following form:

$$\frac{\partial \rho}{\partial t} + \text{div}(\rho u) = 0, \tag{1.1.2}$$

which, with the constant density assumption, reduces to

$$\text{div}\, u = 0. \tag{1.1.3}$$

It is worth mentioning that in dissipative systems, the Laplace operator Δ, also denoted by ∇^2, often represents the energy dissipation, as demonstrated in scientific problems in Chaps. 3–6.

1.1.2 Rayleigh–Bénard Convection

The study of transitions in classical and geophysical fluid dynamics and in climate dynamics has enormous economic and social implications. Classical fluid dynamics has a long history, and models of typical fluid flows have been developed from first principles. The atmosphere and ocean surrounding the Earth are rotating geophysical fluids, and they are two important components of the climate system. The phenomena of the atmosphere and ocean are extremely rich in their organization and complexity, with scales ranging from meters to thousands of kilometers. The modeling of important geophysical fluid flows remains an important part of current and future research.

To demonstrate the basic ideas on the modeling and the mathematical settings, we consider the classical Rayleigh–Bénard convection problem. Convection is the well-known phenomenon of fluid motion induced by buoyancy when a fluid is heated from below. It is familiar as the driving force in atmospheric and oceanic phenomena. The Rayleigh–Bénard convection problem originated in the famous experiments conducted in 1900 by H. Bénard, who investigated a liquid in a horizontal dish with diameter much larger than the height, keeping the upper surface free to interact with the air and heating the dish from below. He noticed that when the lower surface temperature T is less than some critical temperature T_c, the liquid is in a static state, but when T becomes greater than T_c, the liquid suddenly breaks into a somewhat regular cellular pattern of hexagonal convection cells. The cells have the convection roll structure in their vertical section, as illustrated in Fig. 1.1.

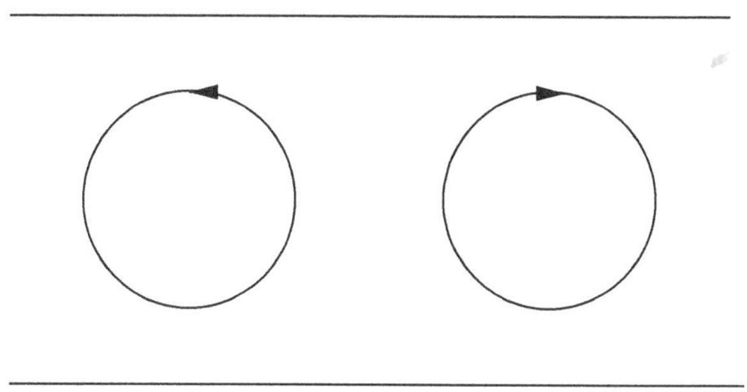

Fig. 1.1 The vertical section of thermal convection

The modeling of the Bénard problem includes the following main ingredients.

FIRST, the energy conservation law is needed to take the effect of the heat into consideration. The basic physical law is the first law of thermodynamics, which is an expression of the more universal physical law of conservation of energy:

$$\frac{\partial T}{\partial t} + (u \cdot \nabla)T - \kappa \Delta T = 0. \tag{1.1.4}$$

Here, as we discussed previously, the term $-\kappa\Delta T$ is due to the energy diffusion, and κ is the diffusion coefficient.

SECOND, an equation of state is needed to provide a relation between state variables: the pressure p, the temperature T, and the density ρ. Boyle's law was perhaps the first expression of an equation of state. For water as well as for many other fluids, an empirical relationship derived using the Taylor expansion near a basic state is often used. In particular, we shall use the following linear approximation of the equation of state in our discussions of Rayleigh–Bénard convection:

$$\rho = \rho_0(1 - \alpha_T(T - \overline{T}_0)), \tag{1.1.5}$$

where $\overline{T}_0$ is the reference temperature, and α_T is the expansion coefficient.

THIRD, the external forcing here is given by $-\rho g k$, where g is the gravitational constant, and $k = (0,0,1)$ is the vertical unit vector. Then using the momentum equation (1.1.1) with $-\rho g k$ as the external forcing and the continuity equation (1.1.2), we obtain the following system of partial differential equations governing the motion and states of the flow:

$$
\begin{aligned}
&\rho\left[\frac{\partial u}{\partial t}+(u\cdot\nabla)u\right]=\mu\Delta u-\nabla p-\rho gk,\\
&\frac{\partial T}{\partial t}+(u\cdot\nabla)T-\kappa\Delta T=0,\\
&\frac{\partial\rho}{\partial t}+\mathrm{div}(\rho u)=0,\\
&\rho=\rho_0(1-\alpha_T(T-\overline{T}_0)).
\end{aligned}
$$

As mentioned before, modeling of a given physical problem consists in the mathematical representation of physical laws, and approximations are made based on the specific characteristics of the underlying physics problem. In this case, we have already made a simplification to the equation of state. Another simplification is often made using the classical Boussinesq assumption, which amounts to saying that density differences are sufficiently small to be neglected, except where they appear in terms multiplied by g, the acceleration due to gravity, and in the equation of state. With this assumption, the above system of equations is reduced to the following equations; see among others (Rayleigh 1916; Drazin and Reid 1981; Chandrasekhar 1981):

$$
\begin{aligned}
&\frac{\partial u}{\partial t}+(u\cdot\nabla)u-\nu\Delta u+\rho_0^{-1}\nabla p=-gk[1-\alpha(T-\overline{T}_0)],\\
&\frac{\partial T}{\partial t}+(u\cdot\nabla)T-\kappa\Delta T=0,\\
&\mathrm{div}u=0,
\end{aligned}
\tag{1.1.6}
$$

where $\nu=\mu/\rho_0$. This system of equations is often called the *Boussinesq equations*.

The spatial domain is usually given by $\Omega=D\times(0,h)$, where $D\subset\mathbb{R}^2$ is a bounded domain in $\mathbb{R}^2$ and h is the height of the container. The system is often supplemented with the following boundary conditions:

$$
\begin{aligned}
&u_3=0, && \frac{\partial(u_1,u_2)}{\partial x_3}=0, T=\overline{T}_0 && \text{at } x_3=0,\\
&u_3=0, && \frac{\partial(u_1,u_2)}{\partial x_3}=0, T=\overline{T}_1 && \text{at } x_3=h,\\
&u_n=0, && \frac{\partial u_\tau}{\partial n}=0,\quad \frac{\partial T}{\partial n}=0 && \text{on } \partial D\times(0,h),
\end{aligned}
\tag{1.1.7}
$$

where n and τ are the respective unit normal and tangent vectors on $\partial D\times[0,h]$, and $u_n=u\cdot n, u_\tau=u\cdot\tau$.

1.1.3 Mathematical Formulation of Physical Problems

As discussed in the previous sections, many problems in the sciences are modeled by partial differential equations. With these models in our disposal, one basic level question is the well-posedness of the model, which consists in the existence, uniqueness, and regularity of solutions, as well as the dependence of the solutions on physical parameters. There is a long history of research in this direction throughout the last century. With well-posedness, we can put a model into the perspective of a dynamical system, either finite- or infinite-dimensional, as follows:

$$\frac{du}{dt} = L_\lambda u + G(u,\lambda), \quad u(0) = \varphi, \tag{1.1.8}$$

where $u : [0,\infty) \to X$ is the unknown function, and X is the phase space equipped with the norm $\|\cdot\|$. Often, the system depends on a control parameter $\lambda \in \mathbb{R}^N$ ($N \geq 1$). In general, $L_\lambda : X_1 \to X$ is a linear operator, and $G(\cdot,\lambda) : X_1 \to X$ is a nonlinear operator of order greater than that of u:

$$G(u,\lambda) = o(\|u\|) \quad \text{(order greater than that of } u), \tag{1.1.9}$$

which amounts to saying that $u = 0$ is always a steady-state solution of (1.1.8), and that it represents the basic solution of the underlying physical system. For a system with nonzero basic state, the function u here always represents the *deviation* from the basic solution, as demonstrated in the example below.

To demonstrate the basic ideas, consider the Boussinesq equations (1.1.6) with the boundary conditions (1.1.7). We proceed as follows.

Basic State and Its Deviation Equations. The problem possesses a basic solution given by

$$\overline{u} = 0, \quad \overline{T} = \overline{T}_0 - \frac{\overline{T}_0 - \overline{T}_1}{h} x_3, \quad \overline{p} = p_0 - \rho_0 g x_3 - \frac{\alpha_T(\overline{T}_0 - \overline{T}_1)}{2h} x_3^2.$$

Consider the deviation (u', T', p') from the basic solution $(0, \overline{T}, \overline{p})$ defined by

$$u = u', \quad T = \overline{T} + T', \quad p = \overline{p} + p'.$$

Then the resulting equations of the deviations are given by

$$\begin{aligned} &\frac{\partial u}{\partial t} + (u\cdot\nabla)u - \nu\Delta u + \rho_0^{-1}\nabla p = gk\alpha_T T, \\ &\frac{\partial T}{\partial t} + (u\cdot\nabla)T - \frac{\overline{T}_0 - \overline{T}_1}{h} u_3 - \kappa\Delta T = 0, \\ &\text{div} u = 0, \end{aligned} \tag{1.1.10}$$

where we have dropped the primes for simplicity.

Nondimensionalization. To study the dynamics of a problem and its physical implications, we need to nondimensionalize. For this purpose, we set

$$\begin{aligned} &(x,t) = (hx', h^2t'/\kappa), && (u,T,p) = (\kappa u'/h, \beta h(T'/\sqrt{R}), \rho_0\kappa^2 p'/h^2),\\ &\mathrm{Pr} = \nu/\kappa, && R = \frac{g\alpha\beta}{\kappa\nu}h^4. \end{aligned}$$

Once again, omitting the primes, (1.1.10) can be rewritten as

$$\begin{aligned} &\frac{1}{\mathrm{Pr}}\left[\frac{\partial u}{\partial t} + (u\cdot\nabla)u + \nabla p\right] - \Delta u - \sqrt{R}Tk = 0,\\ &\frac{\partial T}{\partial t} + (u\cdot\nabla)T - \sqrt{R}u_3 - \Delta T = 0, \qquad (1.1.11)\\ &\mathrm{div}u = 0. \end{aligned}$$

The spatial domain (still denoted by Ω) now becomes $\Omega = D\times(0,1)$, and the boundary conditions are

$$\begin{aligned} &u_3 = 0, && \frac{\partial(u_1,u_2)}{\partial x_3} = 0, T = 0 && \text{at } x_3 = 0,1,\\ &u_n = 0, && \frac{\partial u_\tau}{\partial n} = 0, \quad \frac{\partial T}{\partial n} = 0 && \text{on } \partial D\times(0,1). \end{aligned} \qquad (1.1.12)$$

Mathematical Setup. For the problem (1.1.11) with (1.1.12), we define the function spaces as

$$\begin{aligned} &H = \{(u,T)\in L^2(\Omega)^4 \mid \mathrm{div}u = 0,\ u_n = 0 \text{ on } \partial\Omega\},\\ &H_1 = \{(u,T)\in H^2(\Omega)^4 \mid \mathrm{div}u = 0,\ (u,T) \text{ satisfies (1.1.12)}\}. \end{aligned} \qquad (1.1.13)$$

Also, let $L_\lambda = -A + B_\lambda : H_1 \to H$, and let $G: H_1 \to H$ be defined by

$$\begin{aligned} &A\phi = (-\mathrm{Pr}\,P(\Delta u), -\Delta T),\\ &B_\lambda\phi = \lambda(\mathrm{Pr}\,P[Tk], u_3), \qquad (1.1.14)\\ &G(\phi) = (-P[(u\cdot\nabla)u], -(u\cdot\nabla)T), \end{aligned}$$

for all $\phi = (u,T)\in H_1$. Here $\lambda = \sqrt{R}$, and

$$P: L^2(\Omega)^3 \to \{u\in L^2(\Omega)^3 \mid \text{div } u = 0, \quad u\cdot n|_{\partial\Omega} = 0\}$$

is the Leray projection. Then it is easy to see that the Boussinesq equations (1.1.11) with (1.1.12) can be put into the operator form (1.1.8).

1.2 Introduction to Dynamic Transition Theory

1.2.1 Motivation and Key Philosophy

In order to understand the transitions of a system from one state to another and the stability/robustness of the new states, the authors have recently developed a dynamic transition theory for nonlinear partial differential equations that can be applied to both equilibrium and nonequilibrium phase transitions in the nonlinear sciences. This theory is based on Key Philosophy 2, which states that we seek a full set of transition states representing physical reality after a transition. The set of transition states is often represented by a local attractor.

Dynamic transition theory is developed with this philosophy and the close link to the physics in mind, and the theory is then applied to physical problems to gain new insights into the underlying physical phenomena. An important part of the research presented in this book is the identification and classification of local attractors after a transition.

1.2.2 Principle of Exchange of Stability

It is clear that linear theory plays a crucial role in the study of nonlinear dynamic transitions. Classically, one starts with the following normal-mode solution of the problem (1.1.8):

$$u = e^{\beta t}\phi, \tag{1.2.1}$$

where ϕ is time-independent. Inserting this normal-form solution into (1.1.8) and dropping the nonlinear term, we find that

$$L_\lambda \phi = \beta \phi. \tag{1.2.2}$$

It is then easy to derive the following general statements in many nonlinear problems of the form (1.1.8):

(1) If all eigenvalues of the eigenvalue problem (1.2.2) have negative real parts, then the basic solution $u = 0$ is linearly stable.
(2) If an eigenvalue of (1.2.2) has positive real part, then the basic solution is linearly unstable.
(3) If some eigenvalues of (1.2.2) cross the imaginary axis from left to right as the control parameter λ crosses a critical value λ_0, then the stable basic solution losses its stability, and a dynamic transition likely occurs.

In the dynamic transition theory and its applications presented in this book, we always take a special path in the control parameter space $\lambda \in \mathbb{R}^N$. Namely, when we discuss dynamic transitions, it is understood that the path we take can often be represented as

$$\lambda(s) = (\lambda_1(s), \cdots, \lambda_N(s)) \quad \text{for } s \in \mathbb{R}. \tag{1.2.3}$$

However, for brevity, we abuse notation and simply write $\lambda = (\lambda_1, \cdots, \lambda_N)$.

Physically, the eigenvalue crossing condition in item (3) above is called the *principle of exchange of stability (PES)*; see among others (Chandrasekhar 1981; Drazin and Reid 1981).

Definition 1.2.1 *Let $\{\beta_j(\lambda) \in \mathbb{C} \mid j \in \mathbb{N}\}$ be the eigenvalues (counting multiplicity) of L_λ.*[1] *These eigenvalues satisfy the principle of exchange of stability at λ_0 with multiplicity m if*

$$Re\ \beta_i(\lambda) \begin{cases} < 0 & \text{if } \lambda < \lambda_0, \\ = 0 & \text{if } \lambda = \lambda_0, \\ > 0 & \text{if } \lambda > \lambda_0, \end{cases} \qquad \forall 1 \le i \le m, \tag{1.2.4}$$

$$Re\ \beta_j(\lambda_0) < 0 \qquad \forall j \ge m+1. \tag{1.2.5}$$

It is worth noting that if (1.1.8) represents a dissipative system, then the linear operator $L_\lambda : X_1 \to X$ is often a sectorial operator and is a completely continuous field.[2] In fact, in many applications, the linear operator is often an elliptic operator such as a Laplacian plus low-order differential terms, which is always a sectorial operator. In this case, there is a countable number of eigenvalues for (1.2.2), and the above PES makes sense.

Finally, the dynamic transition theory in the next chapter will be used to identify and classify the dynamic transitions for the nonlinear problem (1.1.8). Applications to various problems in science and engineering are then presented in the subsequent chapters.

1.2.3 Equation of Critical Parameters

The PES and statement (3) above suggest that the critical value $\lambda_0 = (\lambda_1^0, \cdots, \lambda_N^0)$ of the control parameter $\lambda = (\lambda_1, \cdots, \lambda_N)$ should satisfy the following equation:

$$\text{Re}\ \beta_1(\lambda_1, \cdots, \lambda_N) = 0, \tag{1.2.6}$$

which is called the *equation of critical parameters* in the physics literature. Here β_1 is the first eigenvalue.

Equation (1.2.6) defines a hypersurface in the control parameter space $\mathbb{R}^N$, which divides $\mathbb{R}^N$ into several disjoint regions. The graph of (1.2.6) in $\mathbb{R}^N$ is called the phase diagram for (1.1.8). The type of transition as the control parameter crosses the hypersurfaces from one region to another is derived from nonlinear dynamic transition theory.

[1] For brevity, we use here $\lambda \in \mathbb{R}$.

[2] See the beginning of Chap. 2 for the precise definition, and (Ma and Wang 2005b) for more detailed discussion.

1.2.4 Classifications of Dynamic Transitions

The concept of phase transition originated in statistical physics and thermodynamics. In physics, "phase" is usually used as a state of matter. In this book, we use phase to stand for a (stable) state in a system of some nonlinear science including physics, chemistry, biology, ecology, economics, fluid dynamics, and geophysical fluid dynamics. Hence, the concept of phase transition has been endowed with more general significance. In fact, the phase transition dynamics introduced in this book treats a wide variety of topics involving the universal critical phenomena of state changes in nature from a unified mathematical viewpoint.

From statistical physics, the phase transitions of nonlinear dissipative systems can be divided into two types: equilibrium phase transitions and the nonequilibrium phase transitions. Namely, if the system (1.1.8) possesses a gradient-type structure, then the phase transitions are called "equilibrium phase transition"; otherwise, they are called the "nonequilibrium phase transitions."

From the physical point of view, an equilibrium phase transition signifies that a thermodynamic system undergoes a transition at a critical point from one equilibrium state to another, maintaining minimal free energy. We shall see later that equilibrium phase transitions appear extensively in statistical physics, solid-state physics, materials science, thermodynamics, and elastomechanics. Nonequilibrium phase transitions appear mainly in chemical reaction dynamics, biology, ecology, economics, fluid dynamics and geophysical fluid dynamics. In addition, these two types of phase transitions have remarkable differences in their dynamical properties. In particular, an equilibrium phase transition does not lead to chaos, while a nonequilibrium phase transition may.

Equilibrium phase transitions are usually classified by the Ehrenfest classification scheme, which groups phase transitions based on the degree of nonanalyticity involved; see among many others (Reichl 1998; Stanley 1971) for detailed discussion.

One important goal of dynamic transition theory presented in this book is to establish the following dynamic classification scheme for dynamic transitions; see also Definition 2.1.4. This dynamic classification scheme is applicable to both equilibrium and nonequilibrium phase transitions. It was first presented in (Ma and Wang 2007b, 2009b, c, 2008a).

Definition 1.2.2 (Dynamic Classification of Phase Transitions) *Let $\lambda_0 \in \mathbb{R}$ be a critical point of (1.1.8), and suppose that (1.1.8) always undergoes a transition from a basic state Σ^1_λ to a state Σ^2_λ. The transitions can be classified into three types: Type I, Type II, and Type III, as λ crosses λ_0, depending on their dynamic properties.*

Mathematically, these three types of transitions are also respectively called continuous, jump and mixed transitions, as shown in Figs. 1.2–1.4; we refer interested readers to Chap. 2 for the precise definition, and to Chaps. 3–5 for concrete examples.

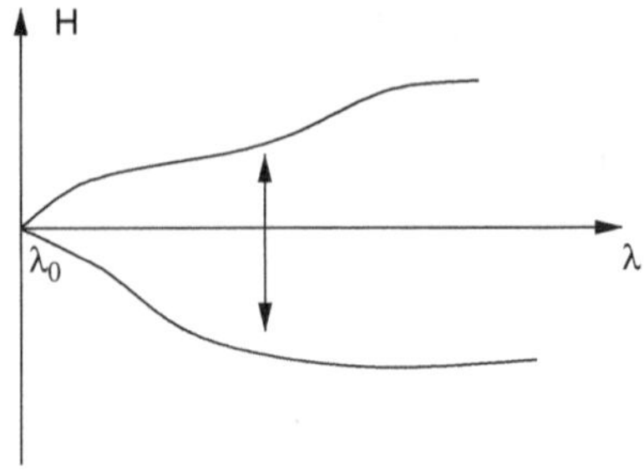

Fig. 1.2 Schematic of Type I transition: The transition states are represented by a local attractor Σ_λ, which attracts a neighborhood of the basic solution

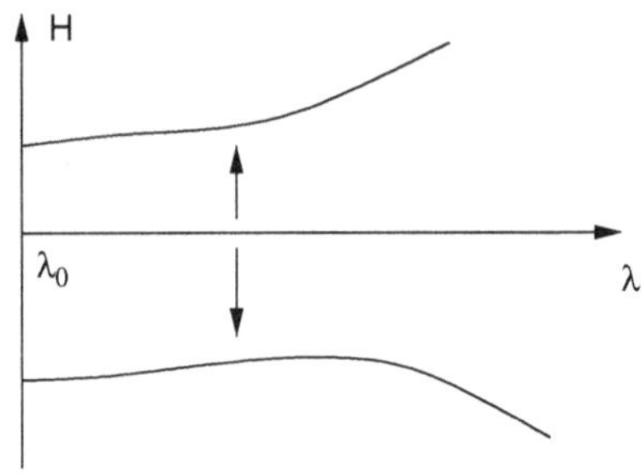

Fig. 1.3 Schematic of Type II transition: The transition states are represented by some local attractors away from the basic state at the critical point λ_0. In particular, the system undergoes a drastic change

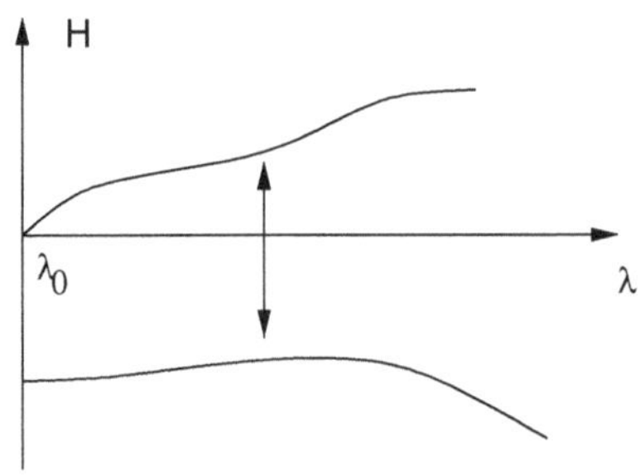

Fig. 1.4 Schematic of Type III transition: The transition states are represented by two local attractors, with one as in a Type II transition, and the other as in a Type I transition

Among other features, the dynamic classification scheme provides a classification of both equilibrium and nonequilibrium phase transitions. It includes the dynamic behavior of the system, and it offers clear information on the transition types in the classical sense for equilibrium transition problems. In equilibrium phase transition problems studied in Chap. 3, Type I transitions correspond to second- or higher-order transitions in terms of the Ehrenfest classification scheme, and transitions of Types II and III correspond to first-order transitions, in which latent heat, metastable states, and hysteresis occur. These phenomena will be explored in detail in Chap. 3, leading to some physical predictions. Hence the dynamic classification scheme provides not only the type of transition, but also the dynamic properties of the system.

1.2.5 Structure and Characterization of Dynamic Transitions

With the key Philosophy 2, linear theory, and dynamic classification scheme in hand, the main objective now becomes the identification and classification of dynamic transitions and transition states. Concrete methods and ideas are developed in detail in Chap. 2, and they are applied to a wide range of scientific problems in Chaps. 3–6. Here we present a few general remarks.

First, the dynamic classification scheme is universal for dissipative dynamic systems, and it is applicable to both equilibrium and nonequilibrium phase transitions. A Type I transition is essentially determined by attractor bifurcation theory (Ma and Wang 2005b), although as we shall see in the next chapter, there exist Type I transitions with no bifurcated attractors. In a dissipative dynamic system, Type II and Type III transitions always give rise to a singular separation, such as a saddle-node bifurcation, and the existence of metastable states.

Second, dynamic transition theory characterizes dynamic transitions both near and far away from the basic equilibrium state. This global nature of the theory is often derived by combining detailed local characterizations as given in Chap. 2 with the global properties of the problem such as the existence of a global attractor. For example, a Type II transition leads to transition states far from the basic solution, and a Type III transition gives rise to two types of transition states: one near and one far from the basic solution. The singular separation is another global feature of Type II and Type III transitions.

Third, one crucial component of dynamic transition theory is center manifold reduction; see Appendix A.1 for details. Under the PES assumption (1.2.4) and (1.2.5), by reducing (1.1.8) to a center manifold, we know that the type of transitions for (1.1.8) at $(0,\lambda_0)$ is completely dictated by its reduced equation near $\lambda=\lambda_0$, which can be expressed as

$$\frac{dx}{dt}=J_\lambda x+PG(x+\Phi(x,\lambda),\lambda) \quad \text{for } x\in\mathbb{R}^m, \tag{1.2.7}$$

where J_λ is the $m\times m$ Jordan matrix corresponding to the eigenvalues given by (2.1.4), $\Phi(x,\lambda)$ is the center manifold function of (2.1.1) near λ_0, $P: X\to E_\lambda$ is the canonical projection, and

$$E_\lambda=\bigcup_{1\leq i\leq m}\bigcup_{k\in N}\{u\in X_1|\ \ (L_\lambda-\beta_i(\lambda))^k u=0\}$$

is the eigenspace of L_λ corresponding to the first m eigenvalues.

In many applications, it is crucial to calculate the center manifold function Φ. For this purpose, a systematic approach is given in Appendix A.1. This approach provides useful formulas that are used in the applications presented in the remaining chapters of this book. For example, if $G(u,\lambda)$ has the Taylor expansion

$$G(u,\lambda)=G_k(u,\lambda)+o(\|u\|^k) \qquad \text{for some } k\geq 2, \tag{1.2.8}$$

where $G_k(u,\lambda)$ is a k-multilinear operator, and J_λ is diagonal near $\lambda=\lambda_0$, then by (A.1.14), we have

$$\begin{aligned} &\Phi(x,\lambda)=\sum_{j=m+1}^{\infty}\Phi_j(x,\lambda)e_j,\\ &\Phi_j(x,\lambda)=\frac{-1}{\beta_j(\lambda)(e_j,e_j^*)}(G_k(x,\lambda),e_j^*)+o(k), \end{aligned} \tag{1.2.9}$$

where $x=\sum_{i=1}^{m}x_ie_i$, and $\beta(\lambda)=(\beta_1(\lambda),\cdots,\beta_m(\lambda))$ is as in (1.2.4). Here e_j and e_j^* $(1\leq j\leq m)$ are the eigenvectors of L_λ and L_λ^* respectively corresponding to the eigenvalues $\beta_j(\lambda)$ as in (1.2.4). Here $(\cdot,\cdot)$ stands for the dual product between the function space X and its dual X^*; see also (1.1.8).

1.2.6 General Features of Dynamic Transitions

In a nutshell, here are some general features of dynamic transitions for dissipative systems.

First, we derive in this book a general principle, Principle 1, that dynamic transitions of all dissipative systems can be classified into three categories: continuous, catastrophic, and random.

Second, when a thermodynamic system undergoes a phase transition at some critical temperature and pressure, it will always reach a new equilibrium state in which its thermodynamic potential reaches its minimum. From the viewpoint of minimal potential energy, equilibrium phase transitions are changes in the point of minimal value of the energy functional at some critical point λ_0 of the system control parameter λ.

Third, in the 1960s, B.B. Belosov and A.M. Zhabotinsky found a class of chemical reactions, now called the Belosov–Zhabotinsky reactions, that lie in nonthermodynamic states. In their study of these reactions, Prigogine and colleagues found that the phase transitions for thermodynamic equilibrium states are universal for nonlinear dissipative systems. Prigogine's self-organization theory suggests that a stable state far from equilibrium needs to be maintained by energy input from the outside.

Fourth, mathematically speaking, both equilibrium and nonequilibrium phase transitions can be put into the perspective of a dynamical system with dissipative structure, which we often call a dissipative system. The dissipation here is associated with dissipation of the generalized energy. The global dynamical behavior of a dissipative system can be described by the existence of an absorbing set and a global attractor.

The dynamic transition theory developed in this book should provide a systematic approach for studying phase transitions and critical phenomena. Some general dynamic features are summarized as follows:

Physical States and Universality of Dynamic Transitions: A dissipative system always lives in a stable state, which is often represented by a local attractor of the system. Dynamic transitions of a dissipative system are a universal phenomenon in nature.

Generalized Energy: In every stable state there is some energy dissipation. Therefore, a dissipative system needs an energy supply in the generalized sense.

Every nonlinear dissipative system occurring in nature has only a finite number of stable states, denoted by $\Sigma_1^\lambda, \cdots, \Sigma_n^\lambda$. The stability of each state Σ_i^λ is sustained in an energy range

$$\Sigma_i^\lambda : \lambda_i^1 < \lambda < \lambda_i^2 \qquad \text{for } 1 \leq i \leq n,$$

where λ represents the generalized supplied energy. Namely, only when the externally supplied generalized energy λ is maintained in the range $\lambda_i^1 < \lambda < \lambda_i^2$ can the system sustain the state Σ_i^λ; otherwise, the system will undergo a transition to another state.

The lowest-energy state Σ_1^λ is a basic state of the system. Its energy range is

$$\Sigma_1^\lambda : \quad \lambda_1^1 \leq \lambda < \lambda_1^2.$$

Namely, Σ_1^λ needs the minimum possible energy to sustain itself.

Metastable States: It is possible for several stable states to coexist at a certain energy level. We call them metastable states, and dynamic transitions among them can occur due to fluctuations and uncertainty.

Chaotic States: A chaotic state of a system may be defined as an attractor $\mathscr{A} \subset U$ that attracts its open neighborhood U and contains no minimal attractors of singular elements (singular points and periodic orbits). It is clear then from this definition that a gradient-type flow does not have a chaotic state.

If a system has a chaotic state, then the chaotic state is Σ_n^λ:

$$\Sigma_n^\lambda : \quad \lambda_n^1 < \lambda < \infty,$$

which needs maximal energy to sustain itself. Moreover, once a chaotic state Σ_n^λ appears, no new ordered states will occur for any $\lambda > \lambda_n^1$ unless the generalized energy drops below the level λ_n^1.

In general, if (1.1.8) undergoes successive transitions at $\lambda_0 < \cdots < \lambda_k$ $(k \geq 1)$ such that the global attractor $\mathscr{A}_\lambda$ of (1.1.8) satisfies

$$\mathscr{A}_\lambda = \{0\} \qquad \text{for } \lambda \leq \lambda_0,$$

and $\mathscr{A}_\lambda$ has no minimal attractors of singular elements on $\lambda > \lambda_k$, then with successive transitions, the system passes through ordered states to a chaotic state.

1.3 Examples of Typical Phase Transition Problems

Hereinafter, we shall present a few typical phase transition problems addressed in this book, demonstrating the need for dynamical transition theory and some of its advantages. The main focus is on physical implications and predictions derived from dynamic transition theory, and the reader is referred to later chapters for further details.

1.3.1 Rayleigh–Bénard Convection

As mentioned earlier, Rayleigh–Bénard convection together with the Taylor problem has become a paradigm of nonequilibrium phase transitions with their instability, bifurcations, pattern formation, and chaotic behavior in fluid dynamics, as well as in physical systems in general.

Consider the nondimensional equations (1.1.11) governing the motion and states of the Bénard convection problem. By solving the linearized eigenvalue problem, as given, for example, by (4.1.12) in Chap. 4, the critical parameter equation (1.2.6) takes the following form:

$$R = R_c, \tag{1.3.1}$$

where R_c is the first critical Rayleigh number.

With dynamic transition theory, we can show (Ma and Wang 2004b) that as the Rayleigh number R crosses the critical parameter R_c, the system always undergoes a continuous (Type I) transition to an $(m-1)$-dimensional sphere $\mathscr{A}_R = S^{m-1}$, where m is the dimension of the eigenspace for the linearized problem corresponding to the critical Rayleigh number R_c.

The structure of the set $\mathscr{A}_R = S^{m-1}$ of transition states and the related pattern formation mechanism depend on the physical boundary conditions and the spatial geometry (of the domain); see (Ma and Wang 2007a) and Sect. 4.1 for more details.

First, when the (nondimensional) spatial domain is given by $\Omega = (0,L)^2 \times (0,1)$ with free slip lateral boundary condition and with L satisfying (4.1.62), the transition attractor $\mathscr{A}_R = S^1$ (i.e., $m = 2$) consists of exactly eight singular points Φ_i $(1 \leq i \leq 8)$ and eight heteroclinic orbits connecting the singular points, as shown in Fig. 1.5, where $\Phi_1, \Phi_3, \Phi_5, \Phi_7$ are minimal attractors, and $\Phi_2, \Phi_4, \Phi_6, \Phi_8$ are saddle points. See Theorem 4.1.14. We note here that the connecting orbits are the transients, which represent most of the states in the transition attractor.

Second, when the problem is equipped with periodic boundary conditions in the horizontal directions with periods L_1 and L_2, the transition attractor $\mathscr{A}_R$ is given by an attractor $\mathscr{A}_R$ homeomorphic to a sphere, i.e.,

$$\mathscr{A}_R = \begin{cases} S^5 & \text{if} \quad L_2 = \sqrt{k^2-1}\, L_1, \quad k = 2,3,\cdots, \\ S^3 & \text{otherwise.} \end{cases}$$

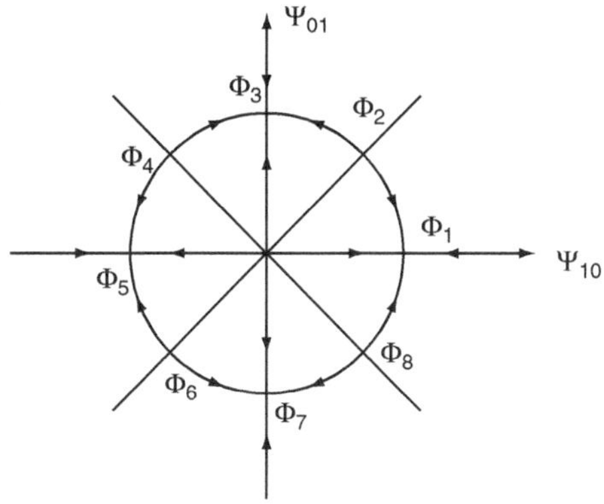

Fig. 1.5 The singular points $\Phi_1, \Phi_3, \Phi_5, \Phi_7$ are the minimal attractors, and the other singular points are the saddles

Due to the translational invariants in the horizontal directions, the high-dimensional attractor $\mathscr{A}_R$ consists of a finite number of two-dimensional tori (containing only steady states) and the transients containing connecting the tori. The collection of the tori of steady states is a set of measure zero in $\mathscr{A}_R$. Namely, as in the previous case, most of the states in $\mathscr{A}_R$ are transients.

It is worth mentioning that Philosophy 2 and Principle 1 play a crucial role in the derivation of the above results, which cannot be directly derived by other existing methods. In fact, without Philosophy 2 and Principle 1, one might not have been able to envision these results, which are global in nature.

Finally, dynamic transition theory opens a new avenue to pattern formation, since the patterns and structure of transition states are dictated by solutions in $\mathscr{A}_R$, which are again determined by eigenvectors of the corresponding linearized problem (with high-order approximations). With this theory, we can, for example, easily examine the formation and stability of different structures such as rolls, rectangles, and hexagons. However, due to the limitation in the scope of this book, we touch issues of pattern formation only occasionally in this book, and we intend to leave further exploration of patterns from the dynamic transition point of view to the future.

1.3.2 El Niño Southern Oscillation

The El Niño Southern Oscillation (ENSO) is a global coupled ocean–atmosphere phenomenon, associated with floods, droughts, and other disturbances around the world. ENSO is the most prominent known source of interannual variability in weather and climate (about three to eight years). In spite of its importance and a long history of research, an understanding of its nature and mechanism is still lacking, and a fundamental examination of the problem is crucial.

ENSO is closely related to the Walker circulation over the Tropics, as shown in Figs. 1.6 and 1.7. When convective activity weakens or reverses, an El Niño phase takes place, causing the ocean surface to be warmer than average, reducing or terminating the upwelling of cold water. On the other hand, a particularly strong Walker circulation causes a La Niña event, resulting in cooler sea-surface temperature (SST) due to stronger upwelling.

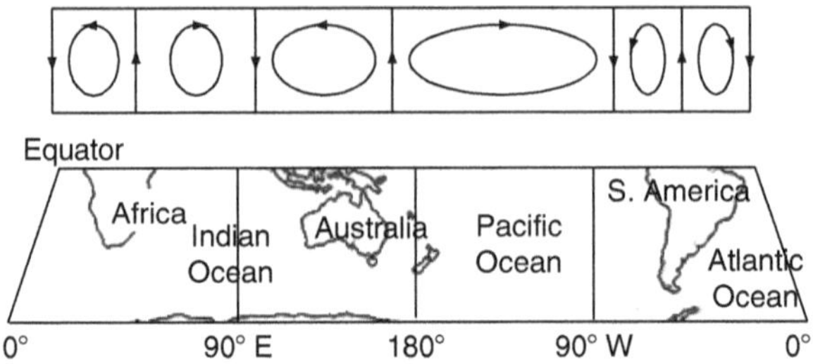

Fig. 1.6 Global atmospheric Walker circulation over the Tropics

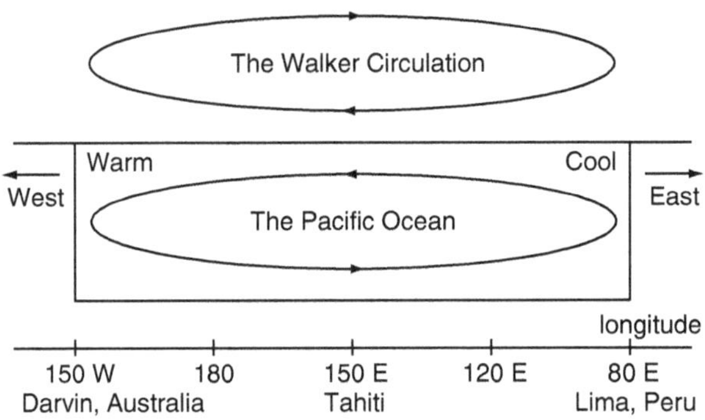

Fig. 1.7 Walker circulation over the tropical Pacific Ocean

An interesting current debate is whether ENSO is best modeled as a stochastic or a chaotic system—linear and noise-forced, or nonlinear oscillatory and unstable (Philander and Fedorov 2003).

With dynamic transition theory and the geometric theory of incompressible flows (Ma and Wang 2005d), we derive a new mechanism of ENSO as a self-organizing and self-excitatory system, with two highly coupled processes:

- oscillation between the two metastable warm (El Niño phase) and cold (La Niña phase) events, and
- spatiotemporal oscillation of the SST field.

The symbiotic interplay between these two processes gives rise to the climate variability associated with ENSO. It leads to ENSO's random and deterministic features and defines a new natural *feedback mechanism* that drives its sporadic oscillation. This new mechanism is schematically illustrated in Fig. 1.8; see also Sect. 5.2 for more details.

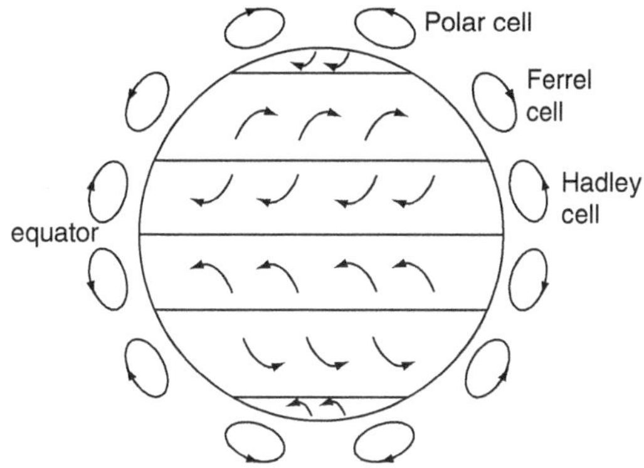

Fig. 1.8 Transition diagram for the equatorial atmospheric circulation, where R^* is the saddle-node bifurcation point and R_c^ε is the first critical Rayleigh number. Here the Rayleigh number R is proportional to SST

We emphasize in particular that the transition behavior of ENSO phenomena depicted here offers a natural explanation of the randomness associated with mixed (Type III) transitions, whence the terminology "random transition." The randomness is due to fluctuations of initial data for deterministic systems.

Another good example is pipe flow. Under some natural conditions, the basic flow is always linearly and nonlinearly stable for all Reynolds numbers (Eckhardt et al. 2007). However, for large Reynolds numbers, both experimental and numerical simulations lead to chaotic behavior. The crucial point of the problem is that the stable basic solution possesses a very narrow basin of attraction, and small fluctuations lead immediately to chaotic motion outside the basin of attraction of the basic flow.

1.3.3 Dynamic Transition Versus Transition in Physical Space

The nature of a flow's boundary-layer separation from the boundary and interior separation plays a fundamental role in many physical problems. Typical examples of boundary layer separation include the vortices forming from the bank of a river and the formation of vortices around an airfoil. The formation of tornadoes is a typical example of interior separation. Basically, in the boundary layer, the shear flow can separate from the boundary, generating a slow backflow and leading to more complicated turbulent behavior; see Fig. 1.9. The interior separation is often connected with spin-offs.

The precise characterization of the boundary layer and interior separations of fluid flows is a long-standing problem in fluid mechanics, going back to the pioneering work of (Prandtl 1904). Progress has been made recently using a systematic geometric theory of incompressible flows (Ma and Wang 2005d).

The main philosophy of studying the flow structures in physical space is to classify the topological structure and its transitions of the *instantaneous* velocity field (i.e., streamlines in Eulerian coordinates), treating the time variable as a parameter as though one were watching a movie. Based on this philosophy, research can progress in two directions:

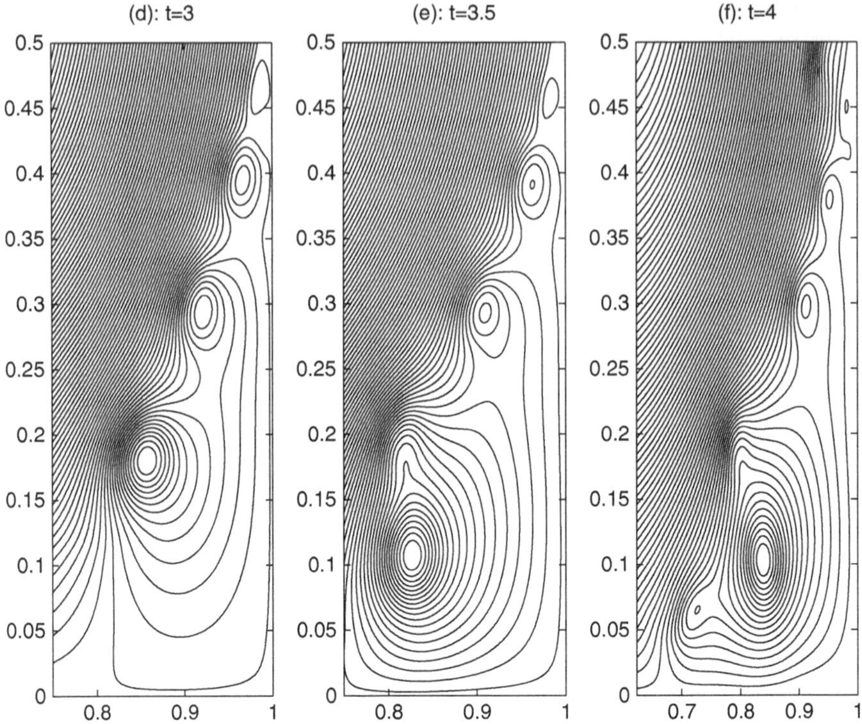

Fig. 1.9 Boundary layer separation of shear flow

(1) the development of a global geometric theory of divergence-free vector fields on general two-dimensional compact manifolds with or without boundaries; and
(2) the connections between the solutions of the Navier–Stokes (or Euler) equations and the dynamics of velocity fields in physical space.

As part of this program, a rigorous characterization of both boundary-layer and interior separations have been derived by the authors, in collaboration in part with Michael Ghil; see, e.g., (Ghil et al. 2001, 2005; Ma and Wang 2001, 2005d).

To demonstrate some of the ideas for this characterization and its applications, we consider the Taylor–Couette flow, a viscous fluid between two coaxial rotating cylinders. For the classical Taylor problem, the basic (Couette) flow becomes unstable as soon as the rotation speed of the inner cylinder exceeds a critical value. This instability gives rise to stationary axisymmetric counterrotating vortices that fill the whole annular region. The associated flow is referred to as Taylor–Couette (TC) flow.

When a through flow driven by a pressure gradient along the rotating axis is added, the base flow consists of a superposition of circular Couette flow and annular Poiseuille flow, called Couette–Poiseuille (CP) flow; see Fig. 1.10. The axial through flow suppresses the basic stationary instability, and as the axial pressure

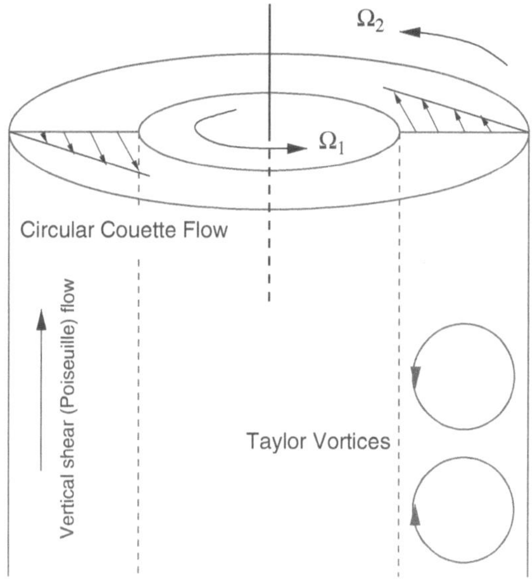

Fig. 1.10 Taylor–Couette–Poiseuille flow

gradient increases while the rotation speed of the inner cylinder is held fixed, the first bifurcation gives rise to a traveling train of axisymmetric Taylor vortices, commonly referred to as propagating Taylor vortices (PTV). Henceforth, the term Taylor–Couette–Poiseuille (TCP) flow is used to refer to all hydrodynamic phenomena pertaining to the open system described above; see among others (Raguin and Georgiadis 2004) and the references therein.

With dynamic transition theory, combined with a characterization of boundary-layer and interior separations (Ma and Wang 2005d), we show that there are two transition processes associated with the formation of propagating Taylor vortices; see (Ma and Wang 2009a):

- The first transition is a dynamic one at the first critical Taylor number, leading to a CP-like flow. In this case, there is no formation of PTV.
- The second transition occurs at another critical Taylor number, T_1. It is the transition of the flow structure in physical space. This transition leads to the formation of PTV, as shown in Fig. 1.11.

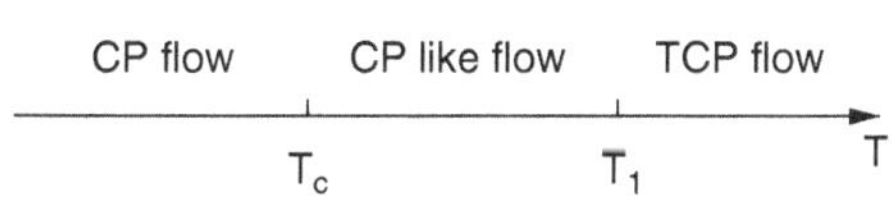

Fig. 1.11 Dynamic and structural transitions leading to the formation of Taylor vortices

In a nutshell, contrary to what is commonly believed, we show that the PTV do not appear after the first dynamic transition and that they appear only when the Taylor number is further increased to cross another critical value so that a structural bifurcation in the physical spaces occurs.

1.3.4 Andrews Critical Point and Third-Order Gas–Liquid Transition

The physical-vapor transport (PVT) system is one of the most basic models employed in the study of equilibrium phase transitions. As we know, a PVT system is composed of one type of molecule, and the interaction between molecules is governed by the van der Waals law. The molecules generally have a repulsive core and a short-range attraction region outside the core. Such systems have a number of phases: gas, liquid, and solid, and a solid can itself appear in a few phases. The most typical example of a PVT system is water. In general, the phase transitions of PVT systems mainly refer to the gas–liquid, gas–solid, and liquid–solid phase transitions. These transitions are all of first order in the Ehrenfest sense (i.e., discontinuous) and are accompanied by a latent heat and a change in density.

A PT-phase diagram of a typical PVT system is schematically illustrated in Fig. 1.12, where point A is the triple point at which the gas, liquid, and solid phases can coexist. Point C is the Andrews critical point, at which the gas–liquid coexistence curve terminates (Reichl 1998; Stanley 1971; Kleman and Lavrentovich 2003; Fisher 1964). The fact that the gas–liquid coexistence curve has a critical point means that if we choose the right path, we can go continuously from a gaseous state to a liquid state without ever meeting an observable phase transition.

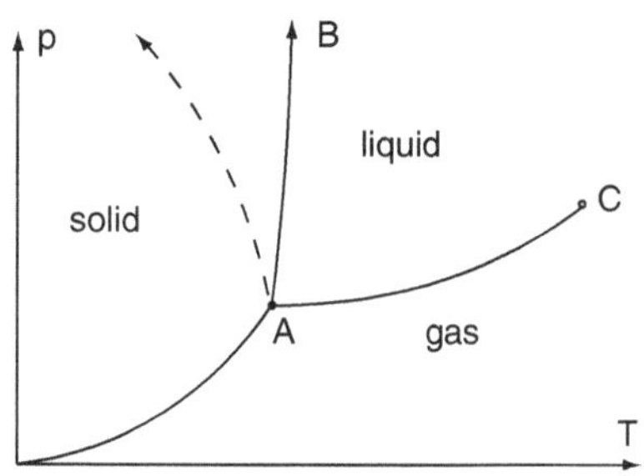

Fig. 1.12 Coexistence curves of a typical PVT system: A is the triple point, C is the critical point, and the dashed cure is a melting line with negative slope

A natural question to ask is why the Andrews critical point exists and what the order of transition is going beyond this critical point. This is a longstanding problem, which is currently solved using dynamic transition theory in (Ma and Wang 2008b, 2011e):

First, a dynamic model for phase transition in a PVT system is established based on (a) the unified dynamic models for equilibrium phase transitions established in Sect. 3.1, (b) the classical phase diagram given in Fig. 1.12, and (c) the van der Waals equations.

Second, with this dynamic model, we are able to derive a theory of the Andrews critical point C:

(1) The Andrews point C is derived using a dynamic approach different from the classical derivation.

(2) The gas–liquid coexistence curve can be extended beyond the Andrews critical point.
(3) The transition is of *first order* before the critical point, *second order* at the critical point, and *third order* beyond the Andrews critical point.

This clearly explains why it is hard to observe the gas–liquid phase transition beyond the Andrews critical point. Furthermore, the analysis leads naturally to the introduction of a general asymmetry principle of fluctuations and the preferred transition mechanism for a thermodynamic system. The theoretical results derived are in agreement with the experimental results obtained in (Nishikawa and Morita 1998). Also, the derived second-order transition at the critical point is consistent with the result obtained in (Fisher 1964).

In addition, a careful examination of the dynamic transitions of the PVT system suggests that there are two possible phase transition behaviors near a critical point, and theoretically, each of them has the same probability of occurring. However, only one of them can appear in reality, which suggests the asymmetry of fluctuations given in the conjecture Physical Principle 3.4.3. Namely, the symmetry of fluctuations for general thermodynamic systems may not be universally true. In other words, in some systems with multiequilibrium states, the fluctuations near a critical point occur only in one basin of attraction of some equilibrium states, which are the ones that can be physically observed.

1.3.5 Binary Systems

Materials consisting of two components A and B, such as binary alloys, binary solutions, and polymers, are called binary systems. Sufficient cooling of a binary system may lead to phase separations. Namely, at the critical temperature, the concentrations of both components A and B with homogeneous distributions undergo changes, leading to heterogeneous spatial distributions (Kleman and Lavrentovich 2003; Reichl 1998; Pismen 2006; Novick-Cohen and Segel 1984; Cahn and Hillard 1957).

A commonly used model for a binary system is the classical Cahn–Hilliard equation, which can also be derived using the unified model presented early in this chapter; see Chap. 3 for more details. Let u_A and u_B $(= 1 - u_A)$ be the concentrations of components A and B respectively. The Helmholtz free energy is given by

$$F(u) = F_0 + \int_\Omega \left[\frac{\mu}{2}|\nabla u_B|^2 + f(u_B)\right] dx, \tag{1.3.2}$$

where the energy density function f can be explicitly given using the Bragg–Williams or Hildebrand theory (Reichl 1998); see (3.5.2):

$$\begin{aligned} f =& \mu_A(1-u_B) + \mu_B u_B + RT(1-u_B)\ln(1-u_B) \\ &+ RTu_B \ln u_B + a u_B(1-u_B), \end{aligned} \tag{1.3.3}$$

where μ_A, μ_B are the chemical potentials of A and B respectively, R the molar gas constant, and $a > 0$ the measure of repulsion between A and B.

In a homogeneous state, u_B is a constant in space and is equal to its mean value, i.e., $u_B = \overline{u}_B$. Let

$$u = u_B - \overline{u}_B, \qquad u_0 = \overline{u}_B.$$

The classical Cahn–Hilliard equation is then derived using Landau mean field theory:

$$\frac{\partial u}{\partial t} = -k\Delta^2 u + \Delta[b_1 u + b_2 u^2 + b_3 u^3], \tag{1.3.4}$$

where the coefficients k, b_1, b_2, and b_3 can be given explicitly in terms of the temperature T, the pressure p, and the homogeneous molar fraction u_0 of component B, using the explicit expression for f; see (3.5.4).

To illustrate the main ideas, for simplicity, we consider here only the case in which the binary solution is contained in a cube $(0, L_d)^3$; see Chap. 3 and (Ma and Wang 2009b,c) for more general cases.

Critical Parameter Equation. First, the critical parameter curve equation is derived by setting the first eigenvalue of the linearized problem to zero, $\beta_1 = 0$, leading to

$$T_c = \frac{u_0(1-u_0)}{RD}\left(a - \frac{k\pi^2}{2L_d^2}\right), \tag{1.3.5}$$

where T_c is the critical temperature.

Order of Phase Transitions and Phase Diagrams. By nonlinear dynamic transition theory, we then derive that the order of transition is completely determined by the sign of the number K_d (the subscript d refers to dimensional form), given by

$$K_d = \frac{26L_d^2}{27\pi^2}\frac{b_2^2}{k} - b_3, \tag{1.3.6}$$

which depends on the temperature T, the pressure p, and the homogeneous molar fraction u_0 of component B. Namely, when $K_d > 0$, the transition is of first order with latent heat, corresponding to a catastrophic transition in the dynamic classification scheme, and when $K_d < 0$, the transition is of second order.

To examine the order of separation, we start with two important formulas for the critical parameters. First, by (3.5.4), solving $K_d = 0$ gives a precise formula for the critical length scale L_d in terms of the critical temperature T_c and the molar fraction u_0:

$$L_d = \frac{3\sqrt{3k}\pi\sqrt{u_0(1-u_0)(1-3u_0+3u_0^3)}}{2\sqrt{26DRT_c}\,|u_0 - \frac{1}{2}|}. \tag{1.3.7}$$

We are ready now to examine the order of separation in terms of the length scale L_d and molar fraction u_0. First, using (1.3.7), we derive the Lu_0-phase diagram given by Fig. 1.13. Then using Fig. 1.13 and (1.3.5), we derive the Tu_0-phase diagram given in Fig. 1.14.

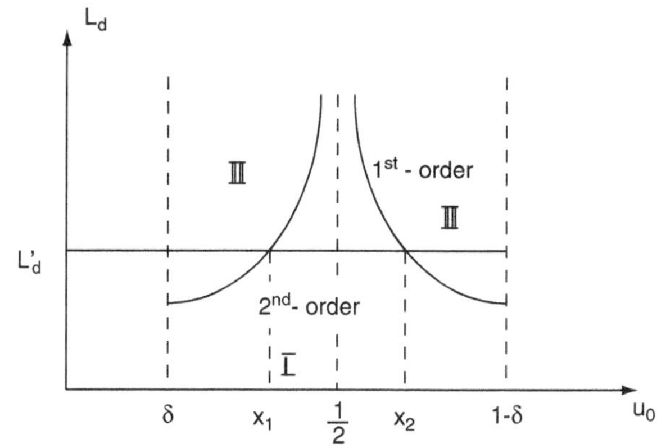

Fig. 1.13 Lu_0-phase diagram derived using (3.5.60) or (3.5.61): region II is the first-order transition region with latent heat, and region I is the second-order transition region

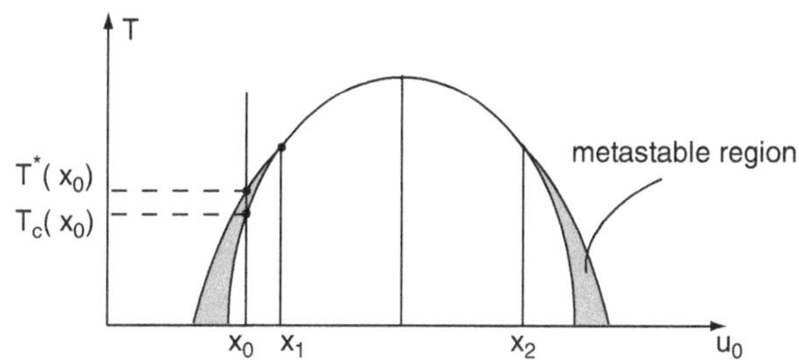

Fig. 1.14 Tu_0-phase diagram for a fixed length scale L', derived using Fig. 1.13 and (1.3.5): the shaded area stands for the metastable region, and $T^*(x_0)$ is a critical temperature corresponding to a saddle-node bifurcation, as shown in Chap. 3, and $T_c(x_0)$ is the critical temperature

From these two phase diagrams, we obtain immediately the following physical conclusions:

(1) For a fixed length scale $L = L'$, there are numbers $x_1 < \frac{1}{2} < x_2$ such that the transition is of second order if the molar fraction satisfies $x_1 < u_0 < x_2$, and the transition is of first order if $u_0 > x_2$ or $u_0 < x_1$. See Fig. 1.13.
(2) The phase diagram of Fig. 1.14 is for this fixed length scale L'. The points x_1 and x_2 are the two molar concentrations where there is no metastable region and no hysteresis phenomena for $x_1 < u_0 < x_2$. In other words,

$$T^*(u_0) = T_c(u_0) \qquad \text{for } x_1 < u_0 < x_2;$$

$T^*(x_0)$ and $T_c(x_0)$ are critical temperatures.

It is worth mentioning that the results derived using the Cahn–Hilliard equation (1.3.4) are also valid for the model using the full Bragg–Williams potential (1.3.3).

Pattern Formation and Structure of Transition Attractor. As indicated earlier, in the case $K_d > 0$, the transition is catastrophic. For the case $K_d < 0$, the transition is of second order, and the local attractor representing the complete transition states $\Sigma_T = S^2$ consists of 26 steady states and the transients connecting these 26 steady states, as shown in Fig. 1.15. This clearly demonstrates again the point mentioned before for the Bénard convection, that dynamic transition theory opens a new avenue toward to pattern formation, since the patterns and structure of transition states are dictated by solutions in Σ_T.

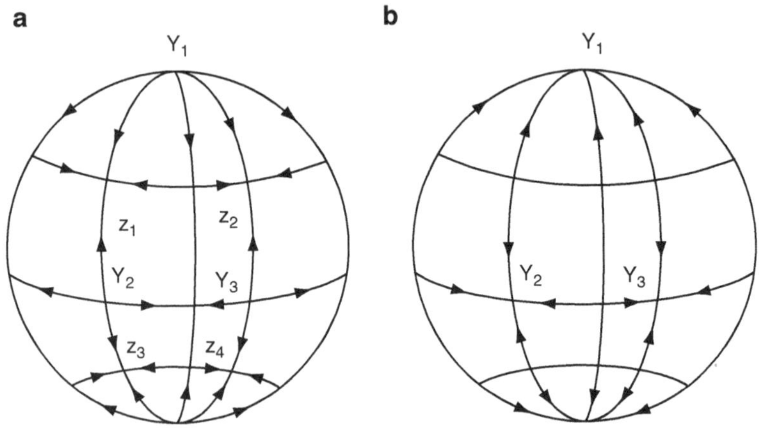

Fig. 1.15 Structure of $\Sigma_T = S^2$ with either that (a) $\pm Z_k$ $(1 \leq k \leq 4)$ are attractors, or (b) $\pm Y_k$ $(1 \leq k \leq 3)$ are attractors, determined by a simple relationship between b_2 and b_3; see Theorem 3.5.5

Chapter 2
Dynamic Transition Theory

This chapter introduces the dynamic transition theory for nonlinear dissipative systems developed recently by the authors. The main focus is the derivation of a general principle, Principle 1, on dynamic transitions for dissipative systems and a study of the types and structure of dynamic transitions. Our work is based on Philosophy 2, which states that we should search for a complete set of transition states.

2.1 General Dynamic Transition Theory

2.1.1 Classification of Dynamic Transitions

Let X and X_1 be two Banach spaces, and $X_1 \subset X$ a compact and dense inclusion. In this chapter, we always consider the following nonlinear evolution equations:

$$\frac{du}{dt} = L_\lambda u + G(u,\lambda), \quad u(0) = \varphi, \tag{2.1.1}$$

where $u : [0,\infty) \to X$ is the unknown function. Here we assume that the system control parameter λ is in $\mathbb{R}^1$. As explained in (1.2.3), in the case $\lambda \in \mathbb{R}^N$, we always take a one-dimensional path for λ, and we can always treat λ as in the case for $N = 1$.

Assume that[1] $L_\lambda : X_1 \to X$ is a parameterized linear completely continuous field depending continuously on λ that satisfies

[1] We follow here the notation used in Ma and Wang (2005b). In particular, a linear operator $L : X_1 \to X$ is called a completely continuous field if $L = -A + B : X_1 \to X$, $A : X_1 \to X$ is a linear homeomorphism, and $B : X_1 \to X$ is a linear compact operator. Also, we refer the interested reader to classical books such as (Kato 1995) for a basic knowledge of linear operators, and to Henry (1981), Pazy (1983) for semigroups of linear operators and sectorial operators.

T. Ma and S. Wang, *Phase Transition Dynamics*, DOI 10.1007/978-1-4614-8963-4_2,

$$\begin{aligned} &L_\lambda = -A + B_\lambda && \text{is a sectorial operator,} \\ &A : X_1 \to X && \text{is a linear homeomorphism, and} \\ &B_\lambda : X_1 \to X && \text{is a linear compact operator.} \end{aligned} \tag{2.1.2}$$

In this case, we can define the fractional-order spaces X_σ for $\sigma \in \mathbb{R}^1$ using interpolation, and we refer the interested reader to Henry (1981), Pazy (1983), and Ma and Wang (2005b) for the precise definitions. Also, we assume that $G(\cdot,\lambda) : X_\alpha \to X$ is a C^r $(r \geq 1)$ bounded mapping for some $0 \leq \alpha < 1$, depending continuously on $\lambda \in \mathbb{R}^1$, and

$$G(u,\lambda) = o(\|u\|_{X_\alpha}) \quad \forall \lambda \in \mathbb{R}^1. \tag{2.1.3}$$

Hereinafter we always assume the conditions (2.1.2) and (2.1.3), which imply that the system (2.1.1) has a dissipative structure.

Definition 2.1.1 *We say that the system (2.1.1) undergoes a (dynamic) transition from $(u,\lambda) = (0,\lambda_0)$ at $\lambda = \lambda_0$ if the following two conditions hold:*

(1) If $\lambda < \lambda_0$, $u = 0$ is locally asymptotically stable for (2.1.1).
(2) If $\lambda > \lambda_0$, then there exists a neighborhood $U \subset X$ of $u = 0$ independent of λ such that for every $\varphi \in U \setminus \Gamma_\lambda$, the solution $u_\lambda(t,\varphi)$ of (2.1.1) satisfies

$$\limsup_{t\to\infty} \|u_\lambda(t,\varphi)\| \geq \delta(\lambda) > 0, \qquad \lim_{\lambda\to\lambda_0} \delta(\lambda) \geq 0,$$

where Γ_λ is the stable manifold of $u = 0$, with codim $\Gamma_\lambda \geq 1$ in X for $\lambda > \lambda_0$.

The above definition amounts to saying that the basic state $u = 0$ is stable for $\lambda < \lambda_0$ and loses its stability completely for $\lambda > \lambda_0$. Obviously, the attractor bifurcation of (2.1.1) is a dynamic transition. However, bifurcation and transition are two different, but related, concepts. Definition 2.1.1 defines the transition of (2.1.1) from a stable equilibrium point to other, not necessarily equilibrium, states. In general, we can define transitions from one attractor to another as follows.

Definition 2.1.2 *Let $\Sigma_\lambda \subset X$ be an invariant set of (2.1.1). We say that (2.1.1) undergoes a transition from $(\Sigma_{\lambda_0},\lambda_0)$ at $\lambda = \lambda_0$ if the following conditions are satisfied:*

(1) When $\lambda < \lambda_0$, then Σ_λ is a local minimal attractor.
(2) When $\lambda > \lambda_0$, there exists a neighborhood $U \subset X$ of Σ_λ independent of λ such that for every $\varphi \in U \setminus (\Gamma_\lambda \cup \Sigma_\lambda)$, the solution $u(t,\varphi)$ of (2.1.1) satisfies

$$\limsup_{t\to\infty} \mathrm{dist}(u(t,\varphi),\Sigma_\lambda) \geq \delta(\lambda) > 0, \qquad \lim_{\lambda\to\lambda_0} \delta(\lambda) = \delta \geq 0,$$

where Γ_λ is the stable manifold of Σ_λ with codim $\Gamma_\lambda \geq 1$.

The main focus of this book is dynamic transitions from equilibrium states. We start with the principle of exchange of stabilities (PES). Let the (generalized) eigenvalues (counting multiplicity) of L_λ be given by $\{\beta_j(\lambda) \in \mathbb{C} \mid j = 1,2,\cdots\}$, and let

$$\text{Re}\,\beta_i(\lambda)\begin{cases} <0 & \text{if } \lambda<\lambda_0, \\ =0 & \text{if } \lambda=\lambda_0, \\ >0 & \text{if } \lambda>\lambda_0 \end{cases} \qquad \forall 1\le i\le m, \tag{2.1.4}$$

$$\text{Re}\,\beta_j(\lambda_0)<0 \qquad \forall j\ge m+1. \tag{2.1.5}$$

A few technical tools for verifying the PES for applications are introduced in Ma and Wang (2008d), by calculating the derivatives of the first eigenvalues that cross the imaginary axis at the critical parameter.

The following theorem is a basic principle of transitions from equilibrium states. It provides sufficient conditions and a basic classification for transitions of nonlinear dissipative systems. The proof of this theorem is given at the end of this subsection.

Theorem 2.1.3 *Let the conditions (2.1.4) and (2.1.5) hold. Then the system (2.1.1) always undergoes a dynamic transition from $(u,\lambda)=(0,\lambda_0)$, and there is a neighborhood $U\subset X$ of $u=0$ such that the transition is one of the following three types:*

(1) **Continuous Transition***: there exists an open and dense set $\widetilde{U}_\lambda\subset U$ such that for every $\varphi\in\widetilde{U}_\lambda$, the solution $u_\lambda(t,\varphi)$ of (2.1.1) satisfies*

$$\lim_{\lambda\to\lambda_0}\limsup_{t\to\infty}\|u_\lambda(t,\varphi)\|=0.$$

(2) **Jump Transition***: for every $\lambda_0<\lambda<\lambda_0+\varepsilon$ with some $\varepsilon>0$, there is an open and dense set $U_\lambda\subset U$ such that for any $\varphi\in U_\lambda$,*

$$\limsup_{t\to\infty}\|u_\lambda(t,\varphi)\|\ge\delta>0,$$

where $\delta>0$ is independent of λ.

(3) **Mixed Transition***: for every $\lambda_0<\lambda<\lambda_0+\varepsilon$ with some $\varepsilon>0$, U can be decomposed into two open (not necessarily connected) sets U_1^λ and U_2^λ: $\overline{U}=\overline{U}_1^\lambda\cup\overline{U}_2^\lambda$, $U_1^\lambda\cap U_2^\lambda=\emptyset$, such that*

$$\lim_{\lambda\to\lambda_0}\limsup_{t\to\infty}\|u(t,\varphi)\|=0 \qquad \forall\varphi\in U_1^\lambda,$$

$$\limsup_{t\to\infty}\|u(t,\varphi)\|\ge\delta>0 \qquad \forall\varphi\in U_2^\lambda,$$

where U_1^λ and U_2^λ are called metastable domains.

With this theorem at our disposal, it is natural to classify all dynamic transitions into the three categories as described in this theorem.

Definition 2.1.4 (Classification of Dynamic Transitions) *The dynamic transitions for (2.1.1) at $\lambda=\lambda_0$ are classified using their dynamic properties: continuous, jump, and mixed, as given in Theorem 2.1.3, which are called Type I, Type II, and Type III respectively.*

If the system (2.1.1) possesses a gradient-type structure, then the dynamic transitions are associated with the equilibrium phase transitions in statistical physics; otherwise, they relate to nonequilibrium phase transitions.

For equilibrium phase transition problems in statistical physics, there are classically several ways to classify phase transitions. The most notable is the classical Ehrenfest classification scheme, whereby phase transitions are labeled by the lowest derivative of the free energy that is discontinuous at the transition.

Proof of Theorem 2.1.3. Using the center manifold reduction outlined in Chap. 1, we have only to consider the reduced equation (1.2.7). By the PES assumption (2.1.4) and (2.1.5), we know that for $\lambda > \lambda_0$, $x = 0$ is a repeller. Namely, there exists a neighborhood $\mathscr{O}_\lambda$ of $x = 0$ such that for every $x_0 \in \mathscr{O}_\lambda$,

$$\liminf_{t\to\infty} \|x(t,x_0)\| > 0.$$

As $\lambda \to \lambda_0^+$, there exists a neighborhood $\mathscr{O}$ of $x = 0$ such that $\mathscr{O} = \overline{\mathscr{O}}_1 \bigcup \overline{\mathscr{O}}_2$ and

$$\begin{aligned}
\mathscr{O}_1 &= \{x_0 \in \mathscr{O} \mid \lim_{\lambda\to\lambda_0} \limsup_{t\to\infty} \|x(t,x_0)\| = 0\},\\
\mathscr{O}_2 &= \{x_0 \in \mathscr{O} \mid \lim_{\lambda\to\lambda_0} \limsup_{t\to\infty} \|x(t,x_0)\| > \delta > 0\},
\end{aligned}$$

where $\delta > 0$ is obtained using the unstable manifold theorem and finite-dimensionality of the reduced problem. Then there can be only the following three cases:

(a) $\mathscr{O}_2$ is nowhere dense and $\mathscr{O}_1$ is open and dense in $\mathscr{O}$,
(b) $\mathscr{O}_1$ is nowhere dense and $\mathscr{O}_2$ is open and dense in $\mathscr{O}$, and
(c) both $\mathscr{O}_1$ and $\mathscr{O}_2$ contains interior points.

We note that by the continuity of orbits, the case whereby both $\mathscr{O}_1$ and $\mathscr{O}_2$ are dense in $\mathscr{O}$ cannot occur. Again by the continuity of orbits, we see that cases (a)–(c) correspond to the three types of transitions respectively for (1.2.7). □

2.1.2 *Characterization of Transition Types*

An important aspect of transition theory is to determine which of the three types of transitions given by Theorem 2.1.3 occurs in a specific problem. The following theorem is useful in distinguishing the transition types of (2.1.1) at $(u,\lambda) = (0,\lambda_0)$, which can be deduced directly from Theorem 2.1.3, the attractor bifurcation theorem, Theorem 2.2.2, and Theorem 2.1.15 in Ma and Wang (2005d) on the stability of extended orbits.

Theorem 2.1.5 *Let the conditions (2.1.4) and (2.1.5) hold. Then the transition type of (2.1.1) at $\lambda = \lambda_0$ is completely dictated by its reduced equation (1.2.7). In particular, let $U \subset \mathbb{R}^m$ be a neighborhood of $x = 0$. Then we have the following assertions:*

(1) *If the transition of (2.1.1) at* $(0,\lambda_0)$ *is continuous, then there is an open and dense set* $\widetilde{U} \subset U$ *such that for every* $x_0 \in \widetilde{U}$*, the solution* $x(t,x_0)$ *of (1.2.7) at* $\lambda = \lambda_0$ *with* $x(0,x_0) = x_0$ *satisfies*

$$\lim_{t\to\infty} x(t,x_0) = 0.$$

(2) *If there exists an open and dense set* $\widetilde{U} \subset U$ *such that for every* $x_0 \in \widetilde{U}$*, the solution* $x(t,x_0)$ *of (1.2.7) at* $\lambda = \lambda_0$ *satisfies*

$$\limsup_{t\to\infty} |x(t,x_0)| \neq 0,$$

then the transition is a jump transition.

(3) *If the transition is mixed, then there exist two open sets* $\widetilde{U}_1, \widetilde{U}_2 \subset U$ *such that*

$$\begin{aligned} &\lim_{t\to\infty} x(t,x_0) = 0 && \forall x_0 \in \widetilde{U}_1, \\ &\limsup_{t\to\infty} |x(t,x_0)| \neq 0 && \forall x_0 \in \widetilde{U}_2. \end{aligned}$$

In general, the conditions in assertions (1)–(3) of Theorem 2.1.5 are not sufficient. They do, however, give sufficient conditions when (2.1.1) has a variational structure. To see this, let (2.1.1) be a gradient-type equation. Under the conditions (2.1.4) and (2.1.5), by Lemmas A.2.7 and A.2.8, in a neighborhood $U \subset X$ of $u = 0$, the center manifold M^c in U at $\lambda = \lambda_0$ consists of three subsets

$$M^c = W^u + W^s + D,$$

where W^s is the stable set, W^u the unstable set, and D the hyperbolic set of (1.2.7).

Theorem 2.1.6 *Let (2.1.1) be a gradient-type equation, and suppose that the conditions (2.1.4) and (2.1.5) hold. If* $u = 0$ *is an isolated singular point of (2.1.1) at* $\lambda = \lambda_0$*, then we have the following assertions:*

(1) *The transition of (2.1.1) at* $(u,\lambda) = (0,\lambda_0)$ *is continuous if and only if* $u = 0$ *is locally asymptotically stable at* $\lambda = \lambda_0$*, i.e., the center manifold is stable:* $M^c = W^s$.

(2) *The transition of (2.1.1) at* $(u,\lambda) = (0,\lambda_0)$ *is a jump transition if and only if there is an open and dense set* $\widetilde{U}$ *in a neighborhood* $U \subset H$ *of* $u = 0$ *such that for every* $\varphi \in \widetilde{U}$*, the solution* $u(t,\varphi)$ *of (2.1.1) at* $\lambda = \lambda_0$ *satisfies*

$$\limsup_{t\to\infty} \|u(t,\varphi)\| \neq 0.$$

Proof. The sufficiency of assertion (1) follows from Theorem 2.2.2. By Lemmas A.2.7 and A.2.8, the hyperbolic set $D \subset M^c$ is open, and if $D = \emptyset$, then the center manifold M^c satisfies

$$M^c = W^s \quad \text{or } M^c = W^u. \tag{2.1.6}$$

Thus, the necessity of assertion (1) follows from assertion (1) of Theorem 2.1.5 and (2.1.6). Assertion (2) follows from assertion (2) of Theorem 2.1.5. The proof is complete. □

2.1.3 Local Topological Structure of Transitions

The study of both the local and global topological structures of transitions is an important topic in phase transition dynamics in the nonlinear sciences. In this subsection we briefly discuss the local topological structure of transitions of (2.1.1) near $(u,\lambda) = (0,\lambda_0)$. Equivalently, we have only to address the topological structure of transitions of the finite-dimensional system (1.2.7) near $(x,\lambda) = (0,\lambda_0)$.

Let $\mathscr{O} \subset \mathbb{R}^m$ be a neighborhood of $x = 0$. By Theorem 2.1.5, for the transition of (2.1.1) at $(u,\lambda) = (0,\lambda_0)$, there are two subsets in $\mathscr{O}$, the stable set M^s and the complementary set $\mathscr{O}_2 = \mathscr{O} \setminus M^s$. These sets are related to transitions as follows:

$$
\begin{aligned}
&\overline{M^s} = \overline{\mathscr{O}}, \quad \mathring{\mathscr{O}}_2 = \emptyset && \Leftarrow \quad \text{Continuous transition},\\
&\mathring{M}^s = \emptyset, \quad \overline{\mathscr{O}}_2 = \overline{\mathscr{O}} && \Rightarrow \quad \text{Jump transition},\\
&\overline{\mathscr{O}} = \overline{M^s} + \overline{\mathscr{O}_2}, \quad \mathring{M}^s \neq \emptyset, \quad \mathring{\mathscr{O}}_2 \neq \emptyset && \Leftarrow \quad \text{Mixed transition},
\end{aligned}
$$

where $\mathring{U}$ stands for the set of interior points of a set U.

We now examine continuous transitions based on the properties of the stable set.

Case That the Stable Set M^s equals $\mathscr{O}$: In this case, the transition is an attractor bifurcation, and the topological structure of transitions in $\mathscr{O}$ is characterized by the attractor, which is an $(m-1)$-dimensional homological sphere.

Case That $\overline{M^s} = \overline{\mathscr{O}}$ and $M^s \neq \mathscr{O}$: In this case, $\mathring{M}^s$ might not be connected, and both the following two cases can occur:

(a) There is an attractor in $\mathring{M}^s$ bifurcated from $(0,\lambda_0)$.
(b) There is no attractor in $\mathring{M}^s$ bifurcated from $(0,\lambda_0)$.

Example 2.1.7. The this example shows that the case (b) does happen:

$$
\begin{aligned}
\frac{dx_1}{dt} &= 4\lambda x_1 + \lambda x_2 + x_1^2 - 4x_1x_2 - x_2^2,\\
\frac{dx_2}{dt} &= -6\lambda x_1 + 6x_1x_2.
\end{aligned}
\tag{2.1.7}
$$

The linear and nonlinear operators in (2.1.7) are given by

$$
A_\lambda = \begin{pmatrix} 4\lambda & \lambda \\ -6\lambda & 0 \end{pmatrix}, \qquad G(x) = \begin{pmatrix} x_1^2 - 4x_1x_2 - x_2^2 \\ 6x_1x_2 \end{pmatrix}.
$$

The matrix A_λ has two eigenvalues, $\beta_1(\lambda) = \lambda(2+i\sqrt{2})$, $\beta_2(\lambda) = \lambda(2-i\sqrt{2})$, which satisfy (2.1.4) with $\lambda_0 = 0$. We can calculate that

$$\text{ind}(G,0) = 2. \tag{2.1.8}$$

To further study the transitions of Eq. (2.1.7), we need to introduce a lemma, called the Poincaré formula, which provides some connections between the local topological structure of singular points of a two-dimensional vector field and the indices of the singular points. This formula will also be useful in later discussions.

Lemma 2.1.8 *Let $v \in C^r(\Omega, \mathbb{R}^2)$ $(r \geq 0)$ be a 2-dimensional vector field, and $v(0) = 0$. Then the following Poincaré index formula holds:*

$$ind(v,0) = 1 + \frac{1}{2}(e-h), \tag{2.1.9}$$

where e is the number of elliptic regions, h is the number of hyperbolic regions, and the elliptic, hyperbolic, and parabolic regions E, H, and P in a neighborhood $U \subset \mathbb{R}^2$ of $x = 0$ are defined by (see Fig. 2.1)

$$\begin{aligned} E &= \{x \in U \mid \lim_{t\to\infty} S(t)x = \lim_{t\to\infty} S(-t)x = 0\}, \\ H &= \{x \in U \mid S(t)x,\ S(-t)x \notin U \text{ as } t > t_0 \text{ for some } t_0 > 0\}, \\ P &= \{x \in U \mid \text{either } \lim_{t\to\infty} S(t)x = 0,\ S(-t)x \notin U\,(t > t_0), \\ &\qquad \text{or } \lim_{t\to\infty} S(-t)x = 0,\ S(t)x \notin U, \\ &\qquad \text{or } S(t)x,\ S(-t)x \in U,\ \forall t \geq 0\}, \end{aligned}$$

where $S(t)$ is the semigroup generated by v.

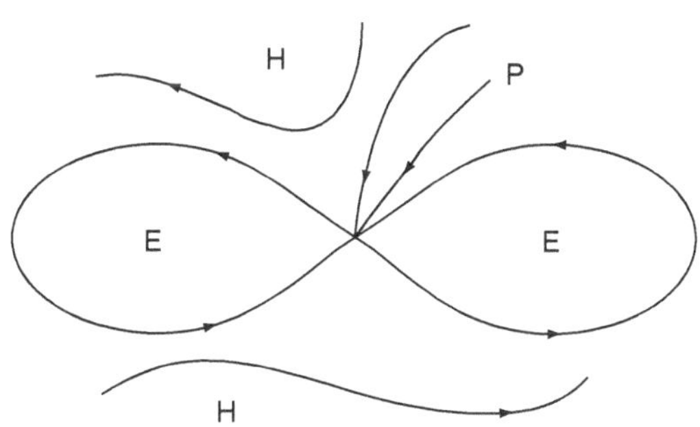

Fig. 2.1 Schematic of elliptic, parabolic, and hyperbolic regions

By the Poincaré index formula (2.1.9), we deduce from (2.1.8) that the vector field $G(x)$ has at least two elliptic regions near $x = 0$. In fact, $G(x)$ has just two elliptic regions. To see this, it is clear that the equation

$$k = \frac{x_2}{x_1} = \frac{G_2(x)}{G_1(x)} = \frac{6x_1x_2}{x_1^2 - 4x_1x_2 - x_2^2} = \frac{6k}{1 - 4k - k^2}$$

has only one real solution $k=0$, which implies that $G(x)$ has only two straight-line orbits on the x_1-axis that reach to $x=0$. Therefore, the topological structure of $G(x)$ is as shown in Fig. 2.2, and for each point $x_0 \in \mathbb{R}^2$ and $x_0 \neq (x_1,0)$ with $x_1>0$, we have

$$\lim_{t\to\infty} S(t)x_0 = 0,$$

where $S(t)$ is the semigroup generated by $G(x)$.

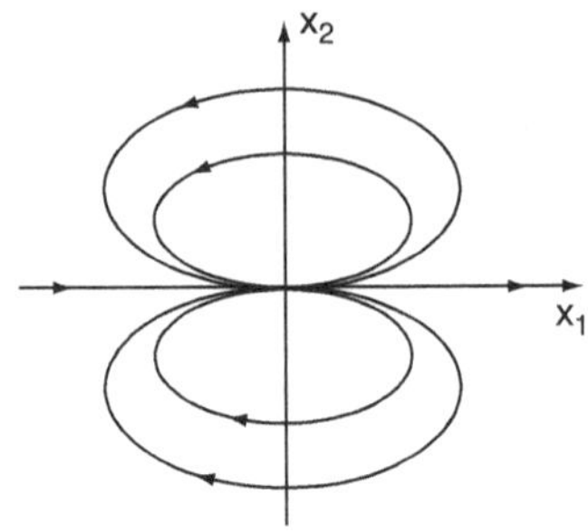

Fig. 2.2 Topological structure of G

When $\lambda > \lambda_0 = 0$, the system (2.1.7) has a continuous transition and bifurcates from $(0,\lambda_0)$ to a unique singular point $x_\lambda = (0,\lambda)$. The topological structure of this transition is as shown in Fig. 2.3. Obviously, there is no attractor bifurcated from $(0,\lambda_0)$ on $\lambda > \lambda_0$.

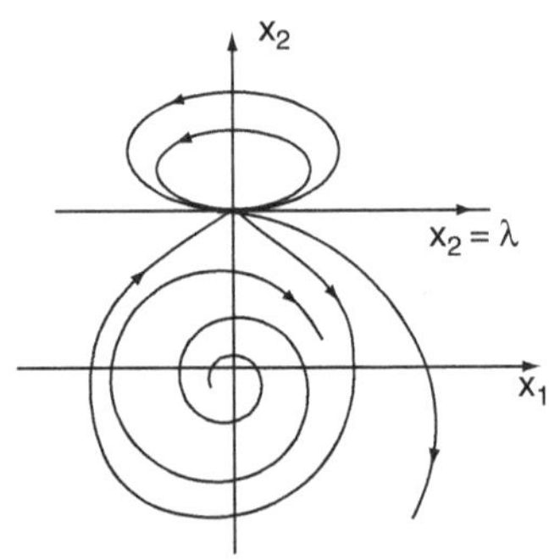

Fig. 2.3 Topological structure after the transition

By Theorem 2.1.5, we know that if the interior of the stable set is empty, i.e., $\mathring{M}^s = \emptyset$, then $\overline{\mathscr{O}}_2 = \overline{\mathscr{O}}$, and the transition of (2.1.1) at $(0,\lambda_0)$ is a jump transition. However, the converse statement is not true. The following example shows that there exist systems that have a jump transition, but their stable sets M_s at the critical state at $\lambda = \lambda_0$ are open and dense in a neighborhood $\mathscr{O} \subset \mathbb{R}^m$ of $x=0$.

Example 2.1.9. Consider the system given by

$$\begin{aligned}\frac{dx_1}{dt} &= 2\lambda x_1 + 3\lambda x_2 + x_1^2 - x_2^2,\\ \frac{dx_2}{dt} &= -\lambda x_1 + x_1 x_2.\end{aligned} \tag{2.1.10}$$

The linear operator in (2.1.10) has the eigenvalues

$$\beta_1(\lambda) = \lambda(1 + i\sqrt{2}), \quad \beta_2(\lambda) = \lambda(1 - i\sqrt{2}).$$

It is easy to check that the nonlinear operator $G(x) = \left(x_1^2 - x_2^2, x_1 x_2\right)^t$ is topologically equivalent to the structure shown in Fig. 2.2, i.e., the stable set M^s of G is open and dense in $\mathbb{R}^2$. However, when $\lambda > \lambda_0 = 0$, the system (2.1.10) bifurcates from $(0, \lambda_0)$ to a unique singular point $x_\lambda = (x_1^\lambda, x_2^\lambda) = (0, 3\lambda)$, and the eigenvalues of the Jacobian matrix at x_λ of the vector field in (2.1.10) are

$$\widetilde{\beta}_1(\lambda) = \lambda(1 + i\sqrt{5}), \quad \widetilde{\beta}_2(\lambda) = \lambda(1 - i\sqrt{5}).$$

This shows that the transition of (2.1.10) at $(0, \lambda_0) = (0, 0)$ on $\lambda > \lambda_0$ is a jump transition, which is schematically illustrated in Fig. 2.4.

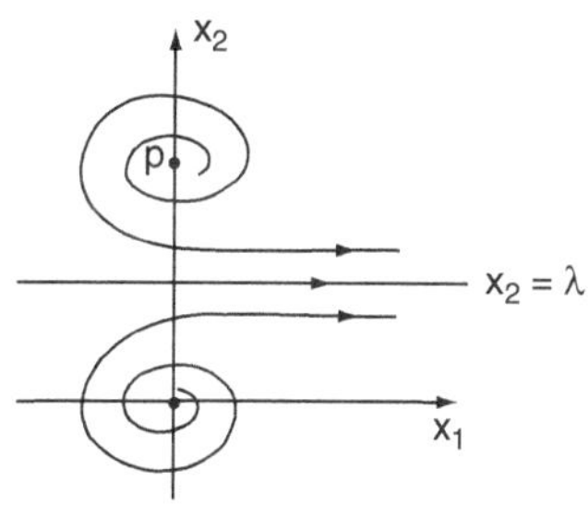

Fig. 2.4 Jump transition for Example 2.1.9, with $p = (0, 3\lambda)$

An important topic related to jump transitions is saddle-node bifurcation, found in phase transition problems in nonlinear sciences. This will be addressed in later sections.

The structural problem of mixed transitions deals mainly with stable and unstable domains and the structure of bifurcated attractors in stable domains. Mathematically, a related interesting topic is the topological structure of bifurcated invariant sets for (2.1.1).

2.2 Continuous Transition

2.2.1 Finite-Dimensional Systems

In this subsection, we study attractor bifurcation for the following finite-dimensional systems:

$$\frac{dx}{dt} = A_\lambda x + G(x,\lambda), \quad \lambda \in \mathbb{R}^1, \quad x \in \mathbb{R}^n \ (n \geq 1), \tag{2.2.1}$$

where A_λ is an $n \times n$ matrix whose entries are continuous functions of λ, and $G : \mathbb{R}^n \times \mathbb{R}^1 \to \mathbb{R}^n$ is $C^r (r \geq 1)$ on $x \in \mathbb{R}^n$ and continuous on $\lambda \in \mathbb{R}^1$:

$$A_\lambda = \begin{pmatrix} a_{11}(\lambda) & \cdots & a_{1n}(\lambda) \\ \vdots & & \vdots \\ a_{n1}(\lambda) & \cdots & a_{nn}(\lambda) \end{pmatrix}, \qquad G(x,\lambda) = o(|x|) \qquad \forall \lambda \in \mathbb{R}^1. \tag{2.2.2}$$

Let all eigenvalues (counting multiplicities) of A_λ be given by $\beta_1(\lambda), \cdots, \beta_n(\lambda)$. It is known that the eigenvalues $\beta_i(\lambda)$ $(1 \leq i \leq n)$ depend continuously on λ.

In the following, we give some definitions of dynamic bifurcation for (2.2.1), which are also suitable for infinite-dimensional systems; see also (Ma and Wang 2005b,c).

Definition 2.2.1

(1) We say that the system (2.2.1) bifurcates from $(x,\lambda) = (0,\lambda_0)$ to an invariant set Σ_λ if there exists a sequence of invariant sets $\{\Sigma_{\lambda_n}\}$ such that

$$0 \notin \Sigma_{\lambda_n}, \qquad \lim_{n\to\infty} \lambda_n = \lambda_0, \qquad \lim_{n\to\infty} d(\Sigma_{\lambda_n}, 0) = \lim_{n\to\infty} \max_{x\in\Sigma_{\lambda_n}} |x| = 0.$$

(2) If the invariant sets Σ_λ are attractors of (2.2.1), then the bifurcation is called an attractor bifurcation.

(3) If Σ_λ are attractors of (2.2.1) that have the same homology as an m-dimensional sphere, then we say that the system has an S^m-attractor bifurcation at $(0,\lambda_0)$, and Σ_λ is called an m-dimensional homological sphere.

For the eigenvalues $\beta_i(\lambda)$ of A_λ, we assume that

$$\text{Re}\beta_i(\lambda) \begin{cases} < 0 & \text{if } \lambda < \lambda_0, \\ = 0 & \text{if } \lambda = \lambda_0, \\ > 0 & \text{if } \lambda > \lambda_0 \end{cases} \qquad \forall 1 \leq i \leq m, \tag{2.2.3}$$

$$\text{Re}\beta_j(\lambda_0) < 0 \qquad \forall m+1 \leq j \leq n. \tag{2.2.4}$$

Let E_0 be the eigenspace of A_λ at λ_0 associated with the first m eigenvalues; then $\dim E_0 = m$.

The main results in this section are the following S^m-attractor bifurcation theorems for the finite dimensional system (2.2.1); see (Ma and Wang 2005b).

Theorem 2.2.2 *Assume that (2.2.2)–(2.2.4) hold, and $x = 0$ is locally asymptotically stable for (2.2.1) at $\lambda = \lambda_0$. Then we have the following assertions.*

(1) The system (2.2.1) bifurcates from $(0, \lambda_0)$ to an attractor Σ_λ for $\lambda > \lambda_0$ with $m - 1 \leq \dim \Sigma_\lambda \leq m$, which is connected when $m \geq 2$.
(2) There exists a neighborhood $U \subset \mathbb{R}^n$ of $x = 0$ such that Σ_λ attracts U/Γ, where Γ is the stable manifold of $x = 0$ with codimension codim $\Gamma = m$ in $\mathbb{R}^n$.
(3) The bifurcated attractor Σ_λ is an $(m-1)$-dimensional homological sphere; more precisely, $\Sigma_\lambda = \bigcap_{k=1}^{\infty} M_k$, where $M_{k+1} \subset M_k$ are m-dimensional annuli. In particular, if Σ_λ is a finite simplicial complex, then Σ_λ is a deformation retract of an m-annulus.
(4) For every $x_\lambda \in \Sigma_\lambda$, x_λ can be expressed as

$$x_\lambda = z_\lambda + o(|z_\lambda|), \quad z_\lambda \in E_0.$$

(5) If the number of singular points of (2.2.1) in Σ_λ is finite, then we have the following index formula:

$$\sum_{x \in \Sigma_\lambda} ind(-(A_\lambda + G), x) = \begin{cases} 2 & \text{if } m = odd, \\ 0 & \text{if } m = even. \end{cases}$$

Theorem 2.2.3 *Assume that (2.2.2)–(2.2.4) hold and that $x = 0$ is globally asymptotically stable for (2.2.1) at $\lambda = \lambda_0$. Then for every bounded open set $U \subset \mathbb{R}^n$ with $0 \in U$, there is an $\varepsilon > 0$ such that if $\lambda_0 < \lambda < \lambda_0 + \varepsilon$, the attractor Σ_λ of (2.2.1) bifurcated from $(0, \lambda_0)$ attracts $U \setminus \Gamma$, where $\Gamma \subset \mathbb{R}^n$ is the stable manifold of $x = 0$ with $\dim \Gamma = n - m$. In particular, if (2.2.1) has global attractors near $\lambda = \lambda_0$, then Σ_λ attracts $\mathbb{R}^n \setminus \Gamma$.*

2.2.2 S^1-Attractor Bifurcation

We now prove that the bifurcated attractor Σ_λ of (2.2.1) from an eigenvalue with multiplicity two is homeomorphic to S^1.

Let $v \in C^r(\Omega, \mathbb{R}^{m+1})(r \geq 1)$ be an n-dimensional vector field given by

$$v_\lambda(x) = \lambda x - G(x, \lambda) \qquad \text{for } x \in \Omega \subset \mathbb{R}^{m+1}, \tag{2.2.5}$$

where $G(x, \lambda) = G_k(x, \lambda) + o(|x|^k)$, and G_k is a k-multilinear field that satisfies

$$c|x|^{k+1} \leq (G_k(x, \lambda), x) \qquad \forall \lambda \in \mathbb{R}^1, \tag{2.2.6}$$

for some constant $c > 0$, and $k = 2p + 1$ with $p \geq 1$.

Theorem 2.2.4 *Let $v_\lambda \in C^r(\Omega, \mathbb{R}^2)$ satisfy (2.2.5) and (2.2.6). Then v_λ bifurcates from $(x, \lambda) = (0, 0)$ on $\lambda > 0$ to an attractor Σ_λ, which is homeomorphic to a circle S^1. Moreover, one and only one of the following is true:*

(1) Σ_λ *is a periodic orbit.*
(2) Σ_λ *contains some arc segments* $\Gamma_i \subset \Sigma_\lambda$ $(1 \leq i \leq l)$ *such that* Γ_i *consist of singular points.*
(3) Σ_λ *contains at most* $2(k+1) = 4(p+1)$ *singular points. Furthermore,* Σ_λ *has an even number, say* $2N$*, saddle points, and* $2N$ *stable node points (possibly degenerate), and* n $(\leq 4(p+1) - 4N)$ *singular points with index zero; see Fig. 2.5 for* $N = 1$ *and* $n = 2$*.*

Proof. We proceed in the following five steps.

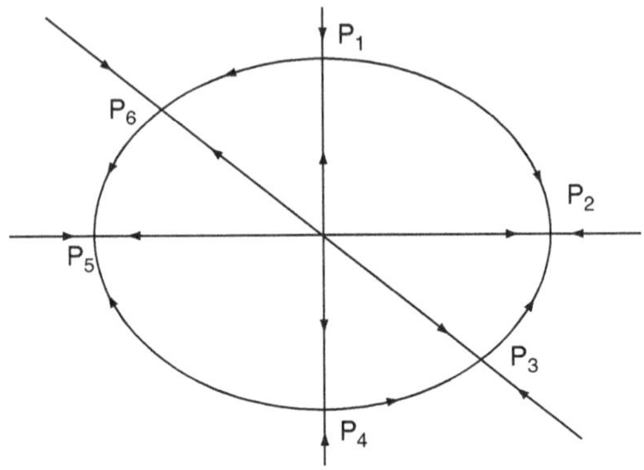

Fig. 2.5 $\Sigma_\lambda = S^1$ has $4N+n$ $(N=1, n=2)$ singular points, where P_1, P_4 are saddles, P_2, P_5 are nodes, P_3, P_6 are singular points with index zero

Step 1. Obviously, (2.2.6) implies that $x = 0$ is asymptotically stable for (2.2.5) at the critical state $\lambda = 0$. Hence, by Theorem 2.2.2, the vector field v_λ bifurcates from $(x,\lambda) = (0,0)$ to an attractor Σ_λ on $\lambda > 0$, which has the homology type of a circle S^1.

Step 2. Let Σ_λ have no singular points. Then Σ_λ must contain at least a periodic orbit. We need to show that Σ_λ contains only one periodic orbit.

In the polar coordinate system $(x_1, x_2) = (r\cos\theta, r\sin\theta)$, the vector field v_λ becomes

$$\frac{dr}{d\theta} = r\frac{\cos\theta v_1 + \sin\theta v_2}{\cos\theta v_2 - \sin\theta v_1}. \tag{2.2.7}$$

We see that

$$\begin{aligned}
\cos\theta v_1 &= \lambda r\cos^2\theta - \cos\theta g_1(r\cos\theta, r\sin\theta, \lambda),\\
\sin\theta v_2 &= \lambda r\sin^2\theta - \sin\theta g_2(r\cos\theta, r\sin\theta, \lambda),\\
\cos\theta v_2 &= \lambda r\cos\theta\sin\theta - \cos\theta g_2(r\cos\theta, r\sin\theta, \lambda),\\
\sin\theta v_1 &= \lambda r\sin\theta\cos\theta - \sin\theta g_1(r\cos\theta, r\sin\theta, \lambda),
\end{aligned}$$

where $G(x,\lambda) = (g_1(x,\lambda), g_2(x,\lambda))$. Let $g_i(x,\lambda) = g_{ki}(x,\lambda) + o(|x|^k)$ for $i = 1,2$. By assumption, $g_{ki}(x,\lambda)$ $(i = 1,2)$ are $k = 2m+1$ multilinear functions. Hence (2.2.7) can be rewritten as

$$\frac{dr}{d\theta} = \frac{\lambda - r^{2m}(\cos\theta g_{k1} + \sin\theta g_{k2}) + o(r^{2m})}{r^{2m-1}(\sin\theta g_{k1} - \cos\theta g_{k2}) + o(r^{2m-1})}. \tag{2.2.8}$$

Based on (2.2.6), we have

$$0 < c \leq \cos\theta g_{k1}(\cos\theta, \sin\theta, \lambda) + \sin\theta g_{k2}(\cos\theta, \sin\theta, \lambda). \tag{2.2.9}$$

On the other hand, by assumption, Σ_λ contains a periodic orbit for every $\lambda > 0$ sufficiently small. Hence there exists $C > 0$ such that

$$0 < C \leq \sin\theta g_{k1}(\cos\theta, \sin\theta, \lambda) - \cos\theta g_{k2}(\cos\theta, \sin\theta, \lambda), \tag{2.2.10}$$

for every $0 \leq \theta \leq 2\pi$ and $\lambda > 0$ small. The condition (2.2.10) amounts to saying that the orbits of v_λ are circular around $x = 0$.

Let $r(\theta, r_0)$ be the solution of (2.2.8) with initial value $r(0, r_0) = r_0$. Then we have the Taylor expansion

$$r^{2m}(\theta, r_0) = r_0^{2m} + R(\theta) \cdot o(r_0^{2m}), \quad R(0) = 0. \tag{2.2.11}$$

It follows from (2.2.8) and (2.2.11) that

$$\begin{aligned}\frac{1}{2m}\left[r^{2m}(2\pi, r_0) - r^{2m}(0, r_0)\right] =& \int_0^{2\pi} \frac{\lambda - r^{2m}\alpha(\theta) + o(r^{2m})}{\beta(\theta) + o(1)} d\theta \\ =& a\lambda - b r_0^{2m} + o(r_0^{2m}),\end{aligned} \tag{2.2.12}$$

where

$$a = \int_0^{2\pi} \frac{1}{\beta(\theta) + o(1)} d\theta, \qquad b = \int_0^{2\pi} \frac{\alpha(\theta)}{\beta(\theta) + o(1)} d\theta,$$
$$\alpha(\theta) = \cos\theta g_{k1} + \sin\theta g_{k2}, \qquad \beta(\theta) = \sin\theta g_{k1} - \cos\theta g_{k2}.$$

From (2.2.12) we see that the periodic solutions of v_λ near $x = 0$ correspond to positive solutions of

$$a\lambda - b r_0^{2m} + o(r_0^{2m}) = 0. \tag{2.2.13}$$

By (2.2.9) and (2.2.10), $a > 0, b > 0$. Therefore, (2.2.13) has a unique positive solution near $r = 0$:

$$r_0 = \left(\frac{a\lambda}{b}\right)^{1/2m} + o(\lambda^{1/2m}),$$

for every $\lambda > 0$ sufficiently small. Thus, Σ_λ consists of a unique periodic orbit.

Step 3. In the case that Σ_λ contains a singular point, we assert that Σ_λ contains either a finite number of singular points or a few arc segments $\Gamma_i \bigcap \Sigma_\lambda$ $(1 \leq i \leq l)$ consisting of singular points, and if it contains a finite number of singular points, then there are at most $2(k+1)$ of them near $x = 0$.

In fact, if

$$\frac{g_1(x,\lambda)}{g_2(x,\lambda)} = \frac{x_1}{x_2} + f(x,\lambda),$$

where $f(x,\lambda)$ are cutoff functions

$$f(r,\theta)\begin{cases} =0 & \forall\theta\in[\theta_i,\theta_{i+1}], \\ \neq 0 & \theta\notin[\theta_i,\theta_{i+1}] \end{cases} \qquad \forall 1\leq i\leq l,$$

where (r,θ) is the polar coordinate system, then Σ_λ contains l $(l\geq 1)$ arc segments Γ_i that consist of singular points. In particular, if $f(x,\lambda)=0$, namely

$$\frac{g_1(x,\lambda)}{g_2(x,\lambda)} = \frac{x_1}{x_2}, \quad \forall x\in\mathbb{R}^2, \tag{2.2.14}$$

then Σ_λ has a cycle of singular points. Otherwise, by (2.2.6), the number of singular points of v_λ is finite. The maximal number of singular points for v_λ is determined by the following equation:

$$\lambda x - G_k(x,\lambda) = 0. \tag{2.2.15}$$

Since G_k is a k-multilinear vector field, the singular points of (2.2.15) must be on the straight lines $x_2 = zx_1$, where z satisfies

$$z = \frac{g_{k2}(x_1,x_2,\lambda)}{g_{k1}(x_1,x_2,\lambda)} = \frac{g_{k2}(1,z,\lambda)}{g_{k1}(1,z,\lambda)}. \tag{2.2.16}$$

The number of solutions of (2.2.16) is at most $k+1$. Since $k =$ odd, the number of solutions of (2.2.15) is at most $2(k+1)$.

Step 4. Let Σ_λ contain a cycle S^1 of singular points, then we shall see that $\Sigma_\lambda = S^1$.

Under the polar coordinate system, we have

$$v_r(\theta,r) = (v_\lambda(x),x) = \lambda r^2 - r^{k+1}\alpha(\theta) + o(r^{k+1}),$$

where $\alpha(\theta)$ is as in (2.2.12). By (2.2.6),

$$0 < c \leq \alpha(\theta) \leq c_1, \quad \forall 0\leq\theta\leq 2\pi.$$

It is clear that for each $\theta(0\leq\theta\leq 2\pi), v_r$ has a unique zero point

$$r_\lambda(\theta) = (\lambda/\alpha(\theta))^{1/k-1} + o(|\lambda|^{1/k-1}),$$

near $r=0$ for $\lambda>0$. Hence, the set

$$\widetilde{\Omega}_\lambda = \{(\theta,r_\lambda(\theta))|\ v_r(\theta,r_\lambda(\theta)) = 0, 0\leq\theta\leq 2\pi\}$$

is homeomorphic to a cycle S^1, and all singular points of v_λ near $x=0$ are in $\widetilde{\Omega}_\lambda$. This implies that $\widetilde{\Omega}_\lambda \subset \Sigma_\lambda$. Let

$$g_{kl} = \sum_{i+j=k} a^l_{ij} x_1^i x_2^j, \quad l = 1,2. \tag{2.2.17}$$

By (2.2.14), we know that $x_2 g_{k1} = x_1 g_{k2}$. Then we infer from (2.2.6) that

$$0 < \alpha^1_{k0} = \alpha^2_{k-11}. \tag{2.2.18}$$

At the singular point $(\widetilde{x}_1, 0) \in \widetilde{\Omega}_\lambda$ of v_λ, for every $\lambda > 0$ sufficiently small, we have

$$\begin{aligned}
\operatorname{div} v_\lambda(\widetilde{x}_1, 0) &= 2\lambda - k\alpha^1_{k0}\widetilde{x}_1^{k-1} - \alpha^2_{k-11}\widetilde{x}_1^{k-1} + o(|\widetilde{x}_1|^{k-1}) \\
&= (\text{by } \alpha^1_{k0}\widetilde{x}_1^{k-1} = \lambda \quad \text{and} \quad (2.2.18)) \\
&= -(k-1)\lambda + o(|\lambda|) \\
&< 0.
\end{aligned}$$

In the same fashion, for any point $\widetilde{x} \in \widetilde{\Omega}_\lambda$, we can take an orthogonal system transformation such that $\widetilde{x}$ is on the x_1-axis, and then we can prove that

$$\operatorname{div} v_\lambda(x) < 0, \quad \forall x \in \widetilde{\Omega}_\lambda,$$

which implies that $\Sigma_\lambda = \widetilde{\Omega}_\lambda = S^1$.

Step 5. Σ_λ contains finite number of singular points. We prove that $\Sigma_\lambda = S^1$. By the Brouwer degree theory, it follows from (2.2.6) that

$$\deg(v_\lambda, \Omega, 0) = 1 \qquad \forall |\lambda| > 0 \text{ sufficiently small},$$

in some neighborhood $\Omega \subset \mathbb{R}^2$ of $x = 0$. It is known that

$$\operatorname{ind}(v_\lambda, 0) = 1, \quad \forall |\lambda| \neq 0.$$

Hence, we have

$$\sum_{z_i \in \Sigma_\lambda} \operatorname{ind}(v_\lambda, z_i) = 0. \tag{2.2.19}$$

Let $z \in \Sigma_\lambda$ be a singular point of v_λ. Without loss of generality, we may take the orthogonal coordinate system such that $z = (x_1, 0)$. Then by (2.2.6) and (2.2.17), the Jacobian matrix of v_λ at z is given by

$$Dv_\lambda(z) = \begin{pmatrix} -(k-1)\lambda + o(\lambda) & * \\ 0 & (1 - \alpha^2_{k-11}/\alpha^1_{k0})\lambda \end{pmatrix},$$

where $\alpha^1_{k0} > 0$. Obviously, $Dv_\lambda(z)$ has an eigenvalue $\beta = -(k-1)\lambda + o(\lambda) < 0$. Hence, for every singular point $z \in \Sigma_\lambda$ of v_λ, the index of v_λ at z can only be one of $1, -1$, and 0. It is easy to see that if the index is 1, then z is a stable node point.

Let the index of v_λ at z be -1:

$$\text{ind}(v_\lambda, z) = -1. \tag{2.2.20}$$

When $\alpha^2_{k-11} \neq \alpha^1_{k0}$, z is nondegenerate. Therefore, v_λ has a unique unstable manifold at z. When $\alpha^2_{k-11} = \alpha^1_{k0}$,

$$\text{div} v_\lambda(z) = -(k-1)\lambda + o(\lambda) < 0. \tag{2.2.21}$$

If the unstable manifold of v_λ at z is not unique, then the local structure of v_λ at z is topologically equivalent to that as shown in Fig. 2.6.

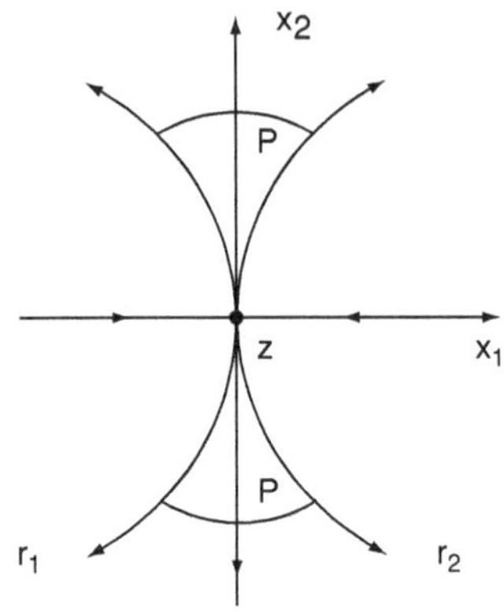

Fig. 2.6 r_1 and r_2 are unstable manifolds of v_λ at z; the flow in P is outward from z

On the other hand, (2.2.21) means that there is a neighborhood $O \subset \mathbb{R}^2$ of $z = 0$ such that

$$\text{div} v_\lambda(x) < 0, \quad \forall x \in O,$$

which implies that for every open set $\widetilde{O} \subset O$, the area of $\widetilde{O}$ satisfies

$$|\widetilde{O}| > |\widetilde{O}_t|, \quad \forall 0 < t < t_0, \tag{2.2.22}$$

where $t_0 > 0$ depends on $\widetilde{O}, \widetilde{O}_t = S(t)\widetilde{O}$, and $S(t)$ is the operator semigroup generated by v_λ.

However, it is clear that for every open set $\widetilde{O} \subset O$ in a domain P as shown in Fig. 2.6, the property (2.2.22) is not true. Hence, the unstable manifold of v_λ at z must be unique.

We can prove in the same fashion that if the index of v_λ at z is zero, then the unstable manifold of v_λ at z is also unique.

By the Poincaré–Bendixson theorem, all unstable manifolds of the singular points of v_λ with indices -1 and 0 are connected to the singular points with indices 1 and 0 as shown in Fig. 2.5. Thus by the uniqueness of unstable manifolds for each singular point with either index -1 or index 0, the set of all singular points and unstable manifolds is a cycle S^1, i.e., $\Sigma_\lambda = S^1$, because Σ_λ is homological to S^1.

Step 6. Finally, if

$$\Sigma_\lambda = \bigcup_{i=1}^{r} \Gamma_i + \Sigma,$$

where Γ_i are arc segments consisting of singular points and Σ contains only a finite number of singular points, we can apply the methods of Steps 4 and 5 to prove that $\Sigma_\lambda = S^1$.

Thus the theorem is proved.□

More generally, consider a two-dimensional vector field given by

$$v(x,\lambda) = \begin{pmatrix} \beta_1(\lambda) & 0 \\ 0 & \beta_2(\lambda) \end{pmatrix} \begin{pmatrix} x_1 \\ x_2 \end{pmatrix} + G(x,\lambda), \qquad G(x,\lambda) = o(|x|), \tag{2.2.23}$$

where

$$\beta_i(\lambda) \begin{cases} < 0 & \text{if } \lambda < \lambda_0, \\ = 0 & \text{if } \lambda = \lambda_0, \\ > 0 & \text{if } \lambda > \lambda_0 \end{cases} \qquad \text{for } i = 1,2. \tag{2.2.24}$$

In polar coordinates (r,θ), the projection v_r of v in the r-direction is expressed as

$$v_r(r,\theta,\lambda) = (v(x,\lambda),\hat{x}) = r(\beta_1(\lambda)\cos^2\theta + \beta_2(\lambda)\sin^2\theta) - g(r,\theta,\lambda),$$

where $\hat{x} = (\cos\theta, \sin\theta)$. Assume that for some $\delta > 0$,

$$(G(x,\lambda),x) = g(x,\lambda) > 0 \quad \forall 0 < |x| < \delta. \tag{2.2.25}$$

If for every $\lambda - \lambda_0 > 0$ sufficiently small and $0 \le \theta \le 2\pi$ there exists a unique $r_\theta > 0$ such that

$$\begin{aligned} & r_\theta(\beta_1\cos^2\theta + \beta_2\sin^2\theta) - g(r_\theta,\theta,\lambda) = 0, \quad 0 \le \theta \le 2\pi, \\ & \frac{\partial g(r_\theta,\theta,\lambda)}{\partial r} > \beta_1(\lambda)\cos^2\theta + \beta_2(\lambda)\sin^2\theta, \end{aligned} \tag{2.2.26}$$

then the vector field $v(x,\lambda)$ given by (2.2.23) bifurcates from $(x,\lambda) = (0,\lambda_0)$ on $\lambda > \lambda_0$ to a circle attractor $\Sigma_\lambda = S^1$.

In fact, by Theorem 2.2.2, it follows from (2.2.24) and (2.2.25) that $v(x,\lambda)$ bifurcates from $(0,\lambda_0)$ to an attractor Σ_λ that is homological to a circle S^1, i.e., Σ_λ may be one of the cases as shown in Fig. 2.7a–c.

Intuitively, the conditions (2.2.24)–(2.2.26) exclude the cases shown in Fig. 2.7b, c.

To see this, we know that the function $r_\theta = r(\theta,\lambda)$ satisfying (2.2.26) represents a circle S^1, i.e., the set

$$\Sigma_\lambda = \{(r(\theta,\lambda),\theta) \in \mathbb{R}^2 | \, 0 \le \theta \le 2\pi\}$$

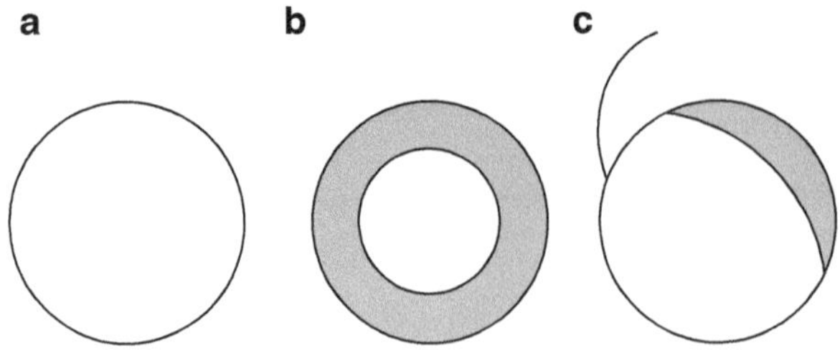

Fig. 2.7 (**a**) A circle S^1, (**b**) an annulus, and (**c**) a set having the homotopical type of S^1

is homeomorphic to S^1, and all singular points of v in Σ_λ are also in Σ_λ, which implies, by (2.2.26), that for every singular point $x_0 \in \Sigma_\lambda$ of v, the Jacobian matrix $Dv(x_0)$ has at least an eigenvalue λ such that $\text{Re}\lambda < 0$. It follows that the unstable manifold of a singular point of v in Σ_λ is unique. Thus, $\Sigma_\lambda = S^1$.

2.2.3 S^m-Attractor Bifurcation

We now consider the attractor bifurcation of an m-dimensional sphere S^m for the vector fields $v \in C^r(\Omega, \mathbb{R}^{m+1})$ given by (2.2.5). Let

$$v(x) = Ax + G(x) \qquad \forall x \in \Omega \subset \mathbb{R}^{m+1},$$

where A is an $(m+1) \times (m+1)$ matrix, and $G(x) = o(|x|)$. The following is the S^m attractor bifurcation theorem; see (Ma and Wang 2007b):

Theorem 2.2.5 *Let $v_\lambda \in C^r(\Omega, \mathbb{R}^{m+1})$ satisfy (2.2.5) and (2.2.6). Then v bifurcates from $(x,\lambda) = (0,0)$ to an attractor Σ_λ on $\lambda > 0$ that is homeomorphic to an m-dimensional sphere S^m.*

2.2.4 Structural Stability of Dynamic Transitions

In this subsection, we introduce the stability of dynamic bifurcation for the parameterized vector field $v(x,\lambda)$ that is not only the stability of bifurcated branches, but also the local structural stability of bifurcated solutions. We shall see later that the stability of dynamic bifurcation is important in the transition theory of the perturbed systems, which is related to phase transition stability in science and engineering.

Let $\Omega \subset \mathbb{R}^n$ be an open set, $x = 0 \in \Omega$, and $\lambda \in I = (a,b) \subset \mathbb{R}^1$. We denote the space of parameterized vector fields by

$$C_0^{k,1}(\Omega \times I, \mathbb{R}^n) = \{v : \Omega \times I \to \mathbb{R}^n | \, v(0,\lambda) = 0, \quad \forall \lambda \in I\}$$

endowed with the norm

$$\|u\|_{C^{k,1}} = \sup_{(x,\lambda)\in\Omega\times I} \left[\sum_{p=0}^{k} |D_x^p v| + |D_\lambda v| + |D_{\lambda x}^2 v| \right].$$

Obviously, if $v \in C_0^{k,1}(\Omega \times I, \mathbb{R}^n)(k \geq 1)$, then v is k-times differentiable at $x \in \Omega$ and differentiable at $\lambda \in I$, and v can be expressed as

$$v(\cdot,\lambda) = A_\lambda + G(\cdot,\lambda), \tag{2.2.27}$$

where A_λ and $G(\cdot,\lambda)$ are the same as in (2.2.1) and (2.2.2).

It is known that the simple real and complex eigenvalues of A_λ are differentiable on λ (Kato 1995), which can be expressed by

$$\beta_i(\lambda) = \alpha_i + \sigma_i\lambda + o(|\lambda|) \qquad \forall 1 \leq i \leq n.$$

In the following, we introduce the definitions of stability of dynamic bifurcation.

Definition 2.2.6 *Let $v_1, v_2 \in C_0^{k,1}(\Omega \times I, \mathbb{R}^n)$, and let $\lambda_i \in I$ $(i = 1,2)$ be a bifurcation point of the invariant set $\Gamma_i(\rho)$ of $v_i(x,\rho)$. We say that both bifurcation points λ_1 and λ_2 have the same structure if v_1 and v_2 are locally topologically equivalent at $\Gamma_1(\rho_1)$ and $\Gamma_2(\rho_2)$ with $\rho_1 - \lambda_1 = \rho_2 - \lambda_2$, i.e., there are neighborhoods $U_i \subset \mathbb{R}^n$ of $\Gamma_i(\rho)$ and a homeomorphism of $\varphi : U_1 \to U_2$ such that φ takes the orbits of v_1 in U_1 to orbits of v_2 in U_2 preserving orientation.*

Definition 2.2.7 *Let $v \in C_0^{k,1}(\Omega \times I, \mathbb{R}^n)$, $\lambda_0 \in I$ be a bifurcation point of $v(x,\lambda)$. We say that λ_0 is stable in $C_0^{k,1}(\Omega \times I, \mathbb{R}^n)$ if for every $\varepsilon > 0$ sufficiently small there is a neighborhood $O \subset C_0^{k,1}(\Omega \times I, \mathbb{R}^n)$ of v such that for every $u \in O$, u has a bifurcation point λ_1 with $|\lambda_1 - \lambda_0| < \varepsilon$, and λ_1 and λ_0 have the same structure. If all bifurcation points of v are stable, we say that v is stable for bifurcation in $C_0^{k,1}(\Omega \times I, \mathbb{R}^n)$.*

Let $v \in C_0^{k,1}(\Omega \times I, \mathbb{R}^n)(k \geq 1)$. The point $\lambda_0 \in I$ is called an eigenparameter with multiplicity m if the eigenvalues $\beta_i(\lambda)$ of $A_\lambda = Dv(0,\lambda)$ satisfy

$$\begin{aligned} &\mathrm{Re}\beta_i(\lambda_0) = 0 && \text{if } 1 \leq i \leq m, \\ &\mathrm{Re}\beta_j(\lambda_0) \neq 0 && \text{if } m+1 \leq j \leq n. \end{aligned} \tag{2.2.28}$$

The eigenparameter $\lambda_0 \in I$ is simple if $m = 1$ for $\beta_1(\lambda_0) = 0$ and $m = 2$ for $\beta_1(\lambda_0) = \overline{\beta}_2(\lambda_0) = i\beta$ $(\beta \neq 0)$. If $\lambda_0 \in I$ is simple and the eigenvalues in (2.2.28) satisfy

$$\frac{d}{d\lambda}\mathrm{Re}\beta_i(\lambda_0) \neq 0 \qquad \forall 1 \leq i \leq m, \tag{2.2.29}$$

then λ_0 is called a regular eigenparameter.

The following two theorems provide sufficient and necessary conditions for structural stability of dynamic bifurcation of parameterized vector fields in $C_0^{3,1}(\Omega \times I, \mathbb{R}^n)$; they are proved in Ma and Wang (2005b).

Theorem 2.2.8 *Let $v \in C_0^{3,1}(\Omega \times I, \mathbb{R}^n)$ and $\lambda_0 \in I$ be a simple eigenparameter of v. Then there exists a number $b(\lambda_0)$, called the bifurcation number of v at λ_0, depending continuously on $\|v\|_{C^{3,1}}$ such that the following assertions hold:*

(1) λ_0 is a stable bifurcation point of v if and only if λ_0 is regular and the bifurcation number satisfies $b(\lambda_0) \neq 0$.
(2) If λ_0 has multiplicity $m = 1$ and $b(\lambda_0) \neq 0$, then $v(x,\lambda)$ bifurcates to a unique branch of singular points $x(\lambda)(\lambda \neq \lambda_0)$ on each side of $\lambda = \lambda_0$ that are hyperbolic, and $x(\lambda)$ has Morse index $k+1$ for $\beta_1(\lambda) < 0$ and Morse index k for $\beta_1(\lambda) > 0$, where k equals the number of eigenvalues with $Re\beta_j(\lambda_0) > 0$.
(3) If λ_0 has multiplicity $m = 2$, then $v(x,\lambda)$ bifurcates to a unique branch of periodic orbits Γ_λ for $Re\beta_1(\lambda) < 0$ as $b(\lambda_0) > 0$ and for $Re\beta_1(\lambda) > 0$ as $b(\lambda_0) < 0$, which are hyperbolic.

Theorem 2.2.9 *A parameterized vector field $v \in C_0^{3,1}(\Omega \times I, \mathbb{R}^n)$ is structurally stable for bifurcation if and only if*

(1) all eigenparameters of v are simple and regular,
(2) each bifurcation number of v is nonzero.

Moreover, the set of all vector fields with stable bifurcation is open and dense in $C_0^{3,1}(\Omega \times I, \mathbb{R}^n)$.

Remark 2.2.10 We note that in $C_0^{1,1}(\Omega \times I, \mathbb{R}^n)$ there are no vector fields that are stable for bifurcation except the vector fields that have no eigenparameters in I. In $C_0^{2,1}(\Omega \times I, \mathbb{R}^n)$, the set of all vector fields with stable bifurcation comprises those whose eigenparameters are regular with multiplicity $m = 1$.

2.2.5 Infinite-Dimensional Systems

For infinite-dimensional systems, consider (2.1.1). Let the eigenvalues of L_λ be given by $\{\beta_j(\lambda) | \ j = 1,2,\cdots\} \subset \mathbb{C}$. Assume that the conditions (2.1.4) and (2.1.5) hold. Let E_0 be the eigenspace of L_λ at $\lambda = \lambda_0$; then dim $E_0 = m$.

The following theorem generalizes Theorems 2.2.2 and 2.2.3 to the infinite-dimensional system (2.1.1). It provides sufficient conditions for continuous transitions and gives a local transition structure.

Theorem 2.2.11 *Under the conditions (2.1.4) and (2.1.5), if $u = 0$ is a locally asymptotically stable equilibrium point of (2.1.1), then we have the following assertions:*

(1) Equation (2.1.1) bifurcates from $(u,\lambda) = (0,\lambda_0)$ on $\lambda > \lambda_0$ to an attractor $\mathscr{A}_\lambda$ with $m - 1 \leq dim \mathscr{A}_\lambda \leq m$, i.e., (2.1.1) has a continuous transition at $\lambda = \lambda_0$, and $\mathscr{A}_\lambda$ is connected if $m \geq 2$.
(2) There exists a neighborhood $U \subset X$ such that $\mathscr{A}_\lambda$ attracts $U \setminus \Gamma$, where Γ is the stable manifold of $u = 0$ with codim $\Gamma = m$.

(3) *$\mathscr{A}_\lambda$ is an $(m-1)$-dimensional homological sphere; in particular, if $\mathscr{A}_\lambda$ is a finite simplicial complex, then $\mathscr{A}_\lambda$ has the homotopy type of the $(m-1)$-dimensional sphere S^{m-1}.*
(4) *For every $u_\lambda \in \mathscr{A}_\lambda$, u_λ can be expressed as*

$$u_\lambda = v_\lambda + o(\|v_\lambda\|), \quad v_\lambda \in E_0.$$

(5) *If the number of singular points of (2.1.1) in $\mathscr{A}_\lambda$ is finite, then we have the index formula*

$$\sum_{u_i \in \Sigma_\lambda} ind(-(L_\lambda + G), u_i) = \begin{cases} 2 & \text{if } m = odd, \\ 0 & \text{if } m = even. \end{cases}$$

(6) *If $u = 0$ is globally asymptotically stable for (2.1.1) at $\lambda = \lambda_0$, then for every bounded open set $U \subset H$, there is an $\varepsilon > 0$ such that when $\lambda_0 < \lambda < \lambda_0 + \varepsilon$, the attractor $\mathscr{A}_\lambda$ attracts $U \setminus \Gamma$ in H. In particular, if (2.1.1) has a global attractor for all λ near λ_0, then $\mathscr{A}_\lambda$ attracts $H \setminus \Gamma$.*

Proof. By the reduction of (2.1.1) on the center manifold, the bifurcation equation of (2.1.1) is given by (1.2.7). By (2.1.3) and the fact that the center manifold function is of high order, i.e., $\Phi(x,\lambda) = o(\|x\|)$, we have

$$PG(x, \Phi(x,\lambda), \lambda) = o(\|x\|).$$

Since $u = 0$ is asymptotically stable for (2.1.1), $x = 0$ is a local asymptotically stable singular point of (1.2.7) at $\lambda = \lambda_0$. Hence, assertions (1)–(5) follow from Theorem 2.2.2, and assertion (6) can be deduced by Theorem 5.4 in Ma and Wang (2005b). The proof is complete. □

When the multiplicity in (2.1.4) and (2.1.5) is $m = 1$, then we have the pitchfork attractor bifurcation theorem for (2.1.1) as follows.

Theorem 2.2.12 *Under the assumptions of Theorem 2.2.11, if $m = 1$ in (2.1.4) and $G(u,\lambda)$ is analytic at $u = 0$, then there exists an open set $U \subset X$ with $0 \in U$ such that (2.1.1) bifurcates from $(u,\lambda) = (0,\lambda_0)$ on $\lambda > \lambda_0$ to exactly two singular points $u_1(\lambda)$ and $u_2(\lambda) \in U$, and U can be decomposed into two open sets U_1^λ and U_2^λ satisfying the following properties:*

(1) *$\overline{U} = \overline{U}_1^\lambda + \overline{U}_2^\lambda, U_1^\lambda \cap U_2^\lambda = \emptyset, \Gamma = \partial U_1^\lambda \cap \partial U_2^\lambda$ is the stable manifold of $u = 0$, and $u_i(\lambda) \in U_i^\lambda$ $(i = 1,2)$.*
(2) *Let $u(t,\varphi)$ be the solution of (2.1.1). Then*

$$\lim_{n\to\infty} \|u(t,\varphi) - u_i(\lambda)\| = 0 \qquad \forall \varphi \in U_i^\lambda \quad (i = 1,2).$$

Proof. Assertions (1) and (2) can be deduced from Theorem 5.5 in Ma and Wang (2005b), and the conclusion that (2.1.1) bifurcates to exactly two singular points can be directly derived from the Krasnosel'skii theorem; see Theorem 1.11 in Ma and Wang (2005b). The proof is complete. □

2.3 Transition from Simple Eigenvalues

2.3.1 *Real Simple Eigenvalues*

In this section, we consider the transition of (2.1.1) from a simple critical eigenvalue. Let the eigenvalues $\beta_j(\lambda)$ of L_λ satisfy (2.1.4) and (2.1.5) with $m = 1$. In this case, $\beta_1(\lambda)$ is a real eigenvalue. Let $e_1(\lambda)$ and $e_1^*(\lambda)$ be the eigenvectors of L_λ and L_λ^* respectively corresponding to $\beta_1(\lambda)$ with

$$L_{\lambda_0}e_1 = 0, \quad L_{\lambda_0}^* e_1^* = 0, \quad (e_1, e_1^*) = 1.$$

Let $\Phi(x,\lambda)$ be the center manifold function of (2.1.1) near $\lambda = \lambda_0$, and assume that

$$(G(xe_1 + \Phi(x,\lambda_0),\lambda_0), e_1^*) = \alpha x^k + o(|x|^k), \tag{2.3.1}$$

where $k \geq 2$ is an integer and $\alpha \neq 0$ a real number.

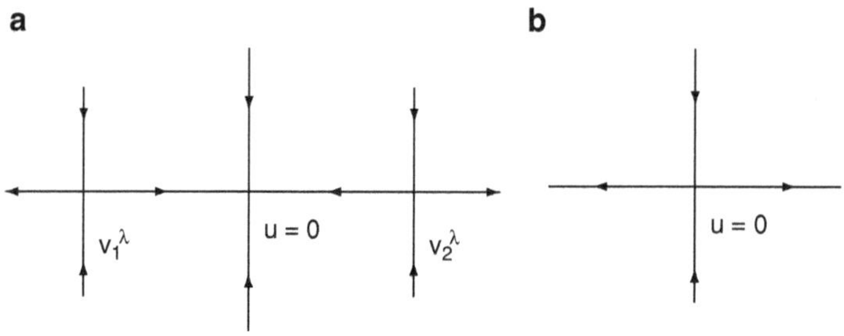

Fig. 2.8 Topological structure of the jump transition of (2.1.1) when k is odd and $\alpha > 0$: (**a**) $\lambda < \lambda_0$; (**b**) $\lambda \geq \lambda_0$. Here the *horizontal line* represents the center manifold

Theorem 2.3.1 *Let (2.1.4) and (2.1.5) with $m = 1$ and (2.3.1) hold. If k is odd and $\alpha \neq 0$ is as in (2.3.1), then the following assertions hold:*

(1) If $\alpha > 0$, then (2.1.1) has a jump transition from $(0,\lambda_0)$ and bifurcates on $\lambda < \lambda_0$ to exactly two saddle points v_1^λ and v_2^λ with Morse index one, as shown in Fig. 2.8.

(2) If $\alpha < 0$, then (2.1.1) has a continuous transition from $(0,\lambda_0)$ that is an attractor bifurcation as described in Theorem 2.2.12 and as shown in Fig. 2.9.

(3) The bifurcated solutions v_1^λ and v_2^λ can be expressed as

$$v_{1,2}^\lambda = \pm|\beta_1(\lambda)/\alpha|^{1/(k-1)}e_1(\lambda) + o(|\beta_1|^{1/(k-1)}).$$

Theorem 2.3.2 *Let (2.1.4) and (2.1.5) with $m = 1$ and (2.3.1) hold. If k is even and $\alpha \neq 0$, then we have the following assertions:*

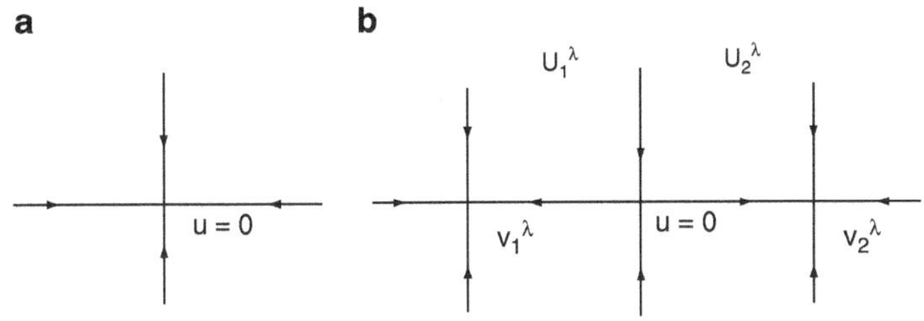

Fig. 2.9 Topological structure of the continuous transition of (2.1.1) when k is odd and $\alpha < 0$: **(a)** $\lambda \leq \lambda_0$; **(b)** $\lambda > \lambda_0$

1. *Equation (2.1.1) has a mixed transition from* $(0, \lambda_0)$. *More precisely, there exists a neighborhood* $U \subset X$ *of* $u = 0$ *such that* U *is separated into two disjoint open sets* U_1^λ *and* U_2^λ *by the stable manifold* Γ_λ *of* $u = 0$ *such that the local transition structure is as shown in Fig. 2.10 and the following hold:*

   a. $U = U_1^\lambda + U_2^\lambda + \Gamma_\lambda$;
   b. *the transition in* U_1^λ *is a jump transition;*
   c. *the transition in* U_2^λ *is continuous.*

2. *Equation (2.1.1) bifurcates in* U_2^λ *to a unique singular point* v^λ *on* $\lambda > \lambda_0$ *that is an attractor such that for every* $\varphi \in U_2^\lambda$,

$$\lim_{t \to \infty} \|u(t, \varphi) - v^\lambda\| = 0.$$

3. *Equation (2.1.1) bifurcates on* $\lambda < \lambda_0$ *to a unique saddle point* v^λ *with Morse index one.*
4. *The bifurcated singular point* v^λ *can be expressed as*

$$v^\lambda = -(\beta_1(\lambda)/\alpha)^{1/(k-1)} e_1 + o(|\beta_1|^{1/(k-1)}).$$

Remark 2.3.3 Theorem 2.3.2 shows that for the mixed transition from real simple eigenvalues, there is a bifurcated attractor in the stable domain, which is a singular point. However, we shall see later that in general, this property does not hold. In other words, a mixed transition may have no bifurcated attractor in its stable domain.

Remark 2.3.4 The local topological structure of the transitions of (2.1.1) from simple eigenvalues is schematically shown in the center manifold in Figs. 2.8 and Remark 2.4.18.

Proof of Theorems 2.3.1 and 2.3.2. Under the assumption, the reduction equation of (2.1.1) on the center manifold is given by

$$\frac{dx}{dt} = \beta_1(\lambda)x + (G(xe_1(\lambda) + \Phi(x, \lambda)), e_1^*(\lambda)). \tag{2.3.2}$$

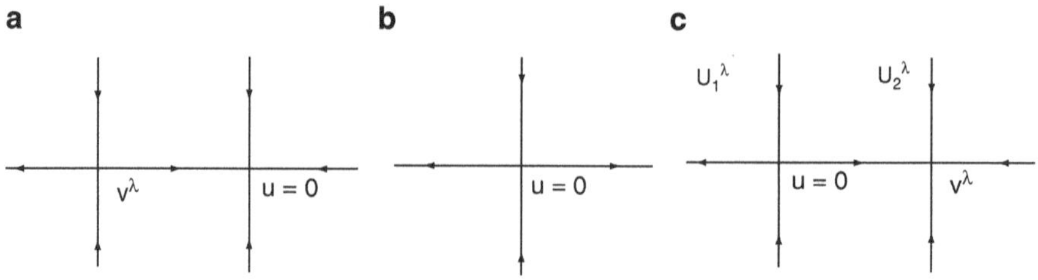

Fig. 2.10 Topological structure of mixed transition of (2.1.1) when k is even and $\alpha \neq 0$: **(a)** $\lambda < \lambda_0$; **(b)** $\lambda = \lambda_0$; **(c)** $\lambda > \lambda_0, U_1^\lambda$ is the unstable domain, and U_2^λ the stable domain

By (2.3.1), (2.3.2) is rewritten as

$$\frac{dx}{dt} = \beta_1(\lambda)x + \alpha_\lambda x^k + o(|x|^k) \qquad \text{for } x \in \mathbb{R}^1, \tag{2.3.3}$$

where $\lim_{\lambda \to \lambda_0} \alpha_\lambda = \alpha \neq 0$.

It is known that the transition of (2.1.1) and its local topological structure are determined completely by (2.3.3). Obviously, when k is odd and $\alpha > 0$, the solutions of (2.3.3) for $\lambda > \lambda_0$ ($\beta_1(\lambda) > 0$) satisfy

$$\limsup_{t \to \infty} |x(t, x_0)| > \delta > 0 \quad \forall 0 < |x_0| < \varepsilon, \tag{2.3.4}$$

where $\delta > 0$ is a constant, and $x(0, x_0) = x_0$. From (2.3.4) we know that the transition of (2.3.3) at $\lambda = \lambda_0$ is a jump transition, which implies that the transition of (2.1.1) from $(0, \lambda_0)$ is a jump transition.

It is clear that (2.3.3) has exactly two bifurcated solutions on $\lambda < \lambda_0$ ($\beta_1(\lambda) < 0$) as follows:

$$x_\pm = \pm|\beta_1(\lambda)/\alpha_\lambda|^{1/(k-1)} + o(|\beta_1(\lambda)|^{1/(k-1)}).$$

Therefore,

$$v_{1,2}^\lambda = x_\pm e_1 + \Phi(x_\pm, \lambda) \tag{2.3.5}$$

is the set of bifurcated singular points of (2.1.1) on $\lambda < \lambda_0$. By Theorem 3.10 in Ma and Wang (2005b), the Morse index of v_i^λ ($i = 1, 2$) is one. Thus assertion (1) of Theorem 2.3.1 is proved, and the expression (2.3.5) is as in assertion (3).

Other conclusions and Theorem 2.3.2 can be derived from (2.3.3) in the same fashion, and we omit the details. The proof is complete.□

We remark that Theorems 2.3.1 and 2.3.2 are very convenient for applications. However, a complete theory of transitions from real simple eigenvalues is given by the following theorem, together with Theorems 2.3.1 and 2.3.2.

Let $U \subset X$ be a neighborhood of $u = 0$. An open set $D \subset U$ is called a connected unstable domain if for every $\varphi \in D$, the solution $u_\lambda(t, \varphi)$ of (2.1.1) satisfies

$$\limsup_{t \to \infty} \|u_\lambda(t, \varphi)\| \geq \delta > 0 \quad \forall \lambda > \lambda_0,$$

and ∂D is an invariant set of (2.1.1) with

$$\lim_{\lambda \to \lambda_0} \lim_{t \to \infty} \|u_\lambda(t, \varphi)\| = 0 \quad \forall \varphi \in \partial D.$$

Theorem 2.3.5 *Let (2.1.4) and (2.1.5) with $m = 1$ hold. The following assertions hold for (2.1.1):*

(1) The transition is continuous if and only if the bifurcation of singular points is on the side of $\lambda > \lambda_0$ and this transition is an attractor bifurcation.
(2) The transition is mixed if and only if there exists a bifurcation of singular points on each side of $\lambda = \lambda_0$, there are exactly one connected unstable domain and one connected stable domain, and (2.1.1) bifurcates on $\lambda > \lambda_0$ to an attractor that attracts the stable domain.
(3) The transition is a jump transition if and only if there is no bifurcation of singular points on the side of $\lambda > \lambda_0$, i.e., the bifurcation can occur only on $\lambda = \lambda_0$ or $\lambda < \lambda_0$, and there are exactly two connected unstable domains.

Theorem 2.3.5 can be directly derived from the reduction equation (2.3.2) of (2.1.1) on the center manifold, and we omit the details here.

2.3.2 Transitions from Complex Simple Eigenvalues

Dynamic Transitions. We now study transitions from a pair of complex eigenvalues. Assume that the eigenvalues of L_λ satisfy

$$\begin{cases} \mathrm{Re}\beta_1(\lambda) = \mathrm{Re}\beta_2(\lambda) \begin{cases} < 0 & \text{if } \lambda < \lambda_0, \\ = 0 & \text{if } \lambda = \lambda_0, \\ > 0 & \text{if } \lambda > \lambda_0, \end{cases} \\ \mathrm{Im}\beta_1(\lambda_0) = -\mathrm{Im}\beta_2(\lambda_0) \neq 0, \end{cases} \tag{2.3.6}$$

$$\mathrm{Re}\beta_j(\lambda_0) < 0 \qquad \forall j \geq 3. \tag{2.3.7}$$

It is known that with the conditions (2.3.6) and (2.3.7), (2.1.1) undergoes a Hopf bifurcation from $(0, \lambda_0)$. The following theorem amounts to saying that the transition of (2.1.1) from $(0, \lambda_0)$ can be one of only two types: continuous and jump.

Theorem 2.3.6 *Under the conditions (2.3.6) and (2.3.7), the transition of (2.1.1) from $(0, \lambda_0)$ is of type either continuous or jump, and the following assertions hold:*

(1) The transition is continuous if and only if (2.1.1) bifurcates to a periodic orbit on $\lambda > \lambda_0$, and this transition is an attractor bifurcation.
(2) The transition is a jump transition if and only if (2.1.1) has no bifurcation to periodic orbits on $\lambda > \lambda_0$, and there is only one connected metastable domain.

Proof. Without loss of generality, let $\mathrm{Im}\beta_1(\lambda) = -\mathrm{Im}\beta_2(\lambda) = 1$. Thus, the reduction equation of (2.1.1) can be rewritten as

$$\frac{dx}{dt} = \begin{pmatrix} \alpha(\lambda) & -1 \\ 1 & \alpha(\lambda) \end{pmatrix} \begin{pmatrix} x_1 \\ x_2 \end{pmatrix} + \begin{pmatrix} g_1(x,\lambda) \\ g_2(x,\lambda) \end{pmatrix}, \tag{2.3.8}$$

where $x = (x_1, x_2)^t \in \mathbb{R}^2$, $g_i(x,\lambda) = o(|x|)$ $(i = 1,2)$, and $\alpha(\lambda) = \text{Re}\beta_1(\lambda)$ satisfies

$$\alpha(\lambda) \begin{cases} < 0 & \text{if } \lambda < \lambda_0, \\ = 0 & \text{if } \lambda = \lambda_0, \\ > 0 & \text{if } \lambda > \lambda_0. \end{cases} \tag{2.3.9}$$

At the critical value $\lambda = \lambda_0$, (2.3.8) is in the form

$$\frac{dx}{dt} = \begin{pmatrix} 0 & -1 \\ 1 & 0 \end{pmatrix} \begin{pmatrix} x_1 \\ x_2 \end{pmatrix} + \begin{pmatrix} g_1(x,\lambda_0) \\ g_2(x,\lambda_0) \end{pmatrix}. \tag{2.3.10}$$

By dynamical system theory, it is known that the singular point $x = 0$ can be only one of the three kinds of singular points of (2.3.10): (a) a stable focus, (b) an unstable focus, (c) a center with an infinite number of periodic orbits in a neighborhood $U \subset \mathbb{R}^2$ of $x = 0$.

For case (a), $x = 0$ is locally asymptotically stable. Hence, by Theorem 2.2.2, the transition is an attractor bifurcation in $\lambda > \lambda_0$.

For cases (b) and (c), the solution $x(t,x_0)$ of (2.3.10) satisfies that for every $0 < |x_0| < \varepsilon$, there is a $\delta_0 > 0$ such that

$$\limsup_{t\to\infty} |x(t,x_0)| \geq \delta_0 > 0, \tag{2.3.11}$$

where $x(0,x_0) = x_0$, and $\varepsilon > 0$ is a constant. By assertion (2) of Theorem 2.1.5, we infer from (2.3.11) that the transition is a jump transition, and (2.1.1) has no bifurcated periodic orbits on $\lambda > \lambda_0$.

Conversely if the transition is an attractor bifurcation, then by assertion (1) of Theorem 2.1.5, $x = 0$ is a stable focus of (2.3.10). □

Transition Types. We give a criterion to determine the transition types of (2.1.1) from complex simple eigenvalues.

Let $e_1(\lambda)$, $e_2(\lambda)$ and $e_1^*(\lambda)$, $e_2^*(\lambda)$ be the eigenvectors of L_λ and L_λ^* respectively corresponding to the complex eigenvalues $\beta_2(\lambda) = \overline{\beta}_1(\lambda) = \alpha(\lambda) + i\sigma(\lambda)$, where α satisfies (2.3.9), $\sigma_0 = \sigma(\lambda_0) \neq 0$, and

$$\begin{cases} L_\lambda e_1(\lambda) = \alpha(\lambda)e_1(\lambda) + \sigma(\lambda)e_2(\lambda), \\ L_\lambda e_2(\lambda) = -\sigma(\lambda)e_1(\lambda) + \alpha(\lambda)e_2(\lambda), \end{cases} \tag{2.3.12}$$

$$\begin{cases} L_\lambda^* e_1^*(\lambda) = \alpha(\lambda)e_1^* - \sigma(\lambda)e_2^*, \\ L_\lambda^* e_2^*(\lambda) = \sigma(\lambda)e_1^* + \alpha(\lambda)e_2^*. \end{cases} \tag{2.3.13}$$

By the spectral theorem, Theorem 3.4 in Ma and Wang (2005b), we can take

$$(e_i(\lambda), e_j^*(\lambda)) = \delta_{ij} \quad \forall 1 \leq i,j \leq 2.$$

Let $\Phi(x,\lambda)$ be the center manifold function of (2.1.1) near $\lambda=\lambda_0$, $x=x_1e_1+x_2e_2$, and $e_i=e_i(\lambda_0)(i=1,2)$. Assume that for $i=1,2$,

$$(G(x+\Phi(x,\lambda_0),\lambda_0),e_i^*)=\sum_{2\leq p+q\leq 3} a^i_{pq}x_1^px_2^q+o(|x|^3). \tag{2.3.14}$$

For (2.3.14) we introduce a number, which is the bifurcation number in Theorem 2.2.8, given by

$$\begin{aligned} b=&\frac{3\pi}{4}(a^1_{30}+a^2_{03})+\frac{\pi}{4}(a^1_{12}+a^2_{21})+\frac{\pi}{2\sigma_0}(a^1_{02}a^2_{02}-a^1_{20}a^2_{20})\\ &+\frac{\pi}{4\sigma_0}(a^1_{11}a^1_{20}+a^1_{11}a^1_{02}-a^2_{11}a^2_{20}-a^2_{11}a^2_{02}). \end{aligned} \tag{2.3.15}$$

The following theorem is essentially the generalization of a theorem of (Andronov et al. 1973) in the finite-dimensional case to the infinite-dimensional case here.

Theorem 2.3.7 *Let the conditions (2.3.6) and (2.3.7) hold.*

1. *If $b<0$, then the transition of (2.1.1) is continuous, and the bifurcated periodic orbit is an attractor. Furthermore, the bifurcated periodic solution can be expressed as*

$$\begin{aligned} u&=x_1e_1+x_2e_2+o(|\alpha(\lambda)|),\\ x_1(t)&=\left(\frac{\alpha(\lambda)}{|b|}\right)^{1/2}\cos\sigma t+o(|\alpha(\lambda)|),\\ x_2(t)&=\left(\frac{\alpha(\lambda)}{|b|}\right)^{1/2}\sin\sigma t+o(|\alpha(\lambda)|), \end{aligned}$$

   where $\alpha(\lambda)$ and $\sigma=\sigma(\lambda)$ are as in (2.3.12).
2. *If $b>0$, then the transition is a jump transition, and (2.1.1) bifurcates on $\lambda<\lambda_0$ to a unique unstable periodic orbit.*

Proof. The reduced equation of (2.1.1) on the center manifold is given by

$$\begin{aligned} \frac{dx_1}{dt}&=\alpha(\lambda)x_1-\sigma(\lambda)x_2+(G(x+\Phi(x,\lambda),\lambda),e_1^*(\lambda)),\\ \frac{dx_2}{dt}&=\sigma(\lambda)x_1+\alpha(\lambda)x_2+(G(x+\Phi(x,\lambda),\lambda),e_2^*(\lambda)), \end{aligned} \tag{2.3.16}$$

where $\alpha(\lambda)$ satisfies (2.3.9). Therefore, by (2.3.14) equations (2.3.16) at $\lambda=\lambda_0$ become

$$\begin{aligned} \frac{dx_1}{dt}&=-\sigma x_2+g_{21}(x)+g_{31}(x)+o(|x|^3),\\ \frac{dx_2}{dt}&=\sigma x_1+g_{22}(x)+g_{32}(x)+o(|x|^3), \end{aligned} \tag{2.3.17}$$

where $\sigma = \sigma(\lambda_0) \neq 0$, and

$$g_{ki}(x) = \sum_{p+q=k} a^i_{pq} x_1^p x_2^q \qquad \text{for } k = 2,3, \quad i = 1,2.$$

Using polar coordinates $(x_1, x_2) = (r\cos\theta, r\sin\theta)$, we obtain from (2.3.17) that

$$\begin{aligned}\frac{dr}{d\theta} &= \frac{r\cos\theta\frac{dx_1}{dt} + r\sin\theta\frac{dx_2}{dt}}{\cos\theta\frac{dx_2}{dt} - \sin\theta\frac{dx_1}{dt}} \\ &= \frac{r^2[\cos\theta(g_{21}+rg_{31}) + \sin\theta(g_{22}+rg_{32})] + o(r^3)}{\sigma + r[\cos\theta(g_{22}+rg_{32}) - \sin\theta(g_{21}+rg_{31})] + o(r^2)},\end{aligned} \tag{2.3.18}$$

where $g_{ki} = \sum_{p+q=k} a^i_{pq}\cos^p\theta\sin^q\theta$ for $k = 2,3$ and $i = 1,2$.
Near $r = 0$, (2.3.18) can be rewritten as

$$\frac{1}{r^2}\frac{dr}{d\theta} = \frac{1}{\sigma}(u_1 + r(u_2 + \frac{1}{\sigma}u_1v_1)) + o(r), \tag{2.3.19}$$

where

$$\begin{aligned}u_1(\theta) &= \cos\theta \sum_{p+q=2} a^1_{pq}\cos^p\theta\sin^q\theta + \sin\theta \sum_{p+q=2} a^2_{pq}\cos^p\theta\sin^q\theta, \\ v_1(\theta) &= \sin\theta \sum_{p+q=2} a^1_{pq}\cos^p\theta\sin^q\theta - \cos\theta \sum_{p+q=2} a^2_{pq}\cos^p\theta\sin^q\theta, \\ u_2(\theta) &= \cos\theta \sum_{p+q=3} a^1_{pq}\cos^p\theta\sin^q\theta + \sin\theta \sum_{p+q=3} a^2_{pq}\cos^p\theta\sin^q\theta.\end{aligned}$$

Let $r(\theta, a)$ be the solution of (2.3.19) with the initial value $r(0,a) = a$. Integrating (2.3.19) on θ, we have

$$\frac{r(\theta,a) - a}{ar(\theta,a)} = \frac{1}{\sigma}\int_0^\theta [u_1 + r(u_2 + \frac{1}{\sigma}u_1v_1) + o(r)]d\theta,$$

which implies that $r(\theta, a)$ has the following Taylor expansion:

$$r(\theta,a) = a + \sum_{k=2} f_k(\theta)a^k = a + o(a^2). \tag{2.3.20}$$

It is easy to see that

$$\int_0^{\pm 2\pi} u_1(\theta)d\theta = 0. \tag{2.3.21}$$

We know that the orbits of (2.3.17) around $x = 0$ rotate clockwise if $\sigma < 0$, and counterclockwise if $\sigma > 0$. Hence, we deduce from (2.3.19)–(2.3.21) that if $\sigma > 0$, then

$$\frac{r(2\pi,a)-r(0,a)}{r(2\pi,a)}=\frac{1}{\sigma}a^2\int_0^{2\pi}(u_2+\frac{1}{\sigma}u_1v_1)d\theta+o(a^2),$$

and if $\sigma<0$, then

$$\begin{aligned}\frac{r(-2\pi,a)-r(0,a)}{r(-2\pi,a)}&=\frac{1}{\sigma}a^2\int_0^{-2\pi}(u_2+\frac{1}{\sigma}u_1v_1)d\theta+o(a^2)\\&=\frac{1}{|\sigma|}a^2\int_0^{2\pi}(u_2+\frac{1}{\sigma}u_1v_1)d\theta+o(a^2).\end{aligned}$$

Namely, for $\sigma\neq 0$, we obtain

$$\frac{r(\pm 2\pi,a)-r(0,a)}{r(\pm 2\pi,a)}=\frac{1}{|\sigma|}ba^2+o(a^2), \tag{2.3.22}$$

where

$$b=\int_0^{2\pi}(u_2(\theta)+\frac{1}{\sigma}u_1(\theta)v_1(\theta))d\theta. \tag{2.3.23}$$

From (2.3.22) we can see that when $b<0, x=0$ is a stable focus of (2.3.17), which implies that the transition of (2.3.16) at $\lambda=\lambda_0$ is continuous, and when $b>0, x=0$ is an unstable focus, and the transition is a jump transition. By direct calculation we can derive that the number b in (2.3.23) equals that given in (2.3.15).

Furthermore, we see that when $b<0$, the bifurcated periodic solution of (2.3.16) is given by

$$(x_1(t),x_2(t))=\left(ae^{\alpha(\lambda)}\cos\sigma t+o(|\alpha(\lambda)|),ae^{\alpha(\lambda)}\sin\sigma t+o(|\alpha(\lambda)|)\right),$$

where a is as in (2.3.22). It is known that $ba^2=-\alpha(\lambda)$.

The proof of the theorem is complete.□

We now consider the special case in which the nonlinear term $G(u,\lambda)$ of (2.1.1) has the following Taylor expansion:

$$G(u,\lambda)=G_k(u,\lambda)+G_{k+1}(u,\lambda)+o(\|u\|_{X_1}^{k+1}), \tag{2.3.24}$$

where $k\geq 3$ is an integer, and $G_r(u,\lambda)$ is the r-multilinear operator for $r=k,k+1$. Let e_1,e_2 and e_1^*,e_2^* be as in (2.3.14). Then for $i=1,2$, we have

$$(G_r(x_1e_1+x_2e_2,\lambda_0),e_i^*)=\sum_{p+q=r}a_{pq}^i x_1^p x_2^q.$$

For the coefficients a_{pq}^i, we define a new number by

$$b_0=\begin{cases}\sum\limits_{l=0}^{(k-1)/2}\alpha_{kl}(a^1_{(k-2l)2l}+a^2_{2l(k-2l)}) & \text{for } k=\text{odd},\\ \sum\limits_{l=0}^{k/2}\alpha_{(k+1)l}(a^1_{(k+1-2l)2l}+a^2_{2l(k+1-2l)}) & \text{for } k=\text{even},\end{cases} \tag{2.3.25}$$

where for $r = k, k+1$,

$$\alpha_{rl} = \int_0^{2\pi} \sin^{r+1-2l}\theta \cos^{2l}\theta d\theta.$$

Theorem 2.3.8 *Under the conditions (2.3.6), (2.3.7), and (2.3.24), the following assertions hold:*

1. *If $b_0 > 0$, the system (2.1.1) has a jump transition at $\lambda = \lambda_0$ and bifurcates on $\lambda < \lambda_0$ to a unique periodic orbit, which is a repeller.*
2. *If $b_0 < 0$, then (2.1.1) has a continuous transition and bifurcates on $\lambda > \lambda_0$ to a unique periodic orbit that is an attractor.*

Proof. Under the condition (2.3.24), the reduction equation (2.3.16) at λ_0 is of the following form:

$$\begin{aligned}\frac{dx_1}{dt} &= -\sigma x_2 + g_k^1(x) + g_{k+1}^1(x) + o(|x|^{k+1}),\\ \frac{\partial x_2}{\partial t} &= \sigma x_1 + g_k^2(x) + g_{k+1}^2(x) + o(|x|^{k+1}),\end{aligned} \tag{2.3.26}$$

where

$$g_r^i(x) = (G_r(x_1e_1 + x_2e_2, \lambda_0), e_i^*) = \sum_{p+q=r} a_{pq}^i x_1^p x_2^q.$$

Thus, in polar coordinates, (2.3.26) can be written as

$$\begin{aligned}\frac{1}{r^k}\frac{dr}{d\theta} &= \frac{\cos\theta g_k^1 + \sin\theta g_k^2 + r(\cos\theta g_{k+1}^1 + \sin\theta g_{k+1}^2) + o(r)}{\sigma - r^{k-1}(\sin\theta g_k^1 - \cos\theta g_k^2) + o(r^{k-1})}\\ &= \begin{cases}\dfrac{1}{\sigma}u_k(\theta) + o(r) & \text{if } k = \text{odd},\\ \dfrac{1}{\sigma}[u_k + ru_{k+1}] + o(r) & \text{if } k = \text{even } (> 2),\end{cases}\end{aligned} \tag{2.3.27}$$

where

$$u_r = \cos\theta \sum_{p+q=r} a_{pq}^1 \cos^p\theta \sin^q\theta + \sin\theta \sum_{p+q=r} a_{pq}^2 \cos^p\theta \sin^q\theta.$$

Note that if either p or q is odd, then

$$\int_0^{2\pi} \cos^p\theta \sin^q\theta d\theta = 0.$$

It follows then from (2.3.27) and (2.3.20) that

$$\frac{r^{k-1}(2\pi,a) - r^{k-1}(0,a)}{r^{k-1}(2\pi,a)} = \begin{cases}\dfrac{1}{\sigma}ba^{k-1} + o(a^{k-1}) & \text{if } k = \text{odd},\\ \dfrac{1}{\sigma}ba^k + o(a^k) & \text{if } k = \text{even},\end{cases} \tag{2.3.28}$$

where we consider only the case $\sigma > 0$, and b is given by

$$b = \begin{cases} \int_0^{2\pi} u_k(\theta)d\theta & \text{if } k = \text{odd}, \\ \int_0^{2\pi} u_{k+1}(\theta)d\theta & \text{if } k = \text{even}. \end{cases} \tag{2.3.29}$$

The number in (2.3.29) is the same as in (2.3.25). It is then routine to derive the theorem from (2.3.28). The proof is complete. □

2.3.3 Computation of b

Theorems 2.3.7 and 2.3.8 give a number b that dictates the type of transitions from a pair of complex eigenvalues. However, it is important to know how to compute this number in specific problems. In the following we give two examples to show the methods of computation.

We first consider a finite-dimensional system given by

$$\frac{dx}{dt} = Ax + G(x) + o(|x|^3) \qquad \text{for } x \in \mathbb{R}^n, \tag{2.3.30}$$

where A is an $n \times n$ matrix with eigenvalues

$$\begin{aligned} &\lambda_1 = \overline{\lambda}_2 = -i\alpha && (\alpha \neq 0), \\ &\lambda_j \neq 0 && \forall j = 2, \cdots, n. \end{aligned}$$

The eigenvectors $e_1, \cdots, e_n$ of A are given by

$$\begin{aligned} &Ae_1 = \alpha e_2, \qquad Ae_2 = -\alpha e_1, \\ &(A - \lambda_j I)^{k_j} e_j = 0 && \forall j = 3, \cdots, n, \quad k_j \geq 1, \end{aligned}$$

and the eigenvectors $e_1^*, \cdots, e_n^*$ of A^* by

$$\begin{aligned} &A^* e_1^* = -\alpha e_2^*, \qquad A^* e_2^* = \alpha e_1^*, \\ &(A^* - \lambda_j I)^{m_j} e_j^* = 0 && \forall j = 3, \cdots, n, \quad m_j \geq 1, \\ &(e_i, e_j^*) = \delta_{ij}. \end{aligned}$$

The nonlinear term $G(x) = (g_1(x), \cdots, g_n(x))$ is a polynomial field

$$g_i(x) = \sum_{2 \leq p_1 + \cdots + p_n \leq 3} C^i_{p_1 \cdots p_n} x_1^{p_1} \cdots x_n^{p_n}. \tag{2.3.31}$$

The system (2.3.30) can be written in the following form:

$$\begin{cases} \dfrac{dz_1}{dt} = -\alpha z_2 + (B^{-1}G(Bz), a_1 e_1^*) + o(|z|^3), \\ \dfrac{dz_2}{dt} = \alpha z_1 + (B^{-1}G(Bz), a_2 e_2^*) + o(|z|^3), \end{cases} \tag{2.3.32}$$

$$\frac{d\widetilde{z}}{dt} = J_{n-2}\widetilde{z} + \begin{pmatrix} (B^{-1}G(Bz), a_3 e_3^*) \\ \vdots \\ (B^{-1}G(Bz), a_n e_n^*) \end{pmatrix} + o(|z|^3), \tag{2.3.33}$$

where $z = (z_1, \cdots, z_n)^T, \widetilde{z} = (z_3, \cdots, z_n), J_{n-2}$ is the Jordan matrix associated with the eigenvalues $\lambda_3, \cdots, \lambda_n$, and

$$Bz = \begin{pmatrix} e_{11} & e_{21} & \cdots & e_{n1} \\ \vdots & \vdots & & \vdots \\ e_{1n} & e_{2n} & \cdots & e_{nn} \end{pmatrix} \begin{pmatrix} z_1 \\ \vdots \\ z_n \end{pmatrix},$$

with $e_i = (e_{i1}, \cdots, e_{in})$ $(1 \leq i \leq n)$ being the eigenvectors of A, and

$$a_k = \frac{1}{(e_k, e_k^*)} \qquad \forall 1 \leq k \leq n.$$

By the formulas for center manifold functions (Theorem A.1.1), the center manifold function $\Phi(z_1, z_2)$ of (2.3.30) is determined by (2.3.33), whose first approximation $\Phi_0(z_1, z_2)$ is given by

$$\widetilde{z}_0 = \Phi_0(z_1, z_2) = \Phi_0^1 + \Phi_0^2 + \Phi_0^3,$$

where

$$\begin{aligned} &(-J_{n-2})\Phi_0^1 = \Gamma_1, \\ &(J_{n-2}^2 + 4\alpha^2)(-J_{n-2})\Phi_0^2 = 2\alpha^2\Gamma_1 - 2\alpha^2\Gamma_2, \\ &(J_{n-2}^2 + 4\alpha^2)(-J_{n-2})\Phi_0^3 = \alpha\Gamma_3 + \alpha\Gamma_4, \end{aligned}$$

and

$$\Gamma_1 = \begin{pmatrix} a_3(B^{-1}G_2(z_1 e_1 + z_2 e_2), e_3^*) \\ \vdots \\ a_n(B^{-1}G_2(z_1 e_1 + z_2 e_2), e_n^*) \end{pmatrix},$$

$$\Gamma_2 = \begin{pmatrix} a_3(B^{-1}G_2(z_1e_2 - z_2e_1), e_3^*) \\ \vdots \\ a_n(B^{-1}G_2(z_1e_2 - z_2e_1), e_n^*) \end{pmatrix},$$

$$\Gamma_3 = \begin{pmatrix} a_3(B^{-1}G_2(z_1e_1 + z_2e_2, z_2e_1 - z_1e_2), e_3^*) \\ \vdots \\ a_n(B^{-1}G_2(z_1e_1 + z_2e_2, z_2e_1 - z_1e_2), e_n^*) \end{pmatrix},$$

$$\Gamma_4 = \begin{pmatrix} a_3(B^{-1}G_2(z_2e_1 - z_1e_2, z_1e_1 + z_2e_2), e_3^*) \\ \vdots \\ a_n(B^{-1}G_2(z_2e_1 - z_1e_2, z_1e_1 + z_2e_2), e_n^*) \end{pmatrix},$$

and G_2 is the bilinear term of G.

Inserting $z = z_1e_1 + z_2e_2 + \Phi_0(z_1, z_2)$ into (2.3.32), we obtain

$$(B^{-1}G(Bz), a_ie_i^*)|_{z=(z_1,z_2,\Phi_0)} = \sum_{2\leq k+l\leq 3} a_{kl}^i z_1^k z_2^l + o(|z_0|^3).$$

Thus, we derive the coefficients a_{kl}^1 and a_{kl}^2 $(2 \leq k+l \leq 3)$ that determine the number b defined in (2.3.15) for the system (2.3.30).

2.4 Transition from Eigenvalues with Multiplicity Two

2.4.1 Index Formula for Second-Order Nondegenerate Singularities

In order to investigate transitions of (2.1.1) from eigenvalues with multiplicity two, it is necessary to discuss the index of the following bilinear vector fields at $x = 0$:

$$F = \begin{pmatrix} a_{11}x_1^2 + a_{12}x_1x_2 + a_{22}x_2^2 \\ b_{11}x_1^2 + b_{12}x_1x_2 + b_{22}x_2^2 \end{pmatrix}. \tag{2.4.1}$$

We assume that the vector field (2.4.1) is of second order and nondegenerate at $x = 0$, which implies that $a_{11}^2 + b_{11}^2 \neq 0$, i.e., $x = 0$ is an isolated singular point of (2.4.1). Without loss of generality, we assume that $a_{11} \neq 0$. Let

$$\Delta = a_{12}^2 - 4a_{11}a_{22}.$$

If $\Delta \geq 0$, we define

$$\alpha_1 = \frac{-a_{12}+\sqrt{\Delta}}{2a_{11}}, \qquad \alpha_2 = \frac{-a_{12}-\sqrt{\Delta}}{2a_{11}},$$
$$\beta_i = b_{11}\alpha_i^2 + b_{12}\alpha_i + b_{22}, \qquad i = 1,2.$$

The following index theorem for (2.4.1) will be useful in studying transitions of (2.1.1) from eigenvalues with multiplicity two.

Theorem 2.4.1 *Let the vector field (2.4.1) be of second order and nondegenerate at $x = 0$, and $a_{11} \neq 0$. Then*

$$ind(F,0) = \begin{cases} 0 & \text{if } \Delta < 0 \text{ or } \beta_1\beta_2 > 0, \\ 2 & \text{if } \beta_1 > 0 \text{ and } \beta_2 < 0, \\ -2 & \text{if } \beta_1 < 0 \text{ and } \beta_2 > 0. \end{cases} \tag{2.4.2}$$

Proof. We proceed in several steps as follows.

Step 1. When $\Delta = a_{12}^2 - 4a_{11}a_{22} < 0$, the following quadratic form is either positive or negative definite:

$$a_{11}x_1^2 + a_{12}x_1x_2 + a_{22}x_2^2 > 0 \quad (\text{or} < 0) \quad \forall x \in \mathbb{R}^2, \quad x \neq 0,$$

depending on the sign of a_{11}. Hence the equations

$$\begin{aligned} &a_{11}x_1^2 + a_{12}x_1x_2 + a_{22}x_2^2 = -\varepsilon^2 \ (\text{or } = \varepsilon^2), \\ &b_{11}x_1^2 + b_{12}x_1x_2 + b_{22}x_2^2 = 0, \end{aligned}$$

have no solutions for any $\varepsilon \neq 0$, which implies that

$$\text{ind}(F,0) = 0 \quad \text{if } \Delta < 0.$$

Step 2. If $\Delta \geq 0$, then the vector field F given in (2.4.1) can be rewritten as

$$F = \begin{pmatrix} a_{11}(x_1 - \alpha_1 x_2)(x_1 - \alpha_2 x_2) \\ b_{11}x_1^2 + b_{12}x_1x_2 + b_{22}x_2^2 \end{pmatrix}. \tag{2.4.3}$$

Since F is of second order and nondegenerate at $x = 0, we have \beta_1 \cdot \beta_2 \neq 0$. By (2.4.3), $F = (0, \pm\varepsilon^2)$ with $\varepsilon \neq 0$ is equivalent to

$$x_1 = \alpha_i x_2, \quad \beta_i x_2^2 = \pm\varepsilon^2 \quad (i = 1,2). \tag{2.4.4}$$

If $\beta_1 \cdot \beta_2 > 0$, then one of the systems in (2.4.4), for either $+\varepsilon^2$ or $-\varepsilon^2$, has no solution, which means that the index of F at $x = 0$ is zero.

Step 3. When $\beta_1 \cdot \beta_2 < 0$, it is easy to see that $\alpha_1 \neq \alpha_2$ and $\Delta > 0$. The vector field $F = (F_1, F_2)$ given in (2.4.3) takes the following form:

$$\begin{aligned}F_1 &= a_{11}(x_1-\alpha_1x_2)(x_1-\alpha_2x_2),\\ F_2 &= \frac{1}{(\alpha_1-\alpha_2)^2}[\beta_1(x_1-\alpha_2x_2)^2+\beta_2(x_1-\alpha_1x_2)^2\\ &\quad +\gamma(x_1-\alpha_1x_2)(x_1-\alpha_2x_2)],\end{aligned} \tag{2.4.5}$$

where $\gamma=-(2b_{11}\alpha_1\alpha_2+b_{12}\alpha_1+b_{12}\alpha_2+2b_{22})$. Let

$$\beta_1>0,\quad \beta_2<0.$$

Then the solutions $x=(x_1,x_2)$ of (2.4.4) for $+\varepsilon^2$ are given by

$$x_1^{\pm}=\alpha_1x_2^{\pm},\quad x_2^{\pm}=\pm\beta_1^{-1/2}\varepsilon. \tag{2.4.6}$$

We take

$$(z_1,z_2)=(x_1-\alpha_1x_2,x_1-\alpha_2x_2). \tag{2.4.7}$$

Then the Jacobian matrix of F is given by

$$DF=\begin{pmatrix}\frac{\partial F_1}{\partial z_1} & \frac{\partial F_1}{\partial z_2}\\ \frac{\partial F_2}{\partial z_1} & \frac{\partial F_2}{\partial z_2}\end{pmatrix}\begin{pmatrix}\frac{\partial z_1}{\partial x_1} & \frac{\partial z_1}{\partial x_2}\\ \frac{\partial z_2}{\partial x_1} & \frac{\partial z_2}{\partial x_2}\end{pmatrix}.$$

By (2.4.7), it is easy to see that

$$\det\begin{pmatrix}\frac{\partial z_1}{\partial x_1} & \frac{\partial z_1}{\partial x_2}\\ \frac{\partial z_2}{\partial x_1} & \frac{\partial z_2}{\partial x_2}\end{pmatrix}=\det\begin{pmatrix}1 & -\alpha_1\\ 1 & -\alpha_2\end{pmatrix}=\alpha_1-\alpha_2=\sqrt{\Delta}/a_{11}. \tag{2.4.8}$$

Hence we find from (2.4.5), (2.4.7), and (2.4.8) that

$$\begin{aligned}\det DF(x) &= (\alpha_1-\alpha_2)\det\begin{pmatrix}a_{11}z_2 & a_{11}z_1\\ \frac{2\beta_2z_1+\gamma z_2}{(\alpha_1-\alpha_2)^2} & \frac{2\beta_1z_2+\gamma z_1}{(\alpha_1-\alpha_2)^2}\end{pmatrix}\\ &= \frac{a_{11}^2}{\sqrt{\Delta}}\det\begin{pmatrix}z_2 & z_1\\ 2\beta_2z_1+\gamma z_2 & 2\beta_1z_2+\gamma z_1\end{pmatrix}.\end{aligned} \tag{2.4.9}$$

On the other hand, by (2.4.7), we have

$$z_1^{\pm}=x_1^{\pm}-\alpha_1x_2^{\pm}=0,\quad z_2^{\pm}=x_1^{\pm}-\alpha_2x_2^{\pm}=\pm(\alpha_1-\alpha_2)\beta_1^{-1/2}\varepsilon.$$

Therefore, from (2.4.9) we get

$$\det DF(x^{\pm})=\frac{a_{11}^2}{\sqrt{\Delta}}2\beta_1(z_2^{\pm})^2. \tag{2.4.10}$$

By the Brouwer degree theory, we know that

$$\operatorname{ind}(F,0) = \deg(F,B_r,x_0), \quad x_0 = (0,\varepsilon^2) \in B_r, \tag{2.4.11}$$

where $B_r = \{x \in \mathbb{R}^2 |\ |x| < r\}$ and $r > 0$ is sufficiently small. It follows from (2.4.10) and (2.4.11) that if $\beta_1 > 0$, then

$$\operatorname{ind}(F,0) = \text{sign det } DF(x^+) + \text{sign det } DF(x^-) = 2.$$

In the same fashion, we can obtain that

$$\operatorname{ind}(F,0) = -2 \quad \text{for } \beta_1 < 0, \quad \beta_2 > 0.$$

Thus the formula (2.4.2) is proved, and the proof is complete. □

Remark 2.4.2 If $a_{11} = 0$ and $b_{11} \neq 0$, we let

$$\widetilde{\Delta} = b_{12}^2 - 4b_{11}b_{22}.$$

If $\widetilde{\Delta} \geq 0$, we define

$$\widetilde{\alpha}_1 = \frac{-b_{12} - \sqrt{\widetilde{\Delta}}}{2b_{11}}, \qquad \widetilde{\alpha}_2 = \frac{-b_{12} + \sqrt{\widetilde{\Delta}}}{2b_{11}},$$

$$\widetilde{\beta}_i = a_{11}\widetilde{\alpha}_i^2 + a_{12}\widetilde{\alpha}_i + a_{22}, \quad i = 1,2.$$

Then formula (2.4.2) can be written as

$$\operatorname{ind}(F,0) = \begin{cases} 0 & \text{if } \widetilde{\Delta} < 0 \text{ or } \widetilde{\beta}_1\widetilde{\beta}_2 > 0, \\ 2 & \text{if } \widetilde{\beta}_1 < 0 \text{ and } \widetilde{\beta}_2 > 0, \quad -2 \quad \text{if } \widetilde{\beta}_1 > 0 \text{ and } \widetilde{\beta}_2 < 0. \end{cases}$$

2.4.2 *Bifurcation at Second-Order Singular Points*

Under the conditions (2.1.4) and (2.1.5), the integers m and r are the algebraic and geometric multiplicities of the eigenvalues $\beta_1(\lambda_0)$ of L_λ at $\lambda = \lambda_0$. Here, we assume that $m = r = 2$, and the operator $L_\lambda + G(\cdot,\lambda)$ is of second order and nondegenerate at $(u,\lambda) = (0,\lambda_0)$, i.e.,

$$G(u,\lambda) = G_2(u,\lambda) + o(\|u\|_{X_1}),$$

where $G_2(u,\lambda)$ is a bilinear operator. We set

$$
\begin{aligned}
a_{11}(\lambda) &= (G_2(e_1(\lambda),e_1(\lambda),\lambda),e_1^*(\lambda)),\\
a_{22}(\lambda) &= (G_2(e_2(\lambda),e_2(\lambda),\lambda),e_1^*(\lambda)),\\
a_{12}(\lambda) &= (G_2(e_1(\lambda),e_2(\lambda),\lambda)+G_2(e_2(\lambda),e_1(\lambda),\lambda),e_1^*(\lambda)),\\
b_{11}(\lambda) &= (G_2(e_1(\lambda),e_1(\lambda),\lambda),e_2^*(\lambda)),\\
b_{22}(\lambda) &= (G_2(e_2(\lambda),e_2(\lambda),\lambda),e_2^*(\lambda)),\\
b_{12}(\lambda) &= (G_2(e_1(\lambda),e_2(\lambda),\lambda)+G_2(e_2(\lambda),e_1(\lambda),\lambda),e_2^*(\lambda)),
\end{aligned}
\tag{2.4.12}
$$

where $e_i(\lambda)$ and $e_i^*(\lambda)$ $(i=1,2)$ are the eigenvectors of L_λ and L_λ^* near $\lambda=\lambda_0$, respectively:

$$L_\lambda e_i(\lambda)=\beta_i(\lambda)e_i(\lambda), \qquad L_\lambda^* e_i^*(\lambda)=\beta_i(\lambda)e_i^*(\lambda) \qquad i=1,2.$$

Thus we derive a vector field

$$F(x,\lambda)=\begin{pmatrix} a_{11}(\lambda)x_1^2+a_{12}(\lambda)x_1x_2+a_{22}(\lambda)x_2^2 \\ b_{11}(\lambda)x_1^2+b_{12}(\lambda)x_1x_2+b_{22}(\lambda)x_2^2 \end{pmatrix}. \tag{2.4.13}$$

Consider the steady-state equation of (2.1.1):

$$L_\lambda u+G(u,\lambda)=0. \tag{2.4.14}$$

The following theorem will be useful later in this section.

Theorem 2.4.3 *Let the conditions (2.1.4), (2.1.5) hold, and set $r=m=2$. If $u=0$ is a second-order nondegenerate singular point of $L_\lambda+G$ at $\lambda=\lambda_0$, i.e., $x=0$ is an isolated zero of $F(x,\lambda_0)$, and the two vectors (a_{11},a_{12},a_{13}) and (b_{11},b_{12},b_{13}), defined by (2.4.12), are linearly independent at $\lambda=\lambda_0$, then we have the following assertions:*

1. *There are at most three branches of (2.4.14) bifurcated from $(0,\lambda_0)$ on each side of $\lambda=\lambda_0$.*
2. *If the bifurcated branches on $\lambda>\lambda_0$ (respectively on $\lambda<\lambda_0$) are regular, then the number of branches on this side is either 1 or 3.*
3. *On any given side, if the number of branches is 3, then these three branches must be regular.*
4. *If the number of branches on one side is 2, then one of them is regular.*

Proof. We proceed in several steps as follows.

Step 1. Since we have $m=r=2$ in (2.1.4) and (2.1.5), the bifurcation equations (2.4.14) can be rewritten as follows:

$$
\begin{aligned}
&\beta_1(\lambda)x_1+a_{11}x_1^2+a_{12}x_1x_2+a_{22}x_2^2+o(|x|^2)=0,\\
&\beta_2(\lambda)x_2+b_{11}x_1^2+b_{12}x_1x_2+b_{22}x_2^2+o(|x|^2)=0.
\end{aligned}
\tag{2.4.15}
$$

Step 2. We now show that there is an $\varepsilon>0$ such that for every C^∞ function $(f_1(x),f_2(x))$ with $|f_1(x)|+|f_2(x)|$ small near $x=0$, the equations

$$a_{11}x_1^2+a_{12}x_1x_2+a_{22}x_2^2+f_1(x)+o(|x|^2)=0, \tag{2.4.16}$$
$$b_{11}x_1^2+b_{12}x_1x_2+b_{22}x_2^2+f_2(x)+o(|x|^2)=0, \tag{2.4.17}$$

have at most four solutions in $|x|<\varepsilon$.

Since the second-order terms in (2.4.16) and (2.4.17) are of second order and nondegenerate, at least one of the coefficients a_{11}, a_{22}, b_{11}, and b_{22} is not zero. Without loss of generality, we assume that $a_{22}\neq 0$. Then we obtain from (2.4.16) that

$$x_2^{\pm}=-\frac{a_{12}}{2a_{22}}x_1\pm\frac{1}{2a_{22}}\sqrt{(a_{12}^2-4a_{11}a_{22})x_1^2-4a_{22}f_1(x)+o(|x|^2)},$$

which, by the implicit function theorem, implies that

$$x_2^{\pm}=-\frac{a_{12}}{2a_{22}}x_1\pm\frac{1}{2a_{22}}\Delta, \tag{2.4.18}$$

where

$$\Delta=\sqrt{(a_{12}^2-4a_{11}a_{22})x_1^2+g(x_1)+o(|x_1|^2)}$$
$$g(x_1)\to 0 \quad \text{if} \quad f_1(x)\to 0.$$

Inserting (2.4.18) into (2.4.17), we have

$$\alpha x_1^2+b_{22}g(x_1)+4a_{22}^2f_2(x)+o(|x|^2)=\mp\beta x_1\Delta, \tag{2.4.19}$$

where

$$\alpha=4a_{22}(a_{22}b_{11}-a_{11}b_{22})+2a_{12}(a_{12}b_{22}-a_{22}b_{12}), \qquad \beta=2(a_{22}b_{12}-a_{12}b_{22}).$$

We infer then from (2.4.18) and (2.4.19) that

$$\rho x_1^4+h_1(x_1)x_1^2+h_2(x_1)+o(|x_1|^4)=0, \tag{2.4.20}$$

where

$$\rho=\alpha^2-\beta^2(a_{11}^2-4a_{11}a_{22}),$$
$$h_1(x_1),h_2(x_1)\to 0 \quad \text{if} \quad f_1(x),f_2(x)\to 0.$$

It is easy to verify that the condition $\rho\neq 0$ is equivalent to the following conditions:

$$b_{22}z_{\pm}^2+b_{11}z_{\pm}+b_{11}\neq 0, \qquad z_{\pm}=\frac{-a_{12}\pm\sqrt{a_{12}^2-4a_{11}a_{22}}}{2a_{22}}. \tag{2.4.21}$$

By the assumption of this theorem, the conditions in (2.4.21) are valid. Therefore, $\rho\neq 0$, which implies that (2.4.20) has at most four real solutions satisfying $x_1\to 0$ as $f_1(x)$, $f_2(x)\to 0$.

On the other hand, each solution of (2.4.20) corresponds to one of the signs $\pm$ in (2.4.19) and (2.4.18). Hence (2.4.16) and (2.4.17) have at most four real solutions near $x = 0$. Assertion (1) is proved.

Step 3. Proof of assertion (2). Let $F_\lambda : \mathbb{R}^2 \to \mathbb{R}^2$ be defined by

$$F_\lambda(x) = \begin{pmatrix} \beta_1(\lambda)x_1 + a_{11}x_1^2 + a_{12}x_1x_2 + a_{22}x_2^2 + o(|x|^2) \\ \beta_2(\lambda)x_2 + b_{11}x_1^2 + b_{12}x_1x_2 + b_{22}x_2^2 + o(|x|^2) \end{pmatrix}.$$

By Theorem 3.9 in Ma and Wang (2005b), the singular points $z(\lambda)$ of (2.4.15) near $x = 0$ are regular if and only if the bifurcated branches of (2.4.14) from $(0, \lambda_0)$ are regular. Hence, we have

$$\mathrm{ind}(F_\lambda, z(\lambda)) = \pm 1 \tag{2.4.22}$$

for $|\lambda - \lambda_0| \neq 0$ sufficiently small. In addition, we know that if $\lambda \neq \lambda_0$, then

$$\mathrm{ind}(F_\lambda, 0) = \mathrm{sign}\beta_1(\lambda)\beta_2(\lambda) = 1. \tag{2.4.23}$$

It is known that the topological degree of an even mapping is even. Hence we have

$$\sum_{|z_i|<\varepsilon} \mathrm{ind}(F_\lambda, z_i(\lambda)) = \text{even}, \tag{2.4.24}$$

for every $\varepsilon > 0$ small. Hence assertion (2) follows from (2.4.22)–(2.4.24) and assertion (1).

Step 4. Proof of assertions (3) and (4). We know from Step 2 that the solutions of (2.4.16) and (2.4.17) are in one-to-one correspondence with those of (2.4.20). Let $x_0 \in \mathbb{R}^1$ be a solution of (2.4.20) near $x = 0$. Then (2.4.20) can be expressed as

$$(x - x_0)^k g(x) = 0, \qquad k \geq 1. \tag{2.4.25}$$

We shall prove that a solution $z_0 \in \mathbb{R}^2$ of (2.4.16) and (2.4.17) corresponding to x_0 is nondegenerate if and only if $k = 1$ in (2.4.25).

Obviously, if $k \geq 2$ in (2.4.25), then under a perturbation there are more solutions of (2.4.16) and (2.4.17) near z_0.

Suppose z_0 is degenerate, i.e.,

$$\det DF(z_0) = 0, \tag{2.4.26}$$

where $DF(z_0)$ stands for the Jacobian matrix of F, and

$$F = \begin{cases} a_{11}x_1^2 + a_{12}x_1x_2 + a_{22}x_2^2 + f_1(x) + o(|x|^2), \\ b_{11}x_1^2 + b_{12}x_1x_2 + b_{22}x_2^2 + f_2(x) + o(|x|^2). \end{cases}$$

Under a suitable coordinate system, $DF(z_0)$ must assume one of the following three forms:

$$DF(z_0) = \begin{pmatrix} 0 & 0 \\ 0 & 0 \end{pmatrix}, \tag{2.4.27}$$

$$DF(z_0) = \begin{pmatrix} \alpha & 0 \\ 0 & 0 \end{pmatrix}, \tag{2.4.28}$$

$$DF(z_0) = \begin{pmatrix} 0 & \alpha \\ 0 & 0 \end{pmatrix}, \tag{2.4.29}$$

for some $\alpha \neq 0$.

Since f_1, f_2 are C^∞ sufficiently small functions, and the quadratic forms in F are of second order and nondegenerate, for both cases of (2.4.27) and (2.4.28) it is known that with a linear perturbation, (2.4.16) and (2.4.17) will yield one more solution near z_0, which implies $k \geq 2$ in (2.4.25). We proceed with the case (2.4.29).

Near z_0, F can be expressed as

$$\widetilde{F}(y) = \begin{cases} \alpha y_2 + c_{11}y_1^2 + c_{12}y_1y_2 + c_{22}y_2^2 + o(|y|^2), \\ \qquad d_{11}y_1^2 + d_{12}y_1y_2 + d_{22}y_2^2 + o(|y|^2), \end{cases} \tag{2.4.30}$$

where $y = x - z_0$.

The zero point $y = 0$ of (2.4.30) stands for the zero point z_0 of F. Now we consider the perturbation equation

$$\alpha y_2 = -(c_{11}y_1^2 + c_{12}y_1y_2 + c_{22}y_2^2) + o(|y|^2), \tag{2.4.31}$$

$$\varepsilon y_1 = -(d_{11}y_1^2 + d_{12}y_1y_2 + d_{22}y_2^2) + o(|y|^2), \tag{2.4.32}$$

for ε small. By the implicit function theorem, we obtain from (2.4.31) a solution $y_2 = f(y_1)$. Then (2.4.32) implies that

$$\varepsilon y_1 = -(d_{11}y_1^2 + d_{12}y_1 f(y_1) + d_{22}f^2(y_1)) + o(|y_1|^2). \tag{2.4.33}$$

By the second-order nondegeneracy of $F(y)$, there exist $a \neq 0$ and $m \geq 2$ such that (2.4.33) becomes

$$\varepsilon y_1 = a y_1^m + o(|y_1|^m). \tag{2.4.34}$$

We infer from (2.4.34) that (2.4.31) and (2.4.32) have two solutions $(0,0)$ and $(\widetilde{y}_1, f(\widetilde{y}_1))$, where

$$\widetilde{y}_1 = (\varepsilon/a)^{1/(m-1)} + o\left(|\varepsilon/a|^{1/(m-1)}\right),$$

and ε is taken such that $\varepsilon \cdot a > 0$.

Thus, we prove that if z_0 is a degenerate zero point of (2.4.15) and (2.4.17), then the exponential k of the corresponding solution x_0 in (2.4.25) is greater than or equal to 2.

We are now in a position to prove assertions (3) and (4). If (2.4.15) has three bifurcated solutions (respectively two bifurcated solutions) from $(0, \lambda_0)$, then the corresponding solutions $x_i \neq 0$, $i = 1,2,3$, (respectively $x_j \neq 0$, $j = 1,2$) of (2.4.20) can be expressed as

$$(x-x_1)(x-x_2)(x-x_3)xg_1(x)=0,$$
$$(\text{respectively } (x-x_1)^2(x-x_2)xg_1(x)=0,$$

which implies assertion (3) (respectively assertion (4)).

The proof of the theorem is complete. □

2.4.3 The Case ind $(F,0)=-2$

The vector field F is of second order and nondegenerate at $x=0$ near $\lambda=\lambda_0$. According to Theorem 2.4.3, under conditions (2.1.4) and (2.1.5) with $m=r=2$, if the two vectors (a_{11},a_{12},a_{13}) and (b_{11},b_{12},b_{13}) are linearly independent near $\lambda=\lambda_0$, then there are at most three and at least one bifurcated singular points of (2.1.1) on each side of $\lambda=\lambda_0$.

By Theorem 2.4.1, the index of F defined by (2.4.13) at $x=0$ takes only three values: $0,\pm 2$. We shall show that the transition of (2.1.1) from $(0,\lambda_0)$ is closely related to the index of F at $x=0$.

The following theorem describes the transition of (2.1.1) from $(0,\lambda_0)$ for the case ind $(F,0)=-2$.

Theorem 2.4.4 *Let the conditions (2.1.4) and (2.1.5) with $m=r=2$ hold, let $L_\lambda+G(\cdot,\lambda)$ be of second order and nondegenerate at $(u,\lambda)=(0,\lambda_0)$, and suppose ind $(F,0)=-2$. Then we have the following assertions:*

(1) The transition of (2.1.1) from $(0,\lambda_0)$ is a jump transition, and it has exactly three connected unstable domains.
(2) Equation (2.1.1) bifurcates from $(0,\lambda_0)$ to exactly three saddle points on each side of $\lambda=\lambda_0$, which have Morse index one.
(3) The local topological structure of this transition is as shown in Fig. 2.11.

Proof. We proceed in several steps as follows.

Step 1. Under the conditions (2.1.4) and (2.1.5) with $m=r=2$, the reduction equations of (2.1.1) are given by

$$\begin{aligned}\frac{dx_1}{dt}&=\beta_1(\lambda)x_1+(G(x_1e_1+x_2e_2+\Phi(x,\lambda)),e_1^*(\lambda)),\\ \frac{dx_2}{dt}&=\beta_2(\lambda)x_2+(G(x_1e_1+x_2e_2+\Phi(x,\lambda)),e_2^*(\lambda)),\end{aligned}\tag{2.4.35}$$

where $\Phi(x,\lambda)$ is the center manifold function satisfying $\Phi(x,\lambda)=o(|x|)$ for $x\in\mathbb{R}^2$. Since F is of second order and nondegenerate at $x=0$, it follows that (2.4.35) can be written as

$$\frac{dx}{dt}=J_\lambda x+F(x,\lambda)+o(|x|^2),\tag{2.4.36}$$

where $F(x,\lambda)$ is as in (2.4.13), and

$$J_\lambda x = \begin{pmatrix} \beta_1(\lambda) & 0 \\ 0 & \beta_2(\lambda) \end{pmatrix} \begin{pmatrix} x_1 \\ x_2 \end{pmatrix} = \begin{pmatrix} \beta_1(\lambda)x_1 \\ \beta_2(\lambda)x_2 \end{pmatrix}.$$

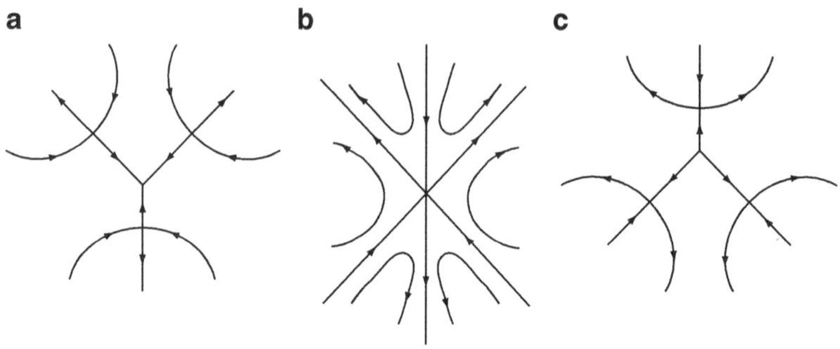

Fig. 2.11 Transition structure from a singular point with index ind $(F,0) = -2$: (**a**) $\lambda < \lambda_0$; (**b**) $\lambda = \lambda_0$; (**c**) $\lambda > \lambda_0$

Since F is of second order and nondegenerate at $(x,\lambda) = (0,\lambda_0)$, the vector field on the right-hand side of (2.4.36) is a perturbation of $J_\lambda + F$ near $x = 0$. Hence, by the local structural stability of nondegenerate singular points, it suffices to prove this theorem for the following system, which is the first-order approximation of (2.4.36):

$$\frac{dx}{dt} = J_\lambda x + F(x,\lambda). \tag{2.4.37}$$

Step 2. Proof of assertion (2). By Theorem 2.4.1, if ind $(F,0) = -2$, then the two vectors (a_{11}, a_{12}, a_{22}) and (b_{11}, b_{12}, b_{22}) are linearly independent. Therefore, it follows from Theorem 2.4.3 that (2.4.37) has at most three bifurcated singular points from $(0,\lambda_0)$. We shall prove that (2.4.37) has just three bifurcated singular points on each side of $\lambda = \lambda_0$.

It is known that

$$\begin{aligned} &\operatorname{ind}(J_\lambda + F, 0) = \operatorname{sign}[\beta_1(\lambda)\beta_2(\lambda)] = 1 \qquad \text{if } \lambda \neq \lambda_0, \\ &\sum_{i=1}^{k} \operatorname{ind}(J_\lambda + F, p_i) + \operatorname{ind}(J_\lambda + F, 0) = \operatorname{ind}(F(\cdot,\lambda_0), 0) = -2, \end{aligned}$$

where p_i $(1 \leq i \leq k)$ are the bifurcated singular points of (2.4.37) from $(0,\lambda_0)$. Hence if $\lambda \neq \lambda_0$, then

$$\sum_{i=1}^{k} \operatorname{ind}(J_\lambda + F, p_i) = -3. \tag{2.4.38}$$

If the number k in (2.4.38) is less than 3, then one of the bifurcated singular points, say p_1, of (2.4.37) satisfies

$$|\text{ind}(J_\lambda + F, p_1)| \geq 2. \tag{2.4.39}$$

By the Brouwer degree theory, if the Jacobian matrix of $J_\lambda + F$ at p_1 is a nonzero matrix, i.e.,

$$D(J_\lambda + F)(p_1) \neq 0,$$

then we have

$$|\text{ind}(J_\lambda + F, p_1)| \leq 1.$$

Therefore, it follows from (2.4.39) that the Jacobian matrix of $J_\lambda + F$ at p_1 is zero:

$$D(J_\lambda + F)(p_1) = J_\lambda + \left(\frac{\partial F_i(p_1)}{\partial x_j}\right) = 0. \tag{2.4.40}$$

Let $p_1 = (z_1, z_2)$. Then we infer from (2.4.40) that

$$\begin{aligned}
&\beta_1 + 2a_{11}z_1 + a_{12}z_2 = 0,\\
&a_{12}z_1 + 2a_{22}z_2 = 0,\\
&\beta_2 + 2b_{22}z_2 + b_{12}z_1 = 0,\\
&b_{12}z_2 + 2b_{11}z_1 = 0,
\end{aligned}$$

which, together with $J_\lambda p_1 + F(p_1, \lambda) = 0$, implies that $p_1 = 0$, a contradiction to $p_1 \neq 0$. Thus, we have shown that $k = 3$. From (2.4.38) and Theorem 2.4.3, we deduce that

$$\text{ind}(J_\lambda + F, p_i) = -1 \quad \text{for } i = 1, 2, 3,$$

which implies that p_i $(1 \leq i \leq 3)$ are saddle points. By Theorem 3.10 in Ma and Wang (2005b), the Morse index of p_i is one. Thus, assertion (2) is proved.

Step 3. Proof of assertion (1). By Theorem 2.1.5 and Lemma 2.1.8, it suffices to prove that the vector field $F(x, \lambda_0)$ has only six hyperbolic regions and has no elliptic or parabolic regions. To this end, we need to introduce a lemma, which is also useful in the later discussion for other cases.

Lemma 2.4.5 *Let $F(x, \lambda_0)$ be of second order and nondegenerate at $x = 0$. Then for every λ near λ_0, the vector field $F(x, \lambda)$ has k straight orbit lines with $1 \leq k \leq 3$:*

$$\alpha_i x_1 + \beta_i x_2 = 0, \quad \alpha_i^2 + \beta_i^2 \neq 0, \quad 1 \leq i \leq k, \tag{2.4.41}$$

where $\sigma_i = -\alpha_i/\beta_i$ or $\sigma_i = -\beta_i/\alpha_i$ are the real solutions of the following algebraic equation:

$$\begin{aligned}
&a_{22}\sigma^3 + (a_{12} - b_{22})\sigma^2 + (a_{11} - b_{12})\sigma - b_{11} = 0,\\
&\text{or}\\
&b_{11}\sigma^3 + (b_{12} - a_{11})\sigma^2 + (b_{22} - a_{12})\sigma - a_{22} = 0.
\end{aligned} \tag{2.4.42}$$

In particular, if $\text{ind}(F, 0) = -2$, then $k = 3$.

Proof. When $F(x,\lambda)$ are of second order and nondegenerate at $x=0$, then $a_{11}^2 + b_{11}^2 \neq 0$. We assume that $a_{11} \neq 0$. By the homogeneity of F, a straight line $x_2 = \sigma x_1$ is an orbit line of $F(x,\lambda)$ if and only if

$$\sigma = \frac{F_2(x,\lambda)}{F_1(x,\lambda)} = \frac{b_{11}x_1^2 + b_{12}x_1x_2 + b_{22}x_2^2}{a_{11}x_1^2 + a_{12}x_1x_2 + a_{22}x_2^2} = \frac{b_{11} + b_{12}\sigma + b_{22}\sigma^2}{a_{11} + a_{12}\sigma + a_{22}\sigma^2}.$$

Hence, the straight lines (2.4.41) satisfying (2.4.42) are orbit lines of $F(x,\lambda)$. Obviously, one of the two equations in (2.4.42) must have at least one real solution and at most three real solutions. Thus, we derive that the number of straight orbit lines is k with $1 \leq k \leq 3$, as shown in Fig. 2.11b.

When ind $(F,0) = -2$, then by assertion (2), the following system will bifurcate exactly to three singular points at $\lambda = \lambda_0$,

$$\frac{dx}{dt} = (\lambda - \lambda_0)x + F(x,\lambda),$$

and each singular point must be on only one straight orbit line of $F(x,\lambda)$. Hence, if ind $(F,0) = -2$, then $k=3$.

Thus the lemma is proved.□

Now we return to prove that $F(x,\lambda_0)$ has only six hyperbolic regions at $x=0$. By Lemma 2.4.5, $F(x,\lambda_0)$ has exactly three straight orbit lines that intersect at $x=0$, and the three straight lines divide the plane $\mathbb{R}^2$ into six invariant regions of $F(x,\lambda_0)$. By the Poincaré index formula (2.1.9), $F(x,\lambda_0)$ possesses at least six hyperbolic regions $H_i \subset \mathbb{R}^2$, and each boundary ∂H_i consists of two rays from $x=0$:

$$\partial H_i = l_i^+ + l_i^- + \{0\},$$

where l_i^+ is an orbit of $F(x,\lambda_0)$ outward to $x=0$, and l^- is an orbit inward to $x=0$, as shown in Fig. 2.12.This shows that $F(x,\lambda_0)$ has exactly six hyperbolic regions, which are separated by the three straight orbit lines. Therefore, F has no elliptic regions, for otherwise, the index ind $(F,0)$ would not equal -2.

Finally, we prove that F has no parabolic regions P at $x=0$. Otherwise, there would have to be a parabolic region $P \subset \mathbb{R}^2$ such that for every $x_0 \in P$, the orbit $x(t,x_0)$ of $F(x,\lambda_0)$ would satisfy

$$\lim_{t\to\infty} x(t,x_0) = 0,$$

where $x(0,x_0) = x_0$, and $x(t,x_0)$ is tangent to some straight orbit as shown in Fig. 2.13.

We shall reach a contradiction. Without loss of generality, we take a straight orbit line of $F(x,\lambda_0)$ as the x_1-axis. Then, in this coordinate system, $F(x,\lambda_0)$ can be expressed as

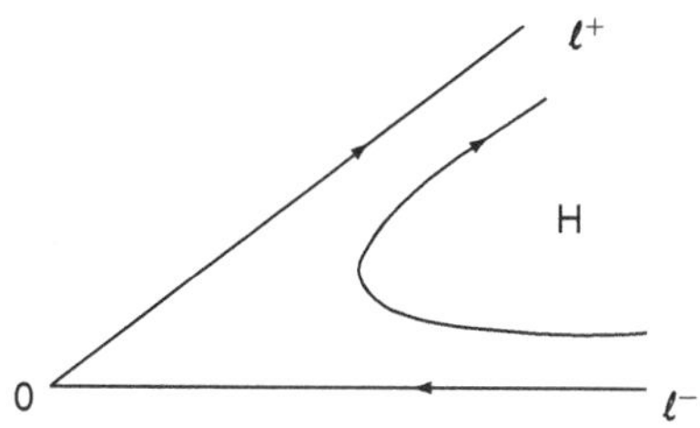

Fig. 2.12 Orbits $l^{\pm}$ of $F(x,\lambda_0)$

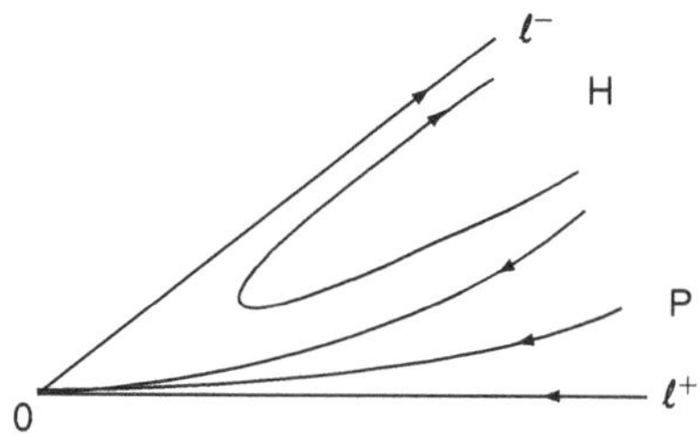

Fig. 2.13 Orbits of x in a sectorial region $S(\theta)$

$$F(x,\lambda_0) = \begin{pmatrix} a_{11}x_1^2 + a_{12}x_1x_2 + a_{22}x_2^2 \\ x_2(b_{12}x_1 + b_{22}x_2) \end{pmatrix}. \tag{2.4.43}$$

Since ind $(F,0) = -2$, we have $b_{12} \neq 0$; otherwise, by Theorem 2.4.1, we would have ind $(F,0) = 0$. Take a coordinate transformation:

$$x_1' = x_1 + \frac{b_{22}}{b_{12}}x_2, \qquad x_2' = x_2.$$

Then the vector field (2.4.43) becomes (the primes are dropped for simplicity)

$$F(x,\lambda_0) = \begin{pmatrix} c_{11}x_1^2 + c_{12}x_1x_2 + c_{22}x_2^2 \\ b_{12}x_1x_2 \end{pmatrix}. \tag{2.4.44}$$

Since the coordinate system transformation preserves the topological structure up to orientation, the vector field (2.4.44) is topologically equivalent to F given by (2.4.43).

By Theorem 2.4.1 and $\text{ind}(F,0) = -2$, we see that

$$b_{12} \cdot \frac{-c_{12} + \sqrt{\Delta}}{2c_{11}} < 0, \qquad b_{12} \cdot \frac{-c_{12} - \sqrt{\Delta}}{2c_{11}} > 0,$$

where $\Delta = c_{12}^2 - 4c_{11}c_{22} > 0$. It follows that

$$\frac{b_{12}}{c_{11}} < 0 \quad (c_{11} \neq 0). \tag{2.4.45}$$

By the above conclusion, for $x_0 \in P$, the orbit $x(t,x_0)$ of (2.4.44) near $x=0$ can be expressed as $x_2 = f(x_1)$ with

$$f(x_1) = o(x_1), \quad x_1 \geq 0. \tag{2.4.46}$$

It follows from (2.4.44) that

$$\frac{df}{dx_1} = \frac{b_{12}x_1x_2}{c_{11}x_1^2 + c_{12}x_1x_2 + c_{22}x_2^2} = \frac{b_{12}}{c_{11}}\frac{f(x_1)}{x_1}\left(1 + \frac{c_{12}}{c_{11}}\frac{f}{x_1} + \frac{c_{22}}{c_{11}}\frac{f^2}{x_1^2}\right)^{-1},$$

which, by (2.4.46), implies that

$$\frac{df}{f} \to \frac{b_{12}}{c_{11}}\frac{dx_1}{x_1} \quad \text{as} \quad x_1 \to 0^+.$$

Namely,

$$f(x_1) \to x_1^{-\sigma} \quad \text{as} \quad x_1 \to 0^+, \tag{2.4.47}$$

where $\sigma = -b_{12}/c_{11} > 0$ (by (2.4.45)).

From (2.4.46) and (2.4.47), we reach a contradiction. Thus we have proved assertion (1) of Theorem 2.4.4, and consequently, we have proved assertion (3). The proof of Theorem 2.4.4 is complete. □

2.4.4 The Case ind $(F,0) = 2$

For the case $\text{ind}(F,0) = 2$, in order to describe the transition structure, we need to introduce some notation. A set $S(\theta) \subset \mathbb{R}^2$ is called a sectorial region with angle $\theta \in [0,2\pi]$, if $S(\theta)$ is enclosed by two curves γ_1 and γ_2 starting with $x=0$ and an arc Γ, and the angle between the tangent lines L_1 and L_2 of γ_1 and γ_2 at $x=0$ is θ; see Fig. 2.14. Let $S_r(\theta)$ be the sectorial region with angle θ and radius $r > 0$ given by

$$S_r(\theta) = \{x \in S(\theta) | \ |x| < r\}.$$

The algebraic equations introduced in (2.4.42) are related to the transition of (2.1.1). It is clear that the number k of real solutions of (2.4.42) that are associated with the straight orbit lines of vector field $F(x,\lambda)$ satisfies $1 \leq k \leq 3$. The following theorem connects the number k with the transition of (2.1.1).

Theorem 2.4.6 *Let ind $(F,0) = 2$, and let the other conditions be as in Theorem 2.4.4. Then for the transition of (2.1.1) from $(0,\lambda_0)$, the following assertions hold:*

(1) When $k > 1$, the transition is mixed, and when $k = 1$, the transition is either continuous or mixed, depending on the terms of order higher than F.
(2) If $\beta_1(\lambda) = \beta_2(\lambda)$ near $\lambda = \lambda_0$, then the stable transition domain is a sectorial region with angle θ ($\pi < \theta \leq 2\pi$) given by

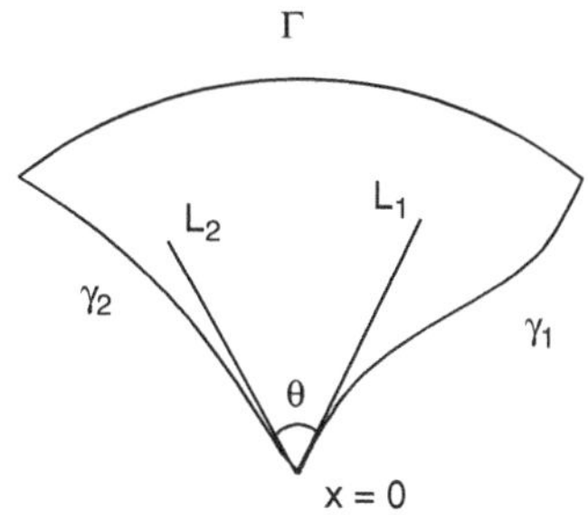

Fig. 2.14 Sectorial region with angle θ

$$D_\theta = \{u \in U \subset X \mid u = x + v, \quad x = x_1 e_1 + x_2 e_2 \in S(\theta)\},$$

where $U \subset X$ is a neighborhood of $u = 0$.

(3) If $\beta_1 = \beta_2$, then (2.1.1) bifurcates on $\lambda > \lambda_0$ to an attractor $\mathscr{A}_\lambda$, which attracts D_θ with dim $\mathscr{A}_\lambda = 0$ or 1 and contains singular points as its minimal attractors.

(4) If $k = 1$, the attractor $\mathscr{A}_\lambda$ consists of a singular point p, and the transition structure is as shown in Fig. 2.15a. If $k = 2$, the attractor $\mathscr{A}_\lambda$ is either a singular point as shown in Fig. 2.15a or two singular points p_1, p_2 and the orbit γ connecting them, as shown in Fig. 2.15b, or three singular points p_0, p_1, p_2 and two orbits γ_1, γ_2 connecting p_0 with p_1, p_2 as shown in Fig. 2.15c. If $k = 3$, then $\mathscr{A}_\lambda$ has the structure shown in Fig. 2.15c.

Remark 2.4.7 If $\beta_1(\lambda) \neq \beta_2(\lambda)$ near λ_0 and $\lambda \neq \lambda_0$, then assertions (3) and (4) in Theorem 2.4.6 are not valid. In this case, the stable transition domain D may contain no attractor. For example, we consider the following equations:

$$\begin{aligned} \frac{dx_1}{dt} &= 4\lambda x_1 + \lambda x_2 + x_1^2 - 4x_1x_2 - x_2^2, \\ \frac{dx_2}{dt} &= -\lambda x_1 + x_1 x_2. \end{aligned} \tag{2.4.48}$$

It is easy to see that the two eigenvalues of (2.4.48) are

$$\beta_1(\lambda) = (2 + \sqrt{3})\lambda, \quad \beta_2(\lambda) = (2 - \sqrt{3})\lambda.$$

Namely, we have $\beta_1(\lambda) \neq \beta_2(\lambda)$, since $\lambda \neq \lambda_0 = 0$. The nonlinear term

$$G(x_1, x_2) = \begin{pmatrix} x_1^2 - 4x_1x_2 - x_2^2 \\ x_1x_2 \end{pmatrix}$$

is of second order and nondegenerate, and by Theorem 2.4.1, we can derive that ind $(G, 0) = 2$. In addition, the algebraic equation (2.4.42) is

$$z^3 + 4z^2 = 0,$$

which has two real solutions $z = 0$ and $z = -4$. However, the system (2.4.48) has no attractors and only one bifurcated singular point $(x_1, x_2) = (0, \lambda)$, the transition with structure as shown in Fig. 2.16.

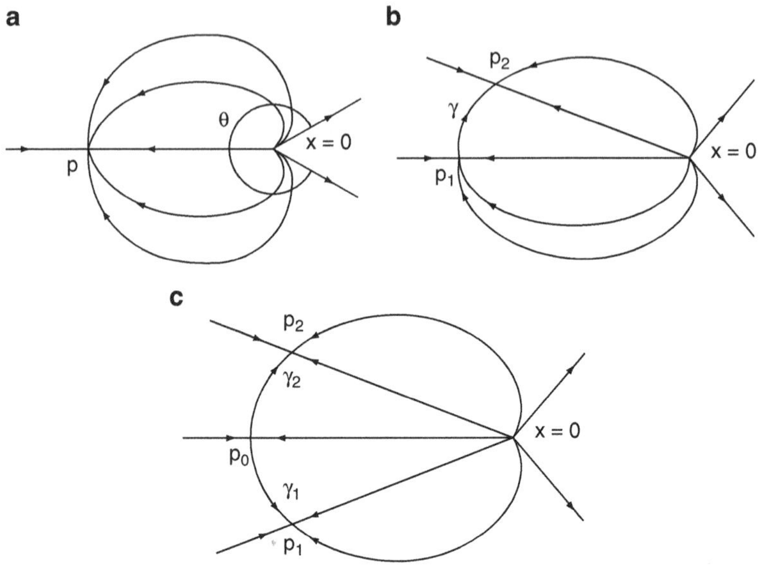

Fig. 2.15 (**a**) When $k = 1$, the attractor $\mathscr{A}_\lambda$ is equal to $\{p\}$. (**b**) When $k = 2, \mathscr{A}_\lambda = \gamma \bigcup \{p_1, p_2\}$. (**c**) When $k = 3, \mathscr{A}_\lambda = \gamma_1 \bigcup \gamma_2 \bigcup \{p_0, p_1, p_2\}$

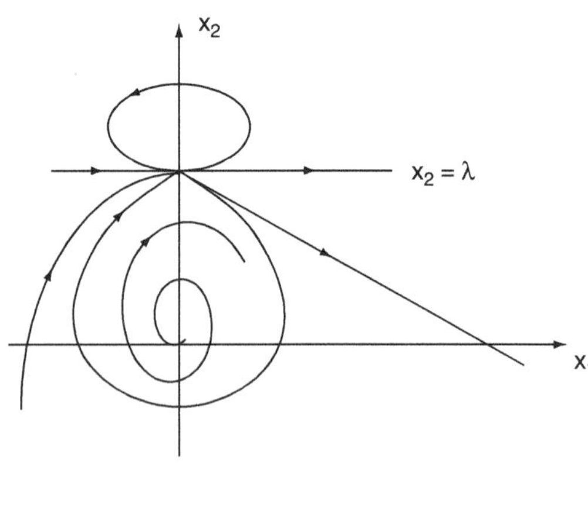

Fig. 2.16 Transition for the example in Remark 2.4.7

Proof of Theorem 2.4.6. As in the proof of Theorem 2.4.4, we need to consider only the system (2.4.37). Theorem 2.4.6 can then be achieved in a few lemmas.

Lemma 2.4.8 *Let ind $(F,0) = 2$, and let k $(1 \leq k \leq 3)$ be the number of real solutions of the algebraic equation (2.4.42). Then we have the following assertions:*

(1) $F(x,\lambda)$ has no hyperbolic regions at $x = 0$.
(2) $F(x,\lambda)$ has exactly two elliptic regions E_1 and E_2.
(3) If $k = 1$, then $F(x,\lambda)$ has no parabolic regions. If $k \geq 2$, then $F(x,\lambda)$ has exactly two parabolic regions P_1 and P_2 such that the orbits in P_1 tend to $x = 0$, and in P_2 are away from $x = 0$.
(4) The elliptic and parabolic regions E and P are sectorial regions $E = S(\theta_1), P = S(\theta_2)$ with $0 < \theta_1, \theta_2 < \pi, \theta_1 + \theta_2 = \pi$, and

$$\theta_2 = \cot^{-1}\sigma_1 - \cot^{-1}\sigma_2,$$

where $\sigma_i = x_2/x_1$ $(i = 0,1,2)$ are the real solutions of (2.4.42) with $\sigma_1 > \sigma_0 \geq \sigma_2$. Moreover, the edges of $S(\theta_1)$ and $S(\theta_2)$ are the straight orbit lines of $F(x,\lambda)$; see Fig. 2.17a–c.

Proof. Based on Lemma 2.4.5, as in the proof of (2.4.44), we can take a coordinate transformation $x' = Ax$ with $\det A > 0$ such that the vector field F can be transformed into the following form, where for brevity, we omit the primes:

$$F(x,\lambda) = \begin{pmatrix} F_1 \\ F_2 \end{pmatrix} = \begin{pmatrix} a(x_1 - \alpha_1 x_2)(x_1 + \alpha_2 x_2) \\ b x_1 x_2 \end{pmatrix}, \tag{2.4.49}$$

where $a \cdot b \neq 0$. Without loss of generality, we assume that $a, b > 0$. Thus by Theorem 2.4.1, we have $\alpha_1, \alpha_2 > 0$.

We can show that (2.4.49) has the topological structure as shown in Fig. 2.17a–c for $a, b > 0$.

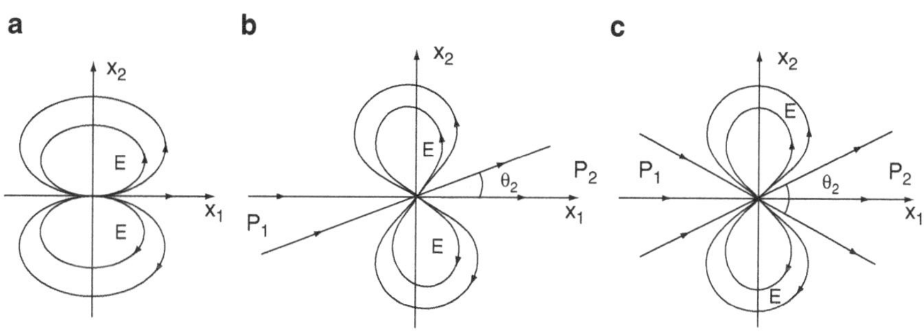

Fig. 2.17 Topological structure of (2.4.49): (**a**) the number of real solutions of (2.4.42) $k = 1$, (**b**) $k = 2$, (**c**) $k = 3$

To derive the topological structure in Fig. 2.17a–c of (2.4.49), let D_1, D_2, D_3, and D_4 be the four open quadrants in $\mathbb{R}^2$, and the two straight lines $x_1 - \alpha_1 x_2 = 0$ and $x_1 + \alpha_2 x_2 = 0$ also divide the plane $\mathbb{R}^2$ into four regions:

$$\begin{aligned}
Q_1 &= \{(x_1,x_2) \in \mathbb{R}^2 |\ x_1 - \alpha_1 x_2 > 0,\ x_1 + \alpha_2 x_2 > 0\},\\
Q_2 &= \{(x_1,x_2) \in \mathbb{R}^2 |\ x_1 - \alpha_1 x_2 < 0,\ x_1 + \alpha_2 x_2 > 0\},\\
Q_3 &= \{(x_1,x_2) \in \mathbb{R}^2 |\ x_1 - \alpha_1 x_2 < 0,\ x_1 + \alpha_2 x_2 < 0\},\\
Q_4 &= \{(x_1,x_2) \in \mathbb{R}^2 |\ x_1 - \alpha_1 x_2 > 0,\ x_1 + \alpha_2 x_2 < 0\}.
\end{aligned}$$

It is easy to see that

$$\begin{aligned}
F_1 &> 0 \quad \text{in } Q_1 \text{ and } Q_3\\
F_1 &< 0 \quad \text{in } Q_2 \text{ and } Q_4,\\
F_2 &> 0 \quad \text{in } D_1 \text{ and } D_3,\\
F_2 &< 0 \quad \text{in } D_2 \text{ and } D_4.
\end{aligned} \tag{2.4.50}$$

The properties (2.4.50) ensure that (2.4.49) has only two elliptic regions E_1 and E_2, with $E_1 \subset \mathbb{R}^2_+ = D_1 + D_2 = \{(x_1,x_2) \in \mathbb{R}^2 |\ x_2 > 0\}$ and $E_2 \subset \mathbb{R}^2_- = D_3 + D_4 = \{(x_1,x_2) \in \mathbb{R}^2 |\ x_2 < 0\}$; see Fig. 2.18.

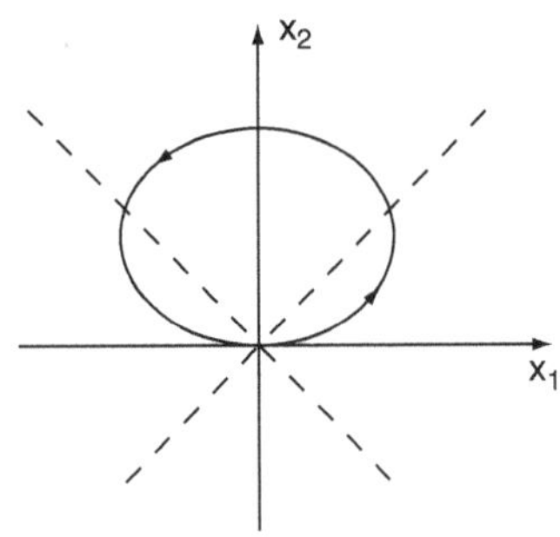

Fig. 2.18 The *dotted straight lines* are $x_1 = -\alpha_2 x_2$ and $x_1 = \alpha_1 x_2$ respectively

Thus, by the Poincaré formula (2.1.9), F has no hyperbolic regions, and assertions (1) and (2) are proved.

By (2.4.42), it follows from (2.4.49) that the straight orbit lines L_i $(i = 0,1,2)$ of $F(x,\lambda)$ are given by

$$L_0 : x_2 = 0, \quad L_1 : x_2 = \sigma_1 x_1, \quad L_2 : x_2 = \sigma_2 x_1,$$

and if σ_1, σ_2 are real, then

$$\sigma_1 = \frac{1}{\alpha_1} - \varepsilon_1, \quad \sigma_2 = -\frac{1}{\alpha_2} + \varepsilon_2,$$

for some real numbers $0 < \varepsilon_1 < 1/\alpha_1$ and $0 < \varepsilon_2 < 1/\alpha_2$. Hence we have

$$L_i \subset Q_1 \bigcup Q_3 \qquad \text{for } i = 0, 1, 2. \tag{2.4.51}$$

Therefore, assertions (3) and (4) follow from (2.4.51) and the radial symmetry of $F(x,\lambda)$. The proof of the lemma is complete. □

Lemma 2.4.9 *If $\beta_1(\lambda) = \beta_2(\lambda)$ and ind $(F,0) = 2$, then (2.4.37) bifurcates to an attractor $\mathscr{A}_\lambda$ on $\lambda > \lambda_0$, which attracts a sectorial region $S_r(\theta)$ with $\pi < \theta \leq 2\pi$. More precisely, $S_r(\theta) \subset (E_1 \bigcup E_2 \bigcup P_1) \bigcap B_r$, where E_1, E_2 are the elliptic regions of F, P_1 is the parabolic region where all orbits of F reach to $x = 0$, $B_r = \{x \in \mathbb{R}^2 | \, |x| < r\}$, $\theta = 2\pi - \theta_0$, and θ_0 is the angle of the parabolic region P_2.*

Proof. We know that under an orthogonal coordinate system transformation, the linear operator

$$J_\lambda = \begin{pmatrix} \beta(\lambda) & 0 \\ 0 & \beta(\lambda) \end{pmatrix}$$

is invariant. Therefore, without loss of generality, we take the vector field F in the form of (2.4.43). By Theorem 2.4.1, $F(x,\lambda)$ can be written as

$$F = \begin{pmatrix} F_1 \\ F_2 \end{pmatrix} = \begin{pmatrix} a(x_1 - \alpha_1 x_2)(x_1 + \alpha_2 x_2) \\ b x_2 (x_1 - \sigma x_2) \end{pmatrix}, \tag{2.4.52}$$

where $a \cdot b \neq 0$, $\alpha_1 > 0$, $\alpha_2 > 0$.

We proceed with the case that $a, b > 0$ and $\sigma \geq 0$. The other case can be proved in the same fashion. By Theorem 2.4.1, we know that $\alpha_1 > \sigma > -\alpha_2$, which implies that the lines $x_1 + \alpha_2 x_2 = 0$, $x_1 - \sigma x_2 = 0$, $x_1 - \alpha_1 x_2 = 0$, and $x_2 = 0$ are alternately positioned in $\mathbb{R}^2$.

Based on the definition of elliptic and parabolic regions, by Lemma 2.4.8, we obtain that

$$\lim_{t \to \infty} S_\lambda(t) x = 0 \qquad \forall x \in E_1 \bigcup E_2 \bigcup P_1, \tag{2.4.53}$$

where $S_\lambda(t)$ is the operator semigroup generated by $F(x,\lambda)$.

On the other hand, we obtain from (2.4.52) that for every $x \in E_1 \bigcup E_2 \bigcup P_1$, there is $t_0(x) \geq 0$ such that

$$S_\lambda(t)x \in D = \{x \in \mathbb{R}^2 | \, x_1 - \sigma x_2 < 0\} \qquad \forall t \geq t_0(x). \tag{2.4.54}$$

It is clear that $P_1 \subset D \subset E_1 \bigcup E_2 \bigcup P_1$. Let

$$D(r) = \{x \in D | \, |x| < r\}, \qquad D(r_1, r_2) = \{x \in D | \, 0 < r_1 < |x| < r_2\},$$

and let $T_\lambda(t)$ be the operator semigroup generated by $J_\lambda + F(\cdot, \lambda)$. It is known that for $\lambda > \lambda_0$, all orbits of $J_\lambda x$ are straight lines issuing outward from $x = 0$. Therefore, by (2.4.54), we deduce that

$$T_\lambda(t)x \in D \qquad \forall t > 0, \quad x \in \partial D, \quad x \neq 0. \tag{2.4.55}$$

Now we prove that for every $\lambda - \lambda_0 > 0$ sufficiently small, there are $r_1, r_2, r_3 > 0$ with $r_1 < r_2 < r_3$ such that

$$T_\lambda(t)x \in D(r_1, r_2) \qquad \forall x \in D(r_3), \quad t > t_x, \tag{2.4.56}$$

for some $t_x \geq 0$.

We know that for $\lambda > \lambda_0$, the singular point $x = 0$ of $J_\lambda + F$ has an unstable manifold M^u with dim $M^u = 2$. We take $r_1 > 0$ such that the ball B_{r_1} is contained in M^u. Then, by (2.4.55), we obtain that

$$T_\lambda(t)x \in D(r_1, r_2) \qquad \forall x \in D(r_1), \quad t > t_x. \tag{2.4.57}$$

If (2.4.56) is not true, then by (2.4.55) and (2.4.57), there exist $\lambda_n \to \lambda_0^+, t_n \to \infty$ and $\{x_n\} \subset D(r_3)$ such that

$$|T_{\lambda_n}(t_n)x_n| \geq r_2 \qquad \forall n \geq 1. \tag{2.4.58}$$

Let $x_n \to x_0 \in D(r_3)$. Then by the stability of extended orbits, Theorem 2.1.15 in Ma and Wang (2005d), and (2.4.58), there is an extended orbit γ of $F(\cdot, \lambda_0)$ with starting point $x_0 \in D(r_3)$ that does not reach $x = 0$. This is a contradiction to (2.4.53).

It follows from (2.4.55) and (2.4.56) that $D(r_1, r_2)$ is an absorbing set in a neighborhood U of $D(r_1, r_2)$. Hence, by the existence theorem of attractors (Theorem 2.2.11), for $\lambda > \lambda_0$, the set $\mathscr{A}_\lambda = \omega(D(r_1, r_2))$, with $0 \notin \mathscr{A}_\lambda$, is an attractor of (2.4.37), which attracts $D(r_3)$.

Applying Theorem 2.1.15 in Ma and Wang (2005d) again, we infer from (2.4.53) that

$$\lim_{\lambda \to \lambda_0} \max_{x \in \mathscr{A}_\lambda} |x| = 0.$$

Thus, $\mathscr{A}_\lambda$ is a bifurcated attractor of (2.4.37) from $(0, \lambda_0)$.

We can deduce from (2.4.54) that $\mathscr{A}_\lambda$ attracts a sectorial region $S_r(\theta) \subset E_1 \bigcup E_2 \bigcup P_1$ with $\theta = 2\pi - \theta_0$, where θ_0 is the angle of the parabolic region P_2. The proof is complete. □

Lemma 2.4.10 *Under the hypotheses of Lemma 2.4.9, for the attractor $\mathscr{A}_\lambda$, we have the following assertions:*

(1) If the number k of real solutions of (2.4.42) is equal to 1, then $\mathscr{A}_\lambda = \{p\}$, where p is a singular point of (2.4.37) with Morse index zero whose structure is as shown in Fig. 2.15a.
(2) If $k = 2$, then $\mathscr{A}_\lambda = \{p_1, p_2\} \bigcup \gamma$, where p_1, p_2 are two singular points of (2.4.37), p_1 is degenerate with index zero, p_2 is nondegenerate with Morse index zero, and γ is the orbit connecting p_1 and p_2, as shown in Fig. 2.15b.
(3) If $k = 3$, then $\mathscr{A}_\lambda = \{p_0, p_1, p_2\} \bigcup \gamma_1 \bigcup \gamma_2$, where p_0, p_1, p_2 are nondegenerate singular points with p_0 a saddle point, p_1, p_2 two attractors, and γ_1, γ_2 the orbits connecting p_0 with p_1 and p_2, as shown in Fig. 2.15c.

Proof. Obviously, the attractor $\mathscr{A}_\lambda$ contains all bifurcated singular points. We first need to prove that $\mathscr{A}_\lambda$ does not contain extended orbits homeomorphic to S^1.

By Lemma 2.4.5, all singular points of (2.4.37) must be in the straight orbit lines of $F(x,\lambda)$. Each straight orbit line is an invariant set of (2.4.37) and consists of orbits and singular points.

Using the method of the proof of Lemma 2.4.9, we see that for any straight orbit line L, we can take an orthogonal coordinate system transformation with L as its x_1-axis. Thus, the vector field $F(x,\lambda)$ takes the form of (2.4.52), and the bifurcated singular point $x_0 = (x_1^0, x_2^0)$ of $L_\lambda + F$ on L is

$$x_0 = (x_1^0, x_2^0) = (-\beta(\lambda)/a, 0),$$

and the Jacobian matrix of $J_\lambda + F$ at x_0 is given by

$$D(J_\lambda + F)(x_0) = \begin{pmatrix} -\beta(\lambda) & * \\ 0 & (1-\frac{b}{a})\beta(\lambda) \end{pmatrix}. \tag{2.4.59}$$

Hence, on each straight orbit line, there is only one singular point x_0 of $J_\lambda + F$, and there are two orbits r_1 and r_2 in L reaching x_0. Moreover, one of r_1 and r_2 connects $x = 0$ to x_0. It follows that the attractor $\mathscr{A}_\lambda$ containing all singular points has no closed extended orbits. By the Poincaré–Bendixon theorem, we obtain that $\dim \mathscr{A}_\lambda \leq 1$, and the number k of bifurcated singular points is the number of straight orbit lines.

When $k = 1$, by ind $(F, 0) = 2$, we have

$$\mathrm{ind}(J_\lambda + F, x_0) = 2 - \mathrm{ind}(J_\lambda + F, 0) = 1.$$

It then follows from (2.4.59) that x_0 is an attractor. We shall prove that x_0 is non-degenerate. To this end, by (2.4.59) it suffices to show that $b/a > 1$. For (2.4.52), (2.4.42) takes the form

$$z = \frac{bz(1-\sigma z)}{a(1-\alpha_1 z)(1+\alpha_2 z)}, \tag{2.4.60}$$

which has a real solution $z = 0$. Then (2.4.60) can be written as

$$\alpha_1\alpha_2 z^2 + \left(\alpha_1 - \frac{b\sigma}{a} - \alpha_2\right) z + \frac{b}{a} - 1 = 0. \tag{2.4.61}$$

Hence, (2.4.60) has no real solutions other than $z = 0$ if and only if the coefficients of (2.4.61) satisfy

$$\left(\frac{\sigma}{a} b + \alpha_2 - \alpha_1\right)^2 < 4\alpha_1\alpha_2 \left(\frac{b}{a} - 1\right).$$

Since $\alpha_1, \alpha_2 > 0$, this implies that $b/a > 1$. Thus, we deduce that when $k = 1$, the singular point x_0 is nondegenerate and has Morse index zero. Therefore, $\mathscr{A}_\lambda = \{x_0\}$ has the structure shown in Fig. 2.15a.

When $k=2$, $J_\lambda+F$ has two singular points p_1 and p_2,

$$\mathrm{ind}(J_\lambda+F,p_1)=0, \quad \mathrm{ind}(J_\lambda+F,p_2)=1,$$

which implies, by Theorem 2.4.3 and (2.4.59), that p_2 is an attractor, and p_1 has exactly two hyperbolic regions. Thus, $\mathscr{A}_\lambda$ has the topological structure shown in Fig. 2.15b.

When $k=3$, $J_\lambda+F$ has three singular points p_0,p_1,p_2, and by Theorem 2.4.3, they are regular, and

$$\mathrm{ind}(J_\lambda+F,p_0)=-1, \quad \mathrm{ind}(J_\lambda+F,p_1)=\mathrm{ind}(J_\lambda+F,p_2)=1. \tag{2.4.62}$$

It follows from (2.4.59) and (2.4.62) that p_0 is a saddle point, and p_1 and p_2 are attractors. In this case, the structure of $\mathscr{A}_\lambda$ is as shown in Fig. 2.15c. The proof of the lemma is complete. □

Now, by Lemmas 2.4.8–2.4.10, we can derive assertions (2)–(4) of Theorem 2.4.6, except in the case $k=2$ in assertion (4). In Lemma 2.4.10, for $k=2$, $J_\lambda+F$ has a singular point p_1 with index zero. Hence, under a perturbation of the terms of higher order than F, near p_1 the system (2.4.36) has zero, one, or two singular points. The case of no singular points is shown in Fig. 2.15a, and the case of one singular points is shown in Fig. 2.15b. For the case of two singular points, by (2.4.59), the indices of these singular points are $+1$ and -1, as shown in Fig. 2.15c. Hence assertion (4) for $k=2$ is valid.

Assertion (1) can be deduced from Lemma 2.4.8 and the Poincaré index formula (2.1.9), which illustrate that (2.4.36) bifurcates on $\lambda>\lambda_0$ to at least one singular point p with index one such that p has at least an elliptic region. Hence, the transition is not a jump transition.

The proof of Theorem 2.4.6 is complete. □

2.4.5 The Case ind $(F,0)=0$

The following theorem gives transitions for the case of index zero.

Theorem 2.4.11 *Let ind* $(F,0)=0$ *and assume that the other conditions are as in Theorem 2.4.4. Then the transition of (2.1.1) from* $(0,\lambda_0)$ *is either a jump or mixed transition. If it is a jump transition, then the structure is as shown in Fig. 2.19a. If it is a mixed transition, then the structure is as shown in Fig. 2.19b or c. In particular, for the number k of real solutions of (2.4.42), we have the following assertions:*

(1) If $k=1$*, the transition is a jump transition.*
(2) If $k=2$*, the transition is either a jump or mixed transition.*
(3) If $k=3$ *and* $\beta_1(\lambda)=\beta_2(\lambda)$ *near* $\lambda=\lambda_0$*, then the transition is mixed, and (2.1.1) bifurcates on* $\lambda>\lambda_0$ *to three singular points* p_0,p_1,p_2 *such that* p_0 *is an attractor that attracts a sectorial region* $D_r(\theta)$ *with* $0<\theta<\pi$*, and* p_1,p_2 *are saddle points, as shown in Fig. 2.19c.*

Proof. As was done in the proof of Lemma 2.4.8, one can prove that if ind $(F,0)=0$, then the topological structure of $F(x,\lambda)$ at $x=0$ near $\lambda=\lambda_0$ is just one of the three types shown in Fig. 2.20a–c, depending on the number k of real solutions of (2.4.42).

On the other hand, the sum of indices of all singular points near $(x,\lambda)=(0,\lambda_0)$ is zero:

$$\sum_{i=1}^{n}\mathrm{ind}(J_\lambda+g,z_i)+\mathrm{ind}(J_\lambda+g,0)=0,$$

where $g(x,\lambda)=F(x,\lambda)+o(|x|^2)$, which implies that

$$\sum_{i=1}^{n}\mathrm{ind}(J_\lambda+g,z_i)=-1. \tag{2.4.63}$$

When $J_\lambda+g$ has just one bifurcated singular point z_1, by (2.4.63), ind $(J_\lambda+g,z_1)=-1$, which implies that z_1 is a saddle point. The structure of $F(x,\lambda_0)$ at $x=0$ determines this transition structure of $J_\lambda+g$ as shown in Fig. 2.19a.

When $J_\lambda+g$ has two bifurcated singular points z_1 and z_2, by Theorem 2.4.3, z_1 is regular and z_2 is degenerate. Hence, by (2.4.63) we can deduce that

$$\mathrm{ind}(J_\lambda+g,z_1)=-1,\quad \mathrm{ind}(J_\lambda+g,z_2)=0,$$

i.e., z_1 is a saddle point, and z_2 has two hyperbolic regions. Thus, the transition structure is as shown in Fig. 2.19b.

When $J_\lambda+g$ has three bifurcated singular points, by Theorem 2.4.3, they are regular. Therefore, by (2.4.63), we have

$$\mathrm{ind}(J_\lambda+g,z_1)=\mathrm{ind}(J_\lambda+g,z_2)=-1,\qquad \mathrm{ind}(J_\lambda+g,z_3)=1,$$

which shows that the transition structure has to be topologically equivalent to that shown in Fig. 2.19c.

Finally, similar to Theorem 2.4.6, we can obtain assertions (1)–(3). Thus, the proof is complete. □

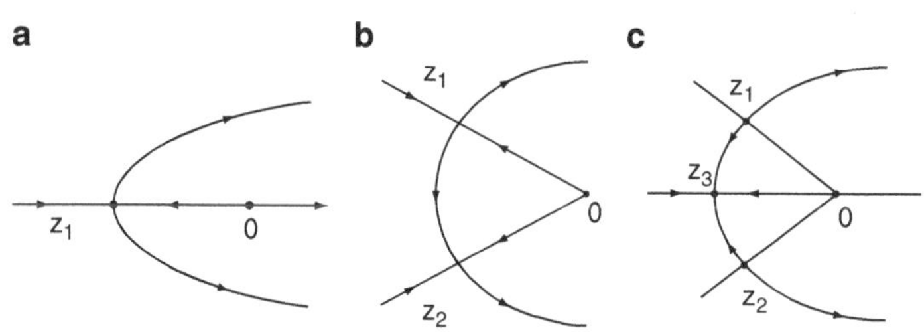

Fig. 2.19 Illustration for the proof of Theorem 2.4.11 for the cases where $J_\lambda+g$ bifurcating to (**a**) one singular point, (**b**) two singular points and (**c**) three singular points

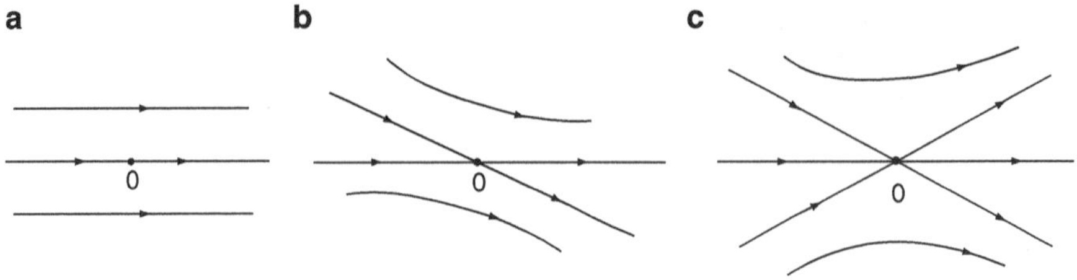

Fig. 2.20 Topological structure of $F(x,\lambda)$ at $x=0$: (**a**) $k=1$; (**b**) $k=2$; (**c**) $k=3$

2.4.6 Indices of kth-order Nondegenerate Singularities

In order to investigate the transition of (2.1.1) from kth-order nondegenerate singular points, it is necessary to study the index of k-homogeneous vector fields at $x=0$. Let

$$F=\begin{pmatrix}F_1\\F_2\end{pmatrix}=\begin{pmatrix}\sum\limits_{i+j=k} a^1_{ij}x_1^i x_2^j\\ \sum\limits_{i+j=k} a^2_{ij}x_1^i x_2^j\end{pmatrix} \tag{2.4.64}$$

be a kth-order homogeneous field, with $k\geq 2$ an integer.

It is known that the index of F at $x=0$ satisfies

$$\mathrm{ind}(F,0)=\begin{cases}\text{even} & \text{if } k=\text{even},\\ \text{odd} & \text{if } k=\text{odd}.\end{cases}$$

Assume that F is kth-order nondegenerate, i.e., $x=0$ is an isolated singular point of F. The components $F_i(x)$ $(i=1,2)$ of F induce two kth-order polynomials as follows:

$$F_i(z,1)=a^i_{k0}z^k+a^i_{k-11}z^{k-1}+\cdots+a^i_{1,k-1}z+a^i_{0k}. \tag{2.4.65}$$

Since F is kth-order nondegenerate, $|a^1_{k0}|+|a^2_{k0}|\neq 0$. Without loss generality, we assume that $a^1_{k0}\neq 0$; otherwise, we can consider the case $a^2_{k0}\neq 0$.

The following theorem provides an index formula for the kth-order nondegenerate vector field (2.4.64), which is a generalized version of the second-order index formula (2.4.2).

Theorem 2.4.12 *Let $z_1\geq\cdots\geq z_N$ be N real zero points of $F_1(z,1)$, i.e., $F_1(z_j,1)=0$, for all $1\leq j\leq N$ (counting multiplicities). Then*

$$ind(F,0)=\begin{cases}\sum\limits_{j=1}^{N}(-1)^{j+1} sign\ F_2(z_j,1) & if\ a^1_{k0}>0,\ N\geq 1,\\ \sum\limits_{j=1}^{N}(-1)^{j} sign\ F_2(z_j,1) & if\ a^1_{k0}<0,\ N\geq 1,\\ 0 & if\ N=0.\end{cases} \tag{2.4.66}$$

Proof. We proceed in several steps as follows.

Step 1. Since $x = 0$ is an isolated singular point of F, for each real zero point z_0 of $F_1(z,1), we have F_2(z_0,1) \neq 0$; otherwise, it is easy to see that all points on the line $x_1 = z_0 x_2$ are singular points of F.

It is clear that when $k = 2$, the formula (2.4.66) is equivalent to (2.4.2).

When $N = 0$, $F_1(x_1,x_2)$ is either positive or negative definite. Hence either

$$(F_1(x_1,x_2),F_2(x_1,x_2)) = (-\varepsilon^2,0) \quad \text{or} \quad (F_1(x_1,x_2),F_2(x_1,x_2)) = (\varepsilon^2,0)$$

has no solutions for any $\varepsilon \neq 0$. Hence we have

$$\text{ind}(F,0) = 0 \qquad \text{for } N = 0.$$

Step 2. Now we need to discuss the case that $N \geq 1$ and $a^1_{k0} > 0$; the case for $a^1_{k0} < 0$ can be proved similarly.

First, we consider the case in which the real zero points $z_1 > \cdots > z_N$ of $F_1(z,1)$, which represent N straight lines, are simple:

$$L_i : x_1 = z_i x_2 \qquad \text{for } 1 \leq i \leq N.$$

The N straight lines L_i $(1 \leq i \leq N)$ divide the region $\mathbb{R}^2_+ = \{(x_1,x_2) \in \mathbb{R}^2 |\ x_2 > 0\}$ into $N+1$ sectorial regions $D_0, D_1, \cdots, D_N$ with

$$\begin{aligned} \partial D_i &= L_i^+ + L_{i+1}^+ \quad \text{for } i \neq 0,N, \\ \partial D_0 &= L_0^+ + L_1^+, \\ \partial D_N &= L_N^+ + L_0^-, \end{aligned} \tag{2.4.67}$$

where

$$\begin{aligned} L_i^+ &= \{(x_1,x_2) \in L_i |\ x_2 \geq 0\} \qquad \text{for } i \neq 0, \\ L_0^+ &= \{(x_1,x_2) |\ x_1 \geq 0, x_2 = 0\}, \\ L_0^- &= \{(x_1,x_2) |\ x_1 \leq 0,\ x_2 = 0\}. \end{aligned}$$

We note, by $a^1_{k0} > 0$, that

$$\begin{aligned} &F_1(x_1,x_2) = 0 && \text{on } L_i : x_1 = z_i x_2, \\ &\text{sign } F_1(x_1,x_2) = (-1)^i && \forall x = (x_1,x_2) \in D_i. \end{aligned} \tag{2.4.68}$$

Let Q_i $(1 \leq i \leq 4)$ be the four quadrants in $\mathbb{R}^2$. For convenience, we assume that $z_i \neq 0$ $(1 \leq i \leq N)$. In fact, $z_i = 0$ corresponds to L_i being the x_2-axis, in which case, the discussion is the same. Thus, there exists l such that

$$\begin{aligned} &D_i \subset Q_1 + Q_4 && \forall 0 \leq i \leq l-1, \\ &D_j \subset Q_2 && \forall l+1 \leq j \leq N, \end{aligned}$$

$$D_l \bigcap Q_1 \neq \emptyset,$$
$$D_l \bigcap Q_2 \neq \emptyset.$$

When $N = 1$, k must be an odd number, and by (2.4.67) and $a_{k0}^1 > 0$, we have

$$F_1(x) \begin{cases} > 0 & \text{if } x \in D_0, \\ < 0 & \text{if } x \in D_1. \end{cases} \tag{2.4.69}$$

By $F(-x) = -F(x)$, it follows from (2.4.69) that $F(x)$ has only a parabolic region at $x = 0$, as shown in Fig. 2.21a,b. The two elliptic and two hyperbolic regions are as shown in Fig. 2.22a (or four hyperbolic regions are as shown in Fig. 2.22b), and

$$F_2(z_1, 1) > 0 \quad (\text{or } F_2(z_1, 1) < 0). \tag{2.4.70}$$

By the Poincaré formula and (2.4.70), we derive that

$$\text{ind}(F, 0) = 1 + \frac{1}{2}(e - h) = \text{sign} F_2(z_1, 1) = \pm 1.$$

Hence, formula (2.4.66) is valid for $N = 1$.

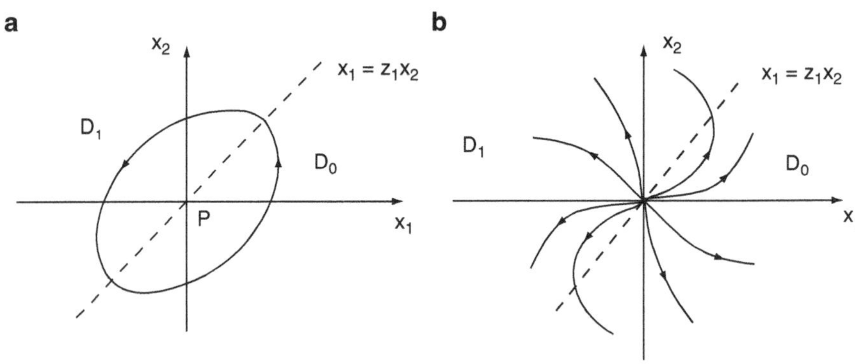

Fig. 2.21 Illustration of Step 2 of the proof of Theorem 2.4.12

Step 3. Let $N = 2$. We consider the structure of F near L_i^+ $(x_1 = z_i\, x_2, x_2 \geq 0)$. We first discuss the case $L_i^+ \subset Q_1$ $(i = 1, 2)$, i.e., $z_1, z_2 > 0$. For the case $L_i^+ \subset Q_2$ $(i = 1, 2)$, the discussion is the same.

First, when $F_1(x) > 0$ in D_0 and $F_2(z_1, 1) > 0, F_2(z_2, 1) < 0$, there is an elliptic region E_1 containing L_1^+, and the adjacent region on the left-hand side of E_1, containing L_2^+, is either an elliptic region or a parabolic region, as shown in Fig. 2.23a or b. In addition, because $N = 2$ and z_1 and z_2 are simple singular points of $F_1(z, 1)$, the number k in (2.4.64) is even. Hence, if the region E_2 containing L_2^+ is elliptic,

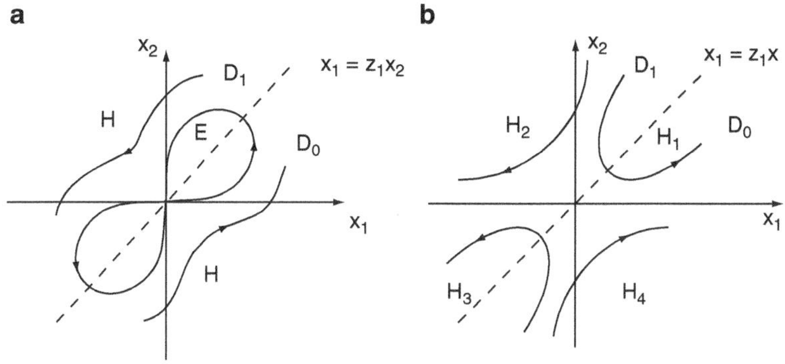

Fig. 2.22 Illustration of Step 2 of the proof of Theorem 2.4.12-continued

then the regions on the left-hand side of E_2 and on the right-hand side of E_1 are hyperbolic; see Fig. 2.23a.

Second, when $F_1(x) > 0$ in D_0, and $F_2(z_1,1) < 0$, $F_2(z_2,1) > 0$, there are exactly three hyperbolic regions H_1, H_2, H_3 of F in $Q_1 + Q_2$, where H_1 contains L_1^+, H_2 contains L_2^+, as shown in Fig. 2.24.

Third, when $F_1(x) > 0$ in D_0, and $F_2(z_1,1) > 0$, $F_2(z_2,1) > 0$, then either there are an elliptic region E_1 and two hyperbolic regions H_2, H_3 of F in $Q_1 + Q_2$ with E_1 containing L_1^+ and H_2 containing L_2^+, as shown in Fig. 2.25a, or there are a parabolic region P containing L_1^+, L_2^+ and a hyperbolic region H, as shown in Fig. 2.25b.

Fourth, when $F_1(x) > 0$ in D_0, and $F_2(z_1,1) < 0$, $F_2(z_2,1) < 0$, then either there are two hyperbolic regions H_0, H_1 and an elliptic region E_2 with H_1 containing L_1^+ and E_2 containing L_2^+, or there are a parabolic region P containing L_1^+, L_2^+ and a hyperbolic region H in $Q_1 + Q_4$, as shown in Fig. 2.26a, b.

Now, we consider the case in which $L_1^+ \subset Q_1$ and $L_2^+ \subset Q_2$, i.e., $D_0 \subset Q_1, D_2 \subset Q_2$ and $D_1 \bigcap Q_1 \neq \emptyset, D_1 \bigcap Q_2 \neq \emptyset$.

Fifth, when $F_1(x) > 0$ in D_0, and $F_2(z_1,1) > 0$, $F_2(z_2,1) < 0$, there is an elliptic region E containing L_1^+ and L_2^+ in $Q_1 + Q_2$, as shown in Fig. 2.27a. When $F_2(z_1,1) < 0$, $F_2(z_2,1) > 0$, there are three hyperbolic regions H_1, H_2, H_3 in $Q_1 + Q_2$, where H_1 contains L_1^+, H_3 contains L_2^+, and H_2 is between L_1^+ and L_2^+; see Fig. 2.27b.

Sixth, when $F_1(x) > 0$ in D_0, and $F_2(z_1,1) > 0$, $F_2(z_2,1) > 0$, there are a parabolic region P containing L_2^+ and a hyperbolic region H containing L_1^+, as shown in Fig. 2.28a, and when $F_2(z_1,1) < 0$, $F_2(z_2,1) < 0$, the structure of F is as shown in Fig. 2.28b.

From the six properties above, it is easy to see that for $N = 2$ with $z_1 > z_2$, the following formula holds:

$$\text{sign} F_2(z_1,1) - \text{sign} F_2(z_2,1) = 1 + \frac{1}{2}(e-h), \tag{2.4.71}$$

where e and h are the numbers of elliptic and hyperbolic regions. Hence the formula (2.4.66) with $N = 2$ follows from (2.4.71) and the Poincaré formula (2.1.9).

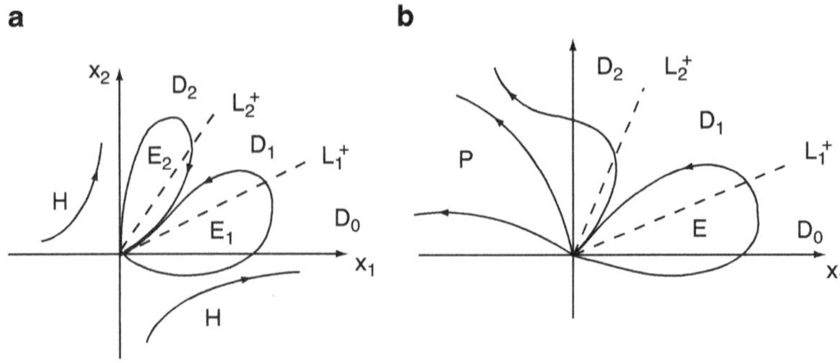

Fig. 2.23 Illustration of Step 3 for the proof of Theorem 2.4.12

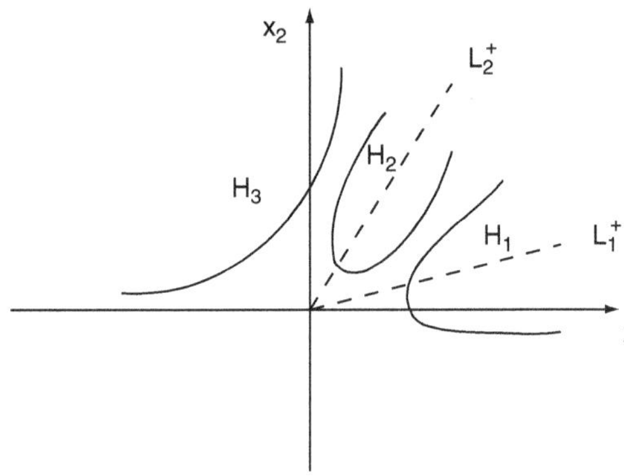

Fig. 2.24 Illustration of Step 3 for the proof of Theorem 2.4.12-continued

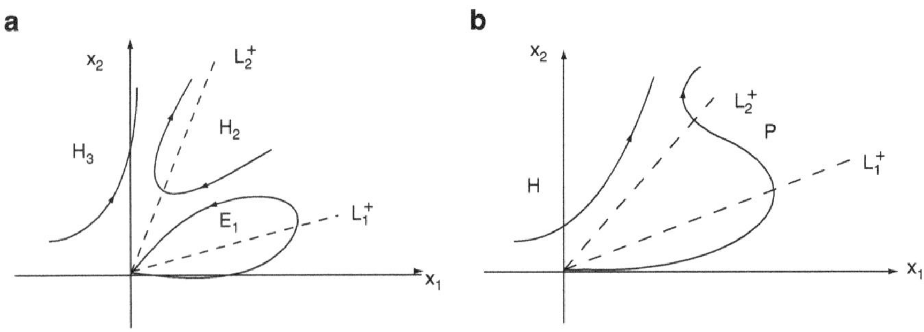

Fig. 2.25 Illustration of Step 3 for the proof of Theorem 2.4.12-continued

Step 4. In the same fashion as above, for $N \geq 3$ we can draw from (2.4.68) the following conclusions:

1. Each elliptic region E intersects with r lines L_{i+j} $(1 \leq j \leq r)$, and the number r satisfies

$$r = \begin{cases} 2m+1 & \text{if } E \subset Q_1 + Q_4 \text{ or } E \subset Q_2 + Q_3, \\ 2m \geq 2 & \text{if } E \bigcap Q_1 \neq \emptyset \text{ and } E \bigcap Q_2 \neq \emptyset. \end{cases}$$

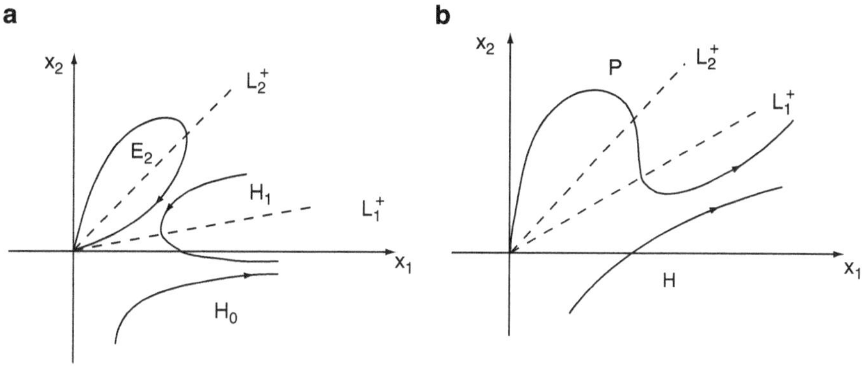

Fig. 2.26 Illustration of Step 3 for the proof of Theorem 2.4.12-continued

In addition, on the lines L_{i+j} $(x_1 = z_{i+j}x_2)$, for $E \subset Q_1 + Q_4$ or $E \subset Q_2 + Q_3$, we have

$$\text{sign}F_2(z_{i+1}, 1) = \cdots = \text{sign}F_2(z_{i+r}, 1),$$

and for $E \bigcap Q_1 \neq \emptyset, E \bigcap Q_2 \neq \emptyset$, the numbers of lines $L^+_{i+s} \subset E \bigcap Q_1$ and $L^+_{i+s} \subset E \bigcap Q_2$ are odd, i.e., l is odd, $r - l$ is odd, and

$$\begin{array}{ll} \text{sign}F_2(z_{i+1}, 1) = \cdots = \text{sign}F_2(z_{i+l}, 1) & \text{for } L^+_{i+s} \subset Q_1, \forall 1 \leq s \leq l, \\ \text{sign}F_2(z_{i+l+1}, 1) = \cdots = \text{sign}F_2(z_{i+r}, 1) & \text{for } L^+_{i+s} \subset Q_2, \forall l+1 \leq s \leq r, \\ \text{sign}F_2(z_{i+1}, 1) = -\text{sign}F_2(z_{i+r}, 1) & \text{for } L^+_{i+1} \subset Q_1, \quad L^+_{i+r} \subset Q_2. \end{array}$$

2. Each hyperbolic region H intersects with r lines L_{i+j} $(1 \leq j \leq r)$, with the number r given by

$$r = \begin{cases} 2m+1 & \text{if } H \subset Q_1 + Q_4 \text{ or } H \subset Q_2 + Q_3, \\ 2m \geq 0 & \text{if } H \bigcap Q_1 \neq \emptyset \text{ and } H \bigcap Q_2 \neq \emptyset. \end{cases}$$

On the lines L_{i+j} $(x_1 = z_{i+j}x_2)$, for $H \subset Q_1 + Q_4$ or $H \subset Q_2 + Q_3$,

$$\text{sign}F_2(z_{i+1}, 1) = \cdots = \text{sign}F_2(z_{i+r}, 1),$$

and for $H \bigcap Q_1 \neq \emptyset, H \bigcap Q_2 \neq \emptyset$, the numbers of lines $L^+_{i+s} \subset E \bigcap Q_1$ and $L^+_{i+s} \subset E \bigcap Q_2$ are even: l is even, $r - l$ is even, and

$$\begin{array}{lll} \text{sign}F_2(z_{i+1}, 1) = \cdots = \text{sign}F_2(z_{i+r}, 1) & \text{for } L^+_{i+s} \subset Q_1, & 1 \leq s \leq l, \\ \text{sign}F_2(z_{i+l+1}, 1) = \cdots = \text{sign}F_2(z_{i+r}, 1) & \text{for } L^+_{i+s} \subset Q_2, & l+1 \leq s \leq r, \\ \text{sign}F_2(z_{i+1}, 1) = -\text{sign}F_2(z_{i+r}, 1) & \text{for } L^+_{i+1} \subset Q_1, & L^+_{i+r} \subset Q_2. \end{array}$$

3. Each parabolic region P intersects with r lines L_{i+j} $(1 \leq j \leq r)$, with

$$r = \begin{cases} 2m \geq 0 & \text{if } P \subset Q_1 + Q_4 \text{ or } P \subset Q_2 + Q_3, \\ 2m+1 & \text{if } P \bigcap Q_1 \neq \emptyset \text{ and } P \bigcap Q_2 \neq \emptyset. \end{cases}$$

Moreover, on the lines L_{i+j} $(x_1 = z_{i+j}x_2)$:

a. If $P \subset Q_1 + Q_4$ or $P \subset Q_2 + Q_3$, then only at an even number of successive points $z_{i+s+1}, \cdots, z_{i+s+2k}$ are the signs of $F_2(z,1)$ the same:

$$\text{sign}F_2(z_{i+s+1}, 1) = \cdots = \text{sign}F_2(z_{i+s+2k}, 1), \quad s + 2k \leq r,$$

b. If $P \bigcap Q_1 \neq \emptyset, P \bigcap Q_2 \neq \emptyset$, then if $P \bigcap Q_1$ (or $P \bigcap Q_2$) contains an odd number of lines $L^+_{i+1}, \cdots, L^+_{i+2k+1}$ (or $L^+_{i+2k+1}, \cdots, L^+_{i+r}$), the signs of $F_2(z,1)$ are the same at the odd number of successive points $z_{i+2k+1} < \cdots < z_{i+2j+1}$ $(j \leq k)$ (or $z_{i+2k+1} > \cdots > z_{i+2j+1}, j \geq k$):

$$\begin{aligned} &\text{sign}F_2(z_{i+2j+1}, 1) = \cdots = \text{sign}F_2(z_{i+2k+1}, 1) && \forall 0 \leq j \leq k, \\ &(\text{or sign}F_2(z_{i+2k+1}, 1) = \cdots = \text{sign}F_2(z_{i+2j+1}, 1) && \forall k \leq j), \end{aligned}$$

and the signs of $F_2(z,1)$ are the same at the remaining even number of successive points of z_{i+j} $(1 \leq j \leq r)$.

4. Let E be an elliptic region containing $r \geq 1$ lines L_{i+j} $(1 \leq j \leq r)$.

a. If

$$\begin{aligned} &\text{sign}F_2(z_{i+r+1}, 1) = -\text{sign}F_2(z_{i+r}, 1) \\ &(\text{respectively sign}F_2(z_i, 1) = -\text{sign}F_2(z_{i+1}, 1)), \end{aligned} \tag{2.4.72}$$

then the adjacent region of E intersecting with L_{i+r+1} (respectively with L_i) that is on the left-hand side (respectively right-hand side) of E is either elliptic or parabolic.

b. If

$$\begin{aligned} &\text{sign}F_2(z_{i+r+1}, 1) = \text{sign}(z_{i+r}, 1) \\ &(\text{respectively sign}F_2(z_i, 1) = \text{sign}(z_{i+1}, 1)), \end{aligned} \tag{2.4.73}$$

then the adjacent region intersecting with L_{i+r+1} (respectively with L_i) is either hyperbolic or parabolic.

5. Let H be a hyperbolic region intersecting with $r(\geq 1)$ lines L_{i+j} $(1 \leq j \leq r)$.

a. If (2.4.72) holds, then the adjacent region of H intersecting with L_{i+r+1} (respectively with L_i) is either hyperbolic or parabolic.

b. If (2.4.73) holds, then the adjacent region intersecting with L_{i+r+1} (respectively with L_i) is either elliptic or parabolic.

6. Let P be a parabolic region intersecting with $r(\geq 0)$ lines L_{i+j} $(1 \leq j \leq r)$. Let $P \subset Q_1$ or $P \subset Q_2$.

a. If the right adjacent region of P, which intersects with L_i, is elliptic, then if

$$\text{sign}F_2(z_i, 1) = -\text{sign}F_2(z_{i+r+1}, 1), \tag{2.4.74}$$

then the left adjacent region of P, which intersects with L_{i+r+1}, must be elliptic, and if

$$\text{sign}F_2(z_i, 1) = \text{sign}F_2(z_{i+r+1}, 1), \tag{2.4.75}$$

then the left adjacent region is hyperbolic.

b. When the right adjacent region of P is hyperbolic, then if (2.4.74) holds, the left adjacent region is also hyperbolic, and if (2.4.75) holds, then the left adjacent region is elliptic.

7. Let P be as in (6), and $P \bigcap Q_1 \neq \emptyset, P \bigcap Q_2 \neq \emptyset$. If the right adjacent region of P is elliptic, then if (2.4.74) holds, the left adjacent region is hyperbolic, and if (2.4.75) holds, then the left adjacent region is elliptic.

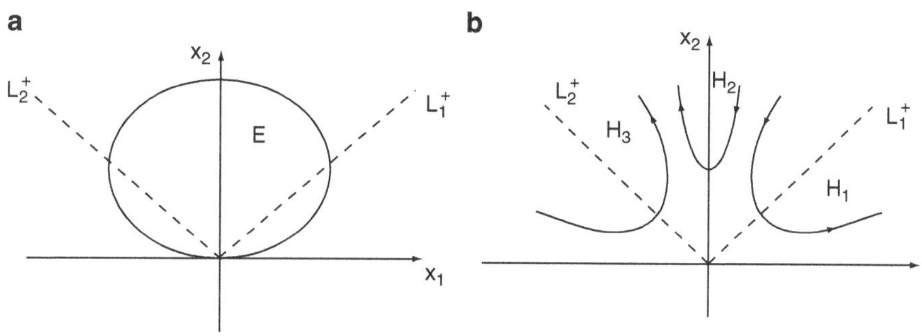

Fig. 2.27 Illustration of Step 3 for the proof of Theorem 2.4.12-continued

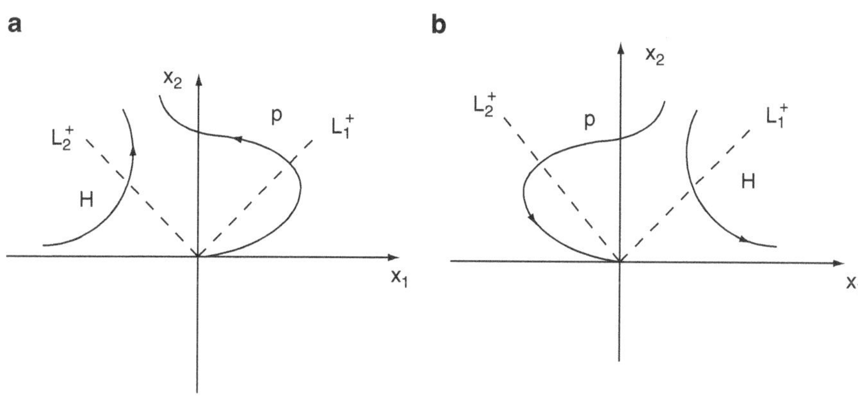

Fig. 2.28 Illustration of Step 3 for the proof of Theorem 2.4.12-continued

For $z_1 > z_2 > \cdots > z_N$ $(N \geq 3)$, from properties 1–7 above, we can see that it suffices to prove (2.4.66) for only the following cases:

(i) Each elliptic region E with $E \subset Q_1$ or $E \subset Q_2$ intersects with only one line L_i $(x_1 = z_i x_2)$.
(ii) The elliptic region E with $E \bigcap Q_1 \neq \emptyset, E \bigcap Q_2 \neq \emptyset$ (if it exists) intersects only with L_l and L_{l+1}, where $L_l \subset Q_1, L_{l+1} \subset Q_2$.
(iii) Each hyperbolic region H with $H \subset Q_1$ or $H \subset Q_2$ intersects with only one line L_i $(1 \leq i \leq N)$.
(iv) The hyperbolic region H with $H \bigcap Q_1 \neq \emptyset, H \bigcap Q_2 \neq \emptyset$ (if it exists) intersects with no line between L_l and L_{l+1}, where $L_l \subset Q_1$ and $L_{l+1} \subset Q_2$.
(v) There are no parabolic regions in Q_1 and Q_2, and if there is a parabolic region P with $P \bigcap Q_1 \neq \emptyset, P \bigcap Q_2 \neq \emptyset$, then P intersects only with either L_l or L_{l+1}, where $L_l \subset Q_1$ and $L_{l+1} \subset Q_2$.

Step 5. Based on properties (i)–(v), we now prove the index formula (2.4.66). Obviously, only one of the elliptic, hyperbolic, and parabolic regions, denoted by R, satisfies $R \bigcap Q_1 \neq \emptyset$ and $R \bigcap Q_2 \neq 0$. Hence, we proceed with the following three cases.

Case 1. There is an elliptic region E_l satisfying $E_l \bigcap Q_1 \neq \emptyset$ and $E_l \bigcap Q_2 \neq \emptyset$. In this case, by property (v), there are no parabolic regions, and all hyperbolic regions are in either Q_1 or Q_2.

For $a^1_{k0} > 0$, using properties (i) and (iii) above, it is easy to verify that the region containing L_1^+ is elliptic if and only if $F_2(z_1, 1) > 0$. Hence, by the previous properties 4 and 5, we deduce that

$$\begin{aligned} (-1)^{j+1}\text{sign}F_2(z_j,1) = 1 &\Leftrightarrow \text{the region containing } L_j^+ \text{ is elliptic,} \\ (-1)^{j+1}\text{sign}F_2(z_j,1) = -1 &\Leftrightarrow \text{the region containing } L_j^+ \text{ is hyperbolic.} \end{aligned} \tag{2.4.76}$$

By properties (i)–(iv), only the elliptic region E_l contains two lines L_l^+ and L_{l+1}^+, and

$$(-1)^k \text{sign}F_2(z_k, 1) = 1, \quad k = l, l+1.$$

Hence, the numbers of elliptic and hyperbolic regions are

$$\begin{aligned} e &= 2\big[-1 + \text{number of } L_j^+ \text{ with } (-1)^{j+1}\text{sign}F_2(z_j,1) = 1\big], \\ h &= 2 \times \text{ number of } L_j^+ \text{ with } (-1)^{j+1}\text{sign}F_2(z_j,1) = -1. \end{aligned}$$

Therefore, by the Poincaré formula (2.1.9), we derive that

$$\text{ind}(F,0) = 1 + \frac{1}{2}(e-h) = \sum_{j=1}^{m}(-1)^{j+1}\text{sign}F_2(z_j,1). \tag{2.4.77}$$

Case 2. There is a hyperbolic region H_l such that $H_l \bigcap Q_1 \neq \emptyset$ and $H_l \bigcap Q_2 \neq \emptyset$.

Since there are no parabolic regions in $\mathbb{R}^2$, the conclusion (2.4.76) is valid. By property (iv), this hyperbolic region H_l is the unique one intersecting with no lines of $L_j (1 \le j \le m)$. Hence, the numbers of hyperbolic and elliptic regions are

$$h = 2\left[1 + \text{number of } L_j^+ \text{ with } (-1)^{j+1}\text{sign}F_2(z_j,1) = -1\right],$$
$$e = 2 \text{ number of } L_j^+ \text{ with } (-1)^{j+1}\text{sign}F_2(z_j,1) = 1.$$

Thus we derive (2.4.77).

Case 3. There is a parabolic region P satisfying $P \bigcap Q_1 \neq \emptyset$ and $P \bigcap Q_2 \neq \emptyset$.

By property (v), the parabolic region P is unique, and P intersects with only one line L_l or L_{l+1}.

We know that the parabolic region $P \subset Q_1 + Q_2$ can be characterized by the orientation as follows:

$$P^+ = \{x \in P | \ S(t)x \to 0, \quad t \to -\infty\},$$
$$P^- = \{x \in P | \ S(t)x \to 0, \quad t \to +\infty\},$$

where $S(t)$ is the semigroup generated by $F(x)$. It is not difficult to obtain (see Fig. 2.28a, b)

$$\begin{cases} F_1(x) > 0 & \forall x \in D_{l-1} \quad \text{if } P = P^+ \text{ contains } L_l^+, \\ F_1(x) > 0 & \forall x \in D_l \quad \text{if } P = P^+ \text{ contains } L_{l+1}^+, \end{cases} \tag{2.4.78}$$

$$\begin{cases} F_1(x) < 0 & \forall x \in D_{l-1} \quad \text{if } P = P^- \text{ contains } L_l^+, \\ F_1(x) < 0, & \forall x \in D_l \quad \text{if } P = P^- \text{ contains } L_{l+1}^+, \end{cases} \tag{2.4.79}$$

where D_j with $\partial D_j = L_j^+ + L_{j+1}^+$ are the domains as in (2.4.67), and

$$\begin{cases} F_2(z_l,1) > 0 & \text{if } P = P^+ \text{ contains } L_l^+, \\ F_2(z_{l+1},1) > 0 & \text{if } P = P^+ \text{ contains } L_{l+1}^+, \end{cases} \tag{2.4.80}$$

$$\begin{cases} F_2(z_l,1) < 0 & \text{if } P = P^- \text{ contains } L_l^+, \\ F_2(z_{l+1},1) < 0 & \text{if } P = P^- \text{ contains } L_{l+1}^+. \end{cases} \tag{2.4.81}$$

On the other hand, by assumption, $a_{k0}^1 > 0$, i.e., (2.4.68) holds. Thus it follows from (2.4.78) and (2.4.79) that

$$l = \begin{cases} \text{odd} & \text{if } P = P^+ \text{ contains } L_l^+, \\ \text{even} & \text{if } P = P^+ \text{ contains } L_{l+1}^+, \end{cases} \tag{2.4.82}$$

$$l = \begin{cases} \text{even} & \text{if } P = P^- \text{ contains } L_l^+, \\ \text{odd} & \text{if } P = P^- \text{ contains } L_{l+1}^+. \end{cases} \tag{2.4.83}$$

Therefore, from (2.4.80)–(2.4.83), we obtain that

$$(-1)^{j+1}\text{sign}F_2(z_j,1)=1 \quad \text{as } P \text{ contains } L_j^+. \tag{2.4.84}$$

Thus, we deduce from (2.3.2) that

$$\begin{aligned} e+2 &= 2\times \text{number of } L_j^+ \text{ with } (-1)^{j+1}\text{sign}F_2(z_j,1)=1,\\ h &= 2\times \text{number of } L_j^+ \text{ with } (-1)^{j+1}\text{sign}F_2(z_j,1)=-1, \end{aligned}$$

which implies that (2.4.77) holds.

Thus, formula (2.4.66) is proved for the case in which $z_1>\cdots>z_N$ are simple singular points of $F_1(z,1)$.

Step 6. Now, we consider the case in which z_j may be multiple singular points of $F_1(z,1)$. Without loss of generality, we assume that the multiplicity of z_1 is $r>1$, and that the others are 1, i.e.,

$$z_1=\cdots=z_r<z_{r+1}<\cdots<z_N. \tag{2.4.85}$$

Thus, we have

$$F_1(x)=a_{k0}^1(x_1-z_1x_2)^r(x_1-z_{r+1}x_2)\cdots(x_1-z_Nx_2)f(x), \tag{2.4.86}$$

where $f(x)$ is a positive definite polynomial of order $(k-m)$.

When $r=$ even, we see that

$$\sum_{j=1}^{N}(-1)^{j+1}\text{sign}F_2(z_j,1)=\sum_{i=1}^{N-r}(-1)^{i+1}\text{sign}F_2(z_{r+i},1).$$

On other hand, by (2.4.86), we see that the lines $L_i: x_1=z_{r+i}x_2 (1\le i\le N-r)$ with $z_{r+1}>\cdots>z_{r+N}$ divide the half-plane $\mathbb{R}^2_+=\{x\in\mathbb{R}^2|\ x_2>0\}$ into $N-r+1$ sectorial domains $D_0,\cdots,D_{N-r}$, as given by (2.4.67), where (2.4.68) is valid except in D_0, and

$$\text{sign}F_1(x)=1 \quad \text{as} \quad x\in D_0/L, \quad L: x_1=z_1x_2.$$

Thus, the proof above is valid for the case (2.4.85) with r even.

When r is odd, we have

$$\sum_{j=1}^{N}(-1)^{j+1}\text{sign}F_2(z_j,1)=\sum_{i=1}^{N-r+1}(-1)^{i+1}\text{sign}F_2(z_{r+i-1},1).$$

By (2.4.86), the lines $L_i: x_1=z_{r+i-1}x_2$ $(1\le i\le m-r+1)$ with $z_r>\cdots>z_N$ divide $\mathbb{R}^2_+$ into $N-r+2$ domains $D_0,\cdots,D_{N-r+1}$ where (2.4.68) is valid. Thus, the proof above is still valid.

The proof of the theorem is complete. □

Remark 2.4.13 When $a_{k0}^1=0$ and $a_{k0}^2\neq 0$, if $z_1\ge\cdots\ge z_N$ are N real zero points of $F_2(z,1)$, then

$$\text{ind}(F,0) = \begin{cases} 0 & \text{if } N = 0, \\ \sum_{j=1}^{N} (-1)^j \text{sign} F_1(z_j, 1) & \text{if } a_{k0}^2 > 0, \quad N \neq 0, \\ \sum_{j=1}^{N} (-1)^{j+1} \text{sign} F_1(z_j, 1) & \text{if } a_{k0}^2 < 0, \quad N \neq 0. \end{cases} \tag{2.4.87}$$

Remark 2.4.14 Formula (2.4.66) shows that for a k-homogeneous vector field (2.4.64), the index of F at $x = 0$ takes values in the following range:

$$\text{ind}(F,0) = \begin{cases} 0, \pm 2, \cdots, \pm k & \text{if } k = \text{even}, \\ \pm 1, \pm 3, \cdots, \pm k & \text{if } k = \text{odd}. \end{cases}$$

2.4.7 Structure of kth-Order Nondegenerate Singularities

In the previous subsections, we studied transitions and their structure from second-order nondegenerate singularities. Now, we shall consider the general kth-order case. For this purpose, in this subsection we investigate the topological structure of a nondegenerate k-homogeneous vector field.

Let $F(x) = (F_1(x), F_2(x))$ be a two-dimensional k-homogeneous nondegenerate vector field given by (2.4.64), and $F_i(z,1)$ the polynomials induced by $F_i(x_1, x_2)$ $(i = 1, 2)$. We introduce another polynomial associated with F:

$$h(z) = a_{k0}^2 z^{k+1} + \sum_{i=1}^{k} (a_{k-ii}^2 - a_{k-i+1i-1}^1) z^{k-i+1} - a_{0k}^1. \tag{2.4.88}$$

The polynomial $h(z)$ is derived by

$$z = \frac{F_1(z,1)}{F_2(z,1)}, \tag{2.4.89}$$

i.e., $h(z) = zF_2(z,1) - F_1(z,1)$. Hence, it is clear that a real zero point z_0 of $h(z)$ represents a straight orbit line L of $F(x)$, given by $L : x_1 = z_0 x_2$.

Hereinafter, we always assume that $a_{k0}^1 \neq 0$, i.e., we take the index formula (2.4.66) in our discussion; otherwise, we take (2.4.87). Thus, the topological structure of F is completely determined by the zero points of $h(z)$ and $F_1(z,1)$.

Because the index of F at $x = 0$ is related to only the odd-multiple zero points of $F_1(z,1)$, we take all zero points of $F_1(z,1)$ with odd multiplicity for $z_1 > \cdots > z_N$, and all real zero points of $h(z)$ for $\alpha_1 > \cdots > \alpha_K$, which correspond to straight lines as follows:

$$\begin{aligned} L_i &: x_1 = z_i x_2 & \forall 1 \leq i \leq N, \\ l_j &: x_1 = \alpha_j x_2 & \forall 1 \leq j \leq K. \end{aligned}$$

Notice that $F(-x) = (-1)^k F(x)$. It suffices to consider the structure of $F(x)$ only in the half-plane

$$\mathbb{R}^2_- = \{x \in \mathbb{R}^2 \,|\, x_1 - \alpha_1 x_2 < 0\}.$$

Thus, the lines l_j $(1 \le j \le K)$ divide $\mathbb{R}^2_-$ into K sectorial domains:

$$S_i = \{(x_1,x_2) \in \mathbb{R}^2_- \,|\, x_1 - \alpha_{i+1}x_2 > 0, \quad x_1 - \alpha_i x_2 < 0\}, \quad 1 \le i \le K-1,$$
$$S_K = \{(x_1,x_2) \in \mathbb{R}^2_- \,|\, x_1 - \alpha_{K-1}x_2 < 0, \quad x_1 - \alpha_1 x_2 < 0\}.$$

Since $F(x)$ is homogeneous, we have the following claim.

Claim A *Each sectorial domain S_j $(1 \le j \le K)$ must be one of the elliptic, hyperbolic, and parabolic regions shown in Fig. 2.29. Hence, if we can be sure of the type of region and its orientation in each sectorial domain $S_j \subset \mathbb{R}^2_-$, then the structure of F in $\mathbb{R}^2$ is completely determined.*

Let S_i be a sectorial domain as above. In the following, we give a method to determine the region type of S_i. By definition,

$$\partial S_i = l_i + l_{i+1}, \quad l_j : x_1 = \alpha_j x_2.$$

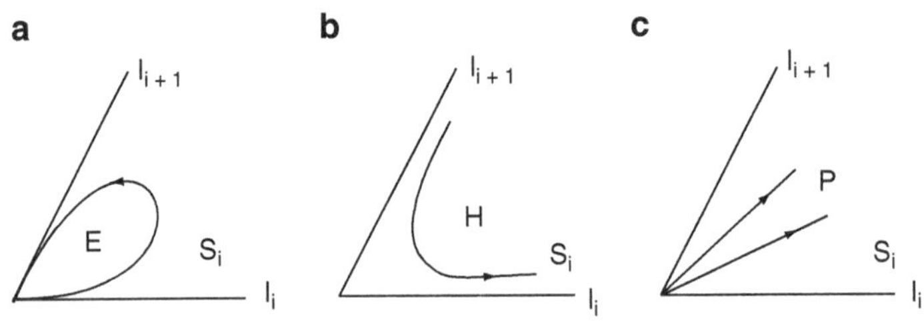

Fig. 2.29 (**a**) S_i is an elliptic region; (**b**) S_i is a hyperbolic region; (**c**) S_i is a parabolic region

We consider only the case $a^2_{k0} \neq 0$, i.e., $F_2(x_1,0) \not\equiv 0$; for the case $a^2_{k0} = 0$, the discussion is the same. Hence, by $\alpha_1 > \alpha_2 > \cdots > \alpha_K$, we have

$$x_1 = \alpha_j x_2 \ (1 \le j \le K) \quad \text{and } (x_1,x_2) \in \mathbb{R}^2_- \Rightarrow x_2 > 0.$$

Let E, H, P be the elliptic, hyperbolic, and parabolic regions. Then

$$\begin{cases} S_i = P \Leftrightarrow \text{sign}F_2(\alpha_i,1) = \text{sign}F_2(\alpha_{i+1},1) & \forall i \neq K, \\ S_K = P \Leftrightarrow \text{sign}F_2(\alpha_K,1) = (-1)^{k+1}\text{sign}F_2(\alpha_1,1). \end{cases} \tag{2.4.90}$$

$$\begin{cases} S_i = E \Leftrightarrow \text{sign}F_2(z_j,1) = \text{sign}F_2(\alpha_i,1) = -\text{sign}F_2(\alpha_{i+1},1) \text{ if } \alpha_i > 0, \\ S_i = E \Leftrightarrow \text{sign}F_2(z_j,1) = \text{sign}F_2(\alpha_{i+1},1) = -\text{sign}F_2(\alpha_i,1) \text{ if } \alpha_i \le 0, \end{cases} \tag{2.4.91}$$

for $i \neq K$, where $z_1 > \cdots > z_K$ are the zero points of $F_1(z,1)$, and z_j satisfies $z_j < \alpha_i < z_{j+1}$ and

$$S_K = E \Leftrightarrow -\text{sign}F_2(z_j,1) = \text{sign}F_2(\alpha_K,1) = (-1)^K \text{sign}F_2(\alpha_1,1). \tag{2.4.92}$$

The hyperbolic region H can be determined by ruling out the parabolic and elliptic cases by (2.4.90)–(2.4.92):

$$S_i = H \Leftrightarrow S_i \neq P \text{ and } S_i \neq E. \tag{2.4.93}$$

The properties (2.4.90)–(2.4.93) completely establish the topological structure of a two-dimensional nondegenerate k-homogeneous vector field F, which is crucial in studying the transition of (2.1.1) from a kth-order nondegenerate singular point.

From Claim A and (2.4.90)–(2.4.93) we obtain the following theorem.

Theorem 2.4.15 *Let F be given by (2.4.64). Then we have the following assertions:*

(1) F has only one parabolic region with $x = 0$ a focus or a center if and only if the polynomial $h(z)$ given by (2.4.90) has no real zero points.
(2) F has only one parabolic region with $x = 0$ a node if and only if for all real zero points α_i $(1 \leq i \leq K)$ of $h(z)$,

$$\text{sign}F_2(\alpha_i,1) = \text{sign}F_2(\alpha_1,1) \qquad \forall 1 \leq i \leq K,$$

and $x = 0$ is stable if $F_2(\alpha_1,1) < 0$, unstable if $F_2(\alpha_1,1) > 0$.
(3) F has exactly $2(k+1)$ hyperbolic regions and no parabolic and elliptic regions if and only if ind $(F,0) = -k$.
(4) F has exactly $2(k-1)$ elliptic regions and no parabolic and hyperbolic regions if and only if ind$(F,0) = k$ and $h(z)$ has only $(k-1)$ real solutions.

2.4.8 Transition from kth-Order Nondegenerate Singularities

We consider the transition of (2.1.1) from the kth-order nondegenerate singular point $u = 0$. Let the nonlinear term $G(u,\lambda)$ of (2.1.1) have Taylor expansion

$$G(u,\lambda) = G_k(u,\lambda) + o(\|u\|_{X_1}^k) \qquad \text{for some } k \geq 2, \tag{2.4.94}$$

where G_k is a k-multilinear operator.

We set

$$F(x,\lambda) = \begin{pmatrix} F_1 \\ F_2 \end{pmatrix} = \begin{pmatrix} (G_k(x_1e_1 + x_2e_2,\lambda), e_1^*) \\ (G_k(x_1e_1 + x_2e_2,\lambda), e_2^*) \end{pmatrix}, \tag{2.4.95}$$

where e_1, e_2 and e_1^*, e_2^* are as in (2.4.13), and

$$(G_k(x_1e_1 + x_2e_2,\lambda), e_l^*) = \sum_{i+j=k} a_{ij}^l x_1^i x_2^j \qquad \text{for } l = 1,2.$$

The polynomials $F_i(z,1)$ induced by (2.4.95) at $\lambda=\lambda_0$ are as in (2.4.65), and $h(z)=h(z,\lambda_0)$ is as in (2.4.88).

Here we always assume that the conditions (2.1.4) and (2.1.5) with $m=r=2$ hold.

Theorem 2.4.16 *Let $F(x,\lambda)$ be given by (2.4.95), which is kth-order nondegenerate. Then we have the following assertions:*

(1) If $h(z)$ has no real zero points and $a_{k0}^2\neq 0$, then (2.1.1) bifurcates from $(u,\lambda)=(0,\lambda_0)$ to a periodic orbit. For the number

$$\alpha=\int_0^{2\pi}\frac{\cos\theta F_1(\cos\theta,\sin\theta)+\sin\theta F_2(\cos\theta,\sin\theta))}{\cos\theta F_2(\cos\theta,\sin\theta)-\sin\theta F_1(\cos\theta,\sin\theta))}d\theta, \tag{2.4.96}$$

if $\alpha<0$, the transition from $(0,\lambda_0)$ is an attractor bifurcation, and if $\alpha\geq 0$, the transition is a jump transition.

(2) If $h(z)$ has real zero points α_i $(1\leq i\leq k)$, and

$$\mathrm{sign}F_2(\alpha_i,1)=\mathrm{sign}F_2(\alpha_1,1)\quad\forall 1\leq i\leq k,$$

then if $F_2(\alpha_1,1)<0$, the transition of (2.1.1) from $(0,\lambda_0)$ is an S^1-attractor bifurcation, and if $F_2(\alpha_1,1)>0$, the transition is a jump transition. Moreover, there are at least four singular points bifurcated on $\lambda>\lambda_0$ for $F_2(\alpha_1,1)<0$ and on $\lambda<\lambda_0$ for $F_2(\alpha_1,1)>0$.

(3) If ind $(F_{\lambda_0},0)\leq 0$, then the transition is either a jump or mixed transition, and there are singular points bifurcated on both sides of λ_0. In particular, if ind$(F_{\lambda_0},0)=-k$, then the transition is a jump transition and there are exactly $(k+1)$ singular points bifurcated on each side of λ_0.

(4) If ind$(F_{\lambda_0},0)=k$ and $h(z)$ has only $(k-1)$ real solutions, then the transition is continuous. In particular, if $\beta_1(\lambda)=\beta_2(\lambda)$ in (2.1.4), the transition is an attractor bifurcation and the bifurcated attractor $\mathcal{A}_\lambda$ consists of $(k-1)$ singular points that are minimal attractors.

(5) If ind$(F_{\lambda_0},0)\geq 2$, the transition is either mixed or continuous, and if $\beta_1(\lambda)=\beta_2(\lambda)$, then (2.1.1) bifurcates in the stable domain to an attractor $\mathcal{A}_\lambda$ with dim $\mathcal{A}_\lambda=0$ or 1, and $\mathcal{A}_\lambda$ attracts at least n sectorial domains $D(\theta_i)$ with $\sum_{i=1}^n\theta_i\leq 2\pi, n=$ ind$(F,0)-1$.

Proof. We need to consider only the reduced equation of (2.1.1) as follows:

$$\begin{aligned}\frac{dx_1}{dt}&=\beta_1(\lambda)x_1+F_1(x,\lambda)+o(|x|^k),\\ \frac{dx_2}{dt}&=\beta_2(\lambda)x_2+F_2(x,\lambda)+o(|x|^k),\end{aligned}\tag{2.4.97}$$

where $F(x,\lambda)=(F_1(x,\lambda),F_2(x,\lambda))$ is given by (2.4.95).

Proof of assertion (1). By Theorem 2.4.15, when $h(z)$ has no real zero points, this implies that at $\lambda=\lambda_0$, $x=0$ is either a focus of (2.4.97) or there are infinite periodic orbits of (2.4.97) in a neighborhood of $x = 0$. In the second case, it means a bifurcation of periodic orbits. Hence, we consider only the first case.

When $x = 0$ is a stable focus, by the attractor bifurcation theorem (Theorem 2.2.2), the system (2.4.97) bifurcates from $(0, \lambda_0)$ to an S^1-attractor. Because $h(z) = zF_2(z,1) - F_1(z,1)$ has no real zero points, (2.4.97) has no singular points bifurcated from $(0, \lambda_0)$. Hence, by Theorem 2.6, the bifurcated S^1-attractor is a periodic orbit. Likewise, when $x = 0$ is an unstable focus, $x = 0$ is a stable focus of $-(F(x) + o(|x|^k))$. Thus, as above, we can prove that the vector field

$$\begin{pmatrix} \beta_1(\lambda) & 0 \\ 0 & \beta_2(\lambda) \end{pmatrix} \begin{pmatrix} x_1 \\ x_2 \end{pmatrix} - F(x,\lambda) - o(|x|^k) \tag{2.4.98}$$

bifurcates from $(0, \lambda_0)$ on $\lambda > \lambda_0$ to a periodic orbit, which implies that (2.4.97) bifurcates on $\lambda < \lambda_0$ to a periodic orbit.

It is clear that the transition type of (2.4.96), continuous or jump, depends on whether $x = 0$ is a stable or unstable focus of $F(x,\lambda)$. Take the polar coordinates $x_1 = r\cos\theta, x_2 = r\sin\theta$. Then the equation $dx/dt = F(x,\lambda_0)$ becomes

$$\frac{dr}{rd\theta} = \frac{\cos\theta F_1(\cos\theta,\sin\theta) + \sin\theta F_2(\cos\theta,\sin\theta)}{\cos\theta F_2(\cos\theta,\sin\theta) - \sin\theta F_1(\cos\theta,\sin\theta)}. \tag{2.4.99}$$

In addition, by assumption, $h(z)$ has no zeros:

$$\cos\theta F_2(\cos\theta,\sin\theta) - \sin\theta F_1(\cos\theta,\sin\theta) \neq 0 \qquad \forall 0 \leq \theta \leq 2\pi.$$

Hence, it follows from (2.4.99) that

$$\ln\frac{r(2\pi)}{r(0)} = \alpha, \tag{2.4.100}$$

where α is given by (2.4.96). Thus, we infer from (2.4.100) that if $\alpha < 0$, then $x = 0$ is a stable focus, and if $\alpha > 0$, then $x = 0$ is unstable. Assertion (1) is proved.

Proof of assertion (2). It follows from Theorems 2.4.15 and 2.2.2 that if $F_2(\alpha_1, 1) < 0$, the system (2.4.97) has an S^1-attractor bifurcation, and if $F_2(\alpha_1, 1) > 0$, then (2.4.97) has a jump transition.

We consider the case in which $x = 0$ is a stable node, while for the case in which $x = 0$ is unstable, the discussion is the same. It is known that the flows of $F(x,\lambda)$ are inward to $x = 0$, and the flows of $(\beta_1(\lambda)x_1, \beta_2(\lambda)x_2)$ are radially outward from $x = 0$. Hence, for any $\lambda - \lambda_0 > 0$, there exists a curve $\gamma(\lambda)$ passing through $x = 0$ such that $F(x,\lambda)$ is parallel to $\beta x = (\beta_1(\lambda)x_1, \beta_2(\lambda)x_2)$ on $\gamma(\lambda)$, i.e.,

$$\begin{aligned} &\beta(\lambda)x = \rho(\lambda,x)F(x,\lambda) && \forall x \in \gamma(\lambda), \\ &\rho(\lambda,x) < 0 && \forall x \in \gamma(\lambda) \text{ and } x \neq 0. \end{aligned} \tag{2.4.101}$$

We note that $F(\alpha x,\lambda) = \alpha^k F(x,\lambda)$ $(k \geq 3)$. Hence, for any given $\lambda - \lambda_0 > 0$ sufficiently small, there exists $r(\lambda) > 0$ such that

$$\begin{aligned}
&\beta(\lambda)|x| > |F(x,\lambda)| && \text{if } 0<|x|<r(\lambda),\\
&\beta(\lambda)|x| < |F(x,\lambda)| && \text{if } |x|>r(\lambda),\\
&\lim_{\lambda\to\lambda_0^+} r(\lambda)=0,
\end{aligned}$$

yielding, by (2.4.101), that

$$\beta(\lambda)x_\lambda + F(x_\lambda,\lambda)=0, \quad x_\lambda\in\gamma(\lambda), \quad |x_\lambda|=r(\lambda). \tag{2.4.102}$$

Since $F(x,\lambda)+o(|x|^k)$ is a perturbation of $F(x,\lambda)$ near $x=0$, the above procedure is also valid for any higher-order perturbation of $F(x,\lambda)$. Hence we have

$$\mathrm{ind}(\beta id + F, x_\lambda)\neq 0.$$

Noting that $\beta id+F$ is odd ($k\geq 3$ is an odd number), we deduce that $\pm x_\lambda$ are the solutions of (2.4.102), and

$$\mathrm{ind}(\beta\mathrm{id}+F,x_\lambda)=\mathrm{ind}(\beta id+F,-x_\lambda)\neq 0. \tag{2.4.103}$$

Let $\pm x_\lambda^k$ $(1\leq k\leq n)$ be all solutions of (2.4.102). Then we have

$$\sum_{k=1}^{n}\mathrm{ind}(\beta id+F,x_\lambda^k)=\mathrm{ind}(F_{\lambda_0},0)=1. \tag{2.4.104}$$

On the other hand, we know that

$$\mathrm{ind}(\beta\mathrm{id}+F,0)=1. \tag{2.4.105}$$

We then infer from (2.4.103)–(2.4.105) that there are at least four bifurcated solutions $\pm x_\lambda^1, \pm x_\lambda^2$ of (2.4.102) such that

$$\mathrm{ind}(\beta id+F,\pm x_\lambda^1)=-\mathrm{ind}(\beta id+F,\pm x_\lambda^2)\neq 0. \tag{2.4.106}$$

By the perturbation invariance of the index, assertion (2) follows from (2.4.105).

The proofs of assertions (3) and (4) are similar to those of Theorems 2.4.4 and 2.4.6.

Proof of assertion (5). By the Poincaré formula (2.1.9), when the index ind $(F_{\lambda_0},0)$ is greater than or equal to 2, there are at least n ($n=\mathrm{ind}(F_{\lambda_0},0)-1$) pairs of elliptic regions such as E_1,E_2 with a parabolic region P (it is possible that $P=\emptyset$), as shown in Fig. 2.30, such that

$$\lim_{t\to\infty}S(t)x=0 \quad \forall x\in E_1\bigcup E_2\bigcup P,$$

where $S(t)$ is the semigroup generated by $F(x,\lambda_0)$. Then, as done for Theorem 2.4.6, one can verify assertion (5).

The proof of the theorem is complete. □

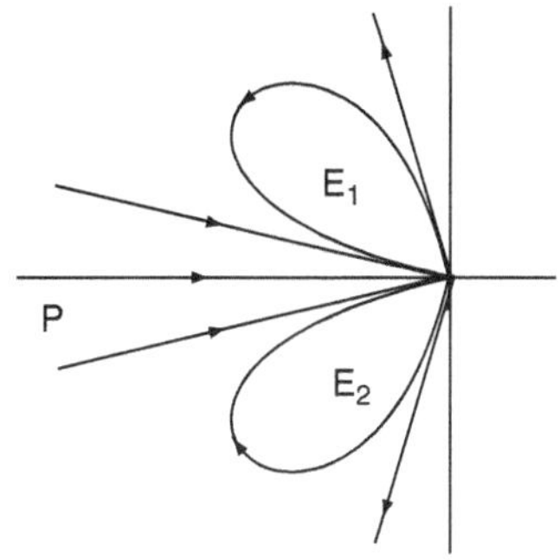

Fig. 2.30 Illustration of proof of assertion 5 for Theorem 2.4.16

2.4.9 Bifurcation to Periodic Orbits

Consider the case $m=2$, $r=1$, and k is odd, $k \geq 3$ in (2.1.4) and (2.1.5), where $r=1$ is the geometric multiplicity of $\beta_1(\lambda)$ at $\lambda=\lambda_0$. In this case, the two eigenvectors e_1 and e_2 of L_λ at $\lambda=\lambda_0$ satisfy

$$L_{\lambda_0}e_1=0, \quad L_{\lambda_0}e_2=ae_1, \qquad L^*_{\lambda_0}e^*_2=0, \quad L^*_{\lambda_0}e^*_1=ae^*_2, \quad a\neq 0,$$

$$(e_i,e^*_j)\begin{cases} >0 & \text{for } i=j, \\ =0 & \text{for } i\neq j. \end{cases}$$

Let $\alpha \in \mathbb{R}$ be defined by

$$\alpha=(G_k(e_1,\lambda_0),e^*_2), \tag{2.4.107}$$

where G_k is as in (2.4.94). Then we have the following transition theorem for the bifurcation of periodic orbits from the real eigenvalues.

Theorem 2.4.17 *Assume the conditions (2.1.4) and (2.1.5) with $m=2$, $r=1$, and k odd, $k\geq 3$. Let α be given by (2.4.107). If $\alpha\cdot a<0$, then the transition of (2.1.1) from $(u,\lambda)=(0,\lambda_0)$ is either a continuous (attractor bifurcation) or jump transition, and (2.1.1) bifurcates from $(0,\lambda_0)$ to a periodic orbit.*

Proof. We proceed in the following four steps.

Step 1. By the center manifold reduction, it suffices to consider the following reduced equations of (2.1.1):

$$\begin{aligned} \frac{dx_1}{dt} &= \beta_1(\lambda)x_1+ax_2+(G(x+h(x,\lambda),\lambda),e^*_1(\lambda)), \\ \frac{dx_2}{dt} &= \beta_2(\lambda)x_2+(G(x+h(x,\lambda),\lambda),e^*_2(\lambda)), \end{aligned} \tag{2.4.108}$$

where $x=x_1e_1(\lambda)+x_2e_2(\lambda)$, $h(x,\lambda)$ is the center manifold function, and

$$L_\lambda e_1(\lambda)=\beta_1(\lambda)e_1(\lambda), \qquad L_\lambda e_2(\lambda)=\beta_2(\lambda)e_2(\lambda)+av_1(\lambda),$$

$$L_\lambda^* e_2^*(\lambda) = \beta_2(\lambda) e_2^*(\lambda), \qquad L_\lambda^* e_1^*(\lambda) = \beta_1(\lambda) e_1^*(\lambda) + a e_2^*(\lambda),$$
$$(e_i(\lambda), e_j^*(\lambda)) = \delta_{ij}, \qquad a \neq 0,$$

where δ_{ij} is the Kronecker symbol.

By (2.4.94) and (2.4.107), we have that (2.4.108) at $\lambda = \lambda_0$ reads

$$\frac{dx}{dt} = F(x) = \begin{pmatrix} F_1(x) \\ F_2(x) \end{pmatrix}, \tag{2.4.109}$$

where

$$F_1(x) = a x_2 + o(|x_1|^k, |x_2|^k),$$
$$F_2(x) = \alpha x_1^k + o(|x_1|^{k+1}, |x_2|^k, |x_2|^{k-1}|x_1|, \cdots, |x_2||x_1|^{k-1}).$$

Since $\alpha \cdot a < 0$ and k is odd, $k \geq 3$, we have

$$\text{ind}(F, 0) = 1. \tag{2.4.110}$$

Step 2. We now prove that the number of elliptic regions of F at $x = 0$ is zero, i.e., $e = 0$. Assume otherwise; then there exists an orbit γ of (2.4.109) connected to $x = 0$, i.e.,

$$\lim_{t \to \infty} S(t)x = 0 \quad \forall x \in \gamma,$$

where $S(t)$ is the operator semigroup generated by (2.4.109). Let γ can be expressed near $x = 0$ as

$$x_2 = f(x_1) \qquad \text{for } (x_1, x_2) \in \gamma.$$

It follows from (2.4.109) that for every $(x_1, x_2) \in \gamma$,

$$\frac{dx_2}{dx_1} = \frac{\alpha x_1^k + o(|x_1|^{k+1}, |f|^k, |x_1||f|^{k-1}, \cdots, |x_1|^{k-1}|f|)}{a f(x) + o(|x_1|^k, |f|^k)}.$$

Thus, we obtain

$$a f(x_1) f'(x_1) + o(|x_1|^k, |f|^k) f' = \alpha x_1^k + o(|x_1|^{k+1}, |x_1|^{k-i}|f|^i), \tag{2.4.111}$$

which implies that

$$f(x) = \beta x^m + o(|x|^m) \qquad \text{for } 2 \leq m = \frac{k+1}{2} < k, \quad \beta \neq 0. \tag{2.4.112}$$

Therefore, from (2.4.111) and (2.4.112), we get

$$\alpha = a m \beta^2,$$

where $a \cdot \alpha < 0$, $m \geq 2$, and $\beta \neq 0$, which is a contradiction. Hence $e = 0$.

Step 3. By (2.4.110) and the Poincaré formula (2.1.9), there are no hyperbolic regions near $x = 0$. Hence, (2.4.109) has only one parabolic region P, and by the proof in Step 2, the orbits in P are circular about $x = 0$. Thus, $x = 0$ is (a) a stable focus or (b) an unstable focus or (c) a singular point having infinite periodic orbits in its neighborhood.

Case (c) implies a bifurcation to periodic orbits for (2.4.108).

For case (a), $x = 0$ is an asymptotically stable singular point of (2.4.109). Then by Theorem 2.2.2, (2.4.108) bifurcates from $(x,\lambda) = (0,\lambda_0)$ to an S^1-attractor Σ_λ on $\lambda > \lambda_0$.

For case (b), $x = 0$ is an asymptotically stable singular point of $-F(x)$. Therefore, the vector field

$$-\begin{pmatrix} \beta_1(\lambda) & 0 \\ 0 & \beta_2(\lambda) \end{pmatrix} x - F(x,\lambda)$$

bifurcates from $(0,\lambda_0)$ to an S^1-attractor Σ_λ on $\lambda > \lambda_0$, which implies that (2.4.108) bifurcates from $(0,\lambda_0)$ on $\lambda < \lambda_0$ to an S^1-repeller Σ_1, which is an invariant set.

Step 4. Now we need to prove that the S^1-invariant set Σ_λ contains no singular points. Consider the following equations:

$$\begin{aligned} &\beta_1(\lambda)x_1 + ax_2 + o(|x_1|^k, |x_2|^k) = 0, \\ &\beta_2(\lambda)x_2 + \alpha x_1^k + o(|x_1|^{k+1}, |x_2|^{k-i}|x_i|^i) = 0. \end{aligned}$$

Hence

$$\begin{aligned} &x_2 = -\beta_1(\lambda)a^{-1}x_1 + o(|x_1|^k, |\beta_1|^k), \\ &a\alpha x_1^{k-1} - \beta_1(\lambda)\beta_2(\lambda) + o(|x_1|^k, |x_1|^k|\beta_1|) = 0. \end{aligned} \tag{2.4.113}$$

For $a\alpha < 0$ and $\beta_1(\lambda)\beta_2(\lambda) > 0$, we see that (2.4.113) has no solutions near $(x,\lambda) = (0,\lambda_0)$. Hence, there are no singular points in Σ_λ, which implies that Σ_λ must contain a periodic orbit.

Finally, as with Theorem 2.3.6, we can prove that the transition of (2.4.108) is either continuous, which is an attractor bifurcation, or a jump transition. Thus the proof is complete. □

Remark 2.4.18 Theorem 2.4.17 can be considered a complement of Theorem 4.3 in Ma and Wang (2005b) for $m = 2$, $r = 1$, $\alpha < 0$ $(a = 1)$, where there are no bifurcations of singular points.

2.4.10 Application to Parabolic Systems

Consider

$$\begin{aligned} &\frac{\partial u_1}{\partial t} = \Delta u_1 + \lambda\alpha_1 u_1 + au_2 + g_1(u,\lambda), \\ &\frac{\partial u_2}{\partial t} = \Delta u_2 + \lambda\alpha_2 u_2 + g_2(u,\lambda), \\ &u|_{\partial\Omega} = 0, \end{aligned} \tag{2.4.114}$$

where $u=(u_1,u_2)^t$, the parameters α_1, α_2, and a are constants, $\Omega\subset\mathbb{R}^n (n\geq 1)$ is a bounded domain, and

$$g_l(u,\lambda)=\sum_{i+j=k} a^l_{ij}u_1^i u_2^j+o(|u|^k) \quad \text{for } l=1,2, \quad k\geq 2. \tag{2.4.115}$$

Let $\lambda_i>0$ and $h_i(x)$ be the ith eigenvalue and eigenfunction of $-\triangle$ on Ω with Dirichlet boundary condition. Let

$$X=L^p(\Omega,\mathbb{R}^2), \qquad X_1=W^{2,p}(\Omega,\mathbb{R}^2)\bigcap W^{1,p}_0(\Omega,\mathbb{R}^2), \qquad \text{for } p\geq n,$$

and define the operators $L_\lambda=-A+B_\lambda : X_1\to X$ and $G: X_1\to X$ by

$$-Au=\begin{pmatrix}\Delta u_1\\ \Delta u_2\end{pmatrix}, \quad B_\lambda u=\begin{pmatrix}\lambda\alpha_1 & a\\ 0 & \lambda\alpha_2\end{pmatrix}\begin{pmatrix}u_1\\ u_2\end{pmatrix}, \quad G(u,\lambda)=\begin{pmatrix}g_1(u,\lambda)\\ g_2(u,\lambda)\end{pmatrix}.$$

Then (2.4.114) can be written in the form (2.1.1), and the conditions (2.1.2) and (2.1.3) are satisfied.

Case 1. $\alpha_2>\alpha_1>0$. Thus, all eigenvalues of L_λ are $\alpha_1\lambda-\lambda_i$ and $\alpha_2\lambda-\lambda_i$ $(1\leq i<\infty)$, which enjoy the properties (2.1.3) and (2.1.4) with $\lambda_0=\alpha_1^{-1}\lambda_1$. The first eigenvectors e_1 and e_1^* of L_λ and L^*_λ are given by

$$e_1=\begin{pmatrix}h_1\\ 0\end{pmatrix}, \quad e_1^*=\begin{pmatrix}h_1\\ -a\alpha_1 h_1/(\alpha_2-\alpha_1)\lambda_1\end{pmatrix}.$$

Let the k-homogeneous term in (2.4.115) be given by

$$G_k(u,\lambda)=\begin{pmatrix}\sum_{i+j=k} a^1_{ij}u_1^i u_2^j\\ \sum_{i+j=k} a^2_{ij}u_1^i u_2^j\end{pmatrix},$$

and set

$$\alpha=(G_k(e_1,\lambda_0),e_1^*)=\left[a^1_{k0}-\frac{\alpha_1 a}{(\alpha_2-\alpha_1)\lambda_1}a^2_{k0}\right]\int_\Omega h_1^{k+1}dx.$$

Let $\alpha\neq 0$, i.e.,

$$a^1_{k0}-\frac{\alpha_1 a}{(\alpha_2-\alpha_1)\lambda_1}a^2_{k0}\neq 0.$$

When k is even, by Theorem 2.3.2, equations (2.4.114) have a mixed transition from $(u,\lambda)=(0,\alpha_1^{-1}\lambda_1)$, and there is a neighborhood $U\subset X$ of $u=0$ that can be decomposed into two open sets

$$U_1=\left\{(u_1,u_2)\in U|\ u_1=zh_1+v,\ z\cdot\alpha>0, \quad \int h_1 v dx=0\right\},$$

$$U_2=\left\{(u_1,u_2)\in U|\ u_1=zh_1+v,\ z\cdot\alpha<0, \quad \int h_1 v dx=0\right\},$$

such that U_1 is an unstable domain, U_2 is a stable domain of (2.4.114), and equations (2.4.114) bifurcate in U_2 to an attractor that is a singular point

$$v_\lambda = -\frac{\beta_1(\lambda)}{\alpha}\begin{pmatrix} h_1(x) \\ 0 \end{pmatrix} + o(|\beta_1|) \in U_2 \qquad \text{for } \lambda > \lambda_0,$$

and $\{v_\lambda\}$ attracts U_2 in the X-norm.

When k is odd, by Theorem 2.3.1, equations (2.4.114) have a jump transition from $(0, \alpha_1^{-1}\lambda_1)$ for $\alpha > 0$, and they have an attractor bifurcation for $\alpha < 0$.

Case 2: $\alpha_1 = \alpha_2 > 0$ **and** $a = 0$**.** We then have $\beta_1(\lambda) = \beta_2(\lambda) = \alpha_1\lambda - \lambda_1$, and the corresponding eigenvectors are given by

$$e_1 = e_1^* = \begin{pmatrix} h_1(x) \\ 0 \end{pmatrix}, \quad e_2 = e_2^* = \begin{pmatrix} 0 \\ h_1(x) \end{pmatrix}. \tag{2.4.116}$$

The k-homogeneous vector field induced by (2.4.114) reads

$$F(x,\lambda) = C\begin{pmatrix} \sum_{i+j=k} a_{ij}^1 u_1^i u_2^j \\ \sum_{i+j=k} a_{ij}^2 u_1^i u_2^j \end{pmatrix},$$

where $C = \int_\Omega h_1^{k+1}(x)dx > 0$.

If $F(x,\lambda)$ is kth-order nondegenerate, then the results in this section are valid for (2.4.114).

Case 3. $\alpha_1 = \alpha_2 > 0, a \neq 0$**, and** k **= odd**≥ 3. In this case, $m = 2$, $r = 1$, and the first eigenvalues are $\beta_1(\lambda) = \beta_2(\lambda) = \alpha_1\lambda - \lambda_1$, and the corresponding eigenvectors are given by (2.4.116), which satisfy

$$\begin{aligned} L_\lambda e_1 &= \beta_1(\lambda)e_1, & L_\lambda e_2 &= \beta_1(\lambda)e_2 + ae_1, \\ L_\lambda^* e_2^* &= \beta_1(\lambda)e_2^*, & L_\lambda^* e_1^* &= \beta_1(\lambda)e_1^* + ae_2^*. \end{aligned}$$

Let

$$\alpha = (G_k(e_1,\lambda_0), e_2^*) = a_{k0}^2 \int_\Omega h_1^{k+1}(x)dx.$$

By Theorem 2.4.17, if $a \cdot a_{k0}^2 < 0$, then equations (2.4.114) bifurcate from $(0, \alpha_1^{-1}\lambda_1)$ to a periodic orbit, and the transition is either a jump or continuous transition.

2.5 Singular Separation

2.5.1 General Principle

In this section, we study an important problem associated with the jump and mixed transitions of (2.1.1), which we call singular separation.

Definition 2.5.1

1. *An invariant set Σ of (2.1.1) is called a singular element if Σ is either a singular point or a periodic orbit.*
2. *Let $\Sigma_1 \subset X$ be a singular element of (2.1.1) and $U \subset X$ a neighborhood of Σ_1. We say that (2.1.1) has a singular separation of Σ at $\lambda = \lambda_1$ if:*

   a. *Equation (2.1.1) has no singular elements in U when $\lambda < \lambda_1$ (or $\lambda > \lambda_1$), and it generates a singular element $\Sigma_1 \subset U$ at $\lambda = \lambda_1$.*
   b. *There are branches of singular elements Σ_λ that are separated from Σ_1 for $\lambda > \lambda_1$ (or $\lambda < \lambda_1$), i.e.,*

$$\lim_{\lambda \to \lambda_1} \max_{x \in \Sigma_\lambda} dist(x, \Sigma_1) = 0.$$

A special case of singular separation is the saddle-node bifurcation:

Definition 2.5.2 *Let $u_1 \in X$ be a singular point of (2.1.1) at $\lambda = \lambda_1$ with $u_1 \neq 0$. We say that (2.1.1) has a saddle-node bifurcation at (u_1, λ_1) if:*

1. *The index of $L_\lambda + G$ at (u_1, λ_1) is zero, i.e., $ind(-(L_{\lambda_1} + G), u_1) = 0$.*
2. *There are two branches $\Gamma_1(\lambda)$ and $\Gamma_2(\lambda)$ of singular points of (2.1.1) that are separated from u_1 for $\lambda > \lambda_1$ (or $\lambda < \lambda_1$), i.e., for every $u_\lambda \in \Gamma_i(\lambda)$ $(i = 1, 2)$, we have*

$$u_\lambda \to u_1 \text{ in } X \quad \text{as } \lambda \to \lambda_1.$$

3. *The indices of $u_\lambda^i \in \Gamma_i(\lambda)$ are as follows:*

$$ind(-(L_\lambda + G), u_\lambda) = \begin{cases} 1 & \text{if } u_\lambda \in \Gamma_2(\lambda), \\ -1 & \text{if } u_\lambda \in \Gamma_1(\lambda). \end{cases}$$

Intuitively, the saddle-node bifurcation is as schematically shown in Fig. 2.31, where the singular points in $\Gamma_1(\lambda)$ are saddle points and in $\Gamma_2(\lambda)$ are nodes, and the singular separation of periodic orbits is as in Fig. 2.32.

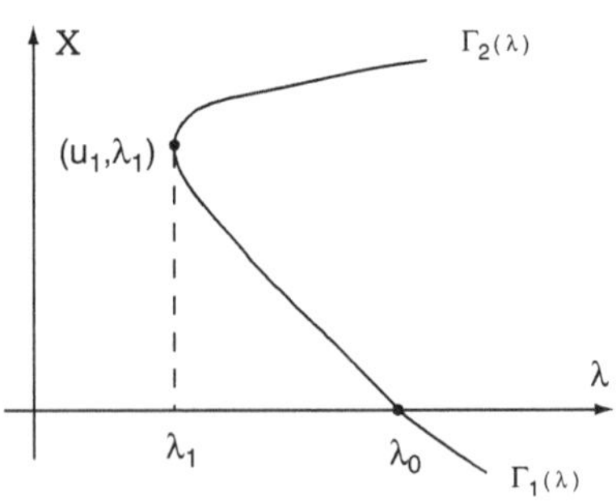

Fig. 2.31 Saddle-node bifurcation

For singular separation, we can give a general principle as follows, which provides a basis for singular separation theory.

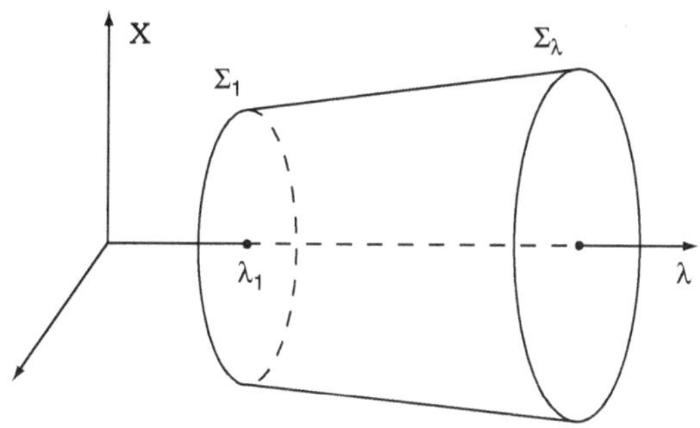

Fig. 2.32 Singular separation of periodic orbits

Theorem 2.5.3 *Let the conditions (2.1.4) and (2.1.5) hold, and suppose*

$$Re\beta_j(\lambda) < 0 \qquad \forall \lambda < \lambda_0, 1 \le j < \infty.$$

Then we have the following assertions:

(1) If (2.1.1) bifurcates from $(u,\lambda) = (0,\lambda_0)$ to a branch Σ_λ of singular elements on $\lambda < \lambda_0$ that is bounded in $X \times (-\infty, \lambda_0)$, then (2.1.1) has a singular separation of singular elements at some $(\Sigma_0, \lambda_1) \subset X \times (-\infty, \lambda_0)$.

(2) If the branch Σ_λ consists of singular points that have index -1, i.e.,

$$ind(-(L_\lambda + G), u_\lambda) = -1 \quad \forall u_\lambda \in E_\lambda, \quad \lambda < \lambda_0,$$

then the singular separation is a saddle-node bifurcation from some $(u_1, \lambda_1) \in X \times (-\infty, \lambda_0)$.

Proof. By assumption, for every $\lambda < \lambda_0$, the eigenvalues $\beta_j(\lambda)$ $(1 \le j < \infty)$ of L_λ have negative real part:

$$\mathrm{Re}\beta_j(\lambda) < 0 \quad \forall \lambda < \lambda_0 \quad 1 \le j < \infty.$$

Hence, no singular elements are bifurcated from $(u,\lambda) = (0,\lambda)$ for all $\lambda < \lambda_0$. This implies that the singular elements of the branch Σ_λ bifurcated from $(0,\lambda_0)$ do not converge to $u = 0$. On the other hand, by assumption, Σ_λ is bounded in $X \times (-\infty, \lambda_0)$, which means that Σ_λ separates from some singular element Σ_0 at $\lambda_1 < \lambda_0$. Assertion (1) is proved. Assertion (2) follows from assertion (1) and Definition 2.5.2. The proof is complete. □

2.5.2 *Saddle-Node Bifurcation*

For a nonlinear dissipative system (2.1.1), singular separation is closely related to jump and mixed transitions, which may lead to chaos. In addition, singular separation can provide important information on the global topological structure of jump and mixed transitions.

We consider (2.1.1) defined on two Hilbert spaces $X = H$ and $X_1 = H_1$. Let $L_\lambda = -A + \lambda B$. For L_λ and $G(\cdot,\lambda) : H_1 \to H$, we assume that $A : H_1 \to H$ is symmetric, and

$$(Au,u) \geq c\|u\|^2_{H_{1/2}}, \tag{2.5.1}$$

$$(Bu,u) \leq c\|u\|^2, \tag{2.5.2}$$

$$(Gu,u) \leq -c_1\|u\|^p + c_2\|u\|^2, \tag{2.5.3}$$

where $p > 2, c, c_1, c_2 > 0$ are constants.

Theorem 2.5.4 *Assume the conditions (2.1.3), (2.1.4), and (2.5.1)–(2.5.3). Then (2.1.1) has a transition at* $(u,\lambda) = (0,\lambda_0)$, *and the following assertions hold:*

(1) If $u = 0$ *is an even-order nondegenerate singular point of* $L_\lambda + G$ *at* $\lambda = \lambda_0$, *then (2.1.1) has a singular separation of singular points at some* $(u_1,\lambda_1) \in H \times (-\infty,\lambda_0)$.
(2) If $m = 1$ *and* G *satisfies (2.3.1) with* $\alpha > 0$ *if* k *is odd and* $\alpha \neq 0$ *if* k *is even, then (2.1.1) has a saddle-node bifurcation at some singular point* (u_1,λ_1) *with* $\lambda_1 < \lambda_0$.

Proof. By Theorem 2.1.3, under the conditions (2.1.3) and (2.1.4), equation (2.1.1) has a transition at $(0,\lambda_0)$. By the conditions (2.5.1)–(2.5.3), we can deduce that for every $\lambda \in \mathbb{R}^1$, there are constants $\alpha_1, \alpha_2 > 0$ such that

$$(L_\lambda u + G(u,\lambda),u) \leq -\alpha_1\|u\|^2_{H_{1/2}} + \alpha_2.$$

Then it is clear that for every $\lambda \in \mathbb{R}^1$, equation (2.1.1) possesses a global attractor in H.

If $u = 0$ is of even order and nondegenerate for $L_\lambda + G$ at $\lambda = \lambda_0$, then by Theorem 4.1 in Ma and Wang (2005b), (2.1.1) bifurcates from $(0,\lambda_0)$ to a branch Γ_λ of singular points on $\lambda < \lambda_0$. Since (2.1.1) has a global attractor, for $u_\lambda \in \Gamma_\lambda$ we have

$$\|u_\lambda\| \leq c_\lambda, \quad c_\lambda > 0 \text{ a constant.} \tag{2.5.4}$$

In addition, it follows from (2.5.1)–(2.5.3) that

$$\begin{aligned}(L_\lambda u + G(u,\lambda),u) &\leq -c\|u\|^2_{H_{1/2}} - c_1\|u\|^p + (c_2 + \lambda c)\|u\|^2,\\ &\leq -\alpha\|u\|^2, \quad \forall\lambda \leq -c_2c^{-1}, \quad u \in H,\end{aligned}$$

which implies that when $\lambda \leq -c_2c^{-1}$, $u = 0$ is a global asymptotically stable singular point of (2.1.1). Hence, it follows from (2.5.4) that Γ_λ is bounded in $H \times (-\infty,\lambda_0)$. Thus, by Theorem 2.5.3, we derive assertion (1). Likewise, assertion (2) follows from Theorems 2.3.1, 2.3.2, and 2.5.3.

The proof is complete. □

The following theorem is motivated by applications in hydrodynamic stability and transition. For the operators $L_\lambda = -A + B_\lambda$ and G, we assume the following:

(A_1) $A: H_1 \to H$ is positive definite (not necessarily symmetric), i.e., the operator A satisfies (2.5.1), and $B_\lambda = 0$ at $\lambda = 0$.

(A_2) $G: H_1 \to H$ is bilinear, satisfying

$$(G(u,\lambda),u) = 0 \quad \forall u \in H_1.$$

(A_3) The conditions (2.1.3) and (2.1.4) with $m = 1$ and $\lambda_0 > 0$ hold, and for the first eigenvectors e_1 and e_1^* of L_λ and L_λ^* at $\lambda = \lambda_0$, we have

$$(G(e_1,\lambda_0),e_1^*) \neq 0.$$

(A_4) For every $\lambda \in \mathbb{R}^1$, (2.1.1) possesses a global attractor in H.

Theorem 2.5.5 *Assume conditions (A_1)–(A_4). Then (2.1.1) has a saddle-node bifurcation at some (u_1,λ_1) with $0 < \lambda_1 < \lambda_0$. Moreover, the two branches $\Gamma_1(\lambda)$ and $\Gamma_2(\lambda)$ separated from (u_1,λ_1), as shown in Fig. 2.31, have the following properties:*

(1) For $u_\lambda \in \Gamma_1(\lambda), u_\lambda \to 0$ as $\lambda \to \lambda_0$, and when $\lambda_1 < \lambda < \lambda_0, u_\lambda$ is a saddle with the Morse index one, and when $\lambda > \lambda_0$, u_λ is an attractor that attracts a sectorial domain $D(\theta)$ with angle $\theta = \pi$.

(2) For $u_\lambda \in \Gamma_2(\lambda), u_{\lambda_0} \neq 0$, and there exists $\lambda^* > \lambda_1$ such that when $\lambda_1 < \lambda < \lambda^*, u_\lambda$ is an attractor.

Proof. By Theorem 2.3.2, it follows from (A_3) that (2.1.1) has a mixing transition at $(0,\lambda_0)$, and there is a branch $\Gamma_1(\lambda)$ bifurcated from $(0,\lambda_0)$ on each side of $\lambda = \lambda_0$, and when $\lambda > \lambda_0, u_\lambda \in \Gamma_1(\lambda)$ is an attractor that attracts a sectorial domain $D(\theta)$ with $\lambda < \lambda_1, then u_\lambda \in \Gamma_1(\lambda)$ is a saddle point with Morse index one.

By Theorem 2.5.3, we need to prove that $\Gamma_1(\lambda)$ is bounded in $H \times (-\infty,\lambda_0)$ and $\Gamma_1(\lambda) = \emptyset$ at $\lambda = 0$.

In fact, by $(A_1), we have that B_\lambda = 0$ at $\lambda = 0$. Hence, (2.1.1) at $\lambda = 0$ is in the following form:

$$\frac{du}{dt} = -Au + G(u,0). \tag{2.5.5}$$

From (2.5.1) and (A_2), we can deduce that $u = 0$ is a global asymptotically stable singular point of (2.5.5). Hence, $\Gamma_1(0) = \emptyset$. By $(A_4), \Gamma_1(\lambda)$ is bounded in $H \times (0,\lambda_0)$. Thus, (2.1.1) has a saddle-node bifurcation at some (u_1,λ_1) with $\lambda_1 < \lambda_0$. By the homotopy invariance of index we derive assertion (1).

Assertion (2) can be proved by the fact that the node $u_\lambda \in \Gamma_2(\lambda)$ separated from (u_1,λ_1) is stable for $\lambda_1 < \lambda < \lambda_1 + \varepsilon$ for some $\varepsilon > 0$. Thus the theorem is proved. □

Remark 2.5.6 The objective of global dynamics of a nonlinear dissipative system is to study the topological structure of global attractors and their transitions when as system parameter λ varies. Essentially, singular separation and transition of (2.1.1) provide key information about the topological structure of global attractors and their transitions. As we shall see later, the theories of singular separation and transition are important tools in understanding the global behavior of phase transitions in the nonlinear sciences. In particular, Type I transitions are essentially characterized by attractor bifurcation, and Type II and Type III transitions often lead to singular separations.

2.5.3 Singular Separation of Periodic Orbits

We consider in this subsection the singular separation of periodic orbits. Let $\beta_1(\lambda) = \overline{\beta_2(\lambda)}$ be a pair of complex eigenvalues of L_λ. Assume that the PES (2.3.6) and (2.3.7) hold.

For $L_\lambda = -A + B_\lambda$ and $G(\cdot,\lambda) : H_1 \to H$, we also assume that

$$\begin{aligned} &(G(u,\lambda),u) = 0 && \forall u \in H_1, \\ &(L_{\lambda^*}u,u) \leq -\alpha\|u\|^2_{H_{1/2}} && \text{for some } \lambda^* < \lambda_0, \end{aligned} \tag{2.5.6}$$

or

$$\begin{aligned} &(G(u,\lambda),u) \leq -c_1\|u\|^p + c_2\|u\|^2, \\ &(L_\lambda u,u) \leq -\alpha\|u\|_{H_{1/2}} + c_\lambda\|u\|^2, \end{aligned} \tag{2.5.7}$$

where $\alpha, c_1, c_2 > 0$ are constants, $p > 2$, $c_{\lambda^*} \leq -c_2$ for some $\lambda^* < \lambda_0$, and λ_0 is as in the PES (2.3.6) and (2.3.7).

Theorem 2.5.7 *Under the PES (2.3.6) and (2.3.7), equation (2.1.1) has a Hopf bifurcation at* $(0,\lambda_0)$*. If the branch* Σ_λ *of bifurcated periodic orbits is on* $\lambda < \lambda_0$*, then we have the following assertions.*

(1) If (2.5.6) holds, and for every λ*, (2.1.1) possesses a global attractor in* H*, then (2.1.1) has a singular separation of periodic orbits at some* $(\Sigma_1,\lambda_1) \subset H \times (\lambda^*,\lambda_0)$*.*
(2) If (2.5.7) holds, then (2.1.1) has a singular separation of periodic orbits at some $(\Sigma_1,\lambda_1) \subset H \times (\lambda^*,\lambda_0)$*.*
(3) The branch Σ_λ *of bifurcated periodic orbits converges to* Σ_1 *as* $\lambda \to \lambda_1$*.*

The proof of Theorem 2.5.7 is similar to those of Theorems 2.5.4 and 2.5.5, and therefore, it is omitted here.

Example 2.5.8. As an example, we consider the stability, transition, and singular separation of the 2D Navier–Stokes equations given by

$$\begin{aligned} &\frac{\partial u}{\partial t} + (u\cdot\nabla)u = \mu\Delta u - \nabla p + \lambda f, \\ &\text{div} u = 0, \\ &u|_{\partial\Omega} = 0, \end{aligned} \tag{2.5.8}$$

where $\Omega \subset \mathbb{R}^2$ is a bounded and open set. It is known that for every $\lambda \in \mathbb{R}$, the steady-state equations of (2.5.8) are given by

$$\begin{aligned} & -\mu \Delta u + (u\cdot\nabla)u + \nabla p = \lambda f, \\ & \mathrm{div} u = 0, \\ & u|_{\partial\Omega} = 0, \end{aligned}$$

and have a solution $v_\lambda(x) \in H^2(\Omega, \mathbb{R}^2)$ that is unique for $|\lambda| < \varepsilon$. Let $u = u' + v_\lambda(x)$. Then (2.5.8) can be rewritten as (for simplicity, we omit the prime)

$$\begin{aligned} & \frac{\partial u}{\partial t} = \mu\Delta u - (v_\lambda\cdot\nabla)u - (u\cdot\nabla)v_\lambda - \nabla p - (u\cdot\nabla)u, \\ & \mathrm{div} u = 0, \\ & u|_{\partial\Omega} = 0. \end{aligned} \tag{2.5.9}$$

It is clear, then, that (2.5.9) can be expressed in the form of (2.1.1). Also, it is well known that (2.5.9) possesses a global attractor; see among others (Foiaş and Temam 1979; Temam 1997).

Obviously, for $\lambda > 0$ sufficiently small, all eigenvalues $\beta_j(\lambda)$ of $L_\lambda = -A + B_\lambda$ satisfy

$$\mathrm{Re}\beta_j(\lambda) < 0 \quad \forall j \geq 1.$$

Based on the theories of transition and singular separation, we can claim that for almost every $f \in L^2(\Omega, \mathbb{R}^2)$, only one of the following three assertions holds:

(1) For every $\lambda \geq 0$, the steady state $v_\lambda(x)$ of (2.5.8) is locally asymptotically stable.
(2) There exists $\lambda_0 > 0$ such that (2.5.8) has an attractor bifurcation from $(v_{\lambda_0}, \lambda_0)$ on $\lambda > \lambda_0$.
(3) Equation (2.5.8) has a singular separation at some $(\Sigma_1, \lambda_1) \subset H \times (0, \lambda_0)$, and there is a branch Σ_λ of singular elements separating from (Σ_1, λ_1) that converges to $(v_{\lambda_0}, \lambda_0)$, i.e.,

$$\lim_{\lambda\to\lambda_0} \max_{u_\lambda \in \Sigma_\lambda} \|u_\lambda - v_{\lambda_0}\| = 0.$$

2.6 Perturbed Systems

2.6.1 General Eigenvalues

We consider the following perturbed equation of (2.1.1):

$$\frac{du}{dt} = (L_\lambda + S_\lambda^\varepsilon)u + G(u,\lambda) + T^\varepsilon(u,\lambda), \tag{2.6.1}$$

where L_λ and G are as in (2.1.1), $S^\varepsilon_\lambda : X_\sigma \to X$ is a linear perturbed operator, $T^\varepsilon : X_\sigma \to X$ is a C^1 nonlinear perturbed operator, and X_σ is the fractional order space, $0 \le \sigma < 1$. We also assume that G_λ, T^ε are C^3 on u, and

$$\|S^\varepsilon_\lambda\| < \varepsilon, \qquad \|T^\varepsilon(\cdot,\lambda)\| < \varepsilon, \qquad T^\varepsilon(u,\lambda) = o(\|u\|_\sigma). \tag{2.6.2}$$

By the continuity of eigenvalues of linear operators, under the conditions (2.1.4) and (2.1.5), for every $\varepsilon > 0$ sufficiently small, there exists λ_0^ε with

$$\lim_{\varepsilon\to 0} \lambda_0^\varepsilon = \lambda_0,$$

such that the eigenvalues $\{\beta_j^\varepsilon(\lambda)\}$ of $L_\lambda + S^\varepsilon_\lambda$ satisfy

$$\mathrm{Re}\beta_i^\varepsilon(\lambda) \begin{cases} <0 & \text{if } \lambda < \lambda_0^\varepsilon, \\ =0 & \text{if } \lambda = \lambda_0^\varepsilon, \\ >0 & \text{if } \lambda > \lambda_0^\varepsilon \end{cases} \qquad \forall 1 \le i \le m_1, \tag{2.6.3}$$

$$\mathrm{Re}\beta_j^\varepsilon(\lambda_0^\varepsilon) < 0 \qquad \forall j \ge m_1 + 1, \tag{2.6.4}$$

where $1 \le m_1 \le m$ and m is as in (2.1.4).

The following theorem determines the transition to an S^m-attractor for the perturbed system (2.6.1) at the critical point $(u,\lambda) = (0,\lambda_0^\varepsilon)$. This theorem is very useful in many nonlinear science problems. In particular, the transition theory for perturbed systems plays an important role in the study of the Taylor problem in fluid dynamics and the Walker circulation in the equatorial atmospheric layer, which will be discussed later. This is essentially Theorem 6.14 in Ma and Wang (2005b), and we omit the proof here.

Theorem 2.6.1 *Under the hypotheses of Theorem 2.2.11, there exist $\varepsilon > 0$ and $\delta > 0$ such that if S^ε_λ and T^ε_λ satisfy (2.6.2) and $\lambda_0^\varepsilon < \lambda < \lambda_0^\varepsilon + \delta$, then the transition of (2.6.1) at $(0,\lambda_0^\varepsilon)$ enjoys the following properties:*

(1) There is a neighborhood $U \subset X$ of $u = 0$ such that (2.6.1) has an attractor $\Sigma^\varepsilon_\lambda \subset U$, $\Sigma^\varepsilon_\lambda$ attracts an open and dense set U, and

$$dim\Sigma^\varepsilon_\lambda \le m, \qquad 0 \notin \Sigma^\varepsilon_\lambda, \qquad \lim_{\varepsilon\to 0} \sup_{x\in\Sigma^\varepsilon_\lambda} dist(x,\Sigma_\lambda) = 0, \tag{2.6.5}$$

where Σ_λ is the attractor of (2.1.1) bifurcated from $(0,\lambda_0)$.

(2) When $\varepsilon > 0$ is sufficiently small, $\Sigma^\varepsilon_\lambda$ is an $(m-1)$-dimensional homological sphere for $\lambda > \lambda_0 + \sigma^\varepsilon$ (σ^ε a small number).

(3) Each $u_\lambda \in \Sigma^\varepsilon_\lambda$ can be expressed as

$$u_\lambda = v_\lambda + w^\varepsilon_\lambda, \quad v_\lambda \in E_0, \qquad \lim_{\lambda\to\lambda_0} v_\lambda = 0, \qquad \lim_{\lambda\to\lambda_0,\varepsilon\to 0} \frac{\|w^\varepsilon_\lambda\|}{\|v_\lambda\|} = 0, \tag{2.6.6}$$

where E_0 is the first eigenspace of L_{λ_0} as in Theorem 2.2.11.

(4) *If $u = 0$ is a globally asymptotically stable equilibrium point of (2.1.1), and (2.6.1) has a global attractor for every λ near λ_0, then $\Sigma_\lambda^\varepsilon$ attracts $X \setminus \Gamma$, where Γ is a set with codimension codim $\Gamma \geq 1$ in X.*

2.6.2 Simple Eigenvalues

In this subsection, we study the transition and singular separation of a perturbed system from simple eigenvalues.

Let the PES (2.1.4) and (2.1.5) with $m = 1$ hold, let $G(u,\lambda) = G_2(u,\lambda) + o(\|u\|_{X_1}^2)$, where $G_2(\cdot,\lambda)$ is a bilinear operator, and let

$$b = (G_2(e,\lambda_0), e^*) \neq 0, \tag{2.6.7}$$

where $e \in X$ and $e^* \in X^*$ are the eigenvectors of L_λ and L_λ^* corresponding to $\beta_1(\lambda)$ at $\lambda = \lambda_0$ respectively.

The following result is a stability theorem for perturbed systems.

Theorem 2.6.2 *Assume the PES (2.1.4) and (2.1.5) with $m = 1$, and (2.6.7). Then there is an $\varepsilon > 0$ such that for every S_λ^ε and T_λ^ε satisfying (2.6.2), the transition of the perturbed system (2.6.1) at $(0, \lambda_0^\varepsilon)$ has the same structure as that of (2.1.1) at $(0, \lambda_0)$.*

This theorem is a direct corollary of the structural stability theorem for dynamic bifurcation. In fact, when (2.1.1) is reduced to the center manifold, the number b given by (2.6.7) is the bifurcation number. Hence, using Theorems 2.2.8 and 2.2.9, one immediately derives Theorem 2.6.2, and we omit the details.

We now consider the transition associated with the saddle-node bifurcation of the perturbed system (2.6.1). Let $h(x,\lambda)$ be the center manifold function of (2.1.1) near $\lambda = \lambda_0$. Assume that

$$(G(xe + h(x,\lambda_0),\lambda_0), e^*) = b_1 x^3 + o(|x|^3), \tag{2.6.8}$$

where $b_1 \neq 0$, and e and e^* are as in (2.6.7).

Then we have the following theorems.

Theorem 2.6.3 *Let the PES (2.1.4) and (2.1.5) with $m = 1$, and (2.6.8) hold, and let $b_1 < 0$. Then there is an $\varepsilon > 0$ such that if S_λ^ε and T_λ^ε satisfy (2.6.2), then the transition of (2.6.1) is either continuous or mixed. If the transition is continuous, then assertions (2) and (3) of Theorem 2.3.1 are valid for (2.6.1). If the transition is mixed, then the following assertions hold:*

(1) *Equation (2.6.1) has a saddle-node bifurcation at some point $(u_1, \lambda_1) \in X \times (-\infty, \lambda_0^\varepsilon)$, and there are exactly two branches*

$$\Gamma_i^\lambda = \{(u_i^\lambda, \lambda) |\ \lambda_1 < \lambda < \lambda_0^\varepsilon + \delta\}, \qquad i = 1, 2,$$

separated from (u_1, λ_1), *as shown in Fig. 2.31, that satisfy*

$$\|u_2^\lambda\| \neq 0 \qquad \forall (u_2^\lambda, \lambda) \in \Gamma_2^\lambda, \quad \lambda_1 < \lambda < \lambda_0^\varepsilon + \delta,$$
$$\lim_{\lambda \to \lambda_0^\varepsilon} \|u_1^\lambda\| = 0 \qquad \forall (u_1^\lambda, \lambda) \in \Gamma_1^\lambda.$$

(2) *There is a neighborhood* $U \subset X$ *of* $u = 0$ *such that for each* λ *with* $\lambda_1 < \lambda < \lambda_0^\varepsilon + \delta$ *and* $\lambda \neq \lambda_0^\varepsilon$, U *contains only two nontrivial singular points* u_1^λ *and* u_2^λ *of (2.6.1).*

(3) *For each* $\lambda_1 < \lambda < \lambda_0^\varepsilon + \delta$, U *can be decomposed into two open sets* $\overline{U} = \overline{U}_1^\lambda + \overline{U}_2^\lambda$ *with* $U_1^\lambda \bigcap U_2^\lambda = \emptyset$ *such that:*

   a. *if* $\lambda_1 < \lambda < \lambda_0^\varepsilon$, *then* $0 \in U_1^\lambda$, $u_2^\lambda \in U_2^\lambda$, $u_1^\lambda \in \partial U_1^\lambda \bigcap \partial U_2^\lambda$, *with* $u = 0$ *and* u_2^λ *attractors that attract* U_1^λ *and* U_2^λ *respectively.*
   b. *If* $\lambda_0^\varepsilon < \lambda < \lambda_0^\varepsilon + \delta$, *then* $u_1^\lambda \in U_1^\lambda$, $u_2^\lambda \in U_2^\lambda$, $0 \in \partial U_1^\lambda \bigcap \partial U_2^\lambda$, *with* u_1^λ *and* u_2^λ *attractors that attract* U_1^λ *and* U_2^λ *respectively.*

(4) *Near* $(u, \lambda) = (0, \lambda_0^\varepsilon)$, u_1^λ *and* u_2^λ *can be expressed as*

$$\begin{aligned} &u_1^\lambda = \alpha_1(\lambda, \varepsilon)e + o(|\alpha_1|), \quad && u_2^\lambda = \alpha_2(\lambda, \varepsilon)e + o(|\alpha_2|), \\ &\lim_{\lambda \to \lambda_0^\varepsilon} \alpha_1(\lambda, \varepsilon) = 0, && \alpha_2(\lambda_0^\varepsilon, \varepsilon) \neq 0, \end{aligned} \tag{2.6.9}$$

where e *is as in (2.6.8).*

Theorem 2.6.4 *Assume the PES (2.1.4) and (2.1.5) with* $m = 1$, *and (2.6.8) with* $b_1 > 0$. *Then there is an* $\varepsilon > 0$ *such that when* S_λ^ε *and* T_λ^ε *satisfy (2.6.2), the transition of (2.6.1) is either a jump or mixed transition. If it is a jump transition, then assertions (1) and (3) of Theorem 2.3.1 are valid for (2.6.1). If it is mixed, then the following assertions hold:*

(1) *Equation (2.6.1) has a saddle-node bifurcation at some point* (u_1, λ_1) *with* $\lambda_0^\varepsilon < \lambda_1$, *and there are exactly two branches*

$$\Gamma_i^\lambda = \{(u_i^\lambda, \lambda) | \, \lambda_0^\varepsilon - \delta < \lambda < \lambda_1\} \qquad (i = 1, 2),$$

separated from (u_1, λ_1), *that satisfy*

$$\|u_2^\lambda\| \neq 0 \qquad \forall (u_2^\lambda, \lambda) \in \Gamma_2^\lambda, \quad \lambda_0^\varepsilon - \varepsilon < \lambda < \lambda_1,$$
$$\lim_{\lambda \to \lambda_0^\varepsilon} \|u_1^\lambda\| = 0 \qquad \forall (u_1^\lambda, \lambda) \in \Gamma_1^\lambda.$$

(2) *There is a neighborhood* $U \subset X$ *of* $u = 0$ *such that for each* λ *with* $\lambda_0^\varepsilon - \delta < \lambda < \lambda_1$, U *contains only two nontrivial singular points* u_1^λ *and* u_2^λ *of (2.6.1).*

(3) *For every* $\lambda_0^\varepsilon - \delta < \lambda < \lambda_1$, U *can be decomposed into three open sets* $\overline{U} = \overline{U}_0 + \overline{U}_1 + \overline{U}_2$ *with* $U_i \bigcap U_j = \emptyset$ $(i \neq j)$ *such that:*

a. *If* $\lambda_0^\varepsilon - \delta < \lambda < \lambda_0^\varepsilon$, *then*

$$u = 0 \in U_0^\lambda, \quad u_i^\lambda \in \partial U_i^\lambda \bigcap \partial U_0^\lambda \quad (i = 1,2),$$

$u = 0$ *is an attractor that attracts* U_0^λ, *and* u_i^λ $(i = 1,2)$ *are two saddle points with Morse index one.*

b. *If* $\lambda_0^\varepsilon < \lambda < \lambda_1$, *then*

$$u_1^\lambda \in U_1^\lambda, \quad u_2^\lambda \in \partial U_2^\lambda \bigcap \partial U_1^\lambda, \quad 0 \in \partial U_1^\lambda \bigcap \partial U_0^\lambda,$$

u_1^λ *is an attractor that attracts* U_1^λ, *and the two points* u_2^λ *and* $u = 0$ *are saddles with Morse index one.*

(4) *Near* $(0, \lambda_\varepsilon^0), u_1^\lambda$ *and* u_2^λ *can be expressed by (2.6.9).*

Remark 2.6.5 Theorems 2.6.3 and 2.6.4 give precisely the relationship between transition and singular separation that is essential in understanding the global structure of a transition.

Proof of Theorems 2.6.3 and 2.6.4. We need to prove only Theorem 2.6.3, since the proof of Theorem 2.6.4 is identical.

By the PES (2.1.4) and (2.1.5) with $m = 1$, and by (2.6.8), the reduced equation of (2.6.1) near $(0, \lambda_0^\varepsilon)$ can be written in the following form:

$$\frac{dx}{dt} = \beta_1^\varepsilon(\lambda)x + b_\lambda^\varepsilon x^3 + g(x,\varepsilon) + o(|x|^3), \tag{2.6.10}$$

where $x \in \mathbb{R}^1, \beta_1^\varepsilon(\lambda)$ is as in (2.6.3) with $m_1 = 1$, $b_\lambda^\varepsilon \to b_1$ as $\lambda \to \lambda_0$ and $\varepsilon \to 0$, and

$$g(x,\varepsilon) = (T^\varepsilon(xe_\varepsilon(\lambda) + h_\varepsilon(x,\lambda),\lambda), e_\varepsilon^*(\lambda)),$$

where $e_\varepsilon(\lambda)$ and $e_\varepsilon^*(\lambda)$ are the eigenvectors of $L_\lambda + S_\lambda^\varepsilon$ and $L_\lambda^* + S_\lambda^{\varepsilon *}$ corresponding to $\beta_1^\varepsilon(\lambda)$ respectively.

Since G_λ and T_λ^ε are C^3 on u, by (2.6.2) we have

$$\begin{aligned} & g(x,\varepsilon) = a_1(\varepsilon)x^2 + a_2(\varepsilon)x^3 + o(|x|^3), \\ & \lim_{\varepsilon \to 0} a_i(\varepsilon) = 0 \qquad\qquad i = 1,2. \end{aligned} \tag{2.6.11}$$

Notice that

$$b_\lambda^\varepsilon + a_2(\varepsilon) \to b_1, \quad \text{as } \varepsilon \to 0, \quad \lambda \to \lambda_0.$$

Therefore, when $\varepsilon > 0$ is small, if $a_1(\varepsilon) \equiv 0$, then (2.6.10) with (2.6.11) satisfies the conditions of Theorem 2.3.1 with $\alpha < 0$. In this case, the transition of (2.6.1) at $(0, \lambda_0^\varepsilon)$ is continuous, and assertions (2) and (3) of Theorem 2.3.1 are valid for (2.6.1).

When $a_1(\varepsilon) \neq 0$, then (2.6.10) with (2.6.11) satisfies the conditions of Theorem 2.3.2. Hence, we only need to prove that the equation

$$\beta_1^\varepsilon(\lambda)x + a_1(\varepsilon)x^2 + (b_\lambda^\varepsilon + a_2(\varepsilon))x^3 + o(|x|^3) = 0 \tag{2.6.12}$$

has exactly two solutions x_1^λ and x_2^λ near λ_0^ε with $\lambda \neq \lambda_0^\varepsilon$ satisfying

$$\begin{aligned} &x_1^\lambda \to 0 \quad \text{as } \lambda \to \lambda_0^\varepsilon, \\ &x_2^\lambda \neq 0 \quad \forall \lambda_0^\varepsilon - \delta < \lambda < \lambda_0^\varepsilon + \delta, \quad \text{for some } \delta > 0, \end{aligned} \tag{2.6.13}$$

and there exists $\lambda_1 < \lambda_0^\varepsilon$ such that if $\lambda > \lambda_1$, then (2.6.12) has only one solution $x^* \neq 0$ near $x = 0$, with

$$\lim_{\lambda \to \lambda_1^+} x_i^\lambda \to x^*, \qquad i = 1, 2, \tag{2.6.14}$$

and if $\lambda < \lambda_1$, then (2.6.12) has no real nontrivial solutions near $x = 0$.

Let $b_\lambda = b_\lambda^\varepsilon + a_2(\varepsilon)$. Then (2.6.12) can be rewritten as

$$b_\lambda x^2 + a_1(\varepsilon)x + \beta_1^\varepsilon(\lambda) + o(|x|^2) = 0. \tag{2.6.15}$$

It is clear that (2.6.15) has only two nonzero solutions near $x = 0$:

$$x_{1,2} = \frac{-a_1(\varepsilon) \mp \sqrt{a_1^2(\varepsilon) - 4b_\lambda \beta_1^\varepsilon(\lambda)}}{2b_\lambda} + o(|x_{1,2}|). \tag{2.6.16}$$

By assumption, $b_1 < 0$, and hence $b_\lambda < 0$. In addition,

$$\beta_1^\varepsilon(\lambda) \begin{cases} < 0 & \text{if } \lambda < \lambda_0^\varepsilon, \\ = 0 & \text{if } \lambda = \lambda_0^\varepsilon, \\ > 0 & \text{if } \lambda > \lambda_0^\varepsilon. \end{cases}$$

Hence, the solutions x_1^λ and x_2^λ given by (2.6.16) satisfy (2.6.13), and for the number $\lambda_1 < \lambda_0^\varepsilon$ satisfying

$$a_1^2(\varepsilon) = 4b_{\lambda_1}\beta_1^\varepsilon(\lambda_1),$$

(2.6.14) holds, where $x^* = -a_1(\varepsilon)/2b_{\lambda_1}$. Moreover, if $\lambda < \lambda_1$, then $a_1^2(\varepsilon) < 4b_\lambda \beta_1^\varepsilon(\lambda)$. Hence, (2.6.15) has no real solutions.

The proof is complete. □

In the following, we consider a special case that can be applied to phase transitions in fluid dynamics. For the perturbed system (2.6.1), we assume that

$$\begin{aligned} &X_1 \text{ and } X && \text{are Hilbert spaces,} \\ &L_\lambda : X_1 \to X && \text{is symmetric,} \\ &G : X_1 \to X && \text{is a bilinear operator,} \\ &(G(u,v),v) = 0 && \forall u, v \in X_1. \end{aligned} \tag{2.6.17}$$

Theorem 2.6.6 *Assume that the PES (2.1.4) and (2.1.5) with $m = 1$, and (2.6.17) hold. Then the number b_1 defined in (2.6.8) is negative, $b_1 < 0$. Hence the assertions of Theorem 2.6.3 hold.*

Proof. Since L_λ is symmetric, the eigenvectors e is equal to e^*. Let $y = h(x,\lambda_0)$ be the center manifold function of (2.1.1) at $\lambda = \lambda_0$. Then by (2.6.17), we obtain

$$\begin{aligned}(G(xe+y),e) &= (G(xe+y,xe+y),e)\\ &= x(G(e,y),e)+(G(y,y),e)\\ &= -x(G(e),y)+(G(y),e).\end{aligned} \tag{2.6.18}$$

Let $\{e_k |\ k = 1,2,\cdots\}$ be the set of all eigenvectors of L_λ with $e_1 = e$, which constitutes an orthogonal basis of X. Hence, y can be expanded as

$$y = \sum_{j=2}^{\infty} y_j e_j.$$

By Theorem A.1.1, the center manifold function $y = y(x)$ can written as

$$\mathscr{L}_\lambda y = -P_2 G(xe) + o(|x|^2) + O(|\beta_1||x|^2), \tag{2.6.19}$$

where $\mathscr{L}_\lambda e_j = \beta_j(\lambda)e_j$ $(j = 2,3,\cdots)$, $P_2 : X \to E_2 = \text{span}\{e_2,e_3,\cdots\}$ is the canonical projection. Thus, we derive from (2.6.19) that

$$y_j = -\beta_j^{-1}(\lambda)\alpha_j x^2 + o(|x|^2, |x||\beta_1|), \tag{2.6.20}$$

where for $j = 2,3,\cdots$,

$$\alpha_j = (G(e),e_j).$$

Putting (2.6.20) in (2.6.18), we obtain

$$(G(xe+y),e) = -\sum_{j=2}^{\infty} \alpha_j x y_j + O(\|y\|^2) = b_\lambda x^3 + o(|x|^3, |x|^2|\beta_1|),$$

where $b_\lambda = \sum_{j=2}^{\infty} \alpha_j^2 \beta_j^{-1}(\lambda)$.

By the PES (2.1.4) and (2.1.5) with $m = 1$, we have $\beta_j(\lambda_0) < 0$ for every $j \geq 2$. Therefore, $b_1 = b_{\lambda_0} < 0$. The theorem is proved. □

2.6.3 Complex Eigenvalues

Let $\beta_1(\lambda) = \overline{\beta_2(\lambda)}$ be a pair of complex eigenvalues of L_λ, let $e_1(\lambda)$, $e_2(\lambda)$, and $e_1^*(\lambda), e_2^*(\lambda)$ be the eigenvectors of L_λ and L_λ^* corresponding to $\beta_1(\lambda), \beta_2(\lambda)$ respectively. Let $h(x,\lambda)$ be the center manifold function of (2.1.1) near $\lambda = \lambda_0$. Under the PES (2.3.6) and (2.3.7), we shall discuss the transition and singular separation of periodic orbits of (2.6.1) at $(0, \lambda_0^\varepsilon)$.

We first consider the perturbation stability of a transition. To this end, we take the expansion

$$(G(x+h(x,\lambda),\lambda),e_l^*)=\sum_{2\leq i+j\leq 3}a_{ij}^l x_1^i x_2^j+o(|x|^2), \tag{2.6.21}$$

for $l=1,2$, $x=x_1e_1+x_2e_2$.

For (2.6.21) and $\beta_1(\lambda)=\alpha+i\sigma$, we can introduce a number b as defined by (2.3.15), which can determine not only the types but also the stability of transitions of (2.1.1) at $(0,\lambda_0)$. The following is the transition stability theorem of perturbed system (2.6.1).

Theorem 2.6.7 *Under the PES (2.3.6) and (2.3.7), and (2.6.21), if the number b given by (2.3.15) is nonzero, then the transition structure of the perturbed system (2.6.1) at $(0,\lambda_0^\varepsilon)$ is the same as that of (2.1.1) at $(0,\lambda_0)$. More precisely, when $b<0$, the transition of (2.6.1) is continuous, and (2.6.1) bifurcates from $(0,\lambda_0^\varepsilon)$ to an attractor that is a periodic orbit; when $b>0$, the transition is a jump transition, and (2.6.1) bifurcates on $\lambda<\lambda_0^\varepsilon$ to a unique periodic orbit.*

Proof. Under the PES (2.3.6) and (2.3.7), the reduced equation of (2.1.1) on the center manifold reads

$$\begin{aligned}\frac{dx_1}{dt}&=\alpha(\lambda)x_1-\sigma(\lambda)x_2+(G(x+h,\lambda),e_1^*),\\ \frac{dx_2}{dt}&=\sigma(\lambda)x_1+\alpha(\lambda)x_2+(G(x+h,\lambda),e_2^*).\end{aligned} \tag{2.6.22}$$

As in the proof of Theorem 2.6.6, from (2.6.21) and (2.6.22) one can derive the number b given by (2.3.15), which is the bifurcation number as in Theorem 2.2.8. Thus, this theorem is deduced from Theorems 2.2.8, 2.2.9, and 2.3.7. The proof is complete. □

In the following, we consider the instability of transitions for the perturbed system (2.6.1), using the singular separation idea introduced in the previous section. As seen in Theorems 2.6.3 and 2.6.4, when the bifurcation number b is equal to 0, the transition structure of (2.6.1) may be different from the original equation (2.1.1).

Let $G(x+h,\lambda)$ have the following Taylor expansion near $\lambda=\lambda_0$:

$$(G(x+h(x,\lambda),\lambda),e_l^*(\lambda))=\sum_{3\leq i+j\leq 5}a_{ij}^l x_1^i x_2^j+o(|x|^5), \tag{2.6.23}$$

for $l=1,2$. For (2.6.23), we introduce two numbers as follows:

$$\begin{aligned}b_1&=\frac{3}{4}\pi(a_{30}^1+a_{03}^2)+\frac{\pi}{4}(a_{12}^1+a_{21}^2),\\ b_2&=\frac{5\pi}{8}(a_{50}^1+a_{05}^2)+\frac{\pi}{8}(a_{32}^1+a_{14}^1+a_{23}^2+a_{41}^2)\end{aligned} \tag{2.6.24}$$

$$
\begin{aligned}
&+\frac{17\pi}{32\sigma}(a^1_{03}a^2_{03}-a^1_{30}a^2_{30})+\frac{3\pi}{32\sigma}(2a^1_{30}a^1_{21}+2a^1_{03}a^1_{12}+a^1_{30}a^2_{30}\\
&+a^1_{03}a^2_{21}+a^1_{12}a^2_{12}+a^1_{21}a^2_{03}-2a^2_{30}a^2_{21}-2a^2_{03}a^2_{12}-a^1_{03}a^2_{03}\\
&-a^1_{30}a^2_{12}-a^1_{21}a^2_{21}-a^1_{12}a^2_{30})+\frac{3\pi}{64\sigma}(2a^1_{30}a^1_{03}+2a^1_{12}a^1_{21}\\
&+a^1_{12}a^2_{30}+a^1_{21}a^2_{21}+a^1_{30}a^2_{12}-2a^2_{30}a^2_{03}-2a^2_{12}a^2_{21}\\
&-a^1_{21}a^2_{03}-a^1_{12}a^2_{12}-a^1_{03}a^2_{21}).
\end{aligned}
\tag{2.6.25}
$$

Theorem 2.6.8 *Let the PES (2.3.6) and (2.3.7), and (2.6.23) hold, the numbers* $b_1 = 0$, *and* $b_2 < 0$. *If the transition at* $(0, \lambda_0^\varepsilon)$ *is continuous, then (2.6.1) bifurcates on* $\lambda_0^\varepsilon < \lambda$ *to a unique periodic orbit that is an attractor. If the transition is a jump transition, then the following assertions hold:*

(1) Equation (2.6.1) has a singular separation of periodic orbits at some $(\Sigma_1, \lambda_1) \subset X \times (-\infty, \lambda_0^\varepsilon)$, *and there are exactly two branches of periodic orbits* Σ_1^λ $(\lambda_1 < \lambda < \lambda_0^\varepsilon)$ *and* Σ_2^λ $(\lambda_1 < \lambda < \lambda_0^\varepsilon + \delta)$, *separated from* (Σ_1, λ_1), *as shown in Fig. 2.33a, that satisfy*

$$
\begin{aligned}
&\lim_{\lambda\to\lambda_0^\varepsilon} \Sigma_1^\lambda = \{0\}, && \lim_{\lambda\to\lambda_1} \Sigma_i^\lambda = \Sigma_1 \qquad i = 1,2,\\
&dist(\Sigma_2^\lambda, 0) > 0 && \forall \lambda_1 < \lambda < \lambda_0^\varepsilon + \delta \ (\delta > 0).
\end{aligned}
$$

(2) There is a neighborhood $U \subset X$ *of* $u = 0$ *such that if* $\lambda_1 < \lambda < \lambda_0^\varepsilon$, *then* U *can be decomposed into two open sets* $\overline{U} = \overline{U}_1^\lambda + \overline{U}_2^\lambda$ *with* $U_1^\lambda \cap U_2^\lambda = \emptyset$ *such that*

$$
\Sigma_2^\lambda \subset U_2^\lambda, \quad 0 \in U_1^\lambda, \quad \Sigma_1^\lambda \subset \partial U_1^\lambda \bigcap \partial U_2^\lambda,
$$

and Σ_2^λ *and* $u = 0$ *are attractors attracting* U_1^λ *and* U_2^λ *respectively.*

(3) If $\lambda_0^\varepsilon < \lambda < \lambda_0^\varepsilon + \delta$, *then* Σ_2^λ *is an attractor attracting* $U \setminus \Gamma$, *where* Γ *is the stable manifold of* $u = 0$ *with codim* $\Gamma = 2$.

Theorem 2.6.9 *Let the PES (2.3.6) and (2.3.7), and (2.6.23) hold, and suppose* $b_1 = 0$, *and* $b_2 > 0$. *If the transition at* $(0, \lambda_0^\varepsilon)$ *is a jump transition, then (2.6.1) bifurcates on* $\lambda < \lambda_0^\varepsilon$ *to a unique periodic orbit that is a repeller. If it is continuous, then the following assertions hold:*

(1) Equation (2.6.1) has a singular separation of periodic orbits at some $(\Sigma_1, \lambda_1) \subset X \times (\lambda_0^\varepsilon, +\infty)$, *and there are exactly two branches of periodic orbits* Σ_1^λ $(\lambda_0^\varepsilon < \lambda < \lambda_1)$ *and* Σ_2^λ $(\lambda_0^\varepsilon - \delta < \lambda < \lambda_1)$, *separated from* (Σ_1, λ_1), *as shown in Fig. 2.33b, that satisfy*

$$
\begin{aligned}
&\lim_{\lambda\to\lambda_0^\varepsilon} \Sigma_1^\lambda = \{0\}, && \lim_{\lambda\to\lambda_1} \Sigma_i^\lambda = \Sigma_1 \quad i = 1,2,\\
&dist(\Sigma_2^\lambda, 0) > 0 && \forall \lambda_0^\varepsilon - \delta < \lambda < \lambda_1.
\end{aligned}
$$

(2) When $\lambda_0^\varepsilon < \lambda < \lambda_1$, then Σ_1^λ is an attractor that attracts $U_\lambda \setminus \Gamma$, where U_λ is a neighborhood of $u = 0$, Γ the stable manifold of $u = 0$ with codim $\Gamma = 2$, and $\Sigma_2^\lambda \subset \partial U_\lambda$ with ∂U_λ a stable manifold of Σ_2^λ.

(3) When $\lambda_0^\varepsilon - \delta < \lambda < \lambda_0^\varepsilon$, then $u = 0$ is an attractor that attracts a neighborhood U_λ of $u = 0$, and $\Sigma_2^\lambda \subset \partial U_\lambda$ with ∂U_λ a stable manifold of Σ_2^λ.

Proof of Theorems 2.6.8 and 2.6.9. We only need to prove Theorem 2.6.8, since the proof of Theorem 2.6.9 is the same.

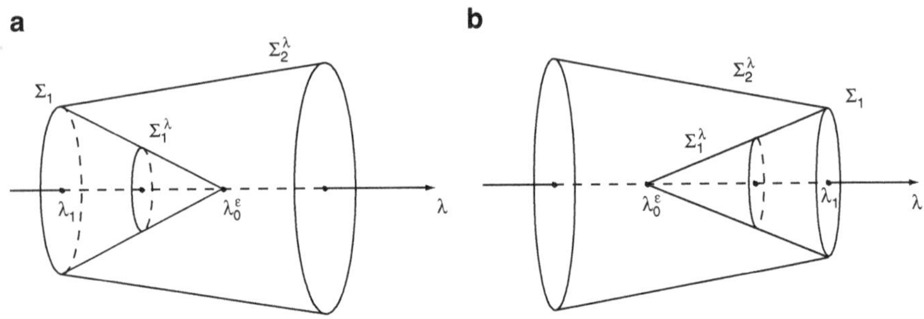

Fig. 2.33 (**a**) The case $b_2 < 0$; (**b**) the case $b_2 > 0$

By (2.6.23), the reduced equation of (2.6.1) on the center manifold near $\lambda = \lambda_0^\varepsilon$ can be written as

$$\begin{aligned} \frac{dx_1}{dt} &= \alpha_\varepsilon(\lambda)x_1 - \sigma_\varepsilon x_2 + \sum_{2 \le i+j \le 5} a_{ij}^1(\varepsilon) x_1^i x_2^j + o(|x|^5), \\ \frac{dx_2}{dt} &= \sigma_\varepsilon x_1 + \alpha_\varepsilon(\lambda)x_1 + \sum_{2 \le i+j \le 5} a_{ij}^2(\varepsilon) x_1^i x_2^j + o(|x|^5), \end{aligned} \tag{2.6.26}$$

where

$$\alpha_\varepsilon(\lambda) \begin{cases} < 0 & \text{if } \lambda < \lambda_0^\varepsilon, \\ = 0 & \text{if } \lambda = \lambda_0^\varepsilon, \\ > 0 & \text{if } \lambda > \lambda_0^\varepsilon, \end{cases} \tag{2.6.27}$$

$$\begin{cases} \lim_{\varepsilon \to 0} \sigma_\varepsilon = \sigma \neq 0, \\ \lim_{\varepsilon \to 0} a_{ij}^l(\varepsilon) = 0 & \text{for } l = 1,2 \text{ and } i+j = 2, \\ \lim_{\varepsilon \to 0} a_{ij}^l(\varepsilon) = a_{ij}^l & \text{for } l = 1,2 \text{ and } 3 \le i+j \le 5. \end{cases} \tag{2.6.28}$$

Taking the polar coordinate $x_1 = r\cos\theta$, $x_2 = r\sin\theta$, we derive then from (2.6.26) that

$$\frac{dr}{d\theta} = \frac{\alpha_\varepsilon(\lambda)r + \sum_{k=2}^5 r^k u_k^\varepsilon(\cos\theta, \sin\theta) + o(r^5)}{\sigma_\varepsilon - \sum_{k=2}^5 r^{k-1} v_k^\varepsilon(\cos\theta, \sin\theta) + o(r^4)}, \tag{2.6.29}$$

where

$$u_k^\varepsilon = \sum_{i+j=k} [a_{ij}^1(\varepsilon)\cos^{i+1}\theta \sin^j \theta + a_{ij}^2(\varepsilon)\cos^i \theta \sin^{j+1}\theta],$$
$$v_k^\varepsilon = \sum_{i+j=k} [a_{ij}^1(\varepsilon)\cos^i \theta \sin^{j+1}\theta - a_{ij}^2(\varepsilon)\cos^{i+1}\theta \sin^j \theta].$$

By $\sigma_\varepsilon \to \sigma \neq 0$ ($\varepsilon \to 0$), near $r = 0$, (2.6.29) can be expressed as

$$\frac{1}{r^2}\frac{dr}{d\theta} = \frac{1}{\sigma_\varepsilon}\left[\frac{\alpha_\varepsilon}{r} + \sum_{k=2}^{5} r^{k-2} f_k^\varepsilon(\cos\theta, \sin\theta) + o(r^3)\right], \tag{2.6.30}$$

where

$$\begin{aligned}
f_2^\varepsilon &= u_2^\varepsilon + \sigma_\varepsilon^{-1}\alpha_\varepsilon v_2^\varepsilon,\\
f_3^\varepsilon &= u_3^\varepsilon + \sigma_\varepsilon^{-1}\alpha_\varepsilon v_3^\varepsilon + \sigma_\varepsilon^{-1}u_2^\varepsilon v_2^\varepsilon + \alpha_\varepsilon\sigma_\varepsilon^{-2}v_2^{\varepsilon 2},\\
f_4^\varepsilon &= u_4^\varepsilon + \sigma_\varepsilon^{-1}\alpha_\varepsilon v_4^\varepsilon + \sigma_\varepsilon^{-1}u_2^\varepsilon v_3^\varepsilon + \sigma_\varepsilon^{-1}u_3^\varepsilon v_2^\varepsilon + 2\alpha_\varepsilon\sigma_\varepsilon^{-2}v_2^\varepsilon v_3^\varepsilon\\
&\quad + \sigma_\varepsilon^{-2}u_2^\varepsilon v_2^{\varepsilon 2} + \alpha_\varepsilon\sigma_\varepsilon^{-3}v_2^{\varepsilon 3},\\
f_5^\varepsilon &= u_5^\varepsilon + \sigma_\varepsilon^{-1}\alpha_\varepsilon v_5^\varepsilon + \sigma_\varepsilon^{-1}u_2^\varepsilon v_4^\varepsilon + \sigma_\varepsilon^{-1}u_3^\varepsilon v_3^\varepsilon + \sigma_\varepsilon^{-1}u_4^\varepsilon v_2^\varepsilon\\
&\quad + 2\alpha_\varepsilon\sigma_\varepsilon^{-2}v_2^\varepsilon v_4^\varepsilon + 2\sigma_\varepsilon^{-2}u_2^\varepsilon v_2^\varepsilon v_3^\varepsilon + \sigma_\varepsilon^{-1}u_3 v_2^{\varepsilon 2}\\
&\quad + 3\alpha_\varepsilon\sigma_\varepsilon^{-3}v_2^{\varepsilon 2}v_3^\varepsilon + \sigma_\varepsilon^{-3}u_2^\varepsilon v_2^{\varepsilon 3} + \alpha_\varepsilon\sigma_\varepsilon^{-4}v_2^{\varepsilon 4}.
\end{aligned}$$

The initial condition is

$$r(0, \lambda, a) = a \qquad (a > 0). \tag{2.6.31}$$

Let $r(\theta, \lambda, a)$ have the Taylor expansion

$$r(\theta, \lambda, a) = a + r_2(\theta, \lambda)a^2 + r_3(\theta, \lambda)a^3 + \alpha_\varepsilon h + o(a^3), \tag{2.6.32}$$

where $\alpha_\varepsilon h(\theta, \lambda, a)$ is contributed by the term containing α_ε in (2.6.30) that vanishes at $\lambda = \lambda_0^\varepsilon$.

Putting (2.6.32) in (2.6.30), we obtain

$$\frac{dr}{d\theta} = \alpha_\varepsilon H(\theta, \lambda, a) + (a + r_2 a^2)^2 f_2^\varepsilon + a^3 f_3^\varepsilon + o(a^3). \tag{2.6.33}$$

Integrating (2.6.33) with respect to θ, by (2.6.31) we get

$$\begin{aligned}
r(\theta, \lambda, a) &= a + a^2 \int_0^\theta f_2^\varepsilon d\theta + a^3 \int_0^\theta (f_3^\varepsilon + 2r_2 f_2^\varepsilon) d\theta\\
&\quad + \alpha_\varepsilon(\lambda) \int_0^\theta H(\theta, \lambda, a) d\theta + o(a^3).
\end{aligned} \tag{2.6.34}$$

In comparing the coefficients of (2.6.32) and (2.6.34), we find that

$$\begin{aligned} r_2(\theta,\lambda) &= \int_0^\theta f_2^\varepsilon(\cos\theta,\sin\theta)d\theta, \\ r_3(\theta,\lambda) &= \int_0^\theta (f_3^\varepsilon(\cos\theta,\sin\theta)+2r_2(\theta,\lambda)f_2^\varepsilon(\cos\theta,\sin\theta)d\theta. \end{aligned} \tag{2.6.35}$$

Inserting (2.6.35) in (2.6.30), and integrating it from $\theta=0$ to $\theta=2\pi$, we deduce that

$$\begin{aligned} \frac{r(2\pi,0)-r(0,a)}{r(2\pi,a)} &= \frac{1}{\sigma_\varepsilon}\Big\{\rho(a,\lambda)\alpha_\varepsilon(\lambda)+a\int_0^{2\pi} f_2^\varepsilon d\theta \\ &+a^2\int_0^{2\pi} f_3^\varepsilon d\theta + a^3\int_0^{2\pi}\left[f_4^\varepsilon+f_3^\varepsilon\left(\int_0^\theta f_2^\varepsilon d\theta\right)\right]d\theta \\ &+a^4\int_0^{2\pi}\Big[f_5^\varepsilon+2f_4^\varepsilon\left(\int_0^\theta f_2^\varepsilon d\theta\right)+f_3^\varepsilon\left(\int_0^\theta f_3^\varepsilon d\theta\right) \\ &+2f_3^\varepsilon\left(\int_0^\theta f_2^\varepsilon\left(\int_0^\theta f_2^\varepsilon d\theta\right)d\theta\right)\Big]d\theta+\alpha_\varepsilon g(\lambda,a)\Big\}+o(a^4), \end{aligned} \tag{2.6.36}$$

where $\rho(a,\lambda)=2\pi+o(a)$ and $g(\lambda,0)=0$.

By a direct computation, we have

$$\begin{aligned} &\int_0^{2\pi} f_2^\varepsilon(\cos\theta,\sin\theta)d\theta=0, \\ &\int_0^{2\pi} f_4^\varepsilon(\cos\theta,\sin\theta)d\theta=0, \\ &\int_0^{2\pi} f_3^\varepsilon\left(\int_0^\theta f_2^\varepsilon d\theta\right)d\theta=\frac{2}{3}a_{02}^2(\varepsilon)\int_0^{2\pi} f_3^\varepsilon d\theta+o(\alpha_\varepsilon), \\ &\int_0^{2\pi} f_3^\varepsilon\left(\int_0^\theta f_3^\varepsilon d\theta\right)d\theta=0. \end{aligned}$$

Hence, (2.6.36) can be rewritten as

$$\frac{r(2\pi,a)-r(0,a)}{r(2\pi,a)}=\varphi(a,\lambda)+\delta(\varepsilon)a^2+\delta_1(\varepsilon)a^3+b_\varepsilon(\lambda)a^4+o(a^5), \tag{2.6.37}$$

where $\varphi(a,\lambda)=\sigma_\varepsilon^{-1}(\rho(a,\lambda)+g(\lambda,a))\alpha_\varepsilon(\lambda)$, by (2.6.27), which satisfies

$$\varphi(\lambda,a)\begin{cases} <0 & \text{if } \lambda<\lambda_0^\varepsilon, \\ =0 & \text{if } \lambda=\lambda_0^\varepsilon, \\ >0 & \text{if } \lambda>\lambda_0^\varepsilon, \end{cases} \tag{2.6.38}$$

and

$$\delta(\varepsilon)=\int_0^{2\pi} f_3^\varepsilon(\cos\theta,\sin\theta)d\theta,$$

$$\delta_1(\varepsilon) = \frac{2}{3} a_{02}^2(\varepsilon)\delta(\varepsilon),$$

$$b_\varepsilon(\lambda) = \int_0^{2\pi} \left[f_5^\varepsilon + 2 f_4^\varepsilon \left(\int_0^\theta f_2^\varepsilon \right) + 2 f_3^\varepsilon \left(\int_0^\theta f_2^\varepsilon \left(\int_0^\theta f_2^\varepsilon \right) \right) \right] d\theta.$$

By (2.6.28), we see that

$$\lim_{\varepsilon \to 0} (u_2^\varepsilon, v_2^\varepsilon, f_2^\varepsilon) = 0,$$
$$\lim_{\varepsilon \to 0} f_3^\varepsilon = f_3 = u_3 + \sigma^{-1}\alpha(\lambda) v_3,$$
$$\lim_{\varepsilon \to 0} f_5^\varepsilon = f_5 = u_5 + \sigma^{-1} u_3 v_3.$$

Hence, we have

$$\lim_{\varepsilon \to 0} \delta(\varepsilon) = b_1(\lambda) = \int_0^{2\pi} [u_3 + \sigma^{-1}\alpha(\lambda) v_3] d\theta,$$
$$\lim_{\varepsilon \to 0} b_\varepsilon(\lambda_0) = b_2 = \int_0^{2\pi} [u_5 + \sigma^{-1} u_3 v_3] d\theta,$$

where $b_1(\lambda_0) = \int_0^{2\pi} u_3 d\theta$ is the number given by (2.6.24), and b_2 is given by (2.6.25). By the assumption $b_1(\lambda_0) = 0$, we have

$$\begin{aligned} &\delta(\varepsilon), \delta_1(\varepsilon) \to 0 && \text{if } \lambda \to \lambda_0, \varepsilon \to 0, \\ &b_\varepsilon(\lambda) < -C && \text{for some } C > 0 \text{ near } \lambda = \lambda_0^\varepsilon. \end{aligned} \tag{2.6.39}$$

It is known that each real singular point $a_0 > 0$ of (2.6.37) is associated with a periodic orbit of (2.6.26), and if

$$r(2\pi, a) - r(0, a) \begin{cases} < 0 & \text{if } a_0 < a, \\ > 0 & \text{if } a_0 > a, \end{cases} \tag{2.6.40}$$

then the periodic orbit associated with a_0 is an attractor. If

$$r(2\pi, a) - r(0, a) \begin{cases} > 0 & \text{if } a_0 < a, \\ < 0 & \text{if } a_0 > a, \end{cases} \tag{2.6.41}$$

then the periodic orbit is a repeller.

Consider the equation

$$\varphi(a, \lambda) + \delta(\varepsilon) a^2 + \delta_1(\varepsilon) a^3 + b_\varepsilon(\lambda) a^4 + o(a^5) = 0. \tag{2.6.42}$$

It follows from (2.6.38) and (2.6.39) that when $\delta(\varepsilon) \leq 0$ then (2.6.42) has a unique real positive solution $a_0 > 0$ near $(a,\lambda) = (0,\lambda_0^\varepsilon)$ on $\lambda > \lambda_0^\varepsilon$ that satisfies (2.6.40), i.e., (2.6.26) bifurcates from $(x,\lambda) = (0,\lambda_0^\varepsilon)$ to an attractor consisting of one periodic orbit. When $\delta(\varepsilon) > 0$ then near $(a,\lambda) = (0,\lambda_0^\varepsilon)$ in $\lambda_1 < \lambda < \lambda_0^\varepsilon$, (2.6.42) has exactly two positive real solutions

$$a_\pm = \left[\frac{\delta(\varepsilon) \pm \sqrt{\delta^2(\varepsilon) - 4|b_\varepsilon|(|\varphi| - \delta_1 a_\pm^3) + o(a_\pm^5)}}{2|b_\varepsilon(\lambda)|}\right]^{1/2},$$

and at $\lambda = \lambda_1$ that satisfies

$$\delta^2(\varepsilon) - 4b_\varepsilon(\lambda_1)(\varphi(a,\lambda_1) + \delta_1(\varepsilon)a^3 + o(a^5)) = 0,$$

(2.6.42) has only one positive real solution

$$a_1 = \left[\frac{-\delta(\varepsilon)}{2b_\varepsilon(\lambda_1)}\right]^{1/2}, \quad (b_\varepsilon < 0),$$

and (2.6.42) has no real solutions on $\lambda < \lambda_1$; however, on $\lambda > \lambda_0^\varepsilon$, (2.6.42) has exactly one positive real solution near $a = 0$:

$$a_+ = \left[\frac{\delta(\varepsilon) + \sqrt{\delta^2(\varepsilon) + 4|b_\varepsilon|(\varphi + \delta_1 a_+^3 + o(a_+^5))}}{2|b_\varepsilon(\lambda)|}\right]^{1/2}.$$

Moreover, (2.6.40) holds for $a_0 = a_+$, and (2.6.41) holds for $a_0 = a_-$. Theorem 2.6.8 is proved. □

Remark 2.6.10 From (2.6.36), we can see that if the condition (2.6.23) is relaxed to

$$(G(x+h(x,\lambda),\lambda),e_l^*) = \sum_{2\leq i+j\leq 5} a_{ij}^l x_1^i x_2^j + o(|x|^5) \qquad \text{for } l = 1,2,$$

i.e., $G(u,\lambda)$ contains a second-order homogeneous term, then we take the numbers $b_1(\lambda)$ and $b_2(\lambda)$ as follows:

$$b_1(\lambda) = \int_0^{2\pi} f_3 d\theta,$$

$$b_2(\lambda) = \int_0^{2\pi} \left[f_5 + 2f_4\left(\int_0^\theta f_2 d\theta\right) + 2f_3\int_0^\theta f_2\left(\int_0^\theta f_2 d\theta\right)d\theta\right]d\theta,$$

where

$$
\begin{aligned}
f_k(\theta) &= u_k + \sum_{i=1}^{k-2} v_{i+1} v_{k-i}, \\
u_k &= \sum_{i+j=k} a_{ij}^1 \cos^{i+1}\theta \sin^j\theta + a_{ij}^2 \cos^i\theta \sin^{j+1}\theta, \\
v_k &= \sum_{i+j=k} a_{ij}^1 \cos^i\theta \sin^{j+1}\theta - a_{ij}^2 \cos^{i+1}\theta \sin^j\theta.
\end{aligned}
$$

In this case, when $b_1(\lambda_0) = 0$, then for $b_2(\lambda_0) < 0$ and $b_2(\lambda_0) > 0$, Theorems 2.6.8 and 2.6.9 hold respectively.

2.7 Notes

Dynamic transition theory has been developed over the last ten years by the authors through a sequence of articles and two books based on Philosophy 2 with strong motivations in applications. The theory is partially presented in Ma and Wang (2004b, 2005a,b,c, 2007b), and this chapter gives a comprehensive account of this theory.

As demonstrated in Chap. 1 and in many examples presented in Chaps. 3–6, dynamic transition theory is a global theory for dynamic transitions of dissipative systems. It is an entirely different theory from the classical bifurcation theory, as demonstrated in the introduction and in Chap. 1.

A combination of mathematical tools used in this chapter include classical dynamical systems system, bifurcation theory, topology, and differential equations. For steady-state bifurcation theory, see among others (Krasnosel'skii 1956; Rabinowitz 1971; Crandall and Rabinowitz 1977; Nirenberg 2001, 1981; Chow and Hale 1982; Ma and Wang 2005b, 2007b). For Hopf bifurcation, see among many others (Hopf 1942; Marsden and McCracken 1976; Ma and Wang 2005b). Although not used in the book, the reader is referred to Golubitsky and Schaeffer (1985), Sattinger (1978, 1979, 1980, 1983), Field (1996) for bifurcation theory with symmetry. For dynamical systems theory, see, for example, (Henry 1981; Perko 1991; Guckenheimer and Holmes 1990; Wiggins 1990; Temam 1997; Hale. 1988; Ma and Wang 2005d).

Chapter 3
Equilibrium Phase Transition in Statistical Physics

A principal objective in the study of equilibrium phase transitions is to capture the transitions from one equilibrium to another and to study the nature or order of such transitions. The study of equilibrium phase transitions presented in this book involves a combination of modeling, mathematical analysis, and physical predictions.

Equilibrium phase transitions are classically studied as time-independent problems by exploring the solutions of the equilibrium equation (3.1.10) and their stability. However, as we shall explain in Sect. 3.1, Le Chatelier's principle and the Landau mean field theory allow us to introduce a unified gradient-type dissipative dynamic model for all equilibrium phase transitions. This unified model defines a quasigradient flow with relaxation time. The time-dependent Ginzburg–Landau (TDGL) model of superconductivity and the Cahn–Hilliard model for binary systems are two such classical models. With the success of these two models, the Landau mean field approach can be considered a well-established method for studying many important features of phase transitions. Of course, there are drawbacks to the method, including the failure to predict the correct critical exponents (Stanley 1971).

With this unified model as a general framework, we introduce two recently developed dynamic models for gas–liquid transitions, ferromagnetic systems, and the superfluidity of liquid helium (helium-3, helium-4, and their admixtures) with close links to known experimental results. This analysis leads to a number of physical predictions.

Gas–liquid transition is a classical example, and it is understood both experimentally and theoretically that the transition before the Andrews critical point is of first order. By examining the transition behavior near and beyond the critical point, we are able to derive a first example of third-order transitions. Classically, it is hard to observe third- or higher-order transitions, since they involve higher-order derivatives of the free energy; see, for example, the experimental studies by Nishikawa and Morita (1998); see also (Fisher 1964). Based on a recent article (Ma and Wang 2011e), we derive in Sect. 3.2 a new mechanism of the Andrews critical point and obtain that the liquid–gas phase transition going beyond the Andrews point is of third order—a first example of a third-order transition.

T. Ma and S. Wang, *Phase Transition Dynamics*, DOI 10.1007/978-1-4614-8963-4_3,

Also, by applying dynamic transition theory and the unified model to liquid helium-3, a new prediction on the existence of a new superfluid phase C is made in Sect. 3.8; see also (Ma and Wang 2008e). Of course, some of these predictions have yet to be verified experimentally, and it is hoped that our predictions will help to provide new insights into both theoretical and experimental studies for a better understanding of the underlying physical problems.

Extensive studies of the TDGL model of superconductivity can be found in the literature. See, among many others, (de Gennes 1966; Tinkham 1996; Aftalion and Du 2002; Ginzburg 2004; Du et al. 1992; Bethuel et al. 1994; del Pino et al. 2000; Rubinstein 2004; Serfaty 2006; Lin 1996; Ma and Wang 2005a; Tang and Wang 1995). Cahn–Hilliard equations are often used to model phase transitions for binary systems as well as in other interface problems; see, e.g., (Cahn and Hillard 1957; Novick-Cohen and Segel 1984; Elliott and Songmu 1986; Bates and Fife 1993; Alikakos and Fusco 1993; Shen and Yang 2010; Desai and Kapral 2009). The results presented here on superconductivity and on binary systems are based mainly on Ma and Wang (2005a, 2009b,c).

In his Nobel lecture (Ginzburg 2004), Ginzburg expressed the hope of developing a Ginzburg–Landau model for superfluidity. The study presented here can be considered an attempt in the direction of a realization of his vision. It is based on Ma and Wang (2008a,e, 2009d). For other attempts, the reader can consult (Onuki 2002) and the references therein. However, we would like to mention that the Ginzburg–Landau theory with only one wave function cannot describe the typical phase transition diagrams for superfluidity, and this demonstrates the necessity for introducing new Ginzburg–Landau models.

Finally, there are many excellent general books on statistical physics, thermodynamics, and condensed-matter physics; see, among others, (Reichl 1998; Stanley 1971; Kleman and Lavrentovich 2003; Chalkin and Lubensky 2000).

3.1 Dynamic Models for Equilibrium Phase Transitions

In this section, we present a unified time-dependent model for equilibrium phase transitions based on Le Chatelier's principle, and on the Laudau mean field theory. This model was first presented in the authors' recent work (Ma and Wang 2009b,d, 2008a,e), and is different from the gradient flows used in the literature; see (Onuki 2002; Pismen 2006). The main difference is that the unified model here is only a *gradient-type* flow with important balance terms Φ_i; see, e.g., (3.1.15). In many cases, these added terms are physically important; see also Appendix A.2 for some mathematical properties of systems of gradient type.

3.1.1 Thermodynamic Potentials

Thermodynamics is a macroscopic, phenomenological theory of heat. A thermodynamic potential is a scalar potential function used to represent the thermodynamic state of a system. There are five commonly used thermodynamic potentials: the internal energy, the Helmholtz free energy, the enthalpy, the Gibbs free energy, and the grand potential (Landau potential). The first main thermodynamic potential is the internal energy, and the other four are derived using the Legendre transforms; see among others (Reichl 1998; Stanley 1971).

Conjugate Pairs of State Variables and Internal Energy. First, we note that all thermodynamic potentials are expressed in terms of conjugate pairs. The most commonly considered conjugate thermodynamic variables are as follows:

1. The temperature T and the entropy S,
2. the generalized force f, the generalized displacement X such as the pressure p and the volume V, and
3. The chemical potential μ and the particle number N.

Considering the momentum condensation of bosons, an additional pair of conjugate variables of state consists of

4. the condensation potential β and the number of condensed bosons N_b.

The fundamental equations for the internal energy U and its differential are then given by

$$
\begin{aligned}
U &= ST + fX + \sum_j \mu_j N_j + \beta N_b, \\
dU &= TdS + fdX + \sum_j \mu_j dN_j + \beta dN_b,
\end{aligned}
$$

where μ_j is the chemical potential of the jth particle type, and N_j is the number of j-type particles in the system. The temperature, pressure, and chemical potential are the generalized forces, which drive the changes in entropy, volume, and particle number respectively.

Helmholtz Energy. For an isothermal process, we need to use the Helmholtz free energy, denoted by

$$F = U - TS, \tag{3.1.1}$$

where U is the internal energy and T, f, μ_i, β are functions of variables S, X. In fact, (3.1.1) is the Legendre transformation such that the Helmholtz free energy F is a function of T and X. According to the stability principle, in an equilibrium state (T^*, X^*), $Z = (S, f, \mu, N, \beta, N_b)$ is a continuous function of (T^*, X^*), i.e., $Z^* = Z(T^*, X^*)$, and F reaches its minimal value at (Z^*, T^*, X^*):

$$\frac{\partial F(Z^*, T^*, X^*)}{\partial Z} = 0. \tag{3.1.2}$$

An important viewpoint in the Landau mean field theory is to regard (T,X) as control parameter, i.e.,

$$\lambda = (T,X),$$

and take a characterized variable $u = (u_1,\cdots,u_m)$, called the order parameter, which is derived from $z = (S,f,\mu,N,\beta,N_b)$. Thus $F(Z,T,X)$ can be rewritten as

$$F = F(u,\lambda), \quad u = (u_1,\cdots,u_m),$$

and by (3.1.2), we get the equations describing the thermal equilibrium phase transition as follows:

$$\nabla_u F(u,\lambda) = \left(\frac{\partial F(u,\lambda)}{\partial u_1},\cdots,\frac{\partial F(u,\lambda)}{\partial u_m}\right) = 0. \tag{3.1.3}$$

Thus, the phase transition problem of a thermodynamic process is defined to be the transitions and stability of the solutions of (3.1.3) as the control parameter crosses a critical threshold $\lambda = \lambda_0$.

Gibbs Free Energy. For the various thermal processes, we need different thermodynamic potentials to describe the phase transitions. As seen above, the Helmholtz free energy is for systems having only thermal exchange with the external environment. When the system has only thermal and mechanical exchanges with the external environment, i.e., the thermal process is at constant temperature T with generalized force f, the corresponding potential is the Gibbs free energy, which is defined by

$$G = F - fX, \quad F \text{ the Helmholtz free energy,} \tag{3.1.4}$$

which is the Legendre transformation with (T,f) its independent variables. In this situation, the equations of phase transition are

$$\frac{\partial G(u,\lambda)}{\partial u_i} = 0 \qquad \forall 1 \le i \le m, \tag{3.1.5}$$

where $\lambda = (T,f)$, and u is the order parameter derived from (S,X,μ,N,β,N_b). In equilibrium phase transitions, the most common thermal processes are under constant temperature and force.

Enthalpy. When a thermodynamic system has only mechanical exchange with the external environment, the potential function is the enthalpy, which is defined by

$$H = U - fX, \quad U \text{ the internal energy.} \tag{3.1.6}$$

Therefore, the independent variables of H are $\lambda = (S,f)$, and the order parameter u is derived from (T,X,μ,N,β,N_b).

Grand Potential. When the thermodynamic system is mechanically and materially coupled to the external environment, the potential function is the grand potential, which is defined by

$$J = U - fX - \sum_i \mu_i N_i - \beta N_b. \tag{3.1.7}$$

The independent variables of J are $\lambda = (S, f, N, N_b)$, and the order parameter u is derived from (T, X, μ, β).

Heterogeneity. Consider a thermal system as its order parameter u changes in $\Omega \subset \mathbb{R}^n$ $(1 \leq n \leq 3)$. In this situation, the free energy of this system is of the form

$$\mathscr{H}(u,\lambda) = \mathscr{H}_0 + \int_\Omega \Big[\frac{1}{2}\sum_{i=1}^m \mu_i |\nabla u_i|^2 + g(u, \nabla u, \lambda)\Big] dx, \tag{3.1.8}$$

where $u = (u_1, \cdots, u_m), \mu_i = \mu_i(\lambda) > 0$, and $g(u, \nabla u, \lambda)$ is a $C^r (r \geq 2)$ function of $(u, \nabla u)$ with Taylor expansion

$$g(u, \nabla u, \lambda) = \sum \alpha_{ijk} u_i D_j u_k + \sum_{|I|=1}^N \alpha_I u^I + o(|u|^N) - fX, \tag{3.1.9}$$

where $I = (i_1, \cdots, i_m)$, $i_k \geq 0$, and $N \geq 3$ are integers, $|I| = \sum_{k=1}^m i_k$, the coefficients α_{ijk} and α_I depend continuously on λ, which are determined by the concrete physical problem, $u^I = u_1^{i_1} \cdots u_m^{i_m}$, and fX is the generalized work.

Thus a thermal equilibrium phase transition is regarded as the transition and stability of the following system:

$$\frac{\delta}{\delta u}\mathscr{H}(u,\lambda) = 0, \tag{3.1.10}$$

supplemented with certain natural boundary conditions such as the Dirichlet or Neumann boundary conditions. Here $\delta/\delta u$ is the variational derivative.

3.1.2 Time-Dependent Equations

A thermal system is controlled by some parameter λ. When λ is on one side of a critical point λ_0, the system lies in a stable equilibrium state Σ_1, and when λ reaches or exceeds λ_0, the state Σ_1 becomes unstable, and the system will undergo a transition from Σ_1 to another stable state Σ_2. The basic principle is that there often exist fluctuations in a system leading to deviations from the equilibrium states, and the phase transition process is a dynamical behavior that should be described by a time-dependent equation.

To derive a general time-dependent model, we first recall that the classical Le Chatelier principle amounts to saying that for a stable equilibrium state Σ of a system, when the system deviates from Σ by a small perturbation or fluctuation, there will be a restoring force to return the system to the stable state Σ.

We now derive the dynamic model governing a thermodynamic system using Le Chatelier's principle. For this purpose, consider a system with thermodynamic potential $\mathscr{H}(u,\lambda)$, where $(u_1,\cdots,u_m)$ is the order parameter, and λ is the control parameter. Assume that the order parameters $(u_1,\cdots,u_m)$ are unconserved variables, i.e., the integers

$$\int_\Omega u_i(x,t)dx = a_i(t) \neq \text{constant}.$$

First, we observe that if $\widetilde{u}$ is a stable equilibrium state, then $\widetilde{u}$ is a local minimum of the free energy:

$$\int_\Omega \mathscr{H}(\widetilde{u},\lambda)dx = \min_{u\in\mathscr{O}} \mathscr{H}(u)dx, \tag{3.1.11}$$

where $\mathscr{O}$ is a neighborhood of $\widetilde{u}$. When the system deviates from $\widetilde{u}$ to u_0 due to fluctuations, the process whereby u returns to $\widetilde{u}$ from u_0 is characterized by a function $u(t,u_0)$ satisfying

$$\lim_{t\to\infty} u(t,u_0) = \widetilde{u}. \tag{3.1.12}$$

In addition, $u(t,u_0)$ should satisfy a dynamic equation as follows:

$$\frac{du}{dt} = G(u,\lambda), \quad u(0) = u_0, \tag{3.1.13}$$

and G is an operator to be determined. We note that

$$\frac{d}{dt}\int_\Omega \mathscr{H}(u(t,u_0),\lambda)dx = \int_\Omega \sum_{i=1}^m \frac{\delta}{\delta u_i}\mathscr{H}\cdot\frac{du_i}{dt} = \int_\Omega \sum_{i=1}^m \frac{\delta}{\delta u_i}\mathscr{H}\cdot G_i(u,\lambda),$$

where $\delta/\delta u_i$ are the variational derivatives. By Le Chatelier's principle, the above energy equation should imply that

$$\frac{d}{dt}\int_\Omega \mathscr{H}(u(t,u_0),\lambda)dx \le -k\int_\Omega \left|\frac{\delta\mathscr{H}}{\delta u}\right|^2,$$

for some $k>0$. This energy inequality shows that the energy $\int_\Omega \mathscr{H}(u,\lambda)dx$ is decreasing for $t\ge 0$ except at the equilibrium state u_0. Hence by the properties of gradient-type operators, we derive that (3.1.13) is a gradient-type flow with the operator G given by

$$G(u,\lambda) = \left(-\beta_1\frac{\delta}{\delta u_1}\mathscr{H}(u,\lambda)+\Phi_1(u,\lambda),\cdots,-\beta_m\frac{\delta}{\delta u_m}\mathscr{H}(u,\lambda)+\Phi_m(u,\lambda)\right),$$

for some $\beta_j>0$ and Φ_j satisfying

$$\int_\Omega \sum_i \Phi_i\frac{\delta}{\delta u_i}\mathscr{H}(u,\lambda)dx \le 0. \tag{3.1.14}$$

Hence we deduce that for a thermodynamic system with thermodynamic potential $\mathscr{H}(u,\lambda)$, with unconserved order parameters $(u_1,\cdots,u_m)$, and with control parameter λ, the time-dependent equations are given by

$$\frac{\partial u_i}{\partial t} = -\beta_i \frac{\delta}{\delta u_i}\mathscr{H}(u,\lambda) + \Phi_i(u,\nabla u,\lambda) \quad \text{for } 1 \le i \le m, \tag{3.1.15}$$

$$\int_\Omega \sum_i \Phi_i \frac{\delta}{\delta u_i}\mathscr{H}(u,\lambda)dx = 0. \tag{3.1.16}$$

The system (3.1.15) and (3.1.16) is the mathematical representation of Le Chatelier's principle. It is usually supplemented with a set of boundary conditions, determined by the underlying physical problem; see Chap. 3 for details.

The condition (3.1.16) is required by Le Chatelier's principle. In a concrete problem, the terms Φ_i can be determined by physical laws and (3.1.16). We remark here that following Le Chatelier's principle, one should have an inequality constraint as in (3.1.14). However, physical systems often obey simplified rules, and many existing models for specific problems are consistent with the equality constraint here. This remark applies to the constraint (3.1.22) below as well.

When the system is an $(m+1)$-component system, and the order parameters are the particle number densities and the system has no material exchange with the external environment, then the m independent variables u_j $(1 \le j \le m)$ are conserved, i.e.,

$$\int_\Omega u_j(x,t)dx = \text{constant}. \tag{3.1.17}$$

This conservation law requires a continuity equation

$$\frac{\partial u_j}{\partial t} = -\nabla \cdot J_j(u,\lambda), \tag{3.1.18}$$

where $J_j(u,\lambda)$ is the flux of component u_j. In addition, the J_j satisfy

$$J_j = -\sum_{i=1}^m L_{ij}\nabla(\mu_i - \mu_{m+1}), \tag{3.1.19}$$

where $L = (L_{ij})_{m\times m}$ is called the transport matrix, and μ_l is the chemical potential of component u_l. By the Onsager reciprocity law, L is symmetric and positive definite, which may depend on $u = (u_1, \cdots, u_m)$. In addition,

$$\mu_j - \mu_{m+1} = \frac{\delta}{\delta u_j}\mathscr{H}(u,\lambda) - \phi_j(u,\nabla u,\lambda), \tag{3.1.20}$$

and $\phi_j(u,\nabla u,\lambda)$ is a function depending on the other components of u. Thus, from (3.1.18)–(3.1.20) we obtain the dynamical equations as follows:

$$\frac{\partial u_j}{\partial t} = \beta_j \nabla\left[\sum_{i=1}^m L_{ij}\nabla\left(\frac{\delta}{\delta u_i}\mathscr{H}(u,\lambda) - \phi_i(u,\nabla u,\lambda)\right)\right], \tag{3.1.21}$$

for $1 \le j \le m$, where $\beta_j > 0$ are constants, and ϕ_j satisfy

$$\int_\Omega \sum_j \Delta\phi_j \cdot \frac{\delta}{\delta u_j}\mathscr{H}(u,\lambda)dx = 0. \tag{3.1.22}$$

When $m = 1$, i.e., the system is a binary system consisting of two components A and B, then the term ϕ_j is equal to 0. The above model covers the classical Cahn–Hilliard model. It is worth mentioning that for multicomponent systems, these ϕ_j play an important rule in deriving time-dependent models.

If the order parameters $(u_1, \cdots, u_k)$ are coupled to the conserved variables $(u_{k+1}, \cdots, u_m)$, then the dynamical equations are

$$\begin{aligned} \frac{\partial u_i}{\partial t} &= -\beta_i \frac{\delta}{\delta u_i}\mathscr{H}(u,\lambda) + \Phi_i(u,\nabla u,\lambda), \\ \frac{\partial u_j}{\partial t} &= \beta_j \nabla\left[\sum_{l=1}^{m} L_{lj}\nabla\left(\frac{\delta}{\delta u_l}\mathscr{H}(u,\lambda) - \phi_l(u,\nabla u,\lambda)\right)\right], \end{aligned} \tag{3.1.23}$$

for $1 \le i \le k$ and $k+1 \le j \le m$. Here Φ_i and ϕ_j satisfy (3.1.16) and (3.1.22), respectively.

The model (3.1.23) we derive here gives a general form of the governing equations to thermodynamic phase transitions, and it will play a crucial role in studying the dynamics of equilibrium phase transitions in statistical physics.

Physically, the initial value condition $u(0) = \varphi$ for (3.1.23) stands for the fluctuation of the system or a perturbation from the external environment. Hence, φ is generally small, representing noise or random fluctuations. However, we cannot exclude the possibility of a bigger φ.

It follows from conditions (3.1.16) and (3.1.22) that a steady-state solution u_0 of (3.1.23) satisfies

$$\begin{aligned} &\Phi_i(u_0,\nabla u_0,\lambda) = 0 \quad \forall 1 \le i \le k, \\ &\Delta\phi_j(u_0,\nabla u_0,\lambda) = 0 \quad \forall k+1 \le j \le m, \end{aligned} \tag{3.1.24}$$

which implies that a stable equilibrium state must reach the minimal value of the thermodynamic potential. In fact, u_0 satisfies

$$\begin{aligned} &\beta_i \frac{\delta}{\delta u_i}\mathscr{H}(u_0,\lambda) - \Phi_i(u_0,\nabla u_0,\lambda) = 0, \\ &\beta_j \nabla\left[\sum_{l=1}^{m} L_{lj}\nabla\left(\frac{\delta}{\delta u_l}\mathscr{H}(u_0,\lambda) - \Delta\phi_l(u_0,\nabla u_0,\lambda)\right)\right] = 0, \end{aligned} \tag{3.1.25}$$

for $1 \le i \le k$ and $k+1 \le j \le m$. Multiplying $\Phi_i(u_0,\nabla u_0,\lambda)$ and $\phi_j(u_0,\nabla u_0,\lambda)$ in the first and the second equations of (3.1.25) respectively and integrating, we infer from (3.1.16) and (3.1.22) that

$$\int_\Omega \sum_i \Phi_i^2(u_0,\nabla u_0,\lambda)dx = 0, \qquad \int_\Omega \sum_j |\nabla\phi_j(u_0,\nabla u_0,\lambda)|^2 dx = 0,$$

which implies that (3.1.24) holds.

In practical applications of the above unified model, the terms Φ_i can be determined by physical laws and (3.1.16); see Sect. 3.5 for deriving the TDGL equations for superconductivity with an applied field.

3.2 Classification of Equilibrium Phase Transitions

Based on statistical physics, the phase transitions of nonlinear dissipative systems can be divided into two types: equilibrium phase transitions and nonequilibrium phase transitions. Namely, if the system (1.1.8) possesses a gradient-type structure, then the phase transitions are called equilibrium phase transition; otherwise, they are called nonequilibrium phase transitions.

From the physical point of view, the occurrence of an equilibrium phase transition means that a thermodynamic system has undergone a transition at a critical point from one equilibrium state to another, maintaining minimal free energy. We shall see later that the equilibrium phase transitions appear extensively in statistical physics, solid-state physics, materials science, thermodynamics, and elastomechanics. The nonequilibrium phase transitions appear mainly in chemical reaction dynamics, biology, ecology, economics, fluid dynamics, and geophysical fluid dynamics. In addition, these two types of phase transitions have remarkable differences in their dynamical properties. In particular, an equilibrium phase transition does not lead to chaos, while a nonequilibrium phase transition may.

Equilibrium phase transitions are usually classified by the Ehrenfest classification scheme, which groups phase transitions based on the degree of nonanalyticity involved; see among many others (Reichl 1998; Stanley 1971) for detailed discussions.

In the physics literature, first-order phase transitions are also called discontinuous, and higher-order phase transitions ($n > 1$) are called continuous. Because energy cannot be instantaneously transferred between a system and its environment, first-order transitions are associated with "mixed-phase regimes," in which some parts of the system have completed the transition and others have not. Many important phase transitions fall in this category, including the solid/liquid/gas transitions and Bose–Einstein condensation. Examples of second-order phase transitions are the ferromagnetic transition and the superfluid transition.

As we know, the dynamic classification in Principle 1 offers a new classification scheme for equilibrium phase transitions, based on the dynamic behavior of the transitions. In particular, Type I transitions correspond to second- or higher-order transitions in terms of the Ehrenfest classification scheme, and Type II and III transitions correspond to first-order transitions, under which latent heat, metastable states, and hysteresis occur. These phenomena will be explored in detail in Chap. 3, leading to some physical predictions.

The main characteristic of Type II phase transitions is that there is a gap between basic state $\Sigma_\lambda^1 = \{(0, \lambda)\}$ and the transition state Σ_λ^2 at the critical point λ_0. In thermodynamics, the metastable states correspond in general to the superheated or supercooled states, which have been found in many physical phenomena.

In a Type I phase transition, the above-mentioned two states Σ_λ^1 and Σ_λ^2 meet at λ_0, i.e., the system undergoes a continuous change from Σ_λ^1 to Σ_λ^2.

In a Type III phase transition, there are at least two different stable states Σ_2^λ and Σ_3^λ at λ_0, and the system undergoes a continuous transition to Σ_2^λ or a jump transition to Σ_3^λ, depending on the fluctuations.

For equilibrium phase transitions, it is clear that a Type II phase transition of a gradient-type system must be discontinuous or of zeroth order, because there is a gap between Σ_λ^1 and Σ_λ^j $(2 \leq j \leq K)$. For a Type I phase transition, the energy functional is continuous, and consequently, it is an nth-order transition in the Ehrenfest sense for some $n \geq 2$. A Type III phase transition is indefinite: it is continuous from Σ_λ^1 to Σ_λ^2, and discontinuous from Σ_λ^1 to Σ_λ^3.

Finally, we recall the physical meaning of the derivatives of $F(u,\lambda)$ on λ in statistical physics. In thermodynamics, the parameter λ generally stands for temperature T and pressure p: $\lambda = (T,p)$. The energy functional $F(u,T,p)$ is the thermodynamic potential, and u is the order parameter.

For a PVT system, the first- and second-order derivatives include:

1. the entropy: $S = -\partial F/\partial T$;
2. the phase volume: $V = \partial F/\partial p$;
3. the heat capacity under constant pressure: $C_p = -T\partial^2 F/\partial T^2$;
4. the compression coefficient $\kappa = -\frac{1}{V}\frac{\partial^2 F}{\partial p^2}$;
5. the thermal expansion coefficient $\alpha = \frac{1}{V}\frac{\partial^2 F}{\partial T \partial p}$.

Thus, for a first-order phase transition, the discontinuity of $\partial F/\partial p$ at a critical point $\lambda_0 = (T_0, p_0)$ implies the discontinuity of the phase volume,

$$\Delta V = V^2 - V^1 = \frac{\partial F^+}{\partial p} - \frac{\partial F^-}{\partial p}, \tag{3.2.1}$$

while the discontinuity of $\partial F/\partial T$ implies that there is a gap between both phase entropies:

$$\Delta S = S^2 - S^1 = -\frac{\partial F^+}{\partial T} + \frac{\partial F^-}{\partial T}. \tag{3.2.2}$$

Physically, ΔS is an unmeasurable quantity, and hence we always use latent heat $\Delta H = T\Delta S$ to determine the first-order phase transition, and ΔH stands for the absorbing heat for a transition from phase Σ_λ^1 to phase Σ_λ^2. In the two-phase coexistence situation, ΔH and ΔV are related by the Clausius–Clapeyron equation

$$\frac{dp}{dT} = \frac{\Delta H}{T\Delta V}.$$

For second-order phase transitions, the variance of heat capacity (or specific heat), compression coefficient, and the thermal expansion coefficient at a critical value are important static properties, and they are measurable physical quantities.

3.3 Third-Order Gas–Liquid Phase Transition

3.3.1 Introduction

A physical-vapor transport (PVT) system is a system composed of one type of molecule, and the interaction between molecules is governed by the van der Waals law. The molecules generally have a repulsive core and a short-range attraction region outside the core. Such systems have a number of phases—gas, liquid, and solid—and a solid can appear in a few phases. The most typical example of a PVT system is water. In general, the phase transitions of PVT systems mainly refer to the gas–liquid, gas–solid, and liquid–solid phase transitions. These transitions are all of first order in the Ehrenfest sense (i.e., discontinuous) and are accompanied by a latent heat and a change in density.

A PT-phase diagram of a typical PVT system is schematically illustrated in Fig. 3.1, where point A is the triple point at which the gas, liquid, and solid phases coexist. Point C is the Andrews critical point at which the gas–liquid coexistence curve terminates (Reichl 1998; Stanley 1971).

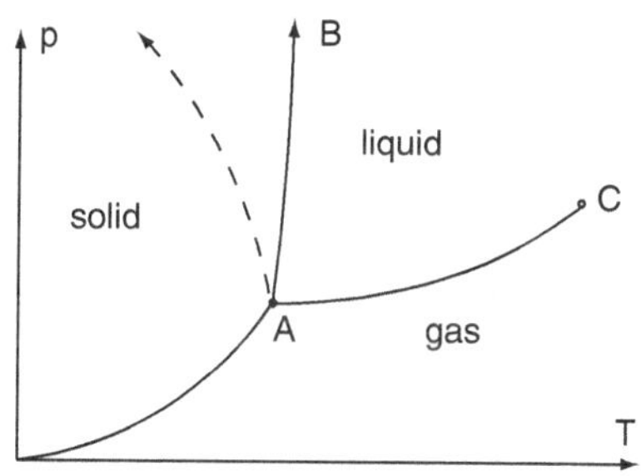

Fig. 3.1 Coexistence curves of a typical PVT system: A is the triple point, C is the critical point, and the *dashed curve* is a melting line with negative slope

The phase diagram in the PV-plane is given in Fig. 3.2, where the dashed curves represent lines of constant temperature. In the region of coexistence of phases, the isotherms (dashed lines) are always flat, indicating that in these regions, the change in volume (density) occurs for constant pressure p and temperature T.

The classical view of the termination of the gas–liquid coexistence curve at a critical point is that the system can pass continuously from a gaseous state to a liquid state without ever meeting an observable phase transition. Note that the critical point in the liquid–solid coexistence curve has never been observed.

The main objective of this section is to study the nature of the Andrews critical point in the gas–liquid transition in a PVT system based on the authors' recent papers (Ma and Wang 2008b, 2011e). In particular, it is shown that (1) the gas–liquid coexistence curve can be extended beyond the Andrews critical point, and (2) the transition is of first order before the critical point, of second order at the critical point, and of third order beyond the Andrews critical point. This clearly explains why it is hard to observe the gas–liquid phase transition beyond the Andrews critical point.

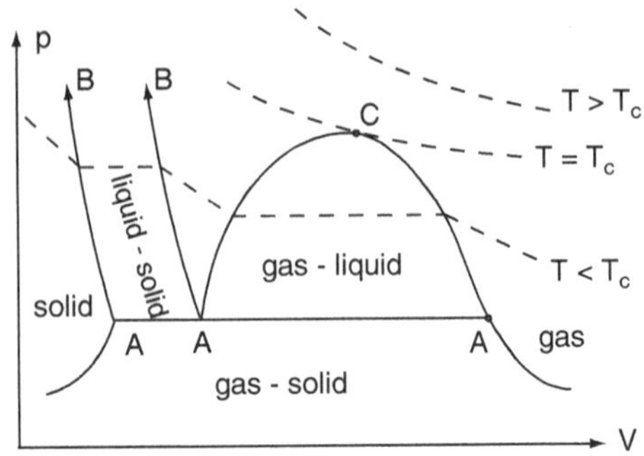

Fig. 3.2 The *PV*-phase diagram with the *dashed line* representing isotherms

The derived theoretical results are in agreement with the experimental discoveries in Nishikawa and Morita (1998) and the references therein. In that article, the density fluctuations of supercritical CO_2 and CF_3H for various thermodynamic states were measured by small-angle X-ray scattering and also calculated using the empirical state equations. For each sample, the ridge of the density fluctuations forms along the extension of the coexistence curve of gas–liquid in the PT-phase diagram and seems to be the locus of a higher-order phase transition. In this section, we shall explicitly calculate the jump of the third-order derivative with respect to the temperature of the Gibbs energy at the critical temperature, and give precise information on the locus of the coexistence curve beyond the Andrews critical point. Also, the derived second-order transition at the critical point is consistent with the result obtained in Fisher (1964).

Finally, the dynamical transition analysis in this section suggests that there are some mathematical states that do not exist in reality. Hence the system appears to avoid the fluctuations in the basin of attraction in these unrealistic states. This leads us to introduce an asymmetry principle of fluctuations and the preferred transition mechanism.

3.3.2 Time-Dependent Models for PVT Systems

Van der Waals Equation and Gibbs Energy. The classical and the simplest equation of state that can exhibit many of the essential features of the gas–liquid phase transition is the van der Waals equation:

$$v^3 - \left(b + \frac{RT}{p}\right) v^2 + \frac{a}{p} v - \frac{ab}{p} = 0, \tag{3.3.1}$$

where v is the molar volume, p is the pressure, T is the temperature, R is the universal gas constant, b is the revised constant of inherent volume, and a is the revised constant of attractive force between molecules. If we adopt the molar density $\rho = 1/v$ to replace v in (3.3.1), then the van der Waals equation becomes

$$-(bp+RT)\rho+a\rho^2-ab\rho^3+p=0. \tag{3.3.2}$$

We shall now apply thermodynamic potentials to investigate the phase transitions of PVT systems, and we shall see later that the van der Waals equation can be derived as an Euler–Lagrange equation for the minimizers of the Gibbs free energy for PVT systems in gaseous states.

Consider an isothermal–isopiestic process. The thermodynamic potential is taken to be the Gibbs free energy. In this case, the order parameters are the molar density ρ and the entropy density S, and the control parameters are the pressure p and temperature T. By (3.1.8), the general form of the Gibbs free energy for PVT systems is given as

$$G(\rho,S,T,p)=\int_\Omega\left[\frac{1}{2}\mu_1|\nabla\rho|^2+\frac{1}{2}\mu_2|\nabla S|^2+g(\rho,S,T,p)-ST-\alpha(\rho,T,p)p\right]dx, \tag{3.3.3}$$

where g and α are differentiable with respect to ρ and S, $\Omega\subset\mathbb{R}^3$ is the container, and αp is the mechanical coupling term in the Gibbs free energy given by

$$\alpha(\rho,T,p)p=\rho p-\frac{1}{2}b\rho^2 p, \tag{3.3.4}$$

where $b=b(T,p)$ depends continuously on T and p.

Based on both physical and mathematical considerations, we take the Taylor expansion of $g(\rho,S,T,p)$ with respec to ρ and S as follows:

$$g=\frac{1}{2}\alpha_1\rho^2+\frac{1}{2}\beta_1 S^2+\beta_2 S\rho^2-\frac{1}{3}\alpha_2\rho^3+\frac{1}{4}\alpha_3\rho^4, \tag{3.3.5}$$

where α_i $(1\le i\le 3)$, β_1, and β_2 depend continuously on T and p, and

$$\beta_1=\beta_1(T,p)>0,\qquad \alpha_i=\alpha_i(T,p)>0,\quad i=2,3. \tag{3.3.6}$$

Dynamic Equations for PVT Systems. In a PVT system, the order parameter is $u=(\rho,S)$,

$$\rho=\rho_1-\rho_0,\quad S=S_1-S_0,$$

where ρ_i and S_i $(i=0,1)$ represent the molar density and entropy density, and ρ_0,S_0 are reference points. Hence the conjugate variables of ρ and S are the pressure p and the temperature T. Thus, by the standard model (3.1.15), we derive from (3.3.3)–(3.3.5) the following general form of the time-dependent equations governing a PVT system:

$$\begin{aligned}\frac{\partial\rho}{\partial t}&=\mu_1\Delta\rho-(\alpha_1+bp)\rho+\alpha_2\rho^2-\alpha_3\rho^3-2\beta_2\rho S+p,\\ \frac{\partial S}{\partial t}&=\mu_2\Delta S-\beta_1 S-\beta_2\rho^2+T.\end{aligned} \tag{3.3.7}$$

Although the domain Ω depends on T and p, we can still take the Neumann boundary condition

$$\frac{\partial\rho}{\partial n}=0,\quad \frac{\partial S}{\partial n}=0 \qquad \text{on } \partial\Omega. \tag{3.3.8}$$

An important special case for PVT systems is that the pressure and the temperature functions are homogeneous in Ω. Thus we can assume that ρ and S are independent of $x\in\Omega$, and the free energy (3.3.3) with (3.3.4) and (3.3.5) can be expressed as

$$G(\rho,S,T,p)=\frac{\alpha_1}{2}\rho^2+\frac{\beta_1}{2}S^2+\beta_2S\rho^2-\frac{\alpha_2}{3}\rho^3+\frac{\alpha_3}{4}\rho^4+\frac{b\rho^2p}{2}-\rho p-ST. \tag{3.3.9}$$

We derive then from (3.3.9) the following dynamic equations:

$$\begin{aligned}\frac{d\rho}{dt}&=-(\alpha_1+bp)\rho+\alpha_2\rho^2-\alpha_3\rho^3-2\beta_2S\rho+p,\\ \frac{dS}{dt}&=-\beta_1S-\beta_2\rho^2+T.\end{aligned} \tag{3.3.10}$$

Because $\beta_1>0$ for all T and p, we can replace the second equation of (3.3.10) by

$$S=\beta_1^{-1}(T-\beta_2\rho^2). \tag{3.3.11}$$

Then equations (3.3.10) are equivalent to the following equation:

$$\frac{d\rho}{dt}=-(\alpha_1+bp+2\beta_1^{-1}\beta_2T)\rho+\alpha_2\rho^2-(\alpha_3-2\beta_2^2\beta_1^{-1})\rho^3+p. \tag{3.3.12}$$

It is clear that the steady-state equation of (3.3.12) should agree with the van der Waals equation. Hence it is natural to choose

$$\alpha_1=0,\qquad 2\beta_1^{-1}\beta_2=R,\qquad \alpha_2=a,\qquad (\alpha_3-2\beta_2^2\beta_1^{-1})=ab.$$

3.3.3 Phase Transition Dynamics for PVT Systems

We now use (3.3.12) to discuss dynamical properties of transitions for PVT systems, and remark that similar results can also be derived using the more general form of the time-dependent model (3.3.7).

Let ρ_0 be a steady-state solution of (3.3.12) near the Andrews point $C=(T_c,p_c)$. We take the transformation

$$\rho=\rho_0+\rho'.$$

Then (3.3.12) becomes (we again drop the prime)

$$\frac{d\rho}{dt}=\lambda\rho+a_2\rho^2-ab\rho^3. \tag{3.3.13}$$

Here

$$\lambda = 2a\rho_0 - 3ab\rho_0^2 - \alpha_1 - bp - RT, \quad a_2 = a(1 - 3b\rho_0), \tag{3.3.14}$$

where α_1 is close to zero. Here we emphasize that ρ_0 and (λ, a_2) are all functions of the control parameter $\lambda = (T, p)$.

These are two important physical parameters that are used to fully characterize the dynamic behavior of a gas–liquid transition near the Andrews critical point. In fact, from the derivation of the model, we obtain immediately the following physical meaning of these two parameters:

$$\lambda(T,p) = \frac{d^2G}{d\rho^2}\Big|_{\rho=\rho_0}, \qquad a_2(T,p) = \frac{1}{2}\frac{d^3G}{d\rho^3}\Big|_{\rho=\rho_0}, \tag{3.3.15}$$

where $\rho_0 = \rho_0(T, p)$ is the equilibrium state.

In the PT-plane, near the Andrews point $C = (T_c, p_c)$, the critical parameter equation

$$\lambda = \lambda(T,p) = 0 \qquad \text{in} \quad |T - T_c| < \delta, \quad |p - p_c| < \delta$$

for some $\delta > 0$, defines a continuous function $T = \phi(p)$ such that

$$\lambda \begin{cases} < 0 & \text{if } T > \phi(p), \\ = 0 & \text{if } T = \phi(p), \\ > 0 & \text{if } T < \phi(p). \end{cases} \tag{3.3.16}$$

Equivalently, this is called the principle of exchange of stabilities (Chandrasekhar 1981), which, as we have shown in Ma and Wang (2008b), is a necessary and sufficient condition for the gas–liquid phase transition.

One important component of our theory is that the Andrews critical point is determined by the system of equations

$$\begin{aligned} &\lambda = 0, \\ &a_2 = 0, \\ &-(bp + RT)\rho_0 + a\rho_0^2 - ab\rho_0^3 + p = 0. \end{aligned} \tag{3.3.17}$$

Here the first equation is the critical parameter equation; the second equation, as we shall see below, determines the switching point where the phase transition switches types; and the last equation is the van der Waals equation, which is also the steady-state equation of the dynamic model.

Then by directly solving the system of three algebraic equations in (3.3.17), we determine that the critical point C is given by

$$(\rho_c, p_c, T_c) = \left(\frac{1}{3b}, \frac{a}{27b^2}, \frac{8a}{27bR}\right). \tag{3.3.18}$$

This is in agreement with the classical work by van der Waals. *Here we obtain the Andrews point using a dynamic approach.*

The main result in this section is the following dynamic transition theorem:

Theorem 3.3.1 *Let $T_0 = \phi(p_0)$ and $a_3 > 0$. Then the system (3.3.13) has a transition at $(T,p) = (T_0,p_0)$, and the following assertions hold:*

1. *If the coefficient $a_2 = a_2(T,p)$ in (3.3.13) is zero at (T_0,p_0), i.e., $a_2(T_0,p_0) = 0$, then the transition is of Type I, as schematically shown in Fig. 3.3.*
2. *If $a_2(T_0,p_0) \neq 0$, then the transition is of Type III (i.e., mixed type), and the following assertions hold:*

   (a) *There are two transition solutions near (T_0,p_0) :*

$$\rho^{\pm}(T,p) = \frac{1}{2a_3}\left(a_2 \pm \sqrt{a_2^2 + 4a_3\lambda}\right). \qquad (3.3.19)$$

   (b) *There is a saddle-node bifurcation at (T_1,p_1), where $T_1 > T_0$ and $p_1 < p_0$ if $\phi'(p_0) > 0$, and $p_0 > p_1$ if $\phi'(p_0) < 0$.*

   (c) *For $\phi'(p_0) > 0$, when $a_2(T_0,p_0) > 0$, the transition diagrams are illustrated by Fig. 3.4, where ρ^+ is stable for all (T,p) near (T_0,p_0), and $\rho = 0$ is stable, ρ^- a saddle for $T_0 < T < T_1$ and $p_1 < p < p_0$, and ρ^- is stable, $\rho = 0$ a saddle for $T < T_0$, $p > p_0$.*

   (d) *When $a_2(T_0,p_0) < 0$, the transition diagrams are illustrated by Fig. 3.5, where ρ^- is stable for all (T,p) near (T_0,p_0), and $\rho = 0$ is stable, ρ^+ a saddle for $T_0 < T < T_1$, $p_1 < p < p_0$, and ρ^+ is stable, $\rho = 0$ a saddle for $T < T_0, p_0 < p$.*

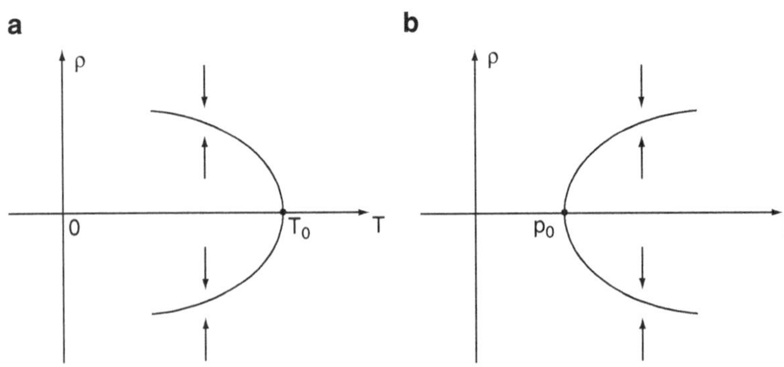

Fig. 3.3 Continuous transition for the case $\phi'(p_0) > 0$

Proof. By (3.3.13), the theorem follows from the transition theorems (Theorems 2.3.1 and 2.3.2) and the singularity separation theorem (Theorem 2.5.4). We omit the detailed routine analysis. □

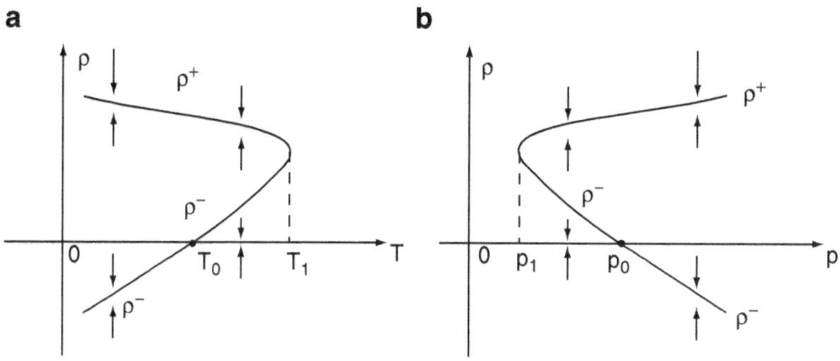

Fig. 3.4 Type III (mixed) transition for $a_2 > 0$ in the (**a**) Tρ-plane, and (**b**) pρ-plane

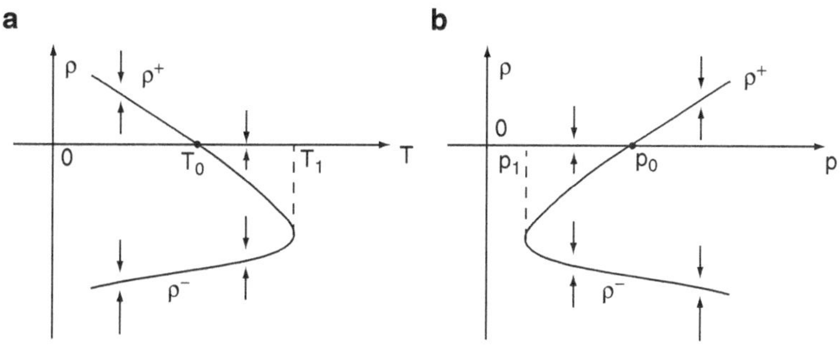

Fig. 3.5 Type III (mixed) transition for $a_2 < 0$

3.3.4 Physical Conclusions

To discuss the physical significance of Theorem 3.3.1, we recall the classical pV-phase diagram given by Fig. 3.2. We take $\rho = 1/V$ to replace volume V. Then the gas–liquid coexistence curve in the ρ, P-plane is illustrated by Fig. 3.6.

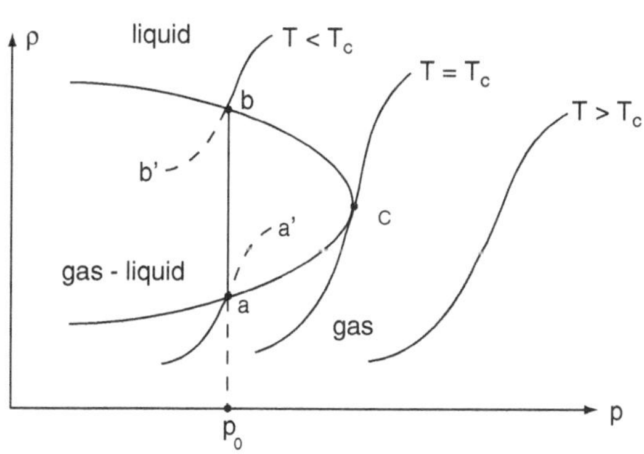

Fig. 3.6 The gas–liquid coexistence curve in the ρ, p-plane: the point C is the Andrews critical point; the *dashed lines* aa' and bb' stand for the metastable states

According to physical experiments, in the gas–liquid coexistence region in the ρ, p phase diagram, there exist metastable states. Mathematically speaking, the metastable states are the attractors that have a small basin of attraction. In Fig. 3.6, the dashed lines aa' and bb' represent the metastable states, and the points in aa' correspond to superheated liquid, while the points in bb' correspond to supercooled gas. The ρ, p phase diagram shows that along the isothermal line $T > T_C$, where $C = (T_C, P_C)$ is the Andrews critical point, when the pressure p increases, the density ρ varies continuously from gaseous to liquid states. However, along the isothermal line $T < T_C$, when the pressure p increases to p_0, the density ρ will undergo an abrupt change, and a transition from the gaseous state a to a liquid state b accompanied by an isothermal exothermal process will occur. Likewise, such processes also occur in the gas–solid and liquid–solid transitions.

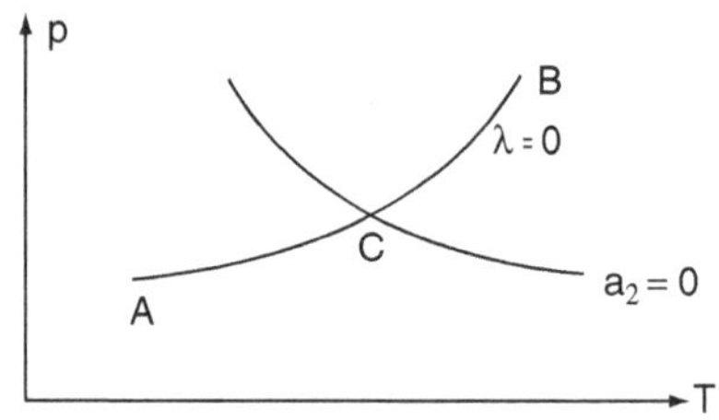

Fig. 3.7 The point $C = (T_c, p_c)$ is the Andrews critical point

We now return to discuss Theorem 3.3.1 and explain the gas–liquid transition near the Andrews critical point C.

First, we have shown in (3.3.18) that at the equilibrium point ρ_0, the two curves given by $\lambda(p,T) = 0$ and $a_2(p,T) = 0$ interact exactly at the critical point C, as shown in Fig. 3.7, and the curve segment AB of $\lambda = 0$ is divided into two parts AC and CB by the point C such that

$$\begin{aligned} a_2(T,p) > 0 \quad &\text{for} \quad (T,p) \in AC, \\ a_2(T,p) < 0 \quad &\text{for} \quad (T,p) \in CB. \end{aligned}$$

Here the curve AC is the classical gas–liquid coexistence curve.

Second, on the curve AC, excluding the critical point C, we have $a_2(T,p) > 0$. The phase transition of the system is of mixed type if we take a path crossing the AC; see Fig. 3.4. In Fig. 3.4, ρ is the deviation from the basic gaseous state ρ_0. We now consider different states given in Fig. 3.4a:

1. For $T > T_1$, the gaseous state ρ_0, corresponding to $\rho = 0$ in the figure, is stable. This is the only stable physical state in this temperature range, and the system is in the gaseous state.
2. For $T_0 < T < T_1$, there are two metastable states given by the gaseous phase ρ_0 and the liquid phase $\rho_0 + \rho^+$.
3. For $T < T_0$, there are three states: the unstable basic gaseous state ρ_0 and the two metastable states: $\rho_0 + \rho^-$ and $\rho_0 + \rho^+$. One important component of our

theory is that the only physical phase here is the liquid phase represented by the metastable state: $\rho_0+\rho^+$. Although mathematically speaking, the gaseous state $\rho_0+\rho^-$ is also metastable, it does not appear in nature. The only possible explanation for this exclusion is the asymmetry principle of fluctuations, to be further explored in the next section.

4. We have shown that as we lower the temperature, the system undergoes a *first-order* transition from a gaseous state to a liquid state with an abrupt change in density. In fact, there is an energy gap between the gaseous and liquid states:

$$\Delta E = G(\rho_0+\rho^+) - G(\rho_0) \sim -\frac{\lambda}{4}(\rho^+)^2 - \frac{a_2}{12}(\rho^+)^3 < 0 \quad \text{for } T < T_1.$$

This energy gap $|\Delta E|$ stands for latent heat, and $\Delta E < 0$ shows that the transition from a gaseous state to a liquid state is an isothermal exothermal process, and that from a liquid state to a gaseous state is an isothermal endothermal process.

Third, at the critical point C, we have $a_2 = 0$. Then the dynamic transition is as shown in Fig. 3.3:

1. As in the previous case, for $T > T_0$, the only physical state is given by the gaseous phase ρ_0, corresponding to zero deviation shown in the figure.
2. As the temperature T is lowered crossing T_0, the gaseous state loses its stability, leading to two metastable states: one is the liquid phase $\rho_0+\rho^+$, and the other is the gaseous phase $\rho_0+\rho^-$. Again, the gaseous phase $\rho_0+\rho^-$ does not appear, and the asymmetry principle of fluctuations is valid in this situation as well.
3. The phase transition here is of second order, since the energy is continuous at T_0. In fact, the energy for the transition liquid state is given by

$$G(\rho_0+\rho^+) = G(\rho_0) - \frac{\alpha^2}{4a_3}(T-T_c)^2 \quad \text{for } T < T_c,$$

for some $\alpha > 0$. Hence the difference of the heat capacity at $T = T_c$ is

$$\Delta C = -T_c\frac{\partial^2}{\partial T^2}\left(G(\rho_0+\rho^+) - G(\rho_0)\right) = \frac{\alpha^2}{2a_3}T_c > 0.$$

Namely, the heat capacity has a finite jump at $T = T_c$, and therefore, the transition at $T = T_c$ is of *second order.*

Fourth, on the curve BC, we have $a_2(T,p) < 0$, and the phase transition diagram is given by Fig. 3.5:

1. For $T > T_1$, the system is in the gaseous phase, which is stable.
2. For $T_0 < T < T_1$, there are two metastable gaseous states given by ρ_0 and $\rho_0+\rho^-$. As before, although it is metastable, the gaseous state $\rho_0+\rho^-$ does not appear.
3. For $T < T_0$, the gaseous phase ρ_0 loses its stability, and the system undergoes a dynamic transition to the metastable liquid state $\rho_0+\rho^+$. Mathematically, the gaseous state $\rho_0+\rho^-$ is also metastable. However, it, too, does not appear, due to the asymmetry principle of fluctuations.

4. The dynamic transition in this case is of third order. In fact, we have

$$G(\rho_0+\rho^+) = G(\rho_0(T)) - \frac{\alpha^3}{6|a_2|^2}(T_0-T)^3 + o(|T_0-T|^3) \text{ for } T < T_0.$$

Namely, the free energy is continuously differentiable up to second order at $T = T_0$, and the transition is of *third order*. This implies that when $(T_0, p_0) \in CB$, the third-order transition at (T_0, p_0) cannot be observed by physical experiments.

In summary, we have obtained a precise characterization of the phase transition behavior near the Andrews point and derived precisely the nature of the Andrews critical point.

3.4 Ferromagnetism

Ferromagnetism refers to the spontaneous magnetization of materials such as iron. For ferromagnetic systems, the magnetization is the first derivative of the free energy with respect to the applied magnetic field strength, and the magnetic susceptibility is the second derivative of the free energy with respect to the field. It has been observed that the magnetization increases continuously from zero as the temperature is lowered below the Curie temperature, and the magnetic susceptibility changes discontinuously. Hence, the ferromagnetic phase transition in materials such as iron is regarded as a second-order phase transition (Kleman and Lavrentovich 2003). However, a theoretical understanding of the transition is still lacking. For example, the classical GL free energy leads to interesting results, in agreement with experiments, for both the steady state and time-dependent models. However, there are obvious discrepancies in the classical theories on both susceptibility and spontaneous magnetization.

The main objective of this section is to identify these discrepancies, to revise the GL free energy to address these issues using dynamic transition theory, and to derive a physical prediction regarding spontaneous magnetization. In addition, our analysis leads naturally to a physical conjecture on asymmetry of fluctuations, which appears in both the ferromagnetic system studied in this section and in PVT systems studied in the previous section; see also (Ma and Wang 2008c, 2011e).

3.4.1 Classical Theory of Ferromagnetism

A ferromagnetic material consists of lattices containing particles with a magnetic moment. When no external field is present and the temperature is above some critical value, called the Curie temperature, the magnetic moments are oriented at random, and there is no net magnetization. However, as the temperature is lowered, magnetic interaction energy between lattice sites becomes more important than the random thermal energy. Below the Curie temperature, the magnetic moments become ordered in space, and a spontaneous magnetization appears. The phase transition from a paramagnetic to a ferromagnetic system takes place at the Curie temperature.

The phase diagrams for magnetic systems are given in Figs. 3.8–3.10. In Fig. 3.8, below the Curie temperature, the magnetization occurs spontaneously, and the zero magnetic field $H = 0$ separates the two possible orientations of magnetization. Figure 3.9 provides a sketch of the isotherms of the magnetic system, and Fig. 3.10 gives the magnetization as a function of temperature; see also (Reichl 1998; Stanley 1971) for details.

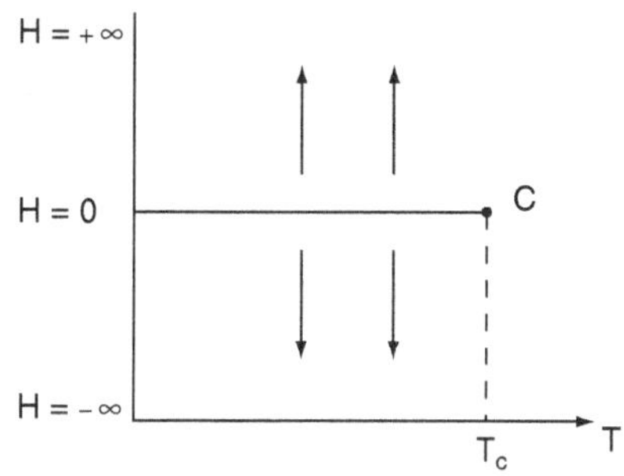

Fig. 3.8 Below the Curie point, magnetization occurs spontaneously; the curve $H = 0$ separates the two possible orientations of magnetization

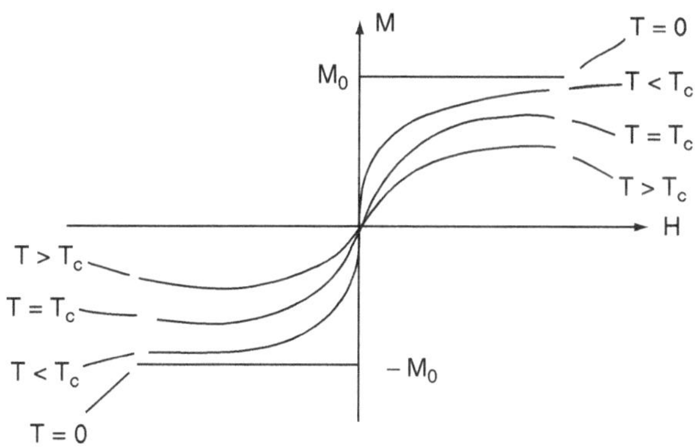

Fig. 3.9 A sketch of the isotherms for a magnetic system

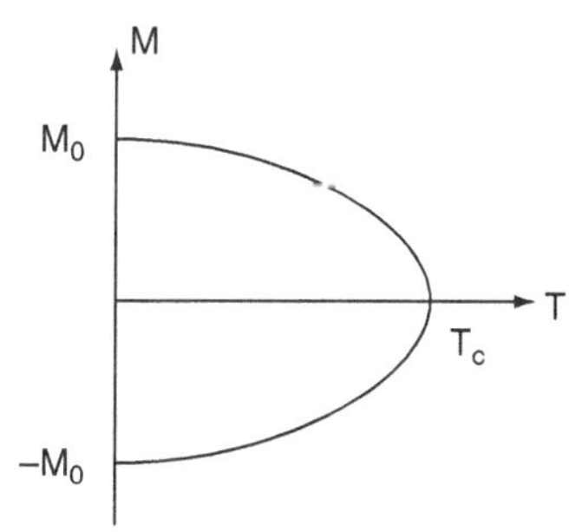

Fig. 3.10 A sketch of the spontaneous magnetization of a magnetic system

Based on the classical Ginzburg–Landau theory, for an isotropic system, the Helmholtz free energy can be expressed as

$$A(M,T) = A_0(T) + \frac{1}{2}\alpha_2(T)|M|^2 + \frac{1}{4}\alpha_4(T)|M|^4 + \cdots,$$

where $A_0(T)$ is a magnetization-independent contribution to the free energy, $|M|^2 = M\cdot M$, and $M = (M_1, M_2, M_3)$ is the magnetization of the system. When an external field H is present, the Gibbs free energy is given by

$$\begin{aligned} G(M,H,T) &= A(M,T) - H\cdot M \\ &= A_0(T) - H\cdot M + \frac{1}{2}\alpha_2(T,H)|M|^2 + \frac{1}{4}\alpha_4(T,H)|M|^4 + \cdots. \end{aligned}$$

For small H, α_2 and α_4 can be considered to be independent of H, and near the Curie point T_c, we have

$$\alpha_2(T) = \alpha_0(T - T_c), \quad \alpha_4(T) > 0, \quad \alpha_0 > 0 \text{ a constant.}$$

Usually, $G(M,H,T)$ is called the Ginzburg–Landau free energy. To omit the terms of order greater than $|M|^4$, it is known that the equilibrium state M of the ferromagnetic system satisfies

$$\frac{\delta}{\delta M}G = \alpha_4|M|^2M + \alpha_2 M - H = 0. \tag{3.4.1}$$

Thus above the Curie point, we obtain from (3.4.1) that

$$M \simeq \frac{1}{\alpha_2}H, \qquad \chi = \frac{\partial M}{\partial H} = \frac{1}{\alpha_2(T)} = \frac{1}{\alpha_0(T - T_c)}, \tag{3.4.2}$$

where χ is the isothermal susceptibility, which is a scalar because the system is isotropic. Below the critical point, for $H = 0$, the magnetization M obeys

$$|M| = \sqrt{\frac{\alpha_0(T_c - T)}{\alpha_4}}, \qquad \frac{\partial |M|}{\partial T} = -\frac{1}{2}\sqrt{\frac{\alpha_0}{\alpha_4(T_c - T)}}. \tag{3.4.3}$$

The heat capacity at $T = T_c$ is

$$\triangle C = -T\frac{\partial^2 G}{\partial T^2} = -T_c\frac{\partial^2}{\partial T^2}\left(\frac{1}{2}\alpha_2|M|^2 + \frac{1}{4}\alpha_4|M|^4\right) = \frac{\alpha_0^2 T_c}{2\alpha_4}. \tag{3.4.4}$$

We infer, then, from (3.4.2)–(3.4.4) the following classical conclusions for an isotropic magnetic system:

1. When an external magnetic field is present, a nonzero magnetization exists above the Curie point T_c, which has the same direction as the applied field H.

2. Near the critical point T_c, the susceptibility χ tends to infinity with the rate $(T-T_c)^{-1}$, i.e., a very small applied field at $T=T_c$ can yield a large effect on the magnetization.
3. In the absence of an external field (i.e., $H=0$), below the critical point a spontaneous magnetization M appears, which depends continuously on T and tends to zero with the rate $(T-T_c)^{1/2}$; namely, the transition is of second order.
4. The heat capacity at $T=T_c$ has a jump at $T=T_c$ with gap $\Delta C=\frac{\alpha_0^2 T_c}{2\alpha_4}$, and the jump has the shape of λ, as shown in Fig. 3.11.

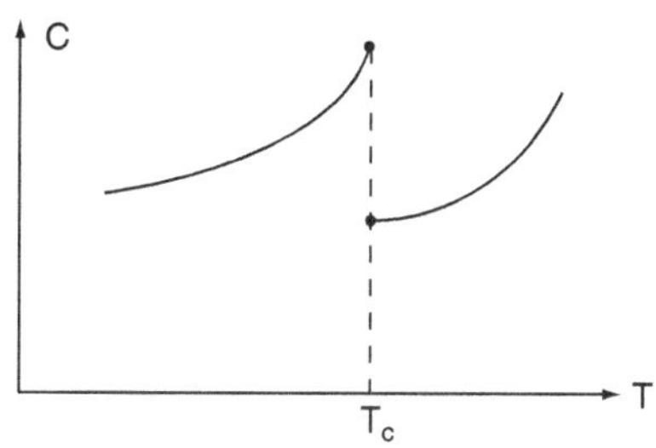

Fig. 3.11 The heat capacity at $T=T_c$ has a jump transition at $T=T_c$

Qualitatively, part of the above conclusions are in agreement with experimental results. However, these conclusions lead to incorrect susceptibility and spontaneous magnetization, whose experimental rates are given by $\chi\propto(T-T_c)^{-r}$ with $r=1.3$, and by $M\propto(T-T_c)^{\beta}$ with $\beta=1/3$.

Free energy G must be a function of the magnetization M. Hence the errors originate from the fact that the expression for G in the Ginzburg–Landau theory is an approximation. It is difficult to derive a precise formula, because G is not analytic on $|M|$; even the differentiability of G on $|M|$ near T_c is very low.

If we study the dynamical properties of ferromagnetic systems using the classical Ginzburg–Landau free energy, we shall see a more serious error when an external field is present.

To see this, note that the dynamic equation of classical theory is given by

$$\frac{dM}{dt}=-\alpha_2 M-\alpha_4|M|^2M+H. \tag{3.4.5}$$

For simplicity, we take $H=(h,0,0)$ with $h>0$, which is equivalent to taking the x_1-axis in the direction of H. Then the equation

$$\alpha_4|M|^2M+\alpha_2 M-H=0$$

has a steady-state solution $M_0=(m_0,0,0)$ for $T\geq 0$, which is the magnetization induced by H. Make the transformation

$$M=M'+M_0.$$

Then (3.4.5) can be rewritten as (we again drop the primes)

$$\begin{aligned}\frac{dM_1}{dt} &= -(\alpha_2+3\alpha_4 m_0^2)M_1 - 2\alpha_4 m_0 M_1^2 - \alpha_4 m_0 |M|^2 - \alpha_4 |M|^2 M_1,\\ \frac{dM_2}{dt} &= -(\alpha_2+\alpha_4 m_0^2)M_2 - 2\alpha_4 m_0 M_1 M_2 - \alpha_4 |M|^2 M_2,\\ \frac{dM_3}{dt} &= -(\alpha_2+\alpha_4 m_0^2)M_3 - 2\alpha_4 m_0 M_1 M_3 - \alpha_4 |M|^2 M_3.\end{aligned} \tag{3.4.6}$$

Comparing the two critical parameter curves

$$\begin{aligned}\alpha_2+3\alpha_4 m_0^2 = 0 &\Rightarrow T_1 = T_c - 3\alpha_4 m_0^2/\alpha_0,\\ \alpha_2+\alpha_4 m_0^2 = 0 &\Rightarrow T_2 = T_c - \alpha_4 m_0^2/\alpha_0,\end{aligned}$$

we find that $T_2 > T_1$. By Theorem 2.1.3, (3.4.6) has the first transition at $T = T_2$, where a new magnetization $M = (M_1, M_2, M_3)$, with $M_2 \neq 0$ and $M_3 \neq 0$, appears. This is unrealistic, because any magnetization M of this system must have the same direction as $H = (h, 0, 0)$; see Fig. 3.8.

In fact, when a magnetic field H is applied to an isotropic system, the direction of H is a favorable one for magnetization. However, in (3.4.5), this point is not manifested. Therefore, to investigate the phase transition dynamics of ferromagnetic systems we need to revise the GL free energy.

3.4.2 Dynamic Transitions in Ferromagnetism

Revised Ginzburg–Landau Free Energy. Let a ferromagnetic system be isotropic. When a magnetic field H is present, we introduce a second-order symmetric tensor

$$A(T,H) = (a_{ij}(T,H)), \quad a_{ij} = a_{ji} \quad \text{for } 1 \leq i,\ j \leq 3,$$

such that $A(T,0) = 0$ and $A(T,H)$ has eigenvalues

$$\lambda_1 = \lambda_1(T,H), \quad \lambda_2 = \lambda_3 = 0 \quad \text{with } \lambda_1(T,H) > 0 \text{ as } H \neq 0,$$

and H is the eigenvector of A corresponding to λ_1:

$$AH = \lambda_1 H \quad (\lambda_1 > 0 \text{ as } H \neq 0, \quad \lambda_1 = 0 \text{ as } H = 0). \tag{3.4.7}$$

It is clear that if we take the coordinate system (x_1, x_2, x_3) with the x_1-axis in the H-direction, then $H = (H_1, 0, 0)$ and

$$A = \begin{pmatrix} \lambda_1 & 0 & 0 \\ 0 & 0 & 0 \\ 0 & 0 & 0 \end{pmatrix}. \tag{3.4.8}$$

Physically, condition (3.4.7) means that H is a favorable direction of magnetization if we add a term

$$-MAM^T = -\sum a_{ij}M_iM_j$$

in the free energy. We also need to consider the nonlinear effect produced by H. To this end, we introduce the term $-|M|^2M\cdot H$ in the free energy.

Thus when the applied field H varies in $\Omega\subset\mathbb{R}^n$ $(n=2,3)$, the GL free energy is in the form

$$\begin{aligned}G(M,T,H) = G_0 + \frac{1}{2}\int_\Omega\Big[&\mu|\nabla M|^2 + \alpha_2|M|^2 + \frac{\alpha_4}{2}|M|^4\\ &-\sum a_{ij}M_iM_j - f(T,H,M)M\cdot H\Big]dx,\end{aligned} \tag{3.4.9}$$

where $G_0 = G_0(T)$ is independent of M and H, and f is a scalar function of T, H, and M, defined by

$$f(T,H,M) = 2(1+\beta|M|^2), \qquad \beta = \beta(T,H) > 0. \tag{3.4.10}$$

For α_2 and α_4, we assume that

$$\alpha_2 = \alpha_0(H)(T - T_0(H)), \quad T_0(0) = T_c, \quad \alpha_0 > 0, \quad \alpha_4 > 0. \tag{3.4.11}$$

By the unified model (3.1.15), we derive from (3.4.9) and (3.4.10) the following dynamical equations:

$$\begin{aligned}\frac{\partial M_i}{\partial t} =&\mu\Delta M_i - \alpha_2 M_i + \sum_{j=1}^{3} a_{ij}M_j - \alpha_4|M|^2M_i\\ &+\beta|M|^2H_i + 2\beta(M\cdot H)M_i + H_i \quad \text{for } 1\le i\le 3,\end{aligned} \tag{3.4.12}$$

supplemented with the Neumann boundary condition $\partial M/\partial n = 0$ on $\partial\Omega$. Obviously, if $H=0$, then equations (3.4.12) coincide with the classical equations. For simplicity, hereinafter we always take

$$H = (h,0,0) \qquad (h>0 \text{ is a constant}). \tag{3.4.13}$$

When H is constant, then in the study of phase transitions of magnetic systems, (3.4.12) can be replaced by a system of ordinary differential equations as follows:

$$\begin{aligned}\frac{dM_1}{dt} &= (\lambda_1 - \alpha_2)M_1 + \beta h|M|^2 + 2\beta hM_1^2 - \alpha_4|M|^2M_1 + h,\\ \frac{dM_2}{dt} &= -\alpha_2M_2 + 2\beta hM_1M_2 - \alpha_4|M|^2M_2,\\ \frac{dM_3}{dt} &= -\alpha_2M_3 + 2\beta hM_1M_3 - \alpha_4|M|^2M_3.\end{aligned} \tag{3.4.14}$$

Equations (3.4.14) have a steady-state solution induced by H:

$$M^* = (m_0, 0, 0) \qquad \text{with } \lim_{h\to 0} m_0 = 0.$$

We see in Fig. 3.9 that the magnetization M^* has an upper bound. Namely there is an M_0 such that

$$\begin{aligned} &|M^*| < |M_0| && \forall h \in \mathbb{R}^1, \quad T \geq 0, \\ &M^* \to M_0 && \text{if } h \to \infty. \end{aligned} \tag{3.4.15}$$

To satisfy (3.4.15), it is necessary to assume that the coefficients α_0 and α_4 as in (3.4.11) possess the properties

$$\alpha_0(H) \to +\infty, \quad \alpha_4(T,H) \to +\infty, \quad \text{as } h \to \infty. \tag{3.4.16}$$

Both conditions (3.4.11) and (3.4.16) are physical.

Dynamic Transitions. To illustrate the main ideas, we shall consider only the case in which H is a constant on Ω. Therefore, we shall study phase transition dynamics of the ferromagnetic systems using equations (3.4.14) for $h > 0$. Analysis of the more general case can be carried out in the same fashion; it will be reported elsewhere.

Take the following transformation in (3.4.14):

$$M = M^* + M'. \tag{3.4.17}$$

Then equations (3.4.14) can be rewritten as (we again drop the primes)

$$\begin{aligned} \frac{dM_1}{dt} &= \beta_1 M_1 - 2a_2 M_1^2 - a_2|M|^2 - \alpha_4|M|^2 M_1, \\ \frac{dM_2}{dt} &= \beta_2 M_2 - 2a_2 M_1 M_2 - \alpha_4|M|^2 M_2, \\ \frac{dM_3}{dt} &= \beta_2 M_3 - 2a_2 M_1 M_3 - \alpha_4|M|^2 M_3, \end{aligned} \tag{3.4.18}$$

where

$$\begin{aligned} &a_2 = \alpha_4 m_0 - \beta h, \\ &\beta_1 = \lambda_1 + 6\beta h m_0 - 3\alpha_4 m_0^2 - \alpha_2, \qquad \beta_2 = 2\beta h m_0 - \alpha_4 m_0^2 - \alpha_2. \end{aligned}$$

The critical parameter curves $\beta_1 = 0$ and $\beta_2 = 0$ are given by

$$\begin{aligned} &\beta_1 = 0 \Rightarrow T_1 = T_0(H) + \frac{1}{\alpha_0}(\lambda_1 + 6\beta h m_0 - 3\alpha_4 m_0^2), \\ &\beta_2 = 0 \Rightarrow T_2 = T_0(H) + \frac{1}{\alpha_0}(2\beta h m_0 - \alpha_4 m_0^2). \end{aligned}$$

It is clear that $T_1 > T_2$ whenever

$$\lambda_1 > 2m_0(\alpha_4 m_0 - 2\beta h), \quad \text{for } h > 0. \tag{3.4.19}$$

Therefore, under condition (3.4.19), equations (3.4.18) have a transition at $T = T_1$ in the space

$$E = \{(M_1, 0, 0)| -\infty < M_1 < \infty\}. \tag{3.4.20}$$

More precisely, we have the following transition theorem.

Theorem 3.4.1 *Assume the condition (3.4.19) and $a_2 \neq 0$. Then (3.4.18) has a Type III (mixed) transition at $T = T_1$, and the transition occurs in the space E. The phase diagram is as shown in Fig. 3.12. Moreover, we have the following assertions:*

(1) There are two stable equilibrium states near $T = T_1$, which are given by

$$M_1^+ = \begin{cases} 0 & \text{if } T > T_1, \\ \dfrac{1}{2\alpha_4}[-3a_2 + \sqrt{9a_2^2 + 4\alpha_4\beta_1}] & \text{if } T < T_1, \end{cases}$$
$$M_1^- = -\frac{1}{2\alpha_4}[3a_2 + \sqrt{9a_2^2 + 4\alpha_4\beta_1}].$$

(2) If $T < T_1$, then M_1^+ is stable in the region $0 < M_1 < \infty$, and M_1^- is stable in $-\infty < M_1 < 0$.
(3) If $T > T_1$, then $M_1^+ = 0$ is stable for $-b < M_1 < \infty$, and M_1^- is stable for $-\infty < M_1 < -b$, where

$$b = \frac{1}{2\alpha_4}\left[3a_2 - \sqrt{9a_2^2 + 4\alpha_4\beta_1}\right] > 0 \qquad \text{for } T > T_1.$$

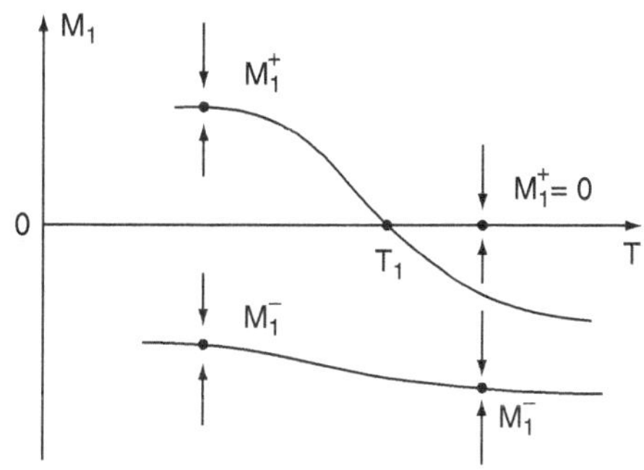

Fig. 3.12 Phase diagram of the transitions given by Theorem 3.4.1

Proof. It is clear that if (3.4.19) holds, then

$$\beta_1(T)\begin{cases} <0 & \text{if } T>T_1,\\ =0 & \text{if } T=T_1,\\ >0 & \text{if } T<T_1,\end{cases} \qquad \beta_2(T_1)<0.$$

Hence by Theorem 2.1.3, the system (3.4.18) has a transition at $T=T_1$. Obviously, the space E defined by (3.4.20) is the center manifold of (3.4.18) near $T=T_1$. Hence, the reduced equation of (3.4.18) on E can be expressed as

$$\frac{dM_1}{dt}=\beta_1 M_1-3a_2M_1^2-\alpha_4M_1^3. \tag{3.4.21}$$

When $a_2\neq 0$, by Theorem 2.3.2, we infer from (3.4.21) that this transition is of type-III, and the transition solutions satisfy

$$\alpha_4M_1^2+3a_2M_1-\beta_1=0.$$

By a direct computation, one obtains assertions (1)–(3). The proof is complete. □

3.4.3 Physical Implications

By (3.4.17), the stable steady states of (3.4.14) near $T=T_1$ are

$$M^{\pm}=(m_0+M_1^{\pm},0,0).$$

From the physical point of view, they should be

$$M_1^+\geq 0,\qquad M_1^-<0,\qquad m_0+M_1^-\geq 0. \tag{3.4.22}$$

The condition (3.4.22) requires that $0<3a_2<\alpha_4m_0$, which is equivalent to

$$\beta h<\alpha_4m_0<\frac{3}{2}\beta h \qquad (h>0), \tag{3.4.23}$$

where $m_0>0$ is a solution of the equation near $T=T_1$:

$$\alpha_4m_0^3-3\beta hm_0^2+(\alpha_2-\lambda_1)m_0-h=0 \qquad (h>0). \tag{3.4.24}$$

Thus, the stable steady states M^+ and M^- of (3.4.14) near $T=T_1$ are physical if the coefficients $\alpha_2(T,h)$, $\alpha_4(T,h)$, $\beta(T,h)$, and $\lambda_1(T,h)$ satisfy (3.4.23) and (3.4.24). In this case, the temperature T_1 is greater than the Curie temperature T_c:

$$T_1(H)>T_c=T_1(0) \qquad \text{for } H\neq 0.$$

The two states M^+ and M^- are mathematically equal, and therefore, by Theorem 3.4.1 alone, we cannot determine the magnetization behavior of ferromagnetic systems near $T = T_1$. However, we see that the magnetization M^+ is stronger than M^-. Physically, this implies that M^+ is favorable in $T < T_1$, and M^- is in $T > T_1$. Thus, from Theorem 3.4.1, there are two possible magnetization behaviors, i.e., two magnetization functions:

$$\mu_1(T) = M^+(T) = (m_0(T) + M_1^+(T), 0, 0),$$
$$\mu_2(T) = \begin{cases} M^+(T) = (m_0(T) + M_1^+(T), 0, 0) & \text{if } T < T_1, \\ M^-(T) = (m_0(T) + M_1^-(T), 0, 0) & \text{if } T \geq T_1. \end{cases}$$

The function $\mu_1(T)$ is continuous on T, as shown in Fig. 3.13a, and its derivative is discontinuous at $T = T_1$:

$$\mu_1'(T_1^-) - \mu_1'(T_1^+) \simeq \left(\frac{1}{3} \frac{d}{dT} \frac{\beta_1(T_1)}{\alpha_2(T_1)}, 0, 0 \right).$$

The function $\mu_2(T)$ has a jump at $T = T_1$, as shown in Fig. 3.13b.

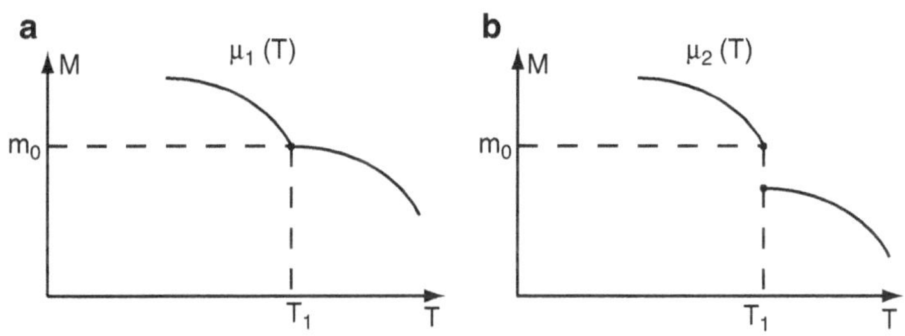

Fig. 3.13 (**a**) The graph of function $\mu_1(T)$; (**b**) the graph of function $\mu_2(T)$

On the other hand, by direct computation, the free energies of M^+ and M^- can be shown to be given by

$$G(M^+) = \begin{cases} G(m_0) & \text{if } T \geq T_1, \\ G(m_0) + \frac{1}{4}(M_1^+)^2(a_2 M_1^+ - \beta_1) & \text{if } T < T_1, \end{cases}$$
$$G(M^-) = G(m_0) + \frac{1}{4}(M_1^-)^2(a_2 M_1^- - \beta_1),$$

where m_0 is the magnetization induced by $H = (h, 0, 0)$ satisfying (3.4.24), and $\beta_1(T_1) = 0$. It is clear that

$$G(M^+) > G(M^-) \quad \text{near } T = T_1. \tag{3.4.25}$$

Hence, it follows from (3.4.25) that the magnetization behavior described by $\mu_2(T)$ is prohibited in the real world because the free energy cannot abruptly increase (or decrease) in a temperature-decreasing (or -increasing) process. Thus, by Theorem 3.4.1 and (3.4.25), we can derive the following physical conclusion:

Physical Conclusion 3.4.2 *When an external field H is present, the magnetization $M_H(T)$ of an isotropic ferromagnetic system is continuous in the temperature T, and there is $T_1(H) > T_c$ (T_c the Curie temperature) with $T_1(H) \to T_c$ as $H \to 0$ such that $M_H(T)$ is not differentiable at $T = T_1$, whose derivative has a finite jump*

$$M'_H(T_1^-) - M'_H(T_1^+) = a > 0 \quad (a < \infty).$$

Moreover, the graph of $M_H(T) = \mu_1(T)$ is as shown in Fig. 3.13a, and $M_H(T) \to M_0(T)$ as $H \to 0$ with

$$M_0(T) = \begin{cases} 0 & \text{if } T \geq T_c, \\ M_s(T) & \text{if } T < T_c, \end{cases}$$

where $M_s(T)$ is the spontaneous magnetization (see Fig. 3.10).

3.4.4 Asymmetry of Fluctuations

The above discussion suggests that for ferromagnetic systems, there are two possible phase transition behaviors near a critical point, and theoretically, each of them has some probability of taking place. However, only one of them can appear in reality. This phenomenon was also observed in the previous section for PVT systems.

One explanation of such phenomena is that the symmetry of fluctuations near a critical point is not generally true in equilibrium phase transitions. To make the statement clearer, we first introduce some concepts.

Let $G(u,\lambda)$ be the free energy of a thermodynamic system, let $u = (u_1,\cdots,u_n)$ be the order parameter, and let $\lambda = (\lambda_1,\cdots,\lambda_N)$ be the control parameter $(n,N \geq 1)$. Assume that u is defined in the function space $L^2(\Omega,\mathbb{R}^n)$ and $\lambda \in \mathbb{R}^N$. Then the space

$$X = \{(u,\lambda) \,|\, u \in L^2(\Omega,\mathbb{R}^n), \lambda \in \mathbb{R}^N\}$$

is called the state space of the system.

Let $(u_0,\lambda_0) \in X$ be a stable equilibrium state of the system; namely, (u_0,λ_0) is a locally minimal state of $G(u,\lambda)$. We say that the system has a fluctuation at (u_0,λ_0) if it deviates randomly from (u_0,λ_0) to $(\widetilde{u},\widetilde{\lambda})$ with

$$\|\widetilde{u} - u_0\| + |\widetilde{\lambda} - \lambda_0| > 0.$$

In this case, $(\widetilde{u},\widetilde{\lambda})$ is called a state of fluctuation.

The so-called symmetry of fluctuation means that for given $r > 0$, all states $(\widetilde{u}, \widetilde{\lambda})$ of fluctuation satisfying

$$\|\widetilde{u} - u_0\| + |\widetilde{\lambda} - \lambda_0| = r, \qquad (\widetilde{u}, \widetilde{\lambda}) \in X,$$

have the same probability of appearing in the real world. Otherwise, we say that the fluctuation is asymmetric.

Observations in both PVT and ferromagnetic systems strongly suggest the following physical conjecture regarding the uniqueness of transition behaviors.

Physical Principle 3.4.3 (Asymmetry of Fluctuations) *The symmetry of fluctuations for general thermodynamic systems may not be universally true. In other words, in some systems with multiequilibrium states, the fluctuations near a critical point occur only in one basin of attraction of some equilibrium states, which are those that can be physically observed.*

3.5 Phase Separation in Binary Systems

Materials compounded by two components A and B, such as binary alloys, binary solutions, and polymers, are called binary systems. Sufficient cooling of a binary system may lead to phase separations, i.e., at a critical temperature, the concentrations of both components A and B with homogeneous distributions undergo changes leading to heterogeneous distributions in space. Phase separation of binary systems observed will be in one of two main ways. The first is by nucleation, whereby sufficiently large nuclei of the second phase appear randomly and grow, and this corresponds to Type II phase transitions. The second is by spinodal decomposition, whereby the systems appear nuclear at once, and a periodic or semiperiodic structure is seen, and this corresponds to second-order phase transitions (Reichl 1998; Pismen 2006).

The Cahn–Hilliard equation describes the process of phase separation, by which the two components of a binary fluid spontaneously separate and form domains pure in each component. There have been extensive studies in the past on the dynamics of the Cahn–Hilliard equations (Cahn and Hillard 1957; Novick-Cohen and Segel 1984; Bates and Fife 1993; Shen and Yang 2010; del Pino et al. 2000).

The main objective of this section is to describe the phase separation mechanism precisely and to make a few physical predictions. The material is based largely on recent articles (Ma and Wang 2009b,c). In particular, four physical predictions on the phase separation of binary systems are derived. First, the order of phase transitions is precisely determined by the sign of a nondimensional parameter K such that if $K > 0$, the transition is of first order with latent heat, and if $K < 0$, the transition is of second order. Here the parameter K is defined in terms of the coefficients in the quadratic and cubic nonlinear terms of the Cahn–Hilliard equation and the typical length scale of the container.

Second, a phase diagram is derived characterizing the order of phase transitions, leading in particular to a prediction that there is only a second-order transition for a molar fraction near $1/2$. This is different from the prediction made by the classical phase diagram. Third, a TL-phase diagram is derived, characterizing the regions of both homogeneous and separation phases and their transitions. Fourth, our research demonstrates again the point, mentioned previously for the Bénard convection, that dynamic transition theory opens a new avenue toward the study of pattern formation, since the patterns and structure of transition states are dictated by solutions in σ_T, which are in turn determined by eigenvectors of the corresponding linearized problem (with high-order approximations).

3.5.1 Modeling

Since binary systems are conserved, the equations describing the Helmholtz process and the Gibbs process are the same. Hence, without distinction we use the term "free energy" to discuss this problem. Let u_A and u_B be the concentrations of components A and B respectively, then $u_B = 1 - u_A$. In a homogeneous state, $u_B = \overline{u}_B$ is a constant. We now consider the case in which the Helmholtz free energy is given by

$$F(u) = F_0 + \int_\Omega \left[\frac{\mu}{2}|\nabla u_B|^2 + f(u_B)\right] dx, \tag{3.5.1}$$

and the density function f is given by the Bragg–Williams potential, also called the Hildebrand theory (Reichl 1998), as follows:

$$f = \mu_A u_A + \mu_B u_B + RT[u_A \ln u_A + u_B \ln u_B] + a u_B u_A, \tag{3.5.2}$$

where μ_A, μ_B are the chemical potentials of A and B respectively, R the molar gas constant, $a > 0$ the measure of repulsion action between A and B.

In a homogeneous state, u_B is a constant in space and is equal to its mean value, i.e., $u_B = \overline{u}_B$. Let

$$u = u_B - \overline{u}_B, \qquad u_0 = \overline{u}_B.$$

Using the Landau mean field theory, we have the following Cahn–Hilliard equation:

$$\frac{\partial u}{\partial t} = -k\Delta^2 u + \Delta[b_1 u^1 + b_2 u^2 + b_3 u^3], \tag{3.5.3}$$

where the coefficients k, b_1, b_2, and b_3 are given by

$$\begin{aligned} k &= \mu D, \\ b_1 &= \frac{D}{2}\frac{d^2 f(u_0)}{du^2} = \left[\frac{RT}{u_0(1-u_0)} - 2a\right]\frac{D}{2} \\ b_2 &= \frac{D}{3!}\frac{d^3 f(u_0)}{du^3} = \frac{2u_0 - 1}{6u_0^2(1-u_0)^2}DRT, \\ b_3 &= \frac{D}{4!}\frac{d^4 f(u_0)}{du^4} = \frac{1-3u_0+3u_0^2}{12u_0^3(1-u_0)^3}DRT, \end{aligned} \tag{3.5.4}$$

where D is the diffusion coefficient.

A physically sound boundary condition for (3.5.3) is either the Neumann boundary condition

$$\frac{\partial u}{\partial n} = \frac{\partial \Delta u}{\partial n} = 0 \qquad \text{on } \partial\Omega, \tag{3.5.5}$$

with $\Omega \subset \mathbb{R}^n$ $(1 \leq n \leq 3)$ a bounded domain, or the periodic boundary condition

$$u(x + KL) = u(x) \tag{3.5.6}$$

with $\Omega = [0,L]^n, K = (k_1, \cdots, k_n), 1 \leq n \leq 3$.

In deriving the model (3.5.3), we use the idea of Landau mean field theory to retain only the terms up to third order. We remark here that the model (3.5.3) is a classical model (Novick-Cohen and Segel 1984), which has been used by many authors. The expansion used here to derive the coefficients is done near critical points of phase transitions.

In the general case, the order parameters are the molar concentration function u_B and the entropy S, and the control parameter contains the temperature T, the pressure p, and u_0. We refer the interested reader to (Ma and Wang 2009b) for the modeling and mathematical analysis of the general case.

3.5.2 Phase Transition in General Domains

We now discuss the Cahn–Hilliard equation from the mathematical point of view. We start with the nondimensional form of the equation. Let

$$\begin{aligned} &x = lx', &&t = \frac{l^4}{k}t', &&u = u_0 u', \\ &\lambda = -\frac{l^2 b_1}{k}, &&\gamma_2 = \frac{l^2 b_2 u_0}{k}, &&\gamma_3 = \frac{l^2 b_3 u_0^2}{k}, \end{aligned}$$

where l is a given length, $u_0 = \overline{u}_B$ is the constant concentration of B, and $\gamma_3 > 0$. Then (3.5.3) can be rewritten as follows (omitting the primes):

$$\begin{aligned}&\frac{\partial u}{\partial t}=-\Delta^2 u-\lambda\Delta u+\Delta(\gamma_2 u^2+\gamma_3 u^3),\\&\int_\Omega u(x,t)dx=0,\\&u(x,0)=\varphi.\end{aligned}\tag{3.5.7}$$

Let

$$H=\left\{u\in L^2(\Omega)\mid\int_\Omega udx=0\right\}.$$

For the Neumann boundary condition (3.5.5), we define

$$H_1=\left\{u\in H^4(\Omega)\bigcap H\,\Big|\,\frac{\partial u}{\partial n}=\frac{\partial\Delta u}{\partial n}=0\text{ on }\partial\Omega\right\},$$

and for the periodic boundary condition (3.5.6), we define

$$H_1=\left\{u\in H^4(\Omega)\bigcap H\,\Big|\,u(x+KL)=u(x)\ \forall K\in\mathbb{Z}^n\right\}.$$

Then we define the operators $L_\lambda=-A+B_\lambda$ and $G:H_1\to H$ by

$$Au=\Delta^2 u,\qquad B_\lambda u=-\lambda\Delta u,\qquad G(u)=\gamma_2\Delta u^2+\gamma_3\Delta u^3.\tag{3.5.8}$$

Thus, (3.5.7) is equivalent to the following operator equation:

$$\frac{du}{dt}=L_\lambda u+G(u),\quad u(0)=\varphi.\tag{3.5.9}$$

It is clear that the conditions (2.1.2) and (2.1.3) hold here.

We first consider the case that $\Omega\subset\mathbb{R}^n$ $(1\le n\le 3)$ is a general bounded and smooth domain. Let ρ_k and e_k be the eigenvalues and eigenfunctions of the following eigenvalue problem:

$$-\Delta e_k=\rho_k e_k,\qquad \frac{\partial e_k}{\partial n}|_{\partial\Omega}=0,\qquad \int_\Omega e_k dx=0.\tag{3.5.10}$$

The eigenvalues of (3.5.10) satisfy $0<\rho_1\le\rho_2\le\cdots\le\rho_k\le\cdots$ and $\lim_{k\to\infty}\rho_k=\infty$. The eigenfunctions $\{e_k\}$ constitute an orthonormal basis of H. Also, $\{e_k\}$ is an orthogonal basis of H_1 with norm

$$\|u\|_1=\left[\int_\Omega|\Delta^2 u|^2dx\right]^{1/2}.$$

We are now in a position to give a phase transition theorem for the problem (3.5.7) with the Neumann boundary condition (3.5.5):

Theorem 3.5.1 *Assume that $\gamma_2 = 0$ and $\gamma_3 > 0$ in (3.5.7). Then the following assertions hold:*

(1) If the first eigenvalue ρ_1 of (3.5.10) has multiplicity $m \geq 1$, then the problem (3.5.7) with (3.5.5) bifurcates from $(u,\lambda) = (0,\rho_1)$ on $\lambda > \rho_1$ to an attractor Σ_λ, homeomorphic to an $(m-1)$-dimensional sphere S^{m-1}, and Σ_λ attracts $H \setminus \Gamma$, where Γ is the stable manifold of $u = 0$ with codimension m.

(2) Σ_λ contains at least 2m singular points. If $m = 1$, thenΣ_λ has exactly two steady states $\pm u_\lambda$, and if $m = 2$, then$\Sigma_\lambda = S^1$ has at most eight singular points.

(3) Each singular point u_λ in Σ_λ can be expressed as

$$u_\lambda = \left(\frac{\lambda - \rho_1}{\gamma_3}\right)^{1/2} e_1 + o(|\lambda - \rho_1|^{1/2}),$$

where e_1 is an eigenfunction corresponding to the first eigenvalue of (3.5.10).

Proof. We proceed in several steps as follows.

Step 1. It is clear that the eigenfunctions $\{e_k\}$ of (3.5.10) are also eigenvectors of the linear operator $L_\lambda = -A + B_\lambda$ defined by (3.5.8), and the eigenvalues of L_λ are given by

$$\beta_k(\lambda) = \rho_k(\lambda - \rho_k), \quad k = 1, 2, \cdots. \tag{3.5.11}$$

It is easy to verify the conditions (2.1.4) and (2.1.5) in our case at $\lambda_0 = \rho_1$. We shall prove this theorem using the attractor bifurcation theory introduced in Sect. 2.2.

We need to verify that $u = 0$ is a global asymptotically stable singular point of (3.5.9) at $\lambda = \rho_1$. By $\gamma_2 = 0$, from the energy integration of (3.5.7) we can obtain

$$\begin{aligned}\frac{1}{2}\frac{d}{dt}\int_\Omega u^2 dx &= \int_\Omega \left[-|\Delta u|^2 + \rho_1 |\nabla u|^2 - 3\gamma_3 u^2 |\nabla u|^2\right] dx \\ &\leq -C\int_\Omega |\Delta v|^2 dx - 3\gamma_3 \int_\Omega u^2 |\nabla u|^2 dx,\end{aligned} \tag{3.5.12}$$

where $C > 0$ is a constant, $u = v + e_1$, and $\int_\Omega v e_1 dx = 0$. It follows from (3.5.12) that $u = 0$ is globally asymptotically stable. Hence for assertion (1), we have only to prove that Σ_λ is homeomorphic to S^{m-1}, since the rest of this assertion follows directly from Theorem 2.2.11.

Step 2. Now we prove that the bifurcated attractor Σ_λ from $(0, \rho_1)$ contains at least $2m$ singular points.

Let $g(u) = -\Delta u - \lambda u + \gamma_3 u^3$. Then the stationary equation of (3.5.7) is given by

$$\Delta g(u) = 0, \qquad \int_\Omega u dx = 0,$$

which is equivalent to

$$-\Delta u - \lambda u + \gamma_3 u^3 = \text{constant}, \qquad \int_\Omega u dx = 0, \qquad \frac{\partial u}{\partial n}|_{\partial\Omega} = 0. \tag{3.5.13}$$

By the Lagrange multiplier theorem, (3.5.13) is the Euler equation of the following functional with zero average constraint:

$$F(u) = \int_\Omega \left[\frac{1}{2}|\nabla u|^2 - \frac{\lambda}{2}u^2 + \frac{\gamma_3}{4}u^4\right] dx, \tag{3.5.14}$$
$$u \in \left\{u \in H^1(\Omega)\bigcap H \;\middle|\; \frac{\partial u}{\partial n}|_{\partial\Omega} = 0 \quad \int_\Omega u dx = 0\right\}.$$

Since F is an even functional, by the classical Krasnosel'skii bifurcation theorem for even functionals, (3.5.14) bifurcates from $\lambda > \rho_1$ at least to $2m$ minimax points, i.e., (3.5.13) has at least $2m$ bifurcated solutions on $\lambda > \rho_1$. Hence, the attractor Σ_λ contains at least $2m$ singular points.

Step 3. To complete the proof, we reduce the equation (3.5.9) to the center manifold near $\lambda = \rho_1$. By center manifold reduction (1.2.9), the reduced equation of (3.5.9) is given by

$$\frac{dx_i}{dt} = \beta_1(\lambda)x_i - \gamma_3\rho_1 \int_\Omega v^3 e_{1i} dx + o(|x|^3) \qquad \text{for } 1 \le i \le m, \tag{3.5.15}$$

where $\beta_1(\lambda) = \rho_1(\lambda - \rho_1)$, $v = \sum_{i=1}^m x_i e_{1i}$, and $\{e_{11}, \cdots, e_{1m}\}$ are the first eigenfunctions of (3.5.10). Equation (3.5.15) can be rewritten as

$$\frac{dx_i}{dt} = \beta_1(\lambda)x_i - \gamma_3\rho_1 \int_\Omega \left(\sum_{j=1}^m x_j e_{1j}\right)^3 e_{1i} dx + o(|x|^3). \tag{3.5.16}$$

Let

$$g(x) = \left(\int_\Omega v^3 e_{11} dx, \cdots, \int_\Omega v^3 e_{1m} dx\right), \qquad v(x) = \sum_{j=1}^m x_j e_{1j}.$$

Then there is a constant $C > 0$ such that for every $x \in \mathbb{R}^m$,

$$(g(x), x) = \sum_{i=1}^m x_i \int_\Omega v^3 e_{1i} dx = \int_\Omega v^4 dx \ge C|x|^4. \tag{3.5.17}$$

Thus by Theorem 2.2.5, it follows from (3.5.16) and (3.5.17) that the attractor Σ_λ is homeomorphic to S^{m-1}. Assertion (1) is proved. Assertions (2) and (3) can then be derived from (3.5.16) and (3.5.17), and the proof is complete. □

Physically, the coefficients γ_2 and γ_3 depend on $u_0 = \overline{u}_B$, the temperature T, and the pressure p:

$$\gamma_k = \gamma_k(u_0, T, p), \qquad k = 2, 3.$$

The set of points satisfying $\gamma_2(u_0, T, p) = 0$ has measure zero in $(u_0, T, p) \in \mathbb{R}^3$. Hence, it is more interesting to consider the case $\gamma_2 \neq 0$.

For this purpose, let the multiplicity of the first eigenvalue ρ_1 of (3.5.10) be $m \geq 1$, and let $\{e_1, \cdots, e_m\}$ be the first eigenfunctions. Let $a_{ij}^k = \int_\Omega e_i e_j e_k dx$ and consider the following quadratic equations:

$$\sum_{i,j=1}^{m} a_{ij}^k x_i x_j = 0 \qquad \text{for } 1 \leq k \leq m. \tag{3.5.18}$$

Theorem 3.5.2 *Let $\gamma_2 \neq 0$, $\gamma_3 > 0$, and let $x = 0$ be an isolated singular point of (3.5.18). Then the phase transition of (3.5.7) and (3.5.5) is of either Type II or Type III. Furthermore, the problem (3.5.7) with (3.5.5) bifurcates to at least one singular point on each side of $\lambda = \rho_1$, and has a saddle-node bifurcation on $\lambda < \rho_1$. In particular, if $m = 1$, then the following assertions hold:*

(1) The phase transition is of Type III, and a neighborhood $U \subset H$ of $u = 0$ can be decomposed into two sectorial regions $\overline{U} = \overline{D}_1(\pi) + \overline{D}_2(\pi)$ such that the phase transition in $D_1(\pi)$ is of first order, and in $D_2(\pi)$ is of order n with $n \geq 3$.
(2) The bifurcated singular point u_λ on $\lambda > \rho_1$ attracts $D_2(\pi)$, which can be expressed as

$$u_\lambda = (\lambda - \rho_1) e_1 / \gamma_2 a + o(|\lambda - \rho_1|), \tag{3.5.19}$$

where by assumption, $a = \int_\Omega e_1^3 dx \neq 0$.
(3) When $|\gamma_2 a| = \varepsilon$ is small, the assertions in the transition perturbation theorems (Theorems 2.6.3 and 2.6.4) hold.

Remark 3.5.3 We shall see later that when $\Omega = \prod_{k=1}^n (0, L_k) \subset \mathbb{R}^n$ is a rectangular domain, then $a = \int_\Omega e_1^3 dx = 0$. However, for almost all nonrectangular domains Ω, the first eigenvalues are simple and $a \neq 0$. Hence, the Type III phase transitions for general domains are generic.

Proof of Theorem 3.5.2. Assertions (1)–(3) can be directly proved using Theorems 2.3.2, 2.6.3, and 2.6.4. By assumption, $u = 0$ is a second-order nondegenerate singular point of (3.5.9) at $\lambda = \rho_1$, which implies that $u = 0$ is not locally asymptotically stable. Hence, it follows from Theorem 2.1.6 and Theorem 4.1 in Ma and Wang (2005b) that the phase transition of (3.5.7) with (3.5.5) is of either Type II or Type III, and there is at least one singular point bifurcated on each side of $\lambda = \rho_1$.

Finally, we shall apply Theorem 2.5.3 to prove that there exists a saddle-node bifurcation on $\lambda < \rho_1$. It is known that

$$\operatorname{ind}(L_\lambda + G, 0) = \begin{cases} \text{even} & \text{if } \lambda = \rho_1, \\ 1 & \text{if } \lambda < \rho_1. \end{cases}$$

Moreover, since $L_\lambda + G$ defined by (3.5.8) is a gradient-type operator, we have

$$\operatorname{ind}(L_{\rho_1} + G, 0) \leq 0.$$

Hence there is a bifurcated branch Σ_λ on $\lambda < \rho_1$ such that

$$\operatorname{ind}(L_\lambda + G, u_\lambda) = -1 \qquad \forall u_\lambda \in \Sigma_\lambda, \quad \lambda < \rho_1.$$

It is clear that the eigenvalues (3.5.11) of L_λ satisfy (2.1.4) and (2.1.5), and for every $\lambda \in \mathbb{R}^1$, (3.5.9) possesses a global attractor. Therefore, for bounded $\lambda, b < \lambda \leq \rho_1$, the bifurcated branch Σ_λ is bounded:

$$\|u_\lambda\| \leq C \qquad \forall u_\lambda \in \Sigma_\lambda, \quad -\infty < b < \lambda \leq \rho_1.$$

We need to prove that there exists $\widetilde{\lambda} < \rho_1$ such that for all $\lambda < \widetilde{\lambda}$, equation (3.5.9) has no nonzero singular points.

By the energy estimates of (3.5.9), for every $\lambda < \widetilde{\lambda} = -\gamma_2^2/2\gamma_3$ and $u \neq 0$ in H, we have

$$\begin{aligned}
&\int_\Omega [|\Delta u|^2 - \lambda|\nabla u|^2 + 2\gamma_2 u|\nabla u|^2 + 3\gamma_3 u^2|\nabla u|^2]\,dx \\
&\geq \int_\Omega |\Delta u|^2 dx + \int_\Omega |\nabla u|^2(-\lambda - 2|\gamma_2 u| + 3\gamma_3 u^2)dx \\
&\geq \int_\Omega |\nabla u|^2 dx + \int_\Omega |\nabla u|^2(-\lambda + \gamma_3 u^2 + 2\gamma_3(u - \frac{\gamma_2}{\gamma_3})^2 - \frac{\gamma_2^2}{2\gamma_3})dx \\
&\geq \int_\Omega |\nabla u|^2 dx + \int_\Omega |\nabla u|^2(-\lambda - \frac{\gamma_2^2}{2\gamma_3})dx \\
&> 0.
\end{aligned}$$

Therefore, when $\lambda < \widetilde{\lambda}$, (3.5.9) has no nontrivial singular points in H. Thus we infer from Theorem 2.5.3 that there exists a saddle-node bifurcation on $\lambda < \rho_1$. The proof is complete. □

3.5.3 *Phase Transition in Rectangular Domains*

The dynamical properties of phase separation of a binary system in a rectangular container are very different from those in a general container. We saw in the previous section that the phase transitions in general domains are of Type III, and we shall show in the following that the phase transitions in rectangular domains are of either Type I or Type II, which are distinguished by a critical size of the domains.

Let $\Omega = \prod_{k=1}^n (0, L_k) \subset \mathbb{R}^n$ $(1 \leq n \leq 3)$ be a rectangular domain. We first consider the case

$$L = L_1 > L_j \qquad \forall 2 \leq j \leq n. \tag{3.5.20}$$

Theorem 3.5.4 *Let $\Omega = \prod_{k=1}^n (0, L_k)$ satisfy (3.5.20). The following assertions hold:*

(1) If

$$\gamma_3 < \frac{2L^2}{9\pi^2}\gamma_2^2,$$

then the phase transition of (3.5.7) and (3.5.5) at $\lambda = \lambda_0 = \pi^2/L^2$ is of Type II. In particular, the problem (3.5.7) with (3.5.5) bifurcates from

$(u,\lambda)=(0,\pi^2/L^2)$ on $\lambda<\pi^2/L^2$ to exactly two equilibrium points, which are saddles, and there are two saddle-node bifurcations on $\lambda<\pi^2/L^2$, as shown in Fig. 3.14.

(2) If

$$\gamma_3 > \frac{2L^2}{9\pi^2}\gamma_2^2,$$

then the transition is of Type I. In particular, the problem bifurcates on $\lambda>\pi^2/L^2$ to exactly two attractors u_1^T and u_2^T, which can be expressed as

$$u_{1,2}^T = \pm\frac{\sqrt{2}(\lambda-\frac{\pi^2}{L^2})^{1/2}}{\sigma^{1/2}}\cos\frac{\pi x_1}{L}+o(|\lambda-\frac{\pi^2}{L^2}|^{1/2}), \tag{3.5.21}$$

where $\sigma=\frac{3\gamma_3}{2}-\frac{L^2\gamma_2^2}{3\pi^2}$.

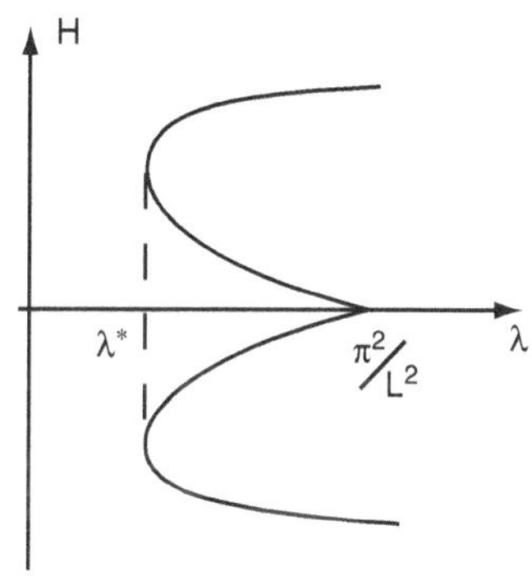

Fig. 3.14 Type II transition as given by Theorem 3.5.4

Proof. With the spatial domain as given, the first eigenvalue and eigenfunction of (3.5.10) are given by

$$\rho_1=\pi^2/L^2, \quad e_1=\cos\frac{\pi x_1}{L}.$$

The eigenvalues and eigenfunctions of $L_\lambda=-A+B_\lambda$ defined by (3.5.8) are as follows:

$$\beta_K=|K|^2(\lambda-|K|^2), \tag{3.5.22}$$

$$e_K=\cos\frac{k_1\pi x_1}{L_1}\cdots\cos\frac{k_n\pi x_n}{L_n}, \tag{3.5.23}$$

where $k_i\in\mathbb{Z}$,

$$K=\left(\frac{k_1\pi}{L_1},\cdots,\frac{k_n\pi}{L_n}\right), \qquad |K|^2=\pi^2\sum_{i=1}^n k_i^2/L_i^2, \qquad |K|^2\neq 0.$$

By Theorem A.1.1, the reduced equation of (3.5.9) to the center manifold is

$$\frac{dy}{dt} = \beta_1(\lambda)y - \frac{2\pi^2}{(L_1\cdots L_n)L_1^2}\int_\Omega \left[\gamma_3 y^3 e_1^4 + \gamma_2(ye_1+\Phi(y))^2 e_1\right]dx, \tag{3.5.24}$$

where $y \in \mathbb{R}^1$, $\Phi(y)$ is the center manifold function, and

$$\beta_1(\lambda) = \frac{\pi^2}{L^2}(\lambda - \frac{\pi^2}{L^2}). \tag{3.5.25}$$

Direct calculation implies that

$$\int_\Omega e_1^3 dx = \int_0^{L_1}\cdots\int_0^{L_n}\cos^3\frac{\pi x_1}{L}dx = 0, \tag{3.5.26}$$

$$\int_\Omega e_1^4 dx = L_2\cdots L_n\int_0^L \cos^4\frac{\pi x_1}{L}dx_1 = \frac{3}{8}L_1\cdots L_n. \tag{3.5.27}$$

By (3.5.26) and $\Phi(y) = O(|y|^2)$, we obtain

$$\int_\Omega (ye_1+\Phi)^2 e_1 dx = 2y\int_\Omega \Phi(y)e_1^2 dx + o(|y|^3). \tag{3.5.28}$$

It follows from Theorem A.1.1 that

$$\Phi(y) = \sum_{|K|^2>\pi^2/L^2}^{\infty} \phi_K(y)e_K + o(|y|^2),$$

$$\phi_K(y) = \frac{\gamma_2 y^2}{-\beta_K\|e_K\|^2}\int_\Omega \Delta e_1^2\cdot e_K dx = \frac{|K|^2\gamma_2 y^2}{\beta_K\|e_K\|^2}\int_\Omega e_K e_1^2 dx.$$

Notice that

$$\int_\Omega e_K e_1^2 dx = \begin{cases} 0 & \forall K \neq \left(\frac{2\pi}{L_1},0,\cdots,0\right), \\ L_1\cdots L_n/4 & \forall K = \left(\frac{2\pi}{L_1},0,\cdots,0\right). \end{cases}$$

Then we have

$$\Phi(y) = \frac{\gamma_2 y^2}{2(\lambda - 4\pi^2/L^2)}\cos\frac{2\pi x_1}{L} + o(|y|^2).$$

Inserting $\Phi(y)$ into (3.5.28), we obtain

$$\begin{aligned}\int_\Omega (ye_1+\Phi)^2 e_1 dx =& \frac{\gamma_2 y^3}{\lambda - \frac{4\pi^2}{L^2}}\int_\Omega \cos\frac{2\pi x_1}{L}e_1^2 dx + o(|y|^3) \\ =& \frac{L_1\cdots L_n}{4}\cdot\frac{\gamma_2}{\lambda - \frac{4\pi^2}{L^2}}y^3 + o(|y|^3). \end{aligned}\tag{3.5.29}$$

Finally, by (3.5.26), (3.5.27), and (3.5.29), we derive from (3.5.24) the following reduced equation of (3.5.9):

$$\frac{dy}{dt} = \beta_1(\lambda)y - \frac{\pi^2}{2L^2}\left(\frac{3\gamma_3}{2} + \frac{\gamma_2^2}{\lambda - \frac{4\pi^2}{L^2}}\right)y^3 + o(|y|^3). \tag{3.5.30}$$

Near the critical point $\lambda_0 = \pi^2/L^2$,

$$\frac{3\gamma_3}{2} + \frac{\gamma_2^2}{\lambda - \frac{4\pi^2}{L^2}} = \frac{3\gamma_3}{2} - \frac{L^2\gamma_2^2}{3\pi^2}.$$

Thus, by Theorem 2.1.5 we derive from (3.5.30) the assertions of the theorem except the claim for the saddle-node bifurcation in assertion (1), which can be proved in the same fashion as used in Theorem 3.5.2. The proof is complete. □

We now consider the case that $\Omega = \prod_{k=1}^n (0, L_k)$ satisfies

$$L = L_1 = \cdots = L_m > L_j \qquad \text{for } 2 \le m \le 3, m < j \le n. \tag{3.5.31}$$

Theorem 3.5.5 *Let $\Omega = \prod_{k=1}^n (0, L_k)$ satisfy (3.5.31). Then the following assertions hold:*

1. If

$$\gamma_3 > \frac{26L^2}{27\pi^2}\gamma_2^2,$$

then the phase transition of the problem (3.5.7) with (3.5.5) at $\lambda_0 = \pi^2/L^2$ is of Type I, satisfying the following properties:

(a) The problem bifurcates on $\lambda > \pi^2/L^2$ to an attractor Σ_λ, containing exactly $3^m - 1$ nondegenerate singular points, and Σ_λ is homeomorphic to an $(m-1)$-dimensional sphere S^{m-1}.

(b) For $m = 2$, the attractor $\Sigma_\lambda = S^1$ contains four minimal attractors, as shown in Fig. 3.15.

(c) For $m = 3$, $\Sigma_\lambda = S^2$ contains eight minimal attractors, as shown in Fig. 3.16a, if

$$\gamma_3 < \frac{22L^2}{9\pi^2}\gamma_2^2,$$

and if contains six minimal attractors, as shown in Fig. 3.16b if

$$\gamma_3 > \frac{22L^2}{9\pi^2}\gamma_2^2.$$

2. *If*

$$\gamma_3 < \frac{26L^2}{27\pi^2}\gamma_2^2,$$

then the transition is of Type II. In particular, the problem has a saddle-node bifurcation on $\lambda < \lambda_0 = \pi^2/L^2$, *and bifurcates on both side of* $\lambda = \lambda_0$ *to exactly* $3^m - 1$ *singular points, which are nondegenerate.*

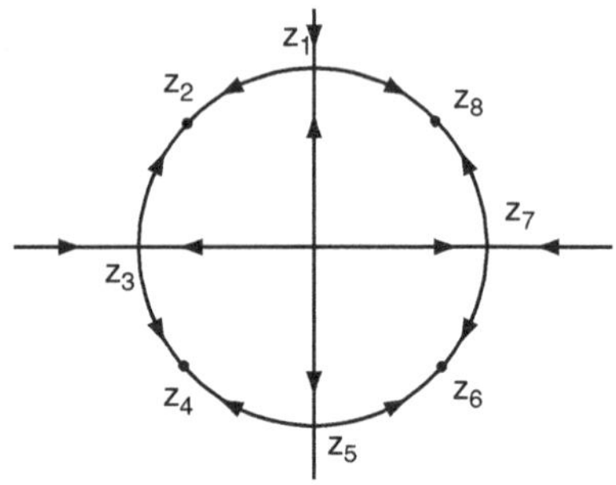

Fig. 3.15 For $m = 2, \Sigma_\lambda = S^1$ and Z_{2k} $(1 \le k \le 4)$ are attractors

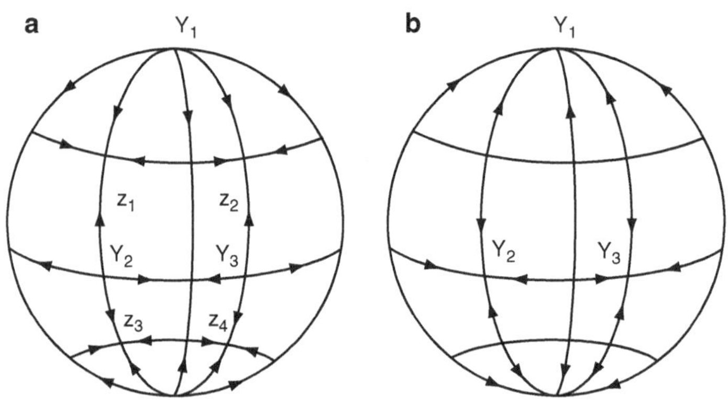

Fig. 3.16 For $m = 3$, $\Sigma_\lambda = S^2$, **(a)** $\pm Z_k$ $(1 \le k \le 4)$ are attractors, **(b)** $\pm Y_k$ $(1 \le k \le 3)$ are attractors

Proof. We proceed in several steps as follows.

Step 1. Consider the center manifold reduction. It is known that the eigenvalues and eigenfunctions of $L_\lambda = -A + B_\lambda$ are given by (3.5.22) and (3.5.23) with $L_1 = \cdots = L_m$. As before, the reduced equations of (3.5.9) are given by

$$\frac{dy}{dt} = \beta_1(\lambda)y + g(y) + o(|y|^3), \tag{3.5.32}$$

where $y = (y_1, \cdots, y_m) \in \mathbb{R}^m$, $\beta_1(\lambda)$ is as in (3.5.25), and

$$\begin{aligned}
g(y) &= \frac{2\pi^2}{L_1\cdots L_n L_1^2}(G_3(y)+G_{23}(y)),\\
G_3(y) &= -\gamma_3\left(\int_\Omega v^3 e_1 dx,\cdots,\int_\Omega v^3 e_m dx\right),\\
G_{23}(y) &= -\gamma_2\left(\int_\Omega u^2 e_1 dx,\cdots,\int_\Omega u^2 e_m dx\right).
\end{aligned}$$

Here $e_i = \cos\pi x_i/L$ for $1\le i\le m$, L is given by (3.5.31), $v=\sum_{i=1}^m y_i e_i$, $u=v+\Phi(y)$, and Φ is the center manifold function. Direct computation shows that

$$\int_\Omega v^2 e_i dx = \int_\Omega \left(\sum_{j=1}^m y_j \cos\pi x_j/L\right)^2 \cos\pi x_i/L dx = 0,$$

$$\begin{aligned}
\int_\Omega v^3 e_i dx &= \int_\Omega \left(\sum_{j=1}^n y_k \cos\pi x_j/L\right)^3 \cos\pi x_i/L dx \\
&= \frac{3}{4}L_1\cdots L_n\left(\frac{1}{2}y_i^3 + y_i\sum_{j\neq i} y_j^2\right),
\end{aligned} \tag{3.5.33}$$

$$\begin{aligned}
\int_\Omega u^2 e_i dx &= \int_\Omega \left(\sum_{j=1}^m y_j e_j + \Phi(y)\right)^2 e_i dx \\
&= 2\sum_{j=1}^m y_j \int_\Omega \Phi(y) e_j e_i dx + \int_\Omega \Phi^2(y) e_i dx.
\end{aligned} \tag{3.5.34}$$

We need to compute the center manifold function $\Phi(y)$. By Theorem A.1.1, we have

$$\Phi(y) = \sum_{|K|>\pi^2/L^2}^\infty \phi_K(y) e_K + o(|y|^2) + O(|y|^2|\beta_1|), \tag{3.5.35}$$

where

$$\begin{aligned}
\phi_K(y) &= \frac{\gamma_2}{-\beta_K(\lambda)(e_K,e_K)}\int_\Omega \Delta v^2 e_K dx \\
&= \frac{|K|^2\gamma_2}{\beta_K(\lambda)(e_K,e_K)}\int_\Omega v^2 e_K dx \\
&= \frac{|K|^2\gamma_2}{\beta_K(e_K,e_K)}\sum_{i,j=1}^m y_i y_i \int_\Omega e_i e_i e_K dx.
\end{aligned}$$

It is clear that

$$\int_\Omega e_i e_j e_K dx = 0 \quad \text{if } K \neq K_i + K_j, \quad K_i = \left(\frac{\pi\delta_{1i}}{L_1},\cdots,\frac{\pi\delta_{ni}}{L_n}\right).$$

By (3.5.22) and (3.5.23), we have

$$\phi_K(y) = \frac{\gamma_2 y_i y_j (2-\delta_{ij})}{(\lambda - |K|^2)(e_K, e_K)} \int_\Omega e_i e_j e_K dx = \begin{cases} \dfrac{\gamma_2 y_i^2}{2(\lambda - \frac{4\pi^2}{L^2})} & \text{if } i = j, \\ \dfrac{2\gamma_2 y_i y_j}{(\lambda - \frac{2\pi^2}{L^2})} & \text{if } i \neq j, \end{cases}$$

for $K = K_i + K_j$. Thus, by (3.5.35), we obtain

$$\Phi(y) = \sum_{i=1}^m \frac{\gamma_2}{2(\lambda - \frac{4\pi^2}{L^2})} y_i^2 \cos \frac{2\pi x_i}{L} + \sum_{l>r}^m \frac{2\gamma_2}{(\lambda - \frac{2\pi^2}{L^2})} y_l y_r \cos \frac{\pi x_l}{L} \cos \frac{\pi x_r}{L}.$$

Inserting $\Phi(y)$ into (3.5.34), we derive

$$\int_\Omega u^2 e_i dx = \frac{\gamma_2}{\lambda - \frac{4\pi^2}{L^2}} \int_\Omega \left[y_i^3 \cos^2 \frac{\pi x_i}{L} \cos \frac{2\pi x_i}{L} + 4 y_i \sum_{j\neq i} y_j^2 \cos^2 \frac{\pi x_j}{L} \cos^2 \frac{\pi x_i}{L} \right] dx$$
$$+ o(|y|^3).$$

Direct computation gives

$$\int_\Omega u^2 e_i dx = \frac{\gamma_2 L_1 \cdots L_n}{4} \left[\frac{y_i^3}{\lambda - \frac{4\pi^2}{L^2}} + \frac{4 y_i}{\lambda - \frac{2\pi^2}{L^2}} \sum_{j\neq i} y_j^2 \right] + o(|y|^3). \tag{3.5.36}$$

Putting (3.5.33) and (3.5.36) into (3.5.32), we obtain the following reduced equations:

$$\frac{dy_i}{dt} = \beta_1(\lambda) y_i - \frac{\pi^2}{2L^2} \left[\sigma_1 y_i^3 + \sigma_2 y_i \sum_{j\neq i} y_j^2 \right] + o(|y|^3) \quad \forall 1 \leq i \leq m, \tag{3.5.37}$$

where

$$\sigma_1 = \frac{3\gamma_3}{2} + \frac{\gamma_2^2}{\lambda - \frac{4\pi^2}{L^2}}, \qquad \sigma_2 = 3\gamma_3 + \frac{4\gamma_2^2}{\lambda - \frac{2\pi^2}{L^2}}. \tag{3.5.38}$$

Step 2. It is known that the transition type of (3.5.9) at the critical point $\lambda_0 = \pi^2/L^2$ is completely determined by (3.5.37), i.e., by the following equations:

$$\frac{dy_i}{dt} = -\frac{\pi^2}{2L^2} \left[\sigma_1^0 y_i^3 + \sigma_2^0 y_i \sum_{k\neq i} y_k^2 \right] \qquad \forall 1 \leq i \leq m, \tag{3.5.39}$$

where

$$\sigma_1^0 = \frac{3\gamma_3}{2} - \frac{L^2\gamma_2^2}{3\pi^2}, \qquad \sigma_2^0 = 3\gamma_3 - \frac{4L^2\gamma_2^2}{\pi^2}.$$

It is easy to see that

$$
\begin{aligned}
\sigma_1^0 + \sigma_2^0 > 0 &\Leftrightarrow \gamma_3 > \frac{26L^2}{27\pi^2}\gamma_2^2, \\
\sigma_1^0 + \sigma_2^0 < 0 &\Leftrightarrow \gamma_3 < \frac{26L^2}{27\pi^2}\gamma_2^2.
\end{aligned}
\tag{3.5.40}
$$

Step 3. We consider the case $m = 2$. Thus, the transition type of (3.5.39) is equivalent to that of the following equations:

$$
\begin{aligned}
\frac{dy_1}{dt} &= -y_1[\sigma_1^0 y_1^2 + \sigma_2^0 y_2^2], \\
\frac{dy_2}{dt} &= -y_2[\sigma_1^0 y_2^2 + \sigma_2^0 y_1^2].
\end{aligned}
\tag{3.5.41}
$$

On the straight lines

$$y_1^2 = y_2^2, \tag{3.5.42}$$

equations (3.5.41) satisfy

$$\frac{dy_2}{dy_1} = \frac{y_2}{y_1} \qquad \text{for } \sigma_1^0 + \sigma_2^0 \neq 0, \quad (y_1, y_2) \neq 0.$$

Hence the straight lines (3.5.42) are orbits of (3.5.41) if $\sigma_1^0 + \sigma_2^0 \neq 0$. Obviously, the straight lines

$$y_1 = 0 \text{ and } y_2 = 0 \tag{3.5.43}$$

are also orbits of (3.5.41).

There are four straight lines determined by (3.5.42) and (3.5.43), and each of them contains two orbits. Hence, the system (3.5.41) has at least eight straight-line orbits. Hence it is not hard to see that the number of straight-line orbits of (3.5.41), if finite, is eight.

Since (3.5.9) is a gradient-type equation, by Lemma A.2.7, there are no elliptic regions at $y = 0$. Hence, when $\sigma_1^0 + \sigma_2^0 > 0$, all the straight-line orbits on (3.5.42) and (3.5.43) tend to $y = 0$, as shown in Fig. 3.17a, which implies that the regions at $y = 0$ are parabolic and stable, and therefore $y = 0$ is asymptotically stable for (3.5.41). Accordingly, by Theorem 2.2.2, the transition of (3.5.37) at $\lambda_0 = \pi^2/L^2$ is of Type I.

When $\sigma_1^0 + \sigma_2^0 < 0$ and $\sigma_1^0 > 0$, namely

$$\frac{2}{9}\frac{L^2\gamma_2^2}{\pi^2} < \gamma_3 < \frac{26}{27}\frac{L^2\gamma_2^2}{\pi^2},$$

the four straight-line orbits on (3.5.42) are outward from $y = 0$, and the other four on (3.5.43) are toward $y = 0$, as shown in Fig. 3.17b, which implies that all regions at $y = 0$ are hyperbolic. Hence, by Theorem 2.1.6, the transition of (3.5.37) at $\lambda_0 = \pi^2/L^2$ is of Type II.

When $\sigma_1^0 \leq 0$, then $\sigma_2^0 < 0$ too. In this case, no orbits of (3.5.41) are toward $y = 0$, as shown in Fig. 3.17c, which implies by Theorem 2.1.6 that the transition is of Type II.

Thus by (3.5.40), for $m=2$ we prove that the transition is of Type I if $\gamma_3 > \frac{26L^2}{27\pi^2}\gamma_2^2$, and of Type II if $\gamma_3 < \frac{26L^2}{27\pi^2}\gamma_2^2$.

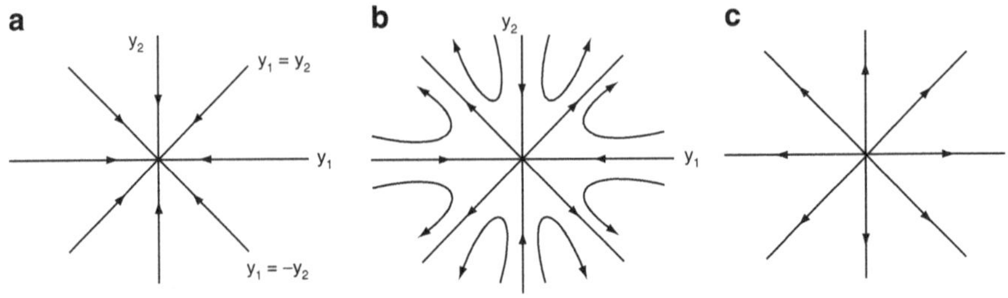

Fig. 3.17 The topological structure of flows of (3.5.37) at $\lambda_0 = \pi^2/L^2$, (**a**) for $\gamma_3 > \frac{26L^2}{27\pi^2}\gamma_2^2$; (**b**) for $\frac{2L^2}{9\pi^2}\gamma_2^2 < \gamma_3 < \frac{26L^2}{27\pi^2}\gamma_2^2$; and (**c**) for $\gamma_3 \leq \frac{2L^2}{9\pi^2}\gamma_2^2$

Step 4. Consider the case $m=3$. Thus, equations (3.5.39) are written as follows:

$$\begin{aligned}\frac{dy_1}{dt} &= -y_1[\sigma_1^0 y_1^2 + \sigma_2^0(y_2^2+y_3^2)],\\ \frac{dy_2}{dt} &= -y_2[\sigma_1^0 y_2^2 + \sigma_2^0(y_1^2+y_3^2)],\\ \frac{dy_3}{dt} &= -y_3[\sigma_1^0 y_3^2 + \sigma_2^0(y_1^2+y_2^2)].\end{aligned} \tag{3.5.44}$$

It is clear that the straight lines

$$y_i = 0, \quad y_j = 0 \qquad \text{for } i \neq j,\ 1 \leq i,\ j \leq 3, \tag{3.5.45}$$

$$\begin{cases} y_i^2 = y_j^2, \quad y_k = 0 & \text{for } i \neq j, i \neq k, j \neq k, 1 \leq i,j,k \leq 3,\\ y_1^2 = y_2^2 = y_3^2, \end{cases} \tag{3.5.46}$$

consist of orbits of (3.5.44). There are altogether 13 straight lines in (3.5.45) and (3.5.46), each of which consists of two orbits. Thus, (3.5.44) has at least 26 straight-line orbits. We shall show that (3.5.44) has just the straight-line orbits given by (3.5.45) and (3.5.46). In fact, we assume that the line

$$y_2 = z_1 y_1, \quad y_3 = z_2 y_1 \quad (z_1, z_2 \text{ are real numbers})$$

is a straight-line orbit of (3.5.44). Then z_1, z_2 satisfy

$$\begin{aligned}\frac{dy_2}{dy_1} &= z_1 = z_1 \frac{\sigma_1^0 z_1^2 + \sigma_2^0(1+z_2^2)}{\sigma_1^0 + \sigma_2^0(z_1^2+z_2^2)},\\ \frac{dy_3}{dy_1} &= z_2 = z_2 \frac{\sigma_1^0 z_2^2 + \sigma_2^0(1+z_1^2)}{\sigma_1^0 + \sigma_2^0(z_1^2+z_2^2)}.\end{aligned} \tag{3.5.47}$$

It is easy to see that when $\sigma_1^0 \neq \sigma_2^0$, the solutions z_1 and z_2 of (3.5.47) take only the values

$$z_1 = 0, \pm 1; \quad z_2 = 0, \pm 1.$$

In the same fashion, we can prove that the straight-line orbits of (3.5.44) given by

$$y_1 = \alpha_1 y_3, \quad y_2 = \alpha_2 y_3, \quad \text{and } y_1 = \beta_1 y_2, \quad y_3 = \beta_2 y_2$$

have to satisfy

$$\alpha_i = 0, \pm 1 \quad \text{and} \quad \beta_i = 0, \pm 1 \quad (i = 1, 2).$$

Thus, we have proved that when $\sigma_1^0 \neq \sigma_2^0$, the number of straight-line orbits of (3.5.44) is exactly 26.

When $\sigma_1^0 = \sigma_2^0$, we have $\gamma_3 = \frac{22}{9}\frac{L^2}{\pi^2}\gamma_2^2$, which implies that $\sigma_1^0 = \sigma_2^0 > 0$. In this case, it is clear that $y = 0$ is an asymptotically stable singular point of (3.5.44). Hence, the transition of (3.5.37) at $\lambda_0 = \pi^2/L^2$ is of Type I.

When $\sigma_1^0 + \sigma_2^0 > 0$ and $\sigma_1^0 \neq \sigma_2^0$, all straight-line orbits of (3.5.44) are toward $y = 0$, which implies that the regions at $y = 0$ are stable, and $y = 0$ is asymptotically stable; see (Ma and Wang 2005b). Therefore, the transition of (3.5.37) is of Type I.

When $\sigma_1^0 + \sigma_2^0 < 0$ with $\sigma_1^0 > 0$, we can see, as in the case of $m = 2$, that the regions at $y = 0$ are hyperbolic, and when $\sigma_1^0 + \sigma_2^0 < 0$ with $\sigma_1 \leq 0$, the regions at $y = 0$ are unstable. Hence, the transition is of Type II.

Step 5. We prove assertion (1). By Steps 3 and 4, if $\gamma_3 > \frac{26L^2}{27\pi^2}\gamma_2^2$, then the reduced equation (3.5.37) bifurcates on $\lambda > \lambda_0 = \pi^2/L^2$ to an attractor Σ_λ. All bifurcated equilibrium points of (3.5.9) are in one-to-one correspondence with the bifurcated singular points of (3.5.37). Therefore, we have only to consider the stationary equations:

$$\beta_1(\lambda) y_i - \frac{\pi^2}{2L^2}[\sigma_1 y_i^3 + \sigma_2 y_i \sum_{j \neq i} y_j^2] + o(|y|^3) = 0, \quad 1 \leq i \leq m, \tag{3.5.48}$$

where σ_1 and σ_2 are as in (3.5.38).

Consider the following approximative equations of (3.5.48):

$$\beta_1(\lambda) y_i - y_i(a_1 y_i^2 + a_2 \sum_{j \neq i} y_j^2) = 0 \qquad \text{for } 1 \leq i \leq m, \tag{3.5.49}$$

where $a_1 = \pi^2 \sigma_1 / 2L^2$, $a_2 = \pi^2 \sigma_2 / 2L^2$. It is clear that each regular bifurcated solution of (3.5.49) corresponds to a regular bifurcated solution of (3.5.48).

We first prove that (3.5.49) has $3^m - 1$ bifurcated solutions on $\lambda > \lambda_0$. For each k $(0 \leq k \leq m-1)$, (3.5.49) has $C_m^k \times 2^{m-k}$ solutions as follows:

$$\begin{aligned} & y_{j_1} = 0, \cdots, y_{j_k} = 0 && \text{for } 1 \leq j_l \leq m, \\ & y_{r_1}^2 = \cdots = y_{r_{m-k}}^2 = \beta_1 (a_1 + (m-k-1) a_2)^{-1} && \text{for } r_i \neq j_l. \end{aligned} \tag{3.5.50}$$

Hence, the number of all bifurcated solutions of (3.5.49) is

$$\sum_{k=0}^{m-1} C_m^k \times 2^{m-k} = (2+1)^m - 1 = 3^m - 1.$$

We need to prove that all bifurcated solutions of (3.5.49) are regular. The Jacobian matrix of (3.5.49) is given by

$$Dv = \begin{pmatrix} \beta_1 - h_1(y) & 2a_2y_1y_2 & \cdots & 2a_2y_1y_m \\ 2a_2y_2y_1 & \beta_1 - h_2(y) & \cdots & 2a_2y_2y_m \\ \vdots & \vdots & & \vdots \\ 2a_2y_my_1 & 2a_2y_my_2 & \cdots & \beta_1 - h_m(y) \end{pmatrix}, \tag{3.5.51}$$

where

$$h_i(y) = 3a_1y_i^2 + a_2\sum_{j\neq i} y_j^2.$$

For the solutions in (3.5.50), without loss of generality, we take

$$\begin{aligned} &y_0 = (y_1^0, \cdots, y_m^0),\\ &y_i^0 = 0 \text{ for } 1 \le i \le k, \qquad y_{k+1}^0 = \cdots = y_m^0 = \beta_1^{1/2}(a_1 + (m-k-1)a_2)^{-1/2}. \end{aligned}$$

Inserting them into (3.5.51), we obtain

$$Dv(y_0) = \begin{pmatrix} \beta I_k & 0 \\ 0 & A_{m-k} \end{pmatrix}, \tag{3.5.52}$$

where

$$\begin{aligned} &\beta = \beta_1\left(1 - \frac{(m-k)a_2}{a_1 + (m-k-1)a_2}\right) = \frac{(\sigma_1 - \sigma_2)\beta_1}{\sigma_1 + (m-k-1)\sigma_2},\\ &A_{m-k} = \begin{pmatrix} \beta_1 - (3a_1 + (m-k-1)a_2)(y_{k+1}^0)^2 & \cdots & 2a_2(y_{k+1}^0)^2 \\ \vdots & & \vdots \\ 2a_2(y_{k+1}^0)^2 & \cdots & \beta_1 - (3a_1 + (m-k-1)a_2)(y_{k+1}^0)^2 \end{pmatrix}. \end{aligned}$$ ■

Direct computation shows that

$$\det A_{m-k} = \frac{\pi^{2m}\beta_1^m}{(a_1 + (m-k-1)a_2)^m L^{2m}} \det \begin{pmatrix} -\sigma_1 & \sigma_2 & \cdots & \sigma_2 \\ \sigma_2 & -\sigma_1 & \cdots & \sigma_2 \\ \vdots & \vdots & & \vdots \\ \sigma_2 & \sigma_2 & \cdots & -\sigma_1 \end{pmatrix},$$

where σ_1 and σ_2 are given by (3.5.38).

Obviously, there are only finitely many $\lambda > \pi^2/L^2$ satisfying

$$\beta(\lambda) = \frac{(\sigma_1(\lambda) - \sigma_2(\lambda))\beta_1(\lambda)}{\sigma_1(\lambda) + (m-k-1)\sigma_2(\lambda)} = 0, \qquad \det A_{m-k}(\lambda) = 0.$$

Hence, for every $\lambda - \pi^2/L^2 > 0$ sufficiently small, the Jacobian matrices (3.5.51) at the singular points (3.5.50) are nondegenerate. Thus, the bifurcated solutions of (3.5.49) are regular.

Since all bifurcated singular points of (3.5.7) with (3.5.5) are nondegenerate, when Σ_λ is restricted to the x_i,x_j-plane $(1 \le i,j \le m)$, the singular points are connected by their stable and unstable manifolds. Hence all singular points in Σ_λ are connected by their stable and unstable manifolds. Therefore, Σ_λ must be homeomorphic to a sphere S^{m-1}.

Assertion (1) is proved.

Step 6. Proof of assertions (2) and (3). When $m = 2$, by Step 5, $\Sigma_\lambda = S^1$ contains eight nondegenerate singular points. By a minimal attractor theorem, namely Theorem 5.6 in Ma and Wang (2005b), four singular points must be attractors and the others are repellers, as shown in Fig. 3.15. When $m = 3$, we take the six singular points

$$\pm Y_1 = (\pm\beta_1 a_1^{-1}, 0, 0), \pm Y_2 = (0, \pm\beta_1 a_1^{-1}, 0), \pm Y_3 = (0, 0, \pm\beta_1 a_1^{-1}).$$

Then the Jacobian matrix (3.5.51) at Y_i $(1 \le i \le 3)$ is

$$Dv(\pm Y_i) = \begin{pmatrix} \rho_1 & 0 & 0 \\ 0 & \rho_2 & 0 \\ 0 & 0 & \rho_3 \end{pmatrix},$$

where $\rho_j = \beta_1\left(1 - \frac{\sigma_2}{\sigma_1}\right)$ if $j \neq i$ and $\rho_i = -2\beta_1$. Obviously, if $\sigma_2 < \sigma_1$, $0 < \rho_j$ $(j \neq i)$ and $\rho_i < 0$. In this case, $\pm Y_k$ $(1 \le k \le 3)$ are repellers in $\Sigma_\lambda = S^2$, which implies that Σ_λ contains eight attractors $\pm z_k$ $(1 \le k \le 4)$, as shown in Fig. 3.16a. If $\sigma_2 > \sigma_1, \rho_j < 0$ $(1 \le j \le 3)$, the six singular points $\pm Y_k$ $(1 \le k \le 3)$ are attractors, which implies that Σ_λ contains only six minimal attractors, as shown in Fig. 3.16b. Thus assertion (2) is proved.

The claim for the saddle-node bifurcation in assertion (3) can be proved by the same method as in the proof of Theorem 3.5.2, and the claim for the singular point bifurcation can be proved by the same fashion as used in Step 5.

The proof of the theorem is complete. □

3.5.4 Spatial Geometry, Transitions, and Pattern Formation

We made a few remarks on the effects of spatial geometry on phase transition and pattern formation.

First, for the domain $\Omega = [0,L]^m \times D \subset \mathbb{R}^n$ $(1 \le m < n)$, where $n \ge 2$ is arbitrary and $D \subset \mathbb{R}^{n-m}$ a bounded open set, Theorems 3.5.4 and 3.5.5 are also valid, provided $\pi^2/L^2 < \lambda_1$, where λ_1 is the first eigenvalue of the equation

$$-\Delta e = \lambda e \quad \text{for } x \in D, \quad \frac{\partial e}{\partial n}|_{\partial D} = 0, \quad \int_D e dx = 0.$$

Second, in Theorem 3.5.5, the minimal attractors in the bifurcated attractor Σ_λ can be expressed as

$$u_\lambda = (\lambda - \pi^2/L^2)^{1/2} e + o(|\lambda - \pi^2/L^2|^{1/2}), \tag{3.5.53}$$

where e is a first eigenfunction of (3.5.10). The expression (3.5.53) can be derived from the reduced equations (3.5.37).

The exponent $\beta = 1/2$ in (3.5.53) is called the critical exponent in physics. It is an important index in phase transition theory in statistical physics. It is interesting to point out that the critical exponent $\beta = 1$ in (3.5.19) is different from the $\beta = 1/2$ appearing in (3.5.21) and (3.5.53). The first occurs when the container $\Omega \subset \mathbb{R}^3$ is a nonrectangular region, and the second one occurs when Ω is a rectangle or a cube.

Third, when the sample or container Ω is a loop, a torus, or a large domain, then periodic boundary conditions are necessary. Let $\Omega = S^1 \times (r_1, r_2) \subset \mathbb{R}^2$ be a loop domain, $0 < r_1 < r_2$. Then the boundary condition is given by

$$\begin{aligned} &u(\theta + 2k\pi, r) = u(\theta, r) \quad &&\text{for } 0 \leq \theta \leq 2\pi,\ r_1 < r < r_2,\ k \in \mathbb{Z}, \\ &\frac{\partial u}{\partial r} = 0, \quad \frac{\partial^3 u}{\partial r^3} = 0 &&\text{at } r = r_1, r_2. \end{aligned} \tag{3.5.54}$$

Assume that the gap $r_2 - r_1$ is small in comparison with the mean radius $r_0 = (r_1 + r_2)/2$, or equivalently, by taking the scaling $r_2 - r_1 = 1$:

$$r_0 = (r_1 + r_2)/2 \gg 1, \qquad r_2 - r_1 = 1. \tag{3.5.55}$$

Then the Laplacian operator can be approximately expressed as

$$\Delta = \frac{\partial^2}{\partial r^2} + \frac{1}{r_0^2} \frac{\partial^2}{\partial \theta^2}. \tag{3.5.56}$$

With the boundary condition (3.5.54) and the operator (3.5.56), the eigenvalues and eigenfunctions of the linear operator $L_\lambda = -A + B_\lambda$ defined by (3.5.8) are given by

$$\begin{aligned} &\beta_K(\lambda) = K^2(\lambda - K^2), && K^2 = \frac{k_1^2}{r_0^2} + k_2^2, \\ &e_K^1 = \cos k_1\theta \cos k_2\pi(r - r_1), && e_K^2 = \sin k_1\theta \cos k_2\pi(r - r_1). \end{aligned}$$

By (3.5.55), the first eigenvalue of L_λ is

$$\beta_1(\lambda) = \frac{1}{r_0^2}(\lambda - \frac{1}{r_0^2}),$$

which has multiplicity 2, and the first eigenfunctions are $e_1^1 = \cos\theta$ and $e_1^2 = \sin\theta$.

Theorem 3.5.6 *Let $\Omega = S^1 \times (r_1, r_2)$ satisfy (3.5.55). Then the following assertions hold:*

1. *If*
$$\gamma_3 > \frac{2r_0^2}{9}\gamma_2^2,$$
then the phase transition of (3.5.7) with (3.5.54) at $\lambda_0 = 1/r_0^2$ is of Type I. Furthermore, the problem (3.5.7) with (3.5.54) bifurcates on $\lambda > \lambda_0 = r_0^{-2}$ to a cycle attractor $\Sigma_\lambda = S^1$ consisting of singular points, as shown in Fig. 3.18, and the singular points in Σ_λ can be expressed as
$$u_\lambda = \sigma^{-1/2}\left(\lambda - \frac{1}{r_0^2}\right)^{1/2}\cos(\theta+\theta_0) + o\left(|\lambda - \frac{1}{r_0^2}|^{1/2}\right), \quad \sigma = \frac{3}{4}\gamma_3 - \frac{1}{6}r_0^2\gamma_2^2,$$
where θ_0 is the angle of u_λ in Σ_λ.
2. *If*
$$\gamma_3 < \frac{2r_0^2}{9}\gamma_2^2,$$
then the transition is of Type II. Moreover, the problem bifurcates on $\lambda < \lambda_0$ to a cycle invariant set $\Gamma_\lambda = S^1$ consisting of singular points, and there is a singularity separation at $\lambda^* < \lambda_0$, where Γ_λ and an S^1 attractor $\Sigma_\lambda = S^1$ are generated such that the system undergoes a transition at $\lambda = \lambda_0$ from $u = 0$ to $u_\lambda \in \Sigma_\lambda$, as shown in Fig. 3.19.

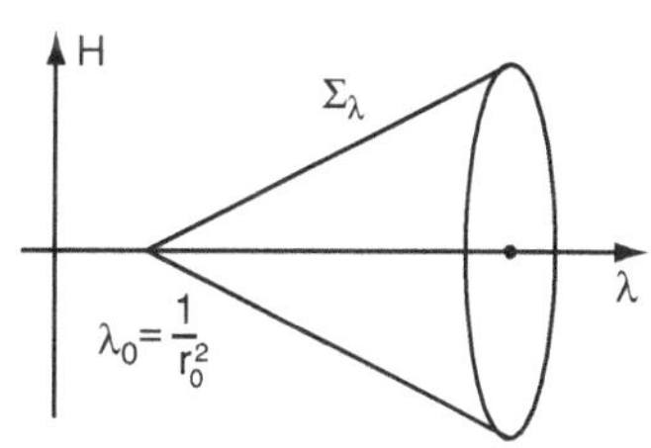

Fig. 3.18 All points in Σ_λ are singular points

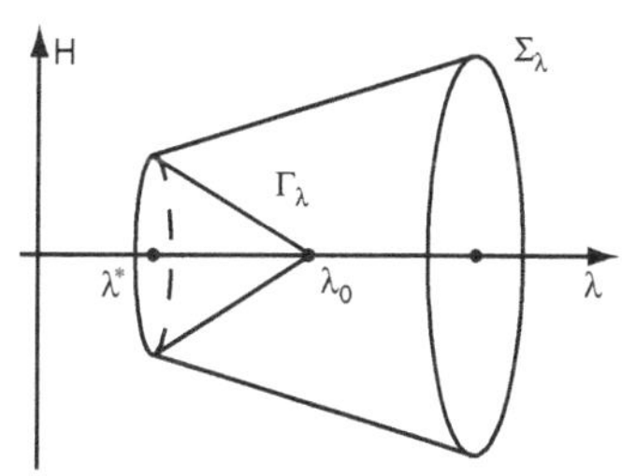

Fig. 3.19 The point λ^* is a singularity separation point from which two invariant sets Σ_λ and Γ_λ are separated with Σ_λ an attractor, and $\Gamma_\lambda \to 0$ as $\lambda \to \lambda_0$

Fourth, consider the problem that the equation (3.5.7) is defined in the whole space $\Omega = \mathbb{R}^n$ $(n \geq 2)$, with the periodic boundary condition

$$u(x+2K\pi) = u(x) \qquad \forall K = (k_1, \cdots, k_n) \in \mathbb{Z}^n. \tag{3.5.57}$$

In this case, the eigenvalues and eigenfunctions of L_λ are given by

$$\begin{aligned} &\beta_K(\lambda) = |K|^2(\lambda - |K|^2), && K = (k_1, \cdots, k_n), \quad |K|^2 = k_1^2 + \cdots + k_n^2, \\ &e_K^1 = \cos(k_1x_1 + \cdots + k_nx_n), && e_K^2 = \sin(k_1x_1 + \cdots + k_nx_n). \end{aligned}$$

It is clear that the first eigenvalue $\beta_1(\lambda) = \lambda - 1$ of L_λ has multiplicity $2n$, and the first eigenfunctions are

$$e_j^1 = \cos x_j, \quad e_j^2 = \sin x_j \qquad \forall 1 \leq j \leq n.$$

Theorem 3.5.7

1. *If*

$$\gamma_3 > \frac{14}{27}\gamma_2^2,$$

then the phase transition of (3.5.7) with (3.5.57) at $\lambda_0 = 1$ is of Type I. Moreover,

(a) *The problem bifurcates from $(u, \lambda) = (0, 1)$ to an attractor Σ_λ homeomorphic to a $(2n-1)$-dimensional sphere S^{2n-1}, and*
(b) *For each k $(0 \leq k \leq n-1)$, the attractor Σ_λ contains C_n^k-dimensional tori $\mathbb{T}^{n-k}$ consisting of singular points.*

2. *If*

$$\gamma_3 < \frac{14}{27}\gamma_2^2,$$

then the transition is of Type II.

3.5.5 Phase Diagrams and Physical Conclusions

Criteria of Separation Order. For simplicity, we consider the case that the spatial domain $\Omega = \prod_{k=1}^3 (0, L_k) \subset \mathbb{R}^3$ is a rectangular domain. Here we use the Neumann boundary condition, which is physically relevant.

Each three-dimensional rectangular domain belongs to one of the following two cases:

$$\begin{aligned} &\text{Case I:} \quad L = L_1 > L_j \qquad \forall j = 2, 3, \\ &\text{Case II:} \quad L = L_1 = L_2 > L_3 \text{ or } L_1 = L_2 = L_3. \end{aligned}$$

We define a nondimensional parameter

$$K = \begin{cases} \dfrac{2L^2}{9\pi^2}\gamma_2^2 - \gamma_3 & \text{for Case I,} \\ \dfrac{26L^2}{27\pi^2}\gamma_2^2 - \gamma_3 & \text{for Case II,} \end{cases} \tag{3.5.58}$$

which is equivalent to the following dimensional parameter:

$$K_d = \begin{cases} \dfrac{2L_d^2}{9\pi^2}\dfrac{b_2^2}{k} - b_3 & \text{for Case I,} \\ \dfrac{26L_d^2}{27\pi^2}\dfrac{b_2^2}{k} - b_3 & \text{for Case II,} \end{cases} \tag{3.5.59}$$

where $L_d = L \cdot l$ is the dimensional length scale.

Theorems 3.5.4 and 3.5.5 give a precise characterization of the order of transitions, which is determined by the sign of this parameter K or K_d. In particular, the following physical conclusion is the direct translation of Theorems 3.5.4 and 3.5.5. Note that the control parameter λ for the transition/bifurcation is naturally linked to the temperature T as

$$\lambda = -\frac{l^2 D}{2k}\left(\frac{RT}{u_0(1-u_0)} - 2a\right).$$

Hence the two critical bifurcation parameters λ^* and λ_c in Theorems 3.5.4 and 3.5.5 lead to two critical temperatures T_c and T^*.

Physical Conclusion 3.5.8 *The order of phase separation is completely determined by the sign of the nondimensional parameter K_d as follows:*

(1) If $K_d < 0$, the separation is of second order, and the dynamic behavior of the Cahn–Hilliard system is as shown in Fig. 3.20.
(2) If $K_d > 0$, the separation is a first-order transition with latent heat. In particular, there are two critical temperatures $T^* > T_c$ such that if the temperature T is greater than T^*, the system is in a homogeneous state, while when $T^* > T > T_c$, the system is in a metastable state accompanied by hysteresis corresponding to a saddle-node bifurcation, and when $T < T_c$, the system is in a phase separation state. In addition, the critical temperatures are functions of u_0 and L: $T^* = T^*(u_0, L), T_c = T_c(u_0, L)$. See Fig. 3.21.

This is in agreement with *part of* the classical phase diagram from the classical thermodynamic theory given in Fig. 3.22; see, among others, (Reichl 1998; Novick-Cohen and Segel 1984; Langer 1971). However, as we shall see below, our result shows that near $u_0 = 1/2$, there is no metastable region; see Fig. 3.23. In other words, the phase diagram given in Fig. 3.23 is different from the one presented in classical studies as shown in Fig. 3.22.**Critical Parameter Equations.** To examine

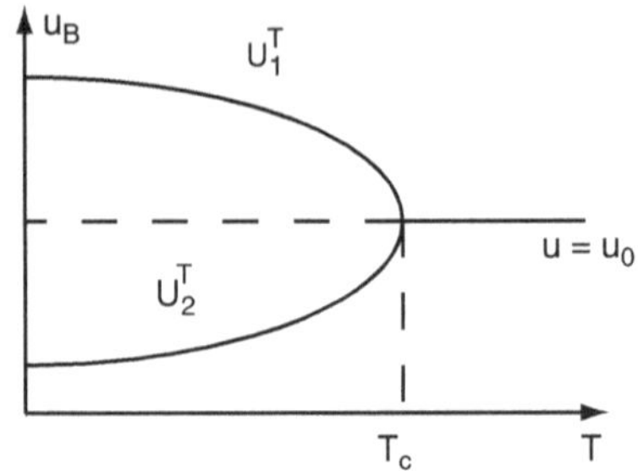

Fig. 3.20 Transition diagram for $K_d < 0$: the state $u_0 = \overline{u}_B$ is stable if $T_c < T$. The state u_0 is unstable, U_1^T and U_2^T are stable if $T < T_c$

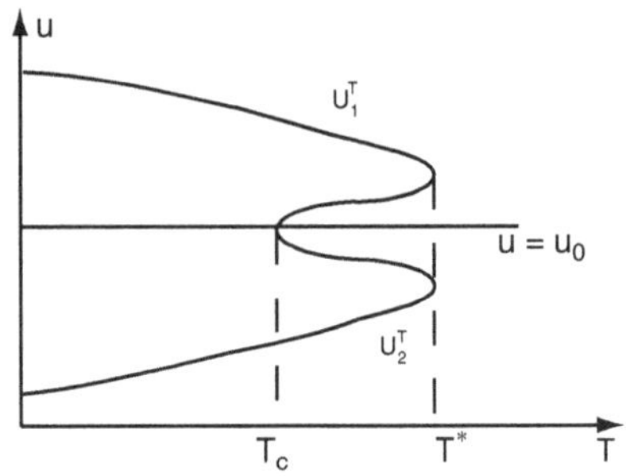

Fig. 3.21 Transition diagram for $K_d > 0$: for fixed u_0 and L, the transition for the case $K_d > 0$ is a first-order separation with latent heat and with hysteresis: U_1^T and U_2^T represent separation states, and u_0 is the homogeneous state. In this case, for $T_c < T < T^*$, all states u_0, U_1^T, U_2^T are metastable states. For $T < T_c$, u_0 is unstable, and U_1^T and U_2^T are stable states

the order of separations, we start with two important formulas for the critical parameters. By (3.5.4), solving $K_d = 0$ gives a precise formula for the critical (dimensional) length scale L_d in terms of the critical temperature T_c and the molar fraction u_0:

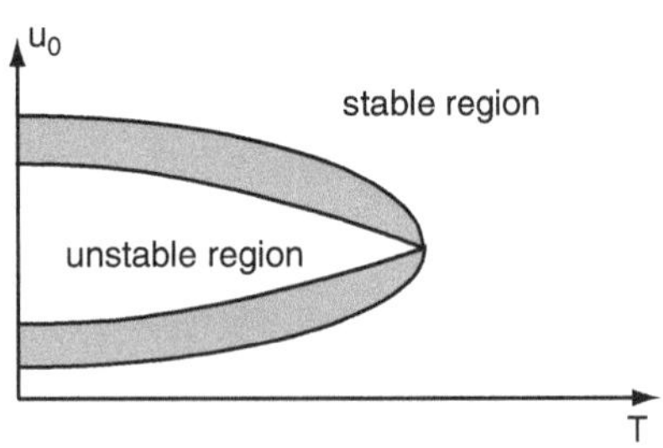

Fig. 3.22 Typical phase diagram from classical thermodynamic theory with the shadowed region metastable

$$L_d = \begin{cases} \dfrac{3\pi\sqrt{k}}{\sqrt{2}} \dfrac{\sqrt{b_3}}{|b_2|} = \dfrac{3\sqrt{3k}\pi\sqrt{u_0(1-u_0)(1-3u_0+3u_0^3)}}{2\sqrt{2DRT_c}\,|u_0-\frac{1}{2}|} & \text{for Case I,} \\[2ex] \dfrac{3\pi\sqrt{3k}}{\sqrt{26}} \dfrac{\sqrt{b_3}}{|b_2|} = \dfrac{3\sqrt{3k}\pi\sqrt{u_0(1-u_0)(1-3u_0+3u_0^3)}}{2\sqrt{26DRT_c}\,|u_0-\frac{1}{2}|} & \text{for Case II.} \end{cases} \tag{3.5.60}$$

It is then easy to derive from (3.5.60) the following asymptotic formula for L_d:

$$L_d = \begin{cases} \dfrac{3\sqrt{3k}\pi}{8\sqrt{2DRT_c}|u_0-\frac{1}{2}|} + O(1) & \text{for Case I,} \\ \dfrac{9\sqrt{k}\pi}{8\sqrt{26DRT_c}|u_0-\frac{1}{2}|} + O(1) & \text{for Case II,} \end{cases} \tag{3.5.61}$$

where T_c is the critical temperature as given in Physical Conclusion 3.5.8.

In fact, the first eigenvalue is given by

$$\beta_1 = -\frac{\pi^2}{L^2}\left(\frac{\pi^2}{L^2} - \lambda\right) = -\frac{\pi^2}{L^2}\left(\frac{\pi^2}{L^2} + \frac{l^2 b_1}{k}\right). \tag{3.5.62}$$

By (3.5.4), the critical parameter curve equation $\beta_1 = 0$ is given by

$$T_c = \frac{u_0(1-u_0)}{RD}\left(a - \frac{k\pi^2}{2l^2L^2}\right) = \frac{u_0(1-u_0)}{RD}\left(a - \frac{k\pi^2}{2L_d^2}\right). \tag{3.5.63}$$

Phase Diagrams. We are ready now to examine the order of separation in terms of the length scale L_d and molar fraction u_0. First, using (3.5.60) or (3.5.61), we derive the Lu_0-phase diagram given by Fig. 3.24. Then using Fig. 3.24 and (3.5.63), we derive the Tu_0-phase diagram given in Fig. 3.23.

Consequently, we have shown the following physical conclusion:

Physical Conclusion 3.5.9

(1) For a fixed length scale $L = L'$, there are numbers $x_1 < \frac{1}{2} < x_2$ such that the transition is of second order if the molar fraction satisfies $x_1 < u_0 < x_2$, and the transition is of first order if $u_0 > x_2$ or $u_0 < x_1$. See Fig. 3.24.

(2) The phase diagram Fig. 3.23 is for this fixed length scale L'. The points x_1 and x_2 are the two molar concentrations where there is no metastable region and no hysteresis phenomenon for $x_1 < u_0 < x_2$. In other words,

$$T^*(u_0) = T_c(u_0) \qquad \text{for } x_1 < u_0 < x_2.$$

We now derive the TL phase diagram. Using (3.5.63), we derive the TL phase diagram given by Fig. 3.25 and the following physical conclusion:

Physical Conclusion 3.5.10 *For a given molar fraction $0 < u_0 < 1$, the critical parameter curve (3.5.63) divides the TL-plane into two regions: the homogeneous region I, and the separation region II such that the following assertions hold:*

(1) If (L,T) is in region I, then the system is in homogeneous phase, and if (L,T) is in region II, the system is in the separation phase.

(2) The phase separation takes place as the control parameter (L,T) crosses the critical curve from region I into region II.

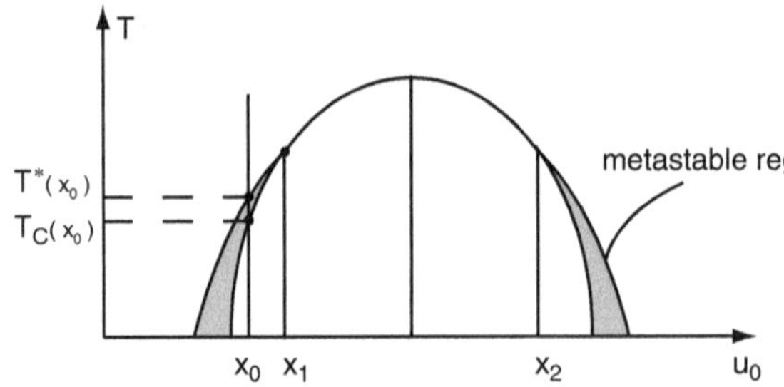

Fig. 3.23 Tu_0-phase diagram for a fixed length scale L', derived using Fig. 3.24 and (3.5.63): the *shadowed region* stands for the metastable region, and $T^*(x_0)$ and $T_c(x_0)$ are critical temperatures as shown in Fig. 3.21

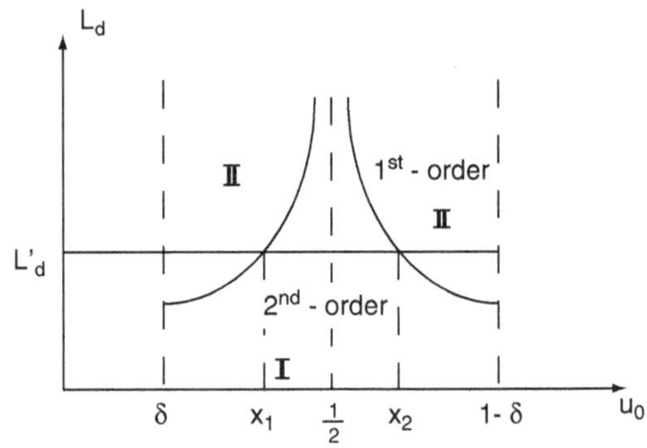

Fig. 3.24 Lu_0-phase diagram derived using (3.5.60) or (3.5.61): region II is the first-order transition region with latent heat, and region I is the second-order transition region

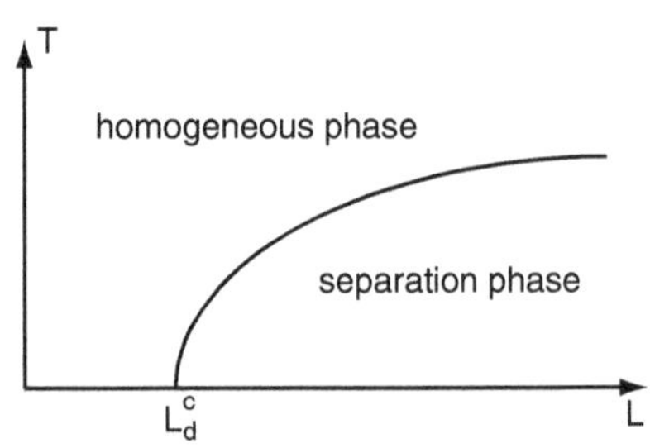

Fig. 3.25 TL phase diagram

In particular, there is a critical (dimensional) length $L_d^c = \sqrt{k\pi^2/2a}$ *such that if* $L_d < L_d^c$*, there is no phase separation for any temperature, and if* $L_d > L_d^c$*, phase separation occurs at the critical temperature* $T = T_c$ *given by (3.5.63).*

Pattern Formation, Symmetry, and Periodic Structure. Physical experiments have shown that in pattern formation via phase separation, a periodic or semiperiodic structure appears. From Theorems 3.5.6 and 3.5.7, we see that for the loop domains and bulk domains, which can be considered $\mathbb{R}^n$ or $\mathbb{R}^m \times D$ $(D \subset \mathbb{R}^{n-m})$, the steady-state solutions of the Cahn–Hilliard equation are periodic, and for rectangular domains they are semiperiodic, and the periodicity is associated with mirror-image symmetry.

For $\Omega = (0,L)^m \times D$, Theorem 3.5.4 is also valid. In fact, the space

$$\widetilde{H} = \left\{ u \in L^2(\Omega) | \ u = \sum_{|K|=1}^{\infty} y_k \cos\frac{k_1\pi}{L}x_1 \cdots \cos\frac{k_m\pi}{L}x_m \right\} \subset H$$

is invariant for the Cahn–Hilliard equation (3.5.7) and (3.5.5). All separated equilibrium states in Theorem 3.5.4 are in $\widetilde{H}$. From the physical viewpoint, all equilibrium states $u(x)$ and their mirror-image states $u(L-x',x'')$ possess the same patterns. Mathematically, under the mirror image transformation

$$x \to (L-x',x''), \quad x' = (x_1,\cdots,x_m), \quad x'' = (x_{m+1},\cdots,x_n),$$

the Cahn–Hilliard equation (3.5.7) with (3.5.5) is invariant. Hence, the steady-state solutions will appear in pairs. In particular, for a Type I phase transition, there is a remarkable mirror-image symmetry. We address this problem as follows.

Let $m=1$ in $\Omega = (0,L)^m \times D$. By (3.5.21), there are two bifurcated stable equilibrium states, and their projections on the first eigenspace are

$$u_{1,2} = \pm y \cos\frac{\pi x_1}{L}, \qquad y = \sqrt{2(\lambda - \pi^2/L^2)} \Big/ \left(\frac{3}{2}\gamma_3 - \frac{L^2}{3\pi^2}\gamma_2^2 \right).$$

It is clear that $u_2(x_1) = u_1(L-x_1)$.

Let $m=2$. By Theorem 3.5.5, the bifurcated attractor Σ_λ contains eight equilibrium states, whose projections are given by

$$u_1^\pm = \pm y_0 e_1, \quad u_2^\pm = \pm y_1(e_1+e_2), \quad u_3^\pm = \pm y_0 e_2, \quad u_4^\pm = \pm y_1(-e_1+e_2),$$

where $e_1 = \left(\cos\frac{\pi x_1}{L}, 0\right)$ and $e_2 = \left(0, \cos\frac{\pi x_2}{L}\right)$ form an orthogonal basis in $\mathbb{R}^2$, and

$$y_0 = \frac{\sqrt{2(\lambda - \pi^2/L^2)}}{\sqrt{\sigma_1}}, \quad y_1 = \frac{\sqrt{2(\lambda - \pi^2/L^2)}}{\sqrt{\sigma_1+\sigma_2}} \quad \text{with } \sigma_1, \sigma_2 \text{ as in (3.5.38).}$$

These eight equilibrium states constitute an octagon in $\mathbb{R}^2$, as shown in Fig. 3.26, and they are divided into two classes $\mathscr{A}_1 = \{u_1^\pm, u_3^\pm\}$ and $\mathscr{A}_3 = \{u_2^\pm, u_4^\pm\}$ by the $\frac{\pi}{2}$-rotation group $G(\frac{\pi}{2})$. Namely, with the action of $G(\frac{\pi}{2})$, $\mathscr{A}_i$ $(i=1,2)$ are invariant:

$$Bu \in \mathscr{A}_i, \quad \forall u \in \mathscr{A}_i \text{ and } B \in G(\frac{\pi}{2}),$$

where $G(\frac{\pi}{2})$ consists of the orthogonal matrices

$$B_1^\pm = \pm\begin{pmatrix} 1 & 0 \\ 0 & 1 \end{pmatrix}, \quad B_2^\pm = \pm\begin{pmatrix} 1 & 0 \\ 0 & -1 \end{pmatrix}, \quad B_3^\pm = \pm\begin{pmatrix} 0 & -1 \\ 1 & 0 \end{pmatrix}, \quad B_4^\pm = \pm\begin{pmatrix} 0 & 1 \\ 1 & 0 \end{pmatrix}.$$

The stability of the equilibrium states u_k is associated with both classes $\mathscr{A}_1$ and $\mathscr{A}_2$, i.e., either the elements in $\mathscr{A}_1$ are stable, or those in $\mathscr{A}_2$ are stable; see Fig. 3.15. By (3.5.52), we can derive the criterion as follows:

$$u_{2k+1}^{\pm} \in \mathscr{A}_1 \text{ are stable} \quad \Leftrightarrow \quad \frac{22}{9}\frac{L^2\gamma_2^2}{\pi^2} > \gamma_3 > \frac{26}{27}\frac{L^2\gamma_2^2}{\pi^2} \quad \text{for } k=0,1,$$
$$u_{2k}^{\pm} \in \mathscr{A}_2 \text{ are stable} \quad \Leftrightarrow \quad \gamma_3 > \frac{22}{9}\frac{L^2\gamma_2^2}{\pi^2} \quad \text{for } k=1,2.$$

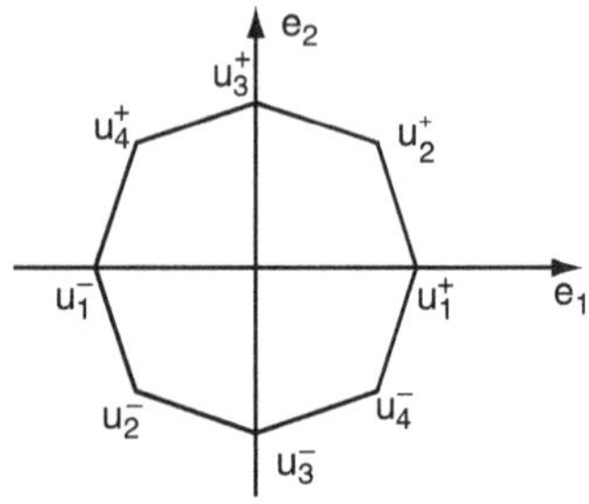

Fig. 3.26 The eight equilibrium states for the case $m=2$ given by Theorem 3.5.5

In Fig. 3.26, we see that elements in $\mathscr{A}_1$ and $\mathscr{A}_2$ have a $\frac{\pi}{4}$ difference in their phase angle. However, in their pattern structure, $u_{2K+1}^{\pm} \in \mathscr{A}_1$ and $u_{2k}^{\pm} \in \mathscr{A}_2$ also have a $\pi/4$ difference at the angle between the lines of $u_{2K+1}^{\pm}=0$ and $u_{2K}^{\pm}=0$. In fact, the lines

$$u_1^+ = -u_1^- = y_0 \cos\frac{\pi x_1}{L} = 0 \qquad \text{and} \qquad u_3^+ = -u_3^- = y_0 \cos\frac{\pi x_2}{L} = 0$$

are given by $x_1 = L/2$ and $x_2 = L/2$ respectively, as shown in Fig. 3.27a, and the lines

$$u_2^+ = -u_2^- = y_1\left(\cos\frac{\pi x_1}{L} + \cos\frac{\pi x_2}{L}\right) = 0, \quad \text{and}$$
$$u_4^+ = -u_4^- = y_1\left(-\cos\frac{\pi x_1}{L} + \cos\frac{\pi x_2}{L}\right) = 0$$

are given by $x_2 = L - x_1$ and $x_2 = x_1$ respectively, as shown in Fig. 3.27b.

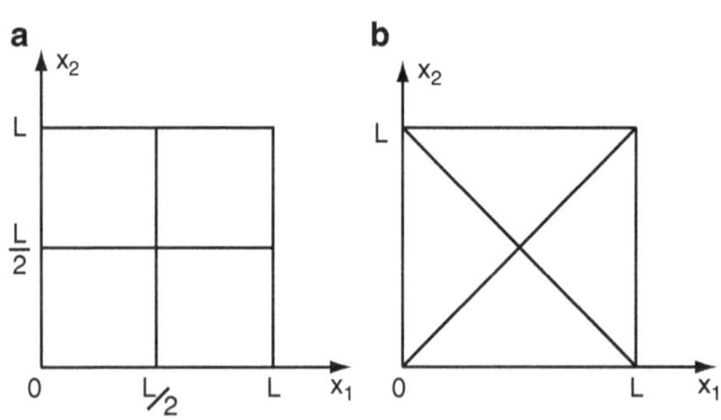

Fig. 3.27 (**a**) Lines for $u_1^{\pm}=0$ and $u_3^{\pm}=0$, and (**b**) lines for $u_2^{\pm}=0$ and $u_4^{\pm}=0$

Let $m = 3$. Then the bifurcated attractor Σ_λ contains 26 equilibrium states, which can be divided into three classes by the three-dimensional $\left(\frac{\pi}{2}, \frac{\pi}{2}, \frac{\pi}{2}\right)$-rotation group $G\left(\frac{\pi}{2}, \frac{\pi}{2}, \frac{\pi}{2}\right)$ as follows:

$$\begin{aligned}
\mathscr{A}_1 &= \left\{u_1^\pm = \pm y_0 e_1, u_2^\pm = \pm y_0 e_2, u_3^\pm = \pm y_0 e_3\right\}, \\
\mathscr{A}_2 &= \left\{u_4^\pm = \pm y_1(e_1+e_2), u_5^\pm = \pm y_1(e_1+e_3), u_6^\pm = \pm y_1(e_2+e_3),\right. \\
&\quad \left. u_7^\pm = \pm y_1(-e_1+e_2), u_8^\pm = \pm y_1(-e_1+e_3), u_9^\pm = \pm y_1(-e_2+e_3)\right\}, \\
\mathscr{A}_3 &= \{u_{10}^\pm = \pm y_1(e_1+e_2+e_3), u_{11}^\pm = \pm y_1(-e_1+e_2+3_3), \\
&\quad u_{12}^\pm = \pm y_1(e_1-e_2+e_3), u_{13}^\pm = \pm y_1(e_1+e_2-e_3)\}.
\end{aligned}$$

Only these elements in $\mathscr{A}_1$ or in $\mathscr{A}_3$ are stable, and they are determined by the following criterion:

$$\begin{aligned}
&\text{elements in } \mathscr{A}_1 \text{ are stable} \Leftrightarrow \frac{22}{9}\frac{L^2\gamma_2^2}{\pi^2} > \gamma_3 > \frac{26}{27}\frac{L^2\gamma_2^2}{\pi^2}, \\
&\text{elements in } \mathscr{A}_3 \text{ are stable} \Leftrightarrow \gamma_3 > \frac{22}{9}\frac{L^2\gamma_2^2}{\pi^2}.
\end{aligned}$$

3.6 Superconductivity

Superconductivity was first discovered in 1911 by H. Kamerlingh Onnes, who found that mercury has zero electric resistance when the temperature decreases below some critical value T_c. Since then, it has been found that a large number of metals and alloys possess the superconducting property. In the superconducting state, once a current is set up in a metal ring, it is expected that no changes in this current will occur over times greater than 10^{10} years; see (Tinkham 1996). In 1933, another important superconducting property, called diamagnetism or the Meissner effect, was discovered by W. Meissner and R. Ochsenfeld. They found not only that a magnetic field is excluded from entering a superconductor, but also that a field in an originally normal sample is expelled as it is cooled below T_c.

One central problem in the theory of superconductivity is to study the phase transition between a normal state, characterized by a complex order parameter ψ that vanishes identically, and a superconducting state, characterized by an order parameter that is not identically zero. The main objective of this section is to address this problem by studying the stability and dynamic transitions of the time-dependent Ginzburg–Landau (TDGL) model of superconductivity.

As we know, the TDGL model for superconductivity was first proposed in (Schmid 1966) and subsequently validated by Gor'kov and Eliashberg (Gor'kov 1968) in the context of the microscopic Bardeen–Cooper–Schrieffer (BCS) theory of superconductivity. Over the years, physical evidence of the importance of the time-dependent models has accumulated over the last few decades.

Analysis of the TDGL leads to some interesting theoretical predictions on the order of transitions and phase diagrams of superconductivity. The material presented in this section is based on Ma and Wang (2005a), and the physical conclusions are presented here for the first time.

3.6.1 Ginzburg–Landau Model

Ginzburg–Landau Free Energy. With the unified model presented in Sect. 3.1, we can derive the model for superconductivity in a much simpler and more transparent fashion.

We start with the Ginzburg–Landau free energy G, corresponding to the notation H used before for the free energy, given by

$$G=\int_\Omega \left[\frac{1}{2m_s}|(-i\hbar\nabla-\frac{e_s}{c}A)\psi|^2+\frac{a}{2}|\psi|^2+\frac{b}{4}|\psi|^4+\frac{H^2}{8\pi}-\frac{HH_a}{4\pi}\right], \tag{3.6.1}$$

where $\Omega\subset\mathbb{R}^n$ $(n=2,3)$ is a bounded domain, the complex-valued function $\psi:\Omega\to\mathbb{C}$ stands for the order parameter, the vector-valued function $H:\Omega\to\mathbb{R}^n$ stands for the magnetic field, H_a is the applied field, $\hbar$ is the Planck constant, e_s and m_s are the charge and mass of a Cooper pair, c is the speed of light, A is the magnetic potential satisfying $H=\text{curl}\,A$, and the parameters $a=a(T), b=b(T)$ are coefficients; see (de Gennes 1966):

$$b=b(T)>0,\quad a=a(T)\begin{cases} >0 & \text{for } T>T_c,\\ <0 & \text{for } T<T_c.\end{cases} \tag{3.6.2}$$

Here T_c is the critical temperature where the incipient superconductivity property may be observed. In the BCS theory, for instance, it is given by (see de Gennes 1966)

$$a(T)=N(0)\frac{T-T_c}{T_c},\qquad b(T)=0.098\frac{N(0)}{(k_bT_c)^2},$$

where k_b is the Boltzmann constant, $N(0)$ is the state density at $T=0$ kelvin.

The order parameter ψ describes the local density n_s of superconducting electrons: $|\psi|^2=n_s$. In addition, ψ is proportional to the energy gap parameter Δ near T_c, which appears in the BCS theory.

Using the Ginzburg–Landau energy, one can derive naturally the steady-state Euler–Lagrange equations, which have been studied by many authors, both physicists and mathematicians, leading to many interesting physical predictions and a rich mathematical theory.

Derivation of the TDGL Model. For the time-dependent problem, because of gauge invariance, in addition to the order parameter and the vector potential, a third variable is needed to complete the description of the physical state of the system in a manner consistent with the gauge invariance.

By (3.6.1), (3.1.15) can be written as

$$k_1\frac{\partial\psi}{\partial t}+\phi_1(\psi,A)=-\frac{1}{2m_s}(i\hbar\nabla+\frac{e_s}{c}A)^2\psi-a\psi-b|\psi|^2\psi, \tag{3.6.3}$$

$$\begin{aligned}k_2\frac{\partial A}{\partial t}+\phi_2(\psi,A)=&-\frac{1}{4\pi}\text{curl}^2A+\frac{1}{4\pi}\text{curl}\,H_a\\&-\frac{e_s^2}{m_sc^2}|\psi|^2A-\frac{e_s\hbar}{2m_sc}i(\psi^*\nabla\psi-\psi\nabla\psi^*),\end{aligned} \tag{3.6.4}$$

where $k_1,k_2>0$ are constants, and $\phi_i(\psi,A)$ are two operators to be determined by physical laws and (3.1.16) in several steps as follows:

First, the Maxwell equations give that

$$\frac{\sigma}{c^2}\frac{\partial A}{\partial t}+\sigma\nabla\phi+\frac{1}{4\pi}\text{curl}^2A-\frac{1}{4\pi}\text{curl}H_a=J_s, \tag{3.6.5}$$

where σ is the conductivity, ϕ is the electric potential, and J_s is the supercurrent density. On the other hand, by quantum mechanics, the supercurrent density J_s is given by

$$J_s=-\frac{e_s^2}{m_sc^2}|\psi|^2A-\frac{e_s\hbar}{2m_sc}i(\psi^*\nabla\psi-\psi\nabla\psi^*). \tag{3.6.6}$$

In comparison with (3.6.4)–(3.6.6), we have

$$\phi_2(\psi,A)=\sigma\nabla\phi. \tag{3.6.7}$$

Second, by direct computation, it is easy to show that by (3.6.3), (3.6.4), and (3.1.16), ϕ_1 is uniquely determined from (3.6.7) by

$$\phi_1(\psi,A)=i\alpha\phi\psi,$$

where α is a coefficient.

Third, according to the physical dimensional balance, also due to P.L. Gork'ov, the coefficients k_1 and α are given by $k_1=h^2/2m_sD$ and $\alpha=he_s/2m_sD$, where D is the diffusion coefficient. Thus, we deduce the following TDGL equations:

$$\begin{aligned}&\frac{\hbar^2}{2m_sD}\left(\frac{\partial}{\partial t}+\frac{ie_s}{\hbar}\phi\right)\psi=-\frac{1}{2m_s}\left(i\hbar\nabla+\frac{e_s}{c}A\right)^2\psi-a\psi-b|\psi|^2\psi,\\&\frac{\sigma}{c^2}\frac{\partial A}{\partial t}+\sigma\nabla\phi=-\frac{1}{4\pi}\text{curl}^2A+\frac{1}{4\pi}\text{curl}H_a-\frac{e_s^2}{m_sc^2}|\psi|^2A\\&\qquad-\frac{e_s\hbar}{2m_sc}i(\psi^*\nabla\psi-\psi\nabla\psi^*).\end{aligned} \tag{3.6.8}$$

Boundary Conditions and Gauge Transformation. Let Ω_d be a spatial domain. A physically sound boundary condition for the order parameter is give by

$$C_1(ih\nabla + \frac{e_s}{c}A)\psi \cdot n = -C_2 ih\psi \quad \text{on} \quad \partial\Omega_d, \tag{3.6.9}$$

which means that no current passes through the boundary, where n is the unit outward normal vector at $\partial\Omega_d$, and $C_1, C_2 \geq 0$ are constants depending on the material with which the contact is made. Physically, they satisfy (see (Tinkham 1996; de Gennes 1966))

$$\begin{aligned} &C_2 = 0, C_1 \neq 0 && \text{for an insulator on } \partial\Omega_d,\\ &C_1 = 0, C_2 \neq 0 && \text{for an magnetic material,}\\ &0 < C_2/C_1 < \infty && \text{for a normal metal.} \end{aligned} \tag{3.6.10}$$

We note that equations (3.6.8) with (3.6.9) are invariant under the following gauge transformation:

$$(\psi, A, \phi) \to \left(\psi e^{i\theta}, A + \frac{hc}{e_s}\nabla\theta, \phi - \frac{h}{e_s}\theta_t \right),$$

where θ is an arbitrary function. If we take θ such that

$$\begin{aligned} &\frac{hc}{e_s}\Delta\theta = \text{div}A && \text{in } \Omega_d,\\ &\frac{hc}{e_s}\frac{\partial\theta}{\partial n} = A \cdot n && \text{on } \partial\Omega_d, \end{aligned}$$

then we obtain an additional equation and a boundary condition as follows (de Gennes 1966; Tang and Wang 1995): □

$$\text{div}A = 0 \quad \text{in } \Omega_d, \tag{3.6.11}$$

$$A_n = A \cdot n = 0 \quad \text{on } \partial\Omega_d. \tag{3.6.12}$$

Another boundary condition often imposed on A is

$$\text{curl}A \times n = H_a \times n \quad \text{on } \partial\Omega_d. \tag{3.6.13}$$

Nondimensional Forms of the TDGL Equations. From both the mathematical and physical points of view, it is convenient to introduce the nondimensional forms of the TDGL equations. We start with the dimensions of various physical quantities. Let m be the mass, L the length scale, t the time, and E the energy. Then we have

$$\begin{aligned} &E : L^2 m/t^2, && h : Et, && D : L^2/t, && e_s^2 : EL,\\ &\sigma : 1/t, && c : L/t, && a : E, && b : EL^3,\\ &\psi : 1/L^{3/2}, && A : (E/L)^{1/2}, && H : (E/L^3)^{1/2}. \end{aligned}$$

We introduce some physical parameters, which are classical and physically important:

$$\begin{aligned}
&\lambda = \lambda(T) = (m_s c^2 b/4\pi e_s^2|a|)^{1/2}, && |\psi_0|^2 = |a|/b, && H_c = (4\pi|a|^2/b)^{1/2},\\
&\xi = \xi(T) = h/(2m_s|a|)^{1/2}, && \kappa = \lambda/\xi, && \eta = 4\pi\sigma D/c^2,\\
&\tau = \lambda^2/D.
\end{aligned}$$

Physically speaking, $|\psi_0|^2$ stands for the equilibrium density, H_c for the thermodynamic critical field, $\lambda = \lambda(T)$ for the penetration depth, $\xi(T)$ for the coherence length, and τ for the relaxation time. The ratio of the two characteristic length scales $\kappa = \lambda/\xi$ is called the Ginzburg–Landau parameter of the substance. When $0 < \kappa < \frac{1}{\sqrt{2}}$, the material is of Type I, and when $\kappa > \frac{1}{\sqrt{2}}$, the material is of Type II.

For the purpose of studying phase transition, the nondimensional form below was first introduced in Ma and Wang (2005b,a); it differs slightly from those often used in the literature. For this purpose, we set

$$\begin{aligned}
&l = \sqrt{b}/e_s, && \tau_0 = hl/e_s^2, && \phi_0 = e_s^2/\sqrt{b},\\
&A_0 = \left(e_s hc^2/D\sqrt{b}\right)^{1/2}, && \alpha = -2a\sqrt{b}m_s D/e_s^3 h, && \mu = hD/\sqrt{b}e_s,\\
&\beta = 2m_s D/h, && \zeta = 4\pi\sigma l e_s^2/c^2 h, && \gamma = 4\pi e_s^2/m_s c^2 l,
\end{aligned}$$

and

$$\begin{aligned}
&x = lx', && t = \tau_0 t', && \psi = l^{-3/2}\psi',\\
&A = A_0 A', && \phi = \phi_0\phi', && H_a = l^{-1}A_0 H_a'.
\end{aligned}$$

Then with the gauge taken such that (3.6.11) and (3.6.12) hold, we obtain the following nondimensional TDGL equations (suppressing the primes):

$$\begin{aligned}
&\frac{\partial\psi}{\partial t} + i\phi\psi = -(i\mu\nabla + A)^2\psi + \alpha\psi - \beta|\psi|^2\psi,\\
&\zeta\left[\frac{\partial A}{\partial t} + \mu\nabla\phi\right] = -\text{curl}^2 A + \text{curl}H_a - \gamma|\psi|^2 A - \frac{\gamma\mu i}{2}(\psi^*\nabla\psi - \psi\nabla\psi^*),\\
&\text{div}A = 0.
\end{aligned} \tag{3.6.14}$$

The nondimensional domain is denoted by Ω. The initial conditions are given by

$$\psi(0) = \psi_0, \quad A(0) = A_0.$$

The boundary conditions can be chosen as one of the following:

Neumann Boundary Condition. For the case in which Ω is enclosed by an insulator:

$$\frac{\partial\psi}{\partial n} = 0, \quad A_n = 0, \quad \text{curl}\, A \times n = H_a \times n \quad \text{on } \partial\Omega.$$

Dirichlet Boundary Condition. For the case in which Ω is enclosed by a magnetic material:

$$\psi = 0, \quad A_n = 0, \quad \text{curl}\, A \times n = H_a \times n \quad \text{on } \partial\Omega.$$

Robin Boundary Condition. For the case in which Ω is enclosed by a normal metal:

$$\frac{\partial \psi}{\partial n} + C\psi = 0, \quad A_n = 0, \quad \text{curl}\, A \times n = H_a \times n \quad \text{on } \partial\Omega.$$

Remark 3.6.1 If the material is a loop or a plate $\Omega = \widetilde{\Omega} \times (0,h)$, with the height h small in comparison to the diameter of $\widetilde{\Omega}$, then it is reasonable to consider the periodic boundary condition either in the x-direction or in the (x,y)-directions.

Remark 3.6.2 We shall see in later discussions that the parameter α plays a key role in the phase transition of superconductivity, which is given in terms of dimensional quantities by

$$\alpha = \alpha(T) = \frac{2\sqrt{b}m_s DN(0)}{e_s^3 h} \cdot \frac{T_c - T}{T_c}. \tag{3.6.15}$$

3.6.2 TGDL as a Gradient-Type System

Mathematical Setting. It is known that for a given applied field H_a with div $H_a = 0$, there exists a field A_a satisfying

$$\begin{aligned} &\text{div} A_a = 0, && \text{curl} A_a = H_a && \text{in } \Omega, \\ &A_a \cdot n = 0 && && \text{on } \partial\Omega. \end{aligned}$$

Let $A = \mathscr{A} + A_a$. Then equations (3.6.14) can be rewritten as

$$\begin{aligned} &\frac{\partial \psi}{\partial t} + i\phi\psi = -(i\mu\nabla + A_a)^2\psi + \alpha\psi - 2A_a \cdot \mathscr{A}\psi \\ &\qquad\qquad - 2i\mu\mathscr{A} \cdot \nabla\psi - |\mathscr{A}|^2\psi - \beta|\psi|^2\psi, \\ &\zeta\left[\frac{\partial \mathscr{A}}{\partial t} + \mu\nabla\phi\right] = -\text{curl}^2\mathscr{A} - \gamma A_a|\psi|^2 - \gamma\mathscr{A}|\psi|^2 \\ &\qquad\qquad - \frac{\gamma\mu}{2} i(\psi^*\nabla\psi - \psi\nabla\psi^*), \\ &\text{div}\mathscr{A} = 0, \end{aligned} \tag{3.6.16}$$

with the following initial and boundary conditions:

$$\psi(0) = \psi_0, \quad \mathscr{A}(0) = \mathscr{A}_0, \tag{3.6.17}$$

$$\mathscr{A}_n|_{\partial\Omega} = 0, \quad \text{curl}\mathscr{A} \times n|_{\partial\Omega} = 0, \tag{3.6.18}$$

and with one of the following three boundary conditions for ψ:

Neumann boundary condition:

$$\frac{\partial\psi}{\partial n}|_{\partial\Omega}=0. \tag{3.6.19}$$

Dirichlet boundary condition:

$$\psi|_{\partial\Omega}=0. \tag{3.6.20}$$

Robin boundary condition:

$$\left(\frac{\partial\psi}{\partial n}+C\psi\right)|_{\partial\Omega}=0,\quad C>0. \tag{3.6.21}$$

Herineafter, we use $H^k(\Omega,\mathbb{C})$ for the Sobolev spaces of complex-valued functions defined on Ω, and $H^k(\Omega,\mathbb{R}^n)(n=2,3)$ for the Sobolev spaces of vector-valued functions defined on Ω. Let

$$\begin{aligned}
&H_B^2(\Omega,\mathbb{C})=\{\psi\in H^2(\Omega,\mathbb{C})|\ \psi \text{ satisfy one of (3.6.19)–(3.6.21)}\},\\
&D^2(\Omega,\mathbb{R}^n)=\{\mathscr{A}\in H^2(\Omega,\mathbb{R}^n)|\ \mathrm{div}\mathscr{A}=0,\mathscr{A}\text{ satisfy (3.6.18)}\},\\
&\mathscr{L}^2(\Omega,\mathbb{R}^n)=\{\mathscr{A}\in L^2(\Omega,\mathbb{R}^n)|\ \mathrm{div}\mathscr{A}=0,\mathscr{A}_n|_{\partial\Omega}=0\}.
\end{aligned}$$

We set

$$H=L^2(\Omega,\mathbb{C})\times\mathscr{L}^2(\Omega,\mathbb{R}^n),\qquad H_1=H_B^2(\Omega,\mathbb{C})\times D^2(\Omega,\mathbb{R}^n).$$

Let $P:L^2(\Omega,\mathbb{R}^n)\to\mathscr{L}^2(\Omega,\mathbb{R}^n)$ be the Leray projection. Then it is known that the function ϕ in (3.6.16) is determined uniquely up to constants by the following equation:

$$\zeta\mu\nabla\phi=(I-P)\left[\frac{\gamma\mu}{2}i(\psi\nabla\psi^*-\psi^*\nabla\psi)-\gamma|\psi|^2(\mathscr{A}+A_a)\right]. \tag{3.6.22}$$

Namely, for every $u=(\psi,\mathscr{A})\in H_1$, there is a unique solution of (3.6.22) up to constants. Therefore, we define a nonlinear operator $\Phi:H_1\to L^2(\Omega)$ by

$$\Phi(u)=\phi=\text{the solution of (3.6.22) with }\int_\Omega\phi dx=0. \tag{3.6.23}$$

We set the mappings $L_\alpha=-K+B_\alpha$ and $G:H_1\to H$ by

$$Ku=\begin{pmatrix}(i\mu\nabla+A_a)^2\psi\\ \zeta^{-1}\mathrm{curl}^2\mathscr{A}\end{pmatrix},\qquad B_\alpha u=\begin{pmatrix}\alpha\psi\\0\end{pmatrix},$$

$$G(u)=-\begin{pmatrix}i\psi\Phi(u)+2A_a\cdot\mathscr{A}\psi+2i\mu\mathscr{A}\cdot\nabla\psi+|\mathscr{A}|^2\psi+\beta|\psi|^2\psi\\ P[\gamma\zeta^{-1}A_a|\psi|^2+\gamma\zeta^{-1}\mathscr{A}|\psi|^2+\frac{\gamma\mu\zeta^{-1}}{2}i(\psi^*\nabla\psi-\psi\nabla\psi^*)\end{pmatrix},$$

where $u = (\psi, \mathscr{A})$, $\Phi(u)$ is defined by (3.6.23), and $P : L^2(\Omega, \mathbb{R}^n) \to \mathscr{L}(\Omega, \mathbb{R}^n)$, the Leray projection. Thus, the problem (3.6.16)–(3.6.18) with one of the boundary conditions (3.6.19)–(3.6.21) can be rewritten in the following operator form:

$$\frac{du}{dt} = L_\alpha u + G(u), \quad u(0) = u_0. \tag{3.6.24}$$

Gradient-Type Structure. Now we show that the Ginzburg–Landau equation has a gradient-type structure. The Ginzburg–Landau energy of (3.6.16) is given by

$$\begin{aligned} E(\psi, \mathscr{A}) =& \frac{1}{2}\int_\Omega \Big[|(i\mu\nabla + \mathscr{A} + A_a)\psi|^2 + \frac{\beta}{2}|\psi|^4 - \alpha|\psi|^2 + \gamma^{-1}|\text{curl}\mathscr{A}|^2\Big]dx \\ & \left(+ \frac{1}{2}\mu^2 C\int_{\partial\Omega}|\psi|^2 ds \text{ for BC (3.6.21)}\right). \end{aligned} \tag{3.6.25}$$

The derivative operator of E is given by

$$DE(u) = \begin{pmatrix} (i\mu\nabla + \mathscr{A} + A_a)^2\psi - \alpha\psi + \beta|\psi|^2\psi \\ \gamma^{-1}\text{curl}^2\mathscr{A} + |\psi|^2(\mathscr{A} + A_a) + \frac{\mu}{2}i(\psi^*\nabla\psi - \psi\nabla\psi^*) \end{pmatrix}.$$

Let

$$\begin{aligned} DE_1 &= (i\mu\nabla + \mathscr{A} + A_a)^2\psi - \alpha\psi + \beta|\psi|^2\psi, \\ DE_2 &= \gamma^{-1}\text{curl}^2\mathscr{A} + |\psi|^2(\mathscr{A} + A_a) + \frac{\mu}{2}i(\psi^*\nabla\psi - \psi\nabla\psi^*), \\ \widetilde{L}u + \widetilde{G}(u) &= -\begin{pmatrix} \zeta DE_1 \\ \gamma DE_2 \end{pmatrix} - \Psi(u), \\ \Psi(u) &= \begin{pmatrix} \zeta i\phi\psi \\ \zeta\mu\nabla\phi \end{pmatrix}. \end{aligned}$$

Let $v = (\zeta\psi, \zeta\mathscr{A})$. Then (3.6.24) can be rewritten as

$$\frac{dv}{dt} = \widetilde{L}v + \widetilde{G}(v), \quad v(0) = \zeta\varphi. \tag{3.6.26}$$

It suffices to prove that (3.6.26) has a gradient-type structure. We see that

$$-(\widetilde{L}u + \widetilde{G}(u), DE(u)) \leq C\|DE\|^2 + (\Psi, DE). \tag{3.6.27}$$

By the boundary conditions and

$$\text{div}\, A_a = 0, \quad \text{div}\, \mathscr{A} = 0, \quad \text{div curl}\, \mathscr{A} = 0,$$

we obtain

$$(DE, \Psi) = \zeta \text{Re}\int_\Omega [((i\mu\nabla + \mathscr{A} + A_a)^2\psi - \alpha\psi + \beta|\psi|^2\psi)\,(i\phi\psi)^*$$

$$
\begin{aligned}
&+\left(|\psi|^2(\mathscr{A}+A_a)+\frac{\mu}{2}i(\psi^*\nabla\psi-\psi\nabla\psi^*)\right)\mu\nabla\phi\Big]\,dx\\
=&\ \zeta\mathrm{Re}\int_\Omega\Big[-i(i\mu\nabla+\mathscr{A}+A_a)^2\psi\cdot\psi^*\phi+\\
&+\left(|\psi|^2(\mathscr{A}+A_a)+\frac{\mu}{2}i(\psi^*\nabla\psi-\psi\nabla\psi^*)\right)\mu\nabla\phi\Big]\,dx\\
=&\ \zeta\mathrm{Re}\int_\Omega\big[i\mu^2\phi\psi^*\Delta\psi+\mu\phi(\mathscr{A}+A_a)\cdot\nabla|\psi|^2\\
&+\mu\nabla\phi\cdot(\mathscr{A}+A_a)|\psi|^2+\mu^2\mathrm{Re}(i\psi^*\nabla\psi)\nabla\phi\big]\,dx\\
=&\ 0.
\end{aligned}
$$

Hence we derive from (3.6.27) that

$$-(\widetilde{L}u+\widetilde{G}(u),DE(u))\le C\|DE(u)\|^2.$$

It is clear that

$$DE(u_0)=0\Leftrightarrow Lu_0+G(u_0)=0.$$

Thus (3.6.26) is a gradient-type equation with energy functional $E(u)$.

It is not hard to establish the following theorem on existence of solutions and global attractors, and we omit the proof. Note also that the existence of strong solutions and a global attractor in H^2-Sobolev space was first derived in Tang and Wang (1995).

Theorem 3.6.3 *Consider the problem (3.6.16)–(3.6.18) with (3.6.19), (3.6.20), or (3.6.21).*

(1) For every initial value $\varphi=(\widetilde{\psi}_0,\widetilde{\mathscr{A}_0})\in H_{1/2}$ and $A_a\in H_1$, the problem has a unique classical solution

$$(\psi,\mathscr{A})\in C^1((0,\infty),H)\bigcap C^0((0,\infty),H_1),\quad \phi\in C^0((0,\infty),H^1(\Omega)).$$

(2) If $\Omega\subset\mathbb{R}^n$ is C^∞ and $A_a\in C^\infty(\Omega,\mathbb{R}^n)$, then the solution is C^∞:

$$
\begin{aligned}
&(\psi,\mathscr{A})\in C^\infty((0,\infty)\times\overline{\Omega},\mathbb{C})\times C^\infty((0,\infty)\times\overline{\Omega},\mathbb{R}^n),\\
&\phi\in C^\infty((0,\infty)\times\overline{\Omega}).
\end{aligned}
$$

(3) For every $k\ge 1/2$, the problem has a global attractor in $H_k\subset H^{2k}(\Omega,\mathbb{C})\times H^{2k}(\Omega,\mathbb{R}^n)$, which attracts H_k in the H^{2k}-norm. Moreover, the set S consisting of all singular points of (3.6.16)–(3.6.18) in the attractor attracts H_k in the H^{2k}-norm.

3.6.3 Phase Transition Theorems

Eigenvalue Problems. In order to describe the dynamical properties of the TDGL equations, it is necessary to consider the eigenvalue problem of the linearized equations of (3.6.16).

Let α_1 be the first eigenvalue of the following equation:

$$(i\mu\nabla + A_a)^2\psi = \alpha\psi \quad \forall x \in \Omega, \tag{3.6.28}$$

with one of the boundary conditions (3.6.19)–(3.6.21). It is clear that (3.6.28) can be equivalently expressed as

$$\begin{aligned} &-\mu^2\Delta\psi_1 + |A_a|^2\psi_1 - 2\mu A_a\cdot\nabla\psi_2 = \alpha\psi_1, \\ &-\mu^2\Delta\psi_2 + |A_a|^2\psi_2 + 2\mu A_a\cdot\nabla\psi_1 = \alpha\psi_2, \end{aligned} \tag{3.6.29}$$

where $\psi = \psi_1 + i\psi_2$.

It is easy to verify that (3.6.29) with one of the boundary conditions (3.6.19)–(3.6.21) is symmetric. Therefore, the eigenvalues of (3.6.28) are real,

$$\alpha_1 \leq \alpha_2 \leq \cdots \leq \alpha_k \leq \cdots, \qquad \lim_{k\to\infty} \alpha_k = +\infty, \tag{3.6.30}$$

and the corresponding eigenvector sequence

$$\{e_k \in H^2_B(\Omega, \mathbb{C}) |\ k = 1, 2, \cdots\} \tag{3.6.31}$$

forms an orthogonal basis of $L^2(\Omega, \mathbb{C})$.

The eigenvalues of (3.6.28) always have even multiplicity, i.e., if ψ is an eigenvector of (3.6.28), then $e^{i\theta}\psi(\theta \in \mathbb{R}^1)$ are also eigenvectors corresponding to the same eigenvalue. Let the first eigenvalue α_1 have multiplicity $2m(m \geq 1)$ with eigenvectors

$$e_{2k-1} = \psi_{k1} + i\psi_{k2}, \quad e_{2k} = -\psi_{k2} + i\psi_{k1} \qquad \forall 1 \leq k \leq m. \tag{3.6.32}$$

We know that α_1 enjoys the following properties:

$$\begin{aligned} &\alpha_1 = \alpha_1(A_a) && \text{depends continuous on } A_a, \\ &\alpha_1(A_a) > 0 && \text{for } A_a \neq 0, \\ &\alpha_1(0) = 0 && \text{for } b.c.\ (3.6.19), \\ &\alpha_1(0) > 0 && \text{for } b.c.\ (3.6.20) \text{ or } (3.6.21). \end{aligned} \tag{3.6.33}$$

Another crucial eigenvalue problem for (3.6.16) is

$$\begin{aligned} &\text{curl}^2 A + \nabla\phi = \rho A, \\ &\text{div} A = 0, \\ &A_n|_{\partial\Omega} = 0, \quad \text{curl} A \times n|_{\partial\Omega} = 0. \end{aligned} \tag{3.6.34}$$

Here we point out that the boundary condition in (3.6.34), i.e., (3.6.18), is the free boundary condition, which can be expressed as

$$A_n|_{\partial\Omega} = 0, \quad \frac{\partial A_\tau}{\partial n}|_{\partial\Omega} = 0, \tag{3.6.35}$$

where τ is the tangent vector on $\partial\Omega$; see (Ma and Wang 2005a).

The eigenvalues of (3.6.34) are real,

$$0 < \rho_1 \leq \rho_2 \leq \cdots \leq \rho_k \leq \cdots, \qquad \lim_{k\to\infty} \rho_k = +\infty, \tag{3.6.36}$$

and the corresponding eigenvectors

$$\{a_k \in D^2(\Omega, \mathbb{R}^n) |\, k = 1, 2, \cdots\} \tag{3.6.37}$$

form an orthogonal basis of $\mathscr{L}^2(\Omega, \mathbb{R}^n)$.

Dynamic Transition Theorems. In superconductivity, the parameter α in (3.6.16) cannot exceed a maximal value $\alpha(T) \leq \alpha(0)$, because $T \geq 0$. Hence, we have to impose a basic hypothesis:

$$\alpha_1 < \alpha(0) = \frac{2\sqrt{b}m_s D N_0}{e_s^3 h}, \tag{3.6.38}$$

where α_1 is the first eigenvalue of (3.6.28), and N_0 is the density of states at the Fermi level.

We mainly consider the case in which the first eigenvalue α_1 of (3.6.28) has multiplicity two, i.e., α_1 is a complex simple eigenvalue. We start with the introduction of a crucial physical parameter, which completely determines the dynamical properties of the phase transition of the Ginzburg–Landau equations.

Let $e \in H^2_B(\Omega, \mathbb{C})$ be a first eigenvector of (3.6.28). Then there is a unique solution to the following equations:

$$\begin{aligned} &\mathrm{curl}^2 \mathscr{A}_0 + \nabla\phi = |e|^2 A_a + \frac{\mu}{2} i(e^* \nabla e - e \nabla e^*), \\ &\mathrm{div} \mathscr{A}_0 = 0, \\ &\mathscr{A}_0 \cdot n|_{\partial\Omega} = 0, \quad \mathrm{curl}\mathscr{A}_0 \times n|_{\partial\Omega} = 0. \end{aligned} \tag{3.6.39}$$

Then we define a physical parameter R as follows:

$$R = -\frac{\beta}{\gamma} + \frac{2\int_\Omega |\mathrm{curl}\mathscr{A}_0|^2 dx}{\int_\Omega |e|^4 dx}. \tag{3.6.40}$$

From (3.6.39) and (3.6.40), it is easy to see that the parameter R is independent of the choice of the first eigenvectors of (3.6.28). $H_0 = \mathrm{curl}\mathscr{A}_0$ given by (3.6.39) depends on the applied magnetic potential A_a and the geometric properties of Ω. The parameter R is essentially a function of $A_a, \Omega, \beta, \gamma, \mu$.

The main results in this section are the following theorems.

Theorem 3.6.4 *Assume that the first eigenvalue α_1 of (3.6.28) with one of the boundary conditions (3.6.19)–(3.6.21) is complex and simple, and that the condition (3.6.38) holds. If the parameter R defined by (3.6.40) satisfies $R < 0$, then the* *phase transition of (3.6.16)–(3.6.18) with one of (3.6.19)–(3.6.21) at $\alpha = \alpha_1$ is of Type I. Moreover, the following assertions hold:*

(1) *The problem bifurcates from $((\psi, \mathscr{A}), \alpha) = (0, \alpha_1)$ to an attractor Σ_α for $\alpha > \alpha_1$, which is homeomorphic to S^1 and consists of steady-state solutions.*
(2) *There is a neighborhood $U \subset H$ of $(\psi, \mathscr{A}) = 0$ such that the attractor Σ_α attracts $U \setminus \Gamma$ in H, where Γ is the stable manifold of $(\psi, \mathscr{A}) = 0$ with codimension two in H.*
(3) *Each $(\psi, \mathscr{A}) \in \Sigma_\alpha$ can be expressed as*

$$\begin{aligned} \psi &= \left|\frac{\alpha - \alpha_1}{R_1}\right|^{1/2} e + o(|\alpha - \alpha_1|^{1/2}), \\ curl^2 \mathscr{A} &= -\gamma \left|\frac{\alpha - \alpha_1}{R_1}\right| \cdot \left[|e|^2 A_a + \mu I_m(e \nabla e^*)\right] + o(|\alpha - \alpha_1|), \end{aligned} \tag{3.6.41}$$

where e is the first eigenvector of (3.6.28) and

$$R_1 = \frac{\gamma R \int_\Omega |e|^4 dx}{\int_\Omega |e|^2 dx}.$$

Theorem 3.6.5 *Assume that the first eigenvalue α_1 of (3.6.28) with one of the boundary conditions (3.6.19)–(3.6.21) is complex simple and that the condition (3.6.38) holds. If $R > 0$, then for the problem (3.6.16)–(3.6.18) with one of (3.6.19)–(3.6.21), we have the following assertions:*

(1) *The phase transition at $\alpha = \alpha_1$ is of Type II.*
(2) *The problem bifurcates from $((\psi, \mathscr{A}), \alpha) = (0, \alpha_1)$ to an invariant set Σ_α on $\alpha < \alpha_1$ and has no bifurcation on $\alpha > \alpha_1$.*
(3) *$\Sigma_\alpha = S^1$ is a circle consisting of steady states, and has a two-dimensional unstable manifold.*

By Theorem 3.6.3, the TDGL equations possess a global attractor. It is clear that $(\psi, \mathscr{A}) = 0$ is a unique minimal-value point of (3.6.25) for $\alpha = 0$, which implies that if $\alpha = 0$, then $(\psi, \mathscr{A}) = 0$ is globally asymptotically stable for (3.6.16)–(3.6.18) with one of the boundary conditions (3.6.19)–(3.6.21). Thus, the following theorem follows from Theorems 2.5.3 and 3.6.5.

Theorem 3.6.6 *Under the assumptions of Theorem 3.6.5, for the case $R > 0$, there exists a saddle-node bifurcation point $\alpha_0 (0 < \alpha_0 < \alpha_1)$ for the TDGL equations such that the following statements hold, which are described schematically by Figs. 3.28 and 3.29:*

(1) *At $\alpha = \alpha_0$, there is a cycle $\Sigma_0 = \Sigma_{\alpha_0}$ of singular points with $0 \notin \Sigma_0$.*
(2) *For $\alpha < \alpha_0$, there are no invariant sets near Σ_0.*
(3) *For $\alpha_0 < \alpha < \alpha_1$, there are two connected branches of cycles Σ_α^1 and Σ_α^2 of singular points, and Σ_α^2 extends to $\alpha \geq \alpha_1$.*
(4) *For each $\alpha > \alpha_0$, Σ_α^2 is an attractor with dist $(\Sigma_\alpha^2, 0) > 0$ at $\alpha = \alpha_1$.*
(5) *For $\alpha_0 < \alpha < \alpha_1$, Σ_α^1 is a repeller with $0 \notin \Sigma_\alpha^1$ and $\Sigma_\alpha^1 \to \{0\}$ as $\alpha \to \alpha_1^-$.*

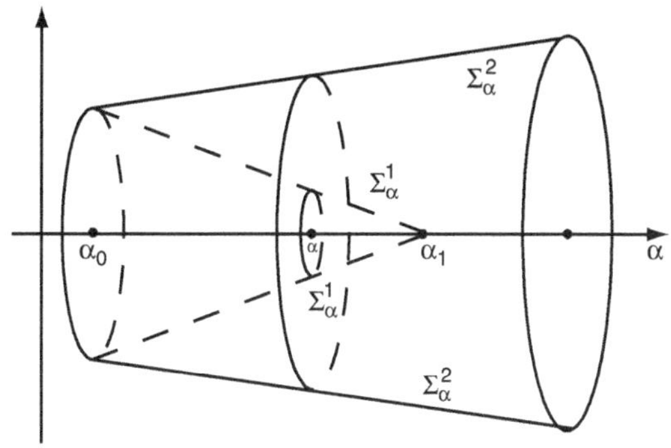

Fig. 3.28 For each $\alpha_0 < \alpha < \alpha_1$, Σ^1_α is a repeller, containing unstable superconducting states. For each $\alpha_0 < \alpha$, Σ^2_α is an attractor, containing stable superconducting states

Remark 3.6.7 The parameter R defined by (3.6.40) can be equivalently expressed as follows:

$$R = -\frac{\beta}{\gamma} + \frac{2\sum_{k=1}^{\infty}\frac{1}{\rho_k}\left[\int_\Omega(|e|^2A_a + 2\mu e_2\nabla e_1)\cdot a_k\right]^2}{\int_\Omega |e|^4 dx}, \tag{3.6.42}$$

where $e = e_1 + ie_2, \rho_k$ are the eigenvalues of (3.6.34) given by (3.6.36), and $\{a_k\}$ are the normalized eigenvectors given by (3.6.37).

Proof of Theorems 3.6.4 and 3.6.5. We proceed in several steps as follows.

Step 1. It is easy to see that $L_\alpha : H_1 \to H$ is a sectorial operator, and the eigenvalues of L_α satisfy

$$\beta_1(\alpha) = \beta_2(\alpha) = \alpha - \alpha_1 \begin{cases} <0 & \text{if } \alpha < \alpha_1, \\ =0 & \text{if } \alpha = \alpha_1, \\ >0 & \text{if } \alpha > \alpha_1, \end{cases}$$

and for $j \geq 3$,

$$\begin{aligned} &\beta_j(\alpha_1) = \alpha_1 - \alpha_k, \quad \text{or } \beta_j(\alpha_1) = -\zeta^{-1}\rho_l, \\ &\beta_j(\alpha_1) < 0, \end{aligned}$$

for all $k > 1$ and $l \geq 1$. Namely, the conditions (2.1.2), (2.1.4), and (2.1.5) are satisfied by (3.6.24). It is clear that the operator $\Phi : H_1 \to L^2(\Omega, C)$ defined by (3.6.23) is C^∞, and it is easy to derive the following estimate:

$$\begin{aligned} \int_\Omega |\Phi(u)\psi|^2 dx &\leq \left[\int_\Omega |\Phi(u)|^3 dx\right]^{2/3} \left[\int_\Omega |\psi|^6 dx\right]^{1/3} \\ &\leq C\left(\|u\|^2_{H_{1/2}} + \|u\|_{H_{1/2}}\|\psi\|^2_{L^4}\right)^2 \|\psi\|^2_{L^6}, \end{aligned}$$

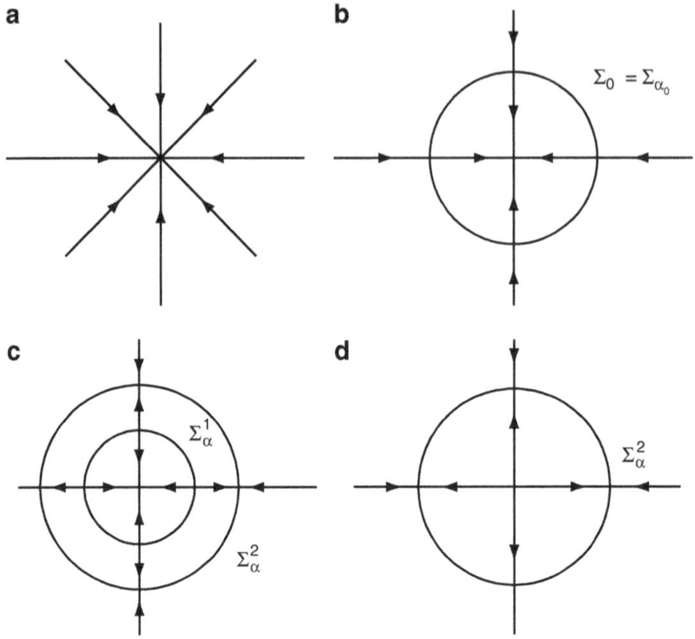

Fig. 3.29 Phase diagram on the center manifold for various α: (**a**) case $\alpha < \alpha_0$, (**b**) case $\alpha = \alpha_0$, (**c**) case $\alpha_0 < \alpha < \alpha_1$, (**d**) case $\alpha_1 \leq \alpha$

where $H_{1/2}$ is the closure of H_1 for the H^1-norm. Hence, it is not difficult to verify that there is a number $1/2 < \sigma < 1$ such that $G : H_\sigma \to H$ is C^∞. Thus, the condition (2.1.3) holds.

Step 2. It is known that the transition of (3.6.24) is determined by its reduced equation to the center manifold. Let

$$\psi_0 \in E = \{ze|\ z \in \mathbb{C},\ e \text{ the first eigenvector of } (3.6.28)\}.$$

Then the reduced equation of (3.6.24) reads

$$\frac{d\psi_0}{dt} = \beta_1(\alpha)\psi_0 - P_1 G(\psi_0 + \widetilde{\psi}(\psi_0), \widetilde{\mathscr{A}}(\psi_0)), \tag{3.6.43}$$

where $P_1 : H \to E_1$ is the canonical projection, and $\widetilde{\Phi}(\psi_0) = (\widetilde{\psi}(\psi_0), \widetilde{\mathscr{A}}(\psi_0)) \in H_1$ is the center manifold function.

The k multilinear operators ($k = 2, 3$) in G are given by

$$G_2(u) = -\begin{pmatrix} 2A_a \cdot \mathscr{A}\psi + 2i\mu \mathscr{A} \cdot \nabla\psi \\ \zeta^{-1}\gamma A_a|\psi|^2 + \frac{\gamma\mu}{2\zeta} i(\psi^*\nabla\psi - \psi\nabla\psi^*) \end{pmatrix},$$

$$G_3(u) = -\begin{pmatrix} i\psi\Phi_2(u) + |\mathscr{A}|^2\psi + \beta|\psi|^2\psi \\ \gamma\zeta^{-1}\mathscr{A}|\psi|^2 \end{pmatrix},$$

where $\Phi_2(u)$ is the bilinear operator in $\Phi(u)$.

By the first approximation of the center manifold reduction, the center manifold function $\widetilde{\Phi} = (\widetilde{\psi}(\psi_0), \widetilde{\mathscr{A}}(\psi_0))$ satisfies

$$\begin{aligned} \text{curl}^2\widetilde{\mathscr{A}} + \mu\nabla\phi &= -\gamma A_a|\psi_0|^2 - \frac{\gamma\mu}{2} i(\psi_0^*\nabla\psi_0 - \psi_0\nabla\psi_0^*) \\ &\quad + o(\|\psi_0\|^2, |\beta_1(\alpha)| \cdot \|\psi_0\|), \qquad (3.6.44) \\ \widetilde{\psi} &= O(\|\widetilde{\mathscr{A}}\| \cdot \|\psi_0\|) = O(\|\psi_0\|^3). \qquad (3.6.45) \end{aligned}$$

Based on (3.6.44) and (3.6.45), (3.6.43) can be expressed as

$$\frac{d\psi_0}{dt} = \beta_1(\alpha)\psi_0 - g_3(\psi_0) + o(\|\psi_0\|^3) + O(\|\psi_0\|^3 \cdot |\beta_1(\alpha)|), \qquad (3.6.46)$$

where

$$g_3(\psi_0) = P_1[\beta|\psi_0|^2\psi_0 + 2A_a \cdot \widetilde{\mathscr{A}_2}\psi_0 + 2i\mu\widetilde{\mathscr{A}_2} \cdot \nabla\psi_0 + i\Phi_2\psi_0], \qquad (3.6.47)$$

$$\begin{cases} \text{curl}^2\widetilde{\mathscr{A}_2} + \nabla\phi = -\gamma A_a|\psi_0|^2 - \dfrac{\gamma\mu}{2} i(\psi_0^*\nabla\psi_0 - \psi_0\nabla\psi_0^*), \\ \text{div}\widetilde{\mathscr{A}_2} = 0, \\ \widetilde{\mathscr{A}_2} \cdot n|_{\partial\Omega} = 0, \quad \text{curl}\widetilde{\mathscr{A}_2} \times n|_{\partial\Omega} = 0. \end{cases} \qquad (3.6.48)$$

The equations (3.6.46)–(3.6.48) are the second-order approximation of the reduction of (3.6.24) to the center manifold.

Step 3. Let $\psi_0 = \psi_1^0 + i\psi_2^0$. We infer from (3.6.47) that

$$\begin{aligned} &(g_3(\psi_0), \psi_0) \\ &= \text{Re}\int_\Omega g_3(\psi_0) \cdot \psi_0^* dx \\ &= \int_\Omega \left[\beta|\psi_0|^4 + 2|\psi_0|^2 A_a \cdot \widetilde{\mathscr{A}_2} + 2\mu\widetilde{\mathscr{A}_2}(\psi_2^0\nabla\psi_1^0 - \psi_1^0\nabla\psi_2^0)\right] dx \\ &= \int_\Omega \left[\beta|\psi_0|^4 + 2|\psi_0|^2 A_a \cdot \widetilde{\mathscr{A}_2} + 4\mu\psi_2^0\widetilde{\mathscr{A}_2} \cdot \nabla\psi_1^0\right] dx. \qquad (3.6.49) \end{aligned}$$

Let $\widetilde{\mathscr{A}_2}$ have the Fourier expansion $\widetilde{\mathscr{A}} = \sum_{k=1}^\infty y_k a_k$ for the basis (3.6.37) of $\mathscr{L}^2(\Omega, \mathbb{R}^n)$. Then for (3.6.48), we can derive the solutions y_k:

$$y_k = -\frac{\gamma}{\rho_k}\int_\Omega \left[|\psi_0|^2 A_a \cdot a_k + 2\mu\psi_2^0\nabla\psi_1^0 \cdot a_k\right] dx.$$

Inserting y_k into (3.6.49), we obtain

$$
\begin{aligned}
(g_3(\psi_0),\psi_0) = \beta \int_\Omega |\psi_0|^4 dx - 2\gamma \sum_{k=1}^{\infty} \frac{1}{\rho_k} \Big[\Big(\int_\Omega |\psi_0|^2 A_a \cdot a_k dx \Big)^2 \\
+ 4\mu \Big(\int_\Omega |\psi_0|^2 A_a \cdot a_k dx \Big) \Big(\int_\Omega \psi_2^0 \nabla \psi_1^0 \cdot a_k dx \Big) \\
+ 4\mu^2 \Big(\int_\Omega \psi_2^0 \nabla \psi_1^0 \cdot a_k dx \Big)^2 \Big].
\end{aligned} \tag{3.6.50}
$$

Let $\psi_0 = x_1 e_1 + x_2 e_2$, where $(x_1, x_2) \in \mathbb{R}^2$, and e_1 and e_2 are as in (3.6.32); then

$$\psi_0 = \psi_1^0 + i\psi_2^0 = (x_1\psi_{11} - x_2\psi_{12}) + i(x_1\psi_{12} + x_2\psi_{11}).$$

Hence

$$
\begin{aligned}
&\int_\Omega |\psi_0|^4 dx = \int_\Omega (|\psi_1^0|^2 + |\psi_2^0|^2)^2 dx = (x_1^2 + x_2^2)^2 \int_\Omega |e_1|^4 dx, \\
&\int_\Omega |\psi_0|^4 A_a \cdot a_k dx = (x_1^2 + x_2^2) \int_\Omega |e_1|^2 A_a \cdot a_k dx, \\
&\int_\Omega \psi_2^0 \nabla \psi_1^0 \cdot a_k dx = (x_1^2 + x_2^2) \int_\Omega \psi_{12} \nabla \psi_{11} \cdot a_k dx.
\end{aligned} \tag{3.6.51}
$$

Putting (3.6.51) into (3.6.50), we obtain

$$(g_3(\psi_0),\psi_0) = -\gamma R (x_1^2 + x_2^2)^2 \int_\Omega |e_1|^4 dx, \tag{3.6.52}$$

where R is as in (3.6.42). It is easy to see that both numbers in (3.6.40) and (3.6.42) are the same.

Step 4. Proof of Theorem 3.6.4. Since the TDGL equations are of gradient-type structure, the Krasnosel'skii bifurcation theorem for the potential operators, Theorem 1.12 in Ma and Wang (2005b), is valid for (3.6.16). Hence the problem (3.6.16)–(3.6.18) with one of the boundary conditions (3.6.19)–(3.6.21) must bifurcate from $((\psi, \mathscr{A}), \alpha) = (0, \alpha_1)$ to at least one steady-state solution.

When $R < 0$, by (3.6.46) and (3.6.52) we infer from Theorem 2.1.6 that the transition at $\alpha = \alpha_1$ is of Type I, and from Theorem 2.2.4 that the Ginzburg–Landau equations bifurcate from $(0, \alpha_1)$ to a cycle of attractor Σ_α for $\alpha > \alpha_1$ that contains at least one singular point. Because of the invariance of the TDGL equations for the gauge transformation $\psi \to \psi e^{i\theta}$ for $\theta \in \mathbb{R}^1$, the steady-state solutions of the Ginzburg–Landau equations appear as a circle S^1. Therefore, the attractor $\Sigma_\alpha = S^1$ consists of steady-state solutions. Assertion (1) is proved.

Assertion (2) follows from Theorem 2.2.11, and assertion (3) can be directly derived from (3.6.46) and (3.6.52). Thus, Theorem 3.6.4 is proved.

Step 5. Proof of Theorem 3.6.5. When $R > 0$, by (3.6.46) and (3.6.52) we infer from Theorem 2.1.6 that the transition is of Type II. In addition, the time-reversed semigroup $S_\alpha(-t)$ generated by (3.6.46) has the same dynamic properties as the following equation:

$$\frac{d\psi_0}{dt} = (\alpha_1 - \alpha)\psi_0 + g_3(\psi_0) + o(\|\psi_0\|^3, |\alpha - \alpha_0|\|\psi_0\|^2). \tag{3.6.53}$$

In the same fashion as used in Step 4, from (3.6.52) and (3.6.53) we deduce that the semigroup $\widetilde{S}_\alpha(t)$ generated by (3.6.53) bifurcates from $(\psi_0, \alpha) = (0, \alpha_1)$ to an S^1 attractor Σ_α for $\alpha < \alpha_1$ that consists of singular points of (3.6.53). Hence assertions (1)–(3) hold for the semigroup $S_\alpha(t)(= \widetilde{S}_\alpha(-t))$ generated by (3.6.46) with $R > 0$.

Theorem 3.6.5 is proved. □

Remark 3.6.8 We know that the first eigenvalue α_1 with complex simplicity is generic. Namely, for each bounded open set $\Omega \subset \mathbb{R}^n$, there is an open and dense set $U \subset L^2(\Omega, \mathbb{R}^n)$ such that for every $A_a \in U$, the first eigenvalue $\alpha_1(A_a)$ of (3.6.28) with one of the boundary conditions (3.6.19)–(3.6.21) is complex and simple. In particular, when A_a is small, $\alpha_1(A_a)$ must be complex and simple.

3.6.4 Model Coupled with Entropy

According to thermodynamic theory, first-order phase transitions (i.e., Type II and Type III phase transitions) release latent heat. Therefore, it is necessary to take entropy into account in the Ginzburg–Landau model. To this end, we revise the Ginzburg–Landau free energy (3.6.25) by

$$\begin{aligned} E = \int_\Omega \Big[& \frac{1}{2}|(i\mu\nabla + \mathscr{A} + A_a)\psi|^2 - \frac{\alpha}{2}|\psi|^2 + \frac{\beta}{4}|\psi|^4 + \frac{\rho}{6}|\psi|^6 \\ & + \frac{1}{2}k_1|\nabla S|^2 + \frac{1}{2}k_2|S|^2 + k_3 S|\psi|^2 + \gamma^{-1}|\text{curl}\mathscr{A}|^2 \Big] dx, \end{aligned} \tag{3.6.54}$$

where $\rho > 0$ and $k_i > 0$ are nondimensional constants, and S is the entropy density deviation. As before, we can derive easily the TDGL corresponding to the free energy (3.6.54). Replace the parameter R defined in (3.6.40) by

$$\widetilde{R} = R + \frac{k_3^2}{\gamma} \frac{\int_\Omega S_1 |e|^2 dx}{\int_\Omega |e|^4 dx}, \tag{3.6.55}$$

where S_0 satisfies the following equation:

$$-k_1 \Delta S_1 + k_2 S_1 = |e|^2, \quad \frac{\partial S_1}{\partial n}|_{\partial\Omega} = 0. \tag{3.6.56}$$

Then we have the following dynamic transition theorem for the TDGL corresponding to the free energy (3.6.54):

Theorem 3.6.9 *If the parameter $\widetilde{R}$ given by (3.6.55) satisfies $\widetilde{R} < 0$, then the statements of Theorem 3.6.4 hold, and if $\widetilde{R} > 0$, then the statements of Theorems 3.6.5 and 3.6.6 hold. In particular, the entropy has a jump at the critical temperature T_c for $\widetilde{R} > 0$:*

$$\delta S = S(T_c^-) - S(T_c^+) = -k_3(-k_1\Delta + k_2)^{-1}|\psi(T_c)|^2,$$

where $\psi(T_c)$ is the stable superconducting state at $T = T_c$.

Physically, the entropy gradient δS implies that the first-order transition from the normal state to superconducting states (respectively from the superconducting states to normal state) is exothermic (respectively is endothermic), with the latent heat given by

$$\delta H = T_c \delta S = -T_c k_3(-k_1\Delta + k_2)^{-1}|\psi(T_c)|^2.$$

3.6.5 Physical Conclusions

General Remarks. The permanent current, called a supercurrent, is expressed in the Ginzburg–Landau equations by (3.6.6). In the steady-state case, the supercurrent in the second type of nondimensional form is written as

$$J_s = J_s(\psi, \mathscr{A}) = -\gamma(A_a + \mathscr{A})|\psi|^2 - \frac{\gamma\mu}{2} i(\psi^* \nabla\psi - \psi\nabla\psi^*). \tag{3.6.57}$$

We take the Meissner effect into account in the Ginzburg–Landau equations. Mathematically speaking, in the normal state, the magnetic field H in a sample should be $H = H_a + \mathscr{H}$, where $H_a = \text{curl}A_a$ is the applied field and $\mathscr{H} = \text{curl}\mathscr{A}$ the nonequilibrium fluctuation, and in the superconducting state, $H = \text{curl}\mathscr{A}$. In the both cases, $\mathscr{A}$ satisfies the Ginzburg–Landau equations (3.6.16) and boundary condition (3.6.18). Namely, we can express the magnetic field H in a sample Ω in the form

$$H = \text{curl}A, \qquad A = \begin{cases} A_a + \mathscr{A} & \text{in the normal state,} \\ \mathscr{A} & \text{in the superconducting state,} \end{cases}$$

and the supercurrent J_s in nondimensional form is given by

$$J_s = \text{curl}^2 \mathscr{A}, \tag{3.6.58}$$

where $\mathscr{A}$ is the superposition of the magnetic potentials associated with supercurrent and nonequilibrium fluctuation, which satisfies (3.6.16) and (3.6.18).

An equilibrium state $(\widetilde{\psi}, \widetilde{\mathscr{A}})$ of the TDGL equations (3.6.16) is said to be in the normal state if $\widetilde{\psi} = 0$, and it is in the superconducting state if $\widetilde{\psi} \neq 0$. A solution $(\psi, \mathscr{A})$ of (3.6.16) is said to be in the normal state if $(\psi, \mathscr{A})$ is in a domain of attraction of a normal equilibrium state; otherwise, $(\psi, \mathscr{A})$ in the superconducting state.

We consider the simplest case, in which the applied field vanishes, $A_a = 0$. In this case, the eigenvalue equation (3.6.28) becomes

$$-\mu\Delta\psi = \alpha\psi. \tag{3.6.59}$$

The first eigenvalue α_1 of (3.6.59) with one of the boundary conditions (3.6.19)–(3.6.21) is complex and simple, and the eigenvector is $e^{i\theta}\psi_0$ with ψ_0 a real function. Therefore, the parameter R defined by (3.6.40) reads

$$R = -\frac{\beta}{\gamma} < 0.$$

By Theorem 3.6.4, when the parameter $\alpha(T)$ is less than or equal to α_1, the solutions $(\psi, \mathscr{A})$ of (3.6.58) are in the normal state, and when $\alpha(T) > \alpha_1$, $(\psi, \mathscr{A})$ with initial values $(\psi_0, \mathscr{A}_0)$ in $U \setminus \Gamma$ are in superconducting states.

When $A_a = 0$, the steady-state solutions $(\widetilde{\psi}, \widetilde{\mathscr{A}})$ of (3.6.58) are real, i.e., $\widetilde{\psi} = e^{i\theta}\psi, Im\psi = 0$. Hence we have

$$\widetilde{\psi}^*\nabla\widetilde{\psi} - \widetilde{\psi}\nabla\widetilde{\psi}^* = 0,$$

which implies that $\widetilde{\mathscr{A}} = 0$. Thus, the supercurrent (3.6.58) vanishes: $J_s = 0$. This shows that with zero applied field $H_a = 0$, there is no supercurrent in a superconductor.

Implications of (3.6.38). For the case with the Neumann boundary condition (3.6.19), e.g., for a sample enclosed by an insulator, $\alpha_1 = 0$ for (3.6.59), which is independent of Ω, the geometry of the sample. Therefore, the condition (3.6.38) always holds.

However, for the Dirichlet and the Robin boundary conditions (3.6.20) and (3.6.21), the situation is different. We know that the first eigenvalue α_1 of (3.6.59) depends on Ω. In particular,

$$\alpha_1 = \alpha_1(\Omega) \to \infty \quad \text{if } |\Omega| \to 0.$$

Hence, the condition (3.6.38) implies that for the cases in which the sample is enclosed by a magnetic material or a normal metal, the volume of a sample must be greater than some critical value $|\Omega| > V_c > 0$. Otherwise, no superconducting state occurs at any temperature. This property also holds for the case in which there is an applied magnetic field H_a present. Of course, in this case, the critical volume V_c depends on H_a as well. In superconductivity, it again shows that it is a universal physical phenomenon that the dynamical properties of phase transitions depend on the geometry of the domain Ω.

Physical theory and experiments show that there is a critical applied magnetic field H_c by which a superconducting state will be destroyed, and H_c satisfies the following approximative equation near the critical temperature T_c, which is given by (3.6.2)

$$H_c(T) = H_0(1 - T^2/T_c^2). \tag{3.6.60}$$

Equation (3.6.60) is an empirical formula. A related equation is the equation of the critical parameter (1.2.6) for superconductivity, which reads

$$\alpha(T) = \alpha_1(H_a, \Omega), \tag{3.6.61}$$

where $\alpha(T)$ is the parameter in (3.6.16) and $\alpha_1 = \alpha_1(H_\alpha, \Omega)$ is the first eigenvalue of (3.6.28), $H_a = \text{curl} A_a$. It is expected that the applied magnetic field H_a satisfying (3.6.61) is the critical field H_c.

Transitions with $R < 0$. As mentioned earlier, there are two phase transitions: Type I and Type II, determined by a simple parameter R defined by (3.6.40). This parameter R links the superconducting behavior with the geometry of the material, the applied field, and the physical parameters.

Let a magnetic field $H_a = \text{curl} A_a$ be applied. By the phase transition theorems, the critical temperature T_c^1 of superconducting transition satisfies $T_c^1 < T_c$, where T_c is as in (3.6.2) and $T = T_c^1$ satisfies (3.6.61). It is known that

$$\alpha_1(H_a, \Omega) \to \infty \quad \text{if } |A_a| \to \infty.$$

This implies that the applied magnetic field H_a cannot be sufficient strong for superconductivity as required by condition (3.6.38).

From Theorems 3.6.4–3.6.6, we see that the number R defined by (3.6.40) is an important parameter for distinguishing two different types of superconducting transitions. We first examine the case $R < 0$.

By Theorem 3.6.4, when $\alpha > \alpha_1$, equations (3.6.16) bifurcate from $((\psi, \mathscr{A}), \alpha) = (0, \alpha_1)$ to a cycle of steady-state solutions $(\psi_\alpha, \mathscr{A}_\alpha)$, which is an attractor attracting an open set $U \setminus \Gamma \subset H$. From the physical point of view, this theorem leads to the following properties of superconducting transitions in the case $R < 0$.

Physical Conclusion 3.6.10 *Let $R < 0$ and $T = T_c^1$ satisfy (3.6.61). Then the following physical properties hold:*

(1) When the control temperature decreases (respectively increases) and crosses the critical temperature T_c^1, there will be a phase transition of the sample from the normal to superconducting state (respectively from the superconducting to normal state).

(2) (Stability.) When the control temperature T is greater than T_c^1, under a fluctuation deviating from the normal state, the sample will soon be restored to the normal state. In addition, when $T < T_c^1$, under a fluctuation deviating from both the normal and superconducting states, the sample will soon be restored to the superconducting state.

(3) In general, the supercurrent given by

$$J_s(\alpha) = -\gamma(A_a + \mathscr{A}_\alpha)|\psi_\alpha|^2 - \frac{\gamma\mu}{2} i(\psi_\alpha^* \nabla \psi_\alpha - \psi_\alpha \nabla \psi_\alpha^*)$$

is nonzero, i.e., $J_s \not\equiv 0$ for $T < T_c^1$ $(\alpha_1 < \alpha)$.

(4) *(Continuity.) The order parameter ψ_α and supercurrent $J_s(\alpha)$ depend continuously on the control temperature T (or the parameter α), namely*

$$\psi_\alpha \to 0, \quad J_s(\alpha) \to 0, \quad \text{if } T \to (T_c^1)^- \quad (\text{or } \alpha \to \alpha_1^+).$$

(5) *The superconducting state of the system is dominated by the lowest-energy eigenfunction of (3.6.28) in the sense given by (3.6.41).*

(6) *The phase transition is of second order in the Ehrenfest sense with the critical exponent $\beta = \frac{1}{2}$, and its phase diagram is as shown in Fig. 3.30.*

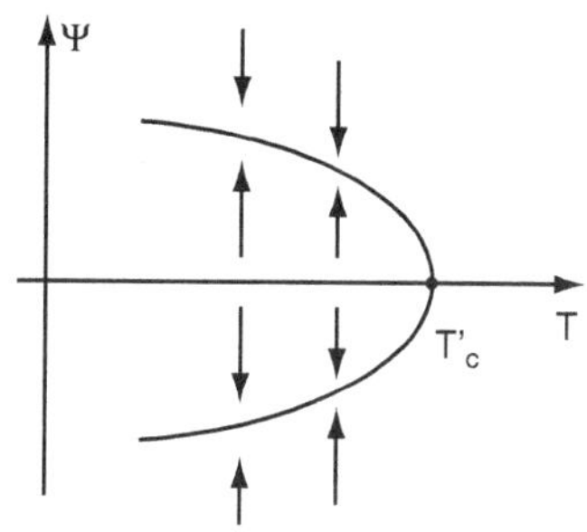

Fig. 3.30 Type I transition n the case $R < 0$

Transitions with $R > 0$. Phase transitions in this case are precisely described by Theorems 3.6.5–3.6.9, as shown in Figs. 3.28 and 3.29. In particular, we have the following conclusions, which recapitulate some superconducting properties obtained in Theorems 3.6.5–3.6.9 in physical terms.

Physical Conclusion 3.6.11 *Consider a material described by the TDGL model with $R > 0$. There are two transition temperatures T_c^0 and T_c^1 ($T_c^0 > T_c^1$) such that*

$$\alpha(T_c^i) = \alpha_i \quad (i = 0, 1) \quad \text{with} \quad \alpha_0 < \alpha_1,$$

and the following phase transition properties hold:

(1) *When the control temperature T decreases and crosses T_c^1 or equivalently α increases and crosses α_1, the stability of the normal state changes from stable to unstable.*

(2) *When $T_c^1 < T < T_c^0$ ($\alpha_0 < \alpha < \alpha_1$), physically observable states consist of the normal state and the superconducting states in Σ_α^2 (see Fig. 3.28). When $T < T_c^1$ ($\alpha_1 < \alpha$), physically observable states are in Σ_α^2.*

(3) *(Instability.) When the control temperature T is in the interval $T_c^1 < T < T_c^0$ (or $\alpha_0 < \alpha < \alpha_1$), the superconducting states in Σ_α^1 are unstable, i.e., with a fluctuation deviating from a superconducting state in Σ_α^1, transition to either the normal state or a superconducting state in Σ_α^2 will occur.*

(4) *(Discontinuity.) At the critical temperature T_c^0 (respectively at T_c^1) of the phase transitions, there is a jump from the superconducting states to the normal state (respectively from the normal state to superconducting states).*

(5) *The other energy-level eigenfunctions possibly have a stronger influence on the superconducting states.*
(6) *In the temperature interval $T_c^1 < T < T_c^0$, phase transitions occur and are accompanied by latent heat.*

Based on the conclusions (1)–(4) above, we can draw the phase diagram in Fig. 3.31, where the superconducting states in Σ_α^2 and the normal state $\psi = 0$ are metastable for $T_c^1 < T < T_c^0$, and Σ_T^2 is stable and $\psi = 0$ is unstable for $T < T_c^1$. Physically speaking, the metastable superconducting phases are the superheated states, and the metastable normal phase is the supercooled state.

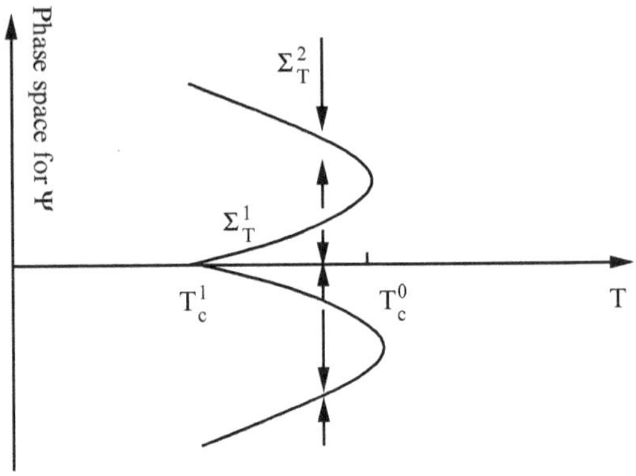

Fig. 3.31 The normal state $\psi = 0$ in $T_c^1 < T < T_c^0$ is supercooled; the superconducting states Σ_T^2 are superheated

Physical Significance of R**.** Physically, it is considered that a phase transition is of second order if the applied magnetic field H_a is equal to 0, and it is of first order if $H_a \neq 0$. However, by Theorems 3.6.4–3.6.6, we know that the two types of continuous and discontinuous transitions should be distinguished by the parameter R. From (3.6.39) we can deduce that for a given material and Ω, there is a critical magnetic field intensity $\widetilde{H}_c^2 > 0$ such that

$$R = R(H_a)\begin{cases} < 0 & \text{if } H_a^2 < \widetilde{H}_c^2, \\ > 0 & \text{if } H_a^2 > \widetilde{H}_c^2. \end{cases} \tag{3.6.62}$$

Hence, when $H_a^2 < \widetilde{H}_c^2$, the phase transition is of second order, and when $H_a^2 > \widetilde{H}_c^2$, it is of zero order, i.e., the Ginzburg–Landau free energy E in (3.6.25) has a jump at the critical temperature.

We now consider the influence of the physical parameters on R. It is noted that the parameter β/γ can be characterized by the Ginzburg–Landau number κ and the parameter μ:

$$\frac{\beta}{\gamma} = \kappa^2\mu^2, \quad \kappa^2 = \frac{m_s^2 C^2}{2\pi e_s^2 h^2} b, \quad \mu^2 = \frac{h^2}{e_s^2}\frac{D^2}{b}. \tag{3.6.63}$$

In the Ginzburg–Landau energy, the term

$$E_0 = \int_\Omega |e|^4 dx \tag{3.6.64}$$

represents the nonlinear part of the energy of the superconducting electrons in the lowest-energy state, and the term

$$E_m = \int_\Omega H_0^2 dx, \quad H_0 = \text{curl}\mathscr{A}_0 \text{ satisfies (3.6.39),;} \tag{3.6.65}$$

is the energy contributed by the magnetic field associated with the supercurrent

$$\text{curl}H_0 = |e|^2 A_a + \frac{\mu}{2} i(e^*\nabla e - e\nabla e^*),$$

which is generated by the applied magnetic potential A_a and the superconducting electrons in the lowest-energy states.

By (3.6.40), we obtain from (3.6.63)–(3.6.65) that

$$R = -\kappa^2\mu^2 + \frac{2E_m}{E_0}.$$

Therefore, the type of superconducting phase transitions among the two described above for a given material depends on the "competition" between the two energies E_0 and E_m:

$$R \begin{cases} < 0 \text{ if } \frac{\kappa^2}{2} E_0 > \frac{1}{\mu^2} E_m, \\ > 0 \text{ if } \frac{\kappa^2}{2} E_0 < \frac{1}{\mu^2} E_m. \end{cases} \tag{3.6.66}$$

According to Abrikosov theory, materials with $\kappa^2 < \frac{1}{2}$ and $\kappa^2 > \frac{1}{2}$ are of types I and II respectively. From (3.6.66), we infer that for a given geometric shape of a sample and applied magnetic field, a Type I material favors a jump phase transition (i.e., the case $(R > 0)$), and a Type II material favors a continuous phase transition (i.e., the case $(R < 0)$).

Finally, we need to consider the action of entropy. By (3.6.55) and (3.6.56), the parameter $\widetilde{R}$ associated with entropy is given by

$$\widetilde{R} = -\kappa^2\mu^2 + \frac{2E_m}{E_0} + \frac{k_3^2}{\gamma} \frac{\int_\Omega \left[|e|^2(-k_1\Delta + k_2)^{-1}|e|^2\right] dx}{E_0}. \tag{3.6.67}$$

If we neglect the term $-k_1\Delta$ in (3.6.67), then we have

$$\widetilde{R} = -\kappa^2\mu^2 + \frac{k_3^2}{\gamma k_2} + \frac{2E_m}{E_0}, \tag{3.6.68}$$

where the parameters $k_2, k_3 > 0$ depend on the materials.

It is known that $E_m = 0$ if $H_a = 0$. For general materials, physical experiments have shown that when $H_a = 0$, the superconducting phase transitions are continuous or of Type I, which implies that for most materials, the parameters in $\widetilde{R}$ satisfy

$$\frac{k_3^2}{\gamma k_2} - \kappa^2\mu^2 \leq 0.$$

However, we cannot exclude the existence of materials in which the parameters satisfy

$$\frac{k_3^2}{\gamma k_2} - \kappa^2\mu^2 > 0. \tag{3.6.69}$$

For materials possessing property (3.6.69), by (3.6.68) and Theorem 3.6.9, the superconducting phase transition is of Type II.

Critical Length Scale and Strength of Applied Field. Let $H_a = KH_a^0 = K\text{curl}A_a^0$, with K the strength of the applied field. Then the parameter R can be written as

$$R = -\kappa^2\mu^2 + 2(K^2q_2 + Kq_1 + q_0). \tag{3.6.70}$$

Here the parameters in (3.6.70) are given by

$$q_2 = \frac{\int_\Omega |\text{curl}\mathscr{A}_{01}|^2 dx}{\int_\Omega |e|^4 dx}, \quad q_1 = \frac{2\int_\Omega \text{curl}\mathscr{A}_{01}\cdot\text{curl}\mathscr{A}_{02} dx}{\int_\Omega |e|^4 dx}, \quad q_0 = \frac{\int_\Omega |\text{curl}\mathscr{A}_{02}|^2 dx}{\int_\Omega |e|^4 dx},$$

and $\mathscr{A}_{01}$ and $\mathscr{A}_{02}$ solve the following Stokes problems:

$$\begin{cases} \text{curl}^2\mathscr{A}_{01} + \nabla\phi = |e|^2 A_a^0, \\ \text{div}\mathscr{A}_{01} = 0, \\ \mathscr{A}_{01}\cdot n|_{\partial\Omega} = 0, \quad \text{curl}\mathscr{A}_{01}\times n|_{\partial\Omega} = 0. \end{cases} \tag{3.6.71}$$

$$\begin{cases} \text{curl}^2\mathscr{A}_{02} + \nabla\phi = \frac{\mu}{2} i(e^*\nabla e - e\nabla e^*), \\ \text{div}\mathscr{A}_{02} = 0, \\ \mathscr{A}_{02}\cdot n|_{\partial\Omega} = 0, \quad \text{curl}\mathscr{A}_{02}\times n|_{\partial\Omega} = 0. \end{cases} \tag{3.6.72}$$

It is easy to see that q_0, q_1, and q_2 satisfy

$$\begin{aligned} &0 < C_1 \leq q_2 \leq C_2, \qquad 0 \leq q_1, q_0 \leq C_2, \\ &q_0 = q_1 = 0 \qquad \text{if } K = 0. \end{aligned}$$

Hence, it is easy to verify that there is a unique solution $K_c > 0$ solving

$$R = -\kappa^2\mu^2 + 2(K^2q_2 + Kq_1 + q_0) = 0.$$

Hence by (3.6.38), we derive the following physical prediction:

Physical Conclusion 3.6.12 (Prediction of Critical Magnetic Strength) *There is a critical magnetic strength K_c such that for $0 \leq K < K_c$, the transition is of second order, and for $K > K_c$, the transition is of first order, provided that*

$$\alpha_1(K_c\widetilde{A}_a) < \frac{2\sqrt{b}m_s D N_0}{e_s^3 h}.$$

Formula of R in Terms of Sample Size. In a case to be defined below, the formula for R given by (3.6.40) reduces to

$$R(L,H) = -\kappa^2\mu^2 + 2L^2(p_3L^4 + p_2L^2 + p_1). \tag{3.6.73}$$

Here the notation is as follows:

First, for $0 < L < \infty$ and $h > 0$, let

$$\begin{aligned}
&\Omega_0 = D_0 \times (0,h) \subset \mathbb{R}^3, \qquad x' = (x_1,x_2) \in D_0, \qquad 0 \in D_0,\\
&\Omega(L) = \{(Lx',x_3)|\ x' \in D_0, 0 < x_3 < h\} = (LD_0) \times (0,h).
\end{aligned}$$

Second, the applied field is $H_a = H_a(x') = (0,0,H(x'))$, where $x' = (x_1,x_2) \in D_0$. Then H_a induces an applied field $\widetilde{H}_a$ on $\Omega(L)$ by $\widetilde{H}(y) = H(y/L)$ for any $y = Lx'$ with $x' \in D_0$. Also, we let $H_a = \text{curl}A_a$, i.e., $H = \frac{\partial A_1}{\partial x_2} - \frac{\partial A_2}{\partial x_1}$, where

$$A_a = (A_1(x'),A_2(x'),0). \tag{3.6.74}$$

Third the parameters in (3.6.73) are given by

$$\begin{cases}
p_3 = \dfrac{\int_{D_0} |\text{curl}A_0|^2 dx'}{\int_{D_0} |e|^4 dx'} > \delta > 0,\\
\text{curl}^2 A_0 + \nabla\phi = |e|^2 A_a \quad \text{in } x' \in D_0,
\end{cases}$$

$$\begin{cases}
p_1 = \dfrac{4\mu^2 \int_{D_0} |\text{curl}B_0|^2 dx'}{\int_{D_0} |e|^4 dx'},\\
\text{curl}^2 B_0 + \nabla\phi = e_{12}\nabla e_{11} \quad \text{in } x' \in D_0,
\end{cases}$$

$$p_2 = \frac{4\mu \int_{D_0} \text{curl}A_0 \cdot \text{curl}B_0 dx'}{\int_{D_0} |e|^4 dx'}.$$

Here $\delta > 0$ is independent of L, and $e(x') = e_{11} + ie_{12}$ is the first eigenfunction of

$$\begin{aligned}
&(i\mu\nabla + L^2 A_a)^2 e = L^2 \alpha e && \text{in } D_0,\\
&\frac{\partial e}{\partial n} = 0 && \text{on } \partial D_0.
\end{aligned} \tag{3.6.75}$$

It is easy to show that there is a unique positive solution L_c such that

$$p_3 L_0^6 + p_2 L_0^4 + p_1 L_0^2 = \frac{1}{2}\kappa^2\mu^2.$$

Hence we derive the following physical prediction:

Physical Conclusion 3.6.13 (Physical Prediction of Critical Length Scale) *Given a superconducting material and an applied field $H_a \neq 0$, there is a critical scale $L_c > 0$ such that the phase transition in $\Omega(L)$ is a second-order transition if $L < L_c$, and is a first-order transition if $L > L_c$.*

Abrikosov Vortices and Topological Structure of Supercurrents. In 1957, A. Abrikosov predicted that in the mixed state of Type II superconductors, there is a square or triangular array of vortices of supercurrents, called the Abrikosov vortices, and this vortex array was confirmed by experiments later. In Abrikosov theory, for Type II superconductors, the first and second critical magnetic fields H_{c_1} and H_{c_2} are given by

$$H_{c_1} = \frac{H_c}{\sqrt{2}\kappa}(\ln\kappa + 0.081), \quad H_{c_2} = \sqrt{2}\kappa H_c,$$

where H_c is the thermodynamic critical magnetic field. When $H_{c_1} < H < H_{c_2}$, the mixed state has two kinds of vortex structures: the square vortex lattice, as shown in Fig. 3.32a, and the triangular vortex lattice, as shown in Fig. 3.32b.

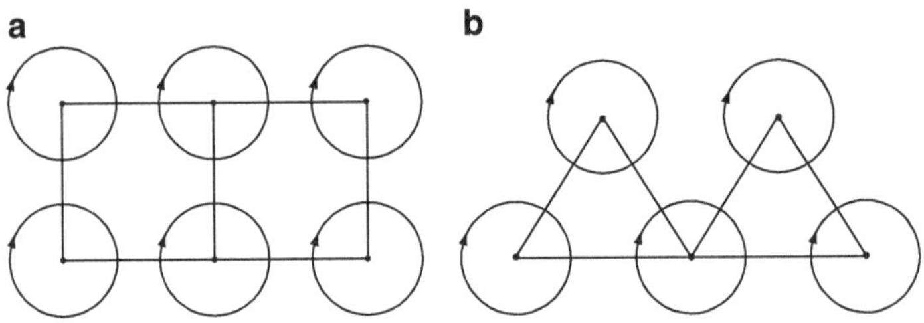

Fig. 3.32 (**a**) The square vortex lattice, (**b**) the triangular vortex lattice

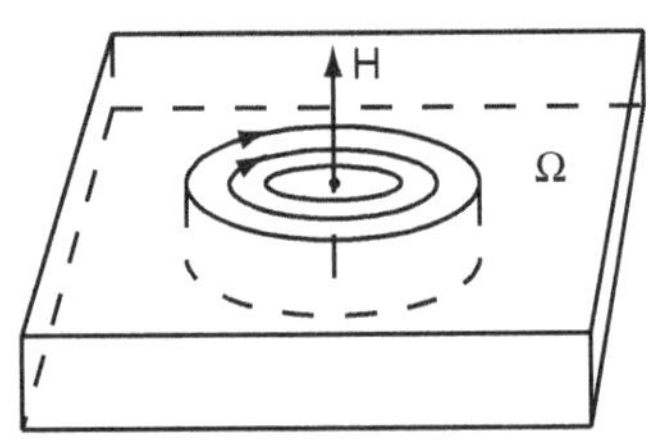

Fig. 3.33 Structure of vortex lines

In each vortex center N, the order parameter vanishes $\psi(N) = 0$, which corresponds to a normal region; the magnetic field H penetrates the sample in a neighborhood of N; and the supercurrents J_s form vortices around N in the reverse direction to that of H, as shown in Fig. 3.33. Each vortex is called a vortex line or a magnetic flux line.

We investigate this problem from the mathematical viewpoint. The supercurrent J_s satisfies

$$J_s = -\gamma|\psi|^2(\mathscr{A} + A_a) - \frac{\gamma\mu}{2}i(\psi^*\nabla\psi - \psi\nabla\psi^*) \tag{3.6.76}$$

$$\text{div} J_s = 0, \quad J_s \cdot n|_{\partial\Omega} = 0. \tag{3.6.77}$$

Based on the geometric theory for two-dimensional incompressible flows introduced in Ma and Wang (2005d), by (3.6.77), the topological structure of J_s corresponding to the square and triangular vortex lattices in mathematics are as shown in Fig. 3.34a, b respectively. However, physically, the supercurrents J_s have the countermagnetic property, which can be seen from the first term in (3.6.76). Thus, the paramagnetic vortices (i.e., the counterclockwise vortices) in Fig. 3.34a, b do not appear in the real world. Therefore, combining this with the physical version, we see that the real topological structure of J_s should be as shown in Fig. 3.35a, b, where the shadowed regions corresponding to the counterclockwise vortex domains in Fig. 3.34 represent the vanishing supercurrent $J_s = 0$, or the surface supercurrents.

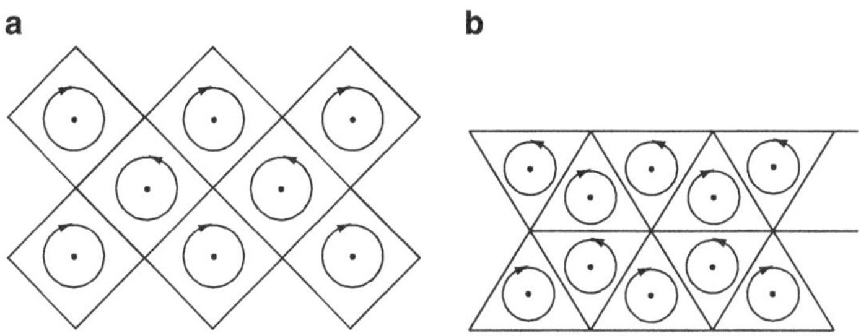

Fig. 3.34 Mathematical vortex structure of J_s

In our previous discussion, we have seen that if the applied magnetic field $H_a = (0,0,H)$ is weak, then the superconductivity is in the Meissner state. In this case, $\psi \simeq$ constant, and the supercurrent J_s in (3.6.76) is approximatively expressed by

$$J_s = -\gamma|\psi|^2 A, \quad (A = \mathscr{A} + A_a).$$

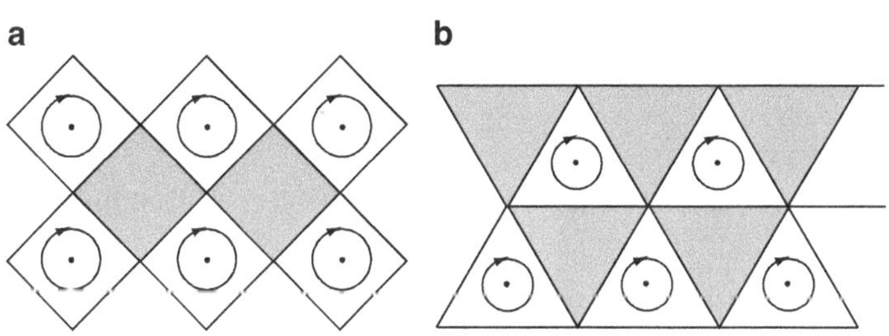

Fig. 3.35 Abrikosov vortex structure of J_s

This supercurrent is completely countermagnetic, which causes the Meissner effect. Furthermore, in the Meissner state, the supercurrent J_s exists only on the surface $\partial\Omega$ of the sample Ω with depth λ, where $\lambda = \lambda(T) = (m_s c^2 b / 4\pi e_s^2 |a|)^{1/2}$ is the penetration depth. Thus, J_s is essentially a two-dimensional incompressible flow, and the index formula (6.1.1) in Ma and Wang (2005d) is valid, i.e.,

$$C - S - \frac{B}{2} = 1 - k,$$

where C is the number of centers, S the number of interior saddles, and B the number of boundary saddles.

In addition, by the structural stability theorems, Theorems 2.1.2 and 2.2.9 in Ma and Wang (2005d), the supercurrents J_s observable in experiments are structurally stable.

3.7 Liquid Helium-4

Superfluidity is a phase of matter in which "unusual" effects are observed when liquids, typically of helium-4 or helium-3, overcome friction by surface interaction when at a stage, known as the "lambda point" for helium-4, at which the liquid's viscosity becomes zero. Also known as a major facet in the study of quantum hydrodynamics, it was discovered by Pyotr Leonidovich Kapitsa, John F. Allen, and Don Misener in 1937 and has been described through phenomenological and microscopic theories.

Atoms of helium have two stable isotopes, ^{4}He and ^{3}He. The former consists of two electrons, two protons, and two neutrons, which are six fermions. Therefore, ^{4}He has an integral spin and obeys the Bose–Einstein statistics. Liquid ^{4}He, called the Bose liquid, displays a direct transition from the normal liquid state (liquid He I) to the superfluid state (liquid He II) at temperature $T = 2.19\,\text{K}$, which can be considered the condensation of particles at a simple quantum state.

In the late 1930s, Ginzburg and Landau proposed a mean field theory of continuous phase transitions. With the successful application of the Ginzburg–Landau theory to superconductivity, it is natural to transfer something similar to the superfluidity case, since the superfluid transitions in liquid ^{3}He and ^{4}He are of similar quantum origin to that of superconductivity. Unfortunately, the classical Ginzburg–Landau free energy is poorly applicable to ^{4}He in a quantitative sense, as described in Ginzburg (2004).

The main objectives of this section are (1) to establish a TDGL model for liquid ^{4}He, and (2) to study its dynamic phase transitions. The results obtained in this section predict:

1. The existence of an unstable region H where both solid and liquid He II states appear randomly depending on fluctuations, and
2. The existence of a switch point M on the λ curve, where the transitions change from first order (Type II with the dynamic classification scheme) to second order (Type I).

An important difference between the dynamic models introduced here and the classical ones is based on the separation of superfluid and normal fluid densities as described in (3.7.1) and their interactions in the Ginzburg–Landau free energies; see among others (Onuki 2002). The results presented in this section were first proved in Ma and Wang (2008a).

3.7.1 Dynamic Model for Liquid Helium-4

Helium 4 was first liquidized at $T = 4.215\,\mathrm{K}$ and $p = 1 \times 10^5(\mathrm{Pa})$ by Karmerlingh Onnes in 1908. In 1938, P.L. Kapitza found that when the temperature T decreases below $T_C = 2.17\,\mathrm{K}$, liquid ^{4}He will transit from a normal liquid state to a superfluid state, in which the fluid possesses zero viscosity, i.e., the viscous coefficient η is equal to 0. Liquids with $\eta = 0$ are called superfluids, and flow without drag is called superfluidity. A superfluid transition is called a λ-phase transition, and its phase diagram is illustrated by Fig. 3.36.

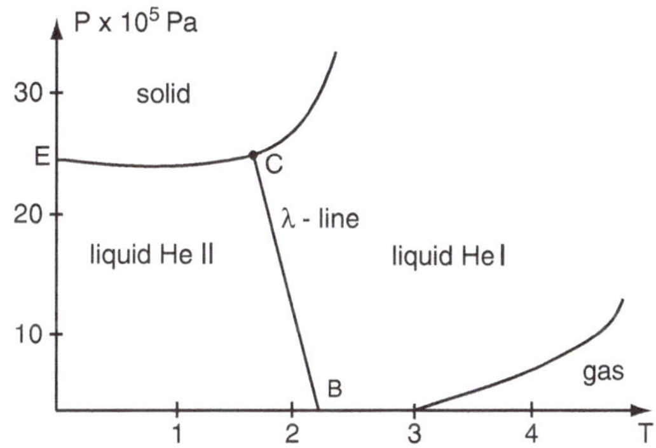

Fig. 3.36 Classical PT phase diagram of ^{4}He

Ginzburg–Landau Free Energy. The order parameter describing superfluidity is characterized by a nonvanishing complexvalued function $\psi : \Omega \to \mathbb{C}$ as in superconductivity, originating from quantum Bose–Einstein condensation. In the two-fluid hydrodynamic theory of superfluidity, the density ρ of ^{4}He is given by

$$\rho = \rho_s + \rho_n, \tag{3.7.1}$$

where ρ_s is the superfluid density and ρ_n the normal fluid density. The square $|\psi|^2$ is proportional to ρ_s, and without loss of generality, we take ψ as

$$|\psi|^2 = \rho_s. \tag{3.7.2}$$

Based on the quantum mechanics, $-ih\nabla\psi$ represents the momentum associated with the Bose–Einstein condensation. Hence, the free energy density contains the term

$$\frac{1}{2m}|-ih\nabla\psi|^2 = \frac{h^2}{2m}|\nabla\psi|^2,$$

where h is Planck's constant, and m the mass of an atom of ^{4}He . Meanwhile, the superfluid state does not obey the classical thermodynamic laws that a normal liquid state obeys. Therefore, in the free energy density, ψ satisfies the Ginzburg–Landau expansion

$$\frac{k_1 h^2}{2m}|\nabla\psi|^2 + \frac{\gamma_1}{2}|\psi|^2 + \frac{\gamma_2}{4}|\psi|^4,$$

and ρ_n has the expansion as in the free energy (3.3.5) for PVT systems. For simplicity, we ignore the entropy, and consider the coupling action of ψ and ρ_n, i.e., we add the term $\frac{1}{2}\mu\rho_n|\psi|^2$ in the free energy density. Thus the Ginzburg–Landau free energy for liquid ^{4}He near the superfluid transition is given by

$$\begin{aligned} G(\psi,\rho_n) = \int_\Omega \Big[&\frac{k_1 h^2}{2m}|\nabla\psi|^2 + \frac{\gamma_1}{2}|\psi|^2 + \frac{\gamma_2}{4}|\psi|^4 + \frac{\gamma_3}{2}\rho_n|\psi|^2 \\ &+ \frac{k_2}{2}|\nabla\rho_n|^2 + \frac{\mu_1}{2}\rho_n^2 + \frac{\mu_2}{3}\rho_n^3 + \frac{\mu_3}{4}\rho_n^4 - p(\rho_n + \frac{\mu_0}{2}\rho_n^2)\Big]dx. \end{aligned} \tag{3.7.3}$$

Dynamic model governing superfluidity. By (3.1.15), we derive from (3.7.3) the following TDGL equations governing the superfluidity of liquid ^{4}He:

$$\begin{aligned} \frac{\partial\psi}{\partial t} &= \frac{k_1 h^2}{m}\Delta\psi - \gamma_1\psi - \gamma_2|\psi|^2\psi - \gamma_3\rho_n\psi, \\ \frac{\partial\rho_n}{\partial t} &= k_2\Delta\rho_n - (\mu_1 - p\mu_0)\rho_n - \mu_2\rho_n^2 - \mu_3\rho_n^3 - \frac{\gamma_3}{2}|\psi|^2 + p. \end{aligned} \tag{3.7.4}$$

It is known that the following problem has a solution $\rho_n^0 \in H^2(\Omega)\bigcap H_0^1(\Omega)$ standing for the density of liquid He I for $p \in L^2(\Omega)$:

$$\begin{aligned} &-k_2\Delta\rho_n^0 + (\mu_1 - p\mu_0)\rho_n^0 + \mu_2(\rho_n^0)^2 + \mu_3(\rho_n^0)^3 = p, \\ &\left.\frac{\partial\rho_n^0}{\partial n}\right|_{\partial\Omega} = 0. \end{aligned} \tag{3.7.5}$$

To derive the nondimensional form of (3.7.4), let

$$\begin{aligned} &(x,t) = (lx', \tau t'), && (\psi,\rho_n) = (\psi_0\psi', \rho_0\rho_n' + \rho_n^0), \\ &\tau = \frac{ml^2}{h^2 k_1}, && \mu = \frac{k_2\tau}{l^2}, \\ &a_1 = \gamma_3\rho_0\tau, && a_2 = \gamma_2|\psi_0|^2\tau, \\ &b_1 = \frac{\gamma_3|\psi_0|^2}{2\rho_0}\tau, && b_2 = \tau\rho_0(3\rho_n^0\mu_3 + \mu_2), \\ &b_3 = \rho_0^2\mu_3\tau, && \lambda_1 = -\tau(\gamma_1 + \gamma_3\rho_n^0), \\ &\lambda_2 = \tau(3(\rho_n^0)^2\mu_3 + 2\rho_n^0\mu_2 + \mu_0 p - \mu_1), \end{aligned}$$

where ρ_n^0 is the solution of (3.7.5).

Thus, suppressing the primes, equations (3.7.4) can be rewritten as

$$\begin{aligned}\frac{\partial \psi}{\partial t} &= \Delta\psi + \lambda_1\psi - a_1\rho_n\psi - a_2|\psi|^2\psi,\\ \frac{\partial \rho_n}{\partial t} &= \mu\Delta\rho_n + \lambda_2\rho_n - b_1|\psi|^2 - b_2\rho_n^2 - b_3\rho_n^3.\end{aligned} \tag{3.7.6}$$

The boundary conditions associated with (3.7.6) are

$$\frac{\partial \psi}{\partial n} = 0, \quad \frac{\partial \rho_n}{\partial n} = 0 \qquad \text{on } \partial\Omega. \tag{3.7.7}$$

When the pressure p is independent of $x \in \Omega$, then the problem (3.7.6) and (3.7.7) can be approximatively replaced by the following systems of ordinary differential equations for superfluid transitions:

$$\begin{aligned}\frac{d\psi}{dt} &= \lambda_1\psi - a_1\rho_n\psi - a_2|\psi|^2\psi,\\ \frac{d\rho_n}{dt} &= \lambda_2\rho_n - b_1|\psi|^2 - b_2\rho_n^2 - b_3\rho_n^3.\end{aligned} \tag{3.7.8}$$

Multiplying by ψ^* both sides of the first equation of (3.7.8) and (3.7.2), equations (3.7.8) are reduced to

$$\begin{aligned}\frac{d\rho_s}{dt} &= \lambda_1\rho_s - a_1\rho_n\rho_s - a_2\rho_s^2,\\ \frac{d\rho_n}{dt} &= \lambda_2\rho_n - b_1\rho_s - b_2\rho_n^2 - b_3\rho_n^3,\\ (\rho_s(0),\rho_n(0)) &= (x_0, y_0),\end{aligned} \tag{3.7.9}$$

for $\rho_s \geq 0$ and $x_0 \geq 0$.

We need to explain the physical properties of the coefficients in (3.7.4) and (3.7.6). It is known that the coefficients γ_i $(1 \leq i \leq 3)$ and μ_j $(0 \leq j \leq 3)$ depend continuously on the temperature T and the pressure p:

$$\gamma_i = \gamma_i(T,p), \quad \mu_j = \mu_j(T,p) \qquad \forall 1 \leq i \leq 3, \quad 0 \leq j \leq 3.$$

From both the mathematical and physical points of view, the following conditions are required:

$$\gamma_2 > 0, \quad \gamma_3 > 0, \quad \mu_3 > 0 \qquad \forall T, p. \tag{3.7.10}$$

In addition, by Landau mean field theory, we have

$$\gamma_1 \begin{cases} > 0 & \text{if } (T,p) \in A_1,\\ < 0 & \text{if } (T,p) \in A_2,\end{cases} \tag{3.7.11}$$

$$\mu_1 \begin{cases} > 0 & \text{if } (T,p) \in B_1,\\ < 0 & \text{if } (T,p) \in B_2,\end{cases} \tag{3.7.12}$$

where A_i, B_i $(i = 1,2)$ are connected open sets such that $\overline{A}_1 + \overline{A}_2 = \overline{B}_1 + \overline{B}_2 = \mathbb{R}^2_+$, and $\overline{A}_1 \bigcap \overline{A}_2, \overline{B}_1 \bigcap \overline{B}_2$ are two simple curves in $\mathbb{R}^2_+$; see Fig. 3.37a, b. In particular, in the PT-plane,

$$\frac{\partial \gamma_1}{\partial T} > 0, \quad \frac{\partial \gamma_1}{\partial p} > 0, \quad \frac{\partial \mu_1}{\partial T} > 0, \quad \frac{\partial \mu_1}{\partial p} > 0. \tag{3.7.13}$$

By (3.7.10), the following nondimensional parameters are positive:

$$a_1 > 0, \quad a_2 > 0, \quad b_1 > 0, \quad b_3 > 0. \tag{3.7.14}$$

By (3.7.11)–(3.7.13), the two critical parameter equations

$$\lambda_1 = \lambda_1(T,p) = 0, \qquad \lambda_2 = \lambda_2(T,p) = 0, \tag{3.7.15}$$

define two simple curves l_1 and l_2 in the PT-plane $\mathbb{R}^2_+$; see Fig. 3.37c, d.

The parameters μ_2 and b_2 depend on the physical properties of the helium atom, and they satisfy the following relations:

$$\begin{aligned} &b_2(T,p) = \tau\rho_0(\mu_2 + 3\rho_n^0\mu_3) < 0 \quad \text{iff } \rho_{sol} > \rho_l \text{ at } \lambda_2(T,p) = 0, \\ &b_2(T,p) = \tau\rho_0(\mu_2 + 3\rho_n^0\mu_3) > 0 \quad \text{iff } \rho_{sol} < \rho_l \text{ at } \lambda_2(T,p) = 0, \end{aligned} \tag{3.7.16}$$

where ρ_{sol} and ρ_l are the densities of the solid and liquid, and ρ_n^0 is the solution of (3.7.5) representing the liquid density. These relations in (3.7.16) can be deduced by the dynamic transition theorem of (2.1.1) from a simple critical eigenvalue.

3.7.2 Dynamic Phase Transition for Liquid ^{4}He

In order to illustrate the main ideas, we discuss only the case in which the pressure p is independent of $x \in \Omega$, i.e., we consider only the equations (3.7.8) and (3.7.9); the general case can be studied in the same fashion and will be reported elsewhere.

***PT*-Phase Diagram.** Based on physical experiments together with (3.7.11)–(3.7.13), the curves of $\gamma_1(T,p) = 0$ and $\mu_1(T,p) = 0$ in the PT-plane are given by Figs. 3.37a, b respectively. By the formulas

$$\begin{aligned} \lambda_1(T,p) &= -\tau(\gamma_1(T,p) + \gamma_3\rho_n^0), \\ \lambda_2(T,p) &= -\tau(\mu_1(T,p) - \mu_0 p - 2\rho_n^0\mu_2 - 3\rho_n^{02}\mu_3), \end{aligned}$$

together with (3.7.10) and (3.7.16), the curves l_1 and l_2 given by (3.7.15) in the PT-plane are illustrated in Fig. 3.37c, d.

Let

$$\begin{aligned} &D_1 = \{(T,p) \in \mathbb{R}^2_+ \mid \lambda_1(T,p) > 0\}, \qquad D_2 = \{(T,p) \in \mathbb{R}^2_+ \mid \lambda_1(T,p) < 0\}, \\ &E_1 = \{(T,p) \in \mathbb{R}^2_+ \mid \lambda_2(T,p) > 0\}, \qquad E_2 = \{(T,p) \in \mathbb{R}^2_+ \mid \lambda_2(T,p) < 0\}. \end{aligned}$$

Let the curve of $\lambda_1(T,p) = 0$ intersect the curve of $\lambda_2(T,p) = 0$ at a point C; see Fig. 3.38.

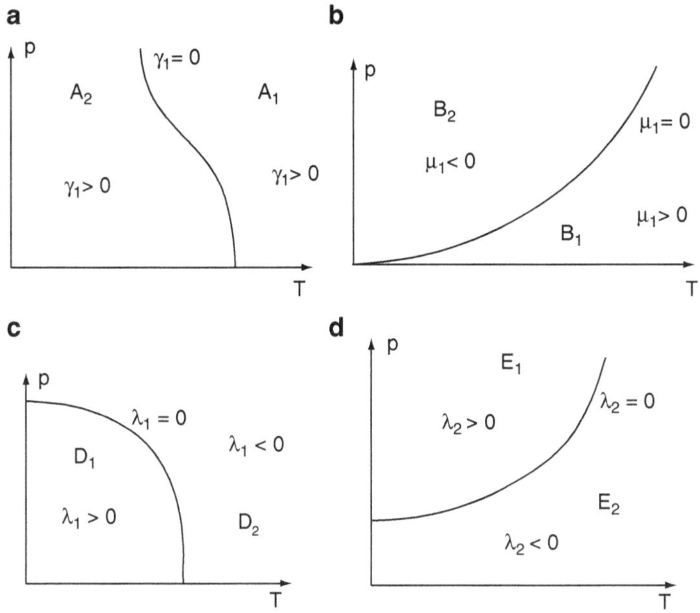

Fig. 3.37 (**a**) The curve of $\gamma_1 = 0$, (**b**) the curve of $\mu_1 = 0$, (**c**) the curve of $\lambda_1 = 0$, (**d**) the curve of $\lambda_2 = 0$

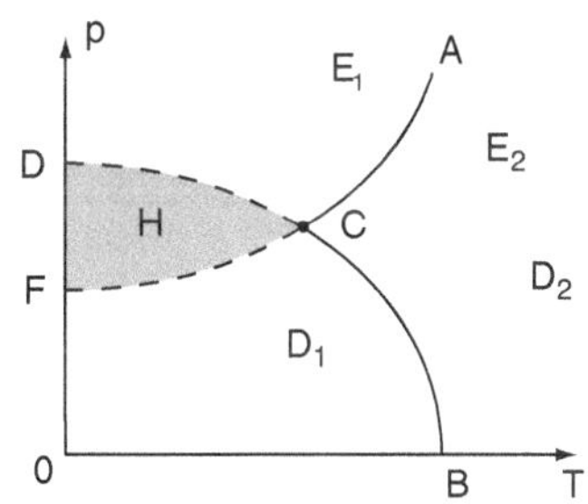

Fig. 3.38 *ACF* is the curve of $\lambda_2 = 0$, and *BCD* is the curve of $\lambda_1 = 0$

When λ_1 crosses the curve segment CB to enter into D_1 from D_2 (see Fig. 3.38), we have $\lambda_2 < 0$, and λ_1 satisfies

$$\lambda_1(T,p) \begin{cases} <0 & \text{if } (T,p) \in D_2, \\ =0 & \text{if } (T,p) \in CB, \\ >0 & \text{if } (T,p) \in D_1. \end{cases}$$

In this case, by Theorem 2.1.3, equations (3.7.9) have a phase change that describes the transition from liquid He-I to liquid He-II, and the second equation of (3.7.9) can be equivalently rewritten as

$$\frac{d\widetilde{\rho}_n}{dt} = (\lambda_2 - 2b_2\rho_n^* - 3b_3(\rho_n^*)^2)\widetilde{\rho}_n - (b_2 + 3b_3\rho_n^*)\widetilde{\rho}_n^2 - b_3\widetilde{\rho}_n^3, \tag{3.7.17}$$

where $\widetilde{\rho}_n = \rho_n - \rho_n^*$, ρ_n^* satisfies the equation

$$\lambda_2 \rho_n - b_2 \rho_n^2 - b_3 \rho_n^3 = b_1 \rho_s,$$

and $\rho_s > 0$ is the transition solution of (3.7.9). Since ρ_n^* is the density deviation of normal liquid, by (3.7.1), we have

$$\rho_n^* = \rho - \rho_n^0 - \rho_s.$$

On the other hand, ρ_n^0 ($\simeq \rho$) represents the density of liquid He I. Thus, $\rho_n^* \simeq -\rho_s < 0$. From the PT-phase diagram of ^{4}He (Fig. 3.36), we see that $\rho_{sol} \geq \rho_l$ for ^{4}He near $T = 0$, and therefore, by (3.7.16), $b_2 \leq 0$. Hence we derive that

$$\lambda_2 - 2b_2 \rho_n^* - 3b_3 (\rho_n^*)^2 < 0, \tag{3.7.18}$$

for (T, p) in the region $D_1 \setminus H$, as shown in Fig. 3.38.

It follows from (3.7.17) and (3.7.18) that when (T, p) is in the region $D_1 \setminus H$, the liquid ^{4}He is in the superfluid state.

When λ_2 crosses the curve segment CA, as shown in Fig. 3.38, to enter into E_1 from E_2, then $\lambda_1 < 0$ and

$$\lambda_2(T, p) \begin{cases} < 0 & \text{if } (T, p) \in E_2, \\ = 0 & \text{if } (T, p) \in CA, \\ > 0 & \text{if } (T, p) \in E_1. \end{cases}$$

In this case, the equations (3.7.9) characterize the liquid–solid phase transition, and the first equation of (3.7.9) is equivalently expressed as

$$\frac{d\rho_s}{dt} = (\lambda_1 - a_1 \widetilde{\rho}_n)\rho_s - a_2 \rho_s^2, \tag{3.7.19}$$

where $\widetilde{\rho}_n > 0$ is the transition state of solid ^{4}He. By (3.7.14), we have

$$\lambda_1(T, p) - a_1 \widetilde{\rho}_n < 0, \tag{3.7.20}$$

for all (T, p) in the region $E_1 \setminus H$, as shown in Fig. 3.38.

From (3.7.19) and (3.7.20), we can derive the conclusion that since (T, p) is in $E_1 \setminus H$, $\rho_s = 0$ is stable, i.e., ^{4}He is in the solid state.

However, the shaded region H in Fig. 3.38 is a metastable domain for the solid and liquid He II states, where any of these two phases may appear depending on random fluctuations. Thus, from the discussion above, we can derive the theoretical PT-phase diagram given by Fig. 3.39, based on equations (3.7.9). In comparison with the experimental PT-phase diagram (Fig. 3.36), a sight difference in Fig. 3.39 is that there exists a metastable region H where it is possible for the solid phase and the He II phase to occur. This metastable region in Fig. 3.39 corresponding to the coexistence curve CE in Fig. 3.36 is a theoretical prediction, which needs to be verified by experiments.

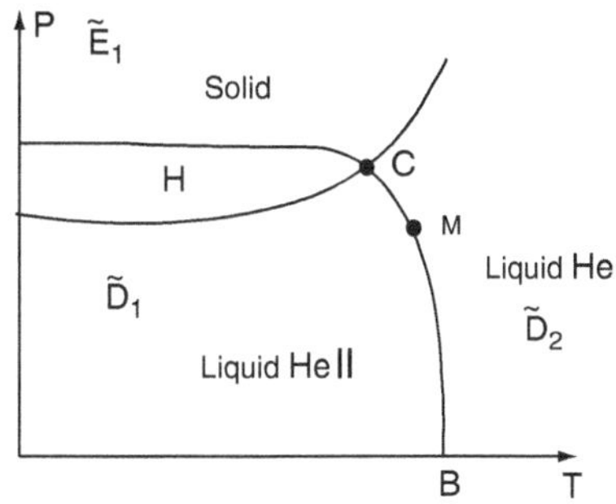

Fig. 3.39 Theoretical *PT*-phase diagram: *H* is a metastable region where both solid and liquid He II states appear randomly depending on fluctuations. The point *M* is a point where the transition between the superfluid state (liquid He II) and the normal fluid state (liquid He I) changes from Type II to Type I; namely, the transition crossing *CM* is of Type II, and the transition crossing *MB* is of Type I

States in the Metastable Region. We consider the dynamical properties of transitions for (3.7.9) in the unstable region. It is clear that at the point $C = (T_C, p_C)$,

$$\lambda_1(T_C, p_C) = 0, \quad \lambda_2(T_C, p_C) = 0, \tag{3.7.21}$$

and the unstable region H satisfies

$$H = \{(T,p) \in \mathbb{R}^2_+ \mid \lambda_1(T,p) > 0,\ \lambda_2(T,p) > 0\}.$$

To study the structure of flows of (3.7.9) for $(T,p) \in H$, it is necessary to consider equations (3.7.9) at the point $C = (T_C, p_C)$, which, by (3.7.21), are given by

$$\begin{aligned} \frac{d\rho_s}{dt} &= -a_1\rho_n\rho_s - a_2\rho_s^2, \\ \frac{d\rho_n}{dt} &= -b_1\rho_s - b_2\rho_n^2 - b_3\rho_n^3. \end{aligned} \tag{3.7.22}$$

Due to (3.7.16) and $\rho_{sol} > \rho_l$, we have

$$b_2 < 0 \quad \text{for} \quad (T,p) \subset H. \tag{3.7.23}$$

Under the condition (3.7.23), equations (3.7.22) have the following two steady-state solutions:

$$Z_1 = (\rho_s, \rho_n) = (0, |b_2|/b_3), \qquad Z_2 = (\rho_s, \rho_n) = \left(\frac{a_1}{a_2}\alpha, -\alpha\right),$$

$$\alpha = \frac{|b_2|}{2b_3}\left(\sqrt{1 + 4a_1b_1b_3/a_2|b_2|^2} - 1\right).$$

By direct computation, we can prove that the eigenvalues of the Jacobian matrices of (3.7.22) at Z_1 and Z_2 are negative. Hence, Z_1 and Z_2 are stable equilibrium points of (3.7.22). Physically, Z_1 stands for the solid state, and Z_2 for the superfluid state.

The topological structure of (3.7.22) is schematically illustrated in Fig. 3.40a, where the two regions R_1 and R_2 divided by the curve AO in Fig. 3.40b are the basins of attraction of Z_1 and Z_2 respectively.

We note that in H, λ_1 and λ_2 are small, i.e.,

$$0 < \lambda_1(T,p), \quad \lambda_2(T,p) \ll 1, \quad \text{for } (T,p) \in H,$$

and (3.7.9) can be considered a perturbed system of (3.7.22).

Thus, for $(T,p) \in H$, the system (3.7.9) has four steady-state solutions $\widetilde{Z}_i = \widetilde{Z}(T,p)$ $(1 \leq i \leq 4)$ such that

$$\lim_{(T,p)\to(T_C,p_C)} (\widetilde{Z}_1(T,p), \widetilde{Z}_2(T,p), \widetilde{Z}_3(T,p), \widetilde{Z}_4(T,p)) = (Z_1, Z_2, 0, 0),$$

and $\widetilde{Z}_1$ and $\widetilde{Z}_2$ are stable, representing the solid and liquid He-II states respectively; $\widetilde{Z}_3$ and $\widetilde{Z}_4$ are two saddle points. The topological structure of (3.7.9) for $(T,p) \in H$ is schematically shown in Fig. 3.40c, and the basins of attraction of $\widetilde{Z}_1$ and $\widetilde{Z}_2$ are $\widetilde{R}_1$ and $\widetilde{R}_2$, as illustrated by Fig. 3.40d.

In summary, with the above analysis and dynamic transition theory, we arrive at the following theorem:

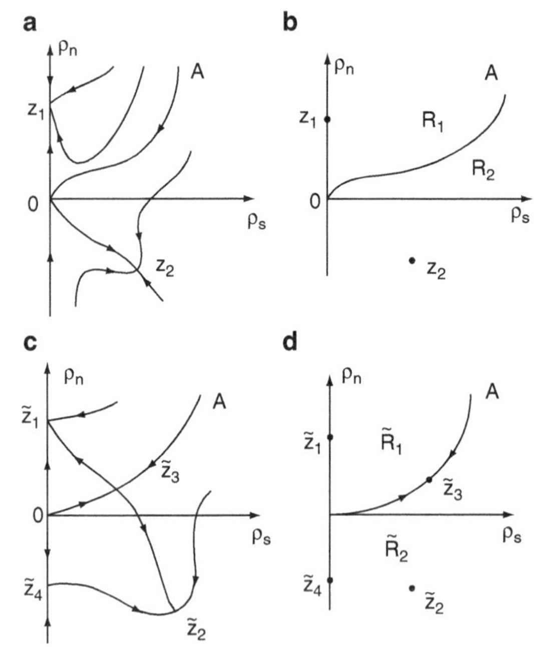

Fig. 3.40 Dynamic behavior of (3.7.9)

Theorem 3.7.1 *There exist four regions $\widetilde{E}_1, \widetilde{D}_1$, $\widetilde{D}_2$, and H in the PT-plane (see Fig. 3.39), which are defined by*

$$\begin{aligned}
\widetilde{E}_1 &= \{(T,p) \in \mathbb{R}^2_+ \mid \lambda_1(T,p) < 0,\ \lambda_2(T,p) > 0\},\\
\widetilde{D}_1 &= \{(T,p) \in \mathbb{R}^2_+ \mid \lambda_1(T,p) > 0,\ \lambda_2(T,p) < 0\},\\
\widetilde{D}_2 &= \{(T,p) \in \mathbb{R}^2_+ \mid \lambda_1(T,p) < 0,\ \lambda_2(T,p) < 0\},\\
H &= \{(T,p) \in \mathbb{R}^2_+ \mid \lambda_1(T,p) > 0,\ \lambda_2(T,p) > 0\},
\end{aligned}$$

such that the following conclusions hold:

(1) If $(T,p) \in \widetilde{E}_1$, the phase of ^4He is in the solid state.
(2) If $(T,p) \in \widetilde{D}_1$, the phase is in the superfluid state.
(3) If $(T,p) \in \widetilde{D}_2$, the phase is in the normal fluid state.
(4) If $(T,p) \in H$, there are two regions $\widetilde{R}_1$ and $\widetilde{R}_2$ in the state space (ρ_s, ρ_n) such that under a fluctuation that is described by the initial value (x_0, y_0) in (3.7.9), if $(x_0, y_0) \in \widetilde{R}_1$, then the phase is in solid the state, and if $(x_0, y_0) \in \widetilde{R}_2$, then it is in the superfluid state.

In general, the observed superfluid transitions of ^{4}He are of second order, i.e., of Type I (continuous) with the dynamic classification. But from (3.7.9), we can prove the following theorem, which shows that for a higher pressure, the superfluid transitions may be of zeroth order, i.e., a Type II (jump) transition with the dynamic classification scheme.

Theorem 3.7.2 *Let (T_0, p_0) satisfy $\lambda_1(T_0, \rho_0) = 0, \lambda_2(T_0, p_0) < 0$. Then equations (3.7.9) have a superfluid transition at (T_0, p_0) from D_2 to D_1. In particular, the following assertions hold:*

(1) Let

$$A = \frac{a_1 b_1}{|\lambda_2|} - a_2 \quad \text{at } (T,p) = (T_0, p_0). \tag{3.7.24}$$

Then if $A < 0$, the superfluid transition is of Type I, and if $A > 0$, then the superfluid transition is of Type II.

(2) The equation

$$A = \frac{a_1 b_1}{|\lambda_2|} - a_2 = 0$$

determines a point M on the CB coexistence curve in Fig. 3.39, where the transition changes from Type II to Type I.

Proof.

Step 1. By the assumption, it is clear that

$$\lambda_1(T,p) = \begin{cases} < 0 & \text{if } (T,p) \in D_2,\\ = 0 & \text{if } (T,p) = (T_0, p_0),\\ > 0 & \text{if } (T,p) \in D_1, \end{cases}$$

where D_1, D_2 are as in Fig. 3.37c. Hence, by Theorem 2.1.3, equations (3.7.9) have a transition at (T_0, p_0). By $\lambda_2(T_0, p_0) < 0$, the first eigenvalue of (3.7.9) is simple, and its eigenvector is given by $e = (e_s, e_n) = (1, 0)$.

The reduced equation of (3.7.9) on the center manifold reads

$$\frac{dx}{dt} = \lambda_1 x - a_1 x h(x) - a_2 x^2, \quad x > 0,$$

where $h(x)$ is the center manifold function. By (3.7.9), h can be expressed as

$$h(x) = \frac{b_1}{\lambda_2} x + o(x^2), \quad x > 0.$$

Thus the reduced equation of (3.7.9) is given by

$$\frac{dx}{dt} = \lambda_1 x + Ax^2 + o(x^2) \qquad \text{for } x > 0, \tag{3.7.25}$$

where A is as in (3.7.24). The theorem follows from (3.7.25) and Theorem 2.1.6.

Step 2. By the nondimensional form, we see that

$$a_1 b_1 = \frac{1}{2} \gamma_3^2 |\psi_0|^2 \tau^2,$$

where γ_3 is the coupled coefficient of ρ_n and ρ_s. Physically, γ_3 is small in comparison with γ_2; namely,

$$0 < a_1 b_1 \ll a_2.$$

On the other hand, we know that

$$\lambda_2(T_0, p_0) \to 0 \quad \text{as} \quad (T_0, p_0) \to (T_C, p_C),$$

where $C = (T_C, p_C)$ is as in (3.7.21). Therefore, we deduce that there exists a pressure p^* $(b < p^* < p_C)$ such that

$$A = \frac{a_1 b_1}{|\lambda_2(T_0, p_0)|} - a_2 \begin{cases} < 0 & \text{if } 0 \leq p_0 < p^*, \\ > 0 & \text{if } p^* < p_0 < p_C. \end{cases}$$

Thus, when the transition pressure p_0 is below some value $p_0 < p^*$, the superfluid transition is of second order, i.e., it is continuous with the dynamic classification scheme, and when $p_0 < p^*$, the superfluid transition is of first order, i.e., it is a jump in the dynamic classification scheme. The proof is complete. □

3.8 Superfluidity of Helium-3

The superfluidity of liquid ^{3}He was discovered in 1971 by D.M. Lee, D.D. Osheroff, and R.C. Richardson, and its transition temperature is $T \simeq 10^{-3}$ K under $p = 1$ atm (10^5 Pa). Helium 3 is a fermion and obeys the Fermi–Dirac statistics , and its atoms have to be paired to form the superfluid phase, leading to the existence of multiple superfluid phases; see Fig. 3.41. In particular, there are two superfluid phases A and B if there is no applied magnetic field.

The main objective of this section is:

1. to establish some TDGL models for ^{3}He with or without applied magnetic fields, and
2. to study superfluid dynamic transitions and their physical significance.

The model is obtained by a careful examination of the classical phase transition diagrams, together with insights from dynamical transition theory and the unified Ginzburg–Landau model presented early in this chapter.

Note that there are four possible pairs for ^{3}He atoms:

$$\uparrow\uparrow, \qquad \downarrow\downarrow, \qquad \uparrow\downarrow, \qquad \downarrow\uparrow, \tag{3.8.1}$$

where $\uparrow$ represents spin $1/2$, and $\downarrow$ stands for spin $-1/2$. Consequently, all paired atoms in (3.8.1) have integral spins and become bosons. Therefore, for the case without an applied magnetic field, the following are possible states for superfluid phases of ^{3}He:

$$\begin{aligned}
&\text{State A:} \quad (a+ib)|\uparrow\uparrow> +(a-ib)|\downarrow\downarrow>,\\
&\text{State B:} \quad (a+ib)|\uparrow\uparrow> +(a-ib)|\downarrow\downarrow> +c[|\uparrow\downarrow> +|\downarrow\uparrow>],\\
&\text{State C:} \quad |\uparrow\downarrow> +|\downarrow\uparrow>.
\end{aligned}$$

State A is called the Anderson–Brinkman–Morel (ABM) state, corresponding to the superfluid phase A. State B is called the Balian–Werthamer (BW) state, corresponding to superfluid phase B. Although physical experiments have illustrated only the existence of phases A and B, the dynamic theory for ^{3}He in this section predicts the existence of a superfluid phase C, characterized by the state C above.

The dynamic theory for superfluidity of ^{3}He in this section includes the following three main ingredients:

First, a dynamic model is established following the main ideas of the unified dynamic model for equilibrium phase transitions. The crucial point for the modeling is the introduction of three wave functions ψ_0, ψ_1, ψ_2 to characterize the superfluidity of ^{3}He, with ψ_0 for the state $|\uparrow\uparrow>$, ψ_1 for the state $|\downarrow\downarrow>$, and ψ_2 for the state $|\uparrow\downarrow> +|\downarrow\uparrow>$. The introduction of these wave functions is based on insights from dynamic transition theory and from the classical phase diagram given in Fig. 3.41; see among many others (Ginzburg 2004; Reichl 1998).

Second, the model is then analyzed using dynamic transition theory, leading to some interesting physical predictions, synthesized in a theoretical PT-phase diagram of ^{3}He given by Fig. 3.42. The analysis of these models leads to some interesting physical predictions. For the case without an applied magnetic field, the main results obtained are synthesized in a theoretical PT-phase diagram of ^{3}He given by Fig. 3.42.

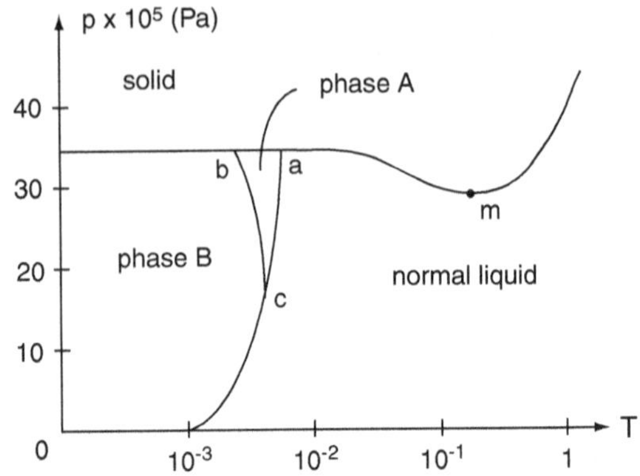

Fig. 3.41 The coexistence curve of ^{3}He without an applied magnetic field

One prediction is the existence of a metastable region $H = H_1 \bigcup H_2$, in which the solid state and the superfluid states A and B appear randomly depending on fluctuations. In particular, in H_1, the phase-A superfluid state and the solid state may appear, and in H_2, the phase-B superfluid state and the solid state may appear.

Another prediction is the possible existence of a phase-C superfluid state, which is characterized by the wave function ψ_2, representing $|\uparrow\downarrow> + |\downarrow\uparrow>$. However, the phase-C region is very narrow, which may be the reason why it is hard to observe experimentally.

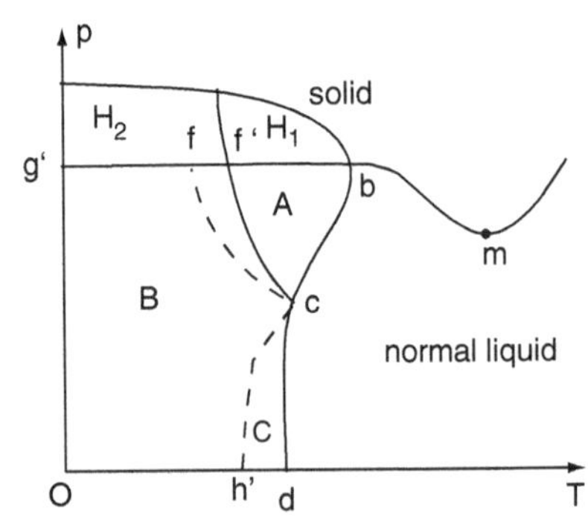

Fig. 3.42 Derived theoretical PT-phase diagram of ^{3}He: The region $H = H_1 \bigcup H_2$ is the metastable domain, where the solid state and the superfluid state appear randomly depending on fluctuations. The curve $\widehat{bcd}$ is the first critical curve where a phase transition between normal fluid and superfluid states occurs

Also, the results predict that near the two triple points b and c, there is a possibility of the existence of two switch points, where the transition on the corresponding coexistence curve switches types at each switch point. The existence of such switch points depends on the physical parameters.

Third, in his Nobel lecture (Ginzburg 2004), Ginzburg expressed a hope to develop a Ginzburg–Landau model for superfluidity. The study presented here can be considered an attempt toward his vision. Our research indicates that on the one hand, in order to have a Ginzburg–Landau theory for superfluidity consistent with the classical phase diagrams, multiple wave functions appear to be crucial, and on the other hand, such a model with multiple wave functions will always lead to new predictions, as outlined above. We are confident that these predictions will be useful for designing better physical experiments and will lead to a better understanding of the physical mechanism of superfluidity, as evidenced by the success of the ideas and methods in other related nonlinear problems.

We would like to mention that the Ginzburg–Landau theory with only one wave function cannot describe this phase transition diagram, and this demonstrates the necessity for introducing the new Ginzburg–Landau models. The results in this section were first derived in Ma and Wang (2008e).

3.8.1 Dynamic Model for Liquid ^{3}He with Zero Applied Field

We introduce three complex-valued functions ψ_0, ψ_1, ψ_2 to characterize the superfluidity of ^{3}He, in which ψ_0 corresponds to the state $|\uparrow\uparrow>$, ψ_1 to the state $|\downarrow\downarrow>$, and ψ_2 to the state $|\uparrow\downarrow> + |\downarrow\uparrow>$. Let ρ_n be the normal fluid density, while ρ_a, ρ_b, and ρ_c represent the densities of superfluid phases A, B, and C respectively. Then we have

$$\begin{aligned} &\rho_a = \tau_0|\psi_0|^2 + \tau_1|\psi_1|^2 && (\tau_0 > 0, \tau_1 > 0),\\ &\rho_b = \tau_2|\psi_0|^2 + \tau_3|\psi_1|^2 + \tau_4|\psi_2|^2 && (\tau_2, \tau_3 > 0, \tau_4 > 0),\\ &\rho_c = \tau_5|\psi_2|^2 && (\tau_5 > 0). \end{aligned}$$

The total density of ^{3}He is given by

$$\rho = \begin{cases} \rho_n & \text{in the normal state,}\\ \rho_n + \rho_a & \text{in the phase A state,}\\ \rho_n + \rho_b & \text{in the phase B state,}\\ \rho_n + \rho_c & \text{in the phase C state.} \end{cases}$$

Physically, the states ψ_0, ψ_1, and ψ_2 are independent, and consequently, there are no coupling terms $|\nabla(\psi_i + \psi_j)|^2$ and $|\psi_i + \psi_j|^{2k}$ $(i \neq j)$ in the free energy density. Since in the case without an applied magnetic field, ψ_0 and ψ_1 are equal, their coefficients in the free energy should be the same. Thus, we have the following Ginzburg–Landau free energy for ^{3}He with $H = 0$:

$$\begin{aligned} G(\psi_0, \psi_1, \psi_2, \rho_n) = \frac{1}{2}\int_\Omega \Big[&\frac{k_1 h^2}{m}|\nabla\psi_0|^2 + \alpha_1|\psi_0|^2 + \alpha_2\rho_n|\psi_0|^2 + \frac{\alpha_3}{2}|\psi_0|^4\\ &+ \frac{k_1 h^2}{m}|\nabla\psi_1|^2 + \alpha_1|\psi_1|^2 + \alpha_2\rho_n|\psi_1|^2 + \frac{\alpha_3}{2}|\psi_1|^4 + \alpha_3|\psi_0|^2|\psi_1|^2\\ &+ \frac{k_2 h^2}{m}|\nabla\psi_2|^2 + \beta_1|\psi_2|^2 + \frac{\beta_2}{4}|\psi_2|^4 + \beta_3|\psi_0|^2|\psi_2|^2 \end{aligned}$$

$$+\beta_3|\psi_1|^2|\psi_2|^2+\beta_4\rho_n|\psi_2|^2+k_3|\nabla\rho_n|^2+\mu_1|\rho_n|^2$$
$$+\frac{2\mu_2}{3}\rho_n^3+\frac{\mu_3}{2}\rho_n^3-p\left(\rho_n+\frac{\mu_0}{2}\rho_n^2\right)\Big]dx, \tag{3.8.2}$$

where the coefficients depend on T and p, and for $1\leq i\leq 3$, $j=2,3,4$,

$$k_i>0,\qquad \beta_j>0,\qquad \alpha_2>0,\qquad \alpha_3>0,\qquad \mu_3>0,\qquad \mu_2<0. \tag{3.8.3}$$

For α_1, β_1, and μ_1, there are regions A_i,B_i,C_i $(i=1,2)$ in the PT-plane $\mathbb{R}^2_+$ such that $\overline{A}_1+\overline{A}_2=\overline{B}_1+\overline{B}_2=\overline{C}_1+\overline{C}_2=\mathbb{R}^2_+$, and

$$\alpha_1=\alpha_1(T,p)\begin{cases}>0 & \text{if } (T,p)\in A_1,\\ <0 & \text{if } (T,p)\in A_2,\end{cases} \tag{3.8.4}$$

$$\beta_1=\beta_1(T,p)\begin{cases}>0 & \text{if } (T,p)\in B_1,\\ <0 & \text{if } (T,p)\in B_2,\end{cases} \tag{3.8.5}$$

$$\mu_1=\mu_1(T,p)\begin{cases}>0 & \text{if } (T,p)\in C_1,\\ <0 & \text{if } (T,p)\in C_2.\end{cases} \tag{3.8.6}$$

It is known that for ^{3}He, $\mu_1=\mu_1(T,p)$ is not monotone on T. In fact, at $T_m=0.318\,\text{K}$, $p_m=29.31\times 10^5$Pa, we have

$$\mu_1(T_m,p_m)=0,\qquad \frac{\partial\mu_1(T_m,p_m)}{\partial T}=0, \tag{3.8.7}$$

where $m=(T_m,p_m)$ is as shown in Fig. 3.41. Near the point m, the famous Pomeranchuk effect takes place, i.e., when pressure increases, the liquid ^{3}He will absorb heat to undergo a transition to solid state.

By the normalized model (3.1.15), we infer from (3.8.2) the following TDGL equations for the superfluidity of liquid ^{3}He:

$$\begin{aligned}
\frac{\partial\psi_0}{\partial t}=&\frac{k_1h^2}{m}\Delta\psi_0-\alpha_1\psi_0-\alpha_2\rho_n\psi_0\\
&-\beta_3|\psi_2|^2\psi_0-\alpha_3|\psi_0|^2\psi_0-\alpha_3|\psi_1|^2\psi_0,\\
\frac{\partial\psi_1}{\partial t}=&\frac{k_1h^2}{m}\Delta\psi_1-\alpha_1\psi_1-\alpha_2\rho_n\psi_1\\
&-\beta_3|\psi_2|^2\psi_1-\alpha_3|\psi_1|^2\psi_1-\alpha_3|\psi_0|^2\psi_1,\\
\frac{\partial\psi_2}{\partial t}=&\frac{k_2h^2}{m}\Delta\psi_2-\beta_1\psi_2-\beta_3|\psi_0|^2\psi_2\\
&-\beta_3|\psi_1|^2\psi_2-\beta_4\rho_n\psi_2-\beta_2|\psi_2|^2\psi_2,\\
\frac{\partial\rho_n}{\partial t}=&k_3\Delta\rho_n-(\mu_1-\mu_0p)\rho_n-\mu_2\rho_n^2\\
&-\mu_3\rho_n^3-\frac{\alpha_2}{2}|\psi_0|^2-\frac{\alpha_2}{2}|\psi_1|^2-\frac{\beta_4}{2}|\psi_2|^2-p.
\end{aligned} \tag{3.8.8}$$

The nondimensional form of (3.8.8) can be written as

$$\begin{aligned}
\frac{\partial \psi_0}{\partial t} &= \Delta \psi_0 + \lambda_1 \psi_0 - a_1 \rho_n \psi_0 - a_2 |\psi_2|^2 \psi_0 - a_3 |\psi_0|^2 \psi_0 - a_3 |\psi_1|^2 \psi_0, \\
\frac{\partial \psi_1}{\partial t} &= \Delta \psi_1 + \lambda_1 \psi_1 - a_1 \rho_n \psi_1 - a_2 |\psi_2|^2 \psi_1 - a_3 |\psi_1|^2 \psi_1 - a_3 |\psi_0|^2 \psi_1, \\
\frac{\partial \psi_2}{\partial t} &= \kappa_1 \Delta \psi_2 + \lambda_2 \psi_2 - b_1 \rho_n \psi_2 - b_2 |\psi_0|^2 \psi_2 - b_2 |\psi_1|^2 \psi_2 - b_3 |\psi_2|^2 \psi_2, \\
\frac{\partial \rho_n}{\partial t} &= \kappa_2 \Delta \rho_n + \lambda_3 \rho_n - c_1 |\psi_0|^2 - c_1 |\psi_1|^2 - c_2 |\psi_2|^2 + c_3 \rho_n^2 - c_4 \rho_n^3,
\end{aligned} \tag{3.8.9}$$

where

$$\begin{aligned}
\lambda_1 &= -\frac{ml^2}{h^2 k_1}(\alpha_1 + \alpha_2 \rho_n^0), \\
\lambda_2 &= -\frac{ml^2}{h^2 k_1}(\beta_1 + b_1 \rho_n^0), \\
\lambda_3 &= -\frac{ml^2}{h^2 k_1}(\mu_1 - \mu_0 p - 2\rho_n^0 \mu_2 - 3(\rho_n^0)^2 \mu_3),
\end{aligned} \tag{3.8.10}$$

and ρ_n^0 is determined by the steady-state solution of the form $(\psi_0, \psi_1, \psi_2, \rho_n) = (0,0,0,\rho_n^0)$ of (3.8.8). By (3.8.3), the coefficients in (3.8.9) satisfy

$$a_i > 0, \quad b_i > 0, \quad c_j > 0 \qquad \forall 1 \le i \le 3, \quad 1 \le j \le 4.$$

The boundary conditions are the Neumann conditions

$$\frac{\partial}{\partial n}(\psi_0, \psi_1, \psi_2, \rho_n) = 0 \qquad \text{on } \partial\Omega. \tag{3.8.11}$$

When p is a constant, the problem (3.8.9) with (3.8.11) can be reduced to the following system of ordinary differential equations:

$$\begin{aligned}
\frac{d\rho_1}{dt} &= \lambda_1 \rho_1 - a_1 \rho_n \rho_1 - a_2 \rho_2 \rho_1 - a_3 \rho_1^2, \\
\frac{d\rho_2}{dt} &= \lambda_2 \rho_2 - b_1 \rho_n \rho_2 - b_2 \rho_1 \rho_2 - b_3 \rho_2^2, \\
\frac{d\rho_n}{dt} &= \lambda_3 \rho_n - c_1 \rho_1 - c_2 \rho_2 + c_3 \rho_n^2 - c_4 \rho_n^3,
\end{aligned} \tag{3.8.12}$$

where $\rho_1 = |\psi_0|^2 + |\psi_1|^2$ and $\rho_2 = |\psi_2|^2$.

3.8.2 Critical Parameter Curves and *PT*-Phase Diagram

Critical Parameter Curves. Critical parameter curves in the PT-plane are given by

$$l_i = \{(T,p) \in \mathbb{R}_+^2 \mid \lambda_i(T,p) = 0\}, \quad i = 1,2,3,$$

where $\lambda_i = \lambda_i(T,p)$ are defined by (3.8.10).

It is clear that the critical parameter curves l_i are associated with the PT-phase diagram of ^{3}He. As in the previous section, if we can determine the critical parameter curves l_i $(1 \leq i \leq 3)$, then we shall obtain the PT-phase diagram.

Phenomenologically, according to the experimental PT-phase diagram (Fig. 3.41), the parameter curves l_i $(i = 1,2,3)$ in the PT-plane should be as schematically illustrated in Fig. 3.43a–c. The combination of the diagrams (a)–(c) in Fig. 3.43 gives Fig. 3.44, in which the real line $\widehat{bm}$ stands for the coexistence curve of the solid and liquid phases, and $\widehat{bcd}$ for the coexistence curve of the superfluid and normal liquid phases.

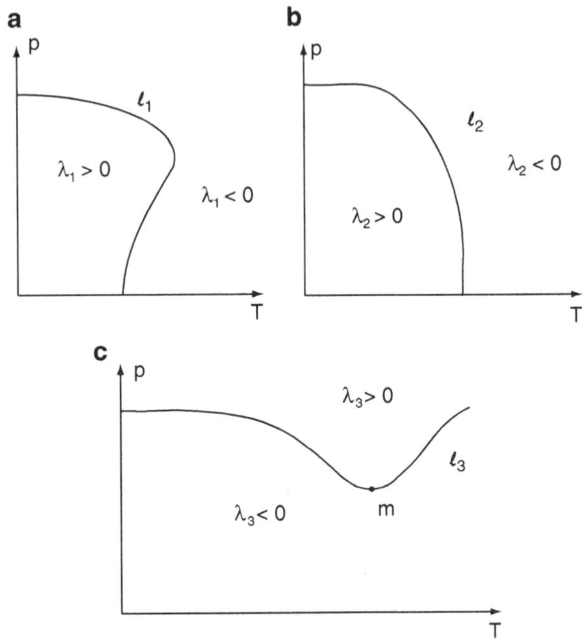

Fig. 3.43 (**a**) The curve l_1 $(\lambda_1 = 0)$, (**b**) the curve l_2 $(\lambda_2 = 0)$, and (**c**) the curve l_3 $(\lambda_3 = 0)$

Now we rigorously examine phase transitions in different regimes determined by the equations and the critical parameter curves.

States in the Metastable Region. We consider the dynamical properties of transitions for (3.8.12) in the metastable region. It is clear that at point $b = (T_b, p_b)$,

$$\lambda_1(T_b, p_b) = \lambda_3(T_b, p_b) = 0, \quad \lambda_2(T_b, p_b) < 0, \tag{3.8.13}$$

and the metastable region H_1 near the triple point b is defined by

$$H_1 = \{(T, p) \in \mathbb{R}^2_+ \mid (\lambda_1, \lambda_2, \lambda_3)(T, p) = (+, -, +)\}.$$

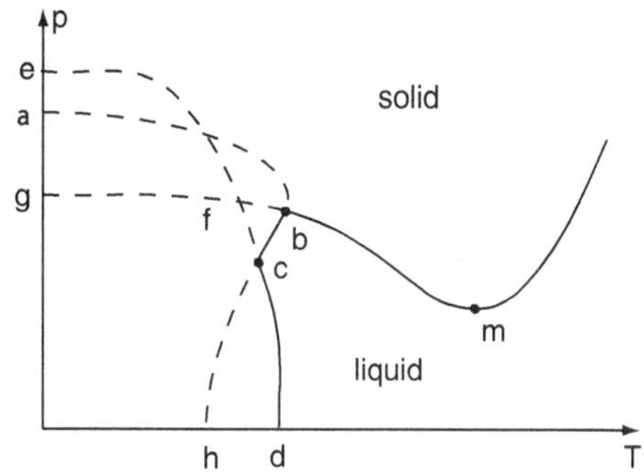

Fig. 3.44 The curve $\widehat{abch}$ is l_1, $\widehat{efcd}$ is l_2, and $\widehat{gfbm}$ is l_3

To study the structure of flows of (3.8.12) for $(T,p) \in H$, it is necessary to consider equations (3.8.12) at the point $b = (T_b, p_b)$, which, by (3.8.13), are given by

$$\begin{aligned} \frac{d\rho_1}{dt} &= -a_1\rho_n\rho_1 - a_3\rho_1^2, \\ \frac{d\rho_n}{dt} &= -c_1\rho_1 + c_3\rho_n^2 - c_4\rho_n^3, \end{aligned} \tag{3.8.14}$$

We know that

$$c_3 > 0 \quad \text{for} \quad (T,p) \subset H_1. \tag{3.8.15}$$

Equations (3.8.14) have the following two steady-state solutions:

$$\begin{aligned} Z_1 &= (\rho_1, \rho_n) = (0, c_3/a_4), \\ Z_2 &= (\rho_1, \rho_n) = \left(\frac{a_1}{a_3}\alpha, -\alpha\right), \\ \alpha &= \frac{c_3}{2c_4}\left(\sqrt{1+\frac{4c_1a_1c_4}{a_3c_3^2}} - 1\right). \end{aligned}$$

By direct computation, we can prove that the eigenvalues of the Jacobian matrices of (3.8.14) at Z_1 and Z_2 are negative. Hence, Z_1 and Z_2 are stable equilibrium points of (3.8.14). Physically, Z_1 stands for the solid state, and Z_2 for the superfluid state. The topological structure of (3.8.14) is schematically illustrated by Fig. 3.45a. The two regions R_1 and R_2 divided by curve AO in Fig. 3.45b are the basins of attraction of Z_1 and Z_2 respectively.

We note that in H_1, λ_1 and λ_3 are small, i.e.,

$$0 < \lambda_1(T,p), \quad \lambda_2(T,p) \ll 1 \quad \text{for } (T,p) \in H,$$

and (3.8.14) can be considered a perturbed system of (3.8.12).

Thus, for $(T,p) \in H_1$, the system (3.8.12) has four steady-state solutions $\widetilde{Z}_i = \widetilde{Z}(T,p)$ $(1 \leq i \leq 4)$ such that

$$\lim_{(T,p)\to(T_C,p_C)} (\widetilde{Z}_1, \widetilde{Z}_2, \widetilde{Z}_3, \widetilde{Z}_4)(T,p) = (Z_1, Z_2, 0, 0),$$

and $\widetilde{Z}_1$ and $\widetilde{Z}_2$ are stable, representing the solid and liquid He-3 states respectively; $\widetilde{Z}_3$ and $\widetilde{Z}_4$ are two saddle points. The topological structure of (3.8.14) for $(T,p) \in H_1$ is schematically shown in Fig. 3.45c, and the basins of attraction of $\widetilde{Z}_1$ and $\widetilde{Z}_2$ are $\widetilde{R}_1$ and $\widetilde{R}_2$, as illustrated by Fig. 3.45d.

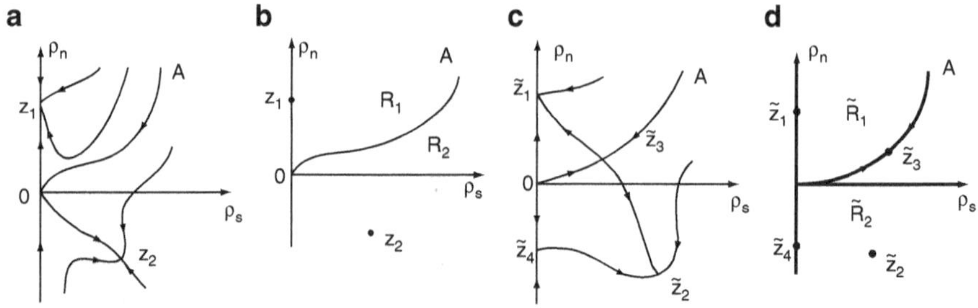

Fig. 3.45 Dynamical properties of transitions for (3.8.12)

First Phase Transition. On the coexistence curve $\widehat{bc}$ in Fig. 3.42,

$$\lambda_1 = 0, \lambda_2 < 0, \lambda_3 < 0.$$

Hence the first phase transition crossing $\widehat{bc}$ is between the normal fluid state and the phase-A superfluid state. On the coexistence curve $\widehat{cd}$,

$$\lambda_1 < 0, \qquad \lambda_2 = 0, \qquad \lambda_3 < 0.$$

In this case, the first phase transition crossing this coexistence curve is between the normal fluid state and the phase-C superfluid state.

Second Phase Transitions. When (T,p) crosses the curve segment $\widehat{bcd}$, (3.8.12) will undergo a second transition. We need to consider two cases.

Case 1. Second transition crossing $\widehat{f'c}$: If (T,p) passes through this curve segment $\widehat{bc}$, then the first transition solution is given by

$$(\rho_1,\rho_2,\rho_n) = (\rho_1^*,0,\rho_n^*), \qquad \rho_1^* = \rho_1^*(T,p) > 0, \qquad \rho_n^* = \rho_n^*(T,p) < 0.$$

Take a transformation

$$\rho_1' = \rho_1 - \rho_1^*, \quad \rho_2' = \rho_2, \quad \rho_n' = \rho_n - \rho_n^*.$$

Then the system (3.8.12) is in the following form (for simplicity, we drop the primes):

$$\begin{aligned}
\frac{d\rho_1}{dt} &= \widetilde{\lambda}_1\rho_1 - a_1\rho_1^*\rho_n - a_2\rho_1^*\rho_2 - a_2\rho_1\rho_2 - a_3\rho_1^3, \\
\frac{d\rho_2}{dt} &= \widetilde{\lambda}_2\rho_2 - b_1\rho_n\rho_2 - b_2\rho_1\rho_2 - b_3\rho_2^2, \\
\frac{d\rho_n}{dt} &= \widetilde{\lambda}_3\rho_n - c_1\rho_1 - c_2\rho_2 + (c_3 - 3c_4\rho_n^*)\rho_n^2 - c_4\rho_n^3,
\end{aligned} \tag{3.8.16}$$

where

$$\begin{aligned}
\widetilde{\lambda}_1 &= \lambda_1 + a_1|\rho_n^*| - 2a_3\rho_1^*, \\
\widetilde{\lambda}_2 &= \lambda_2 + b_1|\rho_n^*| - b_2\rho_1^*, \\
\widetilde{\lambda}_3 &= \lambda_3 - 2c_3|\rho_n^*| - 3c_4\rho_n^{*2}.
\end{aligned} \tag{3.8.17}$$

The linear operator in (3.8.16) reads

$$L = \begin{pmatrix} \widetilde{\lambda}_1 & -a_2\rho_1^* & -a_1\rho_1^* \\ 0 & \widetilde{\lambda}_2 & 0 \\ -c_1 & -c_2 & \widetilde{\lambda}_3 \end{pmatrix}.$$

The three eigenvalues of L are

$$\beta_1 = \widetilde{\lambda}_2, \quad \beta_\pm = \frac{1}{2}\left[\widetilde{\lambda}_1 + \widetilde{\lambda}_3 \pm \sqrt{(\widetilde{\lambda}_3 - \widetilde{\lambda}_1)^2 + 4a_1c_1\rho_1^*}\right]. \tag{3.8.18}$$

It is known that the transition solution $(\rho_1^*, 0, \rho_n^*)$ is stable near $\widehat{bc}$. Therefore, the eigenvalues of L satisfy

$$\beta_1(T,p) < 0, \quad \beta_\pm(T,p) < 0 \text{ for } (T,p) \text{ near } \widehat{bc}.$$

However, near $\widehat{fc}$, there is a curve segment $\widehat{f'c}$ such that

$$\widehat{f'c} = \{(T,p) \in \mathbb{R}_+^2 \mid \beta_1(T,p) = 0\}.$$

Thus, system (3.8.16) has a transition on $\widehat{f'c}$, which is called the second transition of (3.8.12), and $\widehat{f'c}$ is the coexistence curve of phases A and B; see Fig. 3.42.

Case 2. Second transition crossing $\widehat{ch'}$: If (T,p) passes through the curve segment $\widehat{bc}$, then the first transition solution is given by

$$(\rho_1, \rho_2, \rho_n) = (0, \eta_2, \eta_n), \qquad \eta_2 = \eta_2(T,p) > 0, \qquad \eta_n = \eta_n(T,p) < 0.$$

Take a transformation

$$\rho_1' = \rho, \quad \rho_2' = \rho_2 - \eta_2, \quad \rho_n' = \rho_n - \eta_n.$$

Then the system (3.8.12) is in the following form (for simplicity, we drop the primes):

$$\begin{aligned}
\frac{d\rho_1}{dt} &= \widetilde{\widetilde{\lambda}}_1\rho_1 - a_1\rho_1\rho_n - a_2\rho_1\rho_2 - a_3\rho_1^2, \\
\frac{d\rho_2}{dt} &= \widetilde{\widetilde{\lambda}}_2\rho_2 - b_1\eta_2\rho_n - b_2\eta_2\rho_1 - b_1\rho_2\rho_n - b_2\rho_1\rho_2 - b_3\rho_2^2, \\
\frac{d\rho_n}{dt} &= \widetilde{\widetilde{\lambda}}_3\rho_n - c_1\rho_1 - c_2\rho_2 + (c_3 - 3c_4\eta_n)\rho_n^2 - c_4\rho_n^3,
\end{aligned} \tag{3.8.19}$$

where

$$
\begin{aligned}
\widetilde{\widetilde{\lambda}}_1 &= \lambda_1 + a_1|\eta_n| - a_2\eta_2, \\
\widetilde{\widetilde{\lambda}}_2 &= \lambda_2 + b_1|\eta_n| - 2b_2\eta_2, \\
\widetilde{\widetilde{\lambda}}_3 &= \lambda_3 - 2c_3|\eta_n| - 3c_4\eta_n^2.
\end{aligned}
\tag{3.8.20}
$$

The linear operator L in (3.8.19) has three eigenvalues:

$$
\beta_1 = \widetilde{\widetilde{\lambda}}_1, \quad \beta_\pm = \frac{1}{2}\left[\widetilde{\widetilde{\lambda}}_1 + \widetilde{\widetilde{\lambda}}_3 \pm \sqrt{(\widetilde{\widetilde{\lambda}}_3 - \widetilde{\widetilde{\lambda}}_1)^2 + 4b_1c_2\eta_2}\right]. \tag{3.8.21}
$$

It is known that the transition solution $(\rho_1^*, 0, \rho_n^*)$ is stable near $\widehat{bc}$. Therefore, the eigenvalues of L satisfy

$$
\beta_1(T,p) < 0, \quad \beta_\pm(T,p) < 0 \text{ for } (T,p) \text{ near } \widehat{cd}.
$$

However, near $\widehat{hc}$, there is a curve segment $\widehat{h'c}$ such that

$$
\widehat{h'c} = \{(T,p) \in \mathbb{R}_+^2 \mid \beta_1(T,p) = \widetilde{\widetilde{\lambda}}_1 = 0\}.
$$

Thus, system (3.8.19) has a transition on $\widehat{h'c}$, which is called the second transition of (3.8.12), and $\widehat{h'c}$ is the coexistence curve of phases C and B; see Fig. 3.42.

In summary, with the above analysis and dynamic transition theory, we arrive at the following transition theorem:

Theorem 3.8.1 *Define regions in the PT-plane (see Fig. 3.42) by*

$$
\begin{aligned}
&E_1 = \{(T,p) \in \mathbb{R}_+^2 \mid (\lambda_1, \lambda_2, \lambda_3)(T,p) = (-,-,+)\}, \\
&E_2 = \{(T,p) \in \mathbb{R}_+^2 \mid (\lambda_1, \lambda_2, \lambda_3)(T,p) = (-,-,-)\}, \\
&H_1 = \{(T,p) \in \mathbb{R}_+^2 \mid (\lambda_1, \widetilde{\lambda}_2, \lambda_3)(T,p) = (+,-,+)\}, \\
&H_2 = \{(T,p) \in \mathbb{R}_+^2 \mid (\lambda_1, \widetilde{\lambda}_2, \lambda_3)(T,p) = (+,+,+)\}, \\
&\text{Region } f'bc = \{(T,p) \in \mathbb{R}_+^2 \mid (\lambda_1, \lambda_2, \widetilde{\lambda}_2, \lambda_3)(T,p) = (+,-,-,-)\}, \\
&\text{Region } cdh' = \{(T,p) \in \mathbb{R}_+^2 \mid (\lambda_1, \widetilde{\widetilde{\lambda}}_2, \lambda_2, \lambda_3)(T,p) = (-,-,+,-)\},
\end{aligned}
$$

and let the Region $0g'f'ch'$ be the complement of the sum of the above regions. Then the following conclusions hold:

(1) If $(T,p) \in E_1$, the phase of ^{3}He is in the solid state.
(2) If $(T,p) \in E_2$, the phase is in the normal liquid state.
(3) If $(T,p) \in$ Region $f'bc$, the phase is in the phase-A superfluid state.
(4) If $(T,p) \in$ Region cdh', the phase is in the phase-C superfluid state.
(5) If $(T,p) \in$ Region $0g'f'ch'$, the phase is in the phase-B superfluid state.

(6) *If* $(T,p) \in H_1$, *there are two regions* R_1 *and* R_2 *in the state space* (ρ_1,ρ_2,ρ_n) *such that under a fluctuation that is described by the initial value* (x_0,y_0,z_0) *in (3.8.12), if* $(x_0,y_0,z_0) \in R_1$, *then the phase is in the solid state, and if* $(x_0,y_0,z_0) \in R_2$, *then it is in the phase-A superfluid state.*

(7) *If* $(T,p) \in H_2$, *there are two regions* K_1 *and* K_2 *in the state space* (ρ_1,ρ_2,ρ_n) *such that if* $(x_0,y_0,z_0) \in K_1$, *then the phase is in the solid state, and if* $(x_0,y_0,z_0) \in K_2$, *then it is in the phase-B superfluid state.*

3.8.3 Classification of Superfluid Transitions

We now classify the superfluid transitions of (3.8.12) crossing various coexistence curves. First we consider the transitions crossing curve segments $\widehat{cd}$ and $\widehat{bc}$ in Fig. 3.42. Obviously, we have

$$\widehat{cd} = \{(T,p) \in \mathbb{R}^2_+ \mid (\lambda_1,\lambda_2,\lambda_3)(T,p) = (-,0,-)\},$$
$$\widehat{bc} = \{(T,p) \in \mathbb{R}^2_+ \mid (\lambda_1,\lambda_2,\lambda_3)(T,p) = (0,-,-)\}.$$

Let

$$A_1 = a_1c_1 - a_3|\lambda_3|, \qquad A_2 = a_1c_2 - a_2|\lambda_3|,$$
$$B_1 = b_1c_2 - b_3|\lambda_3|, \qquad B_2 = b_1c_1 - b_2|\lambda_3|.$$

In fact, in (3.8.8) and (3.8.9), we see that

$$a_2 = b_2, \quad a_1 = c_1, \quad b_1 = c_2.$$

Therefore $A_2 = B_2$.

Theorem 3.8.2 *For the system (3.8.12), we have the following assertions:*

(1) *If* $(T_0,p_0) \in \widehat{cd}$, *the transition of (3.8.12) at* (T_0,p_0) *is between the phase-B and normal liquid. Furthermore, if* $B_1 \leq 0$, *then it is of Type I, and if* $B_1 > 0$, *then it is of Type II.*

(2) *If* $(T_0,p_0) \in \widehat{bc}$, *the transition is between the phase-A and normal liquid. Moreover, if* $A_1 \leq 0$, *then it is of Type I, and if* $A_1 > 0$, *then it is of Type II.*

Theorem 3.8.2 provides conditions for the first transition of (3.8.12). The following theorem gives sufficient conditions for the second transition near $\widehat{fc}$ in Fig. 3.42.

Obviously, the curve $\widehat{fc}$ is given by

$$\widehat{fc} = \{(T,p) \in \mathbb{R}^2_+ \mid (\lambda_1,\lambda_2,\lambda_3)(T,p) = (+,0,-)\}.$$

To set up the second transition theorem, we need to assume the following conditions. Let $\varepsilon > 0$ be small. Suppose that

$$B_2 > -\varepsilon, \quad b_1a_3 - b_2a_1 > -\varepsilon \quad \text{for } (T,p) \in \widehat{fc}, \tag{3.8.22}$$

and the gap between $\widehat{bc}$ and $\widehat{fc}$ is small, i.e.,

$$|T_2 - T_3| = O(\varepsilon) \quad \forall (T_2, p) \in \widehat{bc}, \ (T_3, p) \in \widehat{fc}. \tag{3.8.23}$$

We also assume that

$$(a_1, c_1) = O(\varepsilon), \quad (b_1, c_2, c_3, a_3, b_3) = O(1), \tag{3.8.24}$$

$$a_3B_1 - b_2A_2 > 0 \qquad \text{in } \widehat{fc} \text{ with } A_1 \leq 0. \tag{3.8.25}$$

Theorem 3.8.3 *Under conditions (3.8.22) and (3.8.23), there exists a curve segment $\widehat{f'c}$ near $\widehat{fc}$ as shown in Fig. 3.42 such that (3.8.12) has the second transition from the first transition solution $(\rho_1^*, 0, \rho_n^*)$, i.e., (3.8.16) has a transition from $(\rho_1, \rho_2, \rho_n) = 0$ in $\widehat{f'c}$, and the transition solutions $(\widetilde{\rho}_1, \widetilde{\rho}_2, \widetilde{\rho}_n)$ satisfy $\widetilde{\rho}_2 > 0$. In addition, if (3.8.24) and (3.8.25) hold, then this transition is of Type II.*

Physical experiments indicate that the superfluid transition of liquid ^{3}He between the normal liquid and superfluid phase B is continuous. Hence, it is necessary to give the conditions of Type I transition of (3.8.12) at the intersection point C of two curves $\lambda_1 = 0$ and $\lambda_2 = 0$.

Theorem 3.8.4 *Let (T_0, p_0) be the point C such that $\lambda_1(T_0, p_0) = 0$ and $\lambda_2(T_0, p_0) = 0$. Then the transition of (3.8.12) at (T_0, p_0) is of Type I if and only if one of the following two conditions holds:*

(i) $A_1 \leq 0, B_1 \leq 0, A_2 = B_2 < 0$,
(ii) $A_1 \leq 0, B_1 \leq 0, A_2 = B_2 \geq 0$ *and* $A_1B_1 > A_2B_2$.

In particular, if the transition is of Type I, then for $\lambda_1 > 0, \lambda_2 > 0$ near (p_0, T_0), there are four types of topological structures of the transition on the center manifold, which are classified as follows:

(1) The transition has the structure shown in Fig. 3.46a if

$$\lambda_1|B_1| + \lambda_2 A_2 > 0 \quad \text{and} \quad \lambda_2|A_1| + \lambda_1 B_2 > 0.$$

(2) The transition has the structure shown in Fig. 3.46b if

$$\lambda_1|B_1| + \lambda_2 A_2 < 0 \quad \text{and} \quad \lambda_2|A_1| + \lambda_1 B_2 < 0.$$

(3) The transitions has the structure shown in Fig. 3.46c if

$$\lambda_1|B_1| + \lambda_2 A_2 < 0 \quad \text{and} \quad \lambda_2|A_1| + \lambda_1 B_2 > 0.$$

(4) The transition has the structure shown in Fig. 3.46d if

$$\lambda_1|B_1| + \lambda_2 A_2 > 0 \quad \text{and} \quad \lambda_2|A_1| + \lambda_1 B_2 < 0.$$

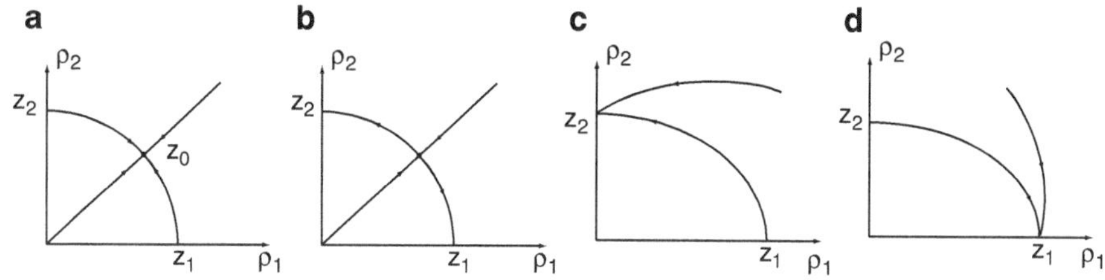

Fig. 3.46 Topological structure of a Type I transition near the intersection point $C = (T_0, p_0)$ of $\lambda_1 = 0$ and $\lambda_2 = 0$. Here ρ_1 and ρ_2 stand for the densities of phases A and C respectively

Remark 3.8.5 Physically, the transition between normal liquid and superfluid for ^{3}He is generally of Type I, the transition between superfluid phases A and B is of Type II, and the region of phase A is narrow. Therefore, under the conditions (3.8.22)–(3.8.25) and

$$\begin{aligned} &B_1 < 0 && \text{in } \widehat{cd}, \\ &B_1 \leq 0, A_1 < 0 && \text{near point } C = \overline{\Gamma}_1 \bigcap \overline{\Gamma}_2 = (T_0, p_0), \end{aligned}$$

the above theorems (Theorems 3.8.2–3.8.4) provide a precise mathematical proof of superfluid transitions of liquid ^{3}He with no applied magnetic field.

By condition (3.8.24), we see that

$$a_1 \cdot c_1 = O(\varepsilon^2), \quad a_3 = O(1).$$

Assertion (1) of Theorem 3.8.2 implies that only in a very small range of (T, p) near the change point of superfluid and solid is the transition between normal liquid and superfluid of Type II, and this range is

$$0 > \lambda_3(T, p) > -\frac{a_1 c_1}{a_3} = -O(\varepsilon^2).$$

Moreover, the superfluid density of phase A near the solid phase is of quantitative order ε^3, i.e.,

$$\rho_1 = \frac{a_1^2 c_1}{a_3^2 c_3} = O(\varepsilon^3).$$

Hence, the difference between Type I and Type II phase transitions in experiments is very small.

Note that Theorem 3.8.3 remains valid if condition (3.8.22) is replaced by

$$\frac{\partial \lambda_2}{\partial T} \gg 1 \quad \text{for } (T, p) \text{ near } \widehat{fc}.$$

Proof of Theorem 3.8.2. If $(T_0, p_0) \in \widehat{bc}, it follows that \lambda_2(T_0, p_0) < 0$ and the space of $(\rho_1, 0, \rho_n)$ is invariant for (3.8.12). Therefore, the transition equations of (3.8.12) at (T_0, p_0) can be expressed in the following form:

$$\begin{aligned}\frac{d\rho_1}{dt} &= \lambda_1\rho_1 - a_1\rho_n\rho_1 - a_3\rho_1^2, \\ \frac{d\rho_n}{dt} &= \lambda_3\rho_n - c_1\rho_1 + c_3\rho_n^2 - c_4\rho_n^3.\end{aligned} \tag{3.8.26}$$

The second-order approximation of the center manifold function ρ_n of (3.8.26) satisfies the equation

$$\lambda_3\rho_n + c_3\rho_n^2 = c_1\rho_1.$$

Its solution is

$$\rho_n = -\frac{c_1\rho_1}{|\lambda_3|} + \frac{c_1^2c_3}{|\lambda_3|^3}\rho_1^2 + o(\rho_1^2).$$

Putting ρ_n in the first equation of (3.8.26) we get the reduced equation of (3.8.12) on the center manifold as follows:

$$\frac{d\rho_1}{dt} = \lambda_1\rho_1 + \frac{1}{|\lambda_3|}A_1\rho_1^2 - \frac{a_1c_1c_3}{|\lambda_3|^3}\rho_1^3 + o(\rho_1^3). \tag{3.8.27}$$

Assertion (2) follows from (3.8.27).

Likewise, if $(T_0, p_0) \in \widehat{cd}$, then $\lambda_1(T_0, p_0) < 0$, and the space of $(0, \rho_2, \rho_n)$ is invariant for (3.8.12). Therefore, in the same fashion as above, we can prove assertion (1). The proof is complete. □

Proof of Theorem 3.8.3. We proceed with the following two cases.

Case 1: $A_1 \leq 0$ **in** $\widehat{bc}$. In this case, by Theorem 3.8.2, the transition of (3.8.12) in $\widehat{bc}$ is of Type I, and the transition solution $(\rho_1^*, 0, \rho_n^*)$ satisfies

$$\rho_n^* = -\frac{c_1}{|\lambda_3|}\rho_1^*.$$

The equations describing the second transition are given by (3.8.16), and the eigenvalues in (3.8.17) and (3.8.18) are rewritten as

$$\begin{aligned}\beta_1 =& \lambda_2 + b_1|\rho_n^*| - b_2\rho_1^* = \lambda_2 + \frac{1}{|\lambda_3|}B_2\rho_1^*, && (3.8.28)\\ \beta_+ =& \frac{1}{2}[\widetilde{\lambda}_1 + \widetilde{\lambda}_3 + \sqrt{(\widetilde{\lambda}_1 - \widetilde{\lambda}_3)^2 + 4a_1c_1\rho_1^*}] && (3.8.29)\\ =& (\text{by (3.8.23) and } \rho_1^*, \widetilde{\lambda}_1 \text{ being small})\\ =& -\lambda_1 + o(|\lambda_1|),\\ \beta_1 <& \beta_+. && (3.8.30)\end{aligned}$$

In addition, we know that

$$\begin{aligned}&\lambda_1 = 0, \quad \lambda_2 < 0 && \text{on } \widehat{bc},\\ &\lambda_2(T - \delta, p) > 0 && \text{for } (T, p) \in \widehat{fc} \text{ and } \delta > 0.\end{aligned}$$

Hence, by assumptions (3.8.22) and (3.8.23), from (3.8.28)–(3.8.30) we can infer that there exists a curve segment $\widehat{f'c}$ near $\widehat{fc}$ such that for $(T_2, p) \in \widehat{bc}$ and $(T_0, p) \in \Gamma_3'$, we have

$$\beta_1(T,p) \begin{cases} < 0 & \text{if } T_0 < T \leq T_2, \\ = 0 & \text{if } T = T_0, \\ > 0 & \text{if } T < T_0, \end{cases} \tag{3.8.31}$$

and $\beta_- < \beta_+ = -\lambda_1 + o(|\lambda_1|) < 0$. Hence, by Theorem 2.1.3, the system (3.8.16) has a transition on Γ_3'.

To determine the transition type, we consider the center manifold function of (3.8.16), which satisfies

$$\begin{pmatrix} \widetilde{\lambda}_1 & -a_1\rho_1^* \\ -c_1 & \widetilde{\lambda}_3 \end{pmatrix} \begin{pmatrix} \rho_1 \\ \rho_n \end{pmatrix} = \begin{pmatrix} a_2\rho_1^*\rho_2 \\ c_2\rho_2 \end{pmatrix} + o(\rho_2). \tag{3.8.32}$$

The solution of (3.8.32) is

$$\rho_1 = \frac{a_2\widetilde{\lambda}_3 + c_2a_1}{\widetilde{\lambda}_3\widetilde{\lambda}_1 - c_1a_1\rho_1^*}\rho_1^*\rho_2 + o(\rho_2), \qquad \rho_n = \frac{\widetilde{\lambda}_1c_2 + c_1a_2\rho_1^*}{\widetilde{\lambda}_3\widetilde{\lambda}_1 - c_1a_1\rho_1^*}\rho_2 + o(\rho_2).$$

From (3.8.17), (3.8.23), and (3.8.24), we can obtain

$$\rho_1 \simeq \frac{A_2\rho_2}{a_3|\lambda_3|}, \quad \rho_n \simeq -\frac{c_2}{|\lambda_3|}\rho_2. \tag{3.8.33}$$

Inserting the center manifold function (3.8.33) into the second equation of (3.8.16), we get the reduced equation as

$$\frac{d\rho_2}{dt} = \widetilde{\lambda}_2\rho_2 + \frac{1}{|\lambda_3|}(B_1 - \frac{b_2A_2}{a_3})\rho_2^2.$$

By (3.8.25), the transition of (3.8.16) is of Type II.

Case 2: $A_1 > 0$ in $\widehat{bc}$. In this case, the transition of (3.8.12) in $\widehat{bc}$ is of Type II, and the transition solution in $\widehat{bc}$ is

$$\rho_1^* = \frac{a_1}{a_3}|\rho_n^*|, \quad \rho_n^* = -\frac{c_3}{2c_4}\left(\sqrt{1 + \frac{4c_4c_1a_1}{a_3c_3^2}} - 1\right).$$

The eigenvalue β_1 in (3.8.18) reads

$$\beta_1 = \lambda_2 + \frac{1}{a_3}(b_1a_3 - b_2a_1)|\rho_n^*|.$$

By (3.8.22) and (3.8.23), this implies that (3.8.31) holds. Hence (3.8.12) has a second transition in Γ_3' for $A_1 > 0$ in $\widehat{bc}$.

Under the condition (3.8.24), we have

$$\rho_n^* \simeq -\frac{a_1c_1}{a_3c_3}, \quad \widetilde{\lambda}_1 \simeq -a_3\rho_1^*, \quad \widetilde{\lambda}_3 \simeq \lambda_3 - \frac{2a_1c_1}{a_3}.$$

By $A_1 > 0$, we get that $|\lambda_3| \leq 0(\varepsilon^2)$. Thus, the solutions of (3.8.32) can be rewritten as

$$\rho_1 \simeq \frac{a_1c_1}{\widetilde{\lambda}_1\widetilde{\lambda}_3 - c_1a_1\rho_1^*}\rho_1^*\rho_2 \simeq \frac{c_2}{c_1}\rho_2, \qquad \rho_2 \simeq -\frac{a_3c_2}{a_1c_1}\rho_2.$$

Putting ρ_1 and ρ_n into the second equation of (3.8.16), we obtain the reduced equation on the center manifold as

$$\frac{d\rho_2}{dt} = \widetilde{\lambda}_2\rho_2 + \left(\frac{b_1c_2a_3}{a_1c_1} - \frac{c_2b_2}{c_1} - b_3\right)\rho_2^2.$$

Due to (3.8.24), we see that

$$\frac{b_1c_2a_3}{a_1c_1} - \frac{c_2b_2}{c_1} - b_3 > 0.$$

Therefore, this transition of (3.8.16) is of Type II.

It is clear that the second transition solutions $(\widetilde{\rho}_1, \widetilde{\rho}_2, \widetilde{\rho}_n)$ satisfy $\widetilde{\rho}_2 > 0$. The theorem is proved. □

Proof of Theorem 3.8.4. At the point $C = (T_0, p_0), \lambda_1(T_0, p_0) = 0$, $\lambda_2(T_0, p_0) = 0$. Hence, the center manifold function of (3.8.12) at (T_0, p_0) reads

$$\rho_n = -\frac{c_1}{|\lambda_3|}\rho_1 - \frac{c_2}{|\lambda_3|}\rho_2 + \frac{c_3}{|\lambda_3|^3}(c_1\rho_1 + c_2\rho_2)^2.$$

Putting ρ_n in the first two equations of (3.8.12), we obtain the reduced equations on the center manifold as

$$\begin{aligned}\frac{d\rho_1}{dt} =& \lambda_1\rho_1 + \frac{1}{|\lambda_3|}(A_1\rho_1^2 + A_2\rho_1\rho_2) - \frac{c_3a_1}{|\lambda_3|^3}(c_1\rho_1 + c_2\rho_2)^2\rho_1,\\ \frac{d\rho_2}{dt} =& \lambda_2\rho_2 + \frac{1}{|\lambda_3|}(B_1\rho_2^2 + B_2\rho_1\rho_2) - \frac{c_3b_1}{|\lambda_3|^3}(c_1\rho_1 + c_2\rho_2)^2\rho_2.\end{aligned} \tag{3.8.34}$$

To verify the Type I transition, by the attractor bifurcation theorem, it suffices to consider the following equations:

$$\begin{aligned}\frac{d\rho_1}{dt} &= A_1\rho_1^2 + A_2\rho_1\rho_2 - \frac{c_3a_1}{|\lambda_3|^2}(c_1\rho_1 + c_2\rho_2)^2\rho_1,\\ \frac{d\rho_2}{dt} &= B_1\rho_2^2 + B_2\rho_1\rho_2 - \frac{c_3b_1}{|\lambda_3|^2}(c_1\rho_1 + c_2\rho_2)^2\rho_2.\end{aligned} \tag{3.8.35}$$

Since equations (3.8.9) have variational structure, the flows of (3.8.35) are of gradient type. Therefore, $(\rho_1,\rho_2)=0$ has no elliptic region for (3.8.35). Hence, in the same fashion as used in Sect. 6.3 in Ma and Wang (2005b), one can prove that the region

$$S=\{(\rho_1,\rho_2)\in\mathbb{R}^2|\ \rho_1>0,\rho_2>0\}$$

is a stable parabolic region. Namely, $(\rho_1,\rho_2)=0$ is an asymptotically stable singular point of (3.8.35) if and only if one of the two conditions (i) and (ii) holds. Thus, we need to prove only assertions (1)–(4).

For a Type I transition at the point $C=(T_0,p_0)$, by conditions (i) and (ii), $A_1<0$ and $B_1<0$. Hence, for $\lambda_1>0,\lambda_2>0$, there are bifurcated solutions of (3.8.34) on the ρ_1-axis and ρ_2-axis as

$$z_1=(\rho_1^*,0)=\left(\frac{|\lambda_3|}{|A_1|}\lambda_1,0\right),\quad z_2=(0,\rho_2^*)=\left(0,\frac{|\lambda_3|}{|B_1|}\lambda_2\right).$$

The Jacobian matrices of (3.8.34) at z_1 and z_2 are given by

$$J(z_1)=\begin{pmatrix}-\lambda_1 & * \\ 0 & \lambda_2+\frac{B_2}{|A_1|}\lambda_1\end{pmatrix},\qquad J(z_2)=\begin{pmatrix}\lambda_1+\frac{A_2}{|B_1|}\lambda_2 & 0 \\ * & -\lambda_2\end{pmatrix}.$$

Since (3.8.34) has at most three bifurcated singular points in the region $\overline{S}$, there are only four types of Type I transitions, as shown in Fig. 3.46a–d, and each type is completely determined by the signs of the eigenvalues of $J(z_1)$ and $J(z_2)$. Thus, from the eigenvalues $\beta(z_1)=\frac{1}{|A_1|}(\lambda_2|A_1|+\lambda_1B_2)$ of $J(z_1)$ and $\beta(z_2)=\frac{1}{|B_1|}(\lambda_1|B_1|+\lambda_2A_2)$ of $J(z_2)$, it is easy to derive assertions (1)–(4). The proof is complete. □

3.8.4 Liquid ^{3}He with Nonzero Applied Field

When liquid ^{3}He is placed in a magnetic field H, the superfluid transition is different from that with $H=0$. Experiments show that when a magnetic field is applied, a new superfluid phase A_1 appears, and the region of phase A can be extended to the bottom at $p=0$. The PT-phase diagram is schematically illustrated in Fig. 3.47.

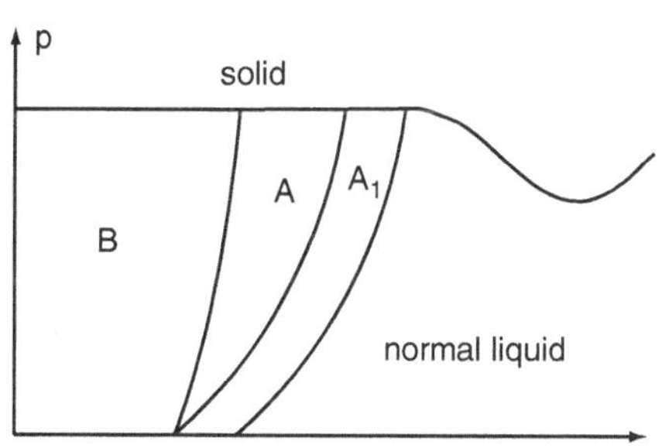

Fig. 3.47 PT-phase diagram of ^{3}He in a magnetic field

When a magnetic field H is applied, a new pairing state appears in which the spin of pairing atoms is parallel to the magnetic field. This new state corresponds to the phase A_1 and is expressed by

$$\sqrt{2}|\Phi>= 2a|\uparrow\uparrow>.$$

We introduce the complex-valued functions ψ_0 to the state $|\uparrow\uparrow>$, ψ_1 to the state $|\downarrow\downarrow>$, and ψ_2 to the state $|\uparrow\downarrow> + |\downarrow\uparrow>$. Let ρ_0, ρ_a, ρ_b stand for the densities of superfluid phases A_1, A, B respectively. Then we have

$$\begin{aligned}
&\rho_0 = |\psi_0|^2, \\
&\rho_a = \tau_0|\psi_0|^2 + \tau_1|\psi_1|^2 && (\tau_0 > 0, \tau_1 > 0), \\
&\rho_b = \tau_2|\psi_0|^2 + \tau_3|\psi_1|^2 + \tau_4|\psi_2|^2 && (\tau_2, \tau_3 \geq 0, \tau_4 > 0),
\end{aligned}$$

and the total density of liquid ^{3}He in a magnetic field is given by

$$\rho = \begin{cases} \rho_n + \rho_0 & \text{at state } A_1, \\ \rho_n + \rho_\alpha & \text{at state } A, \\ \rho_n + \rho_b & \text{at state } B. \end{cases}$$

Thus, similar to (3.8.2), for liquid ^{3}He with $H \neq 0$ we give the Ginzburg–Landau free energy in the following form. For simplicity, we take the nondimensional form:

$$\begin{aligned}
G(\psi_0, \psi_1, \psi_2, \rho_n) = & \frac{1}{2}\int_\Omega \Big[\kappa_0|\nabla\psi_0|^2 - \lambda_0|\psi_0|^2 + \alpha_0\rho_n|\psi_0|^2 \\
& + \alpha_1|\psi_0|^2|\psi_1|^2 + \alpha_2|\psi_0^2||\psi_2|^2 + \frac{\alpha_3}{2}|\psi_0|^4 \\
& + \kappa_1|\nabla\psi_1|^2 - \lambda_1|\psi_1|^2 + a_1\rho_n|\psi_1|^2 \\
& + a_2|\psi_1|^2|\psi_2|^2 + \frac{a_3}{2}|\psi_1|^4 \\
& + \kappa_2|\nabla\psi_2|^2 - \lambda_2|\psi_2|^2 + b_1\rho_1|\psi_2|^2 + \frac{b_3}{2}|\psi_2|^4 \\
& + \kappa_3|\nabla\rho_n|^2 - \lambda_3|\rho_n|^2 - \frac{c_3}{3}\rho_n^3 - \frac{c_4}{4}\rho_n^4\Big] dx. \qquad (3.8.36)
\end{aligned}$$

The equations describing liquid ^{3}He with $H \neq 0$ read

$$\begin{aligned}
\frac{\partial\psi_0}{\partial t} &= \kappa_0\Delta\psi_0 + \lambda_0\psi_0 - \alpha_0\rho_n\psi_0 - \alpha_1|\psi_1|^2\psi_0 - \alpha_2|\psi_2|^2\psi_0 - \alpha_3|\psi_0|^2\psi_0, \\
\frac{\partial\psi_1}{\partial t} &= \kappa_1\Delta\psi_1 + \lambda_1\psi_1 - a_1\rho_n\psi_1 - \alpha_1|\psi_0|^2\psi_1 - a_2|\psi_2|^2\psi_1 - a_3|\psi_1|^2\psi_1, \\
\frac{\partial\psi_2}{\partial t} &= \kappa_2\Delta\psi_2 + \lambda_2\psi_2 - b_1\rho_n\psi_2 - \alpha_2|\psi_0|^2\psi_2 - a_2|\psi_1|^2\psi_2 - b_3|\psi_2|^2\psi_2, \\
\frac{\partial\rho_n}{\partial t} &= \kappa_3\Delta\rho_n + \lambda_3\rho_n - \frac{\alpha_0}{2}|\psi_0|^2 - \frac{a_1}{2}|\psi_1|^2 - \frac{b_1}{2}|\psi_2|^2 + c_2\rho_n^2 - c_3\rho_n^3, \\
\frac{\partial}{\partial n}&(\psi_0, \psi_1, \psi_2, \rho_n) = 0 \qquad \text{on } \partial\Omega,
\end{aligned} \qquad (3.8.37)$$

where the coefficients satisfy that for every $0 \leq i \leq 3$ and $1 \leq j \leq 3$,

$$\alpha_i > 0, \quad a_j > 0, \quad b_1, b_3, c_2, c_3 > 0.$$

Equations (3.8.37) should be the same as (3.8.9) for $H = 0$. Therefore, we assume that when $H = 0$,

$$\kappa_0 = \kappa_1, \quad \lambda_0 = \lambda_1, \quad \alpha_0 = a_1, \quad \alpha_1 = 0, \quad \alpha_2 = a_2, \quad \alpha_3 = a_3. \tag{3.8.38}$$

Based on physical considerations, we also assume that

$$\begin{aligned} &\lambda_0 = \lambda_1(T,p) + \widetilde{\lambda}(T,p,H), \\ &\widetilde{\lambda}(T,p,H) > 0 && \text{if } H \neq 0, \\ &\widetilde{\lambda}(T,p,H) \to 0 && \text{if } H \to 0. \end{aligned} \tag{3.8.39}$$

When the magnetic field H and the pressure p are homogeneous on Ω, the problem (3.8.37) can be reduced to

$$\begin{aligned} \frac{d\rho_0}{dt} &= \lambda_0\rho_0 - \alpha_0\rho_n\rho_0 - \alpha_1\rho_1\rho_0 - \alpha_2\rho_2\rho_0 - \alpha_3\rho_0^2, \\ \frac{d\rho_1}{dt} &= \lambda_1\rho_1 - a_1\rho_n\rho_1 - \alpha_1\rho_0\rho_1 - a_2\rho_2\rho_1 - a_3\rho_1^2, \\ \frac{d\rho_2}{dt} &= \lambda_2\rho_2 - b_1\rho_n\rho_2 - \alpha_2\rho_0\rho_2 - a_2\rho_2\rho_1 - b_3\rho_2^2, \\ \frac{d\rho_n}{dt} &= \lambda_3\rho_n - \frac{\alpha_0}{2}\rho_0 - \frac{a_1}{2}\rho_1 - \frac{b_1}{2}\rho_2 + c_2\rho_n^2 - c_3\rho_n^3, \end{aligned} \tag{3.8.40}$$

where $\rho_i = |\psi_i|^2$ $(i = 0,1,2)$, and λ_i $(i = 1,2,3)$ are as in (3.8.12).

Let λ_1, λ_2, and λ_3 be as shown in Figures 3.43a–c respectively. Then, due to (3.8.39), the curves $\lambda_j(T,p) = 0$ $(0 \leq j \leq 3)$ in the PT-plane are schematically illustrated by Fig. 3.48.

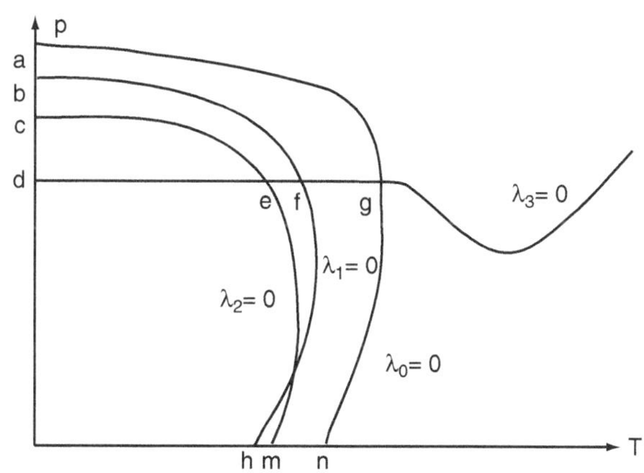

Fig. 3.48 Curve $\widehat{agn}$ is $\lambda_0 = 0, \widehat{bfh}$ is $\lambda_1 = 0, \widehat{cem}$ is $\lambda_2 = 0$, and $\widehat{defg}$ is $\lambda_3 = 0$

Let the applied magnetic field $H \neq 0$ be such that

$$0 < \widetilde{\lambda}(T,p,H) < \varepsilon, \quad \text{for } \varepsilon > 0 \text{ small}, \tag{3.8.41}$$

where $\widetilde{\lambda}$ is as in (3.8.39). Then, under the conditions (3.8.22), (3.8.23), (3.8.38), (3.8.39), and (3.8.41), using the same approach as in Theorems 3.8.2 and 3.8.3, we can prove the following transition theorem for (3.8.40).

Theorem 3.8.6 *Assume the conditions (3.8.22), (3.8.23), (3.8.38), (3.8.39), and (3.8.41). Then for $H \neq 0$, there exist two curve segments $\widehat{f'h'}$ near $\lambda_1 = 0$ and $\widehat{e'm'}$ near $\lambda_2 = 0$ in the PT-plane as shown in Fig. 3.49 such that the following assertions hold:*

(1) The system (3.8.40) has a transition in the curve segment $\lambda_0 = 0$ with $\lambda_3 < 0$ (i.e., the curve segment $\widehat{gn}$ in Fig. 3.49), which is of Type I for $\alpha_0^2 - 2|\lambda_3|\alpha_3 \leq 0$, and is of Type II for $\alpha_0^2 - 2|\lambda_3|\alpha_3 > 0$. The transition solution is given by $(\rho_0^*, 0, 0, \rho_n^*)$ with $\rho_0^* > 0$, $\rho_n^* < 0$.

(2) The system has a second transition from $(\rho_0^*, 0, 0, \rho_n^*)$ in the curve segment $\widehat{f'h'}$ (i.e., $\lambda_1 - a_1\rho_n^* - \alpha_1\rho_0^* = 0$), and the transition solution is $(\rho_0', \rho_1', 0, \rho_n')$ with $\rho_0' > 0$, $\rho_1' > 0$, and $\rho_n' < 0$.

(3) The system has a third transition from $(\rho_0', \rho_1', 0, \rho_n')$ in the curve segment $\widehat{e'm'}$ (i.e., $\lambda_2 - b_1\rho_n' - \alpha_2\rho_0' - a_2\rho_1' = 0$), and the transition solution is $(\rho_0'', \rho_1'', \rho_2'', \rho_n'')$ with $\rho_i'' > 0 (0 \leq i \leq 2)$ and $\rho_n' < 0$.

Remark 3.8.7 The first transition of (3.8.40) in the curve segment $\widehat{gn}$ corresponds to the phase transition of ^{3}He in a magnetic field between normal liquid and superfluid phase A_1, and the second transition in $\widehat{f'h'}$ corresponds to the phase transition between the superfluid phase A_1 and A, and the third transition in $\widehat{e'm'}$ corresponds to the phase transition between superfluid phases A and B; see Fig. 3.49.

Remark 3.8.8 The transition theorems, Theorems 3.7.2–3.8.3 and 3.8.6, provide a theoretical foundation to explain the PT-phase diagrams of superfluidity, while supporting the models of liquid helium based on experiment.

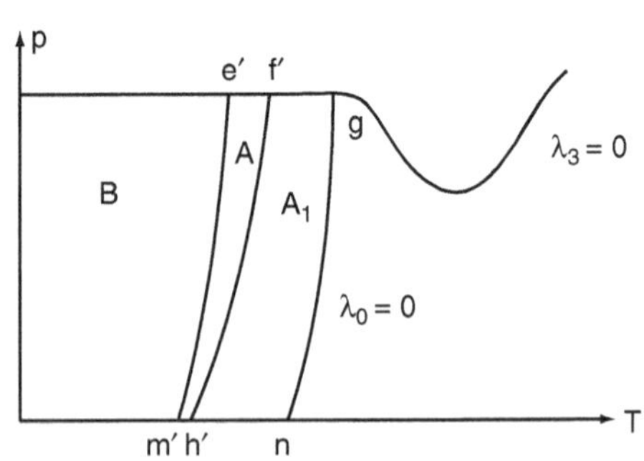

Fig. 3.49 Phase transition diagram

3.8.5 Physical Remarks

By carefully examining the classical phase transition diagrams and with both mathematical and physical insights offered by dynamical transition theory, we have derived two new models for superfluidity of ^{3}He with or without an applied field. A crucial component of these two models is the introduction of three wave functions to represent the Anderson–Brinkman–Morel (ABM) and Balian–Werthamer (BM) states.

We have obtained a theoretical PT-phase diagram of ^{3}He as shown in Fig. 3.42, based on the models and dynamic phase transition analysis. A few main characteristics of the results are as follows.

First, the analysis shows the existence of an unstable region $H = H_1 \bigcup H_2$ in which the solid state and the superfluid states A and B appear randomly depending on fluctuations. In particular, in H_1, the phase-B superfluid state and the solid may appear, and in H_2, the phase-A superfluid state and the solid state may appear.

Second, theoretical analysis suggests the existence of a phase-C superfluid state, which is characterized by the wave function ψ_2, representing $|\uparrow\downarrow> + |\downarrow\uparrow>$. However, the phase-C region is very narrow, which may be why it is hard to observe in experiments.

Third, the curve $\widehat{bcd}$ is the first critical curve where phase transitions between the normal fluid and superfluid states occur. The curve $\widehat{f'c}$ is the coexistence curve between phases A and B. The curve $\widehat{bc}$ is the coexistence curve between the normal fluid state and the phase-A superfluid state; the curve $\widehat{cd}$ is the coexistence curve between the normal fluid state and the phase-C superfluid state; the curve $\widehat{ch'}$ is the coexistence curve between the phase-B and phase-C superfluid states.

Fourth, Theorems 3.8.2–3.8.4 imply that near the two triple points b and c, there is a possibility of the existence of two switch points, where the transition on the corresponding coexistence curve switches types. The existence of such switch points depends on the physical parameters.

In comparison to classical results as shown in Fig. 3.41, our results lead to the predictions of the existence of (1) an unstable region H, (2) a new phase C in a narrow region, and (3) switch points. It is hoped that these predictions will be useful for designing better physical experiments and that they will lead to a better understanding of the physical mechanism of superfluidity.

3.9 Mixture of He-3 and He-4

The main objective of this section is to study λ-phase transitions of liquid helium-4 and phase separations between liquid helium-3 and liquid helium-4 from both the modeling and analysis points of view. This section is based on a recent paper (Ma and Wang 2009d).

The main ingredient in the modeling is that we use an order parameter ψ for the phase transition of liquid ^{4}He between the normal and superfluid states, and the molar fraction u for liquid ^{3}He. Since u is a conserved quantity, a Cahn–Hilliard-type equation is needed for u, and a Ginzburg–Landau-type equation is needed for the order parameter.

The analysis leads to three critical length scales $L_1 < L_2 < L_3$ and the corresponding λ-transition and phase separation diagrams. The derived theoretical phase diagrams based on our analysis agree with classical phase diagrams, as shown, e.g., in Reichl (1998) and Onuki (2002), and it is hoped that the research here will lead to a better understanding of the nature of superfluids.

3.9.1 Model for Liquid Mixture of ^{3}He and ^{4}He

Liquid ^{3}He and ^{4}He can be dissolved into each other. When ^{3}He-atoms are dissolved in liquid ^{4}He and the density of ^{3}He increases, the λ-transition temperature T_λ decreases; see the liquid mixture phase diagram of ^{3}He–^{4}He (Fig. 3.50), where $X = n_3/(n_3+n_4), and n_3$ and n_4 are the atomic numbers of ^{3}He and ^{4}He respectively. When $X = 0, then T_\lambda = 2.17\,\text{K}$, where a λ-phase transition takes place and liquid ^{4}He undergoes a transition to the superfluid phase from the normal liquid phase. When $X = 0.67$ and the temperature decreases to $T = 0.87\,\text{K}$, i.e., at the triple point C in Fig. 3.50, the liquid mixture of ^{3}He–^{4}He has a phase separation.

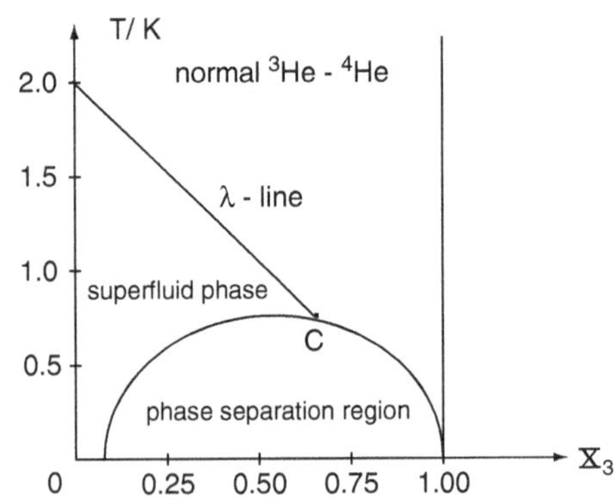

Fig. 3.50 Liquid mixture phase diagram of ^{3}He–^{4}He

Let the complex-valued function $\psi = \psi_1 + i\psi_2$ describe the superfluidity of ^{4}He, and let u be the density of ^{3}He, which is conserved, i.e.,

$$\int_\Omega u dx = c \quad (c > 0 \text{ is a fixed number}). \tag{3.9.1}$$

The Ginzburg–Landau (Gibbs) free energy is taken in the following form:

$$\begin{aligned}G(\psi,\rho_n,u) = \int_\Omega \Big[&\frac{k_1 h^2}{2m}|\nabla\psi|^2 + \frac{\gamma_1}{2}|\psi|^2 + \frac{\gamma_2}{4}|\psi|^4 \\ &+ \frac{k_2}{2}|\nabla u|^2 + \frac{\nu_1}{2}u^2 - \frac{\nu_2}{3}u^3 + \frac{\nu_3}{4}u^4 + \nu_4 u|\psi|^2\Big]dx,\end{aligned} \tag{3.9.2}$$

where h is Planck's constant, m is the mass of a helium-4 atom, and the coefficients satisfy

$$k_1, k_2, \gamma_2, \nu_2, \nu_3, \nu_4 > 0. \tag{3.9.3}$$

When $\psi = 0$, the liquid mixture of helium-3 and helium-4 is a binary system, and $G(0,u)$ stands for the Cahn–Hilliard free energy. By the standard model (3.1.23), from (3.9.1) and (3.9.2), the equations governing liquid mixture of ^{3}He–^{4}He are given as follows:

$$\begin{aligned}\frac{\partial\psi}{\partial t} &= \frac{k_1 h^2}{m}\Delta\psi - \gamma_1\psi - \gamma_2|\psi|^2\psi - 2\nu_4 u\psi, \\ \frac{\partial u}{\partial t} &= -k_2\Delta^2 u + \Delta\left[\nu_1 u - \nu_2 u^2 + \nu_3 u^3 + \nu_4|\psi|^2\right].\end{aligned} \tag{3.9.4}$$

These equations have a physically sound constant steady-state solution given by

$$(\psi, u) = (0, u^0),$$

where $u^0 > 0$ is the density of helium-3 in a homogeneous state. For simplicity, we assume that the total density ρ is equal to 1. Then the system control parameter X, the molar fraction of ^{3}He, becomes

$$X = u^0 \qquad \text{with } 0 \le X \le 1.$$

Now we consider the derivations from this basic state,

$$(\psi, u) = (\psi', X + u'),$$

and we derive the following equations (again, we drop the primes):

$$\begin{aligned}&\frac{\partial\psi}{\partial t} = \mu_1\Delta\psi - \lambda_1\psi - \gamma_2|\psi|^2\psi - 2\nu_4 u\psi, \\ &\frac{\partial u}{\partial t} = -\mu_2\Delta^2 u + \lambda_2\Delta u + \Delta[(-\nu_2 + 3\nu_3 X)u^2 + \nu_3 u^3 + \nu_4|\psi|^2], \\ &\int_\Omega u\,dx = 0,\end{aligned} \tag{3.9.5}$$

with the Neumann boundary condition

$$\frac{\partial}{\partial n}(u, \Delta u, \psi) = 0 \qquad \text{on } \partial\Omega, \tag{3.9.6}$$

where $\mu_1 = k_1 h^2/m$, $\mu_2 = k_2$, and

$$\lambda_1 = \gamma_1 + 2\nu_4 X, \qquad \lambda_2 = \nu_1 - 2\nu_2 X + 3\nu_3 X^2. \tag{3.9.7}$$

3.9.2 Critical Parameter Curves

For simplicity we consider only the case in which $\Omega = (0,L) \times (0,l)^2 \subset \mathbb{R}^3$ is a rectangle with $L > l$, and the control parameters are the temperature T, the molar fraction X, and the length scale L of the container Ω.

Physically, by Hildebrand theory (Reichl 1998), in the lower temperature region and at $p = 1$ atm, the critical parameter curve $\lambda_2 = 0$ is equivalent to

$$T = \frac{2a}{R}(1-X)X - \sigma_0, \tag{3.9.8}$$

where R is the molar gas constant and $a > 0$ is a constant. Here $\sigma_0 > 0$ is a small correction term. The original Hildebrand theory leads to the case $\sigma_0 = 0$. However, as we can see from the classical phase separation of a binary system, Hildebrand theory fails when the molar fraction is near 0 or 1, and the correction term added here agrees with the experimental phase diagram as shown, e.g., in Fig. 4.13 in Reichl (1998).

Furthermore, by (3.9.7) and (3.9.8), we have

$$\nu_1 \simeq \theta_1(T + \sigma_0), \quad 3\nu_3 \simeq 2\nu_2, \quad \frac{2\nu_2}{\theta_1} \simeq \frac{2a}{R}. \tag{3.9.9}$$

Consider the eigenvalue problem of the linear operator in (3.9.5):

$$\begin{pmatrix} \mu_1\Delta - \lambda_1 & 0 & 0 \\ 0 & \mu_1\Delta - \lambda_1 & 0 \\ 0 & 0 & -\Delta(\mu_2\Delta - \lambda_2) \end{pmatrix} \begin{pmatrix} \psi_1 \\ \psi_2 \\ u \end{pmatrix} = \beta \begin{pmatrix} \psi_1 \\ \psi_2 \\ u \end{pmatrix}. \tag{3.9.10}$$

Here $\psi = \psi_1 + \psi_2$. It is known that the first eigenvalue and eigenvector of the Laplacian operator with the Neumann condition and zero average are $\lambda = \pi^2/L^2$ and $u = \cos \pi x_1/L$. Thus, the first two eigenvalues and their corresponding eigenvectors of (3.9.10) are given by

$$\begin{cases} \beta_1 = -\lambda_1 = -(\gamma_1 + 2\nu_4 X), \\ (\psi_1, \psi_2, u) = (1,0,0) \text{ and } (0,1,0), \end{cases} \tag{3.9.11}$$

$$\begin{cases} \beta_2 = -\dfrac{\pi^2}{L^2}\left(\dfrac{\mu_2\pi^2}{L^2} + \lambda_2\right) = -\dfrac{\pi^2}{L^2}\left(\dfrac{\mu_2\pi^2}{L^2} + \nu_1 - 2\nu_2 X + 3\nu_3 X^2\right), \\ (\psi_1, \psi_2, u) = (0,0,\cos \pi x_1/L). \end{cases} \tag{3.9.12}$$

As for v_1, the parameter γ_1 is approximately a linear function of T. Phenomenologically, we take

$$\gamma_1 = -\sigma_1 + \theta_2 T \qquad (\theta_2, \sigma_1 > 0). \tag{3.9.13}$$

Then by (3.9.9) and (3.9.11)–(3.9.13), the critical parameter curves in the TX-plane are as follows:

$$\begin{aligned} l_1 : \beta_1 = 0 \qquad &\Longleftrightarrow \qquad T_{c1} = \frac{\sigma_1 - 2v_4 X}{\theta_2} = \frac{\sigma_1}{\theta_2}(1 - X), \\ l_2 : \beta_2 = 0 \qquad &\Longleftrightarrow \qquad T_{c2} = \frac{2a}{R} X(1 - X) - \sigma_0 - \frac{\mu_2 \pi^2}{\theta_1 L^2}. \end{aligned} \tag{3.9.14}$$

Here we have assumed that $\sigma_1 = 2v_4$. This assumption is true at $X = 1$, as required by $T_{c1} = 0$ at $X = 1$, and is true as well at other values of X if we approximately take σ_1 and v_4 as constants.

The critical parameter curve l_1 ($\beta_1 = 0$) is as shown in Fig. 3.51a. Let

$$L_0 = \sqrt{\frac{\mu_1 \pi^2}{\theta_1 \left(\frac{a}{2R} - \sigma_0\right)}} \qquad \left(\sigma_0 < \frac{a}{2R}\right). \tag{3.9.15}$$

Then when $L > L_0$, the critical parameter curve l_2 ($\beta_2 = 0$) is as shown in Fig. 3.51b.

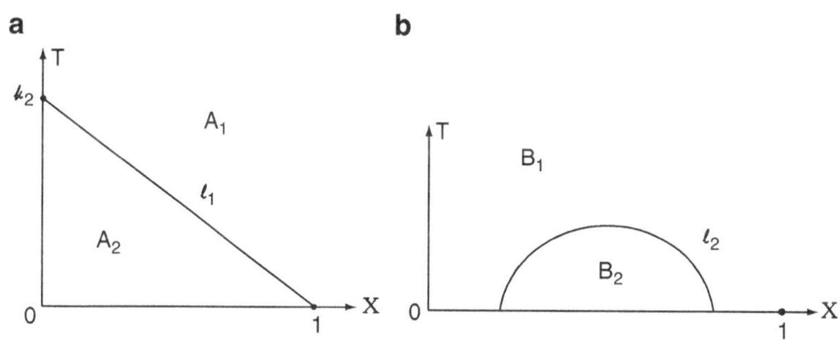

Fig. 3.51 (**a**) $\beta_1 < 0$ in region A_1, and $\beta_1 > 0$ in A_2, (**b**) $\beta_2 < 0$ in region B_1, and $\beta_2 > 0$ in B_2

3.9.3 Transition Theorems

The following is the transition theorem for a liquid mixture ^{3}He–^{4}He. For this purpose, we introduce three length scales:

$$L_1 = \left(\frac{\mu_2 \pi^2}{\theta_1}\right)^{1/2} \cdot \frac{1}{\left(\frac{a}{2R} + \frac{2v_4^2}{\theta_1 \gamma_2} - \sigma_0\right)^{1/2}}, \tag{3.9.16}$$

$$L_2 = \begin{cases} \left(\frac{\mu_2\pi^2}{\theta_1}\right)^{1/2} \cdot \frac{1}{A^{1/2}} & \text{if } A > 0 \\ \infty & \text{otherwise,} \end{cases} \tag{3.9.17}$$

$$L_3 = \begin{cases} \left(\frac{\mu_2\pi^2}{\theta_1}\right)^{1/2} \cdot \frac{1}{B^{1/2}} & \text{if } B > 0 \\ \infty & \text{otherwise,} \end{cases} \tag{3.9.18}$$

where

$$A = \frac{R}{8a}\left(\frac{2a}{R} - \frac{\sigma_1}{\theta_2}\right)^2 + \frac{2\nu_4^2}{\theta_1\gamma_2} - \sigma_0, \qquad B = \frac{R}{8a}\left(\frac{2a}{R} - \frac{\sigma_1}{\theta_2}\right)^2 - \sigma_0.$$

We remark here again that σ_0 is small as a correction term in the Bragg–Williams theory mentioned earlier in (3.9.8).

Theorem 3.9.1 *Let T_{c1} and T_{c2} be given in (3.9.14). For equation (3.9.5), we have the following assertions:*

(1) When $L < L_1$, the system has only the superfluid phase transition (i.e., the λ-phase transition) at $T = T_c^1$, and the TX-phase diagram is as shown in Fig. 3.51a.

(2) Let $\sigma_1/\theta_2 < 4a/R$. Then $L_1 < L_2$, and if $L_1 < L < L_2$, then there are two numbers $0 < X_1 < \frac{1}{2} < X_2 < 1$ given by

$$X_{1,2} = \frac{1}{2}\left[1 \pm \sqrt{1 + \frac{2R}{a}\left(\frac{2\nu_4^2}{\theta_1\gamma_2} - \sigma_0 - \frac{\mu_2\pi^2}{\theta_1 L^2}\right)}\right] \tag{3.9.19}$$

such that the following hold:

(a) If $0 \le X < X_1$ or $X_2 < X < 1$, the system has only the λ-phase transition at $T = T_c^1$.

(b) If $X_1 < X < X_2$, then the system has the λ-phase transition at $T = T_{c1}$, and has phase separation at $T^* = T_{c2} + \frac{2\nu_4^2}{\theta_1\gamma_2} < T_{c1}$. Moreover, the TX-phase diagram is as shown in Fig. 3.52.

(3) Let $\sigma_1/\theta_2 < 4a/R$. For any $L_2 < L < L_3$, the phase diagram is as shown in Fig. 3.53. For any $L_3 < L$, the phase diagram is as shown in Fig. 3.54.

(4) Let $\sigma_1/\theta_2 \ge 4a/R$. Then for $L \ge L_1$, assertion (2) holds.

(5) The λ-phase transition at $T = T_c^1$ is of Type I, which corresponds to a second-order transition.

It is worth mentioning that the order of second transitions crossing $\widetilde{\beta}_1 = 0$ is an interesting open problem.

Proof of Theorem 3.9.1. We proceed in several steps as follows.

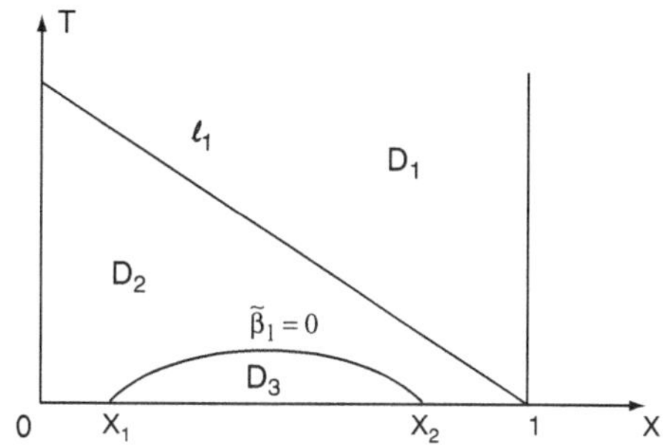

Fig. 3.52 Region D_1 is normal ^3He–^4He, D_2 is the superfluid phase, and D_3 is the phase separation region

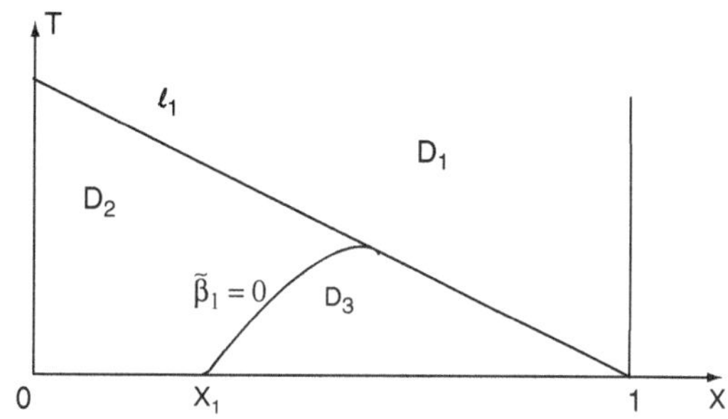

Fig. 3.53 Region D_1 is normal ^3He–^4He, D_2 is the superfluid phase, and D_3 is the phase separation region

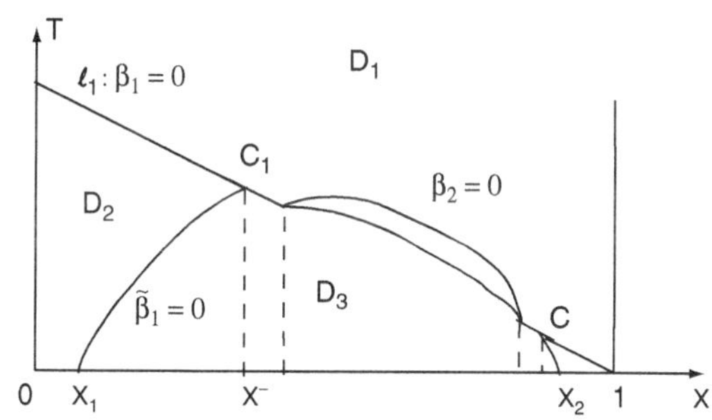

Fig. 3.54 Region D_1 is normal ^3He–^4He, D_2 is the superfluid phase, and D_3 is the phase separation region

Step 1. It is easy to see that the space

$$E = \{(u, \psi) = (0, \psi) |\ \psi \in \mathbb{C}\}$$

is invariant for (3.9.5). Therefore, the transition solutions of (3.9.5) from the critical parameter curve l_1 ($\lambda_1 = 0$) must be in E, which corresponds to a superfluid transition. On the other hand, restricted on E, equations (3.9.4) are equivalent to the following ordinary differential equation:

$$\frac{d\psi}{dt} = -\lambda_1 \psi - \gamma_2 |\psi|^2 \psi. \tag{3.9.20}$$

It is then easy to see that assertion (5) holds, and the bifurcated solutions consist of a circle given by

$$\{\sqrt{-\lambda_1/\gamma_2}\, e^{i\phi} \mid \phi \in [0, 2\pi]\}.$$

Step 2. We now consider the second transition. For this purpose, let the transition solution of (3.9.4) from $T = T_c^1$ (i.e., from $\lambda_1 = 0$) be given by

$$(0, \widetilde{\psi}(T)) = (0, \sqrt{-\lambda_1/\gamma_2}\, e^{i\phi}) \in E. \tag{3.9.21}$$

Take the transition

$$(u, \psi) \to (u', \psi' + \widetilde{\psi}).$$

Then (3.9.5) can be rewritten as (we again drop the primes)

$$\begin{aligned}
&\frac{\partial \psi}{\partial t} = \mu_1 \Delta \psi + \lambda_1 \psi + \lambda_1 \psi^* - 2\nu_4 \sqrt{\frac{-\lambda_1}{\gamma_2}} u - 2\gamma_2 \sqrt{\frac{-\lambda_1}{\gamma_2}} |\psi|^2 - 2\nu_4 u \psi - \gamma_2 |\psi|^2 \psi,\\
&\frac{\partial u}{\partial t} = -\mu_2 \Delta^2 u + \lambda_2 \Delta u + 2\nu_4 \sqrt{\frac{-\lambda_1}{\gamma_2}} \Delta \psi_1 + \Delta[(-\nu_2 + 3\nu_3 X)u^2 + \nu_3 u^3 + \nu_4 |\psi|^2],\\
&\int_\Omega u\,dx = 0,
\end{aligned} \tag{3.9.22}$$

where $\psi = \psi_1 + i\psi_2$.

We consider the transition of (3.9.22) beyond E. The linear operator of (3.9.22) is given by

$$B = \begin{pmatrix} 2\lambda_1 & 0 & -2\nu_4 \sqrt{\frac{-\lambda_1}{\gamma_2}} \\ 0 & \lambda_1 & 0 \\ 2\nu_4 \sqrt{\frac{-\lambda_1}{\gamma_2}} \Delta & 0 & \Delta(-\mu_2 \Delta + \lambda_2) \end{pmatrix}. \tag{3.9.23}$$

Restricted to its first eigenspace

$$E_1 = \left\{ (\psi_1, \psi_2, u) = (y_1, y_2, y_3) \cos \frac{\pi x_1}{L} \mid (y_1, y_2, y_3) \in \mathbb{R}^3 \right\},$$

the linear operator B is given by

$$B|_{E_1} = \begin{pmatrix} 2\lambda_1 & 0 & -2\nu_4 \sqrt{\frac{-\lambda_1}{\gamma_2}} \\ 0 & \lambda_1 & 0 \\ -\frac{2\nu_4 \pi^2}{L^2} \sqrt{\frac{-\lambda_1}{\gamma_2}} & 0 & -\frac{\pi^2}{L^2}\left(\frac{\mu_2 \pi^2}{L^2} + \lambda_2\right) \end{pmatrix}. \tag{3.9.24}$$

The eigenvalues $\widetilde{\beta}_1$, $\widetilde{\beta}_2$, and $\widetilde{\beta}_3$ of $B|_{E_1}$ are given by

$$\begin{aligned}
&\widetilde{\beta}_{1,2} = \lambda_1 - \frac{\pi^2}{2L^2}\left[\frac{\mu_2 \pi^2}{L^2} + \lambda_2\right] \pm \sqrt{\left[\lambda_1 - \frac{\pi^2}{2L^2}\left(\frac{\mu_2 \pi^2}{L^2} + \lambda_2\right)\right]^2 + \frac{2\lambda_1 \pi^2}{L^2}\left(\frac{\mu_2 \pi^2}{L^2} + \lambda_2 - \frac{2\nu_4^2}{\gamma_2}\right)},\\
&\widetilde{\beta}_3 = \lambda_1 < 0.
\end{aligned}$$

We know that

$$\widetilde{\beta}_3 < 0, \quad \widetilde{\beta}_2 < \widetilde{\beta}_1 \qquad \forall T < T_{c1}.$$

By (3.9.9), we have

$$\begin{aligned}\lambda_2 &= \nu_1 - 2\nu_2 X + 3\nu_3 X^2 \\ &= \theta_1 (T + \sigma_0) - 2\nu_2 (1-X)X \\ &= \theta_1 \left(T + \sigma_0 - \frac{2\nu_2}{\theta_1} X(1-X)\right) \\ &= \theta_1 (T - T_{c2}) - \frac{\nu_2 \pi^2}{L^2}.\end{aligned}$$

Hence the transition curve $\widetilde{\beta}_1 = 0$ is given by

$$\frac{\mu_2 \pi^2}{L^2} + \lambda_2 = \frac{2\nu_4^2}{\gamma_2},$$

which is equivalent to

$$\theta_1 (T - T_{c2}) = \frac{2\nu_4^2}{\gamma_2}.$$

Hence the transition curve $\widetilde{\beta}_1 = 0$ is given by

$$T = T_{c2} + \frac{2\nu_4^2}{\theta_1 \gamma_2} = \frac{2a}{R} X(1-X) - \sigma_0 - \frac{\mu_2 \pi^2}{\theta_1 L^2} + \frac{2\nu_4^2}{\theta_1 \gamma_2}. \tag{3.9.25}$$

Then L_1 defined by (3.9.16) is the critical length scale to make the transition curve $\widetilde{\beta}_1 = 0$ achieve its maximum at $(T, X) = (0, 1/2)$.

By definition, it is then easy to see that if $L < L_1$, no phase separation occurs at any temperature, although a phase transition for ^{4}He does occur as the temperature decreases below a certain critical temperature.

Step 3. Now we calculate the length scale L_2 where the critical curves $\beta_1 = 0$ and $\widetilde{\beta}_1 = 0$ are tangent to each other, i.e., where they interact at exactly one point. By definition, we need to solve L such that the equation

$$\frac{\sigma_1}{\theta_2}(1-X) = \frac{2a}{R} X(1-X) - \sigma_0 - \frac{\mu_2 \pi^2}{\theta_1 L^2}$$

has exactly one solution. Hence, it is easy to derive that the formula L_2 is as given by (3.9.17).

By comparing L_1 and L_2, we have

$$L_2 > L_1 \qquad \text{if and only if} \qquad \frac{\sigma_1}{\theta_2} < \frac{4a}{R}. \tag{3.9.26}$$

Now we determine the length scale L_3 where the critical curves $\beta_1 = 0$ and $\beta_2 = 0$ are tangent to each other, i.e., we need to solve L such that the equation

$$\frac{\sigma_1}{\theta_2}(1-X) = \frac{2a}{R}X(1-X) - \sigma_0$$

has exactly one solution. The formula L_3 is as given by (3.9.18).

Step 4. Now we return to prove assertions (1)–(4). First, we consider the case in which $\frac{\sigma_1}{\theta_2} < \frac{4a}{R}$, i.e., $L_2 > L_1$. In this case, for $L_1 < L < L_2$, the second transition curve $\widetilde{\beta}_1 = 0$ is as shown in Fig. 3.52. Namely, as the temperature T entering from region D_2 into D_3 through the curve $\widetilde{\beta}_1 = 0$ decreases, a phase separation occurs.

By (3.9.25), solving

$$\frac{2a}{R}X(1-X) - \sigma_0 - \frac{\mu_2\pi^2}{\theta_1 L^2} + \frac{2\nu_4^2}{\theta_1\gamma_2} = 0$$

gives the smallest and largest molar fractions $X_1 < X_2$ defined by (3.9.19), and assertion (2) follows. Other assertions can be proved in the same fashion. □

3.9.4 Physical Conclusions

We have introduced a dynamical Ginzburg–Landau phase transition/separation model for a mixture of liquid helium-3 and helium-4.

Our analysis of the model established enables us to give a detailed account of the λ-phase transition and the phase separation between liquid ^{3}He and ^{4}He. In particular, we derived three critical length scales $L_1 < L_2 < L_3$ with the following conclusions:

1. For $L < L_1$, there is only a λ-phase transition for ^{4}He and no phase separation between ^{3}He and ^{4}He, as shown in Fig. 3.51a.
2. For $L_1 < L < L_2$, there are no triple points, and phase separation occurs as a second phase transition after the λ-transition when the molar fraction is between two critical values, as shown in Fig. 3.52.
3. For $L_2 < L < L_3$, the λ-transition is always the first transition, as shown in Fig. 3.53.
4. For $L_3 < L$, both the λ-transition and the phase separation can appear as either the first transition or the second transition depending on the molar fractions, as shown in Fig. 3.54. In this case, when the phase separation is the first transition, the separation mechanism is the same as a typical binary system as described in the section on binary systems.

Also, the λ-transition is always of second order. The derived theoretical phase diagram based on our analysis agrees with the classical phase diagram, and it is hoped that the research presented here will lead to a better understanding of the nature of superfluidity. Finally, we remark that the order of the second transition is mathematically more challenging, and will be reported elsewhere.

Chapter 4
Fluid Dynamics

One important source of dynamic transitions and pattern formation is transition and stability problems in fluid dynamics, including in particular Rayleigh–Bénard convection, the Couette–Taylor problem, the Couette–Poiseuille–Taylor problem, and the parallel shear flow problem. The study of these basic problems leads to a better understanding of turbulent behavior in fluid flows, which in turn often leads to new insights and methods in the solution of other problems in science and engineering.

Fluid flow is modeled by Navier–Stokes-type equations, which can be viewed as an infinite-dimensional dissipative system. The dissipative structure of Navier–Stokes equations is mathematically described by global absorbing sets and global attractors; see, among many others, (Foiaş and Temam 1979; Ladyzhenskaya 1982; Vishik 1992; Temam 1997). The existence and properties of solutions of the Navier–Stokes equations are classical topics; see, among others, (Leray 1933; Hopf 1951; Kiselev and Ladyženskaya 1957; Ladyženskaya 1958; Caffarelli et al. 1982a), (Temam 2001, Appendix III), and the references therein.

Also, since Rayleigh's pioneering work on classical Bénard convection, there have been intensive studies of different fluid problems from the point of view of linear analysis and local bifurcation points; see, among many others, (Chandrasekhar 1981; Drazin and Reid 1981; Yudovich 1967a,b; Rabinowitz 1968; Kirchgässner 1975; Busse 1978; Cross and Hohenberg 1993; Palm 1975; Getling 1997; Golubitsky et al. 1988; Swinney 1988; Friedlander et al. 2000; Iooss and Adelmeyer 1998; Koschmieder 1993; Lappa 2009; Ma and Wang 2005d,b).

In addition, the classical point of view on Lagrangian dynamics and its applications has been explored by many authors; see, among many others, (Surana et al. 2006; Samelson and Wiggins 2006; Wiggins 1990; Rogerson et al. 1999; Mezić 2001) and the references therein. For the study of Lagrangian dynamics in conjunction with vortex dynamics, we refer the interested reader to Newton (2001) and the references therein.

The main focus of this chapter is dynamic transitions and pattern-formation mechanisms of classical Bénard convection, the Taylor problem, Taylor–Poiseuille flow, and the rotating convection problem. Structural transition and pattern formation

T. Ma and S. Wang, *Phase Transition Dynamics*, DOI 10.1007/978-1-4614-8963-4_4,

are only briefly touched on, due to the scope and limitations of this book. A crucial component of our study is based on our Principle 1, following Philosophy 2. Namely, we classify transitions into their types—continuous, catastrophic, and random—and determine patterns by searching for a complete set of transition states. The physical implications of our rigorous analysis thereby become transparent. The results presented are based mainly on recent work of the authors (Ma and Wang 2004a,b, 2006, 2007a, 2009a).

For many problems in fluid dynamics, our research involves two types of transitions. The first is the dynamic transition of the solutions of the partial differential equations governing the motion and states of fluids. The second is structural transitions in physical space. Our analysis is based on the dynamic transition theory presented in this book and the geometric theory of incompressible flows synthesized in Ma and Wang (2005d).

The study of structural transitions in physical space was originally motivated by a desire to understand the dynamics of ocean currents. Our approach is to classify the topological structure of the *instantaneous* velocity field (i.e., streamlines in Eulerian coordinates) and its transitions, treating the time variable as a parameter. Based on this approach, our project has two goals: the development of a global geometric theory of divergence-free vector fields on general two-dimensional compact manifolds with or without boundaries, and the establishment of connections between the solutions of the Navier–Stokes (or Euler) equations and the dynamics of velocity fields in physical space. This work is synthesized in Ma and Wang (2005d).

This point of view is explored throughout this chapter. This is a difficult topic, with many potential applications in many areas of nonlinear sciences, including, in particular, pattern formation and climate dynamics; see also Chaps. 5 and 6.

4.1 Rayleigh–Bénard Convection

4.1.1 Bénard Problem

Convection is a well-known phenomenon of fluid motion induced by buoyancy when a fluid is heated from below. It is familiar as the driving force in atmospheric and oceanic phenomena. The Rayleigh–Bénard convection problem originated in the famous experiments conducted in 1900 by H. Bénard, who investigated a liquid in a horizontal dish with diameter much larger than the height, keeping the upper surface free to touch the air and heating the fluid from below. He noticed that while the lower surface temperature T is less than some critical temperature T_c, the liquid is in a static state, and when T is greater than T_c, the liquid suddenly breaks into a rather regular cellular pattern of hexagonal convection cells. The cells have the convection roll structure in their vertical section, as illustrated by Fig. 4.1.

In 1916, Lord Rayleigh, in his paper (Rayleigh 1916), developed a linear theory to interpret the phenomena of the Bénard experiments. He chose the Boussinesq equations with some boundary conditions to model the Bénard experiments, and

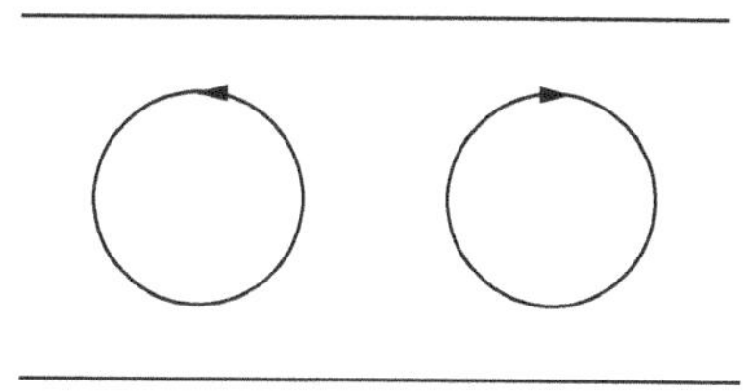

Fig. 4.1 The vertical section of thermal convection

linearized these equations using normal modes. He then showed that convection would occur only when the nondimensional parameter, called the Rayleigh number,

$$R = \frac{g\alpha\beta}{\kappa\nu} h^4 \tag{4.1.1}$$

exceeds a certain critical value, where g is the gravitation constant, α is the coefficient of thermal expansion of the fluid,

$$\beta = |dT/dz| = (\overline{T}_0 - \overline{T}_1)/h$$

is the vertical temperature gradient with $\overline{T}_0$ the temperature on the lower surface and $\overline{T}_1$ on the upper surface, h is the depth of the layer of the fluid, κ is the thermal diffusivity, and ν is the kinematic viscosity.

As mentioned in Sect. 1.3.1, the dynamic transition and patternformation mechanism of the Bénard convection can be thoroughly explored using the dynamic transition theory and the geometric theory of 2-dimensional incompressible flows of Ma and Wang (2005d); see, e.g., (Ma and Wang 2004b, 2007a). This section is devoted to the analysis and physical implications of this study.

4.1.2 Boussinesq Equations

The Bénard problem can be modeled by the Boussinesq equations; see among others (Rayleigh 1916; Drazin and Reid 1981; Chandrasekhar 1981). They read

$$\begin{aligned}
&\frac{\partial u}{\partial t} + (u\cdot\nabla)u - \nu\Delta u + \rho_0^{-1}\nabla p = -gk[1-\alpha(T-\overline{T}_0)],\\
&\frac{\partial T}{\partial t} + (u\cdot\nabla)T - \kappa\Delta T = 0,\\
&\mathrm{div}u = 0,
\end{aligned} \tag{4.1.2}$$

where ν, κ, α, g are the constants defined in (4.1.1), $u = (u_1, u_2, u_3)$ is the velocity field, p is the pressure function, T is the temperature function, $\overline{T}_0$ is a constant representing the lower surface temperature at $x_3 = 0$, and $k = (0,0,1)$ is the unit vector in the x_3-direction.

To make the equations nondimensional, let

$$\begin{aligned}
&(x,t) = (hx', h^2t'/\kappa),\\
&(u,T) = (\kappa u'/h, \beta h(T'/\sqrt{R}) + \overline{T}_0 - \beta h x_3'),\\
&p = \rho_0 \kappa^2 p'/h^2 + p_0 - g\rho_0(hx_3' + \alpha\beta h^2 (x_3')^2/2),\\
&\mathrm{Pr} = \nu/\kappa.
\end{aligned}$$

Here the Rayleigh number R is defined by (4.1.1), β is as in (4.1.1), and Pr is the Prandtl number.

Omitting the primes, Eq. (4.1.2) can be rewritten as

$$\begin{aligned}
&\frac{1}{\mathrm{Pr}}\left[\frac{\partial u}{\partial t} + (u\cdot\nabla)u + \nabla p\right] - \Delta u - \sqrt{R}Tk = 0,\\
&\frac{\partial T}{\partial t} + (u\cdot\nabla)T - \sqrt{R}u_3 - \Delta T = 0, \qquad (4.1.3)\\
&\mathrm{div} u = 0.
\end{aligned}$$

The nondimensional domain is $\Omega = D\times(0,1) \subset \mathbb{R}^3$, and $D\subset\mathbb{R}^2$ is an open set. The coordinate system is given by $x = (x_1, x_2, x_3) \in \Omega$.

The Boussinesq equations (4.1.3) are basic equations for studying the Rayleigh–Bénard problem. They are supplemented with the following initial value conditions:

$$(u,T) = (u_0, T_0) \quad \text{at} \quad t = 0. \qquad (4.1.4)$$

Boundary conditions are needed at the top and bottom boundaries and at the lateral boundary $\partial D\times(0,1)$. At the top and bottom boundaries ($x_3 = 0,1$), either the Dirichlet boundary condition or the free boundary condition is given:

$$(u,T) = 0 \qquad \text{(Dirichlet)}, \qquad (4.1.5)$$

$$T = 0, \quad u_3 = 0, \quad \frac{\partial(u_1,u_2)}{\partial x_3} = 0 \qquad \text{(free)}. \qquad (4.1.6)$$

On the lateral boundary $\partial D\times[0,1]$, one of the following boundary conditions is usually used:

Dirichlet Boundary Condition:

$$u = 0, \quad T = 0 \quad \left(\text{or } \frac{\partial T}{\partial n} = 0\right). \qquad (4.1.7)$$

Free Boundary Condition:

$$u_n = 0, \quad \frac{\partial u_\tau}{\partial n} = 0, \quad T = 0 \quad \left(\text{or } \frac{\partial T}{\partial n} = 0\right), \qquad (4.1.8)$$

where n and τ are the unit normal and tangent vector on $\partial D\times[0,1]$ respectively, and $u_n = u\cdot n, u_\tau = u\cdot\tau$.

Periodic Condition:

$$(u,T)(x_1+k_1L_1,x_2+k_2L_2,x_3)=(u,T)(x_1,x_2,x_3) \qquad \text{for } k_1,k_2\in\mathbb{Z}. \tag{4.1.9}$$

A physically sound boundary condition is a combination of (4.1.5) or (4.1.6) at the top and bottom boundary ($x_3=0,1$) with one of the lateral boundary conditions (4.1.7)–(4.1.9). For simplicity, we study the phase transition problem with the following Dirichlet boundary conditions:

$$(u,T)=0 \quad \text{on } \partial\Omega. \tag{4.1.10}$$

Of course, all results obtained in this section hold as well for other combinations of boundary conditions. For the problem (4.1.3) with (4.1.10), we define the function spaces as

$$\begin{aligned} &H=\{(u,T)\in L^2(\Omega)^4 \mid \text{div}u=0,\ u\cdot n=0 \text{ on } \partial\Omega\},\\ &H_1=\{(u,T)\in H^2(\Omega)^4 \mid \text{div}u=0,\ (u,T)=0 \text{ on } \partial\Omega\}. \end{aligned} \tag{4.1.11}$$

4.1.3 Dynamic Transition Theorems

The linearized equations of (4.1.3) are given by

$$\begin{aligned} &-\Delta u+\nabla p-\sqrt{R}Tk=0,\\ &-\Delta T-\sqrt{R}u_3=0,\\ &\text{div}u=0. \end{aligned} \tag{4.1.12}$$

It is clear that the eigenvalue problem (4.1.12) with (4.1.10) for the Rayleigh number R is symmetric. Hence, we know that all R_k with multiplicities m_k of (4.1.12) with (4.1.10) are real numbers, and

$$0<R_1<R_2<\cdots<R_k<R_{k+1}<\cdots.$$

The first eigenvalue R_1, also denoted by $R_c=R_1$, is called the critical Rayleigh number. Let the multiplicity of R_c be m (≥ 1), and let the first eigenvectors be $\psi_1=(e_1,T_1),\cdots,\psi_m=(e_m,T_m)$, which satisfy

$$(\psi_i,\psi_j)=\int_\Omega [e_i\cdot e_j+T_i\cdot T_j]dx=\delta_{ij},$$

where δ_{ij} is the Kronecker symbol. Let E_0 be the first eigenspace of (4.1.12) with (4.1.10):

$$E_0=\left\{\sum_{k=1}^m \alpha_k\psi_k \ \mid\ \alpha_k\in\mathbb{R}^1,\ 1\leq k\leq m\right\}.$$

The following theorems provide general phase transition results for Rayleigh–Bénard convection.

Theorem 4.1.1 *For the Bénard problem (4.1.3) with (4.1.10), the following assertions hold:*

(1) The problem has a transition at the critical Rayleigh number $R = R_c$, and this transition is of Type I.
(2) When $R < R_c$, the steady state $(u,T) = 0$ is globally asymptotically stable for (4.1.3) and (4.1.10).
(3) When $R > R_c$, the problem bifurcates from $((u,T),R) = (0,R_c)$ to an attractor $\mathscr{A}_R$, which is an $(m-1)$-dimensional homological sphere.
(4) For every $(u,T) \in \mathscr{A}_R$, the velocity field u can be expressed as

$$u = \sum_{k=1}^{m} \alpha_k e_k + o\left(\sum_{k=1}^{m} |\alpha_k|\right), \qquad \lim_{R \to R_c} (\alpha_1, \cdots, \alpha_m) = 0, \tag{4.1.13}$$

where the e_k are the velocity fields of the first eigenvectors in E_0.
(5) The attractor $\mathscr{A}_R$ attracts $H \setminus \Gamma$, where Γ is the stable manifold of $(u,T) = 0$ with codimension m.

Theorem 4.1.2 *If the first eigenvalue of (4.1.12) is simple, i.e., $dimE_0 = 1$, then the following assertions hold:*

1. The bifurcated attractor $\mathscr{A}_R$ of (4.1.3) with (4.1.10) consists of exactly two points $\phi_1, \phi_2 \in H_1$ given by

$$\phi_{1,2} = \pm\alpha \left(\sqrt{R} - \sqrt{R_c}\right)^{1/2} \psi + o(|\sqrt{R} - \sqrt{R_c}|^{1/2}), \tag{4.1.14}$$

where $\alpha > 0$ is a constant, and $\psi \in E_0$ is the first eigenvector.
2. The phase space H can be decomposed into two open sets:

$$H = \overline{U}_1 + \overline{U}_2, \quad U_1 \bigcap U_2 = \emptyset,$$

such that for $i = 1,2$ and $\varphi_i \in U_i$,

$$\lim_{t \to \infty} \|S_R(t)\varphi_0 - \varphi_i\| = 0 \qquad \forall \varphi_0 = (u_0, T_0) \in U_i,$$

where $S_R(t)\varphi_0$ is the solution of (4.1.3) with (4.1.4) and (4.1.10).

A few remarks are now in order.

Remark 4.1.3 As mentioned earlier, both theorems hold for the Boussinesq equations (4.1.3) with different combinations of boundary conditions given by (4.1.5)–(4.1.9).

Remark 4.1.4 As we shall see later, (4.1.13) and (4.1.14) are crucial for studying the topological structure of Rayleigh–Bénard convection in physical space.

Remark 4.1.5 Theorem 4.1.2 corresponds to the classical pitchfork bifurcation, and it is much stronger in the sense that assertion 2 here provides a global asymptotic stability of the bifurcated steady-state solutions, which does not appear to be derived by any other existing methods.

Proof of Theorem 4.1.1. We shall apply Theorem 2.2.11 to prove Theorem 4.1.1, and proceed in the following steps.

Step 1. For simplicity, we assume that the Prandtl number Pr is equal to 1; otherwise, we have only to consider the following form of (4.1.3), and the proof is the same:

$$\begin{aligned}
&\frac{\partial u}{\partial t}+(u\cdot\nabla)u+\nabla p-\mathrm{Pr}\Delta u-\sqrt{R\mathrm{Pr}}\theta k=0,\\
&\frac{\partial \theta}{\partial t}+(u\cdot\nabla)\theta-\sqrt{R\mathrm{Pr}}u_3-\Delta\theta=0,\\
&\mathrm{div} u=0,
\end{aligned}$$

where $\theta=\sqrt{\mathrm{Pr}}T$.

Let H and H_1 be the function space defined by (4.1.11). Then let $L_\lambda=-A+B_\lambda:H_1\to H$, and $G:H_1\to H$ be defined by

$$\begin{aligned}
&A\phi=(-P(\Delta u),-\Delta T), && B_\lambda\phi=\lambda(P[Tk],u_3),\\
&G(\phi,\phi_1)=(-P[(u\cdot\nabla)u_1],-(u\cdot\nabla)T_1), && G(\phi)=G(\phi,\phi)
\end{aligned}\tag{4.1.15}$$

for $\phi=(u,T),\phi_1=(u_1,T_1)\in H_1$. Here $\lambda=\sqrt{R}$, and $P:L^2(\Omega)^3\to H$ is the Leray projection. These operators enjoy the following properties:

(a) The linear operators A,B_λ and L_λ are all symmetric.
(b) The nonlinear operator G is orthogonal, i.e.,

$$(G(\phi_1,\phi),\phi)=-\int_\Omega[(u_1\cdot\nabla)u\cdot u+(u_1\cdot\nabla)T\cdot T]dx=0.\tag{4.1.16}$$

(c) The conditions (2.1.2) and (2.1.3) hold for the operators defined in (4.1.15).

Then the Boussinesq equations (4.1.3) can be rewritten in the following operator form:

$$\frac{d\phi}{dt}=L_\lambda\phi+G(\phi).\tag{4.1.17}$$

Step 2. Verification of conditions (2.1.4) and (2.1.5). Consider the eigenvalue problem

$$L_\lambda\phi=\beta(\lambda)\phi,\tag{4.1.18}$$

which is equivalent to

$$\begin{aligned}
&\Delta u-\nabla p+\lambda Tk=\beta(\lambda)u,\\
&\Delta T+\lambda u_3=\beta(\lambda)T,\\
&\mathrm{div} u=0,
\end{aligned}\tag{4.1.19}$$

supplemented with (4.1.10). For each λ fixed, it is known that the eigenvalues β_k $(k=1,2,\cdots)$ of (4.1.18) are real numbers satisfying

$$\beta_1(\lambda) \geq \beta_2(\lambda) \geq \cdots \geq \beta_k(\lambda) \geq \cdots \to -\infty, \tag{4.1.20}$$

and the first eigenvalue $\beta_1(\lambda)$ of (4.1.18) and the first eigenvalue $\lambda_1 = \sqrt{R_c}$ of (4.1.12) with (4.1.10) have the relation

$$\beta_1(\lambda) \begin{cases} <0 & \text{if } 0 \leq \lambda < \lambda_1, \\ =0 & \text{if } \lambda = \lambda_1. \end{cases} \tag{4.1.21}$$

By Theorem 2.1 in Ma and Wang (2008d) (see also (Ma and Wang 2005b) for a different proof), the first eigenvalue $\beta_1(\lambda)$ is differentiable on λ, and the derivative of $\beta_1(\lambda)$ at $\lambda = \lambda_1$ is given by

$$\frac{d}{d\lambda}\beta_1(\lambda_1) = \int_\Omega [T_1^2 + e_{13}^2]dx > 0, \tag{4.1.22}$$

where $\psi_1 = (e_1, T_1)$ with $e_1 = (e_{11}, e_{12}, e_{13})$ is the eigenvector of (4.1.18) corresponding to $\beta_1(\lambda_1) = 0$.

Thus it follows from (4.1.20)–(4.1.22) that the conditions (2.1.4) and (2.1.5) hold for the eigenvalues of (4.1.18).

Step 3. Finally, in order to use Theorem 2.2.11 to prove Theorem 4.1.1, we need to show that $(u,T)=0$ is a globally asymptotically stable equilibrium point of (4.1.3) with (4.1.10) at the critical Rayleigh number $\lambda_1 = \sqrt{R_c}$. By Theorem 3.16 in Ma and Wang (2005b), it suffices to prove that equations (4.1.3) with (4.1.10) have no invariant sets except the steady state $(u,T)=0$ in the first eigenspace E_0 of (4.1.12).

We know that the Boussinesq equations (4.1.3) with (4.1.10) have a bounded absorbing set in H. Assume that equations (4.1.3) with (4.1.10) have an invariant set $B \subset E_0$ with $B \neq \{0\}$ at $\lambda_1 = \sqrt{R_c}$. Then restricted to B, which contains eigenfunctions of (4.1.18) corresponding to $\beta_1(\lambda_1) = 0$, Eq. (4.1.17) can be rewritten as

$$\frac{d\phi}{dt} = G(\phi) \quad \forall \phi = (u,T) \in B \subset E_0, \tag{4.1.23}$$

which is equivalent to

$$\begin{aligned} &\frac{\partial u}{\partial t} + (u\cdot\nabla)u + \nabla p = 0, \\ &\frac{\partial T}{\partial t} + (u\cdot\nabla)T = 0, \end{aligned} \tag{4.1.24}$$

for every $(u,T) \in B$. It is easy to see that for the solutions $(u,T) \in B$ of (4.1.24), $(\widetilde{u},\widetilde{T}) = \alpha(u(\alpha t), T(\alpha t)) \in \alpha B \subset E_0$ are also solutions of (4.1.24). Namely, for any real number $\alpha \in R$, the set $\alpha B \subset E_0$ is an invariant set of (4.1.24). Thus, we infer that (4.1.3) with (4.1.10) have an unbounded invariant set, which is a contradiction to the existence of bounded absorbing sets. Therefore, the invariant set B consists only of $(u,T)=0$. The proof is complete. □

Proof of Theorem 4.1.2. By Theorem 4.1.1, it suffices to prove that the steady-state equations of (4.1.3) bifurcate to exactly two singular points in H_1 for $R > R_c$. We proceed with the Lyapunov–Schmidt method.

Since the operator $L_\lambda : H_1 \to H$ defined by (4.1.15) is symmetric, all eigenvectors of L_λ,

$$\{\psi_k |\ k = 1, 2, \cdots\} \subset H_1, \tag{4.1.25}$$

constitute an orthogonal basis of H, and H can be decomposed into

$$\begin{aligned}
H &= E_0^\lambda \bigoplus E_1^\lambda,\\
E_0^\lambda &= \{x\psi_1 \mid x \in \mathbb{R}^1,\ \psi_1 \text{ is the first eigenvector of } L_\lambda\},\\
E_1^\lambda &= \left\{\sum_{k=2}^{\infty} y_k\psi_k \mid y_k \in \mathbb{R}^1,\ \psi_k \text{ are as in (4.1.25)}\right\}.
\end{aligned}$$

In addition, E_0^λ and E_1^λ are invariant subspaces of L_λ.

For every $\phi \in H$,

$$\phi = x\psi_1 + \sum_{k=2}^{\infty} y_k\psi_k, \quad x, y_k \in \mathbb{R}^1.$$

Then the equation $L_\lambda\phi + G(\phi) = 0$ can be decomposed into

$$\beta_1(\lambda)x + (G(\phi), \psi_1) = 0, \tag{4.1.26}$$

$$\beta_k(\lambda)y_k + (G(\phi), \psi_k) = 0 \qquad \forall k = 2, 3, \cdots, \tag{4.1.27}$$

where $\beta_k(\lambda)$ are the eigenvalues of L_λ satisfying

$$\beta_1(\lambda) \begin{cases} < 0 & \text{if } \lambda < \lambda_1,\\ = 0 & \text{if } \lambda = \lambda_1,\\ > 0 & \text{if } \lambda > \lambda_1, \end{cases} \tag{4.1.28}$$

$$\beta_k(\lambda_1) < 0 \qquad \forall k \geq 2. \tag{4.1.29}$$

We infer from (4.1.26) and (4.1.16) that

$$\beta_1(\lambda)x - \sum_{k=2}^{\infty} xy_k(G(\psi_1), \psi_k) + O(\|y\|^2) = 0. \tag{4.1.30}$$

We obtain then from (4.1.27) and (4.1.16) that

$$y_k = -\frac{x^2}{\beta_k(\lambda)}(G(\psi_1), \psi_k) + O(|x|^2) \qquad \forall k \geq 2.$$

Putting y_k into (4.1.30), we obtain

$$\beta_1(\lambda)x - ax^3 + o(|x|^3) = 0. \tag{4.1.31}$$

Here, by (4.1.29), we have

$$a=\sum_{k=2}^{\infty}\frac{-1}{\beta_k(\lambda)}|(G(\psi_1),\psi_k)|^2>0 \qquad \text{near } \lambda=\lambda_1.$$

Equation (4.1.31) is the Lyapunov–Schmidt reduction of (4.1.18), which determines the bifurcated singular points of the Boussinesq equations (4.1.3) with (4.1.10). By (4.1.28), Eq. (4.1.31) bifurcates from $(x,\lambda)=(0,\lambda_1)$ to exactly two solutions for $\lambda>\lambda_1$:

$$x_{1,2}=\pm(\beta_1(\lambda)/a)^{1/2}+o(|\beta_1|^{1/2}).$$

Therefore, the problem (4.1.3) with (4.1.10) bifurcates from $(\phi,\lambda)=(0,\lambda_1)$ to exactly two singular points for $\lambda>\lambda_1$ $(R>R_c)$:

$$\phi_1=b\beta_1^{1/2}(\lambda)\psi_1+o(|\beta_1|^{1/2}), \qquad \phi_2=-b\beta^{1/2}(\lambda)\psi_1+o(|\beta_1|^{1/2}), \tag{4.1.32}$$

where $b=a^{-1/2}>0$. By (4.1.32), to prove (4.1.14), it suffices to verify that

$$\beta_1(\lambda)=c(\lambda-\lambda_1)+o(|\lambda-\lambda_1|), \tag{4.1.33}$$

where $\lambda=\sqrt{R}$ and $\lambda_1=\sqrt{R_c}$. The formula (4.1.33) follows from (4.1.22). □

4.1.4 Topological Structure and Pattern Formation

As seen in the expressions (4.1.13) and (4.1.14), the topological structure of the first eigenvectors of the linearized equations plays an important role for studying the onset of Rayleigh–Bénard convection, which essentially determines the thermal convection structure in physical space.

We now examine in detail the first eigenvectors of the linearized problem with the free boundary conditions in a rectangular domain, $\Omega=(0,L_1)\times(0,L_2)\times(0,1)$, and consider the eigenvalue problem (4.1.12) with the free boundary condition:

$$\begin{aligned}
&u_n=0, \quad \frac{\partial u_\tau}{\partial n}=0 && \text{on } \partial\Omega,\\
&T=0 && \text{at } x_3=0,1,\\
&\frac{\partial T}{\partial n}=0 && \text{at } x_1=0,\ L_1 \text{ or } x_2=0,L_2.
\end{aligned} \tag{4.1.34}$$

Solutions of the Eigenvalue Problem. For the eigenvalue problem (4.1.12) with (4.1.34), we take the separation of variables as follows:

$$(u_1,u_2,u_3,T)=\left(\frac{1}{a^2}\nabla_2 f(x_1,x_2)\frac{dH(x_3)}{dx_3},fH(x_3),f\theta(x_3)\right), \tag{4.1.35}$$

where $a^2 > 0$ is an arbitrary constant. It follows from (4.1.12) and (4.1.34) that the functions $f = f(x_1,x_2), H = H(x_3), \theta = \theta(x_3)$ satisfy

$$\begin{cases} -\Delta_2 f = a^2 f, \\ \left.\frac{\partial f}{\partial x_1}\right|_{x_1=0,L_1} = 0, \quad \left.\frac{\partial f}{\partial x_2}\right|_{x_2=0,L_2} = 0, \end{cases} \tag{4.1.36}$$

$$\begin{cases} \left(\frac{d^2}{dx_3^2} - a^2\right)^2 H = a^2 \lambda \theta, \\ \left(\frac{d^2}{dx_3^2} - a^2\right) \theta = -\lambda H, \\ (\theta(x_3), H(x_3), H''(x_3)) = 0 \quad \text{at } x_3 = 0, 1. \end{cases} \tag{4.1.37}$$

It is clear that the solutions of (4.1.36) are given by

$$\begin{aligned} &f(x_1,x_2) = \cos a_1 x_1 \cos a_2 x_2, \\ &(a_1,a_2) = (k_1\pi/L_1, k_2\pi/L_2), \qquad a^2 = a_1^2 + a_2^2 \qquad \forall k_1,k_2 = 0,1,\cdots. \end{aligned} \tag{4.1.38}$$

For each a^2, the first eigenvalue $\lambda_0(a)$ and eigenvectors of (4.1.37) are given by

$$\lambda_0(a) = \frac{(\pi^2+a^2)^{3/2}}{a}, \qquad (H,\theta) = \left(\sin \pi x_3, \frac{1}{a}\sqrt{\pi^2+a^2}\sin \pi x_3\right). \tag{4.1.39}$$

The first eigenvalue $\lambda_1 = \sqrt{R_c}$ of (4.1.12) with (4.1.34) is the minimum of $\lambda_0(a)$:

$$R_c = \min_{a^2=a_1^2+a_2^2} \lambda_0^2(a) = \min_{k_1,k_2\in\mathbb{Z}} \left[\pi^4 \left(1 + \frac{k_1^2}{L_1^2} + \frac{k_2^2}{L_2^2}\right)^3 \Big/ \left(\frac{k_1^2}{L_1^2} + \frac{k_2^2}{L_2^2}\right)\right]. \tag{4.1.40}$$

Thus the first eigenvectors of (4.1.12) with (4.1.34) are immediately derived from (4.1.35), (4.1.38), and (4.1.39):

$$(u,T) = \left(\frac{\pi}{a^2}\nabla_2 f \cdot \cos \pi x_3, f sin\pi x_3, \frac{\sqrt{\pi^2+a^2}}{a} f \sin \pi x_3\right), \tag{4.1.41}$$

where $f = \cos a_1 x_1 \cos a_2 x_2$, and $a^2 = a_1^2 + a_2^2$ achieves the minimum in (4.1.40).

By Theorems 4.1.1 and 4.1.2, the topological structure of the bifurcated solutions of the Bénard problem (4.1.3) with (4.1.34) in a rectangular region is determined by that of (4.1.41), and consequently depends, by (4.1.40), on the horizontal length scales L_1 and L_2. Namely, the pattern of the Rayleigh–Bénard convection depends on the size and form of the containers of the fluid. In fact, the convection pattern in a rectangular region is completely determined by the wave numbers $(k_1,k_2) \in \mathbb{Z}^2$ satisfying (4.1.40), and we now address this issue.

Horizontal Scales and Wave Numbers. By (4.1.40), the wave numbers $(k_1, k_2) \in \mathbb{Z}^2$ of convection depend on the horizontal length scales L_1 and L_2. Let L_1 and L_2 have the (horizontal) aspect ratio α given by

$$L = L_1 = \alpha L_2, \tag{4.1.42}$$

where L is called the horizontal scale, and $\alpha > 0$ is the ratio constant. Thus the critical Rayleigh number (4.1.40) can be rewritten as

$$R_c = \min_{k_1, k_2 \in \mathbb{Z}} \frac{\pi^4 (L^2 + k_1^2 + \alpha^2 k_2^2)^3}{L^4 (k_1^2 + \alpha^2 k_2^2)}. \tag{4.1.43}$$

Let

$$b = k_1^2 + \alpha^2 k_2^2 \tag{4.1.44}$$

achieve its minimum in (4.1.43). Then

$$\frac{dR_c}{db} = \frac{\pi^4}{L^4 b^2} (L^2 + b)^2 (2b - L^2) = 0,$$
$$\frac{dR_c}{dL} = \frac{2\pi^4}{L^5 b} (L^2 + b)^2 (L^2 - 2b) = 0.$$

Hence we derive the formula

$$b = L^2/2. \tag{4.1.45}$$

Namely, the module of wave numbers is proportional to the square of the horizontal scale. Notice that the wave numbers k_1 and k_2 are integers. When the aspect ratio $\alpha > 0$ is fixed, for each given $(k_1, k_2) \in \mathbb{Z}^2$, the corresponding horizontal scale L, with which (k_1, k_2) satisfies (4.1.43), is in an interval.

To see this, let the integer pairs (k_{i1}, k_{i2}) $(i = 1, 2, 3)$ be given so that the corresponding $b_i = k_{i1}^2 + \alpha^2 k_{i2}^2$ satisfy

$$b_1 < b_2 < b_3, \tag{4.1.46}$$

and there are no integers $(k_1, k_2) \in \mathbb{Z}^2$ such that $b = k_1^2 + \alpha^2 k_2^2$ is between b_j and b_{j+1}, i.e., $b \notin (b_j, b_{j+1})$ $(j = 1, 2)$ for all $(k_1, k_2) \in \mathbb{Z}^2$.

The following theorem is a direct consequence of (4.1.43)–(4.1.46).

Theorem 4.1.6 *Let l_1 and l_2 satisfy*

$$b_{i+1} (l_i^2 + b_i)^3 = b_i (l_i^2 + b_{i+1})^3 \qquad \text{for } i = 1, 2. \tag{4.1.47}$$

If the horizontal scale L satisfies

$$l_1 < L < l_2, \tag{4.1.48}$$

then the wave numbers of the Rayleigh–Bénard convection take the integer pair (k_{21}, k_{22}).

This theorem provides a precise relationship between the horizontal scale L of a rectangular container and the wave numbers (k_1,k_2) of thermal convection. In the following, we discuss the convection pattern using Theorem 4.1.6.

Single-Roll Structure. We note that the horizontal scale $L=L_1$ also represents the aspect ratio between the horizontal scale and the vertical scale of the rectangular container for the case $\alpha \geq 1$.

We see from (4.1.44) and (4.1.45) that for $\alpha > 1$, if the aspect ratio L is small, then the wave numbers are given by $(k_1,k_2)=(1,0)$. More precisely, by Theorem 4.1.6, when

$$0 < L^2 < \frac{1+\alpha^2-(1+\alpha^2)^{1/3}}{(1+\alpha^2)^{1/3}-1} \qquad \text{for} \quad \alpha > 1,$$

the convection pattern is a single-roll structure; see Fig. 4.2a, b.

Consider the case $\alpha = 1$, i.e.,

$$L=L_1=L_2 \quad \text{and} \quad 0<L^2<\frac{2-2^{1/3}}{2^{1/3}-1}\simeq 3. \tag{4.1.49}$$

In this case, the wave numbers (k_1,k_2) are given by

$$(k_1,k_2)=(1,0) \quad \text{and} \quad (0,1).$$

Then the multiplicity of the first eigenvalue is two, and the first eigenspace E_0 of (4.1.12) with (4.1.34) is given by

$$E_0=\{\alpha_1\psi_1+\alpha_2\psi_2 \mid \alpha_1,\alpha_2\in\mathbb{R}^1\},$$

where for $i=1,2$, $\psi_i=(e_i,T_i)$ are given by

$$\begin{aligned}
(e_1,T_1)&=\left(-L\sin\frac{\pi x_1}{L}\cos\pi x_3, 0, \cos\frac{\pi x_1}{L}\sin\pi x_3, \sqrt{L^2+1}\cos\frac{\pi x_1}{L}\sin\pi x_3\right),\\
(e_2,T_2)&=\left(0, -L\sin\frac{\pi x_2}{L}\cos\pi x_3, \cos\frac{\pi x_2}{L}\sin\pi x_3, \sqrt{L^2+1}\cos\frac{\pi x_2}{L}\sin\pi x_3\right).
\end{aligned}$$

When $\alpha_1^2+\alpha_2^2=1$ and $\alpha_1\neq\alpha_2$, the topological structure of $u=\alpha_1 e_1+\alpha_2 e_2$ is equivalent to the roll structure as shown in Fig. 4.2a, b.

In addition, the critical Rayleigh number is

$$R_c=\pi^4(1+L^2)^3/L^4. \tag{4.1.50}$$

Remark 4.1.7 The roll structure of $u=\alpha_1 e_1+\alpha_2 e_2 (\alpha_1\neq\alpha_2)$ has a certain stability, although it is not structural stability, i.e., under a perturbation, the roll trait remains invariant. Hence, by Theorem 4.1.1, the solutions $u=\alpha_1 e_1+\alpha_2 e_2+o(|\alpha_1|+|\alpha_2|)$ with $\alpha_1\neq\alpha_2$ in the bifurcated attractor $\mathscr{A}_R$ have also the single-roll structure.

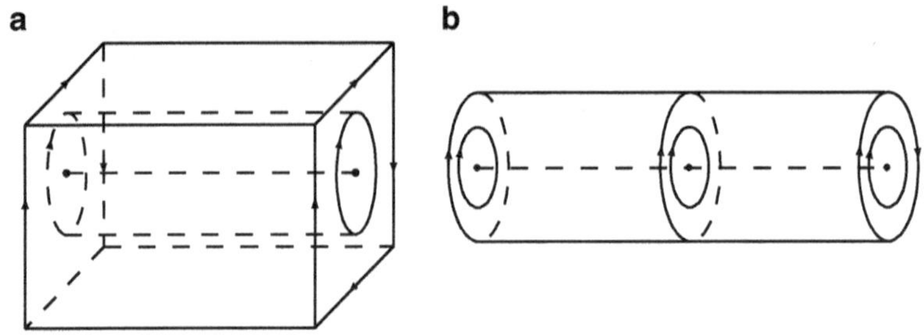

Fig. 4.2 Roll structure: (**a**) an elevation of the flow, and (**b**) flow structure in the interior of the cube

Remark 4.1.8 Since the eigenspace E_0 has dimension two, the classical bifurcation theories cannot be applied to this case for Rayleigh–Bénard convection. In a later subsection, we shall continue to discuss this case using the construction of center manifold functions introduced in Sect. 1.2, and we shall obtain the precise structure of the bifurcated attractor $\mathscr{A}_R$.

Remark 4.1.9 The formula (4.1.50) shows that the critical Rayleigh number R_c depends on the aspect ratio; see also Remark 4.1.10 later.

Double-Roll Structure. We first consider the case

$$L_1 = L_2 = L, \qquad \frac{2-2^{1/3}}{2^{1/3}-1} < L^2 < \frac{2(2-2^{1/3})}{2^{1/3}-1}. \tag{4.1.51}$$

By Theorem 4.1.6, the wave numbers are $(k_1,k_2)=(1,1)$, the eigenvalue is simple, the critical Rayleigh number is

$$R_c = \frac{\pi^4(2+L^4)^3}{2L^4},$$

and the eigenspace is

$$E_0 = \{\beta\psi_1 \mid \beta \in \mathbb{R}^1,\ \psi_1 = (e_1, T_1)\}, \tag{4.1.52}$$

where

$$(e_1,T_1) = \left(\frac{L^2}{2\pi}\nabla_2 f \cos\pi x_3, f\sin\pi x_3, \sqrt{\frac{L^2}{2}+1} f \sin\pi x_3\right), \quad f = \cos\frac{\pi x_1}{L}\cos\frac{\pi x_2}{L}.$$

The topological structure of (4.1.52) consists of two rolls with reverse orientations, and the axes of both rolls are given by (see Fig. 4.3)

$$\left\{\left(\frac{L}{2},x_2,\frac{1}{2}\right) \mid 0 \le x_2 \le \frac{L}{2}\right\} \bigcup \left\{\left(x_1,\frac{L}{2},\frac{1}{2}\right) \mid 0 \le x_1 \le \frac{1}{2}\right\},$$
$$\left\{\left(x_1,\frac{L}{2},\frac{1}{2}\right) \mid \frac{L}{2} \le x_1 \le L\right\} \bigcup \left\{\left(\frac{L}{2},x_2,\frac{1}{2}\right) \mid \frac{L}{2} \le x_2 \le L\right\}.$$

The structure given by Fig. 4.3 is unstable: under a perturbation it will lead to the structures shown in Fig. 4.4.

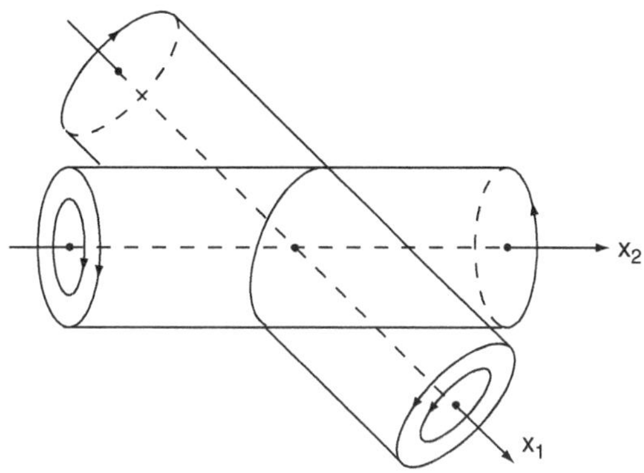

Fig. 4.3 Topological structure of (4.1.52) with (unstable) rolls

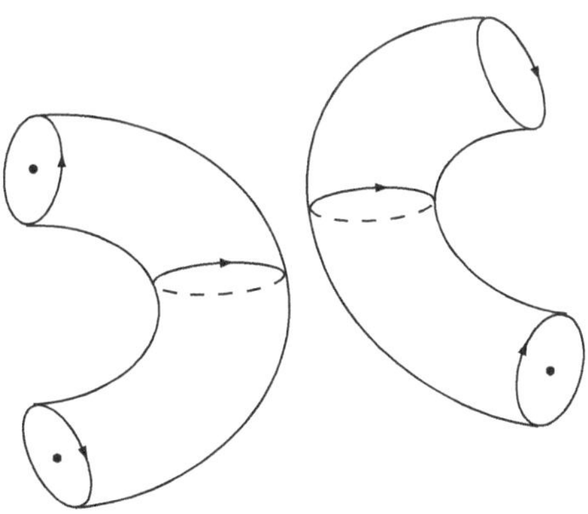

Fig. 4.4 Two stable rolls approximating the structure of (4.1.52)

We now consider the case

$$L_1 = L_2 = L, \qquad \frac{2(2-2^{1/3})}{2^{1/3}-1} \le L^2 \le \frac{5-4(1+\frac{1}{4})^{1/3}}{(1+\frac{1}{4})^{1/3}-1}. \tag{4.1.53}$$

By Theorem 4.1.6, in this case the wave numbers are

$$(k_1,k_2) = (2,0) \quad \text{and} \quad (0,2),$$

and the first eigenspace E_0 is two-dimensional, given by

$$E_0 = \{\alpha_1\psi_1 + \alpha_2\psi_2 |\ \alpha_1, \alpha_2 \in \mathbb{R}^1, \psi_i = (e_i, T_i), i = 1,2\},$$

$$(e_1, T_1) = \left(-\frac{L}{2}\sin\frac{2\pi x_1}{L}\cos\pi x_3, 0, \cos\frac{2\pi x_1}{L}\sin\pi x_3, \frac{\sqrt{4+L^2}}{2}\cos\frac{2\pi x_1}{L}\sin\pi x_3\right),$$

$$(e_2, T_2) = \left(0, -\frac{1}{2}\sin^2\frac{2\pi x_1}{L}\cos\pi x_3, \cos\frac{2\pi x_1}{L}\sin\pi x_3, \frac{1}{\sqrt{4+L^2}}\cos\frac{2\pi x_2}{L}\sin\pi x_3\right),$$

and the critical Rayleigh number is

$$R_c = \frac{\pi^4(4+L^4)^2}{4L^4}.$$

The topological structure of $u = \alpha_1 e_1 + \alpha_2 e_2$ with $\alpha_1 \neq \alpha_2$ is topologically equivalent to the double-roll structure shown in Fig. 4.5.

Remark 4.1.10 These cases of (4.1.49), (4.1.51), and (4.1.53) are consistent with experiments. When we boil water in a container, when the ratio between the diameter and the height is smaller than $\sqrt{3}$ [Condition (4.1.43)], the convection of heating water takes the single-roll pattern as illustrated by Fig. 4.2. When the ratio is between $\sqrt{3}$ and $\sqrt{6}$ [Condition (4.1.51)], the convection takes the double-roll structure as shown in Fig. 4.4. If the ratio is between $\sqrt{6}$ and $\sqrt{8}$ [Condition (4.1.53)], then the convection takes the pattern given by Fig. 4.5.

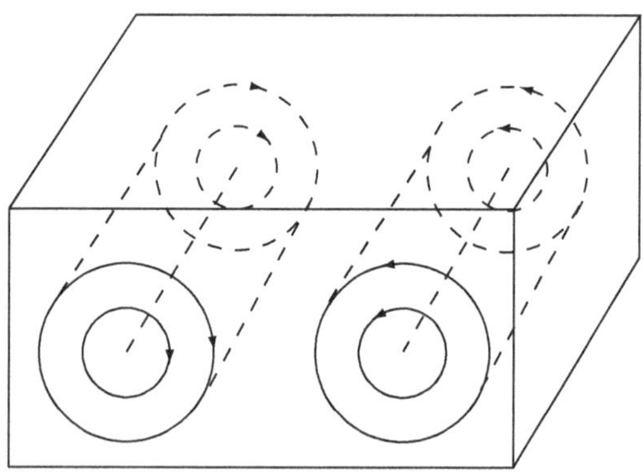

Fig. 4.5 Double-roll structure

Honeycomb Structure. As in the Bénard experiments, if the horizontal length scales L_1 and L_2 are sufficiently large, then it is reasonable to consider the periodic boundary condition in the (x_1, x_2)-plane as follows:

$$\begin{aligned} &(u,T)(x_1+L_1, x_2+L_2, x_3) = (u,T)(x), \\ &T = 0, u_3 = 0, \frac{\partial(u_1,u_2)}{\partial x_3} = 0 \qquad \text{at } x_3 = 0,1. \end{aligned} \tag{4.1.54}$$

In this case, the critical Rayleigh number R_c takes the minimum of $\lambda_0^2(a)$ defined by (4.1.40), which is

$$R_c = \min_a \lambda_0^2(a) = 657.5, \tag{4.1.55}$$

where $a_c = \frac{\pi}{\sqrt{2}}$ is the critical wave number, representing the size of the cells in the Bénard convection. Hence the number $r = 2^{3/2}\pi h/a_c$ can be regarded as the radius of the cells. The first eigenspace E_0 of (4.1.12) with (4.1.54) is generated by eigenvectors of the following type:

$$(e,T) = \left(\frac{\pi}{a_c^2}\nabla_2 f \cos \pi x_3, f(x_1,x_2)\sin \pi x_3, \frac{\sqrt{\pi^2 + a_c^2}}{a_c} f(x_1,x_2)\sin \pi x_3\right), \tag{4.1.56}$$

where $f(x_1,x_2)$ is any one of the following functions:

$$\begin{aligned} &\cos\frac{2\pi k_1 x_1}{L_1}\cos\frac{2\pi k_2 x_2}{L_2}, &&\cos\frac{2\pi k_1 x_1}{L_1}\sin\frac{2\pi k_2 x_2}{L_2},\\ &\sin\frac{2\pi k_1 x_1}{L_1}\cos\frac{2\pi k_2 x_2}{L_2}, &&\sin\frac{2\pi k_1 x_1}{L_1}\sin\frac{2\pi k_2 x_2}{L_2}, \end{aligned}$$

with the periods L_1 and L_2 satisfying

$$4\pi^2\left(\frac{k_1^2}{L_1^2} + \frac{k_2^2}{L_2^2}\right) = a_c^2 = \frac{\pi^2}{2} \qquad \text{for } k_1,k_2 \in \mathbb{Z};$$

namely,

$$\frac{k_1^2}{L_1^2} + \frac{k_2^2}{L_2^2} = \frac{1}{8} \qquad \text{for } k_1,k_2 \in \mathbb{Z}. \tag{4.1.57}$$

It is clear that the dimension of the first eigenspace E_0 is determined by the given periods L_1 and L_2 satisfying (4.1.57), and

$$\dim E_0 = \text{even} \geq 4.$$

Various solutions having the honeycomb structure are found in E_0. For convenience, we list two example as follows.

Square Cells: The solution of (4.1.12) with (4.1.54) given by

$$(e_1,T_1) = \left(\frac{2}{\pi}\nabla_2 f \cos \pi x_3, f \sin \pi x_3, \sqrt{3} f \sin \pi x_3\right), \quad f = \cos\frac{2\pi x_1}{L_1}\cos\frac{2\pi x_2}{L_2},$$

is a rectangular cell with sides of lengths L_1 and L_2, where $\frac{1}{L_1^2} + \frac{1}{L_2^2} = \frac{1}{8}$.

Hexagonal Cells: A solution in E_0 having a hexagonal pattern was found by Christopherson in 1940, and is given in Chandrasekhar (1981):

$$
\begin{aligned}
\psi &= (e,T) = (u_1,u_2,u_3,T),\\
u_1 &= -\frac{2}{\sqrt{6}}\sin\frac{3\pi x_1}{2\sqrt{6}}\cos\frac{\pi x_2}{2\sqrt{2}}\cos\pi x_3,\\
u_2 &= -\frac{2}{3\sqrt{2}}\left(\cos\frac{3\pi x_1}{2\sqrt{6}}+2\cos\frac{\pi x_2}{2\sqrt{2}}\right)\sin\frac{\pi x_2}{2\sqrt{2}}\cos\pi x_3,\\
u_3 &= \frac{1}{3}\left(2\cos\frac{3\pi x_1}{2\sqrt{6}}\cos\frac{\pi x_2}{2\sqrt{2}}+\cos\frac{\pi x_2}{\sqrt{2}}\right)\sin\pi x_3,\\
T &= \frac{1}{\sqrt{3}}\left(2\cos\frac{3\pi x_1}{2\sqrt{6}}\cos\frac{\pi x_2}{2\sqrt{2}}+\cos\frac{\pi x_2}{\sqrt{2}}\right)\sin\pi x_3.
\end{aligned}
$$

This solution is for the case in which the periods are taken as $L_2 = \sqrt{3}L_1$ and $L_1 = 4\sqrt{6}/3$, and the wave numbers are $(k_1,k_2) = (1,1)$ and $(0,1)$.

In summary, for any fixed periods L_1 and L_2, the first eigenspace E_0 of (4.1.12) with (4.1.54) has its dimension determined by

$$
\dim E_0 = \begin{cases} 6 & \text{if } L_2 = \sqrt{k^2-1}\,L_1, \quad k = 2,3,\cdots,\\ 4 & \text{otherwise.}\end{cases}
$$

The attractor bifurcation theorem, Theorem 4.1.1, provides general dynamical properties for Rayleigh–Bénard convection. However, using the construction of center manifold functions, for the problem (4.1.3) with (4.1.54), we can prove the following more refined results:

1. When the Rayleigh number satisfies $R_c < R < R_c + \varepsilon$ [R_c as in (4.1.55)] for some $\varepsilon > 0$, the Bénard problem bifurcates from the trivial state $\phi = 0$ to an attractor $\mathscr{A}_R$ homeomorphic to a sphere, i.e.,

$$
\mathscr{A}_R = \begin{cases} S^5 & \text{if } L_2 = \sqrt{k^2-1}\,L_1, \quad k = 2,3,\cdots,\\ S^3 & \text{otherwise.}\end{cases}
$$

2. The bifurcated attractor $\mathscr{A}_R$ consists of singular points and their, unstable manifold, and the singular points constitute a finite number of tori. We remark here that with the theory for bifurcation problems with symmetry, one can only prove that there is a finite number of two-dimensional tori of steady states, which can never cover all the solutions in a higher-dimensional attractor $\mathscr{A}_R$. In fact, there are many transient states, occupying the same dimensional space as $\mathscr{A}_R$, that are physically important. These transient states cannot be derived by any of the classical bifurcation methods including the theory for bifurcation problems with symmetry.
3. The minimal attractors in $\mathscr{A}_R$ consist of singular points; they are small perturbations of the eigenvectors in E_0, which have the honeycomb structure.
4. $\mathscr{A}_R$ attracts $H \setminus \Gamma$, where Γ is the stable manifold of $\phi = 0$. Consequently, for any initial value $\phi_0 \in H \setminus \Gamma$, the solution $S_R(t)\phi_0$ of (4.1.3) with (4.1.54) converges to the minimal attractors in $\mathscr{A}_R$, which approximates the honeycomb structure.

4.1.5 Asymptotic Structure of Solutions for the Bénard Problem

In this subsection, we study the structural stability of bifurcated solutions and asymptotic structure of solutions for the two-dimensional Bénard problem. It is easy to see that Theorems 4.1.1 and 4.1.2 both hold for the two-dimensional Boussinesq equation with any combination of boundary conditions as described in (4.1.5)–(4.1.9). Hence, we focus on justifying the roll pattern formation generated by the thermal convection as illustrated in Fig. 4.1, and on theoretically verifying the phase transition pattern consistent with the observations in physical experiments.

Technically speaking, we see from (4.1.43) that if $L_2/L_1 = \alpha^{-1}$ is small, then the wave number k_2 is equal to 0. In this case, the three-dimensional Bénard problem is reduced to the two-dimensional one. Furthermore, due to the symmetry in the (x_1,x_2)-plane of the honeycomb structure of the Rayleigh–Bénard convection, from the viewpoint of a cross section, the three-dimensional convection can be well understood from the two-dimensional version.

Let $\Omega = (0,L)\times(0,1)$ with coordinate system $x = (x_1,x_3)$. Thus, the Boussinesq equations take the form

$$\begin{aligned}
&\frac{1}{\Pr}\left[\frac{\partial u_1}{\partial t} + (u\cdot\nabla)u_1 + \frac{\partial p}{\partial x_1}\right] - \Delta u_1 = 0,\\
&\frac{1}{\Pr}\left[\frac{\partial u_3}{\partial t} + (u\cdot\nabla)u_3 + \frac{\partial p}{\partial x_3}\right] - \Delta u_3 - \sqrt{R}T = 0,\\
&\frac{\partial T}{\partial t} + (u\cdot\nabla)T - \sqrt{R}u_3 - \Delta T = 0,\\
&\frac{\partial u_1}{\partial x_1} + \frac{\partial u_3}{\partial x_3} = 0,
\end{aligned} \tag{4.1.58}$$

where $u = (u_1,u_3)$ is the velocity field in the $x = (x_1,x_2)$ coordinate system, and $\nabla = \left(\frac{\partial}{\partial x_1},\frac{\partial}{\partial x_2}\right)$. For simplicity, we consider here only the free boundary conditions:

$$\begin{aligned}
&u_1 = 0, \quad \frac{\partial u_3}{\partial x_1} = 0, \quad \frac{\partial T}{\partial x_1} = 0 \quad \text{at} \quad x_1 = 0, L,\\
&u_3 = 0, \quad \frac{\partial u_1}{\partial x_3} = 0, \quad T = 0 \quad \text{at} \quad x_3 = 0, 1.
\end{aligned} \tag{4.1.59}$$

The initial value condition is

$$(u,T) = \phi_0 = (u_0,T_0) \quad \text{at} \quad t = 0. \tag{4.1.60}$$

In this case, the function space H defined by (4.1.11) is replaced by

$$H = \{(u,T)\in L^2(\Omega)^3 \mid \operatorname{div} u = 0,\ u_3|_{x_3=0,1} = 0, u_1|_{x_1=0,L} = 0\}.$$

By (4.1.41) and (4.1.43), for the problem (4.1.58)–(4.1.60), the wave number k and the critical Rayleigh number are

$$k \simeq L/\sqrt{2} \qquad \text{for } L \text{ large},$$
$$R_c = \pi^4(k^2+L^2)^3/(L^4k^2),$$

and the first eigenspace E_0 is one-dimensional, given by

$$E_0 = \text{span}\{\psi_1 = (e_1, T_1)\},$$
$$(e_1, T_1) = \left(-\frac{L}{k}\sin\frac{k\pi x_1}{L}\cos\pi x_3,\ \cos\frac{k\pi x_1}{L}\sin\pi x_3,\ \frac{\sqrt{L^2+k^2}}{k}\cos\frac{k\pi x_1}{L}\sin\pi x_3\right). \tag{4.1.61}$$

The topological structure of e_1 in (4.1.61) consists of k vortices as shown in Fig. 4.6a, b.

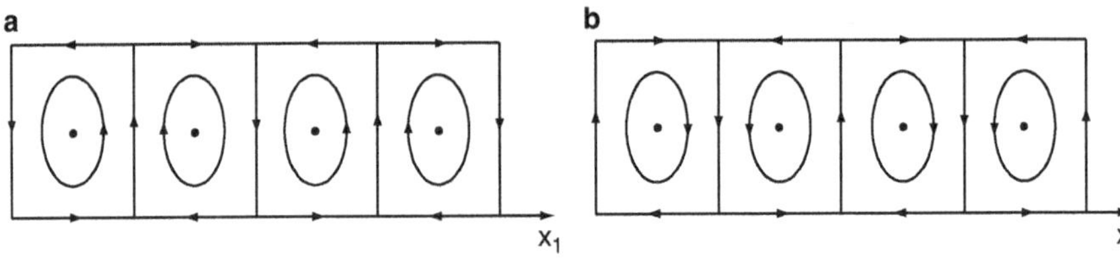

Fig. 4.6 Rolls in (**a**) and (**b**) have reverse orientations

The following theorem shows that the asymptotic structure of solutions of the two-dimensional Bénard problem (4.1.58)–(4.1.60) is topologically equivalent to the roll structure as shown in Fig. 4.6a or b, which is consistent with experimental results.

Theorem 4.1.11 *There is a $\delta > 0$ such that if the Rayleigh number satisfies $R_c < R < R_c + \delta$, then H can be decomposed into two open sets $H = \overline{U}_1 + \overline{U}_2$ with $U_1 \cap U_2 = \emptyset$ such that the following assertions hold:*

(1) For any initial value $\phi_0 = (u_0, T_0) \in U_i$ $(i = 1, 2)$, there exists a time $t_0 > 0$ such that when $t > t_0$, the velocity field $u(t, \phi_0)$ of the solution $S_R(t)\phi_0$ of (4.1.58)–(4.1.60) is topologically equivalent to the structure as shown in either Fig. 4.6a or b, depending on whether $\phi_0 \in U_1$ or $\phi_0 \in U_2$.
(2) When $t > t_0$, the velocity $u(t, \phi_0)$ has k (k is the wave number) vortices. In particular, if L is large, then $k \simeq L/\sqrt{2}$.

Proof. It is easy to check that the velocity fields $\pm e_1$ as in (4.1.61) are of the topological structure as shown in Fig. 4.6a, b, and have k vortices. In particular, by Theorem 4.1.6, if $L \gg 1$ is large, then from the formula

$$(k+1)^2(L^2+k^2)^3 = k^2(L^2+(k+1)^2)^3,$$

one can obtain that the wave number satisfies $k \simeq L/\sqrt{2}$.

Direct computation implies that the singular points of e_1 are nondegenerate. Hence based on the structural stability theorem, Theorem 2.2.9 in Ma and Wang (2005d), the velocity fields $\pm e_1$ are structurally stable.

In addition, by Theorem 4.1.2 and expressions (4.1.14), there exists a $\delta > 0$ such that if $R_c < R < R_c + \delta$, then both velocity fields u_1 and u_2 of the two bifurcated solutions ϕ_1 and ϕ_2 are topologically equivalent to e_1 and $-e_1$ respectively. Furthermore, we know that for every initial value $\phi_0 = (u_0, T_0) \in H$ there is a time $\tau > 0$ such that the solution $S_R(t)\phi_0$ of (4.1.58)–(4.1.60) is C^∞ for $t > \tau$. Therefore, this theorem immediately follows from Theorem 4.1.2 and the following lemma on H^k-uniform boundedness of solutions of (4.1.58)–(4.1.60). Thus, the proof is complete. □

The following lemma is necessary in the proof of the asymptotically structural stability of solutions for the Bénard problem and the Taylor problem discussed in the following section. We refer the interested reader to Ma and Wang (2006) for its proof. Let

$$H_B^k = \{(u,T) \in H^k(\Omega)^3 \bigcap H \mid (u,T) \text{ satisfy } (4.1.59)\}.$$

Lemma 4.1.12 *Let $k \geq 1$ be any integer. If $\phi_0 = (u_0, T_0) \in H_B^k$, then there exists a number $C > 0$ depending on ϕ_0 such that*

$$\|S_R(t)\phi_0\|_{H^k} \leq C \quad \forall t \geq 0,$$

where $S_R(t)\phi_0 = (u(t,\phi_0), T(t,\phi_0))$ is the solution of (4.1.58)–(4.1.60).

Remark 4.1.13 Lemma 4.1.12 and Theorem 4.1.11 hold as well for (4.1.88) with the following boundary conditions:

$$\begin{aligned} & u = T = 0 && \text{at} \quad x_3 = 0, 1, \\ & u_1 = 0, \quad \frac{\partial u_3}{\partial x_1} = 0, \quad \frac{\partial T}{\partial x_1} = 0 && \text{at} \quad x_1 = 0, L. \end{aligned}$$

4.1.6 Structure of Bifurcated Attractors

As an example, we discuss the topological structure of the bifurcated attractor for the Bénard problem (4.1.3) with the free boundary condition (4.1.34) in a rectangular region $\Omega = (0,L)^2 \times (0,1)$. For simplicity, we consider the case

$$0 < L^2 < \frac{2 - 2^{1/3}}{2^{1/3} - 1}. \tag{4.1.62}$$

We proceed by first considering the eigenvalues and eigenvectors of the linearized equations of (4.1.3):

$$\begin{aligned} &\Delta u - \nabla p + \sqrt{R} kT = \beta(R) u, \\ &\Delta T + \sqrt{R} u_3 = \beta(R) T, \\ &\text{div } u = 0, \end{aligned} \tag{4.1.63}$$

supplemented with the boundary condition (4.1.34).

For $\psi = (u_1, u_2, u_3, T)$, we take the separation of variables as follows:

$$\psi = \left(\nabla_2 f\, h'(x_3), a^2 f h(x_3), f\, \theta(x_3)\right), \qquad f = \cos a_1 x_1 \cos a_2 x_2,$$

where as before, $a_i = k_i \pi / L$ for $i = 1, 2$, and $a^2 = a_1^2 + a_2^2$.

We infer from the horizontal velocity equations of (4.1.63) that

$$p = \cos a_1 x_1 \cos a_2 x_2 \left(\left(\frac{d^2}{dx_3^2} - a^2 \right) h' - \beta h' \right).$$

Inserting p into the third component of the velocity equation and the temperature in (4.1.63), we then derive that

$$\begin{aligned} &\left(\frac{d^2}{dx_3^2} - a^2 \right)^2 h - \sqrt{R} \theta = \beta \left(\frac{d^2}{dx_3^2} - a^2 \right) h, \\ &\left(\frac{d^2}{dx_3^2} - a^2 \right) \theta + \sqrt{R} a^2 h = \beta \theta, \end{aligned} \tag{4.1.64}$$

supplemented with the following boundary conditions:

$$(h, h'', \theta) = 0 \quad \text{at} \quad x_3 = 0, 1. \tag{4.1.65}$$

For given Rayleigh number $R > 0$ and a pair of integers k_1 and k_2, the eigenvectors of (4.1.64) and (4.1.65) are given by

$$(h, \theta) = (\sin n\pi x_3, \alpha_{k_1 k_2}^{nm} \sin m\pi x_3) \qquad n \geq 1, \quad m \geq 0, \tag{4.1.66}$$

and the associated eigenvalue $\beta_{k_1 k_2}^{nm}(R)$ and the coefficient $\alpha_{k_1 k_2}^{nm}$ satisfy

$$\begin{aligned} &(n^2\pi^2 + a^2)^2 - \sqrt{R} a^2 \alpha_{k_1 k_2}^{nm} = -\beta_{k_1 k_2}^{nm} (n^2\pi^2 + a^2), \\ &-(m^2\pi^2 + a^2)\alpha_{k_1 k_2}^{nm} + \sqrt{R} = \beta_{k_1 k_2}^{nm} \alpha_{k_1 k_2}^{nm}, \end{aligned}$$

for $a^2 = (k_1^2 \pi^2 + k_2^2 \pi^2)/L^2 \neq 0$. Consequently,

$$\begin{aligned} \alpha_{k_1 k_2}^{nm} &= \frac{1}{a^2 \sqrt{R}} \left[(n^2\pi^2 + a^2)^2 - (m^2\pi^2 + a^2)(n^2\pi^2 + a^2) + \frac{\sqrt{R}(n^2\pi^2 + a^2)}{\alpha_{k_1 k_2}^{nm}} \right], \\ \beta_{k_1 k_2}^{nm} &= -(m^2\pi^2 + a^2) + \sqrt{R}/\alpha_{k_1 k_2}^{nm}, \quad \forall k_1^2 + k_2^2 \geq 1, m \geq 1, n \geq 1. \end{aligned} \tag{4.1.67}$$

When $a^2 = 0$, i.e., $k_1 = k_2 = 0$, the eigenvalues are

$$\beta_0^m = -m^2\pi^2, \quad m = 1, 2, \cdots, \quad n = 0. \tag{4.1.68}$$

Then all eigenvalues of (4.1.63) with (4.1.34) are given by (4.1.67) and (4.1.68), and by (4.1.66) the eigenvectors are

$$\begin{cases} \psi_{k_1k_2nm} = \left(\nabla_2 f_{k_1k_2} \cos n\pi x_3, a^2 f_{k_1k_2} \sin n\pi x_3, \alpha_{k_1k_2}^{nm} f_{k_1k_2} \sin m\pi x_3\right), \\ f_{k_1k_2} = \cos\dfrac{k_1\pi x_2}{L} \cos\dfrac{k_2\pi x_2}{L}, \end{cases} \tag{4.1.69}$$

$$\psi_{0m} = (0, 0, 0, \sin m\pi x_3). \tag{4.1.70}$$

Under the condition (4.1.62), it is known from (4.1.50) that the critical Rayleigh number R_c is equal to $\pi^4(1+L^2)^3/L^4$, and the wave numbers are $(k_1, k_2) = (1, 0)$ and $(0, 1)$. Namely, the first eigenvectors of (4.1.63) with (4.1.34) near $R = R_c$ are given by

$$\psi_{10} = \left(-L\sin\frac{\pi x_1}{L}\cos\pi x_3, 0, \cos\frac{\pi x_1}{L}\sin\pi x_3, \sqrt{1+L^2}\cos\frac{\pi x_1}{L}\sin\pi x_3\right) \tag{4.1.71}$$

$$\psi_{01} = \left(0, -L\sin\frac{\pi x_2}{L}\cos\pi x_3, \cos\frac{\pi x_2}{L}\sin\pi x_3, \sqrt{1+L^2}\cos\frac{\pi x_2}{L}\sin\pi x_3\right), \tag{4.1.72}$$

and by (4.1.67), the first eigenvalue $\beta_1(R)$ of (4.1.63) with (4.1.34) near $R = R_c$ satisfies the principle of exchange of stabilities (PES):

$$\beta_1(R) \begin{cases} < 0 & \text{if } R < R_c, \\ = 0 & \text{if } R = R_c, \\ > 0 & \text{if } R > R_c. \end{cases} \tag{4.1.73}$$

The following is the main result in this subsection.

Theorem 4.1.14 *Assume the condition (4.1.62). Then for the Bénard problem (4.1.3) with (4.1.34), the following assertions hold:*

(1) For $R > R_c (= \pi^4(1+L^2)^3/L^4)$, the problem bifurcates from the trivial solution $((u, T), R) = (0, R_c)$ to an attractor Σ_R, homeomorphic to a cycle S^1.
(2) The attractor Σ_R consists of exactly eight singular points Φ_i $(1 \le i \le 8)$ and eight heteroclinic orbits connecting the singular points, as shown in Fig. 4.7, where $\Phi_1, \Phi_3, \Phi_5, \Phi_7$ are minimal attractors, and $\Phi_2, \Phi_4, \Phi_6, \Phi_8$ are saddle points.
(3) The bifurcated minimal attractors can be expressed as

$$\Phi_{1,5} = \pm 4\beta_1^{1/2}(R)\psi_{10} + o(|\beta_1|), \qquad \Phi_{3,7} = \pm 4\beta_1^{1/2}(R)\psi_{01} + o(|\beta_1|),$$

where β_1 is the first eigenvalue of (4.1.63) with (4.1.34) satisfying (4.1.73), and ψ_{10} and ψ_{01} are the first eigenvectors given by (4.1.71) and (4.1.72) respectively.

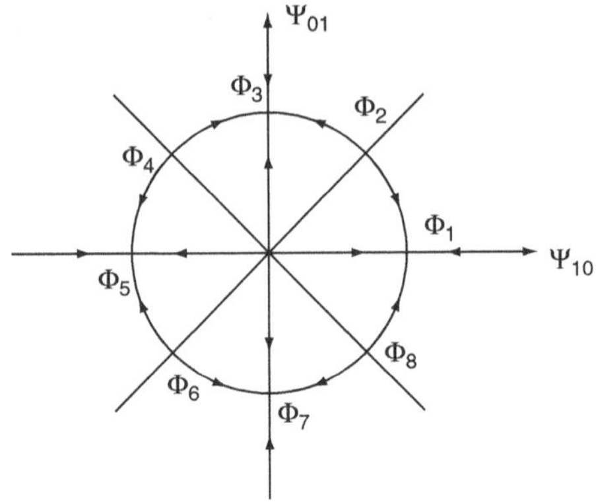

Fig. 4.7 The singular points $\Phi_1, \Phi_3, \Phi_5, \Phi_7$ are the minimal attractors, and the other singular points are the saddles

A few remarks are now in order.

Remark 4.1.15 Theorem 4.1.11 can be proved only by the construction of center manifold functions together with attractor bifurcation theory. The classical theories and methods are unavailable for the Bénard problem (4.1.3) and (4.1.34) in a rectangular region where the first eigenvalue has even multiplicity.

Remark 4.1.16 We know that the multiplicity of the first eigenvalue of (4.1.63) with (4.1.34) in a square region $\Omega = (0,L)^2 \times (0,1)$ is two for many horizontal lengths $L > 0$. In the same fashion, one can prove that assertions (1) and (2) of Theorem 4.1.11 are also valid for these cases.

Remark 4.1.17 It is clear that the velocity fields in ψ_{10} and ψ_{01} have the roll structure as shown in Fig. 4.2. The roll axis of ψ_{10} is the x_2-axis, and that of ψ_{01} is the x_1-axis. The roll structure is of a certain stability, and therefore, the bifurcated minimal attractors $\phi_{2k-1} (1 \leq k \leq 4)$ given byassertion (3) of Theorem 4.1.11 have a similar roll structure as well.

Proof of Theorem 4.1.14. To prove this theorem, we need to reduce (4.1.3) with (4.1.34) to the center manifold.

Let H be the space as defined in (4.1.11). It is known that all eigenvectors (4.1.69) and (4.1.70) constitute an orthogonal basis of H. Hence for every $\psi \in H, \psi$ can be written as

$$\psi = \sum_{K,J} x_{KJ} \psi_{KJ}, \quad K = (k_1, k_2), \quad J = (j_1, j_2).$$

For simplicity, we take $\Pr = 1$ in (4.1.3). Then the eigenvalue problem (4.1.63) with (4.1.34) is written as

$$L_\lambda \psi = \beta \psi, \tag{4.1.74}$$

and β_1 is the first eigenvalue of (4.1.74) satisfying (4.1.73), and ψ_{10} and ψ_{01} are the first eigenvectors given by (4.1.71) and (4.1.72). Let

$$\psi = x\psi_{10} + y\psi_{01} + \Phi,$$
$$\Phi : E_0 \to E_0^\perp \quad \text{the center manifold function,}$$

$$E_0 = \text{span}\{\psi_{10}, \psi_{01}\},$$
$$E_0^\perp = \{\psi \in H \mid (\psi, \psi_{10}) = 0, (\psi, \psi_{01}) = 0\}.$$

The reduced equations of (4.1.3) with (4.1.34) to the center manifold are given by

$$\begin{aligned}\frac{dx}{dt} &= \beta_1 x + \frac{1}{\|\psi_{10}\|^2}(G(x\psi_{10} + y\psi_{01} + \Phi), \psi_{10}),\\ \frac{dy}{dt} &= \beta_1 y + \frac{1}{\|\psi_{01}\|^2}(G(x\psi_{10} + y\psi_{01} + \Phi), \psi_{01}).\end{aligned} \tag{4.1.75}$$

By Theorem A.1.1, the center manifold function Φ satisfies

$$-L_\lambda \Phi = P_2 G(x\psi_{10} + y\psi_{01}) + o(x^2 + y^2), \tag{4.1.76}$$

where $P_2 : H \to E_0^\perp$ is the canonical projection.

For every $\psi_1 = (u, T_1)$ and $\psi_2 = (v, T_2)$, we define

$$G(\psi_1, \psi_2) = (P[-(u \cdot \nabla)v], (u \cdot \nabla)T_2).$$

Then (4.1.76) can be rewritten as

$$\begin{aligned}-L_\lambda \Phi =& P_2[x^2 G(\psi_{10}, \psi_{10}) + y^2 G(\psi_{01}, \psi_{01}) + xyG(\psi_{10}, \psi_{01})\\ &+ xyG(\psi_{01}, \psi_{10})] + o(x^2 + y^2).\end{aligned} \tag{4.1.77}$$

Notice that $\|\psi_{10}\| = \|\psi_{01}\|$. Hence Eqs. (4.1.75) are expressed as

$$\begin{aligned}\frac{dx}{dt} =& \beta_1 x + \frac{1}{\|\psi_{10}\|^2}\Big[x(G(\psi_{10}, \Phi) + G(\Phi, \psi_{10}), \psi_{10})\\ &+ y(G(\psi_{01}, \Phi) + G(\Phi, \psi_{01}), \psi_{10})\Big] + o(|x|^3 + |y|^3),\\ \frac{dy}{dt} =& \beta_1 y + \frac{1}{\|\psi_{10}\|^2}\Big[y(G(\psi_{01}, \Phi) + G(\Phi, \psi_{01}), \psi_{01})\\ &+ x(G(\psi_{10}, \Phi) + G(\Phi, \psi_{10}), \psi_{01})\Big] + o(|x|^3 + |y|^3).\end{aligned} \tag{4.1.78}$$

Using $(G(\psi_1, \psi_2), \psi_3) = -(G(\psi_1, \psi_3), \psi_2)$, we derive from (4.1.78) that

$$\begin{aligned}\frac{dx}{dt} =& \beta_1 x - \frac{1}{\|\psi_{10}\|^2}\Big[x(G(\psi_{10}, \psi_{10}), \Phi) + y(G(\psi_{01}, \psi_{10}), \Phi)\\ &- y(G(\Phi, \psi_{01}), \psi_{10})\Big] + o(|x|^3 + |y|^3),\\ \frac{dy}{dt} =& \beta_1 y - \frac{1}{\|\psi_{10}\|^2}\Big[y(G(\psi_{01}, \psi_{01}), \Phi) + x(G(\psi_{10}, \psi_{01}), \Phi)\\ &- x(G(\Phi, \psi_{01}), \psi_{10})\Big] + o(|x|^3 + |y|^3).\end{aligned} \tag{4.1.79}$$

Direct computation yields

$$G(\psi_{10},\psi_{10}) = P\begin{cases} -L\pi \sin\frac{\pi x_1}{L}\cos\frac{\pi x_1}{L}, \\ 0, \\ -\pi \sin\pi x_3 \cos\pi x_3, \\ -\pi\sqrt{1+L^2}\sin\pi x_3\cos\pi x_3, \end{cases} \tag{4.1.80}$$

$$G(\psi_{01},\psi_{01}) = P\begin{cases} 0, \\ -L\pi \sin\frac{\pi x_2}{L}\cos\frac{\pi x_2}{L}, \\ -\pi \sin\pi x_3 \cos\pi x_3, \\ -\pi\sqrt{1+L^2}\sin\pi x_3\cos\pi x_3, \end{cases} \tag{4.1.81}$$

$$G(\psi_{10},\psi_{01}) = P\begin{cases} 0, \\ -L\pi \cos\frac{\pi x_1}{L}\sin\frac{\pi x_2}{L}\sin^2\pi x_3, \\ -\pi \cos\frac{\pi x_1}{L}\cos\frac{\pi x_2}{L}\sin\pi x_3 \cos\pi x_3, \\ -\pi\sqrt{1+L^2}\cos\frac{\pi x_1}{L}\cos\frac{\pi x_2}{L}\sin\pi x_3\cos\pi x_3, \end{cases} \tag{4.1.82}$$

$$G(\psi_{01},\psi_{10}) = P\begin{cases} -L\pi \sin\frac{\pi x_1}{L}\cos\frac{\pi x_2}{L}\sin^2\pi x_3, \\ 0, \\ -\pi \cos\frac{\pi x_1}{L}\cos\frac{\pi x_2}{L}\sin\pi x_3 \cos\pi x_3, \\ -\pi\sqrt{1+L^2}\cos\frac{\pi x_1}{L}\cos\frac{\pi x_2}{L}\sin\pi x_3\cos\pi x_3. \end{cases} \tag{4.1.83}$$

Under the action of the Leray projection $P : L^2(\Omega,\mathbb{R}^3) \to H$ in G, we find that

$$P\left(\sin\frac{\pi x_1}{L}\cos\frac{\pi x_1}{L}, 0, \sin\pi x_3\cos\pi x_3\right) = 0,$$

$$P\left(0, \sin\frac{\pi x_2}{L}\cos\frac{\pi x_2}{L}, \sin\pi x_3\cos\pi x_3\right) = 0.$$

Hence we have

$$G(\psi_{10},\psi_{10}) = G(\psi_{01},\psi_{01}) = (0,0,0,-\frac{\pi}{2}\sqrt{1+L^2}\sin 2\pi x_3). \tag{4.1.84}$$

Now we return to compute the center manifold function. Let

$$\Phi(x,y) = \sum_{K\neq(1,0),(0,1)} \Phi_{KJ}(x,y)\psi_{KJ} \qquad \text{for } K=(k_1,k_2), J=(j_1,j_2).$$

We infer from (4.1.74) and (4.1.77) that

$$\begin{aligned}-L_\lambda \Phi = &\sum_{K\neq(1,0),(0,1)} \Phi_{KJ}(-\beta_K^J)\psi_{KJ}\\ =&x^2G(\psi_{10},\psi_{10})+y^2G(\psi_{01},\psi_{01})+xyG(\psi_{10},\psi_{01})\\ &+xyG(\psi_{01},\psi_{10})+o(x^2+y^2).\end{aligned} \tag{4.1.85}$$

On the other hand, from (4.1.80)–(4.1.84), we see that

$$\begin{aligned}&(G(\psi_{10},\psi_{10}),\psi_{KJ})=(G(\psi_{01},\psi_{01}),\psi_{KJ})=0 && \forall\psi_{KJ}\neq\psi_{02},\\ &(G(\psi_{10},\psi_{01}),\psi_{KJ})=0 && \forall\psi_{KJ}\neq\psi_{1122},\\ &(G(\psi_{01},\psi_{10}),\psi_{KJ})=0 && \forall\psi_{KJ}\neq\psi_{1122}.\end{aligned}$$

Hence, by (4.1.85), we derive that

$$\Phi(x,y)=\Phi_{02}(x,y)\psi_{02}+\Phi_{1122}(x,y)\psi_{1122}+o(x^2+y^2), \tag{4.1.86}$$

$$\begin{aligned}\Phi_{02}(x,y)&=\frac{x^2+y^2}{-\beta_0^2\|\psi_{02}\|^2}(G(\psi_{10},\psi_{10}),\psi_{02}),\\ \Phi_{1122}(x,y)&=\frac{xy}{-\beta_{11}^{22}\|\psi_{1122}\|^2}(G(\psi_{10},\psi_{01})+G(\psi_{01},\psi_{10}),\psi_{1122})\\ &=\frac{2xy}{-\beta_{11}^{22}\|\psi_{1122}\|^2}(G(\psi_{10},\psi_{01}),\psi_{1122}).\end{aligned}$$

Here ψ_{02} and ψ_{1122} are given by

$$\psi_{02}=(0,0,0,\sin 2\pi x_3),\qquad \psi_{1122}=\begin{cases}-L\sin\frac{\pi x_1}{L}\cos\frac{\pi x_2}{L}\cos 2\pi x_3,\\ -L\cos\frac{\pi x_1}{L}\sin\frac{\pi x_2}{L}\cos 2\pi x_3,\\ \cos\frac{\pi x_1}{L}\cos\frac{\pi x_2}{L}\sin 2\pi x_3,\\ \alpha_{11}^{22}\cos\frac{\pi x_1}{L}\cos\frac{\pi x_2}{L}\sin 2\pi x_3.\end{cases}$$

By (4.1.67), $\alpha_{11}^{22}=\sqrt{1+2L^2}$.

Putting the center manifold function (4.1.86) in (4.1.79), we obtain

$$\begin{aligned}\frac{dx}{dt}&=\beta_1 x-\alpha x(x^2+y^2)-(b+c)xy^2+o(|x|^3+|y|^3),\\ \frac{dy}{dt}&=\beta_1 y-\alpha y(x^2+y^2)-(b+c)yx^2+o(|x|^3+|y|^3),\end{aligned} \tag{4.1.87}$$

where $b>c$ and

$$\alpha=\frac{|(G(\psi_{10},\psi_{10}),\psi_{02})|^2}{-\beta_0^2\|\psi_{10}\|^2\|\psi_{02}\|^2}=\frac{1}{16},$$

$$b = \frac{2|(G(\psi_{10},\psi_{01}),\psi_{1122})|^2}{-\beta_{11}^{22}\|\psi_{1122}\|^2} = \frac{\pi^2L^2\left[1+L^2+\sqrt{(1+L^2)(1+2L^2)}\right]^2}{32(-\beta_{11}^{22})(1+2L^2)},$$

$$c = \frac{2(G(\psi_{1122},\psi_{01}),\psi_{10})(G(\psi_{10},\psi_{01}),\psi_{1122})}{-\beta_{11}^{22}\|\psi_{1122}\|^2}$$

$$= \frac{\pi^2L^2(1+L^2)(1+L^2+\sqrt{(1+L^2)(1+2L^2)})}{32(-\beta_{11}^{22})(1+2L^2)},$$

$$-\beta_{11}^{22} = 2\pi^2(2+\frac{1}{L^2}) - \frac{\sqrt{R}}{\sqrt{1+2L^2}} > 0 \quad \text{at} \quad R = R_c = \frac{\pi^4(1+L^2)^3}{L^4}.$$

Define a vector field $g(x,y)$ by

$$g(x,y) = \begin{pmatrix} g_1 \\ g_2 \end{pmatrix} = \begin{pmatrix} \alpha x(x^2+y^2)+(b+c)xy^2+o(|x|^3+|y|^3) \\ \alpha y(x^2+y^2)+(b+c)yx^2+o(|x|^3+|y|^3) \end{pmatrix}.$$

Then we have

$$xg_1+yg_2 = \alpha(x^2+y^2)^2+2(b+c)x^2y^2+o(|x|^4+|y|^4). \tag{4.1.88}$$

By the S^1-attractor bifurcation theorem (Theorem 2.2.4), it follows from (4.1.73) and (4.1.88) that the reduced equations (4.1.87) bifurcate from $((x,y),R) = (0,R_c)$ to an attractor Σ_R homeomorphic to a cycle S^1, i.e., $\Sigma_R = S^1$. Thus assertion (1) is proved.

In addition, it is easy to see that Eqs. (4.1.87) have eight singular points:

$$\begin{aligned}
&Z_1 = (4\beta_1^{1/2}+o(|\beta_1|^{1/2}),0), && Z_3 = (0,4\beta_1^{1/2}+o(|\beta_1|^{1/2})),\\
&Z_5 = (-4\beta_1^{1/2}+o(|\beta_1|^{1/2}),0), && Z_7 = (0,-4\beta_1^{1/2}+o(|\beta_1|^{1/2})),\\
&Z_2 = (y_0,y_0), && Z_4 = (-y_0,y_0),\\
&Z_6 = (-y_0,-y_0), && Z_8 = (y_0,-y_0),
\end{aligned}$$

where

$$y_0 = 2\sqrt{\frac{1}{1/8+b+c}}\beta_1^{1/2}+o(|\beta_1|^{1/2}).$$

The Jacobian matrices of (4.1.87) at Z_{2k-1} $(1 \le k \le 4)$ are given by

$$J(Z_{1,5}) = \begin{pmatrix} -2\beta_1 & 0 \\ 0 & -1-16(b+c)\beta_1 \end{pmatrix}, \quad J(Z_{3,7}) = \begin{pmatrix} -1-16(b+c)\beta_1 & 0 \\ 0 & -2\beta_1 \end{pmatrix}.$$

Namely, the eigenvalues at Z_{2k-1} $(1\le k\le 4)$ are negative. Therefore, $Z_{2k-1}(1\le k\le 4)$ are the minimal attractors of (4.1.87). Then it follows from Theorem 5.6 in Ma and Wang (2005b) that the other singular points Z_{2k} $(1 \le k \le 4)$ are saddle points, and the topological structure of the bifurcated attractor Σ_R is as shown in Fig. 4.7. Thus assertions (2) and (3) are proved. The proof is complete. □

4.1.7 Physical Remarks

As mentioned earlier, Rayleigh–Bénard convection together with the Taylor problem has become a paradigm of nonequilibrium phase transitions with its instability, bifurcations, pattern formation, and chaotic behaviors in fluid dynamics as well as in physical systems in general.

We have shown that as the Rayleigh number R crosses the critical parameter R_c, the system remains under a continuous (Type I) transition to an $(m-1)$-dimensional sphere $\mathscr{A}_R = S^{m-1}$, where m is the dimension of the eigenspace for the linearized problem, corresponding to the critical Rayleigh number R_c.

The structure of the set $\mathscr{A}_R = S^{m-1}$ of transition states and the related pattern-formation mechanism depend on the physical boundary conditions and the spatial geometry (of the domain). In most cases, the high-dimensional attractor $\mathscr{A}_R$, which has full measure in $\mathscr{A}_R$, consists not only of steady-state solutions, but of transients as well. The Key Philosophy 2 and Principle 1 play a crucial role in deriving these results, which cannot be directly derived by any previously existing methods.

Finally, dynamic transition theory opens a new avenue toward studying pattern formation, since the patterns and structure of transition states are dictated by solutions in $\mathscr{A}_R$, which are again determined by eigenvectors of the corresponding linearized problem (with high-order approximations). With this theory, for example, we can easily examine the formation and stability of different structures such as rolls, rectangles, hexagons, etc.

4.2 Taylor–Couette Flow

We have seen in Rayleigh–Bénard convection that thermal motion is an important cause, leading to instability of fluid flows. In this section, we shall investigate the instability of the Taylor–Couette flow caused by the centrifugal forces of curved fluid flows.

The study of hydrodynamic instability caused by the centrifugal forces originated with the famous experiments conducted by Taylor (1923), in which he observed and studied the stability of an incompressible viscous fluid between two rotating coaxial cylinders. In his experiments, Taylor investigated the case in which the gap between the two cylinders is small in comparison with the mean radius, and the two cylinders rotate in the same direction. He found that when the Taylor number T is smaller than a critical value $T_c > 0$, called the critical Taylor number, the basic flow, called the Couette flow, is stable, and when the Taylor number crosses the critical value, the Couette flow breaks out into a cellular pattern that is radially symmetric; see Fig. 4.8.

The main objective of this section is to explore dynamic transitions and the physical implications of the Taylor problem. The results presented here are based on Ma and Wang (2006, 2010a).

4.2.1 Taylor Problem

Consider an incompressible viscous fluid between two coaxial cylinders. Let r_1 and r_2 $(r_2 > r_1)$ be the radii of the two cylinders, let Ω_1 and Ω_2 be the angular velocities of the inner and the outer cylinders respectively, and set

$$\mu = \Omega_2/\Omega_1, \quad \eta = r_1/r_2. \tag{4.2.1}$$

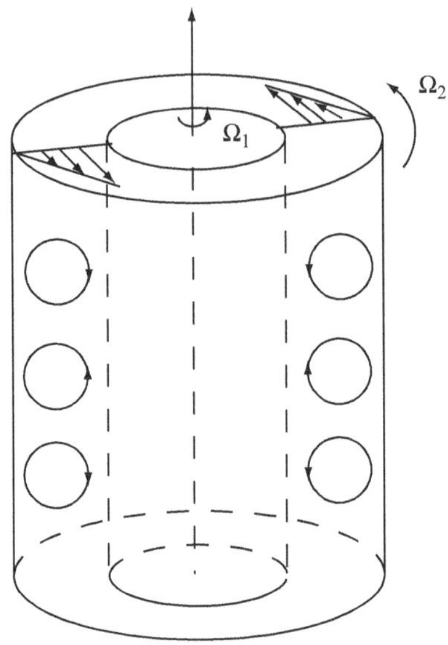

Fig. 4.8 Couette flow and Taylor vortices

The nondimensional Taylor number is defined by

$$T = \frac{4h^4\Omega_1^2}{\nu^2}, \tag{4.2.2}$$

where $\nu > 0$ is the kinematic viscosity and h is the vertical length scale.

There exists a basic steady-state flow, called the Couette flow. In cylindrical polar coordinates (r, θ, z), the Couette flow is defined by

$$(u_r, u_\theta, u_z, p) = \left(0, V(r), 0, \rho \int \frac{1}{r} V^2(r) dr\right), \quad V(r) = ar + \frac{b}{r}, \tag{4.2.3}$$

where (u_r, u_θ, u_z) is the velocity field, p is the pressure, and a, b are constants. It follows from the boundary conditions that

$$V(r_1) = \Omega_1 r_1, \quad V(r_2) = \Omega_2 r_2, \quad a = -\Omega_1 \eta^2 \frac{1-\mu/\eta^2}{1-\eta^2}, \quad b = \Omega_1 \frac{r_1^2(1-\mu)}{1-\eta^2}.$$

Here μ and η are given by (4.2.1).

Based on the Rayleigh criterion, when $\mu > \eta^2$, the Couette flow is always stable at a distribution of angular velocities

$$\Omega(r) = a + b/r^2 \qquad \text{for } r_1 < r < r_2.$$

However, when $\mu < \eta^2$, the situation is different. As in the Taylor experiments, consider the case in which the gap $r_2 - r_1$ is much smaller than the mean radius,

$$r_2 - r_1 \ll r_0 = (r_1 + r_2)/2,$$

and the two cylinders rotate in the same direction. If the Taylor number T in (4.2.2) satisfies $T < T_c$, then the Couette flow (4.2.3) is stable, and if $T_c < T < T_c + \varepsilon$ for some $\varepsilon > 0$, a street of vortices along the z-axis, called the Taylor vortices, emerges abruptly from the basic flow, as shown in Fig. 4.8, and the corresponding flow pattern is radially symmetric and structurally stable.

When the gap $r_2 - r_2$ is not smaller than r_0, or when the cylinders rotate in the opposite directions, the phenomena one observes are much more complex; see (Chandrasekhar 1981) for details. Hence we always assume the condition

$$\eta^2 > \mu \geq 0. \tag{4.2.4}$$

4.2.2 Governing Equations

The hydrodynamic equations governing an incompressible viscous fluid between two coaxial cylinders are the Navier–Stokes equations. In cylindrical polar coordinates (r, θ, z), they are given by

$$\begin{aligned}
&\frac{\partial u_r}{\partial t} + (u \cdot \nabla) u_r - \frac{u_\theta^2}{r} = \nu \left(\Delta u_r - \frac{2}{r^2} \frac{\partial u_\theta}{\partial \theta} - \frac{u_r}{r^2} \right) - \frac{1}{\rho} \frac{\partial p}{\partial r}, \\
&\frac{\partial u_\theta}{\partial t} + (u \cdot \nabla) u_\theta + \frac{u_r u_\theta}{r} = \nu \left(\Delta u_\theta + \frac{2}{r^2} \frac{\partial u_r}{\partial \theta} - \frac{u_\theta}{r^2} \right) - \frac{1}{r\rho} \frac{\partial p}{\partial \theta}, \\
&\frac{\partial u_z}{\partial t} + (u \cdot \nabla) u_z = \nu \Delta u_z - \frac{1}{\rho} \frac{\partial p}{\partial z}, \\
&\frac{\partial (r u_r)}{\partial r} + \frac{\partial u_\theta}{\partial \theta} + \frac{\partial (r u_z)}{\partial z} = 0,
\end{aligned} \tag{4.2.5}$$

where ν is the kinematic viscosity, ρ is the density, $u = (u_r, u_\theta, u_z)$ is the velocity field, p is the pressure function, and

$$u \cdot \nabla = u_r \frac{\partial}{\partial r} + \frac{u_\theta}{r} \frac{\partial}{\partial \theta} + u_z \frac{\partial}{\partial z}, \qquad \Delta = \frac{\partial^2}{\partial r^2} + \frac{1}{r} \frac{\partial}{\partial r} + \frac{1}{r^2} \frac{\partial^2}{\partial \theta^2} + \frac{\partial^2}{\partial z^2}.$$

Then it is easy to see that the Couette flow (4.2.3) is a steady-state solution of (4.2.5). In order to investigate its stability and transitions, we need to consider the perturbed state of (4.2.3):

$$u_r, u_\theta + V(r), u_z, p + \rho \int \frac{1}{r} v^2(r) dr.$$

The perturbed equations read

$$\begin{aligned}
&\frac{\partial u_r}{\partial t} + (u\cdot\nabla)u_r - \frac{u_\theta^2}{r} = \nu\left(\Delta u_r - \frac{2}{r^2}\frac{\partial u_\theta}{\partial \theta} - \frac{u_r}{r^2}\right) - \frac{1}{\rho}\frac{\partial p}{\partial r}\\
&\qquad + \frac{2V(r)}{r}u_\theta - \frac{V(r)}{r}\frac{\partial u_r}{\partial \theta},\\
&\frac{\partial u_\theta}{\partial t} + (u\cdot\nabla)u_\theta + \frac{u_\theta u_r}{r} = \nu\left(\Delta u_\theta + \frac{2}{r^2}\frac{\partial u_r}{\partial \theta} - \frac{u_\theta}{r^2}\right)\\
&\qquad - \frac{1}{r\rho}\frac{\partial p}{\partial \theta} - (V' + \frac{V}{r})u_r - \frac{V}{r}\frac{\partial u_\theta}{\partial \theta},\\
&\frac{\partial u_z}{\partial t} + (u\cdot\nabla)u_z = \nu\Delta u_z - \frac{1}{\rho}\frac{\partial p}{\partial z} - \frac{V}{r}\frac{\partial u_z}{\partial \theta},\\
&\frac{\partial (ru_z)}{\partial z} + \frac{\partial (ru_r)}{\partial r} + \frac{\partial u_\theta}{\partial \theta} = 0.
\end{aligned} \tag{4.2.6}$$

To derive the nondimensional form of Eq. (4.2.6), let

$$\begin{aligned}
&(x,t) = (hx', h^2t'/\nu) && (x = (r, r\theta, z)),\\
&(u,p) = (\nu u'/h, \rho\nu^2 p'/h^2) && (u = (u_r, u_\theta, u_z)).
\end{aligned}$$

Omitting the primes, we obtain the nondimensional form of (4.2.6) as follows:

$$\begin{aligned}
&\frac{\partial u_r}{\partial t} = \Delta u_r - \frac{2}{r^2}\frac{\partial u_\theta}{\partial \theta} - \frac{u_r}{r^2} - (u\cdot\nabla)u_r + \frac{u_\theta^2}{r} - \frac{\partial p}{\partial r}\\
&\qquad - \sqrt{T}\left(\frac{\eta^2-\mu}{1-\eta^2} - \frac{1-\mu}{1-\eta^2}\frac{r_1^2}{r^2}\right)\left(u_\theta - 1/2\frac{\partial u_r}{\partial \theta}\right),\\
&\frac{\partial u_\theta}{\partial t} = \Delta u_\theta + \frac{2}{r^2}\frac{\partial u_r}{\partial \theta} - \frac{u_\theta}{r^2} - (u\cdot\nabla)u_\theta - \frac{u_\theta u_r}{r} - \frac{1}{r}\frac{\partial p}{\partial \theta}\\
&\qquad + \sqrt{T}\frac{\eta^2-\mu}{1-\eta^2}u_r + \frac{\sqrt{T}}{2}\left(\frac{\eta^2-\mu}{1-\eta^2} - \frac{1-\mu}{1-\eta^2}\frac{r_1^2}{r^2}\right)\frac{\partial u_\theta}{\partial \theta},\\
&\frac{\partial u_z}{\partial t} = \Delta u_z - (u\cdot\nabla)u_z - \frac{\partial p}{\partial z} + \frac{\sqrt{T}}{2}\left(\frac{\eta^2-\mu}{1-\eta^2} - \frac{1-\mu}{1-\eta^2}\frac{r_1^2}{r^2}\right)\frac{\partial u_z}{\partial \theta},\\
&\frac{\partial (ru_z)}{\partial z} + \frac{\partial (ru_r)}{\partial r} + \frac{\partial u_\theta}{\partial \theta} = 0,
\end{aligned} \tag{4.2.7}$$

where T is the Taylor number as defined in (4.2.2).

The nondimensional domain for (4.2.7) is

$$\Omega = (l_1, l_2) \times (0, 2\pi) \times (0, L),$$

where $l_i = r_i/h$ $(i = 1, 2)$, and L is the height of the fluid between the two cylinders. The initial value condition for (4.2.7) is given by

$$u(r, \theta, z, 0) = u_0(r, \theta, z). \tag{4.2.8}$$

There is a variety of physically sound boundary conditions. In the θ-direction it is periodic,

$$u(r, \theta + 2k\pi, z) = u(r, \theta, z), \quad \forall k \in \mathbb{Z}. \tag{4.2.9}$$

In the radial direction, there is the rigid boundary condition

$$u = (u_z, u_r, u_\theta) = 0, \quad \text{at} \quad r = l_1, l_2. \tag{4.2.10}$$

At the top and bottom in the z-direction $(z = 0, L)$, the free boundary condition, the rigid boundary condition, or the periodic boundary condition can be used:

Dirichlet Boundary Condition:

$$u = (u_r, u_\theta, u_z) = 0 \quad \text{at} \quad z = 0, L; \tag{4.2.11}$$

Free-Slip Boundary Condition:

$$u_z = 0, \quad \frac{\partial u_r}{\partial z} = \frac{\partial u_\theta}{\partial z} = 0 \quad \text{at} \quad z = 0, L; \tag{4.2.12}$$

Free-Rigid Boundary Condition:

$$\begin{aligned} &u_z = 0, \frac{\partial u_r}{\partial z} = \frac{\partial u_\theta}{\partial z} = 0 &&\text{at } z = L, \\ &u = (u_z, u_r, u_\theta) = 0 &&\text{at } z = 0; \end{aligned} \tag{4.2.13}$$

Periodic Boundary Condition:

$$u(r, \theta, z + 2kL) = u(r, \theta z) \quad \forall k \in \mathbb{Z}. \tag{4.2.14}$$

Remark 4.2.1 From the physical point of view, the free boundary condition is usually used for the case in which the friction on the surface of the boundary is negligible; otherwise, the rigid boundary condition is used. Moreover, if the height L of the fluid is sufficiently large in comparison with the gap $r_2 - r_1$ between the two cylinders, it is reasonable to use the periodic boundary condition.

4.2.3 Narrow-Gap Case with Axisymmetric Perturbations

We consider here the case that the gap $r_2 - r_1$ is small compared to the mean radius $r_0 = (r_1 + r_2)/2$ with $\mu \geq 0$ and with axisymmetric perturbations. This case is the situation investigated by Taylor (1923) in 1923.

We take the length scale $h = r_2 - r_1$. Then the narrow gap condition is given by

$$1 = r_2 - r_1 \ll (r_1 + r_2)/2. \tag{4.2.15}$$

Under the assumption (4.2.15), we can neglect the terms containing r^{-n} $(n \geq 1)$ in (4.2.7). In addition, by (4.2.15) we have

$$\begin{aligned} -\sqrt{T}\left(\frac{\eta^2-\mu}{1-\eta^2} - \frac{1-\mu}{1-\eta^2}\frac{r_1^2}{r^2}\right) &= \sqrt{T}\left(1 - \frac{1-\mu}{1-\eta^2}\frac{r^2-r_1^2}{r^2}\right) \\ &\simeq \sqrt{T}(1-(1-\mu)(r-r_1)). \end{aligned}$$

Let

$$\alpha = \frac{\eta^2-\mu}{1-\eta^2}. \tag{4.2.16}$$

Replacing u_θ by $\sqrt{\alpha}u_\theta$, and assuming that the perturbations are axisymmetric and independent of θ, we obtain from (4.2.7)

$$\begin{cases} \dfrac{\partial u_z}{\partial t} + (\tilde{u}\cdot\nabla)u_z = \Delta u_z - \dfrac{\partial p}{\partial z}, \\ \dfrac{\partial u_r}{\partial t} + (\tilde{u}\cdot\nabla)u_r = \Delta u_r - \dfrac{\partial p}{\partial r} + \sqrt{\alpha T}(1-(1-\mu)(r-r_1))u_\theta, \\ \dfrac{\partial u_\theta}{\partial t} + (\tilde{u}\cdot\nabla)u_\theta = \Delta u_\theta + \sqrt{\alpha T}u_r, \\ \dfrac{\partial u_r}{\partial r} + \dfrac{\partial u_z}{\partial z} = 0, \end{cases} \tag{4.2.17}$$

where

$$\Delta = \frac{\partial^2}{\partial r^2} + \frac{\partial^2}{\partial z^2}, \quad (\tilde{u}\cdot\nabla) = u_r\frac{\partial}{\partial r} + u_z\frac{\partial}{\partial z}.$$

In this case, the spatial domain is $M = (r_1, r_1 + 1) \times (0, L)$. For convenience, we consider here the Dirichlet boundary condition

$$u|_{\partial M} = 0. \tag{4.2.18}$$

The initial value condition is axisymmetric, given by

$$u = u_0(r,z) \quad \text{at} \quad t = 0. \tag{4.2.19}$$

The linearized equations of (4.2.17) read

$$\begin{aligned}
&-\Delta u_z+\frac{\partial p}{\partial z}=0,\\
&-\Delta u_r+\frac{\partial p}{\partial r}=\lambda u_\theta-\lambda(1-\mu)(r-r_1)u_\theta,\\
&-\Delta u_\theta=\lambda u_r,\\
&\frac{\partial u_r}{\partial r}+\frac{\partial u_z}{\partial z}=0,
\end{aligned} \tag{4.2.20}$$

where $\lambda=\sqrt{\alpha T}$, and T is the Taylor number given by (4.2.2).

Let $\lambda_1>0$ be the first eigenvalue of (4.2.20) with (4.2.18). We call

$$T_c=\lambda_1^2/\alpha \tag{4.2.21}$$

the critical Taylor number, where α is given by (4.2.16).

As $\mu\to 1$, Eqs. (4.2.20) are reduced to the following symmetric linear equations:

$$\begin{aligned}
&-\Delta u_z+\frac{\partial p}{\partial z}=0,\\
&-\Delta u_r+\frac{\partial p}{\partial r}=\lambda u_\theta,\\
&-\Delta u_\theta=\lambda u_r,\\
&\frac{\partial u_z}{\partial z}+\frac{\partial u_r}{\partial r}=0.
\end{aligned} \tag{4.2.22}$$

Let the first eigenvalue $\lambda_0>0$ of (4.2.22) with (4.2.18) have multiplicity $m\geq 1$, let the corresponding eigenfunctions be v_i $(i=1,\cdots,m)$, and let the corresponding eigenspace be

$$E_0=\text{span}\{v_i\mid 1\leq i\leq m\}.$$

We remark here that under conditions (4.2.4) and (4.2.15), the condition $\mu\to 1$ can be equivalently replaced by

$$r_1=(2+\delta)/(1-\mu), \tag{4.2.23}$$

for some $\delta>0$. In this case, the parameter α in (4.2.16) is

$$\alpha=(\eta^2-\mu)/(1-\eta^2)\simeq\delta/2.$$

Let $\widetilde{u}=(u_z,u_r)$ and $u=(\widetilde{u},u_\theta)$. For the Taylor problem (4.2.17)∼(4.2.19), we set the function spaces as follows:

$$\begin{aligned}
&H=\{u=(\widetilde{u},u_\theta)\in L^2(M)^3\mid \text{div}\widetilde{u}=0,\widetilde{u}\cdot n|_{\partial M}=0\},\\
&H_1=\{u\in H^2(M)^3\bigcap H\mid u \text{ satisfy (4.2.18)}\}.
\end{aligned}$$

The following theorems provide the nonequilibrium phase transitions at the critical Taylor number T_c [T_c as in (4.2.21)] for the Taylor problem (4.2.17)–(4.2.19) in the narrow-gap case with $\mu>0$ and with axisymmetric perturbations.

Theorem 4.2.2 *Let the conditions (4.2.15) and (4.2.23) hold. Then the Taylor problem (4.2.17)–(4.2.19) has a transition at* $(u,T)=(0,T_c)$. *More precisely, we have the following assertions:*

(1) The problem undergoes a transition to an attractor $\mathscr{A}_T$ *on* $T>T_c$ *with* $0\notin\mathscr{A}_T$, *and* $\mathscr{A}_T$ *is an* $(m-1)$*-dimensional homological sphere.*
(2) The bifurcated attractor $\mathscr{A}_T$ *attracts* $H\setminus\Gamma$, *where* $\Gamma\subset H$ *is the stable manifold of* $u=0$ *with codimension m in H.*
(3) Every $u_T\in\mathscr{A}_T$ *can be expressed as*

$$u_T=v_T+w_T^{\mu},\qquad v_T\in E_0,\qquad \lim_{\mu\to1,\lambda\to\lambda_0}\|w_T^{\mu}\|/\|v_T\|=0, \tag{4.2.24}$$

where $\lambda=\sqrt{\alpha T}$, *and* λ_0 *is the first eigenvalue of (4.2.22) with (4.2.18).*

Theorem 4.2.3 *Under the hypotheses of Theorem 4.2.2, if the first eigenvalue* λ_0 *of (4.2.22) with (4.2.18) is simple, then the transition of (4.2.17)–(4.2.19) at* $(0,T_c)$ *is of either Type I (continuous) or Type III (mixed). Moreover, we have the following assertions:*

(1) The bifurcated attractor $\mathscr{A}_T$ *in Theorem 4.2.2 consists of exactly two equilibrium points of (4.2.17)–(4.2.19), i.e.,* $\mathscr{A}_T=\{u_1^T,u_2^T\}$, *such that*

$$\begin{aligned}&u_1^T=\alpha_1(T,\mu)v_0+w_1(T,\mu), && u_2^T=-\alpha_2(T,\mu)v_0+w_2(T,\mu),\\ &w_i(T,\mu)=o(|\alpha_i|), && v_0\in E_0,\end{aligned} \tag{4.2.25}$$

where $\alpha_i>0$ *and* $\alpha_i(T,\mu)\to0$ *as* $\mu\to1$ *and* $T\to T_0=\sqrt{\lambda_0}/\alpha$.
(2) H can be decomposed into two open sets $H=\overline{U}_1^T+\overline{U}_2^T$ *such that*

$$\begin{aligned}&U_1^T\bigcap U_2^T=\emptyset, && 0\in\partial U_1^T\bigcap\partial U_2^T,\\ &u_i^T\in U_i^T, && \lim_{t\to\infty}\|u(t,u_0)-u_i^T\|=0 \qquad \text{if } u_0\in U_i^T,\end{aligned}$$

for i=1, 2, where $u(t,u_0)$ *is the solution of (4.2.17)–(4.2.19).*
(3) If the transition is of Type III, then the problem (4.2.17)–(4.2.19) has a saddle-node bifurcation at some point (u^*,T^*) *with* $u^*\neq0$ *and* $0<T^*<T_c$, *where* T_c *is given in (4.2.21).*

Remark 4.2.4 For (4.2.17) with other physically sound boundary conditions given in (4.2.10)–(4.2.14), the results of Theorems 4.2.2 and 4.2.3 are also valid.

Remark 4.2.5 The expressions (4.2.24) and (4.2.25) are very useful for the Taylor problem, showing that the asymptotic topological structure of the problem (4.2.17)–(4.2.19) is governed by the first eigenfunctions of the symmetric linearized equations (4.2.22). In the next subsection, we shall see that the first eigenfunctions of (4.2.22) with the free boundary condition have the Taylor vortex type of structure, and these eigenfunctions are structurally stable.

Proof of Theorem 4.2.2. We shall apply Theorem 2.6.1 to prove this theorem, and divide the proof into the following three steps.

Step 1. Let $G, S^{\mu}_{\lambda}, L_{\lambda} = -A + B_{\lambda} : H_1 \to H$ be mappings defined by

$$\begin{aligned} &-Au = \{P\Delta\widetilde{u}, \Delta u_{\theta}\}, \qquad B_{\lambda}u = \{\lambda P(0, u_{\theta}), \lambda u_r\}, \\ &S^{\mu}_{\lambda}u = \{\lambda P(0, -(1-\mu)(r-r_1)u_{\theta}), 0\}, \\ &G(u) = \{-P(\widetilde{u}\cdot\nabla)\widetilde{u}, -(\widetilde{u}\cdot\nabla)u_{\theta}\}, \end{aligned} \tag{4.2.26}$$

where $\lambda = \sqrt{\alpha T}$, $u = (u_z, u_r, u_{\theta}) \in H_1$, and the operator P is the Leray projection. Thus (4.2.17) can be written in the abstract form

$$\frac{du}{dt} = L_{\lambda}u + S^{\mu}_{\lambda}u + G(u). \tag{4.2.27}$$

It is well known that the operators (4.2.26) satisfy the conditions (2.1.2) and (2.1.3), and L_{λ} is symmetric. In addition, the operator S^{μ}_{λ} satisfies

$$\|S^{\mu}_{\lambda}\| \leq \lambda(1-\mu).$$

By (4.2.15) and (4.2.23), $1-\mu \to 0$ as $r_1 \to \infty$. Thus the condition (2.6.2) is satisfied.

Step 2. Consider the eigenvalue problem

$$L_{\lambda}u = \beta(\lambda)u \qquad \text{for } u = (u_z, u_r, u_{\theta}) \in H_1. \tag{4.2.28}$$

By definition, the abstract form (4.2.28) is equivalent to the following eigenvalue equations with boundary condition (4.2.18):

$$\begin{aligned} &\Delta u_z - \frac{\partial p}{\partial z} = \beta(\lambda)u_z, \\ &\Delta u_r - \frac{\partial p}{\partial r} + \lambda u_{\theta} = \beta(\lambda)u_r, \\ &\Delta u_{\theta} + \lambda u_r = \beta(\lambda)u_{\theta}, \\ &\text{div}\, \widetilde{u} = 0, \end{aligned} \tag{4.2.29}$$

where $\widetilde{u} = (u_z, u_r)$. It is known that the eigenvalues β_k $(k = 1, 2, \cdots)$ of (4.2.29) with (4.2.18) are real numbers satisfying

$$\beta_1(\lambda) \geq \beta_2(\lambda) \geq \cdots \geq \beta_k(\lambda) \geq \cdots, \qquad \lim_{k\to\infty} \beta_k(\lambda) = -\infty \tag{4.2.30}$$

At the first eigenvalue λ_0 of (4.2.22) with (4.2.18), the eigenvalues $\beta_k(\lambda)$ $(k = 1, 2, \cdots)$ in (4.2.30) satisfy

$$\begin{aligned} &\beta_i(\lambda_0) = 0 \quad \forall 1 \leq i \leq m, \\ &\beta_j(\lambda_0) < 0 \quad \forall j \geq m+1, \end{aligned} \tag{4.2.31}$$

where $m \geq 1$ is the multiplicity of the first eigenvalue $\beta_1(\lambda_0)$.

By Theorem 2.1 in Ma and Wang (2008d), it is readily verified that the first eigenvalues $\beta_1(\lambda) = \cdots = \beta_m(\lambda)$ are differentiable at $\lambda = \lambda_0$ and have the following Taylor expansions:

$$\beta_i(\lambda) = \beta_i'(\lambda_0)(\lambda - \lambda_0) + o(|\lambda - \lambda_0|), \qquad \beta_i'(\lambda_0) = 2\int_M u_{i\theta} u_{ir} dx > 0, \quad (4.2.32)$$

for $1 \leq i \leq m$, where $u_i = (u_{iz}, u_{ir}, u_{i\theta})$ are the first eigenfunctions.

Hence it follows from (4.2.31) and (4.2.32) that the conditions (2.1.4) and (2.1.5) hold here.

Step 3. It is well known that the Taylor problem (4.2.17)–(4.2.19) possesses a global attractor for all $(\lambda, \mu) \in \mathbb{R}^2$. Hence we need to prove only that $u = 0$ is globally asymptotically stable for the equation

$$\frac{du}{dt} = L_\lambda u + G(u), \tag{4.2.33}$$

at $\lambda = \lambda_0$, which corresponds to

$$\begin{aligned}
&\frac{\partial u_z}{\partial t} + (\widetilde{u} \cdot \nabla) u = \Delta u_z - \frac{\partial p}{\partial z}, \\
&\frac{\partial u_r}{\partial t} + (\widetilde{u} \cdot \nabla) u_r = \Delta u_r - \frac{\partial p}{\partial r} + \lambda u_\theta, \\
&\frac{\partial u_\theta}{\partial t} + (\widetilde{u} \cdot \nabla) u_\theta = \Delta u_\theta + \lambda u_r, \\
&\text{div}\, \widetilde{u} = 0.
\end{aligned}$$

These equations have the same form as that of the two-dimensional Boussinesq equations. We note that L_λ is symmetric, and $G(u)$ is orthogonal. Hence, as in the proof of Theorem 4.1.1, by Theorem 3.16 in Ma and Wang (2005b), we obtain that $u = 0$ is globally asymptotically stable for (4.2.33) at $\lambda = \lambda_0$.

The proof of the theorem is complete. □

Proof of Theorem 4.2.3. The proof follows from Theorem 2.6.6 and the proof of Theorem 4.2.2. □

4.2.4 Asymptotic Structure of Solutions and Taylor Vortices

The main objective of this subsection is to study the structure and its stability in physical space of the solutions associated with the secondary flows for the Taylor problem. In particular, we shall provide a rigorous confirmation of the presence of the Taylor vortex structure as observed by Taylor in his experiments (Taylor 1923).

We mainly consider Eqs. (4.2.17) with one of the following two boundary conditions:

Dirichlet Boundary Condition:

$$\begin{aligned} &u_z = u_r = u_\theta = 0 && \text{at } r = r_1, r_1+1, \\ &u_z = 0, \frac{\partial u_r}{\partial z} = 0, \frac{\partial u_\theta}{\partial z} = 0 && \text{at } z = 0, L; \end{aligned} \tag{4.2.34}$$

Free Boundary Condition:

$$\begin{aligned} &u_r = 0, \frac{\partial u_z}{\partial r} = 0, \frac{\partial u_\theta}{\partial r} = 0 && \text{at } r = r_1, r_1+1, \\ &u_z = 0, \frac{\partial u_r}{\partial z} = 0, \frac{\partial u_\theta}{\partial z} = 0 && \text{at } z = 0, L. \end{aligned} \tag{4.2.35}$$

Eigenvectors of the Linearized Taylor Problem. By the expressions (4.2.24) and (4.2.25) of transition solutions, we see that the first eigenvectors of linearized equations play a crucial role in determining the asymptotic structure of solutions for the Taylor problem. Hence we begin with the computation of the first eigenvalue and eigenvectors of (4.2.22) with either the Dirichlet boundary condition (4.2.34) or the free boundary condition (4.2.35), following (Chandrasekhar 1981).

For the eigenvalue equations (4.2.22), we take the separation of variables as follows:

$$(u_z, u_r, u_\theta) = \left(\frac{1}{a^2}\frac{dh(z)}{dz}\frac{dR(r)}{dr}, h(z)R(r), h(z)\varphi(r)\right), \tag{4.2.36}$$

where $a^2 > 0$ is an arbitrary constant.

By (4.2.22) and the boundary conditions $u_z = 0$ at $z = 0, L$ in (4.2.34), and (4.2.35), we obtain

$$\begin{aligned} &\frac{d^2h}{dz^2} = -a^2 h, \\ &h'(0) = h'(L) = 0. \end{aligned} \tag{4.2.37}$$

Furthermore, the functions R and φ satisfy

$$\begin{aligned} &\left(\frac{d^2}{dr^2} - a^2\right)^2 R = a^2\lambda\varphi, \\ &\left(\frac{d^2}{dr^2} - a^2\right)\varphi = -\lambda R. \end{aligned} \tag{4.2.38}$$

With the Dirichlet boundary condition (4.2.34), we obtain

$$(\varphi(r), R(r), R'(r)) = 0 \qquad \text{at } r = r_1, r_2, \tag{4.2.39}$$

while with the free boundary condition (4.2.35), we have

$$(\varphi(r), R(r), R''(r)) = 0 \qquad \text{at } r = r_1, r_2. \tag{4.2.40}$$

The solutions of (4.2.37) are given by

$$h(z) = \cos az \qquad \text{with } a = k\pi/L, \quad k = 1,2,\cdots. \tag{4.2.41}$$

In the following we discuss the eigenvalue problems of (4.2.38) with one of the two boundary conditions (4.2.39) and (4.2.40).

Dirichlet Boundary Condition Case: The eigenvalue problem of (4.2.38) with (4.2.39) is equivalent to

$$\begin{aligned} &\left(\frac{d^2}{dr^2} - a^2\right)^3 R = -a^2\lambda^2 R, \\ &(R,R') = 0, \quad \left(\frac{d^2}{dr^2} - a^2\right)^2 R = 0 \qquad \text{at } r = r_1, r_2. \end{aligned} \tag{4.2.42}$$

When the height L is large, the first eigenvalue and eigenvector of (4.2.42) are given in Chap. II-15 of Chandrasekhar (1981) by

$$\begin{aligned} &\lambda_0^2 \simeq 1700, \\ &R(r) \simeq \cos\alpha_0\xi - 0.06\cosh(\alpha_1\xi)\cos\alpha_2\xi + 0.1\sinh(\alpha,\xi)\sin\alpha_2\xi, \end{aligned} \tag{4.2.43}$$

where $\xi = r - r_1 - 1/2, \alpha_0 \simeq 3.97, \alpha \simeq 5.2$, and $\alpha_3 \simeq 2.1$. This function is illustrated in Fig. 4.9.

Thus, from (4.2.36), (4.2.38), and (4.2.41) we obtain the first eigenvector of (4.2.22) with (4.2.34) as follows:

$$(u_z, u_r, u_\theta) = \left(-\frac{\sin az}{a}R'(r), \cos azR(r), \frac{\cos az}{a^2\lambda_0}\left[\frac{d^2}{dz^2} - a^2\right]R(r)\right), \tag{4.2.44}$$

where $R(r)$ is given by (4.2.43).

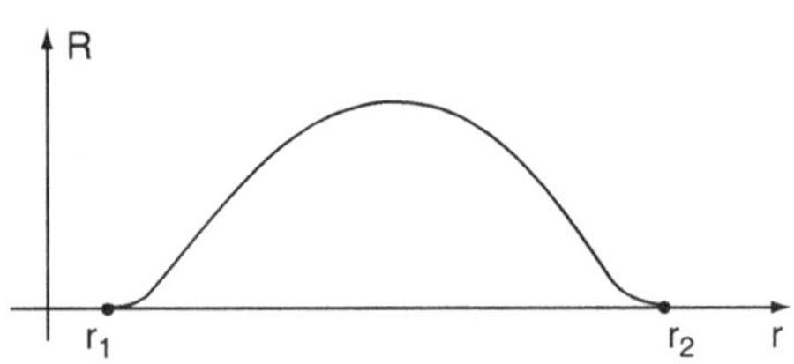

Fig. 4.9 Graph of the function $R(r)$

Free Boundary Condition Case: For each given $a^2 = k^2\pi^2/L^2$, the first eigenvalue $\lambda_0(a)$ and eigenvectors of (4.2.38), with (4.2.40), are given by

$$\lambda_0(a) = \frac{(\pi^2 + a^2)^{3/2}}{a}, \qquad (R,\varphi) = \sin\pi(r - r_1)\left(1, \frac{\sqrt{a^2+\pi^2}}{a}\right). \tag{4.2.45}$$

It is clear that the first eigenvalue λ_0 of (4.2.22) with (4.2.35) is the minimum of $\lambda_0(a)$:

$$\lambda_0^2 = \min_a \lambda_0^2(a) = \min_{k\in\mathbb{N}} \frac{\pi^4 L^2\left(1+\frac{k^2}{L^2}\right)^3}{k^2}. \tag{4.2.46}$$

The corresponding first eigenvectors of (4.2.22) with (4.2.35) are derived from (4.2.36), (4.2.41), and (4.2.45), and are given by

$$\begin{aligned} u_z &= -\frac{\pi}{a}\sin az\cos(r-r_1),\\ u_r &= \cos az\sin\pi(r-r_1),\\ u_\theta &= \frac{1}{a}\sqrt{\pi^2+a^2}\cos az\sin\pi(r-r_1), \end{aligned} \tag{4.2.47}$$

where $a = k\pi/L$ satisfies (4.2.46).

Structural Stability of the First Eigenvectors. It is easy to see that the first eigenvectors $\widetilde{u} = (u_z, u_r)$ given by (4.2.44) and (4.2.47) have the structure of Taylor vortices, and by Theorems 2.1.2 and 2.2.9 in Ma and Wang (2005d), we readily verify the structural stability of the vector fields $\widetilde{u} = (u_z, u_r)$ of (4.2.44) and (4.2.47) in $B^r(M,\mathbb{R}^2)$ $(r \geq 1)$ and $B_0^r(M,\mathbb{R}^2)$ respectively. Here

$$\begin{aligned} B^r(M,\mathbb{R}^2) &= \{\widetilde{u}\in C^r(M,\mathbb{R}^2) \mid \mathrm{div}\widetilde{u} = 0, \widetilde{u} \text{ satisfy (4.2.35)}\},\\ B_0^r(M,\mathbb{R}^2) &= \{\widetilde{u}\in C^r(M,\mathbb{R}^2) \mid \mathrm{div}\widetilde{u} = 0,\ \widetilde{u} \text{ satisfy (4.2.34)}\}. \end{aligned}$$

The following two lemmas establish the structure and its stability for the first eigenvectors given in (4.2.44) and (4.2.47).

Lemma 4.2.6 *For the vector fields given by (4.2.44) and $\alpha \neq 0$,*

$$\widetilde{u} = (u_r, u_z) = \left(\alpha\cos\frac{k\pi z}{L}R(r), -\frac{\alpha L}{k\pi}\sin\frac{k\pi z}{L}R'(r)\right) \tag{4.2.48}$$

are structurally stable in $B_0^r(M,\mathbb{R}^2)$ and are topologically equivalent to the flow structure shown in Fig. 4.10.

Proof. By (4.2.43), a simple calculation shows that

$$R''(r_i) \neq 0 \qquad \text{for } i = 0, 1, 2,$$

where $r_0 = (r_1 + r_2)/2$ satisfies $R'(r_0) = 0$.

It is easy to see that all the interior singular points of the vector field in (4.2.48) are given by

$$(r_0, z_j) = \left(r_0, \frac{jL}{2k}\right) \qquad \forall j = 1, 3, \cdots, 2m-1 \text{ and } m \leq k.$$

By direct computation, we obtain

$$\det\begin{pmatrix} \dfrac{\partial u_z}{\partial z} & \dfrac{\partial u_z}{\partial r} \\ \dfrac{\partial u_r}{\partial z} & \dfrac{\partial u_r}{\partial r} \end{pmatrix}_{(r_0,z_j)} = \frac{\alpha^2 k\pi}{L} R(r_0)R''(r_0) \neq 0.$$

The boundary singular points of (4.2.48) read

$$(r_i,\widetilde{z}_j) = \left(r_i, \frac{jL}{k}\right) \qquad \text{for } i = 1,2 \text{ and } j = 1,2,\cdots,k,$$

and the determinant is given by

$$\det\begin{pmatrix} \dfrac{\partial^2 u_z}{\partial z\partial r} & \dfrac{\partial^2 u_z}{\partial r^2} \\ \dfrac{\partial^2 u_r}{\partial z\partial r} & \dfrac{\partial^2 u_r}{\partial r^2} \end{pmatrix}_{(r_i,\widetilde{z}_j)} = -\alpha^2 R''^2(r_i) \neq 0.$$

Hence, the vector fields (4.2.48) are D-regular for $\alpha \neq 0$. Thus, this lemma follows from Theorem 2.2.9 in Ma and Wang (2005d). The proof is complete. □

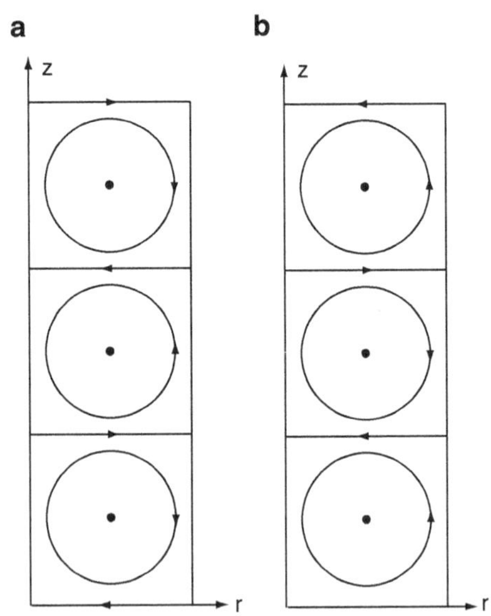

Fig. 4.10 Flow structure of (u_r, u_z) with boundary condition (4.2.34): (**a**) $\alpha < 0$, (**b**) $\alpha > 0$

Lemma 4.2.7 *For the vector fields of (4.2.47) and $\alpha \neq 0$, the vector fields*

$$(u_r, u_z) = \left(\alpha \cos\frac{k\pi z}{L}\sin\pi(r - r_1), -\frac{L\alpha}{k}\sin\frac{k\pi z}{L}\cos\pi(r - r_1)\right) \tag{4.2.49}$$

are structurally stable in $B^k(M,\mathbb{R}^2)$ and are topologically equivalent to the flow structure shown in Fig. 4.11.

Proof. The interior singular points of (4.2.49) are

$$(r_0, z_j) = \left(\frac{r_1+r_2}{2}, \frac{jL}{2k}\right) \qquad \forall j = 1,3,\cdots,2m-1, m \leq k,$$

and the boundary singular points read

$$(r_i, \widetilde{z}_j) = \left(r_i, \frac{jL}{k}\right) \qquad \forall i = 1,2, j = 1,2,\cdots,k.$$

It is clear that the determinants of the Jacobian matrix $D\widetilde{u}$ at the singular points (r_0, z_j) and $(r_i, \widetilde{z}_j)$ are nonzero:

$$\det D\widetilde{u}(r,z) \neq 0 \quad \forall (r,z) = (r_0, z_j) \text{ and } (r_i, \widetilde{z}_j).$$

Hence the vector fields (4.2.49) are regular. Thus this lemma follows from Theorem 2.1.2 in Ma and Wang (2005d). The proof is complete. □

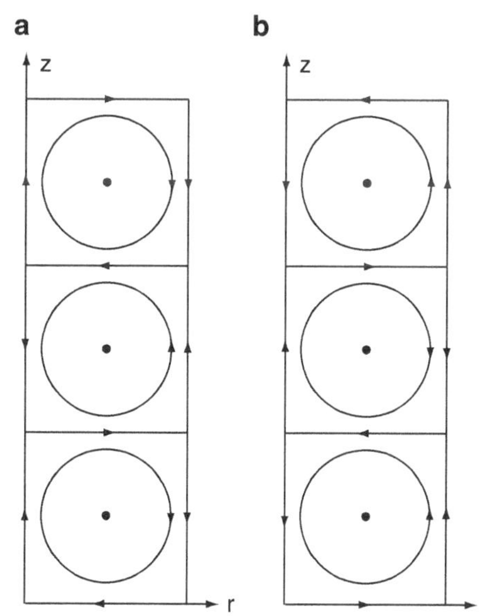

Fig. 4.11 Flow structure of (u_r, u_z) with boundary condition (4.2.35): (**a**) $\alpha < 0$, (**b**) $\alpha > 0$

Taylor Vortex Structure of Solutions. The following theorem shows that the solutions of the Taylor problem with axisymmetric perturbation have Taylor vortices as their asymptotic structure.

Theorem 4.2.8 *There are discrete numbers L_i $(i = 1, 2, \cdots)$ such that for the height $L \neq L_i$, if the Taylor number T satisfies $0 < T - T_c < \delta$ for some $\delta > 0$, where T_c is the critical Taylor number as in (4.2.21), then the following assertions hold:*

(1) The space H can be decomposed into two open sets $H = \overline{U}_1^T + \overline{U}_2^T$ satisfying

$$U_1^T \bigcap U_2^T = \emptyset, \quad 0 \in \partial U_1^T \bigcap \partial U_2^T.$$

(2) For every $\varphi \in U_1^T$, there is a time $t_0 \geq 0$ such that for the solution $u(t, \varphi) = (\widetilde{u}, u_\theta)$ of (4.2.17) and (4.2.34) [respectively of (4.2.17) and (4.2.35)] with $u(0, \varphi) = \varphi$, $\widetilde{u}(t, \varphi) = (u_r, u_z)$ is topologically equivalent to the structure shown in Fig. 4.10a (respectively in Fig. 4.11a) for any $t > t_0$. Namely, the solution $u(t, \varphi)$ of (4.2.17) and (4.2.34) [respectively of (4.2.17) and (4.2.35)] has the Taylor vortex structure for $t > t_0$.
(3) For every $\varphi \in U_2^T$, there is $t_0 \geq 0$ such that the solution $\widetilde{u}(t, \varphi)$ of (4.2.17) and (4.2.34) [respectively of (4.2.17) and (4.2.35)] is topologically equivalent to the structure shown in Fig. 4.10b (respectively in Fig. 4.11b) for any $t > t_0$.

Proof. There are discrete countable numbers L_i $(i = 1, 2, \cdots)$ of the height L such that when $L \neq L_i$, the first eigenvalue of (4.2.22) with (4.2.34) is simple; see (Yudovich 1966; Chandrasekhar 1981). When $L^2 \neq l^2$, where l^2 satisfy that for any integers k_1 and k_2 with $k_1 \neq k_2$,

$$k_2^2(l^2 + k_1^2)^3 = k_1^2(l^2 + k_2^2)^3,$$

we infer from (4.2.46) that the first eigenvalue of (4.2.22) with boundary condition (4.2.35) is also simple. Thus the theorem follows immediately from Theorem 4.2.3, Remark 4.2.4, and Lemmas 4.1.12–4.2.7. The proof is complete. □

Remark 4.2.9 For any height $L > 0$, the multiplicity of the first eigenvalue of (4.2.22) with (4.2.34) or with (4.2.35) is either one or two, and if its is two, then the two corresponding eigenvectors have the respective wave numbers $k_1 = k$ and $k_2 = k + 1$ in (4.2.41) for some $k \geq 1$.

In fact, by the method of construction of the center manifold functions as used in Theorems 4.1.14 and 4.2.12 in the next subsection, we can prove the following results, and so we omit here the details of the proof.

Theorem 4.2.10 *For the Taylor problem (4.2.17) with boundary condition (4.2.34) or with (4.2.35), the following assertions hold:*

(1) The transition of the Taylor problem at $T = T_c$ is of Type I, i.e., the problem has an attractor bifurcation on $T > T_c$.
(2) For any $L > 0$, the assertions of Theorem 4.2.8 hold.
(3) When $L = L_i$ $(i = 1, 2, \cdots)$ such that the first eigenvalue has multiplicity two, the Taylor problem bifurcates from $(u, T) = (0, T_c)$ to an attractor $\mathscr{A}_T$ homeomorphic to S^1, and $\mathscr{A}_T = S^1$ contains exactly eight singular points, four of which are minimal attractors, while the other four are saddle points. The structure of the attractor $\mathscr{A}_T$ is topologically equivalent to that shown in Fig. 4.7.

4.2.5 Taylor Problem with z-Periodic Boundary Condition

We now study the Taylor problem (4.2.7) with the z-periodic boundary condition (4.2.14) and with axisymmetric perturbations. Assuming that Eqs. (4.2.7) are independent of θ, and taking the length scale $h = r_2$ in nondimensional form, we obtain

$$\begin{aligned}
&\frac{\partial u_z}{\partial t} = \Delta u_z - \frac{\partial p}{\partial z} - (\widetilde{u}\cdot\nabla)u_z,\\
&\frac{\partial u_r}{\partial t} = (\Delta - \frac{1}{r^2})u_r + \lambda\left(\frac{1}{r^2} - \kappa\right)u_\theta - \frac{\partial p}{\partial r} + \frac{u_\theta^2}{r} - (\widetilde{u}\cdot\nabla)u_r,\\
&\frac{\partial u_\theta}{\partial t} = \left(\Delta - \frac{1}{r^2}\right)u_\theta + \lambda\kappa u_r - \frac{u_r u_\theta}{r} - (\widetilde{u}\cdot\nabla)u_\theta,\\
&\frac{\partial(ru_z)}{\partial z} + \frac{\partial(ru_r)}{\partial r} = 0,
\end{aligned} \tag{4.2.50}$$

where $(\widetilde{u}\cdot\nabla)$ is as in (4.2.17), $\lambda = \sqrt{T}$, T is the Taylor number, and

$$\begin{aligned}
T &= \frac{4r_2^4\Omega_1^2(1-\mu)^2\eta^4}{\nu^2(1-\eta^2)^2}, && \eta^2 = r_1^2/r_2^2,\\
\kappa &= \frac{1-\mu/\eta^2}{1-\mu}, && \Delta = \frac{\partial^2}{\partial r^2} + \frac{1}{r}\frac{\partial}{\partial r} + \frac{\partial^2}{\partial z^2}.
\end{aligned}$$

The nondimensional domain is $M = (\eta, 1) \times (0, L)$, and the boundary conditions are given by (4.2.10) and (4.2.14), i.e.,

$$\begin{aligned}
&u = (u_z, u_r, u_\theta) = 0 \qquad \text{at } r = \eta, 1,\\
&u \text{ is periodic with period } L \text{ in the } z-\text{direction}.
\end{aligned} \tag{4.2.51}$$

The initial value condition is

$$u = u_0(r,z) \quad \text{at} \quad t = 0. \tag{4.2.52}$$

Functional Setting. For the Taylor problem (4.2.50)–(4.2.52), we set

$$\begin{aligned}
H &= \left\{ u = (\widetilde{u}, u_\theta) \in L^2(M)^3 \;\middle|\; \begin{array}{l} \text{div}(r\widetilde{u}) = 0, u_r = 0 \text{ at } r = \eta, 1, \text{ and} \\ u \text{ is L-periodic in the z-direction} \end{array} \right\}\\
H_1 &= \left\{ u \in H^2(M)^3 \bigcap H \mid u \text{ satisfies } (4.2.51) \right\},
\end{aligned}$$

and the inner product of H is defined by

$$(u, v) = \int_M u \cdot v r dz dr.$$

Let the linear operator $L_\lambda = -A + \lambda B : H_1 \to H$ and nonlinear operator $G : H_1 \to H$ be defined by

$$
\begin{aligned}
Au &= -P\left(\Delta u_z, \left(\Delta - \frac{1}{r^2}\right) u_r, \left(\Delta - \frac{1}{r^2}\right) u_\theta\right),\\
Bu &= P\left(0, \left(\frac{1}{r^2} - \kappa\right) u_\theta, \kappa u_r\right),\\
G(u) &= -P\left((\widetilde{u}\cdot\nabla)u_z, (\widetilde{u}\cdot\nabla)u_r - \frac{u_\theta^2}{r}, (\widetilde{u}\cdot\nabla)u_\theta + \frac{u_\theta u_r}{r}\right),
\end{aligned}
\tag{4.2.53}
$$

where $P : L^2(M)^3 \to H$ is the Leray projection. Thus the Taylor problem (4.2.50)–(4.2.52) can be rewritten in the abstract form

$$
\begin{aligned}
\frac{du}{dt} &= L_\lambda u + G(u),\\
u(0) &= u_0.
\end{aligned}
\tag{4.2.54}
$$

As before, we also use G to denote the corresponding bilinear operator.

Eigenvalue Problem. To study the phase transition of the Taylor problem (4.2.50)–(4.2.52), it is necessary to consider the eigenvalue problem of its linearized equation.

Let L_λ be the linear operator defined by (4.2.53), and L_λ^* its conjugate operator. The associated eigenvalue equation of (4.2.54) reads

$$
L_\lambda u = -Au + \lambda Bu = \beta(\lambda)u, \tag{4.2.55}
$$

and the conjugate equation of (4.2.55) is given by

$$
L_\lambda^* u^* = -A^* u^* + \lambda B^* u^* = \beta(\lambda) u^*. \tag{4.2.56}
$$

The equations corresponding to (4.2.55) are as follows:

$$
\begin{aligned}
&\Delta u_z - \frac{\partial p}{\partial z} = \beta(\lambda) u_z,\\
&\left(\Delta - \frac{1}{r^2}\right) u_r + \lambda\left(\frac{1}{r^2} - \kappa\right) u_\theta - \frac{\partial p}{\partial r} = \beta(\lambda) u_r,\\
&\left(\Delta - \frac{1}{r^2}\right) u_\theta + \lambda\kappa u_r = \beta(\lambda) u_\theta,\\
&\mathrm{div}(r\widetilde{u}) = 0.
\end{aligned}
\tag{4.2.57}
$$

The equations corresponding to (4.2.56) are given by

$$\begin{aligned}
&\Delta u_z^* - \frac{\partial p^*}{\partial z} = \beta(\lambda) u_z^*,\\
&\left(\Delta - \frac{1}{r^2}\right) u_r^* + \lambda \kappa u_\theta^* - \frac{\partial p^*}{\partial r} = \beta(\lambda) u_r^*,\\
&\left(\Delta - \frac{1}{r^2}\right) u_\theta^* + \lambda \left(\frac{1}{r^2} - \kappa\right) u_r^* = \beta(\lambda) u_\theta^*,\\
&\operatorname{div}(r\widetilde{u}^*) = 0.
\end{aligned} \tag{4.2.58}$$

Both (4.2.57) and (4.2.58) are supplemented with the boundary condition (4.2.51).

By Theorem 2.1.3, the phase transition of hydrodynamic equations requires the validity of (2.1.4) and (2.1.5), also called in the physics literature the principle of exchange of stability (PES). In the following, we shall verify the PES for the Taylor problem.

It is known that for each given period L, there is a $\lambda_0^* = \lambda_0(L)$ such that the eigenvalues $\beta_j(\lambda)$ $(j = 1, 2, \cdots)$ of (4.2.57) with (4.2.51) near $\lambda = \lambda_0^*$ satisfy that $\beta_1(\lambda), \cdots, \beta_m(\lambda)$ $(m \geq 1)$ are real, and

$$\begin{aligned}
&\beta_i(\lambda) \begin{cases} < 0 & \text{if } \lambda < \lambda_0^*, \\ = 0 & \text{if } \lambda = \lambda_0^* \end{cases} && \text{for } 1 \leq i \leq m,\\
&\operatorname{Re}\beta_j(\lambda_0^*) < 0 && \text{for } j \geq m+1.
\end{aligned} \tag{4.2.59}$$

In addition, there is a period $L' > 0$ such that

$$\lambda_0 = \lambda_0(L') = \min_{L>0} \lambda_0(L). \tag{4.2.60}$$

Thanks to Yudovich (1966) and Velte (1966), for $\mu = \Omega_1/\Omega_2 \geq 0$, the multiplicity m is equal to 2 in (4.2.59) at $\lambda_0 = \lambda_0(L')$; see also (Kirchgässner 1975; Temam 2001).

In this subsection, we always take L' as the period given by (4.2.60), and we define the following number as the critical Taylor number:

$$T_c = \lambda_0^2(L').$$

For simplicity, omitting the prime, we denote L' by L.

By (4.2.59) and (4.2.60), to verify the PES, it suffices to prove that for $\lambda > \lambda_0$,

$$\beta_i(\lambda) > 0 \qquad \forall 1 \leq i \leq m. \tag{4.2.61}$$

To this end, by Theorem 2.1 in Ma and Wang (2008d), we need to derive the eigenvectors of (4.2.57) and (4.2.58) at $\beta_i(\lambda_0) = 0$ $(i = 1, 2)$.

It is readily checked that the eigenvectors of (4.2.57) with (4.2.51) corresponding to $\beta_i(\lambda_0) = 0$ $(i = 1, 2)$ are given by

$$\psi_1 = (\psi_z, \psi_r, \psi_\theta) = (-\sin az\, D_* h(r), a\cos az\, h(r), \cos az\, \varphi(r)), \tag{4.2.62}$$

$$\widetilde{\psi}_1 = (\widetilde{\psi}_z, \widetilde{\psi}_r, \widetilde{\psi}_\theta) = (\cos az\, D_* h(r), a\sin az\, h(r), \sin az\, \varphi(r)), \tag{4.2.63}$$

where $(h(r),\varphi(r))$ satisfies

$$\begin{aligned}
&(DD_*-a^2)^2h=a^2\lambda_0\left(\frac{1}{r^2}-\kappa\right)\varphi,\\
&(DD_*-a^2)\varphi=-\lambda_0\kappa h, \qquad\qquad (4.2.64)\\
&(h,Dh,\varphi)=0 \qquad\qquad \text{at } r=\eta,1,
\end{aligned}$$

and

$$D=\frac{d}{dr},\quad D_*=\frac{d}{dr}+\frac{1}{r},\quad a=\frac{2\pi}{L}.$$

The dual eigenvectors of (4.2.58) with (4.2.51) read

$$\psi_1^*=(\psi_z^*,\psi_r^*,\psi_\theta^*)=(-\sin az\,D_*h^*(r),a\cos az\,h^*(r),\cos az\,\varphi^*(r)), \qquad (4.2.65)$$
$$\widetilde{\psi}_1^*=(\widetilde{\psi}_z^*,\widetilde{\psi}_r^*,\widetilde{\psi}_\theta^*)=(\cos az\,D_*h^*(r),a\sin az\,h^*(r),\sin az\,\varphi^*(r)), \qquad (4.2.66)$$

where (h^*,φ^*) satisfies

$$\begin{aligned}
&(DD_*-a^2)^2h^*=\lambda_0\kappa\varphi^*,\\
&(DD_*-a^2)\varphi^*=-a^2\lambda_0\left(\frac{1}{r^2}-\kappa\right)h^*, \qquad\qquad (4.2.67)\\
&(h^*,Dh^*,\varphi^*)=0 \qquad\qquad \text{at } r=\eta,1.
\end{aligned}$$

The following lemma shows that the PES is valid for the Taylor problem (4.2.50)–(4.2.52) with $\mu\geq 0$.

Lemma 4.2.11 *If $\mu\geq 0$, then the first eigenvalues $\beta_i(\lambda)$ $(1\leq i\leq m)$ of (4.2.57) are real with multiplicity $m=2$ near $\lambda=\lambda_0=\sqrt{T_c}$, and the first eigenvectors at $\lambda=\lambda_0$ are given by (4.2.62) and (4.2.63). Moreover, the eigenvalues $\beta_j(\lambda)$ $(j=1,2,\cdots)$ satisfy the conditions (2.1.4) and (2.1.5) at $\lambda=\lambda_0$, i.e., the PES holds at the critical Taylor number T_c.*

Proof. We need to prove only (4.2.61). By Theorem 2.1 in Ma and Wang (2008d), it suffices to verify that

$$(B\psi_1,\psi_1^*)\neq 0,\quad (B\widetilde{\psi}_1,\widetilde{\psi}_1^*)\neq 0. \qquad (4.2.68)$$

We infer from (4.2.53), (4.2.62), (4.2.63), (4.2.65), and (4.2.66) that

$$\begin{aligned}
(B\psi_1,\psi_1^*)&=(B\widetilde{\psi}_1,\widetilde{\psi}_1^*)\\
&=\int_0^L\int_\eta^1 r\left[\left(\frac{1}{r^2}-\kappa\right)\psi_\theta\psi_r^*+\kappa\psi_r\psi_\theta^*\right]dzdr\\
&=\frac{La}{2}\int_\eta^1 r\left[\left(\frac{1}{r^2}-\kappa\right)h\varphi^*+\kappa\varphi h^*\right]dr. \qquad (4.2.69)
\end{aligned}$$

Since $\mu \geq 0$, by (4.2.4), we have $0 < \kappa < 1$ and $\frac{1}{r^2} - \kappa > 0$ for $\eta < r < 1$. On the other hand, we know that the first eigenvectors $(h(r), \varphi(r))$ of (4.2.64) and $(h^*(r), \varphi^*(r))$ of (4.2.67) at $\lambda = \lambda_0$ are positive; see (Velte 1966; Kirchgässner 1975; Temam 2001):

$$h(r) > 0, \quad \varphi(r) > 0, \quad h^*(r) > 0, \quad \varphi^*(r) > 0 \quad \forall \eta < r < 1. \tag{4.2.70}$$

Thus (4.2.68) follows from (4.2.69) and (4.2.70). The proof is complete. □

Phase Transition Theorems. Here we always assume that the first eigenvalue of (4.2.57) with (4.2.51) is real with multiplicity $m = 2$, i.e., the first eigenvalue λ_0 of (4.2.64) is simple, and the PES holds. By Lemma 4.2.11, this assumption is valid for all $\mu \geq 0$ and $0 < \eta < 1$.

Let ψ_1 and ψ_1^* be given by (4.2.62) and (4.2.65). We define a number R by

$$R = \frac{1}{(\psi_1, \psi_1^*)} [(G(\Phi, \psi_1), \psi_1^*) + (G(\psi_1, \Phi), \psi_1^*)], \tag{4.2.71}$$

where $\Phi \in H_1$ is defined by

$$(A - \lambda_0 B)\Phi = G(\psi_1, \psi_1). \tag{4.2.72}$$

Here the operators A, B, and G are as in (4.2.53). The solution Φ of (4.2.72) exists because $G(\psi, \psi_1)$ is orthogonal to ψ_1^* and $\widetilde{\psi}_1^*$ in H.

The following results characterize the dynamical properties of phase transitions for the Taylor problem with the z-periodic boundary condition.

Theorem 4.2.12 *If the number $R0$ in (4.2.71) is negative, then the Taylor problem (4.2.50)–(4.2.52) has a Type I (continuous) transition at the critical Taylor number $T = T_c$ or $\lambda = \lambda_0$, and the following assertions hold:*

(1) When the Taylor number T is less than or equal to T_c or $\lambda \leq \lambda_0$, the steady state $u = 0$ is locally asymptotically stable.
(2) The problem bifurcates from $(u, \lambda) = (0, \lambda_0)$ [or from $(u, T) = (0, T_c)$] to an attractor $\mathscr{A}_\lambda$ homeomorphic to a circle S^1 on $\lambda_0 < \lambda$, which consists of steady states of this problem.
(3) Every $u \in \mathscr{A}_\lambda$ can be expressed as

$$u = |\beta_1(\lambda)/R|^{1/2} v + o(|\beta_1|^{1/2}), \qquad v = x\psi_1 + y\widetilde{\psi}_1, \qquad x^2 + y^2 = 1,$$

where $\psi_1, \widetilde{\psi}_1$ are given in (4.2.62) and (4.2.65).
(4) There is an open set $U \subset H$ with $0 \in U$ such that $\mathscr{A}_\lambda$ attracts $U \setminus \Gamma$, where Γ is the stable manifold of $u = 0$ with codimension two in H.
(5) When $1 - \mu > 0$ is small, then for every $u_0 \in U \setminus (\Gamma \bigcup \widetilde{H})$, there exists a time $t_0 \geq 0$ such that for any $t > t_0$, the vector field $\widetilde{u}(t, u_0) = (u_z, u_r)$ is topologically equivalent to one of the patterns shown in Fig. 4.12, where $u = (\widetilde{u}(t, u_0), u_\theta(t, u_0))$ is the solution of (4.2.50)–(4.2.52), and

$$\widetilde{H} = \{u = (u_z, u_r, u_\theta) \in H | \int_0^L \int_\eta^1 r u_z dr dz = 0\}.$$

(6) *When $1 - \mu > 0$ is small, then for every $u_0 \in (U \bigcap \widetilde{H}) \setminus \Gamma$, there exists a time $t_0 \geq 0$ such that for every $t > t_0$, $\widetilde{u}(t, u_0) = (u_z, u_r)$ is topologically equivalent to the structure shown in Fig. 4.13.*

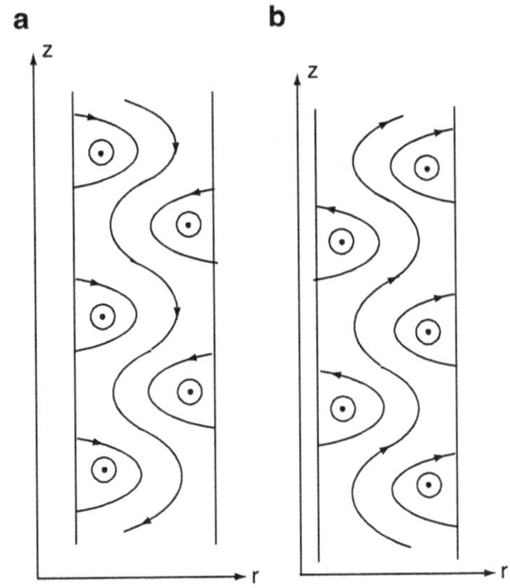

Fig. 4.12 Taylor vortices with a cross-channel flow

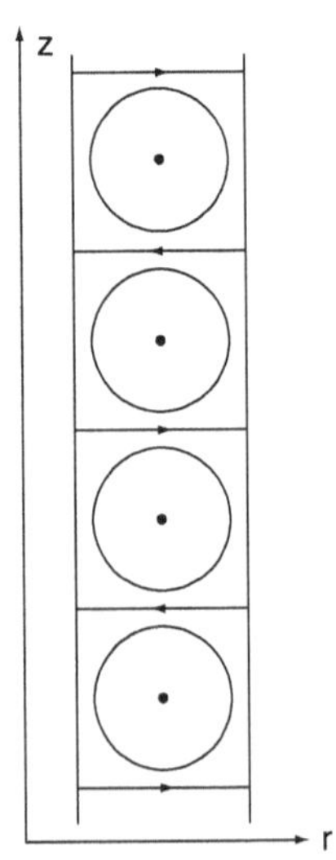

Fig. 4.13 Taylor vortices without a cross-channel flow

Theorem 4.2.13 *For the case $R > 0$, the transition of the Taylor problem (4.2.50)–(4.2.52) at $T = T_c$ is of Type II. Moreover, the Taylor problem has a singularity separation at $T^* < T_c$ ($\lambda^* < \lambda_0$). More precisely, we have the following assertions:*

(1) There exists a number λ^* $(0 < \lambda^* < \lambda_0)$ such that the problem generates a circle $\Sigma^* = S^1$ at $\lambda = \lambda^*$ consisting of singular points, and bifurcates from (Σ^*, λ^*) on $\lambda^* < \lambda$ to at least two branches of circles Σ_1^λ and Σ_2^λ, each consisting of steady states satisfying (see Fig. 4.14):

$$\lim_{\lambda \to \lambda_0} \Sigma_\lambda^1 = \{0\}, \qquad dist\left(\Sigma_2^\lambda, 0\right) = \min_{u \in \Sigma_2^\lambda} \|u\| > 0 \quad at \quad \lambda = \lambda_0.$$

(2) For each $\lambda^* < \lambda < \lambda_0$, the space H can be decomposed into two open sets U_1^λ and U_2^λ : $H = \overline{U}_1^\lambda + \overline{U}_2^\lambda$ with $U_1^\lambda \bigcap U_2^\lambda = \emptyset, \Sigma_1^\lambda \subset \partial U_1^\lambda \bigcap \partial U_2^\lambda$ such that the problem has two disjoint attractors $\mathscr{A}_1^\lambda$ and $\mathscr{A}_2^\lambda$:

$$\mathscr{A}_1^\lambda = \{0\} \subset U_1^\lambda, \quad \Sigma_2^\lambda \subset \mathscr{A}_2^\lambda \subset U_2^\lambda,$$

and $\mathscr{A}_i^\lambda$ attracts U_i^λ $(i = 1, 2)$.

(3) For $\lambda_0 \le \lambda$, the problem has an attractor $\mathscr{A}^\lambda$ satisfying

$$\lim_{\lambda \to \lambda_0} \mathscr{A}_2^\lambda = \mathscr{A}^{\lambda_0}, \quad dist(\mathscr{A}^\lambda, 0) > 0 \quad \forall \lambda \ge \lambda_0,$$

and $\mathscr{A}^\lambda$ attracts $H \setminus \Gamma_\lambda$, where Γ_λ is the stable manifold of $u = 0$ with codimension $m_\lambda \ge 2$ in H.

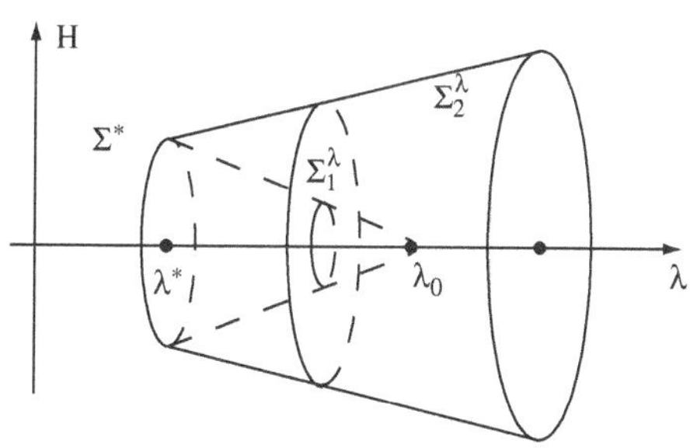

Fig. 4.14 Singularity separation of circles consisting of steady states at $\lambda = \lambda^*$

Proof of Theorem 4.2.12. We proceed as follows.

Step 1. We claim that the problem (4.2.50)–(4.2.52) bifurcates from $(u, \lambda) = (0, \lambda_0)$ to a circle S^1 that consists of steady states.

It is easy to see that the problem (4.2.50) with (4.2.51) is invariant for the transition in the z-direction: $u(z, r, t) \to u(z + z_0, r, t)$ for $z_0 \in \mathbb{R}^1$. Therefore, if u_0 is a steady-state solution of (4.2.50) with (4.2.51), then for every $z_0 \in \mathbb{R}^1$, the function $u_0(z + z_0, r)$ is also a steady-state solution. We can see that the set

$$\Sigma = \{u_0(z+z_0,r) | \, z_0 \in \mathbb{R}^1\}$$

is homeomorphic to a circle S^1 in H_1 for any $u_0 \in H_1$. Hence, the singular points of (4.2.50) with (4.2.51) appear as a circle.

It is shown in Yudovich (1966) and Velte (1966) that there exist singular points bifurcated from $(u,\lambda) = (0,\lambda_0)$. Thus this claim is proved.

Step 2. Reduction to the center manifold. By the spectral theorem [Theorem 3.4 in Ma and Wang (2005b)], the reduced equations of (4.2.54) to the first eigenspace are given by

$$\begin{aligned} \frac{dx}{dt} &= \beta_1(\lambda)x + \frac{(G(u),\psi^*_{1\lambda})}{(\psi_{1\lambda},\psi^*_{1\lambda})}, \\ \frac{dy}{dt} &= \beta_1(\lambda)y + \frac{(G(u),\widetilde{\psi}^*_{1\lambda})}{(\widetilde{\psi}_{1\lambda},\widetilde{\psi}^*_{1\lambda})}, \end{aligned} \tag{4.2.73}$$

where $\psi_{1\lambda}$ and $\widetilde{\psi}_{1\lambda}$ are the eigenvectors of (4.2.57) corresponding to $\beta_1(\lambda)$ near $\lambda = \lambda_0$ with

$$\begin{aligned} &\lim_{\lambda\to\lambda_0} \psi_{1\lambda} = \psi_1 \quad [\psi_1 \text{ as in (4.2.62)}], \\ &\lim_{\lambda\to\lambda_0} \widetilde{\lambda}_{1\lambda} = \widetilde{\psi}_1 \quad [\widetilde{\psi}_1 \text{ as in (4.2.63)}], \end{aligned} \tag{4.2.74}$$

and $\psi^*_{1\lambda}$ and $\widetilde{\psi}^*_{1\lambda}$ are the dual eigenvectors of $\psi_{1\lambda}$ and $\widetilde{\psi}_{1\lambda}$ satisfying

$$\begin{aligned} &\lim_{\lambda\to\lambda_0} \psi^*_{1\lambda} = \psi^*_1 \quad [\psi^*_1 \text{ as in (4.2.65)}], \\ &\lim_{\lambda\to\lambda_0} \widetilde{\psi}^*_{1\lambda} = \widetilde{\psi}^*_1 \quad [\widetilde{\psi}_1 \text{ as in (4.2.66)}]. \end{aligned} \tag{4.2.75}$$

Let $\Psi : E_0 \to E_0^\perp$ be the center manifold function of (4.2.54) at $\lambda = \lambda_0$, where

$$E_0 = \text{span}\{\psi_1,\widetilde{\psi}_1\}, \qquad E_0^\perp = \{u \in H | \, (u,\psi^*_1) = 0, (u,\widetilde{\psi}^*_1) = 0\}.$$

Let $u_0 = x\psi_1 + y\widetilde{\psi}_1 \in E_0$. Then it is easy to check that $G(u_0) \in E_0^\perp$. Hence, by Theorem A.1.1, we find that

$$\begin{aligned} &\Psi = \phi(x,y) + o(|x|^2+|y|^2) + O(\beta_1(\lambda)(|x|^2+|y|^2)), \\ &-L_\lambda\phi = G(\psi_1,\psi_1)x^2 + G(\widetilde{\psi}_1,\widetilde{\psi}_1)y^2 + (G(\psi_1,\widetilde{\psi}_1)+G(\widetilde{\psi}_1,\psi_1))xy. \end{aligned} \tag{4.2.76}$$

On the center manifold, $u = u_0 + \Phi(u_0)$. Therefore, from (4.2.73)–(4.2.76), we obtain the reduced equations of (4.2.54) to the center manifold as follows:

$$
\begin{aligned}
\frac{dx}{dt} =& \beta_1(\lambda)x + \frac{x}{\rho}(G(\phi,\psi_1)+G(\psi_1,\phi),\psi_1^*) \\
&+ \frac{y}{\rho}(G(\phi,\widetilde{\psi}_1)+G(\widetilde{\psi}_1,\phi),\psi_1^*) \\
&+ o(|x|^3+|y|^3)+\varepsilon_1(\lambda)O(|x|^3+|y|^3), \\
\frac{dy}{dt} =& \beta_1(\lambda)y + \frac{x}{\rho}(G(\phi,\psi_1)+G(\psi_1,\phi),\widetilde{\psi}_1^*) \\
&+ \frac{y}{\rho}(G(\phi,\widetilde{\psi}_1)+G(\widetilde{\psi}_1,\phi),\widetilde{\psi}_1^*) \\
&+ o(|x|^3+|y|^3)+\varepsilon_2(\lambda)O(|x|^3+|y|^3),
\end{aligned} \tag{4.2.77}
$$

where $\rho = (\psi_1,\psi_1^*) = (\widetilde{\psi}_1,\widetilde{\psi}_1^*)$, and $\lim_{\lambda\to\lambda_0}\varepsilon_i(\lambda)=0$ for $i=1,2$.

Furthermore, direct calculation shows that

$$
\begin{aligned}
&G(\psi_1,\psi_1) = P\psi_0 + P\psi_2, && G(\widetilde{\psi}_1,\widetilde{\psi}_1) = P\psi_0 - P\psi_2, \\
&G(\psi_1,\widetilde{\psi}_1) = P\widetilde{\psi}_0 + P\widetilde{\psi}_2, && G(\widetilde{\psi}_1,\psi_1) = -P\widetilde{\psi}_0 + P\widetilde{\psi}_2,
\end{aligned}
$$

where $P: L^2(M)^3 \to H$ is the Leray projection, and

$$
\begin{aligned}
\psi_0 &= -\left(0, \frac{1}{2}(a^2hD_*h + a^2hDh - \frac{1}{r}\varphi^2), \frac{a}{2}(\varphi D_*h + hD\varphi + \frac{1}{r}h\varphi)\right), \\
\psi_2 &= -\left(\frac{a}{2}\sin 2az((D_*h)^2 - hDD_*h), \frac{1}{2}\cos 2az(a^2hDh - a^2hD_*h - \frac{1}{r}\varphi^2), \right. \\
&\qquad \left. \frac{a}{2}\cos 2az(hD\varphi - \varphi D_*h + \frac{1}{r}\varphi h)\right), \\
\widetilde{\psi}_0 &= -\left(\frac{a}{2}((D_*h)^2 + hDD_*h), 0, 0\right), \\
\widetilde{\psi}_2 &= -\left(-\frac{a}{2}\cos 2az((D_*h)^2 - hDD_*h), \frac{1}{2}\sin 2az(a^2hDh - a^2hD_*h - \frac{1}{r}\varphi^2), \right. \\
&\qquad \left. \frac{a}{2}\sin 2az(hD\varphi - \varphi D_*h + \frac{1}{r}\varphi h)\right).
\end{aligned}
$$

Thus, (4.2.76) can be rewritten as

$$
(A - \lambda_0 B)\phi = (x^2+y^2)P\psi_0 + (x^2-y^2)P\psi_2 + 2xyP\widetilde{\psi}_2. \tag{4.2.78}
$$

Let

$$
\phi = -[(x^2+y^2)\phi_0 + (x^2-y^2)\phi_2 + 2xy\widetilde{\phi}_2], \tag{4.2.79}
$$

$$
\begin{cases}
\phi_0 = (0,0,\varphi_0), \\
\phi_2 = (-1/2\sin 2az\varphi_z, \cos 2az\varphi_r, \cos 2az\varphi_\theta), \\
\widetilde{\phi}_2 = (1/2\cos 2az\widetilde{\phi}_z, \sin 2az\widetilde{\phi}_r, \sin 2az\widetilde{\phi}_\theta).
\end{cases} \tag{4.2.80}
$$

We then deduce from (4.2.78) and (4.2.79) that

$$\varphi_z = \widetilde{\varphi}_z, \quad \varphi_r = \widetilde{\varphi}_r, \quad \varphi_\theta = \widetilde{\varphi}_\theta,$$

and $(\varphi_z, \varphi_r, \varphi_\theta)$ satisfies

$$\begin{aligned}
&(DD_* - 4a^2)\varphi_r + 4a^2\lambda_0(\frac{1}{r^2} - \kappa)\varphi_\theta = 4a^2H_2 + 2aDH_1,\\
&(DD_* - 4a^2)\varphi_\theta + \lambda_0\kappa\varphi_r = H_3,\\
&\varphi_z = \frac{1}{2a}D_*\varphi_r,\\
&\varphi_z = 0, \varphi_\theta = 0, \varphi_r = D\varphi_r = 0 \quad \text{at} \quad r = \eta, 1,
\end{aligned}$$

where H_1, H_2, and H_3 are given by

$$\begin{aligned}
H_1 &= a((D_*h)^2 - hDD_*h),\\
H_2 &= a^2hDh - a^2hD_*h - \frac{1}{r}\varphi^2,\\
H_3 &= a(hD\varphi - \varphi D_*h + \frac{1}{r}\varphi h).
\end{aligned} \tag{4.2.81}$$

Based on (4.2.80), we obtain

$$\begin{aligned}
&(G(\widetilde{\psi}_1, \phi_i) + G(\phi_i, \widetilde{\psi}_1), \psi_1^*) = 0 \quad \text{for} \quad i = 0, 2,\\
&(G(\psi_1, \phi_i) + G(\phi_i, \psi_1), \widetilde{\psi}_1^*) = 0 \quad \text{for} \quad i = 0, 2,\\
&(G(\widetilde{\psi}_1, \widetilde{\phi}_2) + G(\widetilde{\phi}_2, \widetilde{\psi}_1), \widetilde{\psi}_1^*) = 0,\\
&(G(\psi_1, \widetilde{\phi}_2) + G(\widetilde{\phi}_2, \psi_1), \psi_1^*) = 0.
\end{aligned}$$

Then, putting (4.2.79) into (4.2.77), we deduce that

$$\begin{aligned}
\frac{dx}{dt} = \beta_1 x &- \frac{1}{\rho}x(x^2 + y^2)(G(\phi_0, \psi_1) + G(\psi_1, \phi_0), \psi^*)1)\\
&- \frac{1}{\rho}x(x^2 - y^2)(G(\phi_2, \psi_1) + G(\psi_1, \phi_2), \psi_1^*)\\
&- \frac{2}{\rho}xy^2(G(\widetilde{\phi}_2, \widetilde{\psi}_1) + G(\widetilde{\psi}_1, \widetilde{\phi}_2), \psi_1^*)\\
&+ o(|x|^3 + |y|^3) + \varepsilon_1(\lambda)O(|x|^3 + |y|^3),\\
\frac{dy}{dt} = \beta_1 y &- \frac{1}{\rho}y(x^2 + y^2)(G(\phi_0, \widetilde{\psi}_1) + G(\widetilde{\psi}_1, \phi_0), \widetilde{\psi}_1^*)\\
&- \frac{1}{\rho}y(x^2 - y^2)(G(\phi_2, \widetilde{\psi}_1) + G(\widetilde{\psi}_1, \phi_2), \widetilde{\psi}_1^*)_h\\
&- \frac{2}{\rho}yx^2(G(\widetilde{\phi}_2, \psi_1) + G(\psi_1, \widetilde{\phi}_2), \widetilde{\psi}_1^*)\\
&+ o(|x|^2 + |y|^2) + \varepsilon_2(\lambda)O(|x|^3 + |y|^3).
\end{aligned} \tag{4.2.82}$$

Direct computation yields

$$
\begin{aligned}
(G(\phi_2,\psi_1)+G(\psi_1,\phi_2),\psi_1^*) &= (G(\widetilde{\phi}_2,\widetilde{\psi}_1)+G(\widetilde{\psi}_1,\widetilde{\phi}_2),\psi_1^*) \\
&= (G(\phi_2,\widetilde{\psi}_1)+G(\widetilde{\psi}_1,\phi_2),\widetilde{\psi}_1^*) \\
&= (G(\widetilde{\phi}_2,\psi_1)+G(\psi_1,\widetilde{\phi}_2),\widetilde{\psi}_1^*).
\end{aligned}
$$

Hence (4.2.82) can be rewritten as

$$
\begin{aligned}
\frac{dx}{dt} &= \beta_1 x + Rx(x^2+y^2) + o(|x|^3+|y|^3) + \varepsilon_1(\lambda)O(|x|^3+|y|^3), \\
\frac{dy}{dt} &= \beta_1 y + Ry(x^2+y^2) + o(|x|^3+|y|^3) + \varepsilon_2(\lambda)O(|x|^3+|y|^3),
\end{aligned}
\tag{4.2.83}
$$

where

$$
R = -\frac{1}{\rho}(G(\phi_0+\phi_2,\psi_1)+G(\psi_1,\phi_2+\phi_2),\psi_1^*). \tag{4.2.84}
$$

On the other hand, we infer from (4.2.78) and (4.2.80) that

$$
(A-\lambda_0 B)\phi_i = P\psi_i \qquad \text{for } i = 0,2.
$$

Hence we obtain

$$
\Phi = -(\phi_0+\phi_2), \qquad (A-\lambda_0 B)\Phi = G(\psi_1,\psi_1) = P\psi_0 + P\psi_2.
$$

Thus the number (4.2.84) is the same as that in (4.2.71).

Step 3. Proof of assertions (1)–(4). When $R<0$, $(x,y)=0$ is locally asymptotically stable for (4.2.83) at $\lambda=\lambda_0$. Therefore, $u=0$ is a locally asymptotically stable singular point of (4.2.54). By Theorem 2.2.11, the problem (4.2.50)–(4.2.52) bifurcates from $(u,\lambda)=(0,\lambda_0)$ to an attractor $\mathscr{A}_\lambda$ that attracts an open set $U\setminus\Gamma$, and assertions (1), (3), and (4) hold.

In addition, the nonlinear terms in (4.2.83) satisfy the condition (2.2.6). Hence by the S^1-attractor bifurcation theorem (Theorem 2.2.4) and the conclusion in Step 1, we derive assertion (2).

Step 4. Attraction in C^r-norm. It is known that for every initial value $u_0\in H$, there is a time $t_0>0$ such that the solution $u(t,u_0)$ of (4.2.50)–(4.2.52) is analytic for $t>t_0$, and uniformly bounded in C^r-norm for every $r\geq 1$; see also Lemma 4.1.12. Hence, by assertion (4), for every $u_0\in U\setminus\Gamma$, we have

$$
\lim_{t\to\infty}\min_{v_0\in\mathscr{A}_\lambda}\|u(t,u_0)-v_0\|_{C^r} = 0. \tag{4.2.85}
$$

Step 5. Structure of solutions in $\mathscr{A}_\lambda$. By assertion (3), for every steady-state solution $u_0=(u_z,u_r,u_\theta)\in\mathscr{A}_\lambda$, the vector field $\widetilde{u}=(u_z,u_r)$ of u_0 can be expressed as

$$
\begin{aligned}
u_z &= \gamma\cos a(z+z_0)D_*h(r) + w_1(z,r,\beta_1), \\
u_r &= a\gamma\sin a(z+z_0)h(r) + w_2(z,r,\beta_1),
\end{aligned}
\tag{4.2.86}
$$

for some $z_0 \in \mathbb{R}^1$, where

$$\gamma = |\beta_1(\lambda)/R|^{1/2}, \qquad w_i = o(|\beta_1|^{1/2}) \qquad \text{for } i = 1,2.$$

When $1-\mu$ is small, $r_1 \gg 1$. In this case, we have

$$D_* h(r) \simeq h'(r), \qquad h(r) \simeq h_0(r),$$

where $h_0(r)$, like $R(r)$, satisfies (4.2.38) with (4.2.39). Therefore, as in the proof of Lemma 4.2.6, we deduce that the vector field (4.2.86) is D-regular for all $0 < \lambda - \lambda_0 < \varepsilon$ for some $\varepsilon > 0$. Moreover, the first-order vector field in (4.2.86),

$$(v_z, v_r) = (\gamma \cos a(z+z_0) D_* h(r), a\gamma \sin a(z+z_0) h(r)), \tag{4.2.87}$$

has the topological structure shown in Fig. 4.13.

Furthermore, it is easy to check that the space

$$\widetilde{H} = \left\{ u = (u_z, u_r, u_\theta) \in H \middle| \int_M r u_z dr dz = 0 \right\}$$

is invariant for the operator $L_\lambda + G$ defined by (4.2.53). To see this, since u is z-periodic and $u = 0$ at $r = 1, \eta$, it follows that for every $u \in H$,

$$\begin{aligned} \int_0^L \int_\eta^1 r(\widetilde{u}\cdot\nabla) u_z dr dz &= \int_0^L \int_\eta^1 r u_r \frac{\partial u_z}{\partial r} dr dz \\ &= -\int_0^L \int_\eta^1 \frac{\partial (r u_r)}{\partial r} u_z dr dz = \int_0^L \int_\eta^1 r \frac{\partial u_z}{\partial z} u_z dr dz = 0. \end{aligned}$$

Also,

$$\int_0^L \int_\eta^1 r \Delta u_z dr dz = 0 \qquad \forall u \in \widetilde{H}.$$

Thus we see that $\widetilde{H}$ is invariant for $L_\lambda + G$.

Therefore, for the vector field (4.2.85), $\int_M r u_z dr dz = 0$. By the connection lemma and the orbit-breaking method in Ma and Wang (2005d), the vector field (4.2.86) is topologically equivalent to its first-order field (4.2.87) for $0 < \lambda - \lambda_0 < \varepsilon$.

Step 6. Proof of assertions (5) and (6). For every initial value $u_0 \in U \setminus (\Gamma \bigcup \widetilde{H})$, we have

$$u_0 = \sum_{k=1}^{\infty} \alpha_k e_k + w_0, \tag{4.2.88}$$

where $w_0 \in \widetilde{H}$, and for every $k = 1, 2, \cdots$,

$$e_k = (\widetilde{e}_k(r), 0, 0), \qquad \int_\eta^1 \widetilde{e}_k(r) dr \neq 0,$$

and $\widetilde{e}_k(r)$ satisfies

$$D_*D\widetilde{e}_k = -\rho_k\widetilde{e}_k, \qquad \widetilde{e}_k|_{r=\eta,1} = 0, \qquad 0 < \rho_1 < \rho_2 < \cdots.$$

Let

$$\begin{aligned} &H_1 = E\bigoplus\widetilde{H}_1, && \widetilde{H}_1 = H_1\bigcap\widetilde{H}, \\ &H = E\bigoplus\widetilde{H}, && E = \text{span}\{e_1, e_2, \cdots\}. \end{aligned}$$

Then Eq. (4.2.54) can be decomposed as

$$\begin{aligned} &\frac{de}{dt} = L_\lambda e && e \in E, \\ &\frac{dw}{dt} = L_\lambda w + G(w) && w \in \widetilde{H}_1, \\ &(e(0), w(0)) = (\sum_k \alpha_k e_k, w_0). \end{aligned} \tag{4.2.89}$$

It is obvious that

$$L_\lambda e_k = -\rho_k e_k.$$

Hence for the initial value (4.2.88), the solution $u(t, u_0)$ of (4.2.89) can be expressed as

$$u(t, u_0) = \sum_k \alpha_k e^{-\rho_k t} e_k + w(t, u_0), \qquad \int_\eta^1 e_k(r)dr \neq 0. \tag{4.2.90}$$

By (4.2.85), we have

$$\lim_{t\to\infty} \|w(t, u_0) - v_0\|_{C^r} = 0, \qquad v_0 \in \mathscr{A}_\lambda,$$

which implies by Step 5 that $w(t, u_0)$ is topologically equivalent to (4.2.87) for $t > 0$ sufficiently large, i.e., $w(t, u_0)$ has the topological structure shown in Fig. 4.13.

By the structural stability theorem, Theorem 2.2.9 and Lemmas 2.3.1 and 2.3.3 (connection lemmas) in Ma and Wang (2005d), we infer from (4.2.90) that the vector field in (4.2.90) is topologically equivalent to either the structure shown in Fig. 4.12a or the structure in 4.12(b), dictated by the sign of α_{k_0} in (4.2.88) with $k_0 = \min\{k | \alpha_k \neq 0\}$. Thus assertion (5) is proved.

Assertion (6) can be derived by the invariance of $\widetilde{H}$ under the operator $L_\lambda + G$ and the structural stability theorem with perturbation in $\widetilde{H}$, in the same fashion as in the proof of Theorem 2.2.9 in Ma and Wang (2005d) using the connection lemma.

The proof of Theorem 4.2.12 is complete. □

Proof of Theorem 4.2.13. When $R > 0$, by Theorem 2.1.5 we infer from the reduced equation (4.2.83) that the transition of (4.2.50)–(4.2.52) is of Type II. In the following, we shall use the saddle-node bifurcation theorem, Theorem 2.5.4, to prove this theorem. Let

$$H^* = \{(u_z, u_r, u_\theta) \in H |\ u_z(-z, r) = -u_z(z, r)\}, \qquad H_1^* = H_1\bigcap H^*.$$

It is easy to see that the space H^* is invariant under the action of the operator $L_\lambda + G$ defined by (4.2.53),

$$L_\lambda + G : H_1^* \to H^*, \tag{4.2.91}$$

and the first eigenvalue $\beta_1(\lambda)$ of $L_\lambda : H_1^* \to H^*$ at $\lambda = \lambda_0 (T = T_c)$ is simple, with the first eigenvector ψ_1 given by (4.2.62). Hence, the number R in (4.2.84) is valid for the mapping (4.2.90), i.e.,

$$(G(x\psi_1 + \Phi(x)), \psi_1^*) = Rx^3 + o(|x|^3).$$

Thus, it is readily checked that all conditions in Theorem 2.5.4 are satisfied by the operator (4.2.91). By Step 1 in the proof of Theorem 4.2.12, each singular point of (4.2.91) generates a singularity circle for $L_\lambda + G$ in H. Therefore, Theorem 4.2.13 follows from Theorem 2.5.4.

The proof of Theorem 4.2.13 is complete. □

Explicit Expression of the Parameter R. The parameter R defined by (4.2.71) and (4.2.72) can be explicitly expressed in the following integral formula:

$$\begin{aligned} R = & -\frac{1}{(\psi_1, \psi_1^*)} \Big[\frac{\pi}{2} \int_\eta^1 rh\varphi^* \frac{d\phi_0}{dr} dr \\ & + \int_0^L \int_\eta^1 r \Big((\widetilde{\phi} \cdot \nabla)\psi_z \psi_z^* + (\widetilde{\psi} \cdot \nabla)\phi_z \psi_z^* + (\widetilde{\phi} \cdot \nabla)\psi_r \psi_r^* \\ & + (\widetilde{\psi} \cdot \nabla)\phi_r \psi_r^* + (\widetilde{\phi} \cdot \nabla)\psi_\theta \psi_\theta^* + (\widetilde{\psi} \cdot \nabla)\phi_\theta \psi_\theta^* \\ & + \frac{\phi_\theta \psi_r \psi_\theta^*}{r} + \frac{\psi_\theta \phi_r \psi_\theta^*}{r} - 2\frac{\psi_\theta \phi_\theta \psi_r^*}{r} \Big) drdz \Big], \end{aligned}$$

where $\widetilde{\psi} = (\psi_z, \psi_r)$, $\psi_1 = (\psi_z, \psi_r, \psi_\theta)$, $\psi_1^* = (\psi_z^*, \psi_r^*, \psi_\theta^*)$ are given by (4.2.62) and (4.2.65), and ϕ_0, $\phi = (\phi_z, \phi_r, \phi_\theta)$ satisfy

$$\begin{cases} DD_*\phi_0 = a(\varphi D_* h + hD\varphi + \frac{1}{r}\varphi h), \\ \phi_0|_{r=\eta,1} = 0, \end{cases}$$

$$\begin{cases} -\Delta\phi_z + \frac{\partial p}{\partial z} = -1/2 \sin 2az H_1(r), \\ -\left(\Delta - \frac{1}{r^2}\right)\phi_r - \lambda_0\left(\frac{1}{r^2} - \kappa\right)\phi_\theta + \frac{\partial p}{\partial r} = -1/2 \cos 2az H_2(r), \\ -\left(\Delta - \frac{1}{r^2}\right)\phi_\theta - \lambda_0 \kappa \phi_r = -1/2 \cos 2az H_3(r), \\ \text{div}(r\widetilde{\phi}) = 0, \quad \widetilde{\phi} = (\phi_z, \phi_r), \\ \phi|_{r=\eta,1} = 0. \end{cases}$$

Here $H_i(r)$, $i = 1, 2, 3$, are as in (4.2.81).

Parameter R in the Narrow-Gap Case. When the conditions (4.2.15) and (4.2.23) hold, we ahve $\mu \to 1$ and $r_1 \to \infty$. In this case, equations (4.2.50) are replaced by (4.2.17), and the linearized equations of (4.2.17) reduce to the symmetric linear system (4.2.22). For the approximate problem (4.2.22) with (4.2.51), we use R_0 to denote the number R defined by (4.2.71) and (4.2.72):

$$R_0 = \frac{1}{\|\psi_1\|^2}\left[(G(\Phi,\psi_1),\psi_1) + G(\psi_1,\Phi),\psi_1)\right].$$

Here ψ_1 is given by (4.2.62) with (h,φ) satisfying

$$\begin{aligned}
&(D^2-a^2)^2 h = \lambda_0\varphi,\\
&(D^2-a^2)\varphi = -\lambda_0 h,\\
&h = Dh = 0, \varphi = 0 \qquad \text{at} \quad r = 1,\eta,
\end{aligned}$$

and Φ is defined by

$$(A-\lambda_0 B_0)\Phi = G(\psi_1,\psi_1), \qquad B_0\Phi = P(0,\Phi_\theta,\Phi_r). \tag{4.2.92}$$

We infer from (4.2.92) that

$$R_0 = -\frac{1}{\|\psi_1\|^2}(G(\psi_1,\psi_1),\Phi) = -\frac{1}{\|\psi_1\|^2}((A-\lambda_0 B_0)\Phi,\Phi).$$

We see that $A-\lambda_0 B_0$ is symmetric and positive semidefinite, and

$$G(\psi_1,\psi_1)\perp \mathrm{Ker}(A-\lambda_0 B_0), \quad \Phi\perp \mathrm{Ker}(A-\lambda_0 B_0).$$

Therefore, it follows that

$$R_0 = -\frac{1}{\|\psi_1\|^2}((A-\lambda_0 B_0)^{1/2}\Phi,(A-\lambda_0 B_0)^{1/2}\Phi) < 0.$$

On the other hand, $R(\mu)$ in (4.2.71) is continuous on μ, and $\lim_{\mu to 1} R(\mu) = R_0$. Hence we derive the following conclusion.

Theorem 4.2.14 *For the Taylor problem (4.2.50)–(4.2.52), there exist $\mu_0 < 1$ and $0 < \eta_0 < 1$ such that for every $\mu_0 < \mu < 1$ and $\eta_0 < \eta < 1$ with $\mu < \eta^2$, the parameter $R = R(\mu,\eta)$ defined by (4.2.71) is negative, i.e.,*

$$R(\mu,\eta) < 0, \quad \forall \mu_0 < \mu < 1, \quad \eta_0 < \eta < 1.$$

Consequently, the conclusions in Theorem 4.2.12 hold.

Asymptotic Structure of Solutions. Assertions (5) and (6) in Theorem 4.2.12 provide an asymptotic structure of the solutions in the physical space when the gap $r_2 - r_1$ is small, as observed in experiments. However, for general parameters η and μ, we cannot give the precise theoretical results, and so we present only a qualitative description. Here we consider two general cases as follows.

Case: $\mu \geq 0$**.** Following (Yudovich 1966; Velte 1966), for the eigenvector (h,φ) of (4.2.64), the function h can be taken as positive and having a unique maximum point in the interval $(\eta,1)$. Therefore, for the eigenvectors defined by (4.2.62) and (4.2.63), the vector fields (ψ_z,ψ_r) and $(\widetilde{\psi}_z,\widetilde{\psi}_r)$ are divergence-free and have the topological structure shown in Fig. 4.13. Hence to obtain assertions (5) and (6) in Theorem 4.2.12 for every $\mu \geq 0$, it suffices to prove that

$$h''(r) \neq 0 \quad \text{at} \quad r = r_0, 1, \eta, \tag{4.2.93}$$

where $r_0 \in (\eta,1)$ is the maximum point of h. We conjecture that the property (4.2.93) for the eigenvector (h,φ) of (4.2.64) is valid for all $0 \leq \mu < 1$ and $0 < \eta < 1$.

Case: $\mu < 0$**.** In this case, the situation is different. Numerical results show that the vector field (ψ_z,ψ_r) in (4.2.62) has $k \geq 2$ vortices in the radial direction, called the Taylor vortices. The vector field (ψ_z,ψ_r) has the topological structure shown in Fig. 4.15; see (Chandrasekhar 1981). This type of structure is structurally unstable. However, as discussed in Ma and Wang (2000), under a perturbation either in space H or in

$$\widetilde{H} = \left\{ (u_z,u_r,u_\theta) \in H \Big| \int_0^L \int_\eta^1 r u_z dz dr = 0 \right\},$$

there are only finitely many types of stable structures. In particular, if the vector field (ψ_z,ψ_r) in (4.2.62) is D-regular, i.e., $h(r)$ satisfies (4.2.93), then there is only one class of stable structures regardless of the orientation. For example, when $u_0 \in H/(\Gamma \bigcup \widetilde{H})$, the asymptotic structure of the solution $u(t,u_0)$ of (4.2.50)–(4.2.52) is as shown in Fig. 4.16, and when $u_0 \in \widetilde{H} \setminus \Gamma$, the asymptotic structure of the solution $u(t,u_0)$ is as shown in Fig. 4.17. It is clear that the class of structures illustrated by Fig. 4.16 is different from that illustrated by Fig. 4.17. The first one has a crossing the channel traveling flow in the z-direction, while the second one does not have such a cross the channel flow.

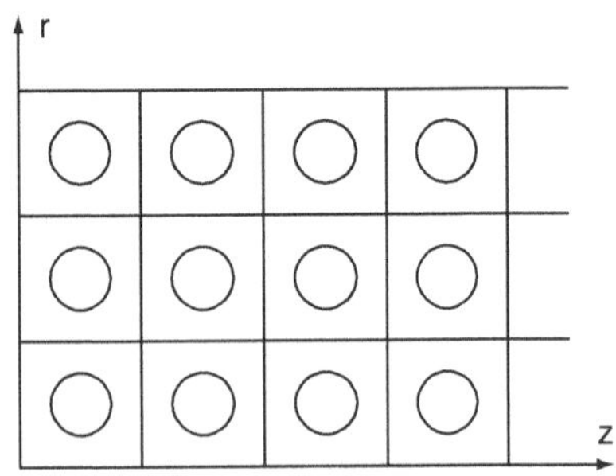

Fig. 4.15 (ψ_z,ψ_r) has k vortices in the r-direction

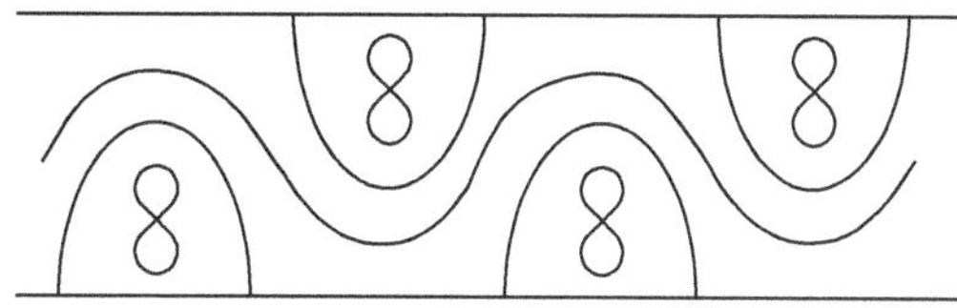

Fig. 4.16 The stable structure with a perturbation in space $H/(\Gamma\bigcup\widetilde{H})$

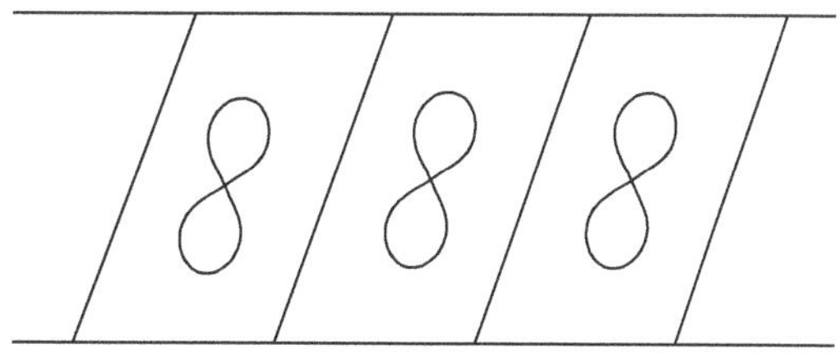

Fig. 4.17 The stable structure with a perturbation in space $\widetilde{H}\setminus\Gamma$

4.2.6 Other Boundary Conditions

Free Boundary Conditions in the z-Direction. We consider the Taylor problem (4.2.50) with the boundary condition (4.2.34), and the results obtained here are also valid for the boundary condition (4.2.35). Let

$$\begin{aligned} H_0 &= \{u=(\widetilde{u},u_\theta)\in L^2(M)^3 \mid \operatorname{div}(r\widetilde{u})=0, u_z|_{z=0,L}=0, u_r|_{r=r_1,r_2}=0\},\\ H_F &= \{u\in H^2(M)^3\bigcap H_0 \mid u \text{ satisfies (4.2.34)}\}. \end{aligned}$$

The mappings $L_\lambda+G: H_F\to H_0$ and $L_\lambda=-A+\lambda B$ are defined as in (4.2.53).

The eigenvalue problem for the free boundary condition is essentially the same as in the case for the periodic boundary condition. Their common characteristic is that the eigenvectors can be obtained by separation of variables.

We consider the following eigenvalue problem:

$$\begin{aligned} &-\Delta u_z+\frac{\partial p}{\partial z}=0,\\ &-\left(\Delta-\frac{1}{r^2}\right)u_r+\frac{\partial p}{\partial r}=\lambda\left(\frac{1}{r^2}-\kappa\right)u_\theta,\\ &-\left(\Delta-\frac{1}{r^2}\right)u_\theta=\lambda\kappa u_r,\\ &\operatorname{div}(r\widetilde{u})=0, \end{aligned} \tag{4.2.94}$$

where $\widetilde{u} = (u_z, u_r)$. The adjoint equations of (4.2.94) are given by

$$\begin{aligned}
&-\Delta u_z^* + \frac{\partial p^*}{\partial z} = 0,\\
&-\left(\Delta - \frac{1}{r^2}\right) u_r^* + \frac{\partial p^*}{\partial z} = \lambda \kappa u_\theta^*,\\
&-\left(\Delta - \frac{1}{r^2}\right) u_\theta^* = \lambda \left(\frac{1}{r^2} - \kappa\right) u_r^*,\\
&\operatorname{div}(r\widetilde{u}^*) = 0.
\end{aligned} \tag{4.2.95}$$

Both (4.2.94) and (4.2.95) are supplemented with the boundary condition (4.2.34).

For the eigenvalue equations (4.2.94) with (4.2.34) we take the separation of variables as follows:

$$\psi = (\psi_z, \psi_r, \psi_\theta) = (-\sin a_k z \, D_* h(r), a_k \cos a_k z \, h(r), \cos a_k z \, \varphi(r)), \tag{4.2.96}$$

where $a_k = k\pi/L$, $k \geq 1$, and $(h(r), \varphi(r))$ satisfies (4.2.64) with a_k replacing a. The eigenvectors of (4.2.95) with (4.2.34) are given by

$$\psi^* = (\psi_z^*, \psi_r^*, \psi_\theta^*) = (-\sin a_k z \, D_* h^*, a_k \cos a_k z \, h^*, \cos a_k z \, \varphi^*), \tag{4.2.97}$$

where $(h^*(r), \varphi^*(r))$ satisfies (4.2.67) with a_k replacing a, and ψ^* is the conjugate eigenvector of ψ in (4.2.96).

When $\mu \geq 0$, it is known that there is a sequence of eigenvalues λ_n $(n = 0, 1, \cdots)$ of (4.2.94) such that λ_0 is real (Temam 2001), and

$$|\lambda_1| < \cdots < |\lambda_n| < |\lambda_{n+1}| < \cdots.$$

Moreover, for all $L > 0$ except a discrete countable set of values, the first eigenvalue λ_0 is simple, and the first eigenvector ψ_0 is given by (4.2.96) with some wave number $k \geq 1$. We can prove in the same fashion as the proof of Lemma 4.2.11 that for the first eigenvalue λ_0, the PES in valid for $\mu \geq 0$. Hence, in this section we always consider the case in which the first eigenvalue λ_0 is simple and the PES holds.

We are now in a position to state and prove the phase transition theorems for the Taylor problem (4.2.50) with (4.2.34).

Let ψ_0 be the first eigenvector of (4.2.94) with (4.2.34), and let ψ_0^* be the conjugate eigenvector. We define a parameter R_0 by

$$R_0 = \frac{1}{(\psi_0, \psi_0^*)} (G(\Phi, \psi) + G(\psi, \Phi), \psi_0^*), \tag{4.2.98}$$

where $\Phi \in H_F$ satisfies

$$(A - \lambda_0 B)\Phi = G(\psi_0, \psi_0),$$

and λ_0 is the first eigenvalue of (4.2.94) with (4.2.34).

Theorem 4.2.15 *When the number R_0 defined by (4.2.98) is negative, i.e., $R_0 < 0$, the Taylor problem (4.2.50) with (4.2.34) has a phase transition of Type I at $\lambda = \lambda_0$. Moreover, we have the following assertions:*

(1) When $\lambda \leq \lambda_0$, the steady state $u = 0$ is locally asymptotically stable.
(2) The problem bifurcates from $(u,\lambda) = (0,\lambda_0)$ to an attractor $\mathscr{A}_\lambda$ consisting of exactly two singular points $\mathscr{A}_\lambda = \{u_1^\lambda, u_2^\lambda\} \subset H_F$ on $\lambda_0 < \lambda$.
(3) There is a neighborhood $U \subset H_0$ of $u = 0$ such that U can be decomposed into two open sets, $\overline{U} = \overline{U}_1^\lambda + \overline{U}_2^\lambda$, satisfying

$$U_1^\lambda \bigcap U_2^\lambda = \emptyset, \quad 0 \in \partial U_1^\lambda \bigcap \partial U_2^\lambda,$$

and for $i = 1,2$,

$$u_i^\lambda \in U_i^\lambda, \qquad \lim_{t\to\infty} \|u(t,u_0) - u_i^\lambda\| = 0 \qquad \forall u_0 \in U_i^\lambda,$$

where $u(t,u_0)$ is the solution of the problem with $u(0,u_0) = u_0$.
(4) If $1-\mu$ is small, then for every $u_0 \in U_1^\lambda + U_2^\lambda$ there is a time $t_0 \geq 0$ such that $\widetilde{u}(t,u_0)$ is topologically equivalent to one of the patterns shown in Fig. 4.10, where $u(t,u_0) = (\widetilde{u}(t,u_0), u_\theta(t,u_0))$ is the solution of the Taylor problem.

Theorem 4.2.16 *If $R_0 > 0$, then the Taylor problem (4.2.50) with (4.2.34) has a phase transition of Type II at $\lambda = \lambda_0$, and the following assertions hold:*

(1) The problem has exactly two saddle-node bifurcation points $(u_1^*,\lambda^*), (u_2^*,\lambda^*) \in H_F$ with $0 < \lambda^* < \lambda_0$, and there are at least four branches $\Gamma_1^\lambda, \Gamma_2^\lambda, \widetilde{\Gamma}_1^\lambda$, and $\widetilde{\Gamma}_2^\lambda$ emanating from (u_1^*,λ^*) and (u_2^*,λ^*) on $\lambda^* < \lambda$; see Fig. 4.18.
(2) The branches Γ_i^λ and $\widetilde{\Gamma}_i^\lambda$ satisfy that for $i = 1,2$,

$$\lim_{\lambda\to\lambda_0} \Gamma_i^\lambda = \{0\}, \qquad dist(\widetilde{\Gamma}_i^\lambda, 0) = \min_{u\in\widetilde{\Gamma}_i^\lambda} \|u\| > 0 \quad at \quad \lambda = \lambda_0.$$

(3) For each $\lambda \in (\lambda^*,\lambda_0)$, the space H_0 can be decomposed into two disjoint open sets $H_0 = \overline{U}_1^\lambda + \overline{U}_2^\lambda$ with $\Gamma_1^\lambda, \Gamma_2^\lambda \subset \partial U_1^\lambda \bigcap \partial U_2^\lambda$, and the problem has two disjoint attractors $\mathscr{A}_1^\lambda$ and $\mathscr{A}_2^\lambda$ such that

$$\mathscr{A}_1^\lambda = \{0\} \subset U_1^\lambda, \quad \widetilde{\Gamma}_1^\lambda, \widetilde{\Gamma}_2^\lambda \subset \mathscr{A}_2^\lambda \subset U_2^\lambda,$$

and $\mathscr{A}_i^\lambda$ attracts U_i^λ $(i = 1,2)$.
(4) When $\lambda_0 \leq \lambda$, the problem has an attractor $\mathscr{A}^\lambda$ such that $0 \notin \mathscr{A}^\lambda$,

$$\lim_{\lambda\to\lambda_0} \mathscr{A}_2^\lambda = \mathscr{A}^{\lambda_0},$$

and $\mathscr{A}^\lambda$ attracts $H_0 \setminus \Gamma_\lambda$, where Γ_λ is the stable manifold of $u = 0$ with codimension $m_\lambda \geq 1$.

Remark 4.2.17 Similar to the case with periodic boundary conditions, if $1-\mu$ is small, the number given by (4.2.98) is negative: $R_c < 0$, as in Theorem 4.2.14. We expect that assertion (4) in Theorem 4.2.15 is valid for every $\mu \geq 0$.

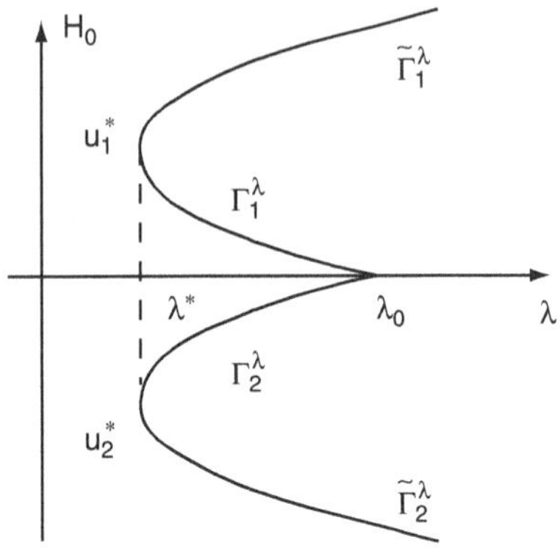

Fig. 4.18 Saddle-node bifurcation and jump transition

Proof of Theorems 4.2.15 and 4.2.16. The proof of Theorem 4.2.15 is similar to that of Theorem 4.2.12, and we omit the details.

Theorem 4.2.16 can be obtained by the same method as used in Theorem 4.2.13 and the fact that the singular points of the Taylor problem (4.2.50) with (4.2.34) appear in pairs. To see this, for $u \in H_F$, we have

$$u = \left(\sum_{k=1}^{\infty} \phi_{zk}(r) \sin \frac{k\pi z}{L}, \sum_{k=1}^{\infty} \phi_{rk}(r) \cos \frac{k\pi z}{L}, \sum_{k=1}^{\infty} \phi_{\theta k}(r) \cos \frac{k\pi z}{L} \right),$$

which implies that the problem is z_2-invariant. Namely, under the transformation: $z \to z+L$, Eqs. (4.2.50) and boundary condition (4.2.34) are invariant. Hence the singular points appear in pairs. □

Rigid and Free-Rigid Boundary Conditions. We know that it is much harder to get explicit expressions for the eigenvectors for both the rigid and the free-rigid boundary conditions. For simplicity, we consider only the rigid boundary condition

$$u|_{\partial M} = 0, \tag{4.2.99}$$

and the discussion holds as well for the free-rigid condition.

When $\mu \geq 0$, it is reasonable to conjecture that there is an open and dense set $S \subset (0,+\infty) \times (0,1)$ such that for every $(L,\eta) \in S$, the first eigenvalue λ_0 of (4.2.94) with (4.2.99) on $M = (0,L) \times (\eta,1)$ is real and simple, and the PES is valid. Moreover, the first eigenvector ψ and its conjugate eigenvector ψ^* satisfy

$$R = (G(\psi,\psi),\psi^*) \neq 0. \tag{4.2.100}$$

Under the assumption above, from Theorem 2.3.2 and 2.5.5 we immediately derive the following theorem.

Theorem 4.2.18 *Let the first eigenvalue λ_0 of (4.2.94) with (4.2.99) be real and simple, and suppose the PES is valid. If (4.2.100) holds, then the Taylor problem (4.2.50) with (4.2.99) undergoes a phase transition of Type III at $\lambda = \lambda_0$. Moreover, we have the following assertions:*

(1) The problem has a saddle-node bifurcation from (u^*, λ^*) with $0 < \lambda^* < \lambda_0$ on $\lambda^* < \lambda$, see Fig. 4.19, and there are two branches $\Gamma_i^\lambda (i = 1,2)$ emanating from (u^*, λ^*) satisfying

$$\lim_{\lambda \to \lambda_0} \Gamma_1^\lambda = \{0\} \quad \text{and} \quad 0 \notin \Gamma_2^{\lambda_0}.$$

(2) When $\lambda^* < \lambda < \lambda_0$, the space H_0 can be decomposed into two disjoint open sets $H_0 = \overline{U}_1^\lambda + \overline{U}_2^\lambda$ with $\Gamma_1^\lambda \subset \partial U_1^\lambda \cap \partial U_2^\lambda$, and the problem has two disjoint attractors $\mathscr{A}_i^\lambda \subset U_i^\lambda$ $(i = 1,2)$ with $\mathscr{A}_1^\lambda = \{0\}, \Gamma_2^\lambda \subset \mathscr{A}_2^\lambda$ such that $\mathscr{A}_i^\lambda$ attracts U_i^λ.

(3) When $\lambda_0 < \lambda < \lambda_0 + \delta$ for some $\delta > 0$, Γ_1^λ consists of a bifurcated singular point $\Gamma_1^\lambda = \{u^\lambda\}$, and H_0 can be decomposed into two disjoint open sets $H_0 = \overline{V}_1^\lambda + \overline{V}_2^\lambda$ with $0 \in \partial V_1^\lambda \cap \partial V_2^\lambda$ such that the problem has two disjoint attractors $\Sigma_i^\lambda \subset V_i^\lambda (i = 1,2)$ with $\Sigma_i^\lambda = \{u^\lambda\}, \Gamma_2^\lambda \subset \Sigma_2^\lambda$, and Σ_i^λ attracts V_i^λ.

(4) The bifurcated attractor u^λ for $\lambda_0 < \lambda < \lambda_0 + \delta$ can be expressed as

$$u^\lambda = -R^{-1}\beta_1(\lambda)\psi + o(|\beta_1(\lambda)|),$$

where R is as in (4.2.100).

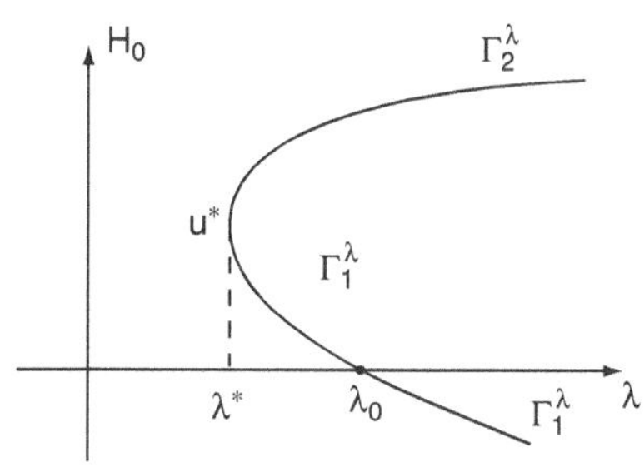

Fig. 4.19 Type III transition

4.2.7 Three-Dimensional Perturbation for the Narrow-Gap Case

In the previous subsections, we always discussed the phase transitions of the Taylor problems under axisymmetric perturbations. In this subsection, we make some remarks on the narrow-gap case with three-dimensional perturbations.

We assume that conditions (4.2.15) and (4.2.23) hold. When the perturbation is three-dimensional and depends on θ, then Eqs. (4.2.7) can be rewritten as

$$\begin{aligned}
\frac{\partial u_r}{\partial t} &= \Delta u_r - (u\cdot\nabla)u_r - \frac{\partial p}{\partial r} + A(r)\frac{\partial u_r}{\partial \theta} \\
&\quad + \lambda(1-(1-\mu)(r-r_1))u_\theta, \\
\frac{\partial u_\theta}{\partial t} &= \Delta u_\theta - (u\cdot\nabla)u_\theta - \frac{1}{r}\frac{\partial p}{\partial \theta} + A(r)\frac{\partial u_\theta}{\partial \theta} + \lambda u_r, \\
\frac{\partial u_z}{\partial t} &= \Delta u_z - (u\cdot\nabla)u_z - \frac{\partial p}{\partial z} + A(r)\frac{\partial u_z}{\partial \theta}, \\
\frac{\partial u_r}{\partial r} &+ \frac{1}{r}\frac{\partial u_\theta}{\partial \theta} + \frac{\partial u_z}{\partial \theta} = 0,
\end{aligned} \tag{4.2.101}$$

where $\lambda = \sqrt{T}$, T is the Taylor number, and

$$\begin{aligned}
T &= \frac{4(r_2-r_1)^4\Omega_1^2(\eta^2-\mu)}{(1-\eta^2)\nu^2}, \qquad & A(r) &= \frac{\sqrt{T}}{2}\left(\frac{\eta^2-\mu}{1-\eta^2} - \frac{1-\mu}{1-\eta^2}\frac{r_1^2}{r^2}\right), \\
\Delta &= \frac{\partial^2}{\partial r^2} + \frac{1}{r^2}\frac{\partial^2}{\partial \theta^2} + \frac{\partial^2}{\partial z^2}, & u\cdot\nabla &= u_r\frac{\partial}{\partial r} + \frac{u_\theta}{r}\frac{\partial}{\partial \theta} + u_z\frac{\partial}{\partial z}.
\end{aligned}$$

The nondimensional domain is $\Omega = (r_1,r_2)\times(0,2\pi)\times(0,L)$ for $x=(r,\theta,z)$. Equations (4.2.101) are supplemented with the following boundary conditions:

$$\begin{aligned}
&u(r,\theta+2k\pi,z) = u(r,\theta,z) \quad \forall k\in\mathbb{Z}, \\
&u = 0 \qquad \text{at } r=r_1,r_2, \quad \text{or} \quad z=0,L.
\end{aligned} \tag{4.2.102}$$

The initial condition is nonaxisymmetric, given by

$$u = u_0(r,\theta,z) \quad \text{at} \quad t=0. \tag{4.2.103}$$

Consider the eigenvalue problem

$$\begin{aligned}
&\Delta u_r - \frac{\partial p}{\partial r} + A(r)\frac{\partial u_r}{\partial \theta} + \lambda(1-(1-\mu)(r-r_1))u_\theta = \beta(\lambda)u_\theta, \\
&\Delta u_\theta - \frac{1}{r}\frac{\partial p}{\partial \theta} + A(r)\frac{\partial u_\theta}{\partial \theta} + \lambda u_r = \beta(\lambda)u_r, \\
&\Delta u_z - \frac{\partial p}{\partial z} + A(r)\frac{\partial u_z}{\partial \theta} = \beta(\lambda)u_z, \\
&\frac{\partial u_r}{\partial r} + \frac{1}{r}\frac{\partial u_\theta}{\partial \theta} + \frac{\partial u_z}{\partial z} = 0,
\end{aligned} \tag{4.2.104}$$

with the boundary condition (4.2.102).

Let all eigenvalues (counting multiplicities) of (4.2.104) be given by

$$\mathrm{Re}\beta_1(\lambda) \geq \mathrm{Re}\beta_2(\lambda) \geq \cdots \geq \mathrm{Re}\beta_k(\lambda) \geq \cdots, \qquad \lim_{k\to\infty}\mathrm{Re}\beta_k(\lambda) = -\infty. \tag{4.2.105}$$

It is clear that if λ^* is an eigenvalue of (4.2.20) with (4.2.102), then there is a real eigenvalue β_k in (4.2.105) satisfying $\beta_k(\lambda^*) = 0$.

Let λ_1 be the first eigenvalue of (4.2.20) with (4.2.102) with multiplicity $m \geq 1$. Then there exists $\beta_{k+1}(\lambda), \cdots, \beta_{k+m}(\lambda)$ for some $k \geq 0$ such that $\beta_{k+j}(\lambda)$ $(1 \leq j \leq m)$ are real near $\lambda = \lambda_1$, and

$$\beta_{k+j}(\lambda) \begin{cases} < 0 & \text{if } \lambda < \lambda_1, \\ = 0 & \text{if } \lambda = \lambda_1, \\ > 0 & \text{if } \lambda > \lambda_1 \end{cases} \qquad \forall 1 \leq j \leq m, \tag{4.2.106}$$

$$\text{Re}\beta_i(\lambda_1) \neq 0 \qquad \forall i \neq k+j, \quad 1 \leq j \leq m. \tag{4.2.107}$$

It is clear that $0 > \text{Re}\beta_1(0)$. Hence it follows from (4.2.105)–(4.2.107) that there exists a λ_0 with $0 < \lambda_0 \leq \lambda_1$ such that for some $m_1 \geq 1$,

$$\text{Re}\beta_i(\lambda) \begin{cases} < 0 & \text{if } \lambda < \lambda_0, \\ = 0 & \text{if } \lambda = \lambda_0, \\ > 0 & \text{if } \lambda > \lambda_0 \end{cases} \qquad \forall 1 \leq i \leq m_1, \tag{4.2.108}$$

$$\text{Re}\beta_j(\lambda_0) < 0 \qquad \forall j \geq m_1 + 1. \tag{4.2.109}$$

Thus by the transition theorem, Theorem 2.1.3, we obtain from (4.2.108) and (4.2.109) the following dynamic transition theorem for the Taylor problem (4.2.101) –(4.2.103):

Theorem 4.2.19 *Let λ_1 be the first eigenvalue of (4.2.20) with (4.2.18). Then there exists a number λ_0 with $0 < \lambda_0 \leq \lambda_1$ such that the properties (4.2.108) and (4.2.109) hold. Therefore, the Taylor problem (4.2.101)–(4.2.103) has a phase transition at $\lambda = \lambda_0$. In particular, if $\lambda_0 = \lambda_1$, then the assertions of Theorems 4.2.2 and 4.2.3 are valid for this problem.*

This theorem is a corollary of Theorems 2.1.3 and 2.6.1 and Theorem 4.1 in Ma and Wang (2008d), and we omit the details of the proof.

In the three-dimensional perturbation problem, we conjecture that the critical parameter λ_0 satisfying (4.2.108) and (4.2.109) is the first eigenvalue λ_1 of (4.2.20) with (4.2.18), i.e., $\lambda_0 = \lambda_1$.

4.2.8 Physical Remarks

Like the Rayleigh–Bénard convection, the Taylor problem has become a paradigm of nonequilibrium phase transitions with its instability, bifurcations, structure formation, and chaotic behaviors in fluid dynamics as well as in physical systems in general.

The classical linear theory explains the instability at the critical Taylor number, and the known nonlinear theory set by Yudovich (1966) and Velte (1966) could deal only with steady-state bifurcations for the z-periodic and the free boundary conditions.

Here in this section we mainly introduce a dynamical theory for the phase transitions of Taylor–Couette flow, using the transition and its classification theorems, the constructive method of center manifold functions, transition perturbation theory, and singularity separation results, together with the geometric theory of incompressible flows established by the authors (Ma and Wang 2005d). In the following, we shall give some remarks on the dynamical characterization for the Taylor problem with various physically sound boundary conditions.

Rigid and Free-Rigid Boundary Conditions. Both the rigid and free-rigid boundary conditions are the most popular conditions used in physics. In this case, Theorem 4.2.18 provides a general description for the phase transitions of the Taylor –Couette flow, although mathematically, Theorem 4.2.18 is not a good result, because it is difficult to check its conditions. However, this theorem is very useful in physics to determine the transition types, because the conditions, in fact, are generic and most realizable in experiments.

Based on Theorems 4.2.2, 4.2.3, and 4.2.18, for the boundary conditions we can provide some interesting and useful physical predictions as follows:

(1) The transition at the critical Taylor number $T_c = \lambda_0^2$ is of Type III, i.e., the transition is mixed.
(2) There are two critical Taylor numbers $T_c^* = \lambda^{*2}$ and $T_c = \lambda_0^2$ with $T_c^* < T_c$ at which the phase transitions may occur.
(3) When $T_c^* < T < T_c$, there is only one type of secondary flow σ_1^T, and the system may have a jump transition between the basic flow σ_0 (the Couette flow) and the secondary flow σ_1^T.
(4) When $T_c < T < T_c + \varepsilon$ for some $\varepsilon > 0$, there are exactly two types of secondary flows σ_1^T and σ_2^T, and the system may have a continuous transition between the basic flow σ_0 and the secondary flow σ_2^λ, or have a jump transition between either σ_0 and σ_1^λ or σ_1^λ and σ_2^λ, depending randomly on the external perturbation.
(5) In the case in which the gap $r_2 - r_1$ is small, we have $T_c^1 \simeq T_c^2$, and the two types of secondary flows σ_1^T and σ_2^T are controlled by the first eigenvectors and cannot be distinguished in experiments.

The z-Periodic and Free Boundary Conditions. As mentioned before, the z-periodic boundary condition is an approximate description for the case in which the ratio $L/(r_2 - r_1)$ between the height L and the gap $r_2 - r_1$ is sufficiently large, and the free boundary condition is suitable in the case that the upper and lower surfaces of the boundary are very smooth, or the viscosity of the fluid is very low.

By Theorems 4.2.2–4.2.16 we obtain the following dynamical properties:

(1) The phase transition at the critical Taylor number $T_c = \lambda_0^2$ is of Type I or Type II.
(2) The two transitions are dictated precisely by the sign of a nondimensional parameter R determined completely by the first eigenvectors and the parameters μ and η.

(3) When $R < 0$, the transition is continuous, i.e., of Type I, and when $R > 0$, it is a jump, i.e., of Type II.

(4) For the continuous transition case, i.e., $R < 0$, the critical exponent of the phase transition, i.e., the exponent in the expression of bifurcated solutions, is $\beta = 1/2$. Moreover, there is only one critical Taylor number T_c such that the secondary flow satisfies $\sigma^T \to 0$ as $T \to T_c$.

(5) When $R > 0$, there are two critical Taylor numbers $T_c = \lambda_0^2$ and $T_c^* = \lambda^{*2}$ with $T_c^* < T_c$, and there is only one type of secondary flow σ^T. When $T_c^* < T < T_c$, the system may have a jump transition between the Couette flow σ_0 and the secondary flow σ^T, and when $T_c \leq T$, the system undergoes a transition from σ_0 to σ^T.

(6) For the narrow-gap case, the parameter R defined by (4.2.71) is negative, $R < 0$, provided the two coaxial cylinders are rotating in the same direction. In fact, when the outer cylinder is static, for the narrow-gap case, the rotation of the inner cylinder induces that the fluid on $r = r_2$ also has a rotation with an angular velocity $\Omega_2 > 0$, and

$$\mu = \Omega_2/\Omega_1 \to 1 \quad \text{as} \quad r_1/(r_2 - r_1) \to \infty,$$

which implies that $R < 0$.

(7) Theoretical analysis shows that a street of vortices appears in the secondary flow for the narrow-gap case with $\mu \to 1$. Thus the theoretical results are consistent with the Taylor experiments.

In summary, the dynamical theory established here provides some interesting and useful physical predictions on the types of phase transitions with various physically sound boundary conditions, on the critical exponent of transitions of Type I, and on the globally dynamical structure of phase transitions. From both the mathematical and physical points of view, the dynamical theories and methods are useful for understanding nonequilibrium phase transitions in fluid dynamics.

4.3 Boundary-Layer and Interior Separations in the Taylor–Couette–Poiseuille Flow

Consider a viscous fluid between two coaxial rotating cylinders. As in the Taylor problem, the basic (Couette) flow becomes unstable as soon as the rotation speed of the inner cylinder exceeds a critical value. This instability gives rise to stationary axisymmetric counterrotating vortices that fill the whole annular region. The associated flow is referred to as Taylor–Couette (TC) flow. When a through flow driven by a pressure gradient along the rotating axis is added, the resulting system can exhibit both convective and absolute instabilities. The base flow consists of a superposition of circular Couette flow and annular Poiseuille flow, called Couette–Poiseuille (CP) flow. The axial throughflow suppresses the basic stationary instability, and as the axial pressure gradient increases while the rotation speed of the inner cylinder is held fixed, the first bifurcation gives rise to a traveling train of axisymmetric Taylor vortices, commonly referred to as propagating Taylor vortices (PTV). Henceforth, the term Taylor–Couette–Poiseuille (TCP) flow is used to refer to all hydrodynamic phenomena pertaining to the open system described above; see among others (Raguin and Georgiadis 2004) and the references therein.

The main objective of this section is to study dynamic transitions of Couette –Poiseuille flow and formation of the PTV. This study involves two types of transitions: the dynamic transition and the structure transition in physical space. The analysis is based on the dynamic transition theory presented in this book and the geometric theory of incompressible flows synthesized in Ma and Wang (2005d).

The results presented here are based on Ma and Wang (2009a). The results show in particular that contrary to what is commonly believed, PTV do not appear after the first dynamical bifurcation, and they appear only when the Taylor number is further increased to cross another critical value so that a structural bifurcation occurs. This structural bifurcation corresponds to the boundary-layer and interior separations of the flow structure in physical spaces.

4.3.1 Model for the Taylor–Couette–Poiseuille Problem

We consider viscous flow between two rotating coaxial cylinders with an axial pressure gradient. The instability of this flow was first discussed by Goldstein (1937) in 1933, and a further investigation was made by Chandrasekhar (1981).

Let r_1 and r_2 $(r_1 < r_2)$ be the radii of two coaxial cylinders, and let Ω_1 and Ω_2 be the angular velocities with which the inner and the outer cylinders rotate respectively. The hydrodynamic equations governing an incompressible viscous fluid between two coaxial cylinders are the Navier–Stokes equations in the cylindrical polar coordinates (r,θ,z), given by (4.2.5). Let a constant pressure gradient

$$p_0 = \partial p/\partial z$$

be applied in the z-direction, let the cylinders rotate, and let there be no radial motion. Under these conditions, (4.2.5) admits a steady-state solution:

$$\mathscr{U}^{cp} = (u_z, u_r, u_\theta) = (W, 0, V), \qquad p = p(r,z), \tag{4.3.1}$$

where

$$\frac{1}{\rho}\frac{dp}{dr} = V^2/r, \tag{4.3.2}$$

$$\nu\frac{d}{dr}\left(\frac{d}{dr} + \frac{1}{r}\right)V = 0, \tag{4.3.3}$$

$$\nu\left(\frac{d}{dr} + \frac{1}{r}\right)\frac{d}{dr}W = \frac{p_0}{p}, \tag{4.3.4}$$

supplemented with the following boundary conditions:

$$\begin{aligned}
&(u_r, u_z) = \left(0, -\frac{p_0}{4\rho\nu} W_0\right) && \text{at } r = r_1, r_2, \\
&u_\theta = r_i \Omega_i && \text{at } r = r_i \text{ (i=1, 2)}.
\end{aligned} \tag{4.3.5}$$

Here $W_0 \geq 0$ is a constant.

We derive then from (4.3.1)–(4.3.5) that

$$V(r) = \frac{\Omega_1}{1-\eta^2}\left(r_1^2(1-\mu)\frac{1}{r} - (\eta^2-\mu)r\right), \tag{4.3.6}$$

$$W(r) = -\frac{p_0}{4\rho\nu}\left(r_1^2 - r^2 + \frac{2r_1 d + d^2}{\ln(1+d/r_1)} \ln\left(\frac{r}{r_1}\right) + W_0\right), \tag{4.3.7}$$

$$p(r,z) = p_0 z + \rho \int \frac{1}{r} V^2(r) dr, \tag{4.3.8}$$

where $d \ (= r_2 - r_1)$ is the gap width, μ and η are as in (4.2.1).

Classically, the flow described by $\mathscr{U}^c = (0,0,V(r))$ is called Couette flow, and the flow described by $\mathscr{U}^p = (W(r),0,0)$ is called Poiseuille flow. Therefore, the flow described by (4.3.1) is a superposition of Couette flow and Poiseuille flow, and is usually called Couette–Poiseuille (CP) flow.

Consider the perturbed state

$$W + u_z, \ u_r, \ V + u_\theta, \qquad p + p_0 z + \rho \int \frac{1}{r} V^2 dr.$$

Assuming that the perturbation is axisymmetric and independent of θ, we derive from (4.2.5) the perturbed equations as follows:

$$\begin{aligned}
&\frac{\partial u_z}{\partial t} + (\widetilde{u}\cdot\nabla)u_z = \nu\Delta u_z - W\frac{\partial u_z}{\partial z} - u_r\frac{dW}{dr} - \frac{1}{\rho}\frac{\partial p}{\partial z}, \\
&\frac{\partial u_r}{\partial t} + (\widetilde{u}\cdot\nabla)u_r - \frac{u_\theta^2}{r} = \nu\left(\Delta u_r - \frac{u_r}{r^2}\right) - \frac{1}{\rho}\frac{\partial p}{\partial r} \\
&\qquad\qquad - W\frac{\partial u_r}{\partial z} + \frac{2V}{r} r_\theta, \\
&\frac{\partial u_\theta}{\partial t} + (\widetilde{u}\cdot\nabla)u_\theta + \frac{u_\theta u_r}{r} = \nu\left(\Delta u_\theta - \frac{u_\theta}{r^2}\right) \\
&\qquad\qquad - \left(V' + \frac{1}{r}V\right)u_r - W\frac{\partial u_\theta}{\partial z}, \\
&\frac{\partial(ru_z)}{\partial z} + \frac{\partial(ru_r)}{\partial r} = 0,
\end{aligned} \tag{4.3.9}$$

where $\widetilde{u} = (u_z, u_r)$, and

$$\Delta = \frac{\partial^2}{\partial r^2} + \frac{1}{r}\frac{\partial}{\partial r} + \frac{\partial^2}{\partial z^2}, \qquad (\widetilde{u}\cdot\nabla) = u_z\frac{\partial}{\partial z} + u_r\frac{\partial}{\partial r}.$$

The spatial domain for (4.3.9) is $M = (0,L)\times(r_1,r_2)$. The physically sound boundary conditions are as follows:

$$\begin{aligned} &u = (u_z,u_r,u_\theta) = 0 && \text{at } r = r_1,\, r_2,\\ &u_z = 0,\ \frac{\partial u_r}{\partial z} = \frac{\partial u_\theta}{\partial z} = 0 && \text{at } z = 0, L. \end{aligned} \tag{4.3.10}$$

We consider the narrow-gap case with $\mu \geq 0$, and assume the condition (4.2.15). Then we deduce from (4.3.6) and (4.3.7) that

$$\begin{aligned} &V' + \frac{1}{r}V = -2\alpha\Omega_1,\\ &\frac{2V}{r} = 2\Omega_1\left(1 - \frac{1-\mu}{1-\eta^2}\frac{r^2-r_1^2}{r^2}\right) \simeq 2\Omega_1\left(1 - \frac{(1-\mu)(r-r_1)}{d}\right),\\ &W \simeq -\frac{p_0}{4\rho\nu}\left[(r-r_1)(r_2-r) + W_0\right], \end{aligned}$$

where α is as in (4.2.16).

Replacing u_θ by $\sqrt{\alpha}u_\theta$ in (4.3.9), we obtain the following approximation equations describing the flow between two cylinders with a narrow gap:

$$\begin{aligned} \frac{\partial u_z}{\partial t} + (\widetilde{u}\cdot\nabla)u_z =& \nu\Delta u_z + \frac{p_0}{4\rho\nu}(d-2\overline{r})u_r\\ &+ \frac{p_0}{4\rho\nu}(W_0 + \overline{r}(d-\overline{r}))\frac{\partial u_z}{\partial z} - \frac{1}{\rho}\frac{\partial p}{\partial z},\\ \frac{\partial u_r}{\partial t} + (\widetilde{u}\cdot\nabla)u_r =& \nu\Delta u_r + 2\sqrt{\alpha}\,\Omega_1\left(1 - \frac{(1-\mu)\overline{r}}{d}\right)u_\theta\\ &+ \frac{p_0}{4\rho\nu}(W_0 + \overline{r}(d-\overline{r}))\frac{\partial u_r}{\partial z} - \frac{1}{p}\frac{\partial p}{\partial r},\\ \frac{\partial u_\theta}{\partial t} + (\widetilde{u}\cdot\nabla)u_\theta =& \nu\Delta u_\theta + 2\sqrt{\alpha}\,\Omega_1 u_r + \frac{p_0}{4\rho\nu}(W_0 + \overline{r}(d-\overline{r}))\frac{\partial u_\theta}{\partial z},\\ \operatorname{div}\widetilde{u} = 0,& \end{aligned} \tag{4.3.11}$$

where $\overline{r} = r - r_1$, $\Delta = \partial^2/\partial z^2 + \partial^2/\partial r^2$.

We need to consider the nondimensionalized form of (4.3.11). To this end, let

$$\begin{aligned} &(x,t) = (x'd, t'd^2/\nu) && (x = (z,r,r\theta)),\\ &(u,p,W_0) = (u'\nu/d, p'\rho\nu^2/d^2, W_0'd^2) && (u = (u_z,u_r,u_\theta)). \end{aligned}$$

Omitting the primes, the equations can be rewritten as

$$
\begin{aligned}
\frac{\partial u_z}{\partial t} =& \Delta u_z + \gamma(1-2\overline{r})u_r \\
& + \gamma(W_0+\overline{r}(1-\overline{r}))\frac{\partial u_z}{\partial z} - \frac{\partial p}{\partial z} - (\widetilde{u}\cdot\nabla)u_z, \\
\frac{\partial u_r}{\partial t} =& \Delta u_r + \lambda(1-(1-\mu)\overline{r})u_\theta \\
& - \gamma(W_0+\overline{r}(1-\overline{r}))\frac{\partial u_r}{\partial z} - \frac{\partial p}{\partial r} - (\widetilde{u}\cdot\nabla)u_r, \\
\frac{\partial u_\theta}{\partial t} =& \Delta u_\theta + \lambda u_r + \gamma(W_0+\overline{r}(1-\overline{r}))\frac{\partial u_\theta}{\partial z} - (\widetilde{u}\cdot\nabla)u_\theta, \\
\text{div}\, \widetilde{u} =& 0,
\end{aligned}
\tag{4.3.12}
$$

where $\lambda = \sqrt{T}$. Here the Taylor number T and the nondimensional parameter γ are given by

$$
T = \frac{4\alpha\Omega_1^2 d^4}{\nu^2}, \qquad \gamma = \frac{p_0 d^3}{4\rho\nu^2}. \tag{4.3.13}
$$

We remark here that the nondimensional parameter γ, proportional to the Reynolds number Re squared times the small gap d, is a small parameter.

When the gap $d = r_2 - r_1$ is small in comparison with the mean radius, we need not distinguish the two equations (4.3.9) and (4.3.12) to discuss their dynamic bifurcation. Therefore, we always consider the problem (4.3.12) instead of (4.3.9). Equations (4.3.12) are supplemented with the following initial condition:

$$
u = \varphi \quad \text{at} \quad t = 0. \tag{4.3.14}
$$

4.3.2 Phase Transition of the TCP Problem

Eigenvalue and Eigenvectors of Linearized Equations. The linearized equations of (4.3.12) with (4.3.10) read

$$
\begin{aligned}
& -\Delta u_z - \gamma(1-2\overline{r})u_r + \gamma(W_0+\overline{r}(1-\overline{r}))\frac{\partial u_z}{\partial z} + \frac{\partial p}{\partial z} = 0, \\
& -\Delta u_r + \lambda(1-\mu)\overline{r}u_\theta + \gamma(W_0+\overline{r}(1-\overline{r}))\frac{\partial u_r}{\partial z} + \frac{\partial p}{\partial r} = \lambda u_\theta, \\
& -\Delta u_\theta - \gamma(W_0+\overline{r}(1-\overline{r}))\frac{\partial u_\theta}{\partial z} = \lambda u_r, \\
& \frac{\partial u_z}{\partial z} + \frac{\partial u_r}{\partial r} = 0,
\end{aligned}
\tag{4.3.15}
$$

with the boundary conditions

$$
\begin{aligned}
& u = 0 && \text{at } r = r_1, r_1+1, \\
& u_z = 0, \quad \frac{\partial u_r}{\partial z} = \frac{\partial u_\theta}{\partial z} = 0 && \text{at } z = 0, L,
\end{aligned}
\tag{4.3.16}
$$

where $\lambda = \sqrt{T}$, T is the Taylor number, T and γ are given by (4.3.13), and μ is defined by (4.2.1). We define

$$\varepsilon = (\gamma, 1-\mu) \in (0,\infty) \times (0,\infty).$$

Let $\lambda_0^\varepsilon > 0$ be the first eigenvalue of (4.3.15) and (4.3.16). The critical Taylor number is given by

$$T_c = (\lambda_0^\varepsilon)^2. \tag{4.3.17}$$

When $\varepsilon \to 0$, (4.3.15) become symmetric as follows::

$$\begin{aligned} -\Delta u_z + \frac{\partial p}{\partial z} &= 0, \\ -\Delta u_r + \frac{\partial p}{\partial r} &= \lambda u_\theta, \\ -\Delta u_\theta &= \lambda u_r, \\ \frac{\partial u_z}{\partial z} + \frac{\partial u_r}{\partial r} &= 0. \end{aligned} \tag{4.3.18}$$

The eigenvalue problem (4.3.18) with (4.3.16) has the first eigenvectors

$$\psi = \left(\frac{1}{a} \sin azR'(r), -\cos azR(r), -\frac{1}{a^2\lambda_0} \cos az \left(\frac{d^2}{dz^2} - a^2 \right)^2 R(r), \right), \tag{4.3.19}$$

where $a = k\pi/L > 0$ is the wavelength for some integer $k \geq 1$ with k depending on L. When L is large, k is taken to be the minimizer of

$$\min_k \left| \frac{k\pi}{L} - 3.117 \right|.$$

The number λ_0 is the first eigenvalue of (4.3.18) with (4.3.16), and when $a \simeq 3.117$, $\lambda_0^2 \simeq 1700$.

The function $R(r)$ in (4.3.19) satisfies

$$\begin{aligned} &\left(\frac{d^2}{dz^2} - a^2 \right)^3 R = -a^2\lambda_0^2 R, \\ &R = 0, \quad R' = 0, \quad \left(\frac{d^2}{dz^2} - a^2 \right)^2 R = 0 \text{ at } r = r_1, r_1 + 1. \end{aligned}$$

Its solution can be expressed as [see Chap. II-15 of Chandrasekhar (1981)]

$$R(r) = \cos\alpha_0 x - \beta_1 \cosh\alpha_1 x + \cos\alpha_2 x + \beta_2 \sinh\alpha_1 x \sin\alpha_2 x, \tag{4.3.20}$$

where $x = \overline{r} - \frac{1}{2}$, $\overline{r} = r - r_1$,

$$\begin{aligned} &\beta_1 = 0.06151664, \quad \beta_2 = 0.10388700, \\ &\alpha_0 = 3.973639, \qquad \alpha_1 = 5.195214, \qquad \alpha_2 = 2.126096. \end{aligned} \tag{4.3.21}$$

In addition, the first eigenvalue λ_0 of (4.3.18) with (4.3.16) is simple except for the following values of L:

$$L_k = \frac{k\pi}{b}, \quad k = 1, 2, \cdots, \quad b \simeq 4.2. \tag{4.3.22}$$

Phase Transition Theorems. We are now in a position to state and prove the phase transition theorems for the Taylor–Couette–Poiseuille problem in the narrow-gap case. For this purpose, we set the function spaces as

$$\begin{aligned} H &= \{u = (\widetilde{u}, u_\theta) \in L^2(M)^3 \mid \text{div}\widetilde{u} = 0, \widetilde{u} \cdot n|_{\partial M} = 0\}, \\ H_1 &= \{u \in H^2(M, \mathbb{R}^3) \bigcap H \mid u \text{ satisfies (4.3.16)}\}. \end{aligned} \tag{4.3.23}$$

Theorem 4.3.1 *Assume the conditions (4.2.15) and (4.2.23). Then there is a $\delta > 0$ such that for every $|\gamma| < \delta$, the problem (4.3.12) with (4.3.10) has a transition at $(u, T) = (0, T_c)$, where $T = \lambda^2$ and T_c is the critical Taylor number given by (4.3.17). Moreover, we have the following:*

(1) The problem undergoes a transition to an attractor $\mathscr{A}_T$ on $T > T_c$ with $0 \notin \mathscr{A}_T$ such that the dimension of $\mathscr{A}_T$ is either zero or one, and $\mathscr{A}_T$ attracts $H \setminus \Gamma$, where Γ is the stable manifold of $u = 0$ with codimension m $(1 \leq m \leq 2)$ in H.
(2) Every $u_T \in \mathscr{A}_T$ can be expressed as

$$u_T = v_T + w_T^\varepsilon, \qquad \lim_{\varepsilon \to 0, T \to T_0} \|w_T^\varepsilon\| / \|v_T\| = 0,$$

where $T_0 = \lambda_0^2, \lambda_0$ and v_T are the first eigenvalue and eigenvector of (4.3.18) with (4.3.16).

Theorem 4.3.2 *Under the hypotheses of Theorem 4.3.1, there exist $\widetilde{L}_K \simeq L_k$, where L_k are given by (4.3.22) such that when $L \neq \widetilde{L}_k$, the following assertions hold:*

(1) The bifurcated attractor $\mathscr{A}_T$ in Theorem 4.3.1 consists of exactly two steady-state solutions $\mathscr{A}_T = \{u_1^T, u_2^T\}$ of (4.3.12) with (4.3.10) such that for $i = 1, 2$, u_i^T can be expressed as

$$\begin{aligned} u_1^T &= \alpha_1(T, \varepsilon)\psi + v_1(T, \varepsilon), \\ u_2^T &= -\alpha_2(T, \varepsilon)\psi + v_2(T, \varepsilon), \\ v_i(\lambda, \varepsilon) &= o(|\alpha_i(\lambda, \varepsilon)|). \end{aligned} \tag{4.3.24}$$

Here ψ is the first eigenvector given by (4.3.19), and

$$\begin{aligned}
&\alpha_{1,2} = \frac{\sqrt{b^2(\varepsilon)+4\beta_1(\lambda,\varepsilon)} \mp |b(\varepsilon)|}{2c} + o(|b|,|\beta_1|),\\
&\beta_1(\lambda,\varepsilon) = \sigma(\varepsilon)(\lambda-\lambda_0^\varepsilon) + o(|\lambda-\lambda_0^\varepsilon|),\\
&\sigma(0)>0,\ \lambda = T^2,\\
&b(\varepsilon) = -\int_M (\widetilde{\psi}^\varepsilon\cdot\nabla)\psi^\varepsilon\cdot\psi^{\varepsilon*}dx,
\end{aligned} \tag{4.3.25}$$

where $\psi^\varepsilon = (\widetilde{\psi}^\varepsilon, \psi_\theta^\varepsilon)$, with $\widetilde{\psi}^\varepsilon = (\psi_z^\varepsilon, \psi_r^\varepsilon)$, is the eigenvector of (4.3.15), and (4.3.16) corresponding to λ_0^ε, $\psi^{\varepsilon*}$ is the dual eigenvector, and $c>0$ a constant.

(2) *The space H can be decomposed into two disjoint open sets $H = \overline{U}_1^T + \overline{U}_2^T$ with $0 \in \Gamma = \partial U_1^T \cap \partial U_2^T$, such that for $i=1,2$, we have $u_i^T \in U_i^T$ and*

$$\lim_{t\to\infty} \|u(t,u_0) - u_i^T\| = 0 \quad \textit{for } u_0 \in U_i^T,$$

where $u(t,u_0)$ is the solution of (4.3.12) with (4.3.10) and (4.3.14).

Remark 4.3.3 It is known that [see (Kato 1995)]

$$\psi^\varepsilon \to \psi, \quad \psi^{\varepsilon*} \to \psi \quad \text{as} \quad \varepsilon \to 0,$$

where ψ is given by (4.3.19). Consequently, the number $b(\varepsilon)$ in (4.3.25) satisfies

$$b(\varepsilon) \to 0 \quad \text{as} \quad \varepsilon \to 0.$$

Remark 4.3.4 The transition theorems (Theorems 4.3.1 and 4.3.2) provide three dynamical ingredients different from any classical bifurcation theories as follows:

(1) The basin of attraction of the bifurcated attractor $\mathscr{A}_T$. As seen in assertion (1) of Theorem 4.3.1, $\mathscr{A}_T$ attracts $H\setminus\Gamma$, i.e., the stable domain of the TCP flow for the external perturbation is almost the entire phase space.
(2) The explicit expressions (4.3.24) and (4.3.25) of the bifurcated solutions. These expressions are crucial to determine, in the next two subsections, the asymptotic structure in physical space of solutions of the problem, leading to the justification of boundary-layer and interior separation for the TCP flow.
(3) Global transition structure. We can see that the explicit forms (4.3.24) and (4.3.25) provide a precise graph of the saddle-node bifurcation. In fact, if we take $\Gamma_i^\lambda = \{u_i^T\}$ with $\lambda = \sqrt{T}$, where u_i^T $(i=1,2)$ are as in (4.3.24) and (4.3.25), then Γ_i^λ $(i=1,2)$ depict in $\lambda - H$ space a graph as shown in Fig. 4.19. Here λ_0^ε represents λ_0, and $\lambda^*(\lambda^* < \lambda_0^\varepsilon)$ satisfies

$$b^2(\varepsilon) + 4\beta_1(\lambda^*,\varepsilon) = 0, \quad \lambda^* = \sqrt{T^*}. \tag{4.3.26}$$

Proof of Theorems 4.3.1 and 4.3.2. We shall apply Theorems 2.6.1 and 2.6.6 to prove these two theorems. Let H and H_1 be defined by (4.3.23). We define the mappings $L_\lambda = A + \lambda B : H_1 \to H$, $S_\lambda^\varepsilon : H_1 \to H$, and $G : H_1 \to H$ by

$$Au = \{P\Delta\widetilde{u}, \Delta u_\theta\}, \qquad Bu = \{P(0,u_\theta), u_r\},$$

$$S^{\varepsilon}_{\lambda}u = \left\{P\left(\gamma(1-2\overline{r})u_r + \gamma(W_0+\overline{r}(1-\overline{r}))\frac{\partial u_z}{\partial z}, \lambda(1-\mu)\overline{r}u_\theta \right.\right.$$
$$\left.\left. + \gamma(W_0+\overline{r}(1-\overline{r}(1-\overline{r}))\frac{\partial u_r}{\partial z}\right), \gamma(W_0+\overline{r}(1-\overline{r}))\frac{\partial u_\theta}{\partial z}\right\},$$

$$G(u) = \{-P(\widetilde{u}\cdot\nabla)\widetilde{u}, -(\widetilde{u}\cdot\nabla)u_\theta\},$$

where $u = (\widetilde{u}, u_\theta) \in H_1$ with $\widetilde{u} = (u_z, u_r)$, and the operator $P : L^2(M)^3 \to H$ is the Leray projection. Thus Eqs. (4.3.12) with (4.3.10) can be written in the abstract form

$$\frac{du}{dt} = L_\lambda u + S^{\varepsilon}_{\lambda}u + G(u).$$

It is well known that the operators $L_\lambda, S^{\varepsilon}_{\lambda}$, and G satisfy the conditions (2.1.2), (2.1.3), and (2.6.2) in Theorems 2.6.1 and 2.6.6. It is clear that L_λ is symmetric, and the condition (2.6.18) in Theorem 2.6.6 holds.

It is clear that the eigenvalues $\beta_k(\lambda)$ $(k = 1, 2, \cdots)$ of the equation

$$L_\lambda u = \beta(\lambda)u$$

satisfy the conditions (2.2.11) and (2.2.12) in Theorems 2.6.1 and 2.6.6. In particular, by Theorem 3.16 in Ma and Wang (2005b), we see that $u = 0$ is a globally asymptotically stable steady-state solution of the following equation:

$$\frac{du}{dt} = L_\lambda u + G(u).$$

Thus Theorem 4.3.1 follows from Theorem 2.6.1, and Theorem 4.3.2 follows from Theorem 2.6.6. We note that the expression (4.3.25) is derived from (2.6.16) in the proof of Theorems 2.6.3 and 2.6.6.

The proofs of Theorems 4.3.1 and 4.3.2 are complete. □

4.3.3 Boundary-Layer Separation from the Couette–Poiseuille Flow

In this and the following subsections, we always assume that the conditions in Theorem 4.3.2 hold. The main objective of these two subsections is to study the asymptotic structure and its transition in physical space in the CP and TCP flow regimes.

We shall see that the results obtained here provide a rigorous characterization on how, when, and where Taylor vortices are generated. In particular, contrary to what is commonly believed, we show that the propagating Taylor vortices (PTV) do not appear after the first phase transition, and they appear only after further increase in the Taylor number, so that boundary-layer or interior separation occurs.

We shall prove that the type of separation (boundary-layer or interior) is dictated by the (vertical) throughflow driven by a pressure gradient along the rotating axis. Hence we consider two cases. The first is the case whereby the throughflow vanishes at the cylinder walls, i.e., $W_0 = 0$ in (4.3.5). This leads to boundary-layer separation, and will be studied in this subsection.

The second is the case $W_0 \neq 0$, leading to interior separations. This case will be addressed in the next subsection.

We now consider the case $W_0 = 0$, and the main results are the following two theorems. The first gives a precise characterization on how, when, and where propagating Taylor vortices (PTV) are generated for the bifurcated solutions, and the second theorem describes time evolution to the PTV for the solutions of the equations.

Theorem 4.3.5 *Assume that the conditions in Theorem 4.3.2 hold, and assume that the constant velocity in boundary condition (4.3.5) is zero, i.e., $W_0 = 0$. Then there is a $\gamma_0 > 0$ such that for every $0 < \gamma < \gamma_0$, where γ is defined by (4.3.13), the following assertions hold for the two (bifurcated) steady-state solutions: $v_i^T = \mathscr{U}^{cp} + u_i^T$ $(i = 1,2)$ of (4.2.5) with (4.3.5), where $\mathscr{U}^{cp}$ and u_i^T are given by (4.3.1) and (4.3.24):*

1. *For v_1^T, there is a number $T_1 > T_c$ [T_c as in (4.3.17)] such that the following hold:*
   - *(a) For $T_c < T < T_1$, the vector field $\widetilde{v}_1^T = (W + u_{1z}^T, u_{1r}^T)$ is topologically equivalent to the vertical shear flow $(W,0)$ as shown in Fig. 4.20.*
   - *(b) For $T_1 < T$, there is a unique center (vortex) of $\widetilde{v}_1^T$ separated from each point $(z_n, r_1) \in \partial M$, where $z_n \simeq (4n+1)\pi/2a$ (respectively $(\overline{z}_n, r_2) \in \partial M, \overline{z}_n = (4n+3)\pi/2a$) for $n = 0,1,\cdots,k_0$, as shown in Fig. 4.21, and $a = k\pi/L$ as in (4.3.19).*

2. *For v_2^T, there is a number $b_0 > 0$ such that only one of the following two assertions holds:*
   - *(a) If $|b(\varepsilon)| < b_0$ [$b(\varepsilon)$ as in (4.3.25)], there is $T_2 > T_c$ such that*
     - *(i) for $T_c < T < T_2$, the vector field $\widetilde{v}_2^T = (W + u_{2z}^T, u_{2r}^T)$ has the topological structure as shown in Fig. 4.20, and*
     - *(ii) for $T_2 < T$, there is a unique center of $\widetilde{v}_2^T$ separated from each point $(\overline{z}_n, r_1) \in \partial M$ (respectively $(z_n, r_2) \in \partial M$) for $n = 0,1,\cdots,k_0$, as shown in Fig. 4.21.*
   - *(b) If $|b(\varepsilon)| > b_0$, then for $T > T_c$, the vector field $\widetilde{v}_2^T$ is topologically equivalent to the structure shown in Fig. 4.22.*

3. *There exists $T_3 > T_1$ $(T_1 \geq T_2)$ such that for $T_i < T < T_3$, the vector field $\widetilde{v}_i^T$ $(i = 1,2)$ is topologically equivalent to the structure shown in Fig. 4.22, and T_3 is independent of γ.*

From Theorems 4.3.2 and 4.3.5, together with the structural stability theorem, Theorem 2.1.2 in Ma and Wang (2005d), we immediately derive the following theorem, which links the dynamics with the structure in physical space.

Theorem 4.3.6 *Assume that the conditions in Theorem 4.3.2 hold, $W_0 = 0$ in (4.3.5), and $0 < \gamma < \gamma_0$. Then there exist $\widetilde{T}_j$ $(1 \le j \le 3)$ with $\widetilde{T}_1 = T_1$, $T^* \le \widetilde{T}_2 \le T_2$, $\widetilde{T}_3 = T_3$, where T_j $(1 \le j \le 3)$ are as in Theorem 4.3.5 and T^* satisfies (4.3.26), such that for each $T < \widetilde{T}_3$, the space H can be decomposed into two disjoint open sets $H = \overline{U}_1^T + \overline{U}_2^T$ with $0 \in \partial U_1^T \bigcap \partial U_2^T$, and the following assertions hold:*

(1) For $u_0 \in U_i^T$, there is a time $t_0 \ge 0$ such that if $T < \widetilde{T}_i$ $(i = 1,2)$ for the solution $u(t,u_0) = (\widetilde{u}, u_\theta)$ of (4.2.5) with (4.3.5) and (4.2.104), then the vector field $\widetilde{u}$ is topologically equivalent to the structure shown in Fig. 4.20, and if $\widetilde{T}_i < T < \widetilde{T}_3$, then $\widetilde{u}$ is topologically equivalent to the structure shown in Fig. 4.22 for all $t > t_0$.

(2) In particular, for the initial value $u_0 \in U_i^T$ near the Couette–Poiseuille flow (4.3.1), if $T > \widetilde{T}_i$, then there is a time $t_0 > 0$ such that the solution $u(t,u_0)$ of (4.2.5) with (4.3.5) and (4.3.14) has a boundary-layer separation at $t = t_0$, i.e., the vector field $\widetilde{u}$ of $u(t,u_0)$ is topologically equivalent to the structure shown in Fig. 4.20 for $t < t_0$, and $\widetilde{u}$ is topologically equivalent to the structure near the boundary ∂M shown in Fig. 4.21 for $t_0 < t$.

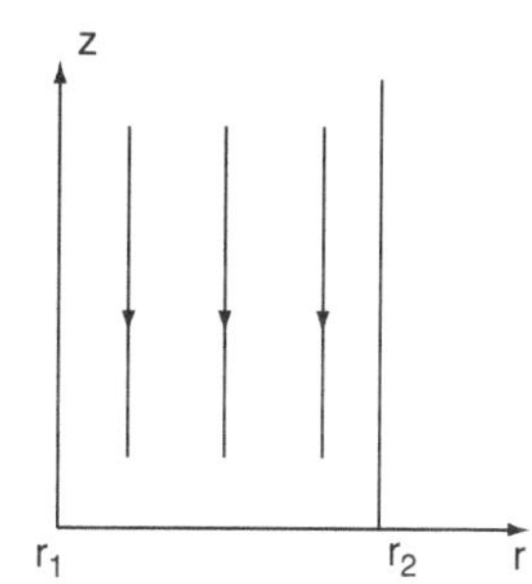

Fig. 4.20 Vertical shear flow

Remark 4.3.7 The critical values $\widetilde{T}_i$ $(i = 1,2)$ in Theorem 4.3.6 that determine the boundary-layer separation for the Taylor–Couette–Poiseuille problem depend continuously on the nondimensional parameter γ given by (4.3.13). The expansion is as follows:

$$\widetilde{T}_i = \alpha_i \gamma + o(\gamma), \quad \alpha_i > 0 \text{ a constant, } i = 1,2.$$

In addition, the condition $0 < \gamma < \gamma_0$ in the above theorems is a consequence of the narrow-gap assumption (i.e., $d = r_2 - r_1$ is small).

Remark 4.3.8 The height L is the product of an integer and the length π/a, i.e., $L = K \cdot \frac{\pi}{a}$ for some integer K. Hence the number k_0 of vortices separated from $r = r_1$ and r_2 in Theorems 4.3.5 and 4.3.6 satisfies

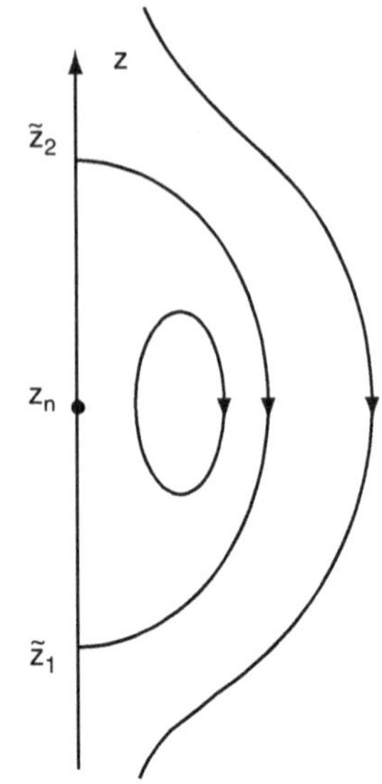

Fig. 4.21 Boundary separation and reattachment

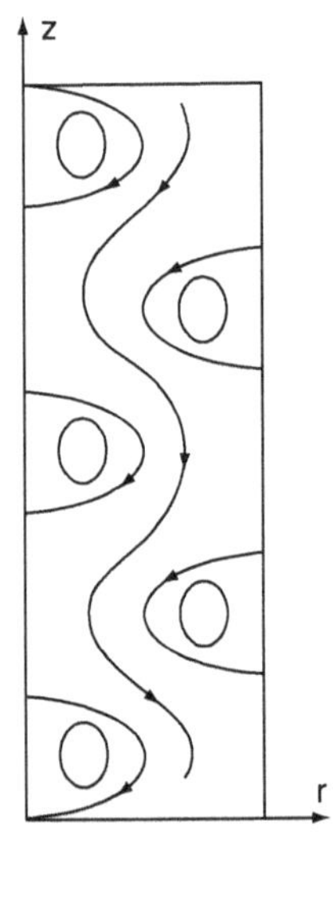

Fig. 4.22 Propagating Taylor vortices after boundary-layer separations

$$k_0 = \begin{cases} \left[\dfrac{K+1}{2}\right] & \text{for } r = r_1 \text{ (or } r = r_2\text{)}, \\ \left[\dfrac{K}{2}\right] & \text{for } r = r_2 \text{ (or } r = r_1\text{)}, \end{cases}$$

where $[\alpha]$ is the integer part of α.

Proof of Theorem 4.3.5. We proceed in a few steps as follows.

Step 1. By Theorem 4.3.2, as the Taylor number $T = \lambda^2$ satisfies $T_c < T$. Equations (4.2.5) with (4.3.5) generate from the basic flow (4.3.1) two steady-state solutions

$$v_i^T = (W(r) + u_{iz}^T, u_{ir}^T, V(r) + u_{i\theta}^T), \quad i = 1,2,$$

which are globally asymptotically stable under axisymmetric perturbation.

As seen in Remark 4.3.3, the number $b(\varepsilon)$ in (4.3.25) satisfies

$$\lim_{\varepsilon \to 0} b(\varepsilon) = 0 \quad (\varepsilon = (1-\mu, r)).$$

Hence, to prove this theorem, it suffices to consider only the following two vector fields:

$$\widetilde{V}_1^T = (W(r), 0) + \frac{\sqrt{b^2 + 4\alpha(\lambda - \lambda_0^\varepsilon)} - b}{2c}\widetilde{u}_0, \tag{4.3.27}$$

$$\widetilde{V}_2^T = (W(r), 0) + \frac{\sqrt{b^2 + 4\alpha(\lambda - \lambda_0^\varepsilon)} + b}{2c}\widetilde{u}_0, \tag{4.3.28}$$

where $\lambda = \sqrt{T}$, $b = |b(\varepsilon)|, \alpha = \alpha(\varepsilon) > 0$, and $c > 0$ is a constant. In addition,

$$\begin{aligned} W(r) &= -\gamma\bar{r}(1-\bar{r}) && (\bar{r} = r - r_1), \\ \widetilde{u}_0 &= \left(\frac{1}{a}\sin azR'(r), -\cos azR(r)\right), \end{aligned}$$

and $R(r)$ is given by (4.3.20).

Step 2. Claim: $\widetilde{V}_1^T$ has a boundary-layer separation on $T_1 < T$ for some $T_1 >$ $T_c = (\lambda_0^\varepsilon)^2$. We shall apply the structural bifurcation theorems, Theorems 5.3.6 and 5.3.7 in Ma and Wang (2005d), to prove this claim. To this end, let

$$\Lambda = \left(\sqrt{b^2 + 4\alpha(\lambda - \lambda_0^\varepsilon)} - b\right)/2ac,$$

which is an increasing function of T with $\Lambda = 0$ at $T_c = (\lambda_0^\varepsilon)^2$. Thus, the vector field (4.3.27) can be rewritten as

$$\widetilde{V}_1^T = (\Lambda \sin azR'(r) - \gamma\bar{r}(1-\bar{r}), -a\Lambda\cos azR(r)). \tag{4.3.29}$$

We discuss the structural bifurcation of (4.3.29) only on the portion $r = r_1$ of ∂M; the bifurcation on the portion $r = r_2$ can be proved in the same fashion.

Since $R(r_1) = R'(r_1) = 0$, the zero points of $\partial\widetilde{V}_1^T/\partial n = \partial\widetilde{V}_1^T/\partial r$ on $r = r_1$ are given by the solutions of the equation

$$\Lambda \sin azR''(r_1) - \gamma = 0. \tag{4.3.30}$$

Let $\Lambda_0 = \gamma/R''(r_1)$. Then we infer from (4.3.30) that for $r = r_1$,

$$\frac{\partial \widetilde{V}_1^T}{\partial n} \begin{cases} \neq 0 & \text{if } \Lambda < \Lambda_0 \\ = 0 & \text{if } \Lambda = \Lambda_0 \text{ and } z_n = \dfrac{(4n+1)}{2a}\pi. \end{cases} \tag{4.3.31}$$

Therefore, $\widetilde{V}_1^T$ has a Taylor expansion at $\Lambda = \Lambda_0$ as follows:

$$\begin{aligned} \widetilde{V}_1^T &= V^0 + (\Lambda - \Lambda_0)V^1, \\ V^0 &= (V_z^0, V_r^0) = (\Lambda_0 \sin azR' - \gamma\overline{r}(1-\overline{r}), -a\Lambda \cos azR) \\ V^1 &= (V_z^1, V_r^1) = (\sin azR', -a\cos azR). \end{aligned} \tag{4.3.32}$$

We derive from (4.3.20) that

$$\begin{aligned} R''(r) = &-\alpha_0^2 \cos\alpha_0 x + \left[\beta_1(\alpha_2^2 - \alpha_1^2) + 2\beta_2\alpha_1\alpha_2\right]\cosh(\alpha_1 x)\cos\alpha_2 x \\ &+ \left[\beta_2(\alpha_1^2 - \alpha_2^2) + 2\beta_1\alpha_1\alpha_2\right]\sinh(\alpha_1 x)\sin\alpha_2 x, \end{aligned} \tag{4.3.33}$$

where $x = \overline{r} - \frac{1}{2}$. Inserting (4.3.21) into (4.3.33), we obtain

$$R''(r_1) \simeq 8 + 1.7e^{5/2} \simeq 28. \tag{4.3.34}$$

Thus we derive from (4.3.31), (4.3.32), and (4.3.34) that at $(z_n, r_1) = (4n + 1/(2a)\pi, r_1)$,

$$\begin{aligned} &\frac{\partial V^0}{\partial n} = \frac{\partial V^0}{\partial r} = 0, \\ &\frac{\partial V^1}{\partial n} = \frac{\partial V^1}{\partial r} \simeq (28, 0), \\ &\frac{\partial^3 V_z^0}{\partial \tau^2 \partial n} = \frac{\partial^3 V_z^0}{\partial z^2 \partial r} \simeq -28a^2\Lambda_0. \end{aligned}$$

Namely, the conditions (5.3.4), (5.3.6), and (5.3.7) in Theorems 5.3.6 and 5.3.7 in Ma and Wang (2005d) are valid.

Now we verify the condition (5.3.5) in (Ma and Wang 2005d), i.e., we need to prove that the index of the vector field $\partial\widetilde{V}_1^T/\partial r$ vanishes at $\Lambda = \Lambda_0$ and (z_n, r_1):

$$\text{ind}\left(\frac{\partial V^{\Lambda_0}}{\partial r}, (z_n, r_1)\right) = 0. \tag{4.3.35}$$

To this end, we consider the equation

$$\frac{\partial V_z^T}{\partial r} = \Lambda \sin azR''(r) - \gamma(1 - 2\overline{r}) = 0. \tag{4.3.36}$$

The function $R''(r)$ has a Taylor expansion at $r = r_1$ as follows:

$$R''(r) = R''(r_1) + R'''(r_1)\overline{r} + o(|\overline{r}|). \tag{4.3.37}$$

We deduce from (4.3.20) and (4.3.21) that

$$R'''(r_1) \simeq -64 \times \frac{1.7}{2} - 10 \times e^{5/2} \simeq -175. \tag{4.3.38}$$

On the other hand, we note that

$$\Lambda = \frac{\gamma\Lambda}{R''(r_1)\Lambda_0} = \frac{\gamma\Lambda}{28\Lambda_0}.$$

Thus, by (4.3.34), (4.3.37), and (4.3.38), Eq. (4.3.36) near $(z,r) = (z_n, r_1)$ can be rewritten in the following form:

$$\left(1 - \frac{\Lambda}{\Lambda_0}\sin az\right) + \left(\frac{175\Lambda \sin az}{28\Lambda_0} - 2\right)\overline{r} + o(\overline{r}) = 0. \tag{4.3.39}$$

Obviously, there is a $\delta > 0$ such that Eq. (4.3.39) has no solutions in $0 < r - r_1 = \overline{r} < \delta, |z - z_n| < \delta$, and $\Lambda < \Lambda_0$. Namely, in a neighborhood of (z_n, r_1), the vector field $\partial \widetilde{V}_1^T/\partial r$ has no singular points for $\Lambda < \Lambda_0$. Hence, by the invariance of the index sum of singular points in a domain, one can derive (4.3.35).

Let T_1 be the number such that $\Lambda(T_1) = \Lambda_0$. It is clear that $T_1 > T_c = (\lambda_0^\varepsilon)^2$. Thus the claim follows from Theorems 5.3.6 and 5.3.7 in Ma and Wang (2005d).

Step 3. The vector field $\widetilde{V}_1^T$ has no singular points in M for $T_c < T < T_1$, and has only one singular point in each domain $(z_k, z_{k+1}) \times (r_1, r_2)$, for $T > T_1$, where $z_k = \frac{k\pi}{a}\ (0 \leq k \leq K = \frac{aL}{\pi} - 1)$.

From (4.3.29), we find that the singular points of $\widetilde{V}_1^T$ must be in the lines $l_k = \{(\frac{k\pi}{a} + \frac{\pi}{2a}, r) \mid r_1 < r < r_2\} \subset M$, and

$$\widetilde{V}_{1r}^T = 0, \quad \widetilde{V}_{1z}^T = (-1)^k \Lambda R'(r) - \gamma\overline{r}(1-\overline{r}) \quad \text{on} \quad l_k.$$

Therefore, to prove this claim we have only to consider

$$(-1)^k \frac{R'(r)}{R''(r_1)} - \frac{\Lambda_0}{\Lambda}\overline{r}(1-\overline{r}) = 0, \quad 0 < \overline{r} < 1. \tag{4.3.40}$$

It is known that $R'(r)$ is odd and $\overline{r}(1-\overline{r})$ is an even function of the variable $x = \overline{r} - \frac{1}{2}$. These functions are illustrated in Fig. 4.23.

Direct computations show that (4.3.40) has no zero points in $0 < \overline{r} < 1$ for $\Lambda \leq \Lambda_0$, and there is a unique zero point $\overline{r}_0 \in (0,1)$ for all $\Lambda_0 < \Lambda$. Furthermore, $0 < \overline{r}_0 < \frac{1}{2}$ if k is even, and $\frac{1}{2} < \overline{r}_0 < 1$ if k is odd. The claim is proved.

Finally, by Steps 1–3 one readily derives assertion (1) of the theorem. For the vector field $\widetilde{V}_2^T$ given by (4.3.28), we can also prove assertion (2) in the same fashion. Obviously, assertion (3) holds.

The proof of the theorem is complete. □

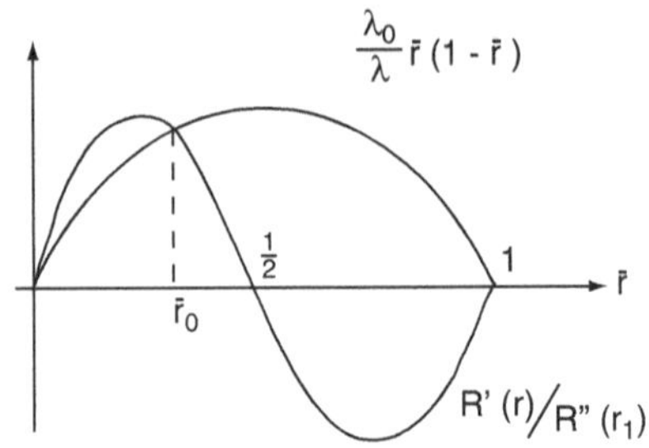

Fig. 4.23 $\bar{r}_0$ is a zero point of (4.3.40)

4.3.4 Interior Separation from the Couette–Poiseuille Flow

In this subsection, we study the structural transition of Couette–Poiseuille flow for the case in which the z-direction boundary velocity W_0 in nonzero in (4.3.5). We show that the secondary flow from the Couette–Poiseuille flow (4.3.1) will have interior separation as the Taylor number T exceeds some critical value $T_1\,(> T_c)$ that is an increasing function of the nondimensional parameter $\gamma = p_0 d^3/4\rho\nu$.

Let the constant velocity W_0 in the boundary condition (4.3.5) be nonzero. Then we have the following interior separation theorems for the TCP flow problem. The results obtained are in agreement with the numerical results obtained by Raguin and Georgiadis; see, e.g., Fig. 8 in Raguin and Georgiadis (2004).

Theorem 4.3.9 *Assume that the conditions in Theorem 4.3.2 hold, and* $W_0 \neq 0$ *in (4.3.5). Then there exists* $\gamma_0 > 0$ *such that if* $0 < \gamma < \gamma_0$, *then for the two (bifurcated) steady-state solutions* $v_i^T = \mathscr{U}^{cp} + u_i^T$ $(i = 1,2)$ *of (4.2.5) with (4.3.5), the following assertions hold:*

(1) For v_1^T, *there is* $T_1 > T_c$ *such that*

(a) for $T_c < T < T_1$, *the vector field* $\widetilde{v}_1^T = (W + u_{1z}^T, u_{1r}^T)$ *is topologically equivalent to the structure shown in Fig. 4.24, and*

(b) for $T_1 < T$, *there is exactly one pair of center and saddle points separated from a point in each domain* $(\widetilde{z}_k, \widetilde{z}_{k+1}) \times (r_1, r_2) \subset M$, *with* $\widetilde{z}_k = \frac{k\pi}{a}$ $(0 \leq k \leq K)$, *as shown in Fig. 4.25.*

(2) For v_2^T, *there is a number* $b_0 > 0$ *such that only one of the following two assertions holds:*

(a) If $|b(\varepsilon)| < b_0$, *where* $b(\varepsilon)$ *is given by (4.3.25), then there is a* $T_2 > T_c$ $(T_1 \geq T_2)$ *such that the same conclusions as assertion (1) hold for* v_2^T *with* T_2 *replacing* T_1.

(b) If $|b(\varepsilon)| > b_0$, *then if* $T > T_c$, *the vector field* $\widetilde{v}_2^T = (W + u_{2z}^T, u_{2r}^T)$ *is topologically equivalent to the structure shown in Fig. 4.26.*

(3) There exists $T_3 > T_1$ $(T_1 \geq T_2)$, *with* T_3 *independent of* γ, *such that if* $T_i < T < T_3$, *the vector field* $\widetilde{v}_i^T$ $(i = 1,2)$ *is topologically equivalent to the structure shown in Fig. 4.26.*

The following theorem is a direct corollary of Theorems 4.3.2 and 4.3.9 and the structural stability theorem, Theorem 2.1.2 in Ma and Wang (2005d). It provides a link between the dynamics and the structure in physical space.

Theorem 4.3.10 *Assume that the conditions in Theorem 4.3.2 hold,* $W_0 \neq 0$ *in (4.3.5), and* $0 < \gamma < \gamma_0$. *Then there exist* $\widetilde{T}_j$ $(1 \leq j \leq 3)$ *with* $\widetilde{T}_1 = T_1, T^* \leq \widetilde{T}_2 \leq T_2, \widetilde{T}_3 = T_3$, *and for each* $T < \widetilde{T}_3$, *the space H can be decomposed into two disjoint open sets* $H = \overline{U}_1^T + \overline{U}_2^T, 0 \in \partial U_1^T \bigcap \partial U_2^T$ *such that the following assertions hold:*

(1) For $u_0 \in U_i^T$, *there is a time* $t_0 > 0$ *such that when* $t > t_0$, *for the solution* $u(t, u_0) = (\widetilde{u}, u_\theta)$ *of (4.2.5) with (4.3.5) and (4.3.14), if* $T < \widetilde{T}_i$, *the vector field* $\widetilde{u}$ *is topologically equivalent to the structure shown in Fig. 4.24, and if* $\widetilde{T}_i < T < \widetilde{T}_3$ $(i = 1, 2)$, $\widetilde{u}$ *is topologically equivalent to the structure shown in Fig. 4.26.*
(2) In particular, for the initial value $u_0 \in U_i^T$ *near the Couette–Poiseuille flow (4.3.1), and for* $T > \widetilde{T}_i$ $(i = 1, 2)$, *there is a time* $t_0 > 0$ *such that for the solution* $u(t, u_0) = (\widetilde{u}, u_\theta)$ *of (4.2.5) with (4.3.5) and (4.3.14),* $\widetilde{u}$ *has an interior separation at* $t = t_0$ *from a point in each domain* $(\widetilde{z}_k, \widetilde{z}_{k+1}) \times (r_1, r_2) \subset M$ *with* $\widetilde{z}_k = k\pi/a$, *i.e.,* $\widetilde{u}$ *is topologically equivalent to the structure shown in Fig. 4.24 for* $t < t_0$, *and equivalent to the structure shown in Fig. 4.26 for* $t > t_0$.

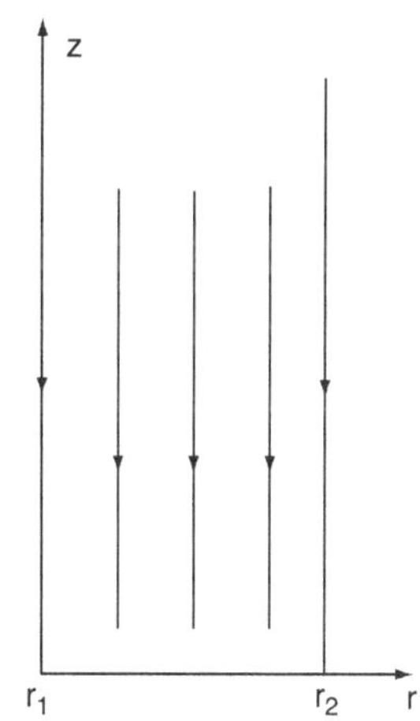

Fig. 4.24 Vertical shear flow with a constant velocity $W_0 \neq 0$ on the boundary

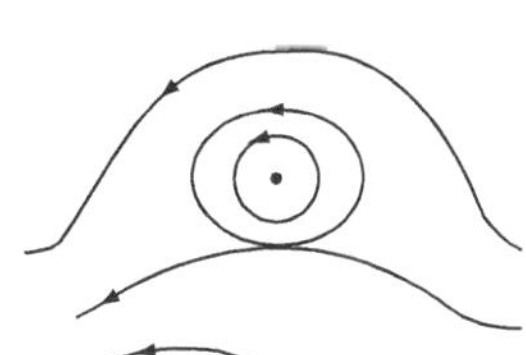

Fig. 4.25 Interior separation

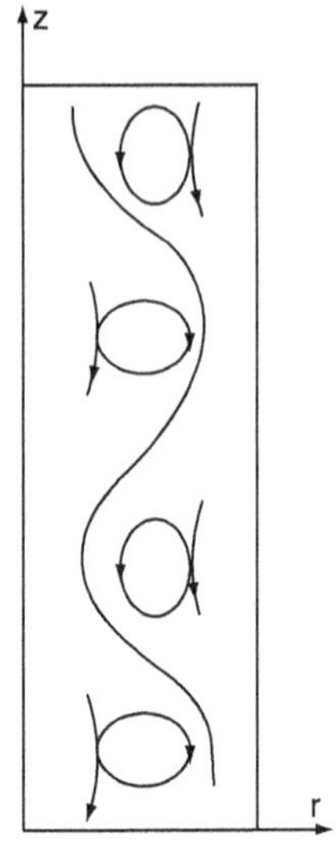

Fig. 4.26 Propagating Taylor vortices separated from the interior

Proof of Theorem 4.3.9. The proof is similar to that of Theorem 4.3.5. Here we need to prove only the interior structural bifurcation in assertion (1) for the following vector field:

$$\widetilde{V}_1^T = (\Lambda \sin azR'(r) - \gamma\overline{r}(1-\overline{r}) - \gamma W_0,\ -a\Lambda \cos azR(r)). \tag{4.3.41}$$

We proceed by applying Theorems 5.6.9 and 5.6.10 in Ma and Wang (2005d). It is clear that the zero points of $\widetilde{V}_1^T$ must be in the lines

$$l_k = \{(z_k, r) |\ z_k = \frac{k\pi}{a} + \frac{\pi}{2a},\ r_1 < r < r_2\}.$$

Hence for the solutions r_0 of the equations

$$\pm \Lambda R'(r) - \gamma\overline{r}(1-\overline{r}) - \gamma W_0 = 0, \tag{4.3.42}$$

we can obtain the singular points $(z_k, r_0) \in l_k$ of $\widetilde{V}_1^T$.

In addition, we see that for given γ $(0 < \gamma < \gamma_0)$ and $W_0 > 0$, there exists $\Lambda_0 > 0$ such that when $\Lambda < \Lambda_0$, (4.3.42) has no solutions, and when $\Lambda = \Lambda_0$, (4.3.42) has exactly two solutions $r_0^{\pm}$. In particular, $r_0^{\pm}$ are the extreme points of the functions

$$f^{\pm}(r) = \pm\Lambda_0 R'(r) - \gamma\overline{r}(1-\overline{r}) - \gamma W_0.$$

Therefore, to find the structural bifurcation points of $\widetilde{V}_1^T$, consider the equations

$$\begin{aligned} &\pm \Lambda R'(r) - \gamma\overline{r}(1-\overline{r}) - \gamma W_0 = 0, \\ &\pm \Lambda R''(r) - \gamma(1-2\overline{r}) = 0. \end{aligned} \tag{4.3.43}$$

By a direct computation, we obtain from (4.3.20) the two unique solutions $(\Lambda_0, r_0^{\pm})$ of (4.3.43) that satisfy

$$r_0^+ - r_1 = r_2 - r_0^-, \quad r_0^+ < r_*^+ < r_*^- < r_0^-,$$

where $r_*^{\pm}$ are the extreme points of $R'(r)$, i.e.,

$$R''(r_*^{\pm}) = 0 \quad (r_*^+ \simeq r_1 + \frac{1}{5}, r_*^- \simeq r_1 + \frac{4}{5}). \tag{4.3.44}$$

The vector field (4.3.41) has the following expression at $\Lambda = \Lambda_0$:

$$\begin{aligned}
\widetilde{V}_1^T &= V^0 + (\Lambda - \Lambda_0)V^1,\\
V^0 &= (V_z^0, V_r^0) = (\Lambda_0 \sin azR' - \gamma\bar{r}(1-\bar{r}) - \gamma W_0, -a\Lambda_0 \cos azR),\\
V^1 &= (V_z^1, V_r^1) = (\sin azR', -a\cos azR).
\end{aligned}$$

It is easy to see that for the solutions $(\Lambda_0, r_0^{\pm})$ of (4.3.38), the points (z_{2k}, r_0^+) and $(z_{2k+1}, r_0^-) \in M$ are singular points of V^0, where $z_m = \frac{m\pi}{a} + \frac{\pi}{2a}$.

For simplicity, we consider the structural bifurcation of $\widetilde{V}_1^T$ only at the singular point $(z_0, r_0) = (\frac{\pi}{2a}, r_0^+)$; the proof for the other singular points is the same.

In the previous discussion, we have seen that if $\Lambda < \Lambda_0$, then the vector field $\widetilde{V}_1^T$ given by (4.3.41) has no singular point in M, and (z_0, r_0) is an isolated singular point of $\widetilde{V}_1^T$ at $\Lambda = \Lambda_0$. Therefore, we have

$$\text{ind}(V^0, (z_0, r_0)) = 0. \tag{4.3.45}$$

In addition, we see that

$$DV^0(z_0, r_0) = \begin{pmatrix} \dfrac{\partial V_z^0}{\partial z} & \dfrac{\partial V_z^0}{\partial r} \\ \dfrac{\partial V_r^0}{\partial z} & \dfrac{\partial V_r^0}{\partial r} \end{pmatrix}_{(z_0, r_0)} = \begin{pmatrix} 0 & 0 \\ a^2\Lambda_0 R(r_0) & 0 \end{pmatrix} \neq 0, \tag{4.3.46}$$

and the eigenvectors e_1 and e_2 satisfying

$$\begin{aligned}
&DV^0(z_0, r_0)e_2 = \alpha e_1 \quad (\alpha = a^2\Lambda_0 R(r_0)),\\
&DV^0(z_0, r_0)e_1 = 0,
\end{aligned}$$

are given by $e_1 = (0,1), \quad e_2 = (1,0)$. Hence we have

$$V^1(z_0, r_0) \cdot e_2 = V_z^1(z_0, r_0) = R'(r_0) \neq 0. \tag{4.3.47}$$

From (4.3.45)–(4.3.47), we find that the conditions (5.6.16)–(5.6.18) in Theorems 5.6.9 and 5.6.10 in Ma and Wang (2005d) hold for the vector field $\widetilde{V}_1^T$ at $\Lambda = \Lambda_0$ and $(z, r) = (z_0, r_0)$.

Finally, we verify condition (5.6.19) in Ma and Wang (2005d). By (4.3.43), we have

$$\left.\frac{\partial^k (V^0 \cdot e_2)}{\partial e_1^k}\right|_{(z_0,r_0)} = \frac{\partial^k V_z^0(z_0,r_0)}{\partial r^k} = 0 \quad \text{for } k=0,1, \tag{4.3.48}$$

$$\frac{\partial^2 V_z^0(z_0,r_0)}{\partial r^2} = \Lambda_0 R'''(r_0) + 2\gamma. \tag{4.3.49}$$

By (4.3.43), we see that

$$\frac{\gamma}{\Lambda_0} = \frac{R''(r_0)}{1-2(r_0-r_1)}. \tag{4.3.50}$$

On the other hand, it is easy to check that

$$-\frac{R'''(r)}{R''(r)} > \frac{2}{1-2(r-r_1)} \quad \forall r_1 \leq r < r_*^+, \tag{4.3.51}$$

where r_*^+ satisfies (4.3.44). It follows from (4.3.49)–(4.3.51) that

$$\frac{\partial^2 V_z^0(z_0,r_0)}{\partial r^2} \neq 0. \tag{4.3.52}$$

Thus we derive from (4.3.48) and (4.3.52) the condition (5.6.19) in Ma and Wang (2005d) with $m=2$. By the interior separation theorem of incompressible flows, Theorem 5.6.10 in Ma and Wang (2005d), the vector field $\widetilde{V}_1^T$ has an interior separation from each point (y_m, r_0) at $\Lambda = \Lambda_0$, where $r_0 = r_0^+$ for $y_m = \frac{2k\pi}{a} + \frac{\pi}{2a}$, and $r_0 = r_0^-$ for $y_m = \frac{(2k+1)\pi}{a} + \frac{\pi}{2a}$.

The proof of the theorem is complete. □

4.3.5 *Nature of Boundary-Layer and Interior Separations*

The boundary-layer and interior separations of incompressible viscous fluid flow play a fundamental role in many hydrodynamic problems, and often lead to more complicated turbulent behaviors. Hence it is very important to understand both the dynamic and kinematic aspects of the nature of the flow's separations from the boundary and the interior of the fluid. These results established in this section give a basic example in fluid dynamics to show when, where, and how the separations occur. Moreover, they also provide a link between the dynamics and the kinematic theories of the boundary-layer and interior separation developed by the authors, in collaboration in part with Ghil; see (Ma and Wang 2005d; Ghil et al. 2001, 2005; Ma and Wang 2001, 2004a,c).

Theorems 4.3.5-4.3.10 reveal the important phenomenon that the structural transitions of a fluid flow undergo in general two bifurcations: the first is the dynamic bifurcation in the hydrodynamic equations, and the second is the structural bifurcation for the bifurcated solutions of the dynamical equations. Meanwhile, the proofs of these theorems also exhibit the mechanism of boundary-layer and interior separations generated by the two types of bifurcations. In the following, we shall briefly illustrate this mechanism in the context of the Navier–Stokes equations.

Consider the Navier–Stokes equations

$$\begin{aligned} &\frac{\partial u}{\partial t}+(u\cdot\nabla)u=\mu\Delta u-\nabla p,\\ &\text{div }u=0, \end{aligned} \tag{4.3.53}$$

on a two-dimensional channel: $M=\{(x_1,x_2)\in\mathbb{R}^2\mid 0<x_2<L\}$ with width $L>0$. Equation (4.3.53) admits the following steady-state solutions, which describe parallel shear flows in M,

$$u_0=(V(x_2),0), \tag{4.3.54}$$

where V satisfies d^2V/dx_2^2 =constant. Therefore, V is given by

$$V_\lambda(x_2)=\lambda x_2^2+\alpha_1(\lambda)x_2+\alpha_2(\lambda).$$

Here $\alpha_1(\lambda)$ and $\alpha_2(\lambda)$ are two arbitrary functions of λ.

Consider the translation

$$u=v+u_0.$$

Equations (4.3.53) become

$$\begin{aligned} &\frac{\partial v_1}{\partial t}+(v\cdot\nabla)v_1=\mu\Delta v_1-(2\lambda x_2+\alpha_1(\lambda))v_2-V_\lambda\frac{\partial v_1}{\partial x_1}-\frac{\partial p}{\partial x_1},\\ &\frac{\partial v_2}{\partial t}+(v\cdot\nabla)v_2=\mu\Delta v_2-V_\lambda\frac{\partial v_2}{\partial x_1}-\frac{\partial p}{\partial x_2},\\ &\text{div}v=0, \end{aligned} \tag{4.3.55}$$

supplemented with the following boundary conditions:

$$\begin{aligned} &v=0 && \text{at } x_2=0,L,\\ &v \text{ is periodic} && \text{in the } x_1-\text{direction}. \end{aligned} \tag{4.3.56}$$

Consider the eigenvalue problem

$$\begin{aligned} &-\mu\Delta u_1+(2\lambda x_2+\alpha_1(\lambda))u_2+V_\lambda\frac{\partial u_1}{\partial x_1}+\frac{\partial p}{\partial x_1}=\beta(\lambda)u_1,\\ &-\mu\Delta u_2+V_\lambda\frac{\partial u_2}{\partial x_1}+\frac{\partial p}{\partial x_2}=\beta(\lambda)u_2,\\ &\text{div}u=0. \end{aligned} \tag{4.3.57}$$

Let the eigenvalues $\beta_j(\lambda)$ of (4.3.57) with (4.3.56) at $\lambda=\lambda_0$ satisfy the PES (1.2.4) and (1.2.5).

By the transition theorem, Theorem 2.1.3, and the existence of global attractors for the Navier–Stokes equations (Temam 1997), the problem (4.3.55) with (4.3.56) undergoes a transition to a minimal attractor Σ_λ. It is clear that each solution $v^\lambda \in \Sigma_\lambda$ of (4.3.55) with (4.3.56) is not a shear flow, whence v^λ has vortices in M, as shown in Fig. 4.27.

Thus, if $\lambda > \lambda_0$, equations (4.3.53) possess new stable states

$$u^*(x,\lambda) = u_0 + v^\lambda = (V_\lambda(x_2) + v_1^\lambda, v_2^\lambda),$$

where $v^\lambda = (v_1^\lambda, v_2^\lambda)$. On the other hand, we can see in Fig. 4.27 that there are curves $l_k \subset \overline{M}$ such that

$$\text{sign} V_\lambda = -\text{sign} v_1^\lambda, \quad \text{and} \quad v_2^\lambda = 0, \quad \text{on } l_k.$$

This implies that if

$$\begin{aligned} &\left|\frac{\partial v_1^{\lambda_0}}{\partial x_2}\right| < \left|\frac{\partial V_{\lambda_0}}{\partial x_2}\right| && \text{at } l_k \bigcap \partial M, \\ &\frac{d}{d\lambda}\left|\frac{\partial v_1^\lambda}{\partial x_2}\right| > \frac{d}{d\lambda}\left|\frac{\partial V_\lambda}{\partial x_2}\right| && \text{at } l_k \bigcap \partial M, \text{ for } \lambda > \lambda_0, \end{aligned} \tag{4.3.58}$$

then there exists a $\lambda_1 > \lambda_0$ such that

$$\begin{aligned} &\frac{\partial}{\partial x_2} u^*(x,\lambda) \neq 0 && \forall x \in \partial M \text{ for } \lambda_0 \leq \lambda < \lambda_1, \\ &\frac{\partial}{\partial x_2} u^*(x_0,\lambda_1) = 0 && \text{for some } x_0 \in l_k \bigcap \partial M. \end{aligned} \tag{4.3.59}$$

Thus, (4.3.58) and (4.3.59) means that

$$\text{ind}\left(\frac{\partial u^*(\cdot,\lambda_1)}{\partial x_2}, x_0\right) = 0, \qquad \frac{\partial}{\partial \lambda}\frac{\partial u^*(x_0,\lambda_1)}{\partial x_2} \neq 0.$$

Hence, by Theorems 5.3.6 and 5.3.7 in Ma and Wang (2005d), a boundary-layer separation at $x_0 \in \partial M$ on $\lambda > \lambda_1$ for $u^*(x,\lambda)$ will occur.

Instead of (4.3.58), if

$$\begin{aligned} &|v_1^{\lambda_0}| < |V_{\lambda_0}| && \text{on } l_k \bigcap M, \\ &\frac{d}{d\lambda}|v_1^\lambda| > \frac{d}{d\lambda}|V_\lambda| && \text{on } l_k \bigcap M \text{ for } \lambda > \lambda_0, \end{aligned} \tag{4.3.60}$$

then there is $\lambda_1 > \lambda_0$ such that

$$\begin{aligned} &u^*(x,\lambda) \neq 0 && \forall x \in M \text{ for } \lambda_0 \leq \lambda < \lambda_1, \\ &u^*(x_0,\lambda_1) = 0 && \text{for some } x_0 \in l_k \bigcap M. \end{aligned} \tag{4.3.61}$$

We infer from (4.3.60) and (4.3.61) that

$$\text{ind}(u^*(\cdot,\lambda_1),x_0)=0, \qquad \frac{\partial}{\partial\lambda}u^*(x_0,\lambda_1)\neq 0.$$

Thus, by Theorems 5.6.9.and 5.6.10 in Ma and Wang (2005d), the stable state $u^*(x,t)$ of (4.3.53) has an interior separation at $x_0 \in M$ on $\lambda > \lambda_1$.

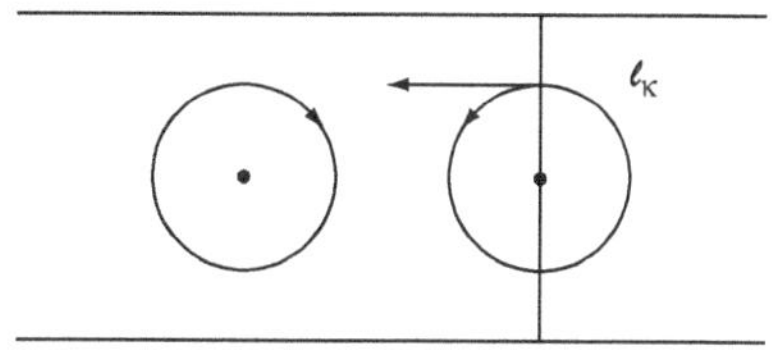

Fig. 4.27 The vortex structure of the transited solution $v^\lambda = (v_1^\lambda, v_2^\lambda)$

We note that there are no external forces in (4.2.5) and (4.3.53) to act directly on the fluid flows. As seen in the previous discussion, in this case, the cause leading to the boundary-layer and interior separation is the transitions between the stable steady-state solutions of the hydrodynamic equations.

However, if there exists an applied force, the mechanism to generate a structural transition is different. To demonstrate this point, in the following we introduce a structural bifurcation theorem for this case, which was proved by Ma and Wang (2004a) and Ma and Wang (2005d).

Consider the following problem:

$$\begin{aligned}&\frac{\partial u}{\partial t}+(u\cdot\nabla)u-\mu\Delta u+\nabla p=\lambda f,\\&\text{div}u=0,\\&u|_{\partial M}=0,\\&u(x,0)=u_0,\end{aligned}\tag{4.3.62}$$

where $u=(u_1,u_2), M\subset\mathbb{R}^2$ is an open set, and $u_0|_{\partial\Omega}=0$.

The following theorem provides a sufficient condition for structural bifurcation near the boundary of the solutions of (4.3.62) generated by the initial data and forcing.

Theorem 4.3.11 *Let $u_0\in C^{2+\alpha}(M,\mathbb{R}^2)$ with $u_0|_{\partial M}=0, f\in C^\alpha(M,\mathbb{R}^2)$ $(0<\alpha<1)$ and $u\in C^1([0,\infty),C^{2+\alpha}(M,\mathbb{R}^2))$ be a solution of (4.3.62). Assume that $\Gamma\subset\partial M$ is a connected component of ∂M, and*

$$\frac{\partial(f\cdot\tau)}{\partial n}\cdot\frac{\partial(u_0\cdot\tau)}{\partial n}<0 \text{ on } \Gamma,\tag{4.3.63}$$

where n and τ are the normal and tangent vectors at ∂M. Then there exists $\lambda_0>0$, depending on $\|f\|_{C^1}, \|u_0\|_{C^3}$ and $\mu>0$, such that for every $\lambda>\lambda_0$, there exist a time $t_0>0$ and point $\overline{x}\in\Gamma$ such that the solution $u(x,t,u_0)$ has a structural bifurcation at $(\overline{x},t_0)$ in its global structure and the index $ind(u(\cdot,t_0),\overline{x})=-k$, $k\geq 0$, is an integer.

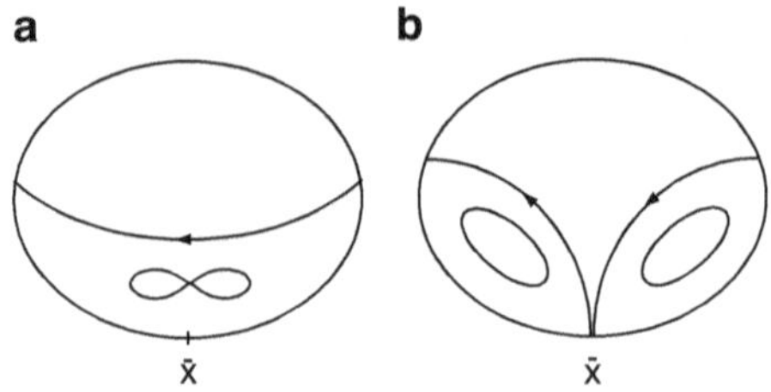

Fig. 4.28 An example of flow transitions for the case with index -1

Remark 4.3.12 Condition (4.3.63) implies that the direction of the external force f is inverse to that of the initial velocity u_0 near the boundary Γ. In fact, only two kinds of structural bifurcations of two-dimensional divergence-free vector fields are generic: the indices are 0 and -1. Namely, in the real world, we can observe only the structural bifurcation of fluid flows in which the index is either 0 or -1; see (Ma and Wang 2005d). The structural bifurcation from index zero is the well-known boundary-layer separation, and from index -1 it has two types of flow transitions, which are shown respectively in Figs. 4.28 and 4.29.

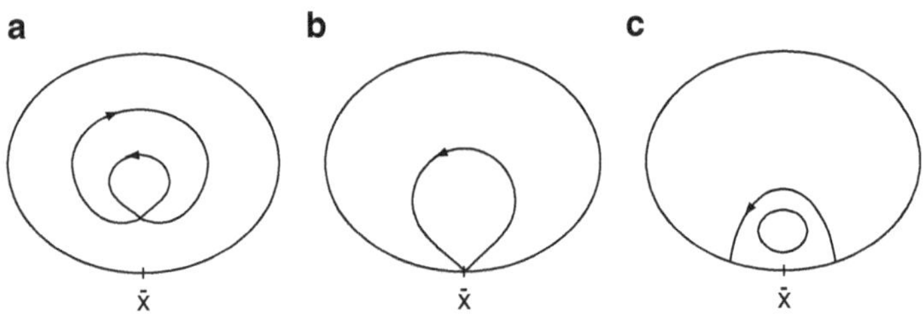

Fig. 4.29 An example of flow transitions for the case with index -1

4.4 Rotating Convection Problem

The stability and transition of a rotating flow are of basic importance in geophysical fluid dynamics, and in particular, in the study of atmospheric and oceanic circulations (Pedlosky 1987; Ghil and Childress 1987; Salby 1996). As we know, both the atmospheric and oceanic flows are flows under the rotation of the Earth. In fact, fast rotation and small aspect ratio are two main characteristics of the large-scale atmospheric and oceanic motions.

In this section, we address the thermal instability of a rotating fluid, and study the stability and transition of basic flows governed by the rotating Boussinesq equations. This study is based on recent articles (Hsia et al. 2007, 2010).

We show in particular that there are only two types of transitions, Type I (continuous) and Type II (jump) transitions, as the Rayleigh number crosses the first real or complex eigenvalues. Specific criteria are given to determine the type of transitions as well. The detailed analysis is based mainly on (1) the study of the eigenvalue problem for the linearized equations around the basic state, and (2) the reduction to the center manifold. In comparison to the classical Rayleigh–Bénard convection (Ma and Wang 2004b, 2007a), the linear problem here is non-self-adjoint, leading to a much more complicated spectrum and more complicated dynamical behaviors. We derive in particular two critical Rayleigh numbers R_{c_1} and R_{c_2} such that R_{c_1} leads to the onset of the transition of steady-state solutions, and R_{c_2} leads to the onset of the transition of periodic solutions (Hopf bifurcation). Both parameters are explicitly given in terms of the physical parameters. In addition, we derive two parameters b_1 and b_2 characterized by the physical parameters to determine the types of transitions of the steady-state solutions and the periodic solutions respectively. It is proved that the transition of steady-state solutions is of Type I (continuous) if $b_1 < 0$, and is of Type II (jump) if $b_1 > 0$. The transition of periodic solutions is of Type I if $b_2 < 0$, and is of Type II (jump) if $b_2 > 0$.

4.4.1 Rotating Boussinesq Equations

The rotating Boussinesq equations are basic equations in the geophysical fluid dynamics, and their nondimensional form is given by

$$\begin{aligned}
&\frac{\partial u}{\partial t} = \Pr \Delta u + \Pr RTk - \omega \bar{k} \times u - (u \cdot \nabla)u - \Pr \nabla p, \\
&\frac{\partial T}{\partial t} = \Delta T + u_3 - (u \cdot \nabla)T, \\
&\operatorname{div} u = 0,
\end{aligned} \tag{4.4.1}$$

for (x_1, x_2, x_3) in the nondimensional domain $\Omega = \mathbb{R}^2 \times (0,1)$, where $u = (u_1, u_2, u_3)$ is the velocity field, $k = (0,0,1)$ the unit vector in the x_3-direction, Pr the Prandtl number, R the Rayleigh number given by (4.1.1), ω a constant proportional to the rotating angular velocity, and T the temperature. In particular, the term $\omega k \times u$ stands for the Coriolis force in the atmosphere–ocean dynamics.

At the top and bottom boundaries, we impose the free–free boundary conditions:

$$(T, u_3) = 0, \qquad \frac{\partial u}{\partial x_3} = \frac{\partial u_2}{\partial x_3} = 0 \quad \text{at } x_3 = 0, 1. \tag{4.4.2}$$

In the x_1 and x_2 directions, we use the periodic boundary condition

$$(u, T)(x_1, x_2, x_3) = (u, T)(x_1 + 2jL_1\pi, x_2 + 2kL\pi, x_3) \qquad \forall j, k \in \mathbb{Z}. \tag{4.4.3}$$

For the problem (4.4.1)–(4.4.3), one of the following two constrained conditions is adopted:

$$\int_\Omega u_1 dx = \int_\Omega u_2 dx = 0, \quad \text{and} \tag{4.4.4}$$

$$u_1(0,0,x_3) = u_2(0,0,x_3) = 0. \tag{4.4.5}$$

The constraint (4.4.5) means that the fluid flow rises or descends along the rotating axis, which is a natural condition in many fluid-dynamical problems. Physically, this condition indicates that the rotation center lies at the point $x = 0$. In particular, we can see this characteristic in the polar cell of the latitudinal circulation in atmospheric physics. For (4.4.4), we set the spaces

$$H = \{(u,T) \in L^2(\Omega,\mathbb{R}^4) \mid \text{div}u = 0, u|_{x_3=0,1} = 0, u \text{ satisfies (4.4.3)}\},$$
$$H_1 = \{(u,T) \in H^2(\Omega,\mathbb{R}^4) \bigcap H \mid (u,T) \text{ satisfies (4.4.2)–(4.4.4)}\},$$

and for the condition (4.4.5), let

$$\widetilde{H} = \{(u,T) \in H \mid (u_1,u_2,u_3,T)(-x_1,-x_2,x_3) = (-u_1,-u_2,u_3,T)(x)\},$$
$$\widetilde{H}_1 = H_1 \bigcap \widetilde{H}.$$

It is clear that $(u,T) \in H$ satisfies (4.4.5) if and only if $(u,T) \in \widetilde{H}$.

Let $L_R = A + B_R : H_1 \to H$ (respectively, $\widetilde{H}_1 \to \widetilde{H}$) and $G : H_1 \to H$ (respectively, $\widetilde{H}_1 \to \widetilde{H}$) be defined by

$$A\psi = P(\Pr \Delta u, \Delta T), \qquad B_R\psi = P(\Pr RTk - \omega k \times u, u_3),$$
$$G(\psi,\widetilde{\psi}) = P(-(u\cdot\nabla)\widetilde{u}, -(u\cdot\nabla)\widetilde{T}), \qquad G(\psi) = G(\psi,\psi),$$

for any $\psi = (u,T) \in H_1$ (respectively $\widetilde{H}_1$), where $P : L^2(\Omega,\mathbb{R}^4) \to H$ is the Leray projection.

We note that $\widetilde{H}_1$ and $\widetilde{H}$ are invariant under the bilinear operator G. Hence, $\widetilde{H}_1$ and $\widetilde{H}$ are invariant under the operator $L_R + G$. Thus, the problem (4.4.1)–(4.4.3) with (4.4.4) [or with (4.4.5)] can be written in the following operator form:

$$\frac{d\psi}{dt} = L_R\psi + G(\psi) \qquad \text{for } \psi = (u,T) \in H_1 \ (\text{or } \widetilde{H}_1). \tag{4.4.6}$$

The existence and basic properties of solutions for rotating Boussinesq equations are known; see (Lions et al. 1992b).

4.4.2 Eigenvalue Problem

To study the transition of (4.4.1)–(4.4.3), consider the following eigenvalue problem:

$$L_R\psi = \beta(R)\psi \qquad \text{for } \psi \in H_1 \ (\text{or } \widetilde{H}_1). \tag{4.4.7}$$

The equations associated with (4.4.7) are given by

$$\begin{aligned}
&\Pr \Delta u + \Pr RTk - \omega k \times u - \Pr \nabla p = \beta u, \\
&\Delta T + u_3 = \beta T, \\
&\operatorname{div} u = 0,
\end{aligned} \tag{4.4.8}$$

supplemented with (4.4.2) and (4.4.3).

Under the constraint (4.4.5), for $\psi = (u_1, u_2, u_3, T) \in \widetilde{H_1}$, the eigenvectors of (10.67) have the separation of variables as follows:

$$\psi = \begin{cases} u_{jk}(x_3)\sin(j\alpha_1 x_1 + k\alpha_2 x_2), \\ v_{jk}(x_3)\sin(j\alpha_1 x_1 + k\alpha_2 x_2), \\ w_{jk}(x_3)\cos(j\alpha_1 x_1 + k\alpha_2 x_2), \\ T_{jk}(x_3)\cos(j\alpha_1 x_1 + k\alpha_2 x_2) \end{cases} \quad \text{for } j,k \in \mathbb{Z} \tag{4.4.9}$$

$$p = p_{jk}(x_3)\cos(j\alpha_1 x_1 + k\alpha_2 x_2) \quad \text{for } j,k \in \mathbb{Z} \tag{4.4.10}$$

$$\alpha_1 = L_1^{-1}, \quad \alpha_2 = L_2^{-1}. \tag{4.4.11}$$

If the condition (4.4.5) is replaced by (4.4.4), then for each $(j,k) \in \mathbb{Z}^2$ with $(j,k) \neq (0,0)$, we have another eigenvector given by

$$\widetilde{\psi} = \begin{cases} -u_{jk}(x_3)\cos(j\alpha_1 x_1 + k\alpha_2 x_2), \\ -v_{jk}(x_3)\cos(j\alpha_1 x_1 + k\alpha_2 x_2), \\ w_{jk}(x_3)\sin(j\alpha_1 x_1 + k\alpha_2 x_2), \\ T_{jk}(x_3)\sin(j\alpha_1 x_1 + k\alpha_2 x_2), \end{cases} \tag{4.4.12}$$

$$\widetilde{p} = p_{jk}\sin(j\alpha_1 x_1 + k\alpha_2 x_2). \tag{4.4.13}$$

Putting (ψ, p) and $(\widetilde{\psi}, \widetilde{p})$ into (4.4.8) respectively, we deduce from (4.4.9) to (4.4.3) that $(u_{jk}, v_{jk}, w_{jk}, T_{jk})$ satisfies the following system of ordinary differential equations:

$$\begin{aligned}
&\Pr(D_{jk}u_{jk} + j\alpha_1 p_{jk}) + \omega v_{jk} = \beta u_{jk}, \\
&\Pr(D_{jk}v_{jk} + k\alpha_2 p_{jk}) - \omega u_{jk} = \beta v_{jk}, \\
&\Pr(D_{jk}w_{jk} - Dp_{jk} + RT_{jk}) = \beta w_{jk}, \\
&D_{jk}T_{jk} + w_{jk} = \beta T_{jk}, \\
&j\alpha_1 u_{jk} + k\alpha_2 v_{jk} + Dw_{jk} = 0, \\
&Du_{jk} = Dv_{jk} = w_{jk} = T_{jk} = 0 \qquad \text{at} \quad x_3 = 0, 1,
\end{aligned} \tag{4.4.14}$$

for all $j,k \in \mathbb{Z}$ with $(j,k) \neq 0$, where

$$D = d/dx_3, \quad D_{jk} = D^2 - \alpha_{jk}^2, \quad \alpha_{jk}^2 = j^2\alpha_1^2 + k^2\alpha_2^2.$$

We infer from (4.4.14) that w_{jk} and $\widetilde{w}_{jk}$ satisfy the equation

$$\Big[(D_{jk}-\beta)(\Pr D_{jk}-\beta)^2 D_{jk}+\omega^2(D_{jk}-\beta)D^2 + \Pr R\alpha_{jk}^2(\Pr D_{jk}-\beta)\Big]w_{jk}=0 \tag{4.4.15}$$

$$w_{jk}=D^2w_{jk}=D^4w_{jk}=D^6w_{jk}=0 \quad \text{at} \quad x_3=0,1. \tag{4.4.16}$$

By (4.4.16), w_{jk} are the sine functions

$$w_{jk}(x_3)=\sin l\pi x_3 \qquad \text{for } l=1,2,\cdots. \tag{4.4.17}$$

In addition, we can deduce from (4.4.14) that

$$(\Pr D_{jk}-\beta)u_{jk}=\frac{-j\alpha_1}{\alpha_{jk}^2}(\Pr D_{jk}-\beta-\frac{k\alpha_2}{j\alpha_1}\omega)Dw_{jk}, \tag{4.4.18}$$

$$(\Pr D_{jk}-\beta)v_{jk}=\frac{-k\alpha_2}{\alpha_{jk}^2}(\Pr D_{jk}-\beta+\frac{j\alpha_1}{k\alpha_2}\omega)Dw_{jk}, \tag{4.4.19}$$

$$(D_{jk}-\beta)T_{jk}=-w_{jk}. \tag{4.4.20}$$

Substituting (4.4.17) into (4.4.15), we see that the eigenvalues β of (4.4.7) satisfy the cubic equations

$$\gamma_{jkl}^2(\beta+\gamma_{jkl}^2)(\beta+Pr\gamma_{jkl}^2)^2+l^2\pi^2\omega^2(\beta+\gamma_{jkl}^2)-\Pr R\alpha_{jk}^2(\beta+\Pr\gamma_{jkl}^2)=0, \tag{4.4.21}$$

for $(j,k,l)\in\mathbb{Z}^2\times\mathbb{N}$ with $(j,k)\neq(0,0)$, where

$$\gamma_{jkl}^2=\alpha_{jk}^2+l^2\pi^2=j^2\alpha_1^2+k^2\alpha_2^2+l^2\pi^2.$$

Consider the following index sets:

$$\begin{aligned}
I_1&=\{(j,k,l)\mid(j,k)\in\mathbb{Z}^2, j\geq 0,(j,k)\neq(0,0),l=1,2,\cdots,\},\\
I_2&=\{(j,k,0)\mid(j,k)\in\mathbb{Z}^2, j\geq 0,(j,k)\neq(0,0)\},\\
I_3&=\{(0,0,l)\mid l=1,2,\cdots,\}.
\end{aligned}$$

First, for $(j,k,0)\in I_2$, we have the following eigenvalues and eigenvectors of (4.4.8) with (4.4.2) and (4.4.3):

$$\begin{aligned}
\beta_{jk0}&=-\Pr\alpha_{jk}^2=-\Pr(j^2\alpha_1^2+k^2\alpha_2^2),\\
\psi_{jk0}&=(k\alpha_2\sin(j\alpha_1x_1+k\alpha_2x_2),-j\alpha_1\sin(j\alpha_1x_1+k\alpha_2x_2),0,0\},\\
\widetilde{\psi}_{jk0}&=(-k\alpha_2\cos(j\alpha_1x_1+k\alpha_2x_2),j\alpha_1\cos(j\alpha_1x_1+k\alpha_2x_2),0,0\};
\end{aligned}$$

namely,

$$L_R\psi_{jk0}=\beta_{jk0}\psi_{jk0},\quad L_R\widetilde{\psi}_{jk0}=\beta_{jk0}\widetilde{\psi}_{jk0}.$$

We define

$$\mathscr{B}_2=\{\beta_{jk0}=-\Pr\alpha_{jk}\mid(j,k,0)\in I_2\},$$

$$E_2 = \{\psi_{jk0}, \widetilde{\psi}_{jk0} |\ (j,k,0) \in I_2\},$$
$$\widetilde{E}_2 = \{\psi_{jk0} |\ (j,k,0) \in I_2\}.$$

Then $\mathscr{B}_2$ is the set of all eigenvalues corresponding to index set I_2, E_2 is the set of all eigenvectors associated with $\mathscr{B}_2$ for the condition (4.4.4), and $\widetilde{E}_2$ is the set of eigenvectors for the condition (4.4.5).

Second, for $(0,0,l) \in I_3$, the eigenvalues and eigenvectors are given by

$$\begin{aligned}
&\beta^1_{00l} = -l^2\pi^2,\ \beta^2_{00l} = -\mathrm{Pr}\ l^2\pi^2 + i\omega,\ \beta^3_{00l} = -\mathrm{Pr}\ l^2\pi^2 - i\omega,\\
&\psi^1_{00l} = (0,0,0,\sin l\pi x_3),\\
&\psi^2_{00l} = (\cos l\pi x_3, 0,0,0),\\
&\psi^3_{00l} = (0, \cos l\pi x_3, 0,0),
\end{aligned}$$

which satisfy

$$\begin{aligned}
L_R\psi^1_{00l} &= \beta^1_{00l}\psi^1_{00l},\\
L_R\psi^2_{00l} &= -\mathrm{Pr}\ l^2\pi^2\psi^2_{00l} + \omega\psi^3_{00l},\\
L_R\psi^3_{00l} &= -\omega\psi^2_{00l} - \mathrm{Pr}\ l^2\pi^2\psi^3_{00l}.
\end{aligned}$$

We denote the sets of eigenvalues and eigenvectors corresponding to I_3 for the conditions (4.4.4) and (4.4.5) respectively as follows:

$$\begin{aligned}
&\mathscr{B}_3 = \{\beta^1_{00l}, \beta^2_{00l}, \beta^3_{00l} |\ l = 1,2,\cdots\}, && E_3 = \{\psi^1_{00l}, \psi^2_{00l}, \psi^3_{00l} |\ l = 1,2,\cdots\},\\
&\widetilde{\mathscr{B}}_3 = \{\beta^1_{00l} |\ l = 1,2,\cdots\}, && \widetilde{E}_3 = \{\psi^1_{00l} |\ l = 1,2,\cdots\}.
\end{aligned}$$

Third, for $(j,k,l) \in I_1$, it is known that all solutions β_{jkl} of the cubic equations (4.4.21) are eigenvalues of (4.4.7). Let $\beta^i_{jkl}(R)$ $(1 \leq i \leq 3)$ be three zeros of (4.4.21) with

$$Re\beta^1_{jkl} \geq Re\beta^2_{jkl} \geq Re\beta^3_{jkl}.$$

Then all eigenvalues corresponding to I_1 are given by

$$\mathscr{B}_1 = \{\beta^i_{jkl}(R) \in D |\ \beta^i_{jkl} \text{ satisfies (4.4.21)}, (j,k,l) \in I_1, 1 \leq i \leq 3\}.$$

Let $\psi^i_{jkl}, \widetilde{\psi}^i_{jkl} \in H_1$ $(\psi^i_{jkl} \in \widetilde{H}_1)$ be the eigenvectors corresponding to $\beta^i_{jkl} \in \mathscr{B}_1$. By (4.4.9)–(4.4.13) and (4.4.17)–(4.4.20) with β being replaced by β^i_{jkl}, we obtain expressions for ψ^i_{jkl} and $\widetilde{\psi}^i_{jkl}$ $(1 \leq i \leq 3)$ as follows:

$$\psi^i_{jkl} = \begin{cases}
u^i_{jkl}\sin(j\alpha_1 x_1 + k\alpha_2 x_2)\cos l\pi x_3,\\
v^i_{jkl}\sin(j\alpha_1 x_1 + k\alpha_2 x_2)\cos l\pi x_3,\\
w^i_{jkl}\cos(j\alpha_1 x_1 + k\alpha_2 x_2)\sin l\pi x_3,\\
T^i_{jkl}\cos(j\alpha_1 x_1 + k\alpha_2 x_2)\sin l\pi x_3,
\end{cases} \tag{4.4.22}$$

$$\widetilde{\psi}^i_{jkl}=\begin{cases} -u^i_{jkl}\cos(j\alpha_1x_1+k\alpha_2x_2)\cos l\pi x_3,\\ -v^i_{jkl}\cos(j\alpha_1x_1+k\alpha_2x_2)\cos l\pi x_3,\\ w^i_{jkl}\sin(j\alpha_1x_1+k\alpha_2x_2)\sin l\pi x_3,\\ T^i_{jkl}\sin(j\alpha_1x_1+k\alpha_2x_2)\sin l\pi x_3,\end{cases} \tag{4.4.23}$$

where

$$\begin{aligned} &u^i_{jkl}=-\frac{l\pi}{\alpha^2_{jk}}(j\alpha_1+k\alpha_2\omega A_1(\beta^i_{jkl})), && v^i_{jkl}=-\frac{l\pi}{\alpha^2_{jk}}(k\alpha_2-j\alpha_1\omega A_1(\beta^i_{jkl})),\\ &w^i_{jkl}=1, && T^i_{jkl}=A_2(\beta^i_{jkl}),\\ &A_1(\beta)=\frac{1}{\Pr\gamma^2_{jkl}+\beta}, && A_2(\beta)=\frac{1}{\gamma^2_{jkl}+\beta}.\end{aligned}$$

When $\beta^q_{jkl}=\overline{\beta}^{q+1}_{jkl}=\sigma+i\rho$ are complex numbers, ψ^i_{jkl} and $\psi^i_{jkl}(i=q)$ are complex functions. Therefore, the eigenvectors $\phi^q_{jkl},\phi^{q+1}_{jkl}$ and $\widetilde{\phi}^q_{jkl},\widetilde{q}^{q+1}_{jkl}$ corresponding to β^q_{jkl} and β^{q+1}_{jkl} can be expressed as

$$\begin{aligned} &\phi^q_{jkl}=Re\psi^i_{jkl}, && \phi^{q+1}_{jkl}=I_m\psi^i_{jkl},\\ &\widetilde{\phi}^q_{jkl}=Re\widetilde{\psi}^i_{jkl}, && \widetilde{\phi}^{q+1}_{jkl}=I_m\widetilde{\psi}^i_{jkl},\end{aligned} \tag{4.4.24}$$

for $i=q$. In this case, ϕ^q_{jkl} and ϕ^{q+1}_{jkl} satisfy

$$\begin{aligned} L_r\phi^q_{jkl}&=\sigma\phi^q_{jkl}-\rho\phi^{q+1}_{jkl},\\ L_R\phi^{q+1}_{jkl}&=\rho\phi^q_{jkl}+\sigma\phi^{q+1}_{jkl}.\end{aligned} \tag{4.4.25}$$

Thus, the sets of all eigenvectors corresponding to I_1 are given by

$$\begin{aligned} E_1&=\{\psi^i_{jkl},\widetilde{\psi}^i_{jkl}|\ (j,k,l)\in I_1,\ 1\le i\le 3\},\\ \widetilde{E}_1&=\{\psi^i_{jkl}|\ (j,k,l)\in I_1,\ 1\le i\le 3\}.\end{aligned}$$

In view of the Fourier expansion, it is clear that $E=E_1\bigcup E_2\bigcup E_3$ is a basis of H, and $\widetilde{E}=\widetilde{E}_1\bigcup\widetilde{E}_2\bigcup\widetilde{E}_3$ is a basis of $\widetilde{H}$, and we have the following lemma:

Lemma 4.4.1

(a) The set $\mathscr{B}=\mathscr{B}_1\bigcup\mathscr{B}_2\bigcup\mathscr{B}_3$ consists of all eigenvalues of (4.4.8) with (4.4.2)–(4.4.4), and the (generalized) eigenvectors in E form a basis of H.

(b) The set $\widetilde{\mathscr{B}}=\mathscr{B}_1\bigcup\mathscr{B}_2\bigcup\widetilde{\mathscr{B}}_3$ consists of all eigenvalues of (4.4.8) with (4.4.2), (4.4.3), and (4.4.5), and the (generalized) eigenvectors in $\widetilde{E}$ form a basis of $\widetilde{H}$.

(c) $Re\beta<0$ for each $\beta\in\mathscr{B}_2\bigcup\mathscr{B}_3$.

(d) Only the eigenvalues $\beta\in\mathscr{B}_1$ depend on the Rayleigh number.

Now we are ready to examine the dual eigenvectors. For $(j,k,l) \in I_1$, the dual eigenvectors ψ^{i*}_{jkl} and $\widetilde{\psi}^{i*}_{jkl}$ of ψ^i_{jkl} and $\widetilde{\psi}^i_{jkl}$ are given by

$$\psi^{i*}_{jkl} = \begin{cases} u^{i*}_{jkl} \sin(j\alpha_1 x_1 + k\alpha_2 x_2)\cos l\pi x_3, \\ v^{i*}_{jkl} \sin(j\alpha_1 x_1 + k\alpha_2 x_2)\cos l\pi x_3, \\ w^{i*}_{jkl} \cos(j\alpha_1 x_1 + k\alpha_2 x_2)\sin l\pi x_3, \\ T^{i*}_{jkl} \cos(j\alpha_1 x_1 + k\alpha_2 x_2)\sin l\pi x_3, \end{cases} \tag{4.4.26}$$

$$\widetilde{\psi}^{i*}_{jkl} = \begin{cases} -u^{i*}_{jkl} \cos(j\alpha_1 x_1 + k\alpha_2 x_2)\cos l\pi x_3, \\ -v^{i*}_{jkl} \cos(j\alpha_1 x_1 + k\alpha_2 x_2)\cos l\pi x_3, \\ w^{i*}_{jkl} \sin(j\alpha_1 x_1 + k\alpha_2 x_2)\sin l\pi x_3, \\ T^{i*}_{jkl} \sin(j\alpha_1 x_1 + k\alpha_2 x_2)\sin l\pi x_3, \end{cases} \tag{4.4.27}$$

where

$$u^{i*}_{jkl} = -\frac{l\pi}{\alpha^2_{jk}}(j\alpha_1 - k\alpha_2\omega A_1(\beta^i_{jkl})), \qquad v^{i*}_{jkl} = -\frac{l\pi}{\alpha^2_{jk}}(k\alpha_2 + j\alpha_1\omega A_1(\beta^i_{jkl})),$$
$$w^{i*}_{jkl} = 1, \qquad T^{i*}_{jkl} = \Pr RA_2(\beta^i_{jkl}).$$

By the spectral theorem, Theorem 3.4 in Ma and Wang (2005b), the dual eigenvectors satisfy

$$(\psi^q_{jkl}, \psi^{p*}_{jkl}) = 0, \qquad (\widetilde{\psi}^q_{jkl}, \widetilde{\psi}^{p*}_{jkl}) = 0 \quad \text{for } q \neq p. \tag{4.4.28}$$

In particular, if $\beta^q_{jkl} = \overline{\beta}^{q+1}_{jkl} = \sigma + i\rho$, then

$$\phi^{q*}_{jkl} = \mathrm{Re}\,\psi^{i*}_{jkl}(\overline{\beta}^q_{jkl}), \qquad \phi^{q+1*}_{jkl} = \mathrm{Im}\,\psi^{i*}_{jkl}(\overline{\beta}^q_{jkl}) \qquad \text{for } i = q, \tag{4.4.29}$$

and for every $\psi^i_J \neq \phi^q_{jkl}, \phi^{q+1}_{jkl}$,

$$L^*_R\phi^{q*}_{jkl} = \sigma\phi^{q*}_{jkl} + \rho\phi^{q+1*}_{jkl}, \qquad L^*_R\phi^{q+1*}_{jkl} = -\rho\phi^{q*}_{jkl} + \sigma\phi^{q+1*}_{jkl}, \tag{4.4.30}$$
$$(\psi^i_J, \phi^{q*}_{jkl}) = (\psi^i_J, \phi^{q+1*}_{jkl}) = 0. \tag{4.4.31}$$

4.4.3 Principle of Exchange of Stabilities

It is known that only the eigenvalues in $\mathscr{B}_1$ depend on the Rayleigh number R. Hence to study the critical crossing of the first eigenvalues of (4.4.7), it suffices to focus this problem on the set $\mathscr{B}_1$, i.e., on the solutions β of (4.4.21), which can be equivalently written in the following form:

$$\beta^3 + (2\mathrm{Pr} + 1)\gamma_{jkl}^2\beta^2 + \left[(\mathrm{Pr}^2 + 2\mathrm{Pr})\gamma_{jkl}^4 + \frac{l^2\pi^2\omega}{\gamma_{jlk}^2} - \mathrm{Pr}\, R\frac{\alpha_{jlk}^2}{\gamma_{jkl}^2}\right]\beta +$$
$$+ \mathrm{Pr}^2\gamma_{jkl}^6 - \mathrm{Pr}^2 R\alpha_{jk}^2 + l^2\pi^2\omega^2 = 0. \tag{4.4.32}$$

Case 1. $\beta = 0$ is a zero of (4.4.32) if and only if

$$\mathrm{Pr}^2\gamma_{jkl}^6 - \mathrm{Pr}^2 R\alpha_{jk}^2 + l^2\pi^2\omega^2 = 0.$$

In this case, we have

$$R = \frac{\gamma_{jkl}^6}{\alpha_{jkl}^2} + \frac{l^2\pi^2\omega^2}{\alpha_{jlk}^2\mathrm{Pr}^2} \geq \frac{(\alpha_{jkl}^2 + \pi^2)^3}{\alpha_{jkl}^2} + \frac{\pi^2\omega^2}{\alpha_{jkl}^2\mathrm{Pr}^2}.$$

Hence the critical Rayleigh number R_{c_1} associated with real eigenvalues is given by

$$R_{c_1} = \min_{(j,k,l)\in I_1}\left(\frac{\gamma_{jkl}^6}{\alpha_{jkl}^2} + \frac{l^2\pi^2\omega^2}{\alpha_{jlk}^2\mathrm{Pr}^2}\right) = \frac{(\alpha_{j_1k_1}^2 + \pi^2)^3}{\alpha_{j_1k_1}^2} + \frac{\pi^2\omega^2}{\alpha_{j_1k_1}^2\mathrm{Pr}^2}, \tag{4.4.33}$$

for some $(j_1, k_1, 1) \in I_1$.

Case 2. Let a purely imaginary number $\beta = i\rho$ ($\rho \neq 0$) be a zero of (4.4.32). Then

$$\begin{aligned} \rho^2 &= (\mathrm{Pr}^2 + 2\mathrm{Pr})\gamma_{jkl}^4 + \frac{l^2\pi^2\omega^2}{\gamma_{jkl}^2} - \mathrm{Pr}\, R\frac{\alpha_{jk}^2}{\gamma_{jkl}^2}, \\ \rho^2 &= \frac{1}{(2\mathrm{Pr} + 1)\gamma_{jkl}^2}\left[\mathrm{Pr}^2\gamma_{jkl}^6 - \mathrm{Pr}^2 R\alpha_{jk}^2 + l^2\pi^2\omega^2\right]. \end{aligned} \tag{4.4.34}$$

We infer from (4.4.34) that Eq. (4.4.32) has a pair of purely imaginary solutions $\beta = \pm i\rho$ if and only if the following conditions hold:

$$\begin{aligned} &(\mathrm{Pr}^2 + 2\mathrm{Pr})\gamma_{jkl}^4 + \frac{l^2\pi^2\omega^2}{\gamma_{jkl}^2} - \mathrm{Pr}\, R\frac{\alpha_{jk}^2}{\gamma_{jkl}^2} > 0, \\ &(2\mathrm{Pr} + 1)\gamma_{jkl}^2\left[(\mathrm{Pr}^2 + 2\mathrm{Pr})\gamma_{jkl}^4 + \frac{l^2\pi^2\omega^2}{\gamma_{jkl}^2} - \mathrm{Pr}\, R\frac{\alpha_{jk}^2}{\gamma_{jkl}^2}\right] \\ &= \mathrm{Pr}^2\gamma_{jkl}^2 - \mathrm{Pr}^2 R\alpha_{jk}^2 + l^2\pi^2\omega^2. \end{aligned}$$

In this case, we have

$$R = \frac{2(\mathrm{Pr} + 1)\gamma_{jkl}^6}{\alpha_{jk}^2} + \frac{2l^2\pi^2\omega^2}{(\mathrm{Pr} + 1)\alpha_{jk}^2}, \tag{4.4.35}$$

$$R < \frac{(\mathrm{Pr} + 2)\gamma_{jkl}^6}{\alpha_{jk}^2} + \frac{l^2\pi^2\omega^2}{\mathrm{Pr}\,\alpha_{jk}^2}. \tag{4.4.36}$$

We derive, from (4.4.35) and (4.4.36), a lower bound for the rotating angular velocity ω:

$$\omega^2 > \frac{\Pr^2(1+\Pr)\gamma_{jkl}^6}{(1-\Pr)l^2\pi^2}, \tag{4.4.37}$$

which can hold only when the Prandtl number Pr is less than 1.

As in Case 1, the minimum of the right-hand side of (4.4.35) is always obtained at $l = 1$. Thus the critical Rayleigh number R_{c_2} associated with complex eigenvalues is given by

$$\begin{aligned} R_{c_2} &= \min_{(j,k,l)\in\Lambda} \left[\frac{2(\Pr+1)\gamma_{jkl}^6}{\alpha_{jk}^2} + \frac{2l^2\pi^2\omega^2}{(\Pr+1)\alpha_{jk}^2} \right] \\ &= \frac{2(\Pr+1)(\alpha_{j_2k_2}^2+\pi^2)^3}{\alpha_{j_2k_2}^2} + \frac{2\pi^2\omega^2}{(\Pr+1)\alpha_{j_2k_2}^2}, \end{aligned} \tag{4.4.38}$$

for some $(j_2,k_2,1) \in \Lambda \subset I_1$, where Λ is the set of (j,k,l) satisfying (4.4.37).

Lemma 4.4.2 *Let* $\alpha_0 = \min\{\alpha_1,\alpha_2\}$. *Assume either* $\Pr \geq 1$ *or*

$$\frac{(1-Pr)\pi^2}{Pr^2(1+Pr)}\omega^2 < (\pi^2+\alpha_0^2)^3 \qquad \textit{for } Pr < 1. \tag{4.4.39}$$

Let $(j,k,l) = (j_1,k_1,1) \in I_1$ *minimize the right-hand side of (4.4.33). Then the zero* $\beta_{j_1k_11}^1(R)$ *of (4.4.32) is a real eigenvalue of (4.4.7) near* R_{c_1}, *and it satisfies*

$$\beta_{j_1k_11}^1(R) \begin{cases} <0 & \text{if } R < R_{c_1}, \\ =0 & \text{if } R = R_{c_1}, \\ >0 & \text{if } R > R_{c_1}, \end{cases} \tag{4.4.40}$$

$$Re\beta(R_{c_1}) < 0 \quad \forall \beta(R) \in \mathscr{B}_1 \bigcup \mathscr{B}_2 \bigcup \mathscr{B}_3 \text{ with } \beta(R_{c_1}) \neq 0. \tag{4.4.41}$$

Proof. We first prove that there is a $\delta > 0$ such that

$$\mathrm{Re}\beta_{jkl}^q(R) < 0 \quad \forall \beta_{jkl}^q \in \mathscr{B}_1, \quad 0 \leq R < \delta. \tag{4.4.42}$$

In fact, for each $(j,k,l) \in I_1$, for $R = 0$, the solutions of (4.4.21) are

$$\beta_{jkl}^1 = -\gamma_{jkl}^2, \quad \beta_{jkl}^2 = \overline{\beta}_{jkl}^3 = -\Pr\gamma_{jkl}^2 + i\frac{l\pi\omega}{\gamma_{jkl}}.$$

Namely, $\mathrm{Re}\beta_{jkl}^q(0) < 0$, which implies that (4.4.42) holds because $\beta_{jkl}^q(R)$ are continuous on R.

By the above discussion, we need to show only that the first eigenvalue crosses the imaginary axis. Let

$$g_{jkl}(\beta) = \gamma_{jkl}^2(\beta+\gamma_{jkl}^2)\left[(\beta+Pr\gamma_{jkl}^2)^2 + l^2\pi^2\omega^2\right],$$

$$h_{jkl}(\beta) = PrR\alpha_{jk}^2(\beta + Pr\gamma_{jkl}^2),$$
$$f_{jkl}(\beta) = g_{jkl}(\beta) - h_{jkl}(\beta).$$

Thus, Eq. (4.4.21) is equivalent to $f_{jkl}(\beta) = 0$.

We note that $f_{j_1k_11}(\beta) = 0$ is equivalent to $g_{j_1k_11}(\beta) = h_{j_1k_11}(\beta)$, i.e.,

$$\gamma_{j_1k_11}^2(\beta + \gamma_{j_1k_11}^2)\left[(\beta + \mathrm{Pr}\ \gamma_{j_1k_11}^2)^2 + l^2\pi^2\omega^2\right] = \mathrm{Pr}\ R\alpha_{j_1k_1}^2(\beta + \mathrm{Pr}\ \gamma_{j_1k_11}^2). \quad (4.4.43)$$

When Pr > 1, we see that both $g_{j_1k_11}(\beta)$ and $h_{j_1k_11}(\beta)$ are strictly increasing for $\beta > -\gamma_{j_1k_11}^2$. Let Γ_1 be the graph of $\eta = g_{j_1k_11}(\beta)$ and Γ_2 the graph of $\eta = h_{j_1k_11}(\beta)$, as shown in Fig. 4.30. When $R = R_{c_1}$, the point S_0, the intersecting point of Γ_1 and Γ_2 corresponding to $\beta_{j_1k_11}(R)$ (i.e., the β coordinate of S_0 is $\beta_{j_1k_11}(R)$), is on the η-axis. When R increases (respectively, decreases), so does S_1 (respectively, S_2). This proves (4.4.40).

When Pr $= 1$, the solution $\beta_{j_1k_11}^1(R)$ of (4.4.43) is given by

$$\beta_{j_1k_11}^1(R) = -\gamma_{j_1k_11}^2 + \gamma_{j_1k_11}^{-1}\sqrt{R\alpha_{j_1k_1}^2 - \pi^2\omega^2},$$

which satisfies (4.4.40).

When Pr < 1, by (4.4.37), the condition (4.4.39) implies that no complex eigenvalues $\beta_{jkl}^q(R)$ satisfy

$$\lim_{R\to R_{c_1}} \mathrm{Re}\beta_{jkl}^q(R) = 0, \quad \lim_{R\to R_{c_1}} \mathrm{Im}\beta_{jkl}^q(R) = 0.$$

In addition, we note that R_{c_1} is the first critical Rayleigh number. Hence, when $R = R_{c_1}$, along the β-increasing direction, Γ_2 crosses Γ_1 at the intersection point S_0 from the left-hand side to the right-hand side of Γ_2, as shown in Fig. 4.31. As proved by (4.4.43) and Fig. 4.30, this implies that (4.4.40) is valid. □

Remark 4.4.3 In the proof of Lemma 4.4.2, as shown by (4.4.43) and Figs. 4.30 and 4.31, we see that near $R = R_{c_1}$, the first eigenvalue $\beta_{j_1k_11}^1(R)$ is a simple zero of $f_{j_1k_11}(\beta)$. We have seen in (4.4.22) and (4.4.23) that there are two eigenvectors $\psi_{j_1k_11}^1$ and $\widetilde{\psi}_{j_1k_11}^1 \in H_1$ corresponding to $\beta_{j_1k_1}^1$. Therefore, the multiplicity of the first eigenvalue of $L_R|_{H_1}$ (respectively $L_R|_{\widetilde{H}_1}$) is $m_{H_1} = 2m$ (respectively $m_{\widetilde{H}_1} = m$), where m is the number of $(j,k,1)'s (\in I_1)$ satisfying $\alpha_{jk}^2 = \alpha_{j_1k_1}^2$.

Remark 4.4.4 For the classical Bénard problem without rotation, the second term on the right-hand side of (4.4.33) is not present. Therefore, the first critical Rayleigh number of the Bénard problem depends only on the aspect of ratio, while the first critical Rayleigh number of the rotating problem depends on the aspect ratio $\alpha = L_2/L_1$, the Prandtl number, and the rotating angular velocity. And it is clear that the first critical Rayleigh number of rotating flows is larger than that of the Bénard problem. This indicates that rotating flows are much more stable than the nonrotating flows; see Sect. 4.1.

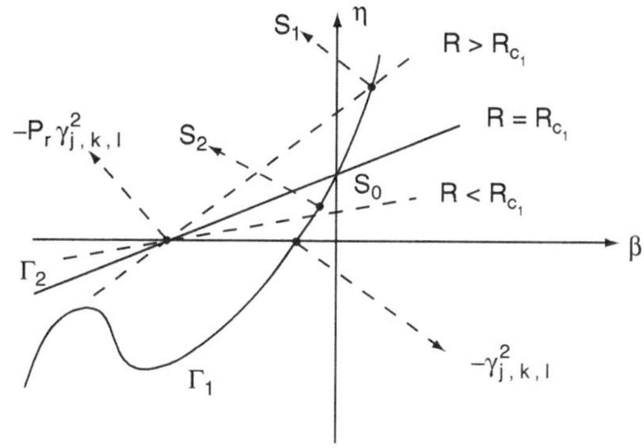

Fig. 4.30 There are at most three intersections of Γ_1 and Γ_2, and if Pr > 1, Γ_2 must cross Γ_1 at S_0 from the left-hand side to the right-hand side of Γ_2 along the β-increasing direction

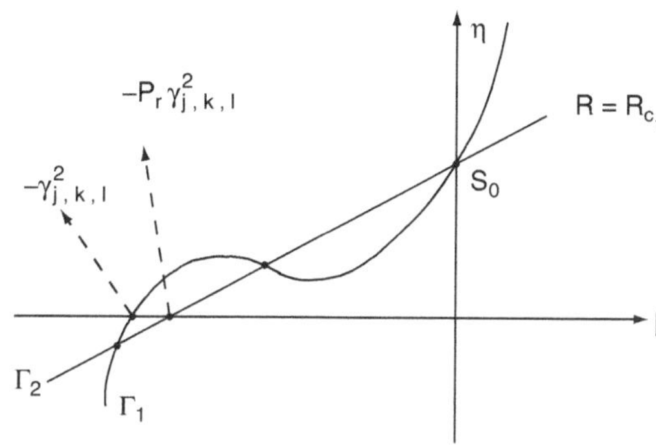

Fig. 4.31 If Pr < 1, the condition (4.4.39) implies that the straight line Γ_2 at $R = R_{c_1}$ must cross Γ_1 at S_0 from the left-hand side to the right-hand side of Γ_1 along the β-increasing direction

When the condition (4.4.39) is not valid, (4.4.37) holds for some indices $(j,k,l) \in I_2$, and there are complex eigenvalues with real part crossing the imaginary axis. In this case, the smaller of the two the critical Rayleigh numbers R_{c_1} and R_{c_2} defined respectively by (4.4.33) and (4.4.38) is the first critical Rayleigh number.

Let $(j_1,k_1,1)$ and $(j_2,k_2,1)$ satisfy (4.4.33) and (4.4.38) respectively. It is readily checked that if

$$\frac{(1-\mathrm{Pr})\pi^2}{\mathrm{Pr}^2(1+\mathrm{Pr})}\omega^2 > \max\{(\alpha^2_{j_1k_1}+\pi^2)^3, (\alpha^2_{j_2k_2}+\pi^2)^3\}, \tag{4.4.44}$$

then R_{c_2} is less than R_{c_1}, and therefore, R_{c_2} is the first critical Rayleigh number. However, if

$$(\alpha^2_{j_2k_2}+\pi^2)^3 < \frac{(1-\mathrm{Pr})\pi^2}{\mathrm{Pr}^2(1+\mathrm{Pr})}\omega^2 < (\alpha^2_{j_1k_1}+\pi^2)^3, \tag{4.4.45}$$

then we need additional methods to identify them.

For $(j,k,l) \in I_1$, there are at most three zeros of $f_{jkl}(\beta)$. Therefore, each pair of complex eigenvalues $\beta^q_{jkl} = \overline{\beta}^{q+1}_{jkl}$ are simple zeros of $f_{jkl}(\beta)$. The following lemma shows that the complex eigenvalues $\beta^q_{j_2k_21}(R) = \overline{\beta}^{q+1}_{j_2k_21}(R)$ cross the imaginary axis at the critical Rayleigh number R_{c_2} regardless of whether these two complex eigenvalues are the first ones.

Lemma 4.4.5 *Let $Pr < 1$ and $\omega^2 > Pr^2(1+Pr)(\pi^2+\alpha^2_{j_2k_2})^3/(1-Pr)\pi^2$. Then the pair of complex eigenvalues $\beta^1_{j_2k_21}(R) = \overline{\beta}^2_{j_2k_21}(R)$ are critical-crossing at the critical Rayleigh number R_{c_2}, i.e.,*

$$Re\beta^1_{j_2k_1l}(R)\begin{cases} <0 & \text{if } R<R_{c_2},\\ =0 & \text{if } R=R_{c_2},\\ >0 & \text{if } R>R_{c_2}.\end{cases}$$

Proof. Since $\beta^1_{j_2k_2l}(R)$, $\beta^2_{j_2k_2l}(R)$, and $\beta^3_{j_2k_2l}(R)$ (real) are zeros of (4.4.32), we obtain

$$\beta^3_{j_2k_2l}(R) = -2Re\beta^1_{j_2k_2l}(R) - -(2\Pr+1)(\alpha^2_{j_2k_2}+\pi^2).$$

Hence, the critical crossing at R_{c_2} is equivalent to

$$\beta^3_{j_2k_2l}(R)\begin{cases} >-(2\Pr+1)(\alpha^2_{j_2k_2}+\pi^2) & \text{if } R<R_{c_2},\\ =-(2\Pr+1)(\alpha^2_{j_2k_2}+\pi^2) & \text{if } R=R_{c_2},\\ <-(2\Pr+1)(\alpha^2_{j_2k_2}+\pi^2) & \text{if } R>R_{c_2},\end{cases}$$

which is true, as shown in Fig. 4.32. The proof is complete. □

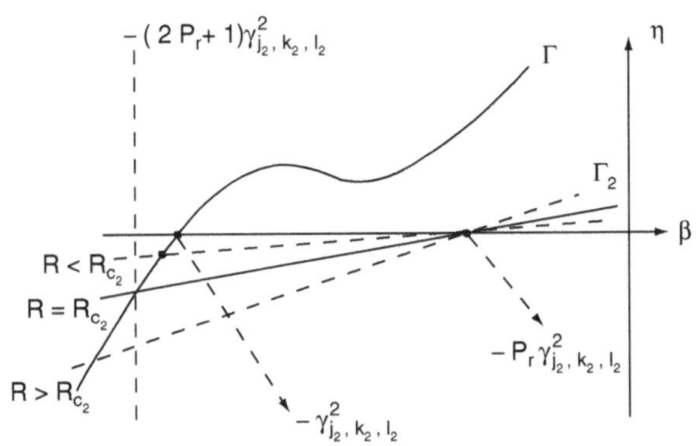

Fig. 4.32 Critical crossing for a complex pair of eigenvalues based on the third real eigenvalue

In fact, for the case (4.4.45), if R_{c_1} is the first critical Rayleigh number, then (4.4.40) and (4.4.41) are also valid. Thus, in view of Lemmas 4.4.2 and 4.4.5, we have the following critical-crossing theorem for the first eigenvalues of (4.4.7).

Theorem 4.4.6 *Let $(j_1,k_1,1)$ and $(j_2,k_2,1)$ satisfy (4.4.33) and (4.4.38) respectively. Assume that*

$$\frac{(1-Pr)\pi^2}{Pr^2(1+Pr)}\omega^2 \neq \begin{cases}(\alpha^2_{j_1k_1}+\pi^2)^3, & \\ (\alpha^2_{j_2k_2}+\pi^2)^3 & \text{if } R_{c_2}<R_{c_1}.\end{cases} \tag{4.4.46}$$

Then if R_{c_1} is the first critical Rayleigh number, if (4.4.40) and (4.4.41) hold, and if $R_{c_2} < R_{c_1}$, we have

$$Re\beta^1_{j_2k_2l}(R) = Re\beta^2_{j_2k_2l}(R) \begin{cases} <0 & \text{if } R < R_{c_2}, \\ =0 & \text{if } R = R_{c_2}, \\ >0 & \text{if } R > R_{c_2}. \end{cases} \tag{4.4.47}$$

$$Re\beta(R_{c_2}) < 0 \quad \forall \beta \in \mathscr{B}_1 \bigcup \mathscr{B}_2 \bigcup \mathscr{B}_3 \text{ with } Re\beta(R_{c_2}) \neq 0. \tag{4.4.48}$$

In particular, if $\Pr \geq 1$ *or (4.4.39) holds, then R_{c_1} is the first critical Rayleigh number, and if (4.4.44) holds, then $R_{c_2} < R_{c_1}$.*

In the following, we consider the multiplicity of the first critical-crossing eigenvalues. It is easy to see that the first eigenvalues of (4.4.7) depend on the periods L_1 and L_2 [i.e., $\alpha^2_{jk} = j^2\alpha_1^2 + k^2\alpha_2^2, \alpha_1$ and α_2 are as in (4.4.11)]. Let

$$L = L_1 = \alpha L_2. \tag{4.4.49}$$

Then by (4.4.33) and (4.4.38), we have

$$\begin{aligned} R_{c_1} &= \min_{(j,k,1)\in I_1} \left(\frac{(\alpha^2_{jk} + \pi^2)^3}{\alpha^2_{jk}} + \frac{\pi^2\omega^2}{\Pr^2 \alpha^2_{jk}} \right) \\ &= \frac{(j_1^2 + k_1^2\alpha^2 + \pi^2L^2)^3}{L^4(j_1^2 + k_1^2\alpha^2)} + \frac{\pi^2\omega^2L^2}{\Pr^2(j_1^2 + k_1^2\alpha^2)}, \end{aligned} \tag{4.4.50}$$

$$\begin{aligned} R_{c_2} &= \min_{(j,k,1)\in \Lambda} 2(\Pr + 1) \left[\frac{(\alpha^2_{jk} + \pi^2)^3}{\alpha^2_{jk}} + \frac{\pi^2\omega^2}{(\Pr + 1)^2 \alpha^2_{jk}} \right] \\ &= 2(\Pr + 1) \left[\frac{(j_2^2 + k_2^2\alpha^2 + \pi^2L^2)^3}{L^4(j_2^2 + k_2^2\alpha^2)} + \frac{\pi^2\omega^2L^2}{(\Pr + 1)^2(j_2^2 + k_2^2\alpha^2)} \right], \end{aligned} \tag{4.4.51}$$

where $(j,k,1) \in \Lambda$ satisfies (4.4.37).

It is known that the multiplicity of the first eigenvalues $\beta^1_{j_1k_1l}(R)$ of $L_R|_{H_1}$ (respectively $L_R|_{\widetilde{H}_1}$) is $m_{H_1} = 2n$ (respectively $m_{\widetilde{H}_1} = n$), where n is the number of $(j,k,l) \in I_1$ satisfying

$$j^2 + k^2\alpha^2 = j_1^2 + k_1^2\alpha^2,$$

which yields

$$\alpha^2 = \frac{j^2 - j_1^2}{k_1^2 - k^2}, \quad (j,k) \neq (0,0). \tag{4.4.52}$$

Likewise, the multiplicity of the first complex eigenvalues $\beta^1_{j_2k_2l} = \overline{\beta}^2_{j_2k_2l}$ is $m_{H_1} = 4n$, and $m_{\widetilde{H}_1} = 2n$.

By (4.4.52), we see that if α^2 is an irrational number, then $n = 1$. Namely, we have the following lemma.

Lemma 4.4.7 *Let α, defined by (4.4.49), be an irrational number. Then the first real eigenvalue $\beta^1_{j_1k_1l}(R)$ at R_{c_1} has multiplicity $m=2$ in H_1 and $m=1$ in $\widetilde{H}_1$, and the first complex eigenvalues $\beta^1_{j_2k_2l}(R)=\beta^2_{j_2k_2l}(R)$ at R_{c_2} have multiplicity $m=4$ in H_1 and $m=2$ in $\widetilde{H}_1$.*

Now we look for conditions to make R_{c_2} the first critical Rayleigh number. For $x>0, b>0$, we define

$$f_b(x)=\frac{(x+\pi^2)^3+b}{x}.$$

Let $x=\alpha^2_{jk}$. Then the right-hand side of (4.4.33) can be expressed as $f_{b_1}(x)$ with $b_1=\pi^2\omega^2/\mathrm{Pr}^2$, and the second line of (4.4.38) can be expressed as $2(\mathrm{Pr}+1)f_{b_2}(x)$ with $b_2=\pi^2\omega^2/(\mathrm{Pr}+1)^2$. Consider

$$f'_b(x)=\frac{(2x-\pi^2)(x+\pi^2)^2-b}{x^2}. \tag{4.4.53}$$

Let x_i be the zero of $f'_{b_i}(x)$ $(i=1,2)$. Then we infer from (4.4.53) that

$$x_1\sim\left(\frac{1}{\mathrm{Pr}}\right)^{2/3}\to\infty,\qquad x_2<\frac{\pi^2}{2}(1+\omega^2)\quad \text{for } \mathrm{Pr}\to 0,$$

which implies that

$$R_{c_1}\sim f_{b_1}(x_1)\to+\infty,\qquad R_{c_2}\sim f_{b_2}(x_2)<+\infty\qquad \text{if } \mathrm{Pr}\to 0.$$

Hence the following lemma holds.

Lemma 4.4.8 *Given $L,\alpha,\omega>0$, there exists a number $\delta=\delta(L,\alpha,\omega)\in(0,1)$ such that if $0<Pr<\delta$, then R_{c_2} is the first critical Rayleigh number, i.e., $R_{c_2}<R_{c_1}$.*

4.4.4 Transition from First Real Eigenvalues

In this section, we consider the transition of (4.4.1)–(4.4.3) from the first real eigenvalues of (4.4.7) with multiplicities $m=2$ in H_1 and $m=1$ in $\widetilde{H}_1$, which are called (generalized) simple eigenvalues. By Lemma 4.4.7, for almost all $(L_1,L_2)=(L,\alpha L)\in\mathbb{R}^2_+$, the first eigenvalues of (4.4.7) are simple. Here we always assume that the number R_{c_1} defined by (4.4.33) is the first critical Rayleigh number, and the first eigenvalues of (4.4.7) are simple near R_{c_1}.

Theorem 4.4.9 *If the Prandtl number Pr and the rotating angular velocity ω satisfy*

$$b_1=\mathrm{Pr}^4(\alpha^2_{j_1k_1}+\pi^2)^3\alpha^2_{j_1k_1}-\pi^2(\pi^2-\mathrm{Pr}^2\alpha^2_{j_1k_1})\omega^2>0, \tag{4.4.54}$$

where (j_1,k_1) satisfies (4.4.50), then the transition of problem (4.4.1)–(4.4.3) at $R=R_{c_1}$ is of Type I (continuous), and the following assertions hold:

(1) *For the constraint (4.4.4), this problem bifurcates from* $((u,T),R) = (0,R_{c_1})$ *to an attractor* $\Sigma_R = S^1 \subset H_1$ *consisting of only steady-state solutions.*
(2) *For the constraint (4.4.5), the problem bifurcates to two steady-state solutions* $\Sigma_R = \{(u_1^R, T_1^R), (u_2^R, T_2^R)\} \subset \widetilde{H}_1$, *which are attractors.*
(3) *The steady-state solutions* $(u,T) \in \Sigma_R = S^1 \subset H_1$ *can be expressed as*

$$(u,T) = a(\beta^1_{j_1k_11}(R))^{1/2}(x\psi^1_{j_1k_11} + y\widetilde{\psi}^1_{j_1k_11}) + o(|\beta^1_{j_1k_11}|^{1/2}),$$

where $a > 0$ *is a constant,* $\psi^1_{j_1k_11}$, *and* $\widetilde{\psi}^1_{j_1k_11}$ *are the first eigenvectors, given by (4.4.22) and (4.4.23), and* $x^2 + y^2 = 1$.
(4) *The two bifurcated steady-state solutions* (u_i^R, T_i^R) $(i = 1,2)$ *in* $\widetilde{H}_1$ *can be written as*

$$(u_i^R, T_i^R) = (-1)^{i+1}a(\beta^1_{j_1k_11}(R))^{1/2}\psi^1_{j_1k_11} + o(|\beta^1_{j_1k_11}|^{1/2}).$$

Theorem 4.4.10 *If* Pr *and* ω *satisfy*

$$b_1 = \mathrm{Pr}^4(\alpha^2_{j_1k_1} + \pi^2)^3\alpha^2_{j_1k_1} - \pi^2(\pi^2 - \mathrm{Pr}^2\alpha^2_{j_1k_1})\omega^2 < 0, \tag{4.4.55}$$

then the transition of problem (4.4.1)–(4.4.3) at $R = R_{c_1}$ *is a Type II (jump) transition. Moreover, we have the following assertions:*

(1) *Under the constraint (4.4.4), there is a singular separation of a cycle of steady-state solutions of* (Σ^*, R^*) *with* $\Sigma^* = S^1, R^* < R_{c_1}$, *i.e., there are two branches of cycles* Σ_R^1 *and* Σ_R^2 *of steady-state solutions bifurcated from* Σ^* *on* $R^* < R$ *such that*

$$\lim_{R\to R_{c_1}} \Sigma_R^1 = \{0\}, \quad \Sigma^2_{R_{c_1}} \neq \{0\} \quad (= S^1),$$

and $\Sigma_R^2 = S^1$ *is an attractor for all* $R^* < R < R_{c_1} + \delta$ *for some* $\delta > 1$.
(2) *Under the constraint (4.4.5), there is a saddle-node bifurcation of steady-state solutions from* (ψ_0^*, R^*) *and* (ψ_1^*, R^*) *with* $\psi_0^*, \psi_1^* \neq 0, R^* < R_{c_1}$, *i.e., there are four branches of steady-state solutions* ψ_R^i $(1 \leq i \leq 4)$ *with* ψ_R^1 *and* ψ_R^2 *bifurcated from* ψ_0^*, ψ_R^3 *and* ψ_R^4 *from* ψ_1^* *on* $R^* < R$, *such that*

$$\lim_{R\to R_{c_1}} \psi_R^1 = 0, \quad \lim_{R\to R_{c_1}} \psi_R^3 = 0, \quad \psi^2_{R_{c_1}}, \psi^4_{R_{c_1}} \neq 0,$$

and ψ_R^2 *and* ψ_R^4 *are attractors for* $R^* < R < R_{c_1} + \delta$ *with some* $\delta > 0$.

Remark 4.4.11 Theorems 4.4.9 and 4.4.10 show that the transition of problem (4.4.1)–(4.4.3) is possibly a jump (Type-II) only under the condition $\pi^2 > \mathrm{Pr}^2\alpha^2_{j_1k_1}$, which provides an upper bound for the Prandtl number Pr leading to the jump transition at R_{c_1}.

Proof of Theorems 4.4.9 and 4.4.10. We first reduce the equations of (4.4.1)–(4.4.3) with (4.4.5) to the center manifold in $\widetilde{H}$.

Let $J_1 = (j_1, k_1, 1)$, $\psi_{J_1} = \psi^1_{j_1 k_1 1}$. The reduced equations of (4.4.1)–(4.4.3) with (4.4.5) in $\widetilde{H}$ read

$$\frac{dx}{dt} = \beta^1_{J_1}(R)x + \frac{1}{(\psi_{J_1}, \psi^*_{J_1})}(G(\psi, \psi), \psi^*_{J_1}), \tag{4.4.56}$$

where $\psi \in \widetilde{H}_1$ is written as

$$\psi = x\psi_{J_1} + \Phi, \tag{4.4.57}$$

Φ is the center manifold function, and $G(\psi_1, \psi_2) = -P((U_1 \cdot \nabla)U_2, (U_1 \cdot \nabla)T_2)$ for $\psi_i = (U_i, T_i)$ $(i = 1, 2)$.

Hereinafter, we make the following convention:

$$o(m) = o(|x|^m) + O(|\beta^1_J| \cdot |x|^m) \quad \text{for } x \in \mathbb{R}^n \text{ near } 0.$$

By $\Phi(x) = O(2)$, inserting (4.4.57) into (4.4.56), we obtain

$$\frac{dx}{dt} = \beta^1_{J_1}(R)x - \frac{x}{(\psi_{J_1}, \psi^*_{J_1})}[(G(\psi_{J_1}, \psi^*_{J_1}), \Phi) + (G(\Phi, \psi^*_{J_1}), \psi_{J_1})] + o(3). \tag{4.4.58}$$

Let the center manifold function be in the form

$$\Phi = \sum_{(J,i) \neq (J_1,1)} y^i_J \psi^i_J. \tag{4.4.59}$$

Then by the approximation formula for center manifold functions in Theorem A.1.1, we obtain

$$L_R \Phi = \sum_{(J,i) \neq (J_1,1)} y^i_J \beta^i_J \psi^i_J = -x^2 G(\psi_{J_1}, \psi_{J_1}) + o(2).$$

It follows that

$$y^i_J = \frac{x^2}{-\beta^i_J(\psi^i_J, \psi^{i*}_J)}(G(\psi_{J_1}, \psi_{J_1}), \psi^{i*}_J) + o(2). \tag{4.4.60}$$

Putting (4.4.59) and (4.4.60) into (4.4.58), we deduce that

$$\frac{dx}{dt} = \beta^1_{J_1}(R)x - \frac{\delta x^3}{(\psi_{J_1}, \psi^*_{J_1})} + o(3), \tag{4.4.61}$$

where

$$\delta = \sum_{(J,i) \neq (J_1,1)} \Big[\frac{(G(\psi_{J_1}, \psi_{J_1}), \psi^{i*}_J)(G(\psi_{J_1}, \psi^*_{J_1}), \psi^i_J)}{-\beta^i_J(\psi^i_J, \psi^{i*}_J)} + \frac{(G(\psi_{J_1}, \psi_{J_1}), \psi^{i*}_J)(G(\psi^i_J, \psi^*_{J_1}), \psi_{J_1})}{-\beta^i_J(\psi^i_J, \psi^{i*}_J)} \Big].$$

By (4.4.22) and (4.4.26),

$$\psi_{J_1}=\begin{cases}u\sin(j_1\alpha_1x_1+k_1\alpha_2x_2)\cos\pi x_3,\\ v\sin(j_1\alpha_1x_1+k_1\alpha_2x_2)\cos\pi x_3,\\ \cos(j_1\alpha_1x_1+k_1\alpha_2x_2)\sin\pi x_3,\\ T\cos(j_1\alpha_1x_1+k_1\alpha_2x_2)\sin\pi x_3,\end{cases}$$

$$\psi^*_{J_1}=\begin{cases}u^*\sin(j_1\alpha_1x_1+k_1\alpha_2x_2)\cos\pi x_3,\\ v^*\sin(j_1\alpha_1x_1+k_1\alpha_2x_2)\cos\pi x_3,\\ \cos(j_1\alpha_1x_1+k_1\alpha_2x_2)\sin\pi x_3,\\ T^*\cos(j_1\alpha_1x_1+k_1\alpha_2x_2)\sin\pi x_3,\end{cases}$$

where

$$u=-\frac{\pi}{\alpha^2_{j_1k_1}}(j_1\alpha_1+k_1\alpha_2\omega A_1),\quad v=-\frac{\pi}{\alpha^2_{j_1k_1}}(k_1\alpha_2-j_1\alpha_1\omega A_1),\quad T=A_2,$$
$$u^*=-\frac{\pi}{\alpha^2_{j_1k_1}}(j_1\alpha_1-k_1\alpha_2\omega A_1),\quad v^*=-\frac{\pi}{\alpha^2_{j_1k_1}}(k_1\alpha_2+j_1\alpha_1\omega A_1),\quad T^*=\mathrm{Pr}\,RA,$$
$$A_1=A_1(\beta^1_{J_1}),\qquad A_2=A_2(\beta^1_{J_1}).$$

Direct calculation shows that

$$G(\psi_{J_1},\psi_{J_1})=\frac{\pi}{2}P(u\sin 2\phi,v\sin 2\phi,0,0)-\frac{\pi}{2}(0,0,0,T\sin 2\pi x_3), \tag{4.4.62}$$
$$G(\psi_{J_1},\psi^*_{J_1})=\frac{\pi}{2}P(u^*\sin 2\phi,v^*\sin 2\phi,0,0)-\frac{\pi}{2}(0,0,0,T^*\sin 2\pi x_3), \tag{4.4.63}$$

where $\phi=j_1\alpha_1x_1+k_1\alpha_2x_2$, and $P:L^2(\Omega)^4\to\widetilde{H}$ is the Leray projection. Also, we know that

$$G(\psi^1_{002},\psi^*_{J_1})=0,\qquad G(\psi_{2(j_1,k_1,0)},\psi^*_{J_1})=0. \tag{4.4.64}$$

It follows from (4.4.62)–(4.4.64) that

$$\begin{aligned}&(G(\psi_{J_1},\psi_{J_1}),\psi^{i*}_J)=0 &&\forall\psi^{i*}_J\neq\psi^*_{2(j_1,k_1,0)},\psi^{1*}_{002},\\ &(G(\psi_{J_1},\psi^*_{J_1}),\psi^i_J)=0 &&\forall\psi^i_J\neq\psi_{2(j_1,k_1,0)},\psi^1_{002},\\ &(G(\psi^i_J,\psi^*_{J_1}),\psi_{J_1})=0 &&\text{for }\psi^i_J=\psi^1_{002}\text{ or }\psi_{2(j_1,k_1,0)}.\end{aligned}$$

Thus, we derive

$$\delta=\frac{(G(\psi_{J_1},\psi_{J_1}),\psi^*_{2(j_1,k_1,0)})(G(\psi_{J_1},\psi^*_{J_1}),\psi_{2(j_1,k_1,0)})}{-\beta_{2(j_1,k_1,0)}(\psi_{2(j_1,k_1,0)},\psi^*_{2(j_1,k_1,0)})}+\frac{(G(\psi_{J_1},\psi_{J_1}),\psi^{1*}_{002})(G(\psi_{J_1},\psi^*_{J_1}),\psi^1_{002})}{-\beta^1_{002}(\psi^1_{002},\psi^{1*}_{002})}.$$

Note that $-\beta^1_{002}=4\pi^2,-\beta_{2(j_1,k_1,0)}=4\mathrm{Pr}\,\alpha^2_{j_1k_1}$, and

$$\psi_{002}^1 = \psi_{002}^{1*} = (0,0,0,\sin 2\pi x_3),\cdots,$$
$$\psi_{2(j_1,k_1,0)} = \psi_{2(j_1,k_1,0)}^* = (k_1\alpha_2 \sin 2\phi, -j_1\alpha_1 \sin 2\phi, 0, 0).$$

By calculation, we get

$$\begin{aligned}\delta &= \frac{L_1L_2\pi^2}{8}\left[\Pr RA_2^2(\beta_{J_1}^1) - \frac{\pi^4\omega^2A_1^2(\beta_{J_1}^1)}{Pr\alpha_{j_1k_1}^4}\right]\\ &= \frac{L_1L_2\pi^2}{8}\left[\frac{\Pr R}{(\alpha_{j_1k_1}^2+\pi^2+\beta_{J_1}^1)^2} - \frac{\pi^4\omega^2}{\Pr\alpha_{j_1k_1}^4(\Pr\alpha_{j_1k_1}^2+\Pr\pi^2+\beta_{J_1}^1)^2}\right].\end{aligned} \tag{4.4.65}$$

We know that

$$\beta_{J_1}^1(R_{c_1}) = 0, \quad R_{c_1} = \frac{(\alpha_{j_1k_1}^2+\pi^2)^3}{\alpha_{j_1k_1}^2} + \frac{\pi^2\omega^2}{\Pr^2\alpha_{j_1k_1}^2}.$$

Hence, by (4.4.65) at $R = R_{c_1}$, we have

$$\delta(R_{c_1}) = \frac{L_1L_2\pi^2}{8(\alpha_{j_1k_1}^2+\pi^2)^2\alpha_{j_1k_1}^2}\left[\Pr(\alpha_{j_1k_1}^2+\pi^2)^3 + \frac{\pi^2\omega^2}{\Pr} - \frac{\pi^4\omega^2}{P_r^3\alpha_{j_1k_1}^2}\right]. \tag{4.4.66}$$

Thus, it follows from (4.4.66) that

$$\begin{aligned}\delta(R_{c_1}) > 0 &\Longleftrightarrow P_r^4(\alpha_{j_1k_1}^2+\pi^2)^3\alpha_{j_1k_1}^2 > \pi^2(\pi^2-\Pr^2\alpha_{j_1k_1}^2)\omega^2,\\ \delta(R_{c_1}) < 0 &\Longleftrightarrow P_r^4(\alpha_{j_1k_1}^2+\pi^2)^3\alpha_{j_1k_1}^2 < \pi^2(\pi^2-\Pr^2\alpha_{j_1k_1}^2)\omega^2.\end{aligned} \tag{4.4.67}$$

In addition, at $R = R_{c_1}$ we see that

$$\begin{aligned}(\psi_{J_1},\psi_{J_1}^*) =& \pi^4\left[\frac{1}{\alpha_{j_1k_1}^2} - \frac{\omega^2A_1}{\alpha_{j_1k_1}^2} + 1 + \Pr R_{c_1}A_2^2\right]\\ =& \frac{\pi^2}{\alpha_{j_1k_1}^2(\alpha_{j_1k_1}^2+\pi^2)^2}\Big[(\alpha_{j_1k_1}^2+\pi^2)^2(\alpha_{j_1k_1}^2+1)\\ &+\frac{\pi^2\omega^2}{\Pr} + \Pr(\alpha_{j_1k_1}^2+\pi^2)^3 - \frac{\omega^2}{\Pr^2}\Big].\end{aligned}$$

It is clear that for $\Pr > \frac{1}{\pi^2}$,

$$(\psi_{J_1},\psi_{J_1}^*) > 0. \tag{4.4.68}$$

In addition, by the spectral theorem, Theorem 3.4 in Ma and Wang (2005b), $(\psi_{J_1},\psi_{J_1}^*) \neq 0$, which is continuous in Pr , ω, and $\alpha_{j_1k_1}$. Therefore, (4.4.68) holds if $\beta_{J_1}^1$ is real and simple near $R = R_{c_1}$.

Hence, by Theorems 2.3.1 and 2.5.3, from (4.4.61), (4.4.67), and (4.4.68), we derive the conclusions of Theorems 4.4.9 and 4.4.10 in $\widetilde{H}$ for the condition (4.4.5).

Under the condition (4.4.4), the reduced equations of (4.4.1)–(4.4.3) on the center manifold in H are given by

$$\begin{aligned}\frac{dx}{dt} &= \beta_{J_1}^1(R)x + \frac{1}{(\psi_{J_1},\psi_{J_1}^*)}(G(\psi,\psi),\psi_{J_1}^*),\\ \frac{dy}{dt} &= \beta_{J_1}^1(R)y + \frac{1}{(\widetilde{\psi}_{J_1},\widetilde{\psi}_{J_1}^*)}(G(\psi,\psi),\widetilde{\psi}_{J_1}^*),\end{aligned} \tag{4.4.69}$$

where $\psi = x\psi_{J_1} + y\widetilde{\psi}_{J_1} + \Phi \in H$. It is easy to verify that

$$\begin{aligned}&(G(x\psi_{J_1} + y\widetilde{\psi}_{J_1}, x\psi_{J_1} + y\widetilde{\psi}_{J_1}),\psi_{J_1}^*) = 0,\\ &(G(x\psi_{J_1} + y\widetilde{\psi}_{J_1}, x\psi_{J_1} + y\widetilde{\psi}_{J_1}),\widetilde{\psi}_{J_1}^*) = 0.\end{aligned}$$

Hence, (4.4.69) can be rewritten as

$$\begin{aligned}\frac{dx}{dt} &= \beta_{J_1}^1 x - \frac{x}{(\psi_{J_1},\psi_{J_1}^*)}\left[(G(\psi_{J_1},\psi_{J_1}^*),\Phi) + (G(\Phi,\psi_{J_1}^*),\psi_{J_1})\right],\\ &\quad - \frac{y}{(\psi_{J_1},\psi_{J_1})}\left[(G(\widetilde{\psi}_{J_1},\psi_{J_1}^*),\Phi) + (G(\Phi,\psi_{J_1}^*),\widetilde{\psi}_{J_1})\right] + o(3),\\ \frac{dy}{dt} &= \beta_{J_1}^1 y - \frac{x}{(\widetilde{\psi}_{J_1},\widetilde{\psi}_{J_1}^*)}\left[(G(\psi_{J_1},\widetilde{\psi}_{J_1}^*),\Phi) + (G(\Phi,\widetilde{\psi}_{J_1}^*),\psi_{J_1})\right]\\ &\quad - \frac{y}{(\widetilde{\psi}_{J_1},\widetilde{\psi}_{J_1})}\left[(G(\widetilde{\psi}_{J_1},\widetilde{\psi}_{J_1}^*),\Phi) + (G(\Phi,\widetilde{\psi}_{J_1}^*),\widetilde{\psi}_{J_1})\right] + o(3).\end{aligned} \tag{4.4.70}$$

Based on the approximation formula for center manifold functions in Theorem A.1.1, we have

$$L_R\Phi = -x^2G(\psi_{J_1},\psi_{J_1}) - xyG(\psi_{J_1},\widetilde{\psi}_{J_1}) - xyG(\widetilde{\psi}_{J_1},\psi_{J_1}) - y^2G(\widetilde{\psi}_{J_1},\widetilde{\psi}_{J_1}) + o(2).$$

Let $\Phi = \Sigma\varphi_J(x,y)\psi_J^i$. Then, by $L_R\Phi = \Sigma\varphi_J\beta_J^i\psi_J^i$, we get

$$\begin{aligned}\varphi_J = &\frac{1}{-\beta_J^i(\psi_J^i,\psi_J^{i*})}\left[x^2(G(\psi_{J_1},\psi_{J_1}),\psi_J^{i*}) + y^2(G(\widetilde{\psi}_{J_1},\widetilde{\psi}_{J_1}),\psi_J^{i*})\right.\\ &\left.+xy(G(\psi_{J_1},\widetilde{\psi}_{J_1}) + G(\widetilde{\psi}_{J_1},\psi_{J_1}),\psi_J^{i*})\right] + o(2).\end{aligned}$$

By a direct calculation, we obtain

$$\begin{aligned}&\Phi = \varphi_1(x,y)\psi_{2(j_1,k_1,0)} + \varphi_2(x,y)\widetilde{\psi}_{2(j_1,k_1,0)} + \varphi_3(x,y)\psi_{002}^1 + o(2), \qquad (4.4.71)\\ &\varphi_1(x,y) = (x^2-y^2)\frac{(G(\psi_{J_1},\psi_{J_1}),\psi_{2(j_1,k_1,0)}^*)}{-\beta_{2(j_1,k_1,0)}(\psi_{2(j_1,k_1,0)},\psi_{2(j_1,k_1,0)}^*)},\\ &\varphi_2(x,y) = 2xy\frac{(G(\psi_{J_1},\widetilde{\psi}_{J_1}),\widetilde{\psi}_{2(j_1,k_1,0)}^*)}{-\beta_{2(j_1,k_1,0)}(\widetilde{\psi}_{2(j_1,k_1,0)},\widetilde{\psi}_{2(j_1,k_1,0)}^*)},\end{aligned}$$

$$\varphi_3(x,y) = (x^2+y^2)\frac{(G(\psi_{J_1},\psi_{J_1}),\psi_{002}^{1*})}{-\beta_{002}^1(\psi_{002}^1,\psi_{002}^{1*})}.$$

Likewise, we obtain

$$\begin{aligned}(G(\Phi,\psi_{J_1}^*),\psi_{J_1}) &= (G(\Phi,\psi_{J_1}^*),\widetilde{\psi}_{J_1}) = o(2),\\ (G(\Phi,\widetilde{\psi}_{J_1}^*),\psi_{J_1}) &= (G(\Phi,\widetilde{\psi}_{J_1}^*),\widetilde{\psi}_{J_1}) = o(2).\end{aligned}$$

Thus, from (4.4.70) and (4.4.71), we derive

$$\begin{aligned}\frac{dx}{dt} &= \beta_{J_1}^1(R)x - \frac{\delta x(x^2+y^2)}{(\psi_{J_1},\psi_{J_1}^*)} + o(3),\\ \frac{dy}{dt} &= \beta_{J_1}^1(R)y - \frac{\delta y(x^2+y^2)}{(\psi_{J_1},\psi_{J_1}^*)} + o(3),\end{aligned} \tag{4.4.72}$$

where δ is as in (4.4.65). Therefore, the conclusions of Theorems 4.4.9 and 4.4.10 in H for the condition (4.4.4) follow from (4.4.72), (4.4.67), and the S^1 attractor bifurcation theorem, Theorem 2.2.4. The proof of the theorems is complete. □

4.4.5 Transition from First Complex Eigenvalues

We now consider the transition of (4.4.1)–(4.4.3) with the condition (4.4.5) from the first complex eigenvalues of (4.4.7). In this section, we assume that the number R_{c_2} defined by (4.4.38) is the first critical Rayleigh number, and the first eigenvalues of (4.4.7) near R_{c_2} are complex and simple, i.e., the multiplicity is $m=2$ in $\widetilde{H}$. By Lemma 4.4.7, for almost all $(L_1,L_2)\in\mathbb{R}_+^2$, the first eigenvalues of (4.4.7) are simple in $\widetilde{H}$.

To determine the transition type, we need to introduce the following parameter:

$$\begin{aligned}b_2 =&\pi^3\omega^2(R_1B_1 - I_1B_2)C_1 + \pi^3\omega^2(I_1B_1+R_1B_2)C_2\\ &+ \Pr R(R_2B_1 - I_2B_2)C_3 + \Pr R(I_2B_1+R_2B_2)C_4,\end{aligned} \tag{4.4.73}$$

where

$$\begin{aligned}R_1 &= \mathrm{Re}\,A_1(\beta_{J_2}^1(R_{c_2})) = \frac{\Pr(\alpha^2+\pi^2)}{\Pr^2(\alpha^2+\pi^2)^2+\rho^2},\\ R_2 &= \mathrm{Re}A_2(\beta_{J_2}^1(R_{c_2})) = \frac{\alpha^2+\pi^2}{(\alpha^2+\pi^2)^2+\rho^2},\\ I_1 &= \mathrm{Im}A_1(\beta_{J_2}^1(R_{c_2})) = \frac{-\rho}{\Pr^2(\alpha^2+\pi^2)^2+\rho^2},\\ I_2 &= \mathrm{Im}A_2(\beta_{J_2}^1(R_{c_2})) = \frac{-\rho}{(\alpha^2+\pi^2)^2+\rho^2},\end{aligned}$$

$$
\begin{aligned}
B_1 &= \pi^2\omega^2 I_1^2 - \Pr R\alpha^2 I_2^2,\\
B_2 &= \pi^2\omega^2 R_1 I_1 - \Pr R\alpha^2 R_2 I_2,\\
C_1 &= \frac{3R_1}{2\Pr\alpha^4} - \frac{\rho^2 R_1}{2\Pr\alpha^4(4\Pr^2\alpha^4+\rho^2)} + \frac{\rho I_1}{\alpha^2(4\Pr^2\alpha^4+\rho^2)},\\
C_2 &= -\frac{I_1}{2\Pr\alpha^4} + \frac{\rho^2 I_1}{2\Pr\alpha^4(4\Pr^2\alpha^4+\rho^2)} + \frac{\rho R_1}{\alpha^2(4\Pr^2\alpha^4+\rho^2)},\\
C_3 &= -\frac{3R_2}{2\pi} + \frac{\rho^2 R_2}{2\pi(4\pi^4+\rho^2)} - \frac{\rho\pi I_2}{4\pi^4+\rho^2},\\
C_4 &= \frac{I_2}{2\pi} - \frac{\rho^2 I_2}{2\pi(4\pi^4+\rho^2)} - \frac{\rho\pi R_2}{4\pi^4+\rho^2},\\
\rho^2 &= \frac{1}{\alpha^2+\pi^2}\left[\frac{1-\Pr}{1+\Pr}\pi^2\omega^2 - \Pr^2(\alpha^2+\pi^2)^3\right],\\
R = R_{c_2} &= \frac{2(\Pr+1)(\alpha^2+\pi^2)^3}{\alpha^2} + \frac{2\pi^2\omega^2}{(\Pr+1)\alpha^2},
\end{aligned}
$$

and $\alpha^2 = \alpha^2_{j_2k_2}$ with (j_2, k_2) satisfying (4.4.38).

We then have the following transition theorem, which provides a criterion for the supercritical and subcritical Hopf bifurcations according to the sign of the number b_2 in (4.4.73).

Theorem 4.4.12 *Let b be the number given by (4.4.73). Then for the problem (4.4.1)–(4.4.3) with (4.4.5), we have the following assertions:*

(1) If $b_2 < 0$, then the problem has a continuous (Type I) transition at $R = R_{c_2}$, and bifurcates to only one periodic solution on $R > R_{c_2}$, which is an attractor.
(2) If $b_2 > 0$, then this problem has a jump (Type II) transition at $R = R_{c_2}$, and bifurcates to a periodic solution on $R < R_{c_2}$, which is a repeller. Moreover, there exists a singular separation of periodic orbits at some R^* with $0 < R^* < R_{c_2}$.

Proof. Let $J = (j_2, k_2, 1), \alpha^2 = \alpha^2_{j_2k_2}$, and let $\beta_J = \beta_{J_2k_2l} = \sigma + i\rho$ be the first eigenvalue of (4.4.7) near $R = R_{c_2}$ in $\widetilde{H}_1$. By (4.4.24), the first eigenvectors $\psi_J = \psi_J^1 + i\psi_J^2$ are given by

$$
\begin{aligned}
\psi_J^1 &= (u_1 \sin\phi\cos\pi x_3, v_1\sin\phi\cos\pi x_3, \cos\phi\sin\pi x_3, T_1\cos\phi\sin\pi x_3),\\
\psi_J^2 &= (u_2 \sin\phi\cos\pi x_3, v_2\sin\phi\cos\pi x_3, 0, T_2\cos\phi\sin\pi x_3),
\end{aligned}
$$

where $\phi = j_2\alpha_1 x_1 + k_2\alpha_2 x_2$, and

$$
(u_1, v_1, T_1) = \left(-\frac{\pi j_2\alpha_1}{\alpha^2} - \frac{\pi k_2\alpha_2}{\alpha^2}\omega \mathrm{Re}A_1(\beta_J), -\frac{\pi k_2\alpha_2}{\alpha^2} + \frac{\pi j_2\alpha_1}{\alpha^2}\omega\mathrm{Re}A_1(\beta_J), R_eA_2(\beta_J)\right),
$$

$$
(u_2, v_2, T_2) = \left(-\frac{\pi k_2\alpha_2}{\alpha^2}\omega\mathrm{Im}A_1(\beta_J), \frac{\pi j_2\alpha_1}{\alpha^2}\omega\mathrm{Im}\,A_1(\beta_J), \mathrm{Im}A_2(\beta_J)\right).
$$

By (4.4.29), the first conjugate eigenvectors $\psi_J^* = \psi_J^{1*} + i\psi_J^{2*}$ are given by

$$\psi_J^{1*} = (u_1^* \sin\phi \cos\pi x_3, v_1^* \sin\phi\cos\pi x_3, \cos\phi \sin\pi x_3, T_1^* \cos\phi \sin\pi x_3),$$
$$\psi_J^{2*} = (u_2^* \sin\phi \cos\pi x_3, v_2^* \sin\phi\cos\pi x_3, 0, T_2^* \cos\phi \sin\pi x_3),$$

where

$$u_1^* = -\frac{\pi j_2\alpha_1}{\alpha^2} + \frac{\pi k_2\alpha_2}{\alpha^2}\omega \mathrm{Re} A_1(\overline{\beta}_J), \qquad v_1^* = -\frac{\pi k_2\alpha_2}{\alpha^2} - \frac{\pi j_2\alpha_1}{\alpha^2}\omega \mathrm{Re} A_1(\overline{\beta}_J),$$
$$T_1^* = \Pr R \mathrm{Re} A_2(\overline{\beta}_J), \qquad u_2^* = \frac{\pi k_2\alpha_2}{\alpha^2}\omega \mathrm{Im} A_1(\overline{\beta}_J),$$
$$v_2^* = -\frac{\pi j_2\alpha_1}{\alpha^2}\omega \mathrm{Im} A_1(\overline{\beta}_J), \qquad T_2^* = \Pr R I_m A_2(\overline{\beta}_J).$$

Note that

$$\beta_J(\psi_J, \psi_J^*) = (L_R\psi_J, \psi_J^*) = (\psi_J, L_R^*\psi_J^*) = \overline{\beta}_J(\psi_J, \psi_J^*).$$

Hence

$$(\psi_J^1, \psi_J^{1*}) = (\psi_J^2, \psi_J^{2*}), \quad (\psi_J^1, \psi_J^{2*}) = -(\psi_J^2, \psi_J^{1*}). \tag{4.4.74}$$

It is clear that for every real number α, the functions $\Phi_J^* = \Phi_J^{1*} + i\Phi_J^{2*}$, with

$$\Phi_J^{1*} = \psi_J^{1*} + \alpha\psi_J^{2*}, \quad \Phi_J^{2*} = -\alpha\psi_J^{1*} + \psi_J^{2*},$$

are also the first conjugate eigenvectors of L_R^*. By (4.4.74), they satisfy

$$\begin{aligned} &(\psi_J^1, \Phi_J^{1*}) = (\psi_J^2, \Phi_J^{2*}) \neq 0,\\ &(\psi_J^1, \Phi_J^{2*}) = (\psi_J^2, \Phi_J^{1*}) = 0,\\ &(\psi_{J_k}^i, \Phi_J^*) = 0, \quad \forall \psi_{J_k}^i \neq \Phi_J^{1*}, \Phi_J^{2*}, \end{aligned} \tag{4.4.75}$$

provided

$$\alpha = \frac{(\psi_J^1, \psi_J^{2*})}{(\psi_J^1, \psi_J^{1*})} = -\frac{(\psi_J^2, \psi_J^{1*})}{(\psi_J^2, \psi_J^{2*})}.$$

The two eigenvectors Φ_J^{1*} and Φ_J^{2*} satisfying (4.4.75) can be expressed as

$$\begin{aligned} \Phi_J^{1*} &= \frac{1}{(\psi_J^1, \psi_J^{1*})}[(\psi_J^1, \psi_J^{1*})\psi_J^{1*} + (\psi_J^1, \psi_J^{2*})\psi_J^{2*}],\\ \Phi_J^{2*} &= \frac{1}{(\psi_J^1, \psi_J^{1*})}[(\psi_J^2, \psi_J^{1*})\psi_J^{1*} + (\psi_J^2, \psi_J^{2*})\psi_J^{2*}]. \end{aligned} \tag{4.4.76}$$

By (4.4.75), the reduced equations of (4.4.1)–(4.4.3) read

$$\begin{aligned} \frac{dx}{dt} &= \sigma x + \rho y + \frac{1}{(\psi_J^1, \Phi_J^{1*})}(G(\psi,\psi), \Phi_J^{1*}),\\ \frac{dy}{dt} &= -\rho x + \sigma y + \frac{1}{(\psi_J^2, \Phi_J^{2*})}(G(\psi,\psi), \Phi_J^{2*}), \end{aligned} \tag{4.4.77}$$

where $\psi \in \widetilde{H}_1$ is written as

$$\psi = x\psi_J^1 + y\psi_J^2 + \Phi(x,y), \tag{4.4.78}$$

and Φ is the center manifold function.

Inserting (4.4.78) into (4.4.77), we obtain

$$\begin{aligned}
\frac{dx}{dt} &= \sigma x + \rho y - \frac{1}{(\psi_J^1, \Phi_J^{1*})}[x(G(\psi_J^1, \Phi_J^{1*}), \Phi) \\
&\quad + y(G(\psi_J^2, \Phi_J^{1*}), \Phi) + x(G(\Phi, \Phi_J^{1*}), \psi_J^1) \\
&\quad + y(G(\Phi, \Phi_J^{1*}), \psi_J^2)] + o(3), \\
\frac{dy}{dt} &= -\rho x + \sigma y - \frac{1}{(\psi_J^2, \Phi_J^{2*})}[x(G(\psi_J^1, \Phi_J^{2*}), \Phi) \\
&\quad + y(G(\psi_J^2, \Phi_J^{2*}), \Phi) + x(G(\Phi, \Phi_J^{2*}), \psi_J^1) \\
&\quad + y(G(\Phi, \Phi_J^{2*}), \psi_J^2)] + o(3).
\end{aligned} \tag{4.4.79}$$

As before, we have

$$\begin{aligned}
G(\psi_J^1, \psi_J^i) &= \frac{\pi}{2}P(u_i \sin 2\phi, v_i \sin 2\phi, 0, 0) + \frac{\pi}{2}(0,0,0,T_i \sin 2\pi x_3), \\
G(\psi_J^1, \psi_J^{i*}) &= \frac{\pi}{2}P(u_i^* \sin 2\phi, v_i^* \sin 2\phi, 0, 0) + \frac{\pi}{2}(0,0,0,T_i \sin 2\pi x_3), \\
G(\psi_J^2, \psi_J^i) &= G(\psi_J^2, \psi_J^{i*}) = 0,
\end{aligned} \tag{4.4.80}$$

for $i = 1,2$. By (A.1.19) and (4.4.80) with ρ replaced by $-\rho$ in (A.1.19), the center manifold function Φ satisfies

$$\begin{aligned}
&(L_R^2 + 4\rho^2)L_R\Phi = -(L_R^2 + 4\rho^2)(x^2 G_{11} + xyG_{12}) \\
&+ 2\rho^2((x^2 - y^2)G_{11} + 2xyG_{12}) - \rho L_R((y^2 - x^2)G_{12} + 2xyG_{11}),
\end{aligned} \tag{4.4.81}$$

where $G_{ij} = G(\psi_J^i, \psi_J^j)$. Based on (4.4.81), we see that

$$\Phi = \Phi_1 + \Phi_2 + \Phi_3,$$

and

$$\begin{aligned}
&L_R\Phi_1 = -(x^2 G_{11} + xyG_{12}), \\
&(L_R^2 + 4\rho^2)L_R\Phi_2 = 2\rho^2((x^2 - y^2)G_{11} + 2xyG_{12}), \\
&(L_R^2 + 4\rho^2)\Phi_3 = -\rho((y^2 - x^2)G_{12} + 2xyG_{11}).
\end{aligned} \tag{4.4.82}$$

By (4.4.80), G_{1i} $(i = 1,2)$ are linear combinations of the eigenvectors $\psi_k = \psi_{2(j_2k_20)}$ and ψ_{002}^1 of L_R. Hence, we have

$$\Phi_j = \varphi_{j1}\psi_k + \varphi_{j2}\psi_{002}^1, \quad j = 1,2,3.$$

From (4.4.80) and (4.4.82) we derive that

$$\begin{aligned}\Phi_1 =& -\frac{\pi^2\omega}{8\mathrm{Pr}\ \alpha^4}(R_1x^2+I_1xy)\psi_k+\frac{1}{8\pi}(R_2x^2+I_2xy)\psi^1_{002},\\ \Phi_2 =& \frac{\pi^2\omega\rho^2}{16\mathrm{Pr}\ \alpha^4(4\mathrm{Pr}^2\alpha^4+\rho^2)}(R_1(x^2-y^2)+2I_1xy)\psi_k\\ &-\frac{\rho^2}{16\pi(4\pi^2+\rho^2)}(R_2(x^2-y^2)+2I_2xy)\psi^1_{002},\\ \Phi_3 =& \frac{\pi^2\omega\rho}{8\alpha^2(4\mathrm{Pr}^2\alpha^4+\rho^2)}(I_1(y^2-x^2)+2R_1xy)\psi_k\\ &-\frac{\pi\rho}{8(4\pi^2+\rho^2)}(I_2(y^2-x^2)+2R_2xy)\psi^1_{002},\end{aligned}$$

where R_j and $I_j (j=1,2)$ are as in (4.4.73). Thus, we get the center manifold function Φ as follows:

$$\Phi=(E_1x^2+E_2xy+E_3y^2)\psi_k+(D_1x^2D_2xy+D_3y^2)\psi^1_{002}, \tag{4.4.83}$$

where

$$\begin{aligned}E_1 &= -\frac{\pi^2\omega R_1}{8\mathrm{Pr}\ \alpha^4}+\frac{\pi^2\omega\rho^2R_1}{16\mathrm{Pr}\ \alpha^4(4\mathrm{Pr}^2\alpha^4+\rho^2)}-\frac{\pi^2\omega\rho I_1}{8\alpha^2(4\mathrm{Pr}^2\alpha^4+\rho^2)},\\ E_2 &= -\frac{\pi^2\omega I_1}{8\mathrm{Pr}\ \alpha^4}+\frac{\pi^2\omega\rho^2 I_1}{8\mathrm{Pr}\ \alpha^4(4\mathrm{Pr}^2\alpha^4+\rho^2)}+\frac{\pi^2\omega\rho R_1}{4\alpha^2(4\mathrm{Pr}^2\alpha^4+\rho^2)},\\ E_3 &= -\frac{\pi^2\omega\rho^2R_1}{16\mathrm{Pr}\ \alpha^4(4\mathrm{Pr}^2\alpha^4+\rho^2)}+\frac{\pi^2\omega\rho I_1}{8\alpha^2(4\mathrm{Pr}^2\alpha^4+\rho^2)},\\ D_1 &= \frac{R_2}{8\pi}-\frac{\rho^2R_2}{16\pi(4\pi^4+\rho^2)}+\frac{\rho\pi I_2}{4(4\pi^4+\rho^2)},\\ D_2 &= \frac{I_2}{8\pi}-\frac{\rho^2I_2}{8\pi(4\pi^4+\rho^2)}-\frac{\rho\pi R_2}{4(4\pi^4+\rho^2)},\\ D_3 &= \frac{\rho^2R_2}{16\pi(4\pi^4+\rho^2)}-\frac{\rho\pi I_2}{8(4\pi^4+\rho^2)}.\end{aligned}$$

It follows from (4.4.80) and (4.4.83) that

$$(G(\psi^2_J,\psi^{i*}_J),\Phi)=0, \qquad (G(\Phi,\psi^{i*}_J),\psi^j_J)=0,$$

for $i,j=1,2$. By (4.4.74) and (4.4.76), (4.4.79) can be rewritten as

$$\begin{aligned}\frac{dx}{dt} =& \sigma x+\rho y-\frac{x}{D^2}[(\psi^2_J,\psi^{2*}_J)(G(\psi^1_J,\psi^{1*}_J),\Phi)\\ &+(\psi^1_J,\psi^{2*}_J)(G(\psi^1_J,\psi^{2*}_J),\Phi)]+o(3),\end{aligned} \tag{4.4.84}$$

$$\frac{dy}{dt} = -\rho x + \sigma y - \frac{x}{D^2}[(\psi_J^2, \psi_J^{1*})(G(\psi_J^1, \psi_J^{1*}), \Phi) + (\psi_J^2, \psi_J^{2*})(G(\psi_J^1, \psi_J^{2*}), \Phi)] + o(3),$$

where $D^2 = (\psi_J^1, \psi_J^{1*})^2 + (\psi_J^1, \psi_J^{2*})^2$.

Inserting (4.4.83) into (4.4.84), we obtain

$$\begin{aligned} \frac{dx}{dt} &= \sigma x + \rho y - \frac{\pi^5}{\alpha^2 D_2}[a_1 x^3 + a_2 x^2 y + a_3 x y^2] + o(3), \\ \frac{dy}{dt} &= -\rho x + \sigma y - \frac{\pi^5}{\alpha^2 D^2}[b_1 x^3 + b_2 x^2 y + b_3 x y^2] + o(3), \end{aligned} \tag{4.4.85}$$

where

$$\begin{aligned} a_1 &= \pi\omega(R_1 B_1 - I_1 B_2)E_1 + \Pr R(R_2 B_1 - I_2 B_2)D_1, \\ a_3 &= \pi\omega(R_1 B_1 - I_1 B_2)E_3 + \Pr R(R_2 B_1 - I_2 B_2)D_3, \\ b_2 &= -\pi\omega(I_1 B_1 + R_1 B_2)E_2 - \Pr R(I_2 B_1 + R_2 B_2)D_2, \end{aligned}$$

and B_1, B_2 are as in (4.4.73).

Thus, the number b in Theorem 2.3.7 reads

$$\begin{aligned} b &= -(3a_1 + a_3 + b_2) \\ &= -\pi\omega(R_1 B_1 - I_1 B_2)(3E_1 + E_3) \\ &\quad - \Pr R(R_2 B_1 - I_2 B_2)(3D_1 + D_3) + \pi\omega(I_1 B_1 + R_1 B_2)E_2 \\ &\quad + \Pr R(I_2 B_1 + R_2 B_2)D_2 \\ &= \frac{1}{4}[\pi^3\omega^2(R_1 B_1 - I_1 B_2)C_1 + \Pr R(R_2 B_1 - I_2 B_2)C_2 \\ &\quad + \pi^3\omega^2(I_1 B_1 + R_1 B_2)C_3 + \Pr R(I_2 B_1 + R_2 B_2)C_4] \\ &= \frac{1}{4}b_2, \end{aligned}$$

where b_2 is as in (4.4.73).

By Theorem 2.3.7, if $b < 0$, the transition of (4.4.85) at $R = R_{c_2}$ is continuous, i.e., the transition is an attractor bifurcation that is a supercritical Hopf bifurcation, and if $b > 0$, the transition is a jump transition, i.e., the transition is a subcritical Hopf bifurcation. Based on Theorem 2.5.7, if the transition is a jump, then there exists a singular separation at some $R^* < R_{c_2}$. Thus the theorem is proved. □

4.4.6 Physical Remarks

We begin with a few remarks on the physical parameters appearing in the model. First, the periods $L_1\pi$ and $L_2\pi$ in (4.4.3) generally represent the horizontal size of the containers in physical experiments, and in the atmospheric and oceanic circulations, they characterize the horizontal motion scale of circulation cells.

Second, the nondimensional number ω is proportional to the rotating angular velocity, which is given by

$$\omega = \Omega L^2/\kappa,$$

where Ω is the rotating angular velocity, L is the length scale, and κ is the thermal diffusivity. In the mid-latitude atmospheric and oceanic circulations, ω is the reciprocal of the Rossby number Ro:

$$\omega = \frac{1}{R_0}.$$

The transition theorems obtained in the previous sections give precise criteria for distinguishing the Type I and Type II transitions. Hereinafter, we shall discuss them in detail.

Transition Types of Steady-State Solutions. Let $\alpha_0 = \min\{\alpha_1, \alpha_2\}, \alpha_i = L_i^{-1}$ $(i=1,2)$. By Lemma 4.4.2, if ω^2/Pr^2 satisfies

$$\lambda \overset{\text{def}}{=} \frac{\omega^2}{\mathrm{Pr}^2} \begin{cases} \geq 0 & \text{for } \mathrm{Pr}^2 \geq 1, \\ < \dfrac{(\pi^2+\alpha_0^2)^3}{\pi^2} & \text{for } \mathrm{Pr}^2 < 1, \end{cases} \tag{4.4.86}$$

then R_{c_1} is the first critical Rayleigh number. Consider the number

$$\begin{aligned} &\mathrm{Pr}^4(\alpha_{j_1k_1}^2+\pi^2)^3\alpha_{j_1k_1}^2 - \pi^2(\pi^2-\mathrm{Pr}^2\alpha_{j_1k_1}^2)\omega^2 \\ &= \mathrm{Pr}^2[\mathrm{Pr}^2\alpha_{j_1k_1}^2(\alpha_{j_1k_1}^2+\pi^2)^3 - \pi^2(\pi^2-\mathrm{Pr}^2\alpha_{j_1k_1}^2)\lambda]. \end{aligned}$$

It is clear that

$$\text{Condition (4.4.54) holds} \Longleftrightarrow \mathrm{Pr}^2 > \frac{\pi^4\lambda}{\alpha_{j_1k_1}^2(\pi^2\lambda+(\alpha_{j_1k_1}^2+\pi^2)^3)}, \tag{4.4.87}$$

$$\text{Condition (4.4.55) holds} \Longleftrightarrow \mathrm{Pr}^2 < \frac{\pi^4\lambda}{\alpha_{j_1k_1}^2(\pi^2\lambda+(\alpha_{j_1k_1}^2+\pi^2)^3)}. \tag{4.4.88}$$

We know that the function

$$f(\lambda) = \frac{\pi^4\lambda}{\alpha_{j_1k_1}^2(\pi^2\lambda+(\alpha_{j_1k_1}^2+\pi^2)^3)}$$

is strictly increasing on λ. Therefore, (4.4.87) and (4.4.88) show that if $\pi^2 > \mathrm{Pr}^2\alpha_{j_1k_1}^2$, then small values of $\lambda = \omega^2/\mathrm{Pr}^2$ lead to a Type I (continuous) transition at $R = R_{c_1}$, and a large value of λ gives rise to a Type II (jump) transition.

Transition Types of Periodic Solutions. By (4.4.37), when R_{c_2} is the first critical Rayleigh number, the number λ defined by (4.4.86) satisfies

$$\lambda > \frac{(1+\mathrm{Pr})(\alpha^2+\pi^2)^3}{(1-\mathrm{Pr})\pi^2} \qquad \text{for } 0 < \mathrm{Pr} < 1,$$

where $\alpha^2 = \alpha_{j_2k_2}^2$ as in (4.4.38). Let

$$\gamma^2 = \alpha_{j_2k_2}^2 + \pi^2.$$

In the following, we show that both Type I and Type II transitions to the periodic solutions, dictated by the sign of b_2, can appear in different physical parameter ranges.

First, we assume that Pr $(0 < \text{Pr} < 1)$ is given, and

$$\omega^2 \gg \gamma^6 \gg 1. \tag{4.4.89}$$

Then we have

$$\begin{aligned}
&R = R_{c_2} \cong \frac{2\pi^2\omega^2}{(\text{Pr}+1)\alpha^2}, && \rho^2 \cong \frac{1}{\gamma^2}\frac{1-\text{Pr}}{1+\text{Pr}}\pi^2\omega^2, \\
&R_1 \cong \frac{\text{Pr}\,\gamma^2}{\rho^2}, && R_2 \cong \frac{\gamma^2}{\rho^2}, \\
&I_1 \cong -\frac{1}{\rho}, && I_2 \cong -\frac{1}{\rho}, \\
&C_1 \cong \frac{R_1}{\text{Pr}\,\alpha^4} - \frac{1}{\alpha^2\rho^2} = \frac{\pi^2}{\alpha^4}\frac{1}{\rho^2}, && C_2 \cong \frac{\text{Pr}(3\alpha^2+\pi^2)}{\alpha^2\rho^3}, \\
&C_3 \cong -\frac{\alpha^2}{\pi\rho^2}, && C_4 \cong -\frac{\pi(\gamma^2+4\pi^2)}{\rho^3}, \\
&B_1 \cong \frac{(1-\text{Pr})\pi^2\omega^2}{\rho^2}, && B_2 \cong \frac{(1-\text{Pr})\text{Pr}\,\gamma^2\pi^2\omega^2}{\rho^3}.
\end{aligned} \tag{4.4.90}$$

By (4.4.89) and (4.4.90), the number b_2 defined by (4.4.73) is approximated by

$$\begin{aligned}
b_2 \cong & \frac{\text{Pr}\,(1-\text{Pr})\pi^3\omega^4}{\alpha^4\rho^6}\Big[2\pi^2 + \pi^4\alpha^2 + \\
& + \frac{2\alpha^4\pi^2 + 10\alpha^2\pi^4}{\text{Pr}+1} - 3\alpha^4\pi^2 - 2\alpha^6 - 2\alpha^4\pi^2\Big] + o(\rho^{-7})\omega^4.
\end{aligned}$$

By (4.4.89), we see that $\rho^{-1} \cong 0$. Hence, for $\alpha^2 \leq \pi^2$, we deduce that $b_2 > 0$. Thus, by Theorem 4.4.12, we claim that when ω^2 is much larger than $\gamma^6 = (\alpha^2+\pi^2)^3$ and $\alpha^2 \geq \frac{\pi}{2}$, the transition of periodic solutions at $R = R_{c_2}$ is of Type II.

Next, we assume that $\rho > 0$ is small enough, i.e.,

$$1 \gg \rho > 0. \tag{4.4.91}$$

In this case, the number b_2 can be approximated by

$$\begin{aligned}
b_2 =& (\pi^4\omega^2(3R_1^2 - I_1^2) + \text{Pr}^2 R\alpha^4(I_2^2 - 3R_2^2))B_1 \\
& - 4(\pi^4\omega^2 R_1 I_1 - \text{Pr}^2 R\alpha^4 R_2 I_2)B_2.
\end{aligned} \tag{4.4.92}$$

By (4.4.91), we see that

$$\begin{aligned}
&\gamma^6 \cong \frac{1-\mathrm{Pr}}{\mathrm{Pr}^2(1+\mathrm{Pr}\,)}\pi^2\omega^2, && R = R_{c2} \cong \frac{2\pi^2\omega^2}{\alpha^2\mathrm{Pr}^2(1+\mathrm{Pr}\,)},\\
&R_1 \cong \frac{1}{\mathrm{Pr}\,\gamma^2}, && R_2 \cong \frac{1}{\gamma^2},\\
&I_1 \cong -\frac{\rho}{\mathrm{Pr}^2\gamma^4}, && I_2 \cong -\frac{\rho}{\gamma^4},\\
&B_1 \cong \frac{\pi^2\omega^2 I_2^2}{\mathrm{Pr}}\left(\frac{1}{\mathrm{Pr}^3}-\frac{2}{1+\mathrm{Pr}}\right), && B_2 \cong \frac{\pi^2\omega^2 R_2 I_2}{\mathrm{Pr}}\left(\frac{1}{\mathrm{Pr}^2}-\frac{2}{1+\mathrm{Pr}}\right).
\end{aligned} \tag{4.4.93}$$

From (4.4.92) and (4.4.93), we derive

$$\begin{aligned}
b_2 \cong \frac{\pi^4\omega^4 I_2^2 R_2^2}{\mathrm{Pr}}&\Big[3\Big(\frac{\pi^2}{\mathrm{Pr}^2}-\frac{2}{\alpha^2(1+\mathrm{Pr}\,)}\Big)\Big(\frac{1}{P_r^3}-\frac{2}{1+\mathrm{Pr}}\Big)\\
&-4\Big(\frac{\pi^2}{P_r^3}-\frac{2}{\alpha^2(1+\mathrm{Pr}\,)}\Big)\Big(\frac{1}{\mathrm{Pr}^2}-\frac{2}{1+\mathrm{Pr}}\Big)\Big].
\end{aligned}$$

It is clear that

$$b_2 < 0 \qquad \text{if } \mathrm{Pr} < \delta_0 \text{ for some } \frac{3}{5} < \delta_0 < 1.$$

Thus, by Theorem 4.4.12, under condition (4.4.91), if Pr $< \delta_0$, then the transition of periodic solutions at $R = R_{c_2}$ is of Type II.

In summary, we have shown that (1) both continuous and jump transitions occur in different flow regimes and (2) the jump transitions lead to hysteresis phenomena and a singularity (steady-state or periodic orbit) separation, giving rise to more complex (possibly chaotic) flow regimes. The transitions are precisely quantified by specific computable parameters.

4.5 Convection Scale Theory

The convection scale law presented in this section was observed recently by the authors and has been used in Ma and Wang (2011b), Ma and Wang (2010c), and Ma and Wang (2010b). This is associated with a phenomenon appearing in convection problems. It is known that thermal convection appears when the Rayleigh number R crosses the critical Rayleigh number R_c. In (4.1.1), the Rayleigh number is given by

$$R = \frac{g\alpha_T(T_0 - T_1)}{\kappa\nu}h^3 \quad (h \text{ the height of the fluid}). \tag{4.5.1}$$

For the free boundary condition, by (4.1.38), the critical Rayleigh number is

$$R_c = \min_{\alpha^2} \frac{(\alpha^2+\pi^2)^3}{\alpha^2} = \frac{(\alpha_c^2+\pi^2)^3}{\alpha_c^2} = \frac{27\pi^4}{4}, \tag{4.5.2}$$

where $\alpha_c^2 = \frac{\pi^2}{2}$, and for the Dirichlet (rigid) boundary condition, by Chandrasekhar (1981), R_c is

$$R_c = 1700. \tag{4.5.3}$$

According to (4.5.1)–(4.5.3), when the horizontal scale L of the fluid is much larger than its height h, i.e., $L \gg h$, thermal convection appears at the temperature difference

$$T_0 - T_1 = \begin{cases} \dfrac{27\kappa\nu\pi^4}{4g\alpha_T h^3} & \text{for the free boundary condition,} \\ \dfrac{1700\kappa\nu}{g\alpha_T h^3} & \text{for the Dirichlet boundary condition.} \end{cases} \tag{4.5.4}$$

From (4.5.4), we see that the critical temperature difference satisfies

$$\Delta T_c = T_0 - T_1 \to \infty \qquad \text{if } h \to 0, \tag{4.5.5}$$

$$\Delta T_c = T_0 - T_1 \to 0 \qquad \text{if } h \to \infty. \tag{4.5.6}$$

Physically, (4.5.5) amounts to saying that for a given fluid, there is a minimal size $h_0 > 0$ such that when $h < h_0$, no convection occurs for any $\Delta T = T_0 - T_1 > 0$. This is in agreement with the fact that in the real world, when h is small, it is hard to maintain a high temperature gradient to generate the vertical convection.

However, (4.5.6) has certain physical discrepancies with the energy conservation of the system. In fact, as h increases, it needs more energy to overcome the frictional force to drive the fluid to move. With ΔT representing this driving force, (4.5.6) shows that when h increases, ΔT decreases.

Here we present a method to resolve this issue. Note that the critical value $\alpha^2 = \pi^2/L^2$ defines the horizontal convection scale L (h scales to $h=1$): $L^2 = \pi^2/\alpha^2$. Hence, at the critical value $\alpha_c^2 = \frac{\pi^2}{2}$ in (4.5.2), the convective cell size is

$$L_c^2 = 2. \tag{4.5.7}$$

When h is not large, this value (4.5.7) is a good approximation, and we need to modify the model to accommodate the case in which h is large.

Instead of (4.5.7), we consider the convection scale L_c as an increasing function of h:

$$L_c^2 = \psi(h), \quad \text{such that } \psi(0) = 2, \ \psi'(h) > 0. \tag{4.5.8}$$

Thus, the critical Rayleigh number R_c reaches its minimal value at

$$\alpha_c^2 = \frac{\pi^2}{\psi(h)}, \tag{4.5.9}$$

rather than at $\alpha_c^2 = \pi^2/2$. The convection scale law of (4.5.8) and (4.5.9) should be reflected by the hydrodynamic equations. In other words, we need to revise the Boussinesq equations leading to new critical values given by (4.5.8) and (4.5.9). The method we propose here is to add a term

$$F = -(\sigma_0 u_1, \sigma_0 u_2, \sigma_1 u_3) \tag{4.5.10}$$

to the right-hand side of the first equations of (4.1.2), where σ_i depends only on h satisfying

$$\sigma_i(0) = 0, \quad \sigma_i'(h) \geq 0, \quad i = 0, 1. \tag{4.5.11}$$

Physically, (4.5.10) and (4.5.11) stand for the added resistance generated by pressure, called the damping terms.

In nondimensional form, the revised Boussinesq equations read

$$\begin{aligned} &\frac{\partial u}{\partial t} + (u \cdot \nabla)u = \mathrm{Pr}\,[\Delta u - \nabla p - f + RTk], \\ &\frac{\partial T}{\partial t} + (u \cdot \nabla)T = \Delta T + u_3, \\ &\mathrm{div} u = 0, \end{aligned} \tag{4.5.12}$$

where $f = (\delta_0 u_1, \delta_0 u_2, \delta_1 u_3)$, and

$$\delta_i(h) = \frac{h^2}{\nu} \sigma_i(h), \quad (i = 0, 1). \tag{4.5.13}$$

For the revised equations (4.5.12) with the free boundary conditions (4.1.34), the critical Rayleigh number is

$$R_c = \min_{\alpha^2} \left[(\alpha^2 + \pi^2)\delta_1 + \frac{(\alpha^2 + \pi^2)^3 + \pi^2(\alpha^2 + \pi^2)\delta_0}{\alpha^2} \right]. \tag{4.5.14}$$

Let $g(x) = (x + \pi^2)\delta_1 + ((x + \pi^2)^3 + \pi^2(x + \pi^2)\delta_0)/x$. Then $x_c = \alpha_c^2$ satisfies $g'(x) = 0$. Thus we get

$$\delta_1 \alpha_c^4 - (\alpha_c^2 + \pi^2)^3 - \pi^2(\alpha_c^2 + \pi^2)\delta_0 + 3\alpha_c^2(\alpha_c^2 + \pi^2)^2 + \pi^2 \alpha_c^2 \delta_0 = 0. \tag{4.5.15}$$

We assume that $\delta_1 > \delta_0$ for $h \neq 0$. By $\alpha_c^2 = \pi^2/L_c^2$, using (4.5.10), (4.5.11), and (4.5.15), we derive the convection scale law of (4.5.8) and (4.5.9). In particular, when h is large, $\delta_1 \gg 1$ and $\delta_1 \gg \delta_0$, we derive from (4.5.14) and (4.5.15) that

$$\alpha_c^4 \cong \frac{\pi^4(\pi^2 + \delta_0)}{\delta_1}, \quad L_c^2 \cong \delta_1^{1/2}/(\pi^2 + \delta_0)^{1/2}, \quad R_c \cong \pi^2\delta_1 + \pi^2\delta_0 L_c^2. \tag{4.5.16}$$

These formulas are very useful in studying large-scale convection motion for the height of fluid $h \geq 10m$. It can resolve the contradiction caused by (4.5.6). Indeed, by (4.5.1), (4.5.13), and (4.5.16), we obtain the critical temperature difference

$$\Delta T_c = \frac{\pi^2 \kappa \sigma_1(h)}{g\alpha_T h} \quad \text{for } h \gg 1m.$$

This suggests that $\sigma_1(h) \cong c \cdot h^2$, which results in the critical temperature gradient being independent of h.

In summary, by (4.5.13), we propose to take

$$\delta_0 = C_0 h^4/\nu, \qquad \delta_1 = C_1 h^4/\nu, \tag{4.5.17}$$

where C_0 and C_1 are constants.

For the rotating convection problem, the revised rotating Boussinesq equations are expressed as

$$\begin{aligned} &\frac{\partial u}{\partial t} + (u \cdot \nabla)u = \text{Pr}\left[\Delta u - \nabla p - f + RTk\right] - \omega k \times u, \\ &\frac{\partial T}{\partial t} + (u \cdot \nabla)T = \Delta T + u_3, \\ &\text{div} u = 0. \end{aligned} \tag{4.5.18}$$

For the boundary condition (4.4.2)–(4.4.3), similar to (4.4.21), the eigenvalues β of the linearized equations of (4.5.18) satisfy

$$\begin{aligned} &\alpha^2(\text{Pr}\,\gamma^2 + \text{Pr}\,\delta_0 + \beta)(\text{Pr}\,\gamma^2 + \text{Pr}\,\delta_1 + \beta)(\gamma^2 + \beta) \\ &+(\text{Pr}\,\gamma^2 + \text{Pr}\,\delta_0 + \beta)^2(\gamma^2 + \beta)\pi^2 + \omega^2(\gamma^2 + \beta)\pi^2 \\ &-\alpha^2 \text{Pr}\,R(\text{Pr}\,\gamma^2 + \text{Pr}\,\delta_0 + \beta) = 0, \end{aligned} \tag{4.5.19}$$

where $\gamma^2 = \alpha^2 + \pi^2, \alpha^2 = \pi^2(j^2 L_1^{-2} + k^2 L_2^{-2})$.

As with (4.4.33), let $\beta = 0$ in (4.5.19). We get the first critical Rayleigh number as follows:

$$R_{c_1} = \min_{\alpha^2}\left[(\alpha^2 + \pi^2)\delta_1 + \frac{(\alpha^2 + \pi^2)^3 + \pi^2(\alpha^2 + \pi^2)\delta_0}{\alpha^2} + \frac{\pi^2\omega^2(\alpha^2 + \pi^2)}{\alpha^2 \text{Pr}^2(\alpha^2 + \pi^2 + \delta_0)}\right]. \tag{4.5.20}$$

When h is large, $\delta_1 \gg 1$ and $\delta_1 \gg \delta_0$; from (4.5.20) we obtain the following scale formulas of large-scale convection motion:

$$\alpha_c^4 \cong \frac{\pi^4}{\delta_1}\left(\pi^2 + \delta_0 + \frac{\omega^2}{\text{Pr}^2(\pi^2 + \delta_0)}\right), \tag{4.5.21}$$

$$\delta_1 \cong (\pi^2 + \delta_0)L_c^4 + \frac{\omega^2}{\text{Pr}^2(\pi^2 + \delta_0)}L_c^4, \tag{4.5.22}$$

$$R_{c_1} \cong \pi^2\delta_1 + \frac{\pi^6 + \pi^4\delta_0}{\alpha_c^2} + \frac{\pi^4\omega^2}{\alpha_c^2 \text{Pr}^2(\pi^2 + \delta_0)}. \tag{4.5.23}$$

Let $\beta = i\rho(\rho \neq 0)$ in (4.5.19). Then we get the second critical Rayleigh number R_{c_2} given by

$$\begin{aligned} R_{c_2} = \min_{\alpha^2} & \frac{1}{\alpha^2 \text{Pr}\,[(\text{Pr}\,+1)\gamma^4 + \text{Pr}\,(\alpha^2\delta_1 + \pi^2\delta_0)]} \\ & \times [\alpha^2\text{Pr}^2(\alpha^2\delta_1 + (\gamma^2+\pi^2)\delta_0) \times B + 2\text{Pr}\,\pi^2\omega^2\gamma^4 \\ & + \alpha^2\text{Pr}\,\gamma^4(2\text{Pr}^2 r_0^2 r_1^2 + (2\text{Pr}\,+1)(r_1^2+\gamma_0^2)r^2) \\ & + 2\pi^2\text{Pr}\,r^4(\text{Pr}^2 r_0^4 + (2\text{Pr}\,+1)\gamma_0^2 r^2)], \end{aligned} \tag{4.5.24}$$

where

$$\begin{aligned} & r_0^2 = \alpha^2 + \pi^2 + \delta_0, \\ & r_1^2 = \alpha^2 + \pi^2 + \delta_1, \\ & B = \alpha^2\text{Pr}\,(\text{Pr}\,r_0^2 r_1^2 + r_1^2 r^2 + r_0^2 r^2) + \pi^2\text{Pr}\,(\text{Pr}\,r_0^4 + 2r_0^2 r^2) + \pi^2\omega^2. \end{aligned}$$

As we shall see later, these two sets of formulas (4.5.16) and (4.5.21)–(4.5.23) provide correct zonal and meridional circulation scales.

Chapter 5
Geophysical Fluid Dynamics and Climate Dynamics

Our Earth's atmosphere and oceans are rotating geophysical fluids that are two important components of the planet's climate system. The atmosphere and the oceans are extremely rich in their organization and complexity, and many phenomena that they exhibit, involving a broad range of temporal and spatial scales (Charney 1948), cannot be reproduced in the laboratory. An understanding of the complex scientific issues of geophysical fluid dynamics requires the combined efforts of scientists in many fields.

The behavior of the atmosphere, the ocean, or the coupled atmosphere and ocean can be viewed as an initial and boundary value problem (Bjerknes 1904; Rossby 1926; Phillips 1956). The ideas of dynamical systems theory were pioneered in (Lorenz 1963a,b; Stommel 1961; Veronis 1963, 1966), among others, whose authors explored the bifurcation structure of low-order models of atmospheric and oceanic flows. The formulation of the primitive equations of atmosphere and ocean as an infinite-dimensional dissipative dynamical system were initiated in (Lions et al. 1992a,b, 1993).

According to (von Neumann 1960), the motion of the atmosphere can be divided into three categories, depending on the time scale of the phenomenon under consideration, or equivalently, on how far into the future one is attempting to make predictions about atmospheric conditions. It is convenient, then, to consider motions corresponding to short-, medium-, and long-term behavior of the atmosphere. Short-term weather forecasting, for example, clearly requires the efficient solution of an initial and boundary value problem of the atmospheric and oceanic systems.

Thanks to the pioneering work (Lorenz 1963a), the medium- and long-term behavior of the atmosphere is often referred to as climate dynamics, and it is governed by the dynamical behavior of infinite-dimensional climate dynamical systems. Great advances in numerical and observational studies have enabled researchers in climate dynamics to set as one of their primary goals to document, through careful theoretical and numerical studies, the presence of low-frequency climate variability, to verify the robustness of this variability's characteristics under changes in the model's parameters, and to explain its physical mechanisms. Obtaining a thorough understanding of this variability is a challenging problem with important practical

T. Ma and S. Wang, *Phase Transition Dynamics*, DOI 10.1007/978-1-4614-8963-4_5,

implications for geophysical efforts to quantify predictability, analyze error growth in dynamical models, and develop efficient forecasting methods.

In mathematical terms, this primary goal is best described by dynamical transitions associated with typical sources of atmospheric and oceanic variability, including, for example, the meridional overturning oceanic circulation and the tropical atmosphere–ocean modes associated with the El Niño–Southern Oscillations (ENSO).

Based on the authors' recent work (Ma and Wang 2010b,c, 2011b), the main objective of this chapter is to initiate a study of dynamical transitions and the stability of large-scale atmospheric and oceanic circulations, focusing on a few typical sources of climate variability. Such variability, independently and interactively, may play a significant role in past and future climate change.

Again, we reiterate that the philosophy of dynamical transition theory of geophysical flows is completely different from that of classical bifurcation studies. A full classification of transition states and their dynamical behavior, including, in particular, that of the transients, is crucial in climate studies, as indicated by existing numerical and observational data.

5.1 Modeling and General Characteristics of Geophysical Flows

The hydrodynamic equations governing large-scale atmospheric and oceanic flows are the Navier–Stokes equations with the Coriolis force generated by the Earth's rotation, coupled with the first law of thermodynamics and with the diffusion equation for humidity and salinity. There are many excellent books in this area; see among others (Pedlosky 1987; Lorenz 1967; Ghil and Childress 1987; Dijkstra 2000a; Salby 1996).

The general system of equations was basically the equations used by L.F. Richardson in his pioneering work (Richardson 1922). However, they are in general too complicated to use in conducting theoretical analysis. As practiced by earlier workers such as J. Charney, and from the lessons learned by the failure of Richardson's pioneering work, one tries to be satisfied with simplified models approximating the actual motions to a greater or lesser degree instead of attempting to deal with the atmosphere in all its complexity. By starting with models incorporating only what are thought to be the most important atmospheric influences, and by gradually bringing in others, one is able to proceed inductively and thereby avoid the pitfalls inevitably encountered when a great many poorly understood factors are introduced all at once. The simplifications are usually done by taking into consideration some of the main features of the large-scale atmosphere.

Hereinafter, we briefly mention a few general features of the large-scale atmosphere and the ocean.

First, the most important feature of the large-scale atmospheric circulation is the smallness of the aspect ratio between the vertical and horizontal scales, leading to a hydrostatic equation replacing the vertical momentum equation. The resulting system of equations is called the system of primitive equations (PEs); see among others (Lions et al. 1992a; Salby 1996).

More precisely, the PEs are obtained from the general equations of hydrodynamics and thermodynamics of the compressible atmosphere by approximating the momentum equation in the vertical direction with the hydrostatic equation:

$$\frac{\partial p}{\partial z} = -\rho g. \tag{5.1.1}$$

This hydrostatic equation is based on the ratio between the vertical and horizontal scale being small. Here ρ is the density, g the gravitational constant, $z = r - a$ the height above the sea level, r the radial distance, and a the mean radius of the Earth.

Equation (5.1.1) expresses the fact that p is a decreasing function along the vertical, so that one can use p instead of z as the vertical variable. Motivated by this hydrostatic approximation, we can introduce a coordinate transformation from (θ, φ, z) to (θ, φ, p).

Second, another important feature of the large-scale geophysical flows is that the Earth rotates much faster than the wind velocity or the velocity of the seawater. The simplification taking the fast rotation into consideration leads to the classical quasi-geostrophic equations (Charney 1947) and many balanced equations.

Third, as discussed in Sect. 4.5, for large-scale atmospheric and oceanic circulations, the critical convection scales of the classical theory are often too small. By carefully examining the Bénard convection, we show that in order to have a correct scaling for the large-scale circulations, we need to add to the basic equations an anisotropic turbulent friction term $(\delta_0 u_1, \delta_0 u_2, \delta_1 u_3)$, where u_1 and u_2 stand for the horizontal velocity components, u_3 stands for the vertical velocity component, and

$$\delta_0 = C_0 h^4/\nu, \qquad \delta_1 = C_1 h^4/\nu, \tag{5.1.2}$$

where C_0 and C_1 are constants, and h is the vertical length scale. We can think about this added friction term as due to the turbulent averaging process, although the mathematical derivation of this scaling law is based on the analysis of the dynamic transitions of convection problems.

Fourth, the large-scale atmospheric circulation can essentially be regarded ad a thermal convection process caused by the temperature gradient between the Earth's surface and the tropopause. There are two types of convective cells: the meridional circulation and the zonal circulation. The meridional circulation is characterized by the polar, the Ferrel, and the Hadley cells, as schematically illustrated in Fig. 5.1. These cells and their interactions are major players in global heat transport.

The large-scale zonal circulation consists of six convective cells over the equator, as shown in Fig. 5.2. The overall atmospheric motion is known as the zonal overturning circulation, and also called the Walker circulation (Salby 1996).

One remarkable feature of the global atmospheric circulation is that the equatorial Walker circulation divides the whole Earth into three invariant regions of atmospheric flow: the northern hemisphere, the southern hemisphere, and the equatorial zone. Also, note that although the large-scale structure of the zonal overturning circulation varies from year to year, as seen, e.g., in the El Niño–Southern Oscillation

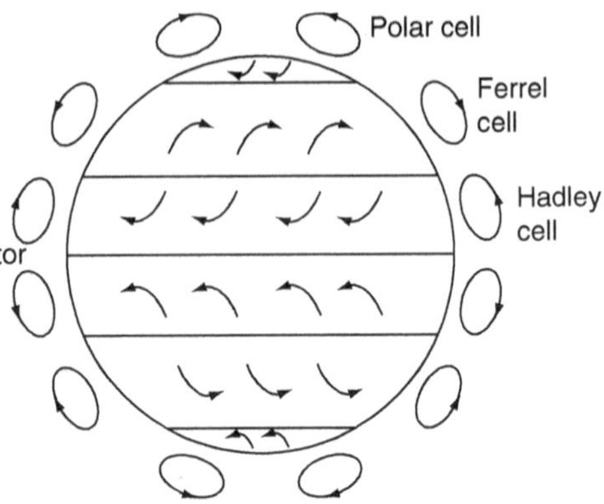

Fig. 5.1 Three large-scale cells in the meridional direction: the polar cell, the Ferrel cell, and the Hadley cell

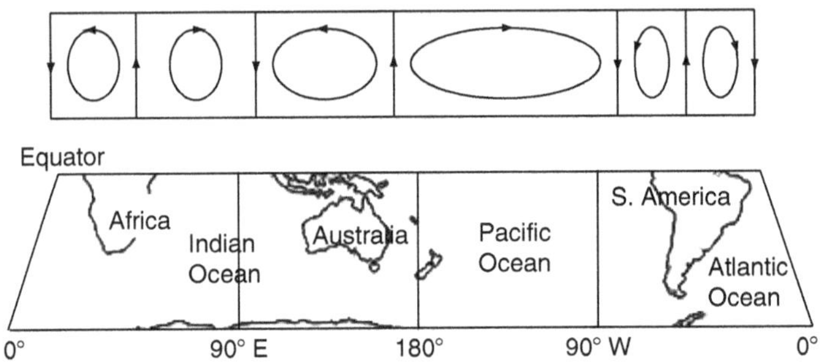

Fig. 5.2 The equatorial Walker circulation

phenomenon over the equatorial Pacific, its basic structure remains fairly stable. Hence to study the formation and dynamics of large-scale atmospheric circulations, it is appropriate to consider the following atmospheric zone hypothesis:

Circulation Zone Hypothesis of Large-Scale Atmosphere. *The atmospheric circulation has three invariant regions: the northern hemisphere domain* $(0 < \theta \leq \frac{\pi}{2})$, *the southern hemisphere domain* $(-\frac{\pi}{2} \leq \theta < 0)$, *and the equatorial zone* $(\theta = 0)$. *Namely, the large-scale circulations in these invariant regions can act alone with a lesser influence on the others. In particular, the velocity field* $u = (u_\varphi, u_\theta, u_z)$ *of the atmospheric circulation has a vanishing latitudinal component in a narrow equatorial zone, i.e,* $u_\theta = 0$ *for* $-\varepsilon < \theta < \varepsilon$, *where* $\varepsilon > 0$ *is a small number.*

5.2 El Niño–Southern Oscillation

The El Niño–Southern Oscillation (ENSO) is the strongest known interannual climate variability associated with strong atmosphere–ocean coupling, which has significant impact on the global climate. ENSO has global impacts on temperature, precipitation, tropical cyclones, human health, and even armed conflict. ENSO is in fact a phenomenon associated with interannual oscillation between warm events (El Niño phase) and cold events (La Niña phase) in the equatorial eastern Pacific sea-surface temperature (SST) anomaly. It is convenient and understandable to employ simplified coupled dynamical models to investigate some essential behaviors of ENSO dynamics; see among many others (Schopf and Suarez 1987; Battisti and Hirst 1989; Jin 1996; Zebiak and Cane 1987; Philander and Fedorov 2003; Jin et al. 1996; Neelin et al. 1998; Neelin 1990a,b; Ghil 2000; Schneider et al. 2003; Sardeshmukh et al. 2000; Samelson 2008).

An interesting current debate is whether ENSO is best modeled as a stochastic or chaotic system—linear and noise-forced, or nonlinear oscillatory and unstable (Philander and Fedorov 2003). It is obvious that a careful fundamental examination of the problem is crucial. The main objective of this section is an attempt to address this fundamental question.

By establishing a rigorous mathematical theory of the formation of the Walker circulation over the tropics, we present in this section a mechanism of the ENSO, first derived in (Ma and Wang 2010c, 2011b). Namely, ENSO is a self-organizing and self-excitation system, with two highly coupled processes:

- the oscillation between the two metastable warm (El Niño phase) and cold (La Niña phase) events, and
- the spatiotemporal oscillation of the SST field.

The symbiotic interplay between these two processes gives rise to the climate variability associated with the ENSO, leads to both the random and deterministic features of the ENSO, and defines a new natural *feedback* mechanism, which drives the sporadic oscillation of the ENSO. On the one hand, the oscillation of metastable states has direct influence on the SST, leading to (1) the intensification or weakening of the oceanic upwelling, and (2) the variation of the atmospheric Rayleigh number, consequently forcing the system to adjust to a different metastable phase, and to intensify/weaken the corresponding event. On the other hand, the oscillation of the SST has a direct influence on the atmospheric Rayleigh number, leading to different El Niño, La Niña, and normal conditions.

5.2.1 Walker Circulation and ENSO

Upwelling and zonal circulation over the tropics contains six convective cells, as shown in Fig. 5.2. Usually, the Walker circulation refers to the Pacific cell over the equator, responsible for creating ocean upwelling off the coasts of Peru and Ecuador. It was discovered by Jacob Bjerknes in 1969 and was named after the English physicist Gilbert Walker, an early-twentieth-century director of British observatories in India, who discovered an indisputable link between periodic pressure variations in the Indian Ocean and the Pacific, which he termed the "Southern Oscillation."

The Walker cell is of such importance that its behavior is a key factor giving rise to the El Niño (more precisely, ENSO) phenomenon. When the convective activity weakens or reverses, an El Niño phase takes place, causing the ocean surface to be warmer than average, reducing or terminating the upwelling of cold water. A particularly strong Walker circulation causes a La Niña event, resulting in cooler SST due to stronger upwelling; see Fig. 5.3.

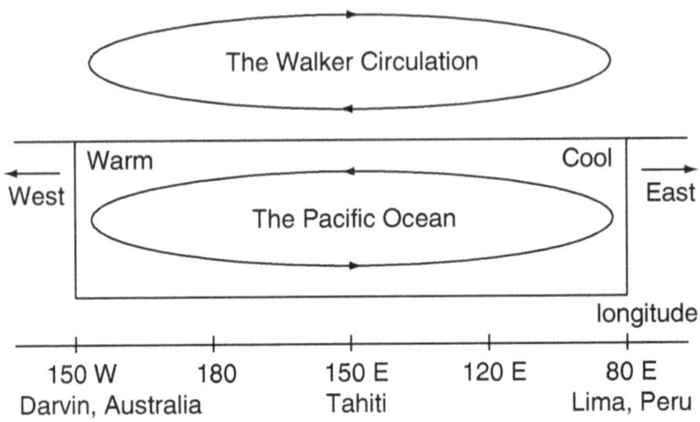

Fig. 5.3 A schematic diagram of the Walker circulation at the equatorial Pacific Ocean, a main factor in the El Niño and the la Niña phenomena

The atmospheric motion associated with the Walker circulation affects the loops on either side. In normal years, the weather behaves as expected. But every few years, winters become unusually warm or unusually cold, or the frequency of hurricanes increases or decreases. This entirely ocean-based cell is seen at the lower surface as easterly trade winds that drive the seawater and air warmed by sun moving westward. The western side of the equatorial Pacific is characterized by warm, wet low-pressure weather, and the eastern side is characterized by cool, dry high-pressure weather. The ocean is about 60 cm higher in the western Pacific as a result of this circulation. The air is returned at the upper surface to the east, and it now becomes much cooler and drier. An El Niño phase is characterized by a breakdown of this water and air cycle, resulting in relatively warm water and moist air in the eastern Pacific. Meanwhile, a La Niña phase is characterized by the strengthening of this cycle, resulting in much cooler water and drier air than in the normal phase. During an El Niño or La Niña event, which occurs about every three to six years, the associated weather sets in for an indeterminate period.

It is then clear that the ENSO is a coupled atmospheric–ocean phenomenon. The main objective of this section is to address the mechanism of atmospheric Walker circulations over the tropics from the dynamic transition point of view.

5.2.2 Equatorial Circulation Equations

The atmospheric motion equations over the tropics are the Boussinesq equations restricted to $\theta = 0$, where the meridional velocity component u_θ is set to zero, and the effect of the turbulent friction is taken into consideration using the scaling law introduced in Sect. 4.5:

$$
\begin{aligned}
\frac{\partial u_\varphi}{\partial t} &= -(u\cdot\nabla)u_\varphi - \frac{u_\varphi u_z}{r_0} + \nu\left(\Delta u_\varphi + \frac{2}{r_0^2}\frac{\partial u_z}{\partial\varphi} - \frac{2u_\varphi}{r_0^2}\right) \\
&\quad - \sigma_0 u_\varphi - 2\Omega u_z - \frac{1}{\rho_0 r_0}\frac{\partial p}{\partial\varphi}, \\
\frac{\partial u_z}{\partial t} &= -(u\cdot\nabla)u_z + \frac{u_\varphi^2}{r_0} + \nu\left(\Delta u_z - \frac{2}{r_0^2}\frac{\partial u_\varphi}{\partial\varphi} - \frac{2u_z}{r_0^2}\right) \\
&\quad - \sigma_1 u_z + 2\Omega u_\varphi - \frac{1}{\rho_0}\frac{\partial p}{\partial z} - (1-\alpha_T(T-T_0))g, \\
\frac{\partial T}{\partial t} &= -(u\cdot\nabla)T + \kappa\Delta T + Q, \\
\frac{1}{r_0}\frac{\partial u_\varphi}{\partial\varphi} &+ \frac{\partial u_z}{\partial z} = 0.
\end{aligned} \tag{5.2.1}
$$

Here the $\sigma_i = C_i h^2$ $(i = 0,1)$ represent the turbulent friction, r_0 is the radius of the Earth, the spatial domain is taken as $M = S^1_{r_0} \times (r_0, r_0+h)$ with $S^1_{r_0}$ the one-dimensional circle with radius r_0, and

$$
\begin{aligned}
(u\cdot\nabla) &= \frac{u_\varphi}{r_0}\frac{\partial}{\partial\varphi} + u_z\frac{\partial}{\partial z}, & \Delta &= \frac{1}{r_0^2}\frac{\partial^2}{\partial\varphi^2} + \frac{\partial^2}{\partial z^2}, \\
(x_1,x_2) &= (r_0\varphi, z), & (u_1,u_2) &= (u_\varphi, u_z).
\end{aligned}
$$

The temperature T_1 at the tropopause $z = r_0 + h$ is assumed to be constant. We take T_0 as the average on the lower surface $z = r_0$. Also, let

$$
\begin{aligned}
& x = hx', \qquad t = h^2 t'/\kappa, \qquad u = \kappa u'/h, \\
& T = (T_0 - T_1)T' + T_0 - (T - T_0)x_2', \\
& p = \kappa\nu\rho_0 p'/h^2 - g\rho_0(hx_2' + \alpha_T(T_0 - T_1)h(x_2')^2/2).
\end{aligned}
$$

Also, we define the Rayleigh number, the Prandtl number, and the scaling laws by

$$
R = \frac{\alpha_T g(T_0 - T_1)h^3}{\kappa\nu}, \qquad \Pr = \frac{\nu}{\kappa}, \qquad \delta_i = \frac{C_i h^4}{\nu} \quad (i = 0,1). \tag{5.2.2}
$$

Omitting the primes, the nondimensional form of (5.2.1) reads

$$
\begin{aligned}
\frac{\partial u_1}{\partial t} &= \Pr\left[\Delta u_1 + \frac{2}{r_0}\frac{\partial u_2}{\partial x_1} - \frac{2}{r_0}u_1 - \delta_0 u_1 - \frac{\partial p}{\partial x_1}\right] \\
&\quad - \omega u_2 - (u\cdot\nabla)u_1 - \frac{1}{r_0}u_1 u_2, \\
\frac{\partial u_2}{\partial t} &= \Pr\left[\Delta u_2 - \frac{2}{r_0}\frac{\partial u_1}{\partial x_1} - \frac{2}{r_0}u_2 - \delta_1 u_2 + RT - \frac{\partial p}{\partial x_2}\right] \\
&\quad + \omega u_1 - (u\cdot\nabla)u_2 + \frac{1}{r_0}u_1^2,
\end{aligned} \tag{5.2.3}
$$

$$\frac{\partial T}{\partial t} = \Delta T + u_2 - (u\cdot\nabla)T + Q,$$
$$\frac{\partial u_1}{\partial x_1} + \frac{\partial u_2}{\partial x_2} = 0,$$

where $(x_1,x_2) \in M = (0,2\pi r_0)\times(r_0,r_0+1)$, δ_0 and δ_1 are as in (5.2.2), $(u\cdot\nabla)$, and Δ are the usual differential operators, and

$$\omega = \frac{2\Omega h^2}{\kappa}. \tag{5.2.4}$$

The problem is supplemented with the natural periodic boundary condition in the x_1-direction, and the free-slip boundary condition on the top and bottom boundaries:

$$(u,T)(x_1+2\pi r_0,x_2,t) = (u,T)(x_1,x_2,t), \tag{5.2.5}$$

$$\begin{cases} u_2 = 0, \ \dfrac{\partial u_1}{\partial x_2} = 0, T = \varphi(x_1) & \text{at } x_2 = r_0, \\ u_2 = 0, \ \dfrac{\partial u_1}{\partial x_2} = 0, \ T = 0 & \text{at } x_2 = r_0 + 1. \end{cases} \tag{5.2.6}$$

Here $\varphi(x_1)$ is the temperature deviation from the average T_0 on the equatorial surface and is periodic:

$$\int_0^{2\pi r_0} \varphi(x_1)dx_1 = 0 \quad \text{and} \quad \varphi(x_1) = \varphi(x_1+2\pi r_0).$$

The deviation $\varphi(x_1)$ is caused mainly by a difference in the specific heat capacities between seawater and land.

5.2.3 Walker Circulation Under Idealized Conditions

In the idealized case, the temperature deviation φ vanishes. In this case, the study of transition of (5.2.3) is of special importance in understanding the longitudinal circulation. Here we address the dynamic bifurcation of (5.2.3), the Walker cell structure of bifurcated solutions, and the convection scale under the following idealized conditions:

$$\varphi(x_1) = 0 \quad \text{for any } 0 \le x_1 \le 2\pi r_0, \qquad Q = 0 \quad \text{in (5.2.3)}. \tag{5.2.7}$$

For the problem (5.2.3) with (5.2.5) and (5.2.6), let

$$H = \{(u,T) \in L^2(M)^3 \mid \text{div} u = 0, (u,T) \text{ satisfies (5.2.5)}\},$$
$$H_1 = \{(u,T) \in H^2(M)^3 \textstyle\bigcap H \mid (u,T) \text{ satisfies (5.2.5) and (5.2.6)}\}.$$

Then define the operators $L_R = A + B_R$ and $G : H_1 \to H$ by

$$A\psi = P\left(\Pr\,(\Delta u_1 + \frac{2}{r_0}\frac{\partial u_2}{\partial x_1} - \delta_0' u_1), \Pr\,(\Delta u_2 - \frac{2}{r_0}\frac{\partial u_1}{\partial x_1} - \delta_1' u_2), \Delta T\right),$$
$$B_R\psi = P(-\omega u_2, \omega u_1 + \Pr\, RT, u_2), \tag{5.2.8}$$
$$G(\psi) = -P\left((u\cdot\nabla)u_1 + \frac{u_1 u_2}{r_0}, (u\cdot\nabla)u_2 - \frac{u_1^2}{r_0}, (u\cdot\nabla)T\right),$$

where $\psi = (u, T) \in H_1$, $P : L^2(M)^3 \to H$ is the Leray projection, and

$$\delta_i' = \frac{2}{r_0^2} + \delta_i \qquad (i = 0, 1).$$

Then the problem (5.2.3) with (5.2.5) and (5.2.6) is equivalent to

$$\frac{d\psi}{dt} = L_R\psi + G(\psi). \tag{5.2.9}$$

Consider the eigenvalue problem $L_R\psi = \beta(R)\psi$, which is equivalent to

$$\begin{aligned}
&\Delta u_1 + \frac{2}{r_0}\frac{\partial u_2}{\partial x_1} - \delta_0' u_1 - \Pr^{-1}\omega u_1 - \frac{\partial p}{\partial x_1} = \Pr^{-1}\beta u_1,\\
&\Delta u_2 - \frac{2}{r_0}\frac{\partial u_1}{\partial x_1} - \delta_1' u_2 + \Pr^{-1}\omega u_2 + RT - \frac{\partial p}{\partial x_2} = \Pr^{-1}\beta u_2,\\
&\Delta T + u_2 = \beta T,\\
&\mathrm{div} u = 0,
\end{aligned} \tag{5.2.10}$$

with the boundary conditions (5.2.5) and (5.2.6). We shall see later that equations (5.2.10) are symmetric, which implies, in particular, that all eigenvalues $\beta_j(R)$ are real, and there exists a number R_c, called the first critical Rayleigh number, such that

$$\beta_i(R)\begin{cases} <0 & \text{if } R < R_c,\\ =0 & \text{if } R = R_c,\\ >0 & \text{if } R > R_c \end{cases} \qquad \text{for } i = 1, 2, \tag{5.2.11}$$

$$\beta_j(R_c) < 0 \qquad \text{for } j > 2. \tag{5.2.12}$$

The following theorem provides a theoretical basis for understanding the equatorial Walker circulation.

Theorem 5.2.1 *Under the idealized condition (5.2.7), the problem (5.2.3) with (5.2.5) and (5.2.6) undergoes a Type I transition at the critical Rayleigh number $R = R_c$. More precisely, the following statements hold:*

(1) When the Rayleigh number R is less than or equal to R_c, the equilibrium solution $(u, T) = 0$ is globally stable in the phase space H.

(2) *When $R_c < R < R_c + \delta$ for some $\delta > 0$, this problem bifurcates from $((u,T),R) = (0,R_c)$ to an attractor $\mathscr{A}_R = S^1$, consisting entirely of steady-state solutions, that attracts $H \setminus \Gamma$, where Γ is the stable manifold of $(u,T) = 0$ with codimension two.*

(3) *For each steady-state solution $\psi_R = (u_R, T_R) \in \mathscr{A}_R$, u_R is topologically equivalent to the structure shown in Fig. 5.4.*

(4) *For every initial value $\psi_0 = (u_0, T_0) \in H \setminus (\Gamma \bigcup E)$, there exists a time $t_0 \geq 0$ such that for every $t > t_0$, the velocity field $u(t, \psi_0)$ is topologically equivalent to the structure shown in Fig. 5.5, either (a) or (b), where $\psi = (u(t,\psi_0), T(t,\psi_0))$ is the solution of the problem with $\psi(0) = \psi_0$, and*

$$E = \{(u,T) \in H | \int_{r_0}^{r_0+1} u_1 dx_2 = 0\}.$$

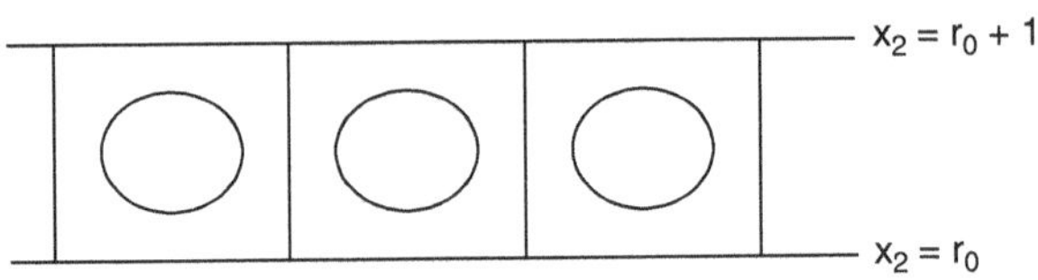

Fig. 5.4 The cell structure of the steady-state solutions in the bifurcated attractor $\mathscr{A}_R$

A few remarks are now in order:

First, in a more realistic case in which $T = T_0 + \varphi(x_1)$ at $x_2 = r_0$, the temperature deviation $\varphi(x_1)$ is small in comparison with the average temperature gradient $T_0 - T_1$, and $Q \neq 0$ is small as well. Therefore, the more realistic case can be considered a perturbation of the idealized case.

Second, mathematically, the conclusion that $\mathscr{A}_R = S^1$ consists of steady-state solutions is derived from the invariance of (5.2.3) under a translation $x_1 \to x_1 + \alpha$.

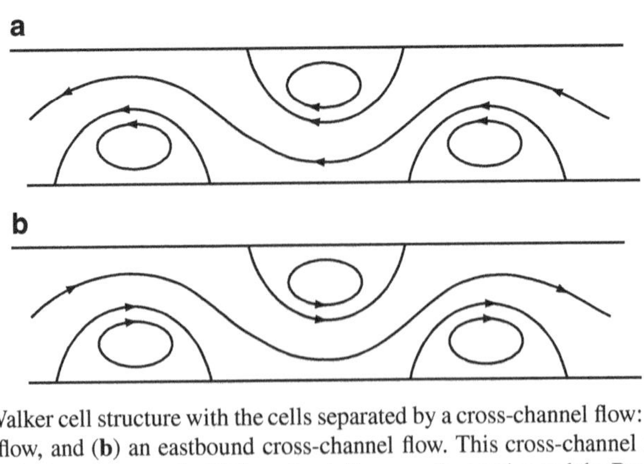

Fig. 5.5 The Walker cell structure with the cells separated by a cross-channel flow: (**a**) a westbound cross-channel flow, and (**b**) an eastbound cross-channel flow. This cross-channel flow pattern has the same topological structure as the Walker circulation over the tropics and the Branstator–Kushnir waves in atmospheric dynamics (Branstator 1987; Kushnir 1987)

It implies that for any two solutions $\psi_1, \psi_2 \in \mathscr{A}_R$, they are the same up to a phase angle α in the longitude, namely

$$\psi_2(x_1,x_2) = \psi_1(x_1+\alpha,x_2) \quad \text{for some } \alpha \in [0,2\pi r_0].$$

For the idealized case, this conclusion is natural, since the equator is treated as homogeneous.

In addition, assertion (2) amounts to saying that as the initial value ψ_0 varies, i.e., as an external perturbation is applied, the rolls of the Walker circulation will be translated by a phase angle α. This behavior is termed translation oscillation, which can be used to explain the ENSO phenomenon in the idealized circumstance.

Third, when a deviation $\varphi(x_1)$ is present and $Q \neq 0$ in (5.2.3), the translation symmetry is broken, and consequently, the bifurcated attractor $\mathscr{A}_R$ consists of isolated equilibrium solutions instead of a circle S^1. Thus, the mechanism of the ENSO will be explained as state exchanges between these equilibrium solutions in $\mathscr{A}_R$, which are metastable. We shall address this point in detail later.

Fourth, by the structural stability theorems in (Ma and Wang 2005d), we see from assertion (4) that the roll structures illustrated by Fig. 5.5 (a) and (b) are structurally stable. Hence, for a deviation perturbation $\varphi(x_1)$, these structures are not destroyed, providing a realistic characterization of the Walker circulation.

Fifth, in assertion (3), these rolls in the bifurcated solutions are closed. However, under some external perturbations, a cross-channel flow appears that separates the rolls and globally transports heat between them.

Sixth, using classical methods as in (Ma and Wang 2007a, 2005b; Chandrasekhar 1981; Drazin and Reid 1981), we can show that the critical Rayleigh number for the revised equation (6.1.2) with (5.2.5) and (5.2.6) is given by

$$R_c = \min_{\alpha^2}\left[(\alpha^2+\pi^2)\delta_1' + \frac{(\alpha^2+\pi^2)^3+\pi^2(\alpha^2+\pi^2)\delta_0'}{\alpha^2}\right]. \tag{5.2.13}$$

Let $g(x) = (x+\pi^2)\delta_1' + ((x+\pi^2)^3+\pi^2(x+\pi^2)\delta_0')/x$. Then $x_c = \alpha_c^2$ satisfies $g'(x) = 0$. Thus we get

$$\delta_1'\alpha_c^4 - (\alpha_c^2+\pi^2)^3 - \pi^2(\alpha_c^2+\pi^2)\delta_0' + 3\alpha_c^2(\alpha_c^2+\pi^2)^2 + \pi^2\alpha_c^2\delta_0' = 0. \tag{5.2.14}$$

We assume that $\delta_1' > \delta_0'$ for $h \neq 0$. By $\alpha_c^2 = \pi^2/L_c^2$, when h is large, $\delta_1' \gg 1$ and $\delta_1' \gg \delta_0'$. Here

$$\delta_i' = \delta_i + \frac{2}{r_0^2}.$$

We infer from (5.2.13) and (5.2.14) that

$$\alpha_c^4 \cong \frac{\pi^4(\pi^2+\delta_0')}{\delta_1'}, \qquad L_c^2 \cong \frac{(\delta_1')^{1/2}}{(\pi^2+\delta_0')^{1/2}}, \qquad R_c \cong \pi^2\delta_1' + \pi^2\delta_0' L_c^2, \tag{5.2.15}$$

which are useful in studying large-scale convections, as shown in the next section.

Proof of Theorem 5.2.1. We proceed in the following four steps.

Step 1. We show that equations (5.2.3) have a form equivalent to that of the classical two-dimensional Bénard problem.

Since the velocity field u defined on $M = S^1_{r_0} \times (r_0, r_0+1)$ is divergence-free, there exists a stream function φ such that

$$u_1 = \frac{\partial \varphi}{\partial x_2}, \quad u_2 = -\frac{\partial \varphi}{\partial x_1},$$

satisfying the given boundary conditions. Therefore, the two vector fields

$$(-\omega u_2, \omega u_1) = \omega \nabla \varphi, \quad (\frac{2}{r_0}\frac{\partial u_2}{\partial x_1}, -\frac{2}{r_0}\frac{\partial u_1}{\partial x_1}) = -\frac{2}{r_0}\nabla \frac{\partial \varphi}{\partial x_1}, \tag{5.2.16}$$

are gradient fields, which can be balanced by ∇p in (5.2.3). Hence, equations (5.2.3) are equivalent to

$$\begin{aligned}
&\frac{\partial u_1}{\partial t} + (u\cdot\nabla)u_1 + \frac{u_1u_2}{r_0} = \mathrm{Pr}\left[\Delta u_1 - \delta_0' u_1 - \frac{\partial p}{\partial x_1}\right],\\
&\frac{\partial u_2}{\partial t} + (u\cdot\nabla)u_2 - \frac{u_1^2}{r_0} = \mathrm{Pr}\left[\Delta u_2 - \delta_1' u_2 + RT - \frac{\partial p}{\partial x_2}\right],\\
&\frac{\partial T}{\partial t} + (u\cdot\nabla)T = \Delta T + u_2,\\
&\mathrm{div}\, u = 0.
\end{aligned} \tag{5.2.17}$$

Therefore, Eq. (5.2.9) is also an abstract form of (5.2.17). Thus the eigenvalue problem (5.2.10) is symmetric. It is clear that the operator G defined in (5.2.8) is orthogonal. Hence Theorem 4.1.1 for the Rayleigh–Bénard convection proved in (Ma and Wang 2004b) is also valid for the problem (5.2.3) with (5.2.5), (5.2.6). Thus, this transition of the problem is of Type I, and assertion (1) is proved.

Step 2. Proof of assertion (2). We have only to verify that the attractor $\mathscr{A}_R$ is homeomorphic to a circle and that it consists of singular points of (5.2.9). This proof is similar to that of Theorem 4.2.12.

We consider the first eigenvectors of (5.2.10) at $R = R_c$. By (5.2.16), equations (5.2.10) at $R = R_c$ are equivalent to the form

$$\begin{aligned}
&\Delta u_1 - \delta_0' u_1 - \frac{\partial p}{\partial x_1} = 0, \\
&\Delta u_2 - \delta_1' u_2 + R_c T - \frac{\partial p}{\partial x_2} = 0, \\
&\Delta T + u_2 = 0, \\
&\mathrm{div} u = 0,
\end{aligned} \tag{5.2.18}$$

with the boundary conditions (5.2.5) and (5.2.6).

As in the eigenvalue problem (4.2.57), we see that the multiplicity of R_c in (5.2.18) is $m = 2$, and the corresponding eigenvectors are given by

$$\psi_1 = (u_1, u_2, T) = \left(-\sin \frac{kx_1}{r_0} H'(x_2), \frac{k}{r_0} \cos \frac{kx_1}{r_0} H(x_2), \cos \frac{kx_1}{r_0} \Phi(x_2) \right), \tag{5.2.19}$$

$$\widetilde{\psi}_1 = (\widetilde{u}_1, \widetilde{u}_2, \widetilde{T}) = \left(\cos \frac{kx_1}{r_0} H'(x_2), \frac{k}{r_0} \sin \frac{kx_1}{r_0} H(x_2), \sin \frac{kx_1}{r_0} \Phi(x_2) \right), \tag{5.2.20}$$

where the functions $H(x_2)$ and $\Phi(x_2)$ satisfy

$$\begin{cases} (D^4 - (\alpha^2 + \alpha_0^2) D^2 + \alpha^2 \alpha_1^2) H = \alpha R_c \Phi, \\ -(D^2 - \alpha^2) \Phi = \alpha H, \end{cases} \tag{5.2.21}$$

$$H = H'' = 0, \quad \Phi = 0 \qquad \text{at } x_2 = r_0, r_0 + 1, \tag{5.2.22}$$

where $D = d/dx_2, \alpha = k/r_0, \alpha_i^2 = \alpha^2 + \delta_i' \ (i = 0, 1)$.

Since equations (5.2.21) with (5.2.22) are symmetric, the number defined by (4.2.71) and (4.2.72) is given by

$$\gamma = \frac{1}{\|\psi_1\|^2} (G(\Psi, \psi_1) + G(\psi_1, \Psi), \psi_1), \tag{5.2.23}$$

$$-L_{R_c} \Psi = G(\psi_1, \psi_1), \tag{5.2.24}$$

L_R is defined by (5.2.8), and $G(\widetilde{\psi}, \psi)$ is defined by

$$G(\widetilde{\psi}, \psi) = P\left((\widetilde{u} \cdot \nabla) u_1 + \frac{\widetilde{u}_1 u_2}{r_0}, (\widetilde{u} \cdot \nabla) u_2 + \frac{\widetilde{u}_1 u_1}{r_0}, (\widetilde{u} \cdot \nabla) T \right),$$

for every $\widetilde{\psi} = (\widetilde{u}, \widetilde{T}), \psi = (u, T) \in H_1$. We derive from (5.2.23) and (5.2.24) that

$$\gamma = -\frac{1}{\|\psi_1\|^2} (G(\psi_1, \psi_1), \Psi) = -\frac{1}{\|\psi_1\|^2} (-L_{R_c} \Psi, \Psi).$$

Due to (5.2.16), $-L_R$ is a symmetric sectorial operator. Thus, we have

$$\gamma = -\frac{1}{\|\psi_1\|^2} ((-L_{R_c})^{1/2} \Psi, (-L_{R_c})^{1/2} \Psi) < 0.$$

As in Theorem 4.2.12, assertion (2) follows.

Step 3. Proof of assertion (3). By Step 1 and the attractor bifurcation theorem, each steady-state solution $\varphi \in \mathscr{A}_R$ can be expressed as

$$\varphi = \frac{|\beta_1(R)}{\gamma|^{1/2}}(x\psi_1 + y\widetilde{\psi}_1) + o(\beta_1(R)), \tag{5.2.25}$$

where $\psi_1, \widetilde{\psi}_1$ are given by (5.2.19) and (5.2.20), $x^2 + y^2 = 1$, and $\beta_1(R)$ is the first eigenvalue of (5.2.10) satisfying (5.2.11).

Let $\varphi = (e, T)$. Then by (5.2.25), the velocity field e can be written as

$$e = \left(r(R) \cos \frac{k}{r_0}(x_1 + \theta)H'(x_2), \frac{k}{r_0} r(R) \sin \frac{k}{r_0}(x_1 + \theta)H(x_2) \right) + o(r(R)),$$

where $r(R) = |\beta_1(R)/\alpha|^{1/2}, (x,y) = (\cos\theta, \sin\theta)$.

By (5.2.21) and (5.2.22), $H(x_2) = \sin j\pi(x_2 - r_0), j \in \mathbb{Z}$. But for the first eigenvectors ψ_1 and $\widetilde{\psi}_1$, we have $H(x_2) = \sin\pi(x_2 - r_0)$. Thus, e can be expressed as

$$\begin{aligned} &e = e_0 + o(r(R)), \\ &e_0 = \left(\pi r(R) \cos \frac{k}{r_0}(x_1 + \theta) \cos\pi(x_2 - r_0), \frac{k}{r_0} r(R) \sin \frac{k}{r_0}(x_1 + \theta) \sin\pi(x_2 - r_0) \right). \end{aligned} \tag{5.2.26}$$

It is easy to check that e_0 is regular. Therefore, e is also regular if $R_c < R < R_c + \delta$ for some $\delta > 0$. Obviously, the vector field e_0 is topologically equivalent to the structure shown in Fig. 5.4.

By the structural stability theorem, Theorem 2.12 in (Ma and Wang 2005d), the vector field e_0 is not structurally stable in H_1, because the boundary saddle points on $x_2 = r_0$ are connected to saddle points on $x_2 = r_0 + 1$, a different connected component. However, e_0 is structurally stable in the space

$$\widetilde{H} = \{(u,T) \in H | \int_0^{2\pi r_0} \int_{r_0}^{r_0+1} u dx = 0\}.$$

To see this, we know that if $u \in \widetilde{H}$, then $u = (u_1, u_2) \neq$constant, and u_1 has the Fourier expansion

$$u_1 = \sum_{k,j} (x_{kj} \cos \frac{kx_1}{r_0} + y_{kj} \sin \frac{kx_1}{r_0}) \cos j\pi(x_2 - r_0).$$

It follows that

$$\int_{r_0}^{r_0+1} u_1 dx_2 = 0 \quad \forall u = (u_1, u_2) \in \widetilde{H}. \tag{5.2.27}$$

According to the connection lemma, Lemma 2.3.1 in (Ma and Wang 2005d), we derive from (5.2.27) that a vector field $u \in \widetilde{H}$ is structurally stable if and only if u is regular, and all interior saddle points of u are self-connected. Therefore, e_0 is structurally stable in $\widetilde{H}$.

If we can prove that the vector field e given by (5.2.26) is in $\widetilde{H}$, then if $R_c < R < R_c + \delta$, e is topologically equivalent to e_0. Hence, to prove assertion (3), it suffices to verify that $e \in \widetilde{H}$.

Obviously, $\widetilde{H}$ is an invariant space for the operator $L_R + G$ defined by (5.2.8), and the orthogonal complement $\widetilde{H}^{\perp}$ of $\widetilde{H}$ in H is

$$\widetilde{H}^{\perp} = \{(u,T) \in H |\ u = (c,0), T = 0, c \in \mathbb{R}^1\}.$$

It is readily shown that all steady-state solutions of (5.2.9) are in $\widetilde{H}$. Thus, assertion (3) is proved.

Step 4. Proof of assertion (4). For every initial value $\psi_0 = (u_0, T_0) \in H \setminus \widetilde{H}$, ψ_0 can be written as

$$\psi_0 = c_0 + \Phi_0, \Phi_0 \in \widetilde{H}, \qquad c_0 = (\alpha_0, 0, 0) \in \widetilde{H}^{\perp}, \quad \alpha_0 \neq 0.$$

Since $G(\psi) \in \widetilde{H}, \forall \psi \in H$, the solution $\psi(t, \psi_0)$ of (5.2.9) with $\psi(0, \psi_0) = \psi_0$ must be of the form

$$\psi(t, \psi_0) = C_0 e^{-\delta_0' t} + \Phi(t, \psi_0), \quad \Phi \in \widetilde{H}.$$

In addition, we know that for each $\psi_0 \in H/\Gamma$, there is a steady-state solution $\varphi = (e, T) \in \mathscr{A}_R$ such that

$$\lim_{t \to \infty} \| \psi(t, \psi_0) - \varphi \|_{C^r} = 0 \quad \text{for any } r \geq 1.$$

Then for every $\psi_0 \in H \setminus (\widetilde{H} \bigcup \Gamma)$, it follows that there exists a time $t_0 > 0$ such that for every $t > t_0$, the velocity field $u(t, \psi_0)$ of $\psi(t_0, \psi_0)$ is topologically equivalent to the following vector field:

$$\widetilde{u} = e + c_0 e^{-\delta_0' t} = e + (\alpha_0 e^{-\delta_0' t}, 0, 0),$$

where e is as in (5.2.26).

The vector field e has the roll structure shown in Fig. 5.4, and $c_0 e^{-\delta_0' t}$ is a parallel channel flow. Hence, it is easy to show that $\widetilde{u} = e + c_0 e^{-\delta_0' t}$ is topologically equivalent to the structure shown in Fig. 5.5 (a) for $\alpha_0 < 0$, and to the structure shown in Fig. 5.5 (b) for $\alpha_0 > 0$, for $t > t_0$ large. Assertion (4) is proved. □

5.2.4 Walker Circulation Under Natural Conditions

Physical Parameters and Effects of the Turbulent Frictional Terms. It is known that atmospheric properties vary with temperature and pressure. Table 5.1 lists some common properties of air: density, kinematic viscosity, thermal diffusivity, expansion coefficient, and the Prandtl number for temperatures between $-100°$C and $100°$C.

Table 5.1 Some physical parameters

Temperature t (°C)	Density ρ (kg/m^3)	Kinematic viscosity ν (m^2/s) $\times 10^{-6}$	Thermal diffusivity κ (m^2/s) $\times 10^{-6}$	Expansion coefficient a (1/k) $\times 10^{-3}$	Prandtl number Pr
−100	1.980	5.95	8.4	5.82	0.74
−50	1.534	9.55	13.17	4.51	0.725
0	1.293	13.30	18.60	3.67	0.715
20	1.205	15.11	21.19	3.40	0.713
40	1.127	16.97	23.87	3.20	0.711
60	1.067	18.90	26.66	3.00	0.709
80	1.000	20.94	29.58	2.83	0.708
100	0.946	23.06	32.8	2.68	0.703

The average temperature over the equator is about 20 °C to 30 °C. Based on the data in Table 5.1, we take a set of physical parameters of air for the Walker circulation as follows:

$$\nu = 1.6\times 10^{-5}\,\mathrm{m}^2/\mathrm{s},\quad \kappa = 2.25\times 10^{-5}\,\mathrm{m}^2/\mathrm{s},\quad \alpha_T = 3.3\times 10^{-3}/^\circ C,\ \ \mathrm{Pr} = 0.71.$$

The height h of the troposphere is taken to be $h = 8\times 10^3$ m. Thus, the Rayleigh number for the equatorial atmosphere is

$$R = \frac{g\alpha_T(T_0-T_1)}{\kappa\nu}h^3 = 3.6\times 10^{19}(T_0-T_1)/^\circ C. \tag{5.2.28}$$

In the classical theory, we have the critical Rayleigh number and the convection scale

$$R_c = \min_{\alpha^2}\frac{(\alpha^2+\pi^2)^3}{\alpha^2} = \frac{27}{4}\pi^4 = 675, \qquad L_c^2 = \pi^2/\alpha_c^2 = 2. \tag{5.2.29}$$

Namely, the diameter of a convective roll is $d = L_c\times h = 20$ km. However, based on atmospheric observations, there are six cells over the tropics, and this shows that the real convective scale is about

$$L = \frac{a\pi}{3} = 6600\,\mathrm{km}. \tag{5.2.30}$$

As a comparison, the values (5.2.29) from the classical theory are too small to match the realistic data given in (5.2.28) and (5.2.30).

We use (4.5.16) to discuss this problem. With the values in (5.2.28) and (5.2.30), since r_0 is large, we take

$$\delta_1' \simeq \delta_1 = 2.7\times 10^{20},\quad \delta_0' \simeq \delta_0 = 3.5\times 10^8. \tag{5.2.31}$$

In comparison with (5.2.2), for air, the constants C_0 and C_1 take the following values:

$$C_0 = 1.37 \times 10^{-12}\,\mathrm{m}^{-2}\cdot\mathrm{s}^{-1}, \qquad C_1 = 1.05\,\mathrm{m}^{-2}\cdot\mathrm{s}^{-1}.$$

Then the convective scale L_c in (4.5.16) takes the value

$$L_c = [\delta_1/\delta_0]^{1/4} h = 7500\,\mathrm{km}, \tag{5.2.32}$$

which is slightly greater than the realistic value $L \cong 6300\,\mathrm{km}$. And, the critical Rayleigh number in (4.5.16) is

$$R_c \cong \pi^2\delta_1 + \pi^2\delta_0 L_c^2 \cong 2.7 \times 10^{21}.$$

Thus, the critical temperature gradient is

$$\Delta T_c \cong 60^\circ C, \tag{5.2.33}$$

which is realistic, although it is slightly smaller than the realistic temperature gradient over the equator.

However, with the classical theory with no frictional terms, the critical values would be $L_c \cong 20\,\mathrm{km}$ and $\Delta T_c \cong 10^{-17\circ}C$, which are certainly unrealistic.

The relation between L_c and the wave number k in the first eigenvectors (5.2.19) and (5.2.20) is

$$k = \text{ the closest even integer to } \frac{2\pi r_0}{L_c},$$

and in this case, we have $k = 6$.

Transition in Natural Conditions. We now return to the natural conditions

$$\varphi(x_1) \neq 0, \qquad Q \neq 0.$$

In this case, Eq. (5.2.3) admits a steady-state solution

$$\widetilde{\psi} = (V, J), \quad (V = (V_1, V_2)). \tag{5.2.34}$$

Consider the deviation from this basic state:

$$u \to u + V, \quad T \to T + J.$$

Then (5.2.3) becomes

$$\begin{aligned}
&\frac{\partial u_1}{\partial t} + (u\cdot\nabla)u_1 + \frac{u_1 u_2}{r_0} = \Pr\,[\Delta u_1 + \frac{2}{r_0}\frac{\partial u_2}{\partial x_1} - \delta_0' u_1 - \frac{\partial p}{\partial x_1}]\\
&\quad -\omega u_2 - (V\cdot\nabla)u_1 - (u\cdot\nabla)V_1 - \frac{V_1}{r_0}u_2 - \frac{V_2}{r_0}u_1,\\
&\frac{\partial u_2}{\partial t} + (u\cdot\nabla)u_2 - \frac{u_1^2}{r_0} = \Pr\,[\Delta u_2 - \frac{2}{r_0}\frac{\partial u_1}{\partial x_1} - \delta_1' u_2 + RT - \frac{\partial p}{\partial x_2}]\\
&\quad +\omega u_1 - (V\cdot\nabla)u_2 - (u\cdot\nabla)V_2 + \frac{2V_1}{r_0}u_1,
\end{aligned} \tag{5.2.35}$$

$$\frac{\partial T}{\partial t} + (u\cdot\nabla)T = \Delta T + u_2 - (u\cdot\nabla)J - (V\cdot\nabla)T,$$
$$\text{div}u = 0.$$

The boundary conditions are the free–free boundary conditions given by

$$\begin{aligned}&(u,T) \text{ is periodic in } x_1 - \text{direction},\\ &T=0, u_2=0, \frac{\partial u_1}{\partial x_2} = 0 \text{ at } x_2 = r_0, r_0+1.\end{aligned} \tag{5.2.36}$$

Let the operators $L_R = A + B_R$ and $G : H_1 \to H$ be as in (5.2.8), and let $\mathscr{L}^\varepsilon : H_1 \to H$ be defined by

$$\begin{aligned}\mathscr{L}^\varepsilon \psi = P(&-(V\cdot\nabla)u_1 - (u\cdot\nabla)V_1 - \frac{V_1}{r_0}u_2 - \frac{V_2}{r_0}u_1,\\ &-(V\cdot\nabla)u_2 - (u\cdot\nabla)V_2 + \frac{2V_1}{r_0}u_1, -(u\cdot\nabla)J - (V\cdot\nabla)T).\end{aligned}$$

Then the problem (5.2.35) and (5.2.36) is equivalent to the abstract form

$$\frac{d\psi}{dt} = L_R\psi + \mathscr{L}^\varepsilon\psi + G(\psi). \tag{5.2.37}$$

Consider the eigenvalue problem

$$L_R\psi + \mathscr{L}^\varepsilon\psi = \beta^\varepsilon(R)\psi. \tag{5.2.38}$$

It is known that with $\Delta T = T_0 - T_1 \cong 100^\circ C$, $|\varphi(x_1)|$ and Q are small. Hence, the steady-state solution (V,J) is also small:

$$\|(V,J)\|_{L^2} = \varepsilon \ll 1.$$

Thus, (5.2.35) is a perturbed equation of (5.2.3).

Since perturbation terms involving (V,J) are not invariant under zonal translation (in the x_1-direction), for general small functions $\varphi(x_1) \neq 0$, the first eigenvalues of (5.2.35) are (real or complex) simple, and by the perturbation theorems in (Ma and Wang 2005b), all eigenvalues of the linearized equation of (5.2.35) satisfy the following principle of exchange of stability (PES):

$$\begin{aligned}&\text{Re}\beta_i^\varepsilon(R)\begin{cases} <0 & \text{for } R < R_c^\varepsilon,\\ =0 & \text{for } R = R_c^\varepsilon,\\ >0 & \text{for } R > R_c^\varepsilon,\end{cases} && \text{for any } 1 \leq i \leq m,\\ &\text{Re}\beta_j^e(R_c^\varepsilon) < 0 && \text{for any } j \geq m+1,\end{aligned}$$

where $m = 1$ when $\beta_1^\varepsilon(R)$ is real, $m = 2$ if $\beta_1^\varepsilon(R)$ is complex near R_c^ε, and R_c^ε is the critical Rayleigh number of the perturbed system (5.2.35).

When $\beta_1^\varepsilon(R)$ near $R = R_c^\varepsilon$ is real, by Theorem 2.6.3, we derive immediately the following transition theorem.

Theorem 5.2.2 *Let $\beta_1^\varepsilon(R)$ near $R = R_c^\varepsilon$ be a simple real eigenvalue. Then the system (5.2.35) has a transition at $R = R_c^\varepsilon$, which is either mixed (Type III) or continuous (Type I), depending on the temperature deviation $\varphi(x_1)$. Moreover, we have the following assertions:*

(1) If the transition is of Type I, then if $R_c^\varepsilon < R < R_c^\varepsilon + \delta$ for some $\delta > 0$, the system bifurcates at R_c^ε to exactly two steady-state solutions ψ_1 and ψ_2 in H, which are attractors. In particular, the space H can be decomposed into two open sets U_1, U_2,

$$H = \overline{U}_1 + \overline{U}_2, \quad U_1 \bigcap U_2 = \emptyset, \quad \psi = 0 \in \partial U_1 \bigcap \partial U_2,$$

such that $\psi_i \in U_i$ $(i = 1, 2)$ and ψ_i attracts U_i.

(2) If the transition is of Type III, then there is a saddle-node bifurcation at $R = R^*$ with $R^* < R_c^\varepsilon$ such that the following statements hold:

(a) If $R^* < R < R_c^\varepsilon + \delta$ with $R \neq R_c^\varepsilon$, the system has two steady-state solutions ψ_R^+ and ψ_R^-, which are attractors, as shown in Fig. 5.6, such that

$$\psi_R^+ = 0 \qquad \text{for } R^* < R < R_c^\varepsilon.$$

(b) There is an open set $U \subset H$ with $0 \in U$ that can be decomposed into two disjoint open sets $\overline{U} = \overline{U}_+^R + \overline{U}_-^R$ with $\psi_R^\pm \in U_\pm^R$, and $\psi_R^\pm$ attracts $U_\pm^R$.

(3) For every initial value $\psi_0 = (u_0, T_0) \in U_-^R$ for $R > R^*$ or $\psi_0 = (u_0, T_0) \in U_+^R$ for $R > R_c^\varepsilon$, there exists a time $t_0 \geq 0$ such that for every $t > t_0$, the velocity field $u(t, \psi_0)$ is topologically equivalent to the structure shown in Fig. 5.5, either (a) or (b), where $\psi = (u(t, \psi_0), T(t, \psi_0))$ is the solution set of the problem with $\psi(0) = \psi_0$.

When $\beta_1^\varepsilon(R)$ is a complex eigenvalue, we obtain by Theorem 2.6.7 the following transition theorem of periodic solutions.

Theorem 5.2.3 *Let $\beta_1^\varepsilon(R)$ be complex near $R = R_c^\varepsilon$. Then the system (5.2.35) bifurcates from $(\psi, R) = (0, R_c^\varepsilon)$ to a periodic solution $\psi_R(t)$ on $R_c^\varepsilon < R$, which is an attractor, and $\psi_R(t)$ can be expressed as*

$$\psi_R(t) = A_R(\cos \rho t \psi_1 + \sin \rho t \widetilde{\psi}_1) + o(|A_R|, \varepsilon), \tag{5.2.39}$$

where $A_R = \alpha(R - R_c^\varepsilon)$, α and ρ are constants depending on $\varphi(x_1)$, and $\psi_1, \widetilde{\psi}_1$ are the first eigenfunctions of the linearized equations of (5.2.3).

A few remarks are now in order.

First, Theorems 5.2.1, 5.2.2, and 5.2.3 provide the possible dynamical behaviors for the zonal atmospheric circulation over the tropics. Theorem 5.2.1 describes the translation oscillation and does not represent a realistic explanation of the ENSO.

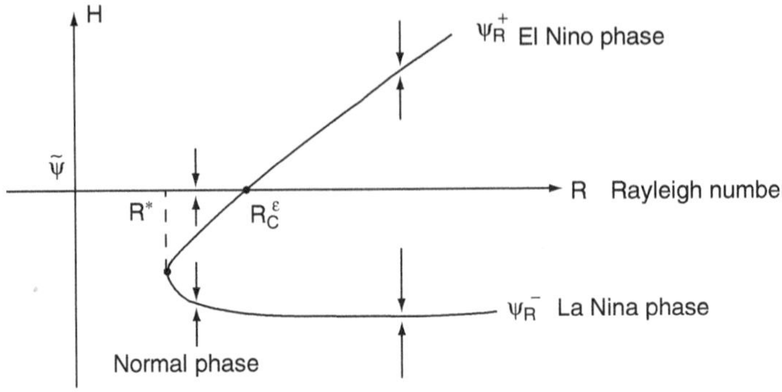

Fig. 5.6 The transition diagram for the equatorial motion equations (5.2.3), R^* is the saddle-node bifurcation point, and R_c^ε is the first critical Rayleigh number

Second, the periodic solution (5.2.39) characterizes a roll pattern translating with a constant velocity over the equator, eastward or westward, as shown in Fig. 5.7 (a) and (b) The time-periodic oscillation obtained in Theorem 5.2.3 does not represent the typical oscillation in the ENSO phenomena, since the Walker circulation does not obey this zonal translational oscillation.

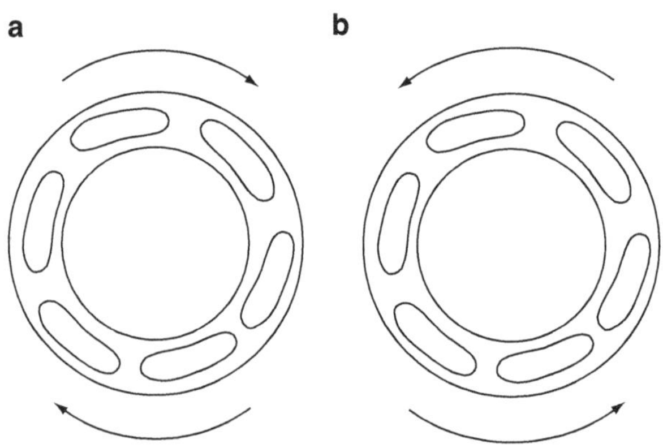

Fig. 5.7 Time-periodic translation of the Walker circulation pattern; (**a**) an eastward translation, (**b**) a westward translation

Third, Theorem 5.2.2 characterizes the oscillation between metastable states. The oscillation between the metastable states in Theorem 5.2.2 can be of either Type I or Type III, depending on the number

$$b = (G(\Psi_1), \Psi_1^*),$$

where Ψ_1 is the first eigenvector of the linearized equation of (5.2.35) at $R = R_c^\varepsilon$, Ψ^* is the corresponding first eigenvector of the adjoint linearized problem, and G represents the nonlinear terms in (5.2.35).

Namely, if $b \neq 0$, the transition is of Type III and if $b = 0$, the transition is of Type I. From the mathematical viewpoint, for almost all functions φ and $Q \neq 0$, $b \neq 0$. Hence, the case $b \neq 0$, and consequently the Type III transition in Theorem 5.2.2, is generic. In other words, the Type III transition derived in this theorem provides a correct oscillation mechanism of the ENSO between two metastable El Niño and La Niña events, and we shall explore this point of view in the next section in detail.

5.2.5 ENSO: Metastable Oscillation Theory

It is well known that the Walker cell at the equatorial Pacific Ocean is closely related to the ENSO phenomenon. Its behavior is the key to understanding ENSO.

Southern Oscillation Phenomenon. The Southern Oscillation Index (SOI) gives a simple measure of the strength and phase of the Southern Oscillation, and it indicates the state of the Walker circulation. The SOI is defined by the pressure difference between Tahiti and Darwin. When the Walker circulation enters its El Niño phase, the SOI is strongly negative; when it is in the La Niña phase, the SOI is strongly positive; and in the normal state, the SOI is small; see Fig. 5.8.

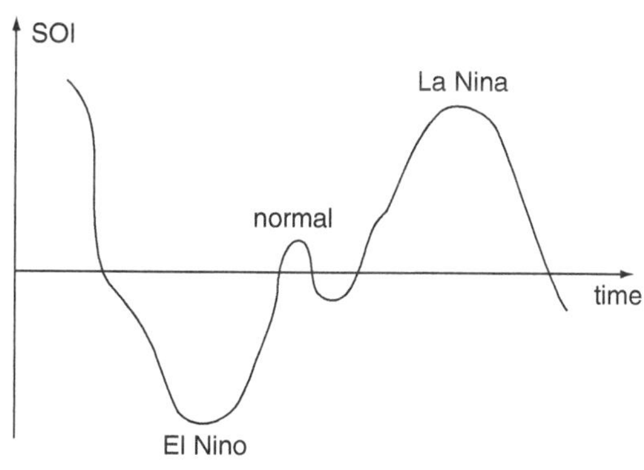

Fig. 5.8 SOI indicates the state of the Walker circulation

Also, the observational data of SOI as shown in Fig. 5.9 provide a clearer picture of both the spatiotemporal and random characteristics of ENSO.

El Niño and La Niña States. The inhomogeneous temperature distribution shows that the locations of the cells over the tropics are relatively fixed. This suggests that the homogeneous case as described by Theorem 5.2.1 is not a valid description of the realistic situation. In addition, the translation oscillation formation suggested in Theorem 5.2.3 also fails to describe the ENSO.

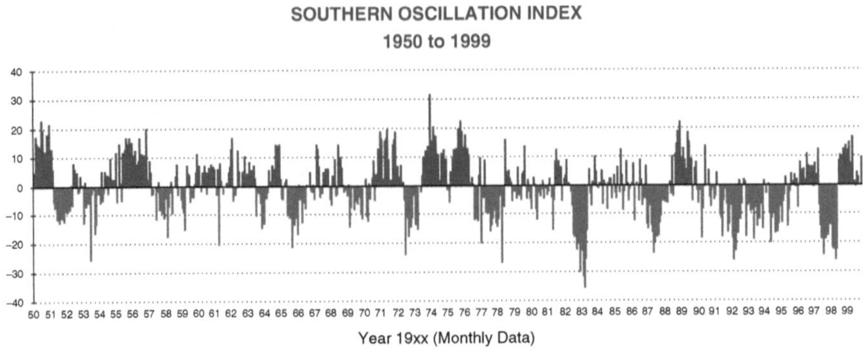

Fig. 5.9 SOI diagram, adopted from from the Australian Bureau of Meteorology website: BOM-Monitoring Climate–Climate variability and El Niño

Hence a natural theory for the atmospheric circulation over the tropics associated with the Walker circulation and the ENSO is given by Theorem 5.2.2. It is known that the steady-state solutions of (5.2.3) can be written as

$$\widetilde{\psi}+\psi_R^{\pm}, \tag{5.2.40}$$

where $\widetilde{\psi}=(V,J)$ is the steady-state solution given by (5.2.34), and $\psi_R^{\pm}$ are the stable equilibria of (5.2.35), derived in Theorem 5.2.2. In fact, the ENSO phenomenon can be explained as the transition between the two metastable steady-state solutions $\widetilde{\psi}+\psi_R^{\pm}$.

Theorem 5.2.2 amounts to saying that for a mixed transition, there are two critical Rayleigh numbers R^* and R_c^{ε}, with $R^* < R_c^{\varepsilon}$, where $R^* \cong R_c \cong 2.7\times 10^{21}$. Equation (5.2.35) has a saddle-node bifurcation at $R=R^*$, and has a mixed transition at $R=R_c^{\varepsilon}$, as shown in Fig. 5.6. When the Rayleigh number R is less than R^*, i.e., $R<R^*$, the system (5.2.3) has only one stable steady-state solution $\widetilde{\psi}=(V,J)$ as in (5.2.34) that attracts H. When $R^*<R<R_c^{\varepsilon}$, this system has two stable equilibria

$$\widetilde{\psi}\in U_+^R, \quad \text{and} \quad \widetilde{\psi}+\psi_R^-\in U_-^R, \quad \text{for} \quad R^*<R<R_c^{\varepsilon}, \tag{5.2.41}$$

where U_+^R, $U_-^R\subset H$ are basins of attraction of $\widetilde{\psi}$ and $\widetilde{\psi}+\psi_R^-$. And when $R_c^{\varepsilon}<R$, this system has as well the two stable equilibria

$$\widetilde{\psi}+\psi_R^-\in U_-^R \quad \text{and} \quad \widetilde{\psi}+\psi_R^+\in U_+^R \quad \text{for } R_c^{\varepsilon}<R, \tag{5.2.42}$$

with U_-^R and U_+^R as their basins of attraction.

Since the problem (5.2.3) with a natural boundary condition is a perturbation for the idealized condition, $\widetilde{\psi}=(V,J)$ is small, and the velocity field V is almost zero. Hence the equilibrium

$$\Psi_R^+=\begin{cases}\widetilde{\psi} & \text{for } R<R_c^{\varepsilon},\\ \widetilde{\psi}+\psi_R^+ & \text{for } R>R_c^{\varepsilon},\end{cases}$$

represents the El Niño phase, and the equilibrium

$$\Psi_R^- = \widetilde{\psi} + \psi_R^-$$

represents the normal phase for $R^* < R < R_c^\varepsilon$ and the La Niña phase for $R_c^\varepsilon < R$. It is clear that both $\Psi_R^\pm$ are metastable, with $U_R^\pm$ as their respective basins of attractions.

Oscillation Mechanism of ENSO. The above theoretical studies suggest that ENSO is the *interplay* between two oscillation processes. The first is the oscillation between the metastable warm event (El Niño phase, represented by Ψ_R^+) and cold event (La Niña phase) represented by Ψ_R^-). The second oscillation is the oscillation of the Rayleigh number caused essentially by the spatiotemporal oscillation of the SST field.

Here we present a brief schematic description of the interplay between these two oscillation processes of the ENSO, based on the saddle-node transition diagram in Fig. 5.6, rigorously proved in Theorem 5.2.2.

We begin with three physical conclusions:

- We observe that as R decreases, the normal phase and La Niña phase Ψ_R^- weaken and their basin of attraction shrinks (to zero as R approaches to R^*).
- As R increases in the interval $R_c^\varepsilon < R$, the strength of the El Niño phase increases. As R increases in (R^*, R_c^ε), the basin of attraction of the El Niño phase shrinks. In particular, near R_c^ε, the El Niño phase is close to the intersection of the two basins of attraction of the El Niño and La Niña phases, forcing the transition from the El Niño phase to the La Niña phase.
- Also, we see that as the El Niño event Ψ_R^+ intensifies, the SST increases, and as the normal and La Niña event Ψ_R^- intensifies, the SST decreases.

From the above three physical conclusions, we obtain the following new mechanism of the ENSO oscillation process:

- When $R^* < R < R_c^\varepsilon$, Ψ_R^- represents the normal condition, and the corresponding upwelling near Peru leads to a decrease of the SST and to R approaching R^*. On the other hand, near R^*, the basin of attraction of the normal condition Ψ_R^- shrinks, and due to the uncertainty of the initial data, the system can undergo a dynamic transition near R^* toward the El Niño phase Ψ_R^+.
- For the El Niño phase $\Psi_R^+ = \widetilde{\psi}$ near R^*, however, the velocity V is small, and SST will increase due to solar heating, which cannot be transported away by weak oceanic currents without upwelling. Hence the corresponding Rayleigh number R increases, and the El Niño phase intensifies. When the Rayleigh number approaches the critical value R_c^ε, then with high probability, the system undergoes a metastable transition from the El Niño phase to either the normal phase or the La Niña phase, depending on the Rayleigh number R and the strength of the phase Ψ_R^-.
- With delaying effect, strong El Niño and La Niña occur at Rayleigh number R larger than the critical number R_c^ε, and the system will repeat the process described in the above two items.

In summary, this proposed new mechanism for ENSO as an interplay of the two processes leads to both the random and deterministic features of the ENSO and defines a new natural *feedback* mechanism that drives the sporadic oscillation of the ENSO. The randomness is closely related to the uncertainty of fluctuations of the initial data between the narrow basins of attraction of the corresponding metastable events, and the deterministic feature is represented by a deterministic coupled atmospheric and oceanic model predicting the basins of attraction and the SST.

5.3 Thermohaline Ocean Circulation

Oceanic circulation is one of key sources of internal climate variability. One important source of such variability is the thermohaline circulation (THC). Physically speaking, the buoyancy fluxes at the ocean surface give rise to gradients in temperature and salinity, which produce, in return, density gradients. These gradients are, overall, sharper along the vertical than the horizontal and are associated therefore with an overturning, or THC.

THC is the global density-driven circulation of the oceans, which is so named because it involves both heat (thermo), and salt (haline). These two attributes, temperature and salinity, together determine the density of seawater, and the gradient in density between the water masses in the oceans causes the water to flow.

THC is also called the great ocean conveyer, the ocean conveyer belt, and the global conveyer belt. The great ocean conveyer produces the greatest oceanic current on the planet. It works in a fashion similar to a conveyer belt transporting an enormous volume of cold, salty water from the North Atlantic to the North Pacific, and bringing warmer, fresher water in return. Figure 5.10 gives a much simplified schematic of the great ocean conveyer, and Fig. 5.11 gives a diagram of oceanic currents associated with THC.

In oceanography, the conveyer is usually described as follows. Under and near the polar-region sea ice in the North Atlantic, the warm and salty water that has been transported northward from the tropical regions is cooled to form frigid water in vast quantities, leading to denser seawater. Note that unlike fresh water, salty water does not have a density maximum at $4°C$, and gets denser as it cools all the way to its freezing point of approximatively $-1.8°C$. When this seawater freezes, the salt is separated (sea ice contains almost no salt), increasing the salinity of the remaining, unfrozen, water. This salinity makes the water denser again. The dense water then sinks into the deep basin of the sea to form the North Atlantic Deep Water (NADW), and drives today's oceanic THC.

The sinking cold and salty water in the North Atlantic flows very slowly southward into the deep abyssal plains of the Atlantic. The deep water then moves through the South Atlantic around South Africa, where it is split into two routes: one into the Indian Ocean and one past Australia into the Pacific. As it continues on its submarine migration, the current mixes with warmer fresh water, and slowly becomes warmer and fresher. Finally, in the North Pacific, the warmer and fresher water up-

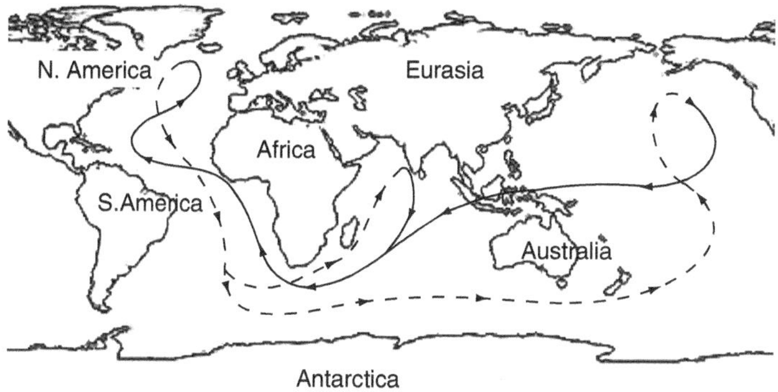

Fig. 5.10 Diagram of the thermohaline circulation. The dotted line represents deep-water currents, while the solid line represents shallow-water currents

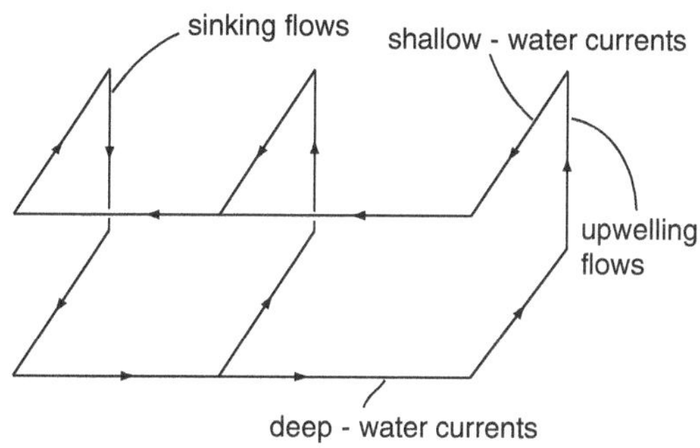

Fig. 5.11 Schematic diagram of oceanic currents of the global thermohaline circulation

wells, while a shallow-water countercurrent is generated. This countercurrent moves southward and westward, through the Indian Ocean, still heading west around southern Africa, then crosses through the South Atlantic, still on the surface (though it extends a kilometer and a half deep). It then moves up along the east coast of North America and crosses the coast of Scandinavia. When this warmer, less salty water reaches high northern latitudes, it chills, and naturally becomes North Atlantic Deep Water, completing a circuit.

THC varies on time scales of decades or longer, and there have been extensive observational, physical, and numerical studies. We refer the interested reader to (Dijkstra and Ghil 2005; Dijkstra 2000a) for an extensive review of the topics; see also, among others, (Stommel 1961; Rooth 1982; Held and Suarez 1974; Ghil 1976; Quon and Ghil 1992, 1995; Thual and McWilliams 1992; Tziperman 1997; Tziperman et al. 1994; Dijkstra and Molemaker 1997, 1999; Dijkstra and Neelin 1999, 2000).

The main objective of this section is to study dynamic stability and transitions in large-scale oceanic circulations associated with THC. The main results presented here were first obtained by (Ma and Wang 2010b).

5.3.1 Boussinesq Equations

For simplicity, we regard the region of the Pacific, Indian and Atlantic Oceans that the ocean conveyer belt occupies as a rectangular domain: $\Omega = (0,l_1) \times (0,l_2) \times (0,h)$, where l_1 stands for the length of the conveyer belt, l_2 for the width, and h for the depth of the ocean.

The basic equations governing the thermohaline circulations are the Boussinesq type of hydrodynamic equations coupled with diffusion equations for salinity and temperature. As we see from Fig. 5.11, the dominated THC is in the zonal direction, leading to the approximation of the Coriolis force in the flat geometry by $(-2\Omega u_3, 0, 2\Omega u_1)$, which is balanced by a gradient force $\nabla\phi$. Hence, we can ignore the rotational effect of the Earth, and the motion and states of the large-scale ocean are governed by the following Boussinesq equations; see among others (Pedlosky 1987; Ghil and Childress 1987; Stern 1960; Malkus and Veronis 1958; Lions et al. 1992b):

$$\begin{aligned}
&\frac{\partial u}{\partial t} + (u\cdot\nabla)u = \nu\Delta u - \frac{1}{\rho_0}(\nabla p + \rho g k),\\
&\frac{\partial T}{\partial t} + (u\cdot\nabla)T = \kappa_T \Delta T,\\
&\frac{\partial S}{\partial t} + (u\cdot\nabla)S = \kappa_S \Delta S,\\
&\mathrm{div} u = 0,
\end{aligned} \tag{5.3.1}$$

where $u = (u_1,u_2,u_3)$ is the velocity field, T is the temperature function, S is the salinity, $k = (0,0,1)$, $\kappa_T > 0$ is the thermal diffusivity, $\kappa_S > 0$ is the salt diffusivity, $\rho_0 > 0$ is the fluid density at the lower surface $x_3 = 0$ for $(x_1,x_2,x_3) \in \Omega$, and ρ is the fluid density given by the equation of state

$$\rho = \rho_0[1 - \alpha_T(T - T_0) + \alpha_S(S - S_0)]. \tag{5.3.2}$$

Here α_T and α_S are assumed to be positive constants. Moreover, the lower boundary $(x_3 = 0)$ is maintained at a constant temperature T_0 and a constant salinity S_0, while the upper boundary $(x_3 = h)$ is maintained at a constant temperature T_1 and a constant salinity S_1. The case in which $T_0 < T_1$ and $S_0 > S_1$ often leads to thermal and solute stabilities, while each of the following three cases may give rise to instability:

$$T_0 > T_1, \quad S_0 < S_1, \tag{5.3.3}$$

$$T_0 > T_1, \quad S_0 > S_1, \tag{5.3.4}$$

$$T_0 < T_1, \quad S_0 < S_1. \tag{5.3.5}$$

The conditions (5.3.3)–(5.3.5) are satisfied over various regions of the oceans. In particular, in the high-latitude ocean regions, the condition (5.3.3) or (5.3.4) is satisfied, and in the tropical ocean regions, the case (5.3.5) occurs. It is the properties (5.3.3)–(5.3.5) that give rise to the global thermohaline circulation, and the theoretical results derived in this section also support this viewpoint.

The trivial steady-state solution of (5.3.1)–(5.3.2) is given by

$$(u^0, T^0, S^0) = (0, T_0 - (T_0 - T_1)x_3/h, S_0 - (S_0 - S_1)x_3/h),$$
$$p^0 = p_0 - g\rho_0[x_3 + \frac{\alpha_T}{2}(T_0 - T_1)x_3^2/h - \frac{\alpha_S}{2}(S_0 - S_1)x_3^2/h],$$

where p_0 is a constant. To make the equations nondimensional, we consider the perturbation of the solution from the trivial steady state

$$u'' = u - u^0, \quad T'' = T - T^0, \quad S'' = S - S^0, \quad p'' = p - p^0.$$

Then we set

$$\begin{aligned} &x = hx', && t = h^2 t'/\kappa_T, \\ &u'' = \kappa_T u'/h, && T'' = (T_0 - T_1)T', \quad S'' = |S_0 - S_1|S', \quad p'' = \rho_0 \nu \kappa_T p'/h^2. \end{aligned}$$

Omitting the primes, equation (5.3.1) can be written as

$$\begin{aligned} &\frac{\partial u}{\partial t} = \mathrm{Pr}\,(\Delta u - \nabla p) + \mathrm{Pr}\,(RT - \mathrm{sign}(S_0 - S_1)\widetilde{R}S)k - (u\cdot\nabla)u \\ &\frac{\partial T}{\partial t} = \Delta T + u_3 - (u\cdot\nabla)T, \\ &\frac{\partial S}{\partial t} = \mathrm{Le}\,\Delta S + \mathrm{sign}(S_0 - S_1)u_3 - (u\cdot\nabla)S, \\ &\mathrm{div}u = 0, \end{aligned} \tag{5.3.6}$$

for $x = (x_1, x_2, x_3)$ in the nondimensional domain $\Omega = (0, L_1) \times (0, L_2) \times (0, 1)$, where $L_i = l_i/h$ $(i = 1, 2)$, and the nondimensional parameters are the thermal Rayleigh number R, the saline Rayleigh number $\widetilde{R}$, the Prandtl number Pr, and the Lewis number Le:

$$R = \frac{\alpha_T g(T_0 - T_1)h^3}{\kappa_T \nu}, \quad \widetilde{R} = \frac{\alpha_S g(S_0 - S_1)h^3}{\kappa_T \nu}, \quad \mathrm{Pr} = \frac{\nu}{\kappa_T}, \quad \mathrm{Le} = \frac{\kappa_S}{\kappa_T}. \tag{5.3.7}$$

We consider the free boundary condition

$$\begin{aligned} &u_1 = 0, \ \frac{\partial u_2}{\partial x_1} = \frac{\partial u_3}{\partial x_1} = \frac{\partial T}{\partial x_1} = \frac{\partial S}{\partial x_1} = 0 && \text{at } x_1 = 0, L_1, \\ &u_2 = 0, \ \frac{\partial u_1}{\partial x_2} = \frac{\partial u_3}{\partial x_2} = \frac{\partial T}{\partial x_2} = \frac{\partial S}{\partial x_2} = 0 && \text{at } x_2 = 0, L_2, \\ &u_3 = T = S = 0, \ \frac{\partial u_1}{\partial x_3} = \frac{\partial u_2}{\partial x_3} = 0 && \text{at } x_3 = 0, 1. \end{aligned} \tag{5.3.8}$$

The initial value conditions are given by

$$(u, T, S) = (\widetilde{u}, \widetilde{T}, \widetilde{S}) \quad \text{at} \quad t = 0. \tag{5.3.9}$$

For the problem (5.3.6)–(5.3.9), we set the spaces

$$\begin{aligned}
&H = \{(u,T,S) \in L^2(\Omega)^5 \mid \mathrm{div} u = 0, u \cdot n|_{\partial\Omega} = 0\},\\
&H_1 = \{(u,T,S) \in H^2(\Omega)^5 \bigcap H \mid (u,T,S) \text{ satisfies (5.3.8) } \}.
\end{aligned}$$

Let $L_\lambda = A + B_\lambda : H_1 \to H$ and $G : H_1 \to H$ be defined by

$$\begin{aligned}
&A\psi = P(\mathrm{Pr}\ \Delta u, \Delta T, \mathrm{Le}\ \Delta S),\\
&B_x\psi = P(\mathrm{Pr}\ (RT - \widetilde{R}S\mathrm{sign}(S_0 - S_1))k, u_3, \mathrm{sign}(S_0 - S_1)u_3),\\
&G(\psi) = P((u \cdot \nabla)u, (u \cdot \nabla)T, (u \cdot \nabla)S),
\end{aligned}$$

for $\psi = (u,T,S) \in H_1, \lambda = (R, \widetilde{R})$. Here $P : L^2(\Omega, \mathbb{R}^5) \to H$ is the Leray projection. Then the problem (5.3.6)–(5.3.9) can be written as

$$\frac{d\psi}{dt} = L_\lambda \psi + G(\psi), \qquad \psi(0) = \psi_0. \tag{5.3.10}$$

5.3.2 Linear Analysis

Linear Eigenvalue Problem. To understand the dynamic transitions of the problem, we need to study the following eigenvalue problem for the linearized equations of (5.3.6)–(5.3.8):

$$\begin{aligned}
&\mathrm{Pr}\ (\Delta u - \nabla p) + \mathrm{Pr}\ (RT - \widetilde{R}S\mathrm{sign}(S_0 - S_1))k = \beta u,\\
&\Delta T + u_3 = \beta T,\\
&\mathrm{Le}\ \Delta S + \mathrm{sign}(S_0 - S_1)u_3 = \beta S,\\
&\mathrm{div} u = 0,
\end{aligned} \tag{5.3.11}$$

supplemented with (5.3.8).

We proceed with the separation of variables. By the boundary condition (5.3.8), $\psi = (u,T,S)$ can be expressed in the following form:

$$\begin{aligned}
&u_1 = u_{jk}(x_3) \sin j\alpha_1 \pi x_1 \cos k\alpha_2 \pi x_2,\\
&u_2 = v_{jk}(x_3) \cos j\alpha_1 \pi x_1 \sin k\alpha_2 \pi x_2,\\
&u_3 = w_{jk}(x_3) \cos j\alpha_1 \pi x_1 \cos k\alpha_2 \pi x_2,\\
&T = T_{jk}(x_3) \cos j\alpha_1 \pi x_1 \cos k\alpha_2 \pi x_2,\\
&S = S_{jk}(x_3) \cos j\alpha_1 \pi x_1 \cos k\alpha_2 \pi x_2,\\
&p = p_{jk}(x_3) \cos j\alpha_1 \pi x_1 \cos k\alpha_2 \pi x_2,
\end{aligned} \tag{5.3.12}$$

for integers j and k, where $\alpha_i = L_i^{-1}$ $(i = 1,2)$. Plugging (5.3.12) into (5.3.11), we obtain the following sets of systems of ordinary differential equations:

$$
\begin{aligned}
&\Pr D_{jk}u_{jk} + \Pr j\alpha_1\pi p_{jk} = \beta u_{jk},\\
&\Pr D_{jk}v_{jk} + \Pr k\alpha_2\pi p_{jk} = \beta v_{jk},\\
&\Pr D_{jk}w_{jk} - \Pr Dp_{jk} + \Pr RT_{jk} - \Pr \widetilde{R}\text{sign}(S_0 - S_1)S_{jk} = \beta W_{jk},\\
&D_{jk}T + w_{jk} = \beta T_{jk},\\
&LeD_{jk}S_{jk} + \text{sign}(S_0 - S_1)w_{jk} = \beta S_{jk}\\
&j\alpha_1\pi u_{jk} + k\alpha_2\pi v_{jk} + Dw_{jk} = 0,\\
&Du_{jk} = Dv_{jk} = w_{jk} = T_{jk} = S_{jk} = 0 \quad \text{at} \quad x_3 = 0,1,
\end{aligned} \tag{5.3.13}
$$

where

$$
D = \frac{d}{dx_3}, \quad D_{jk} = \frac{d^2}{dx_2^2} - \alpha_{jk}^2, \quad \alpha_{jk}^2 = \pi^2(j^2\alpha_1^2 + k^2\alpha_2^2).
$$

If $w_{jk} \neq 0$, then Eq. (5.3.13) can be reduced to a single equation

$$
\begin{aligned}
&[(D_{jk} - \beta)(\text{Le}\, D_{jk} - \beta)(\Pr D_{jk} - \beta)D_{jk}\\
&+ \Pr R\alpha_{jk}^2(\text{Le}\, D_{jk} - \beta) - \Pr \widetilde{R}\alpha_{jk}^2(D_{jk} - \beta)]w_{jk} = 0,\\
&w_{jk} = D^2 w_{jk} = D^4 w_{jk} = D^6 w_{jk} = 0 \quad \text{at} \quad x_3 = 0,1.
\end{aligned} \tag{5.3.14}
$$

It is clear that the solutions of (5.3.14) are sine functions

$$
w_{jk} = \sin l\pi x_3 \quad \text{for } (j,k,l) \in \mathbb{Z}^2 \times \mathbb{N}. \tag{5.3.15}
$$

Substituting (5.3.15) into (5.3.14), we see that the corresponding eigenvalues β of problem (5.3.11) satisfy the cubic equation

$$
\begin{aligned}
&\gamma_{jkl}^2(\gamma_{jkl}^2 + \beta)(\text{Le}\, \gamma_{jkl}^2 + \beta)(\Pr \gamma_{jkl}^2 + \beta)\\
&- \Pr R\alpha_{jk}^2(\text{Le}\, \gamma_{jkl}^2 + \beta) + \Pr \widetilde{R}\alpha_{jk}^2(\gamma_{jkl}^2 + \beta) = 0,
\end{aligned} \tag{5.3.16}
$$

for all $(j,k,l) \in \mathbb{Z}^2 \times \mathbb{N}$, where $\gamma_{jkl}^2 = \alpha_{jk}^2 + l^2\pi^2$.

Moreover, to determine $u_{jk}(x_3), v_{jk}(x_3), T_{jk}(x_3)$, and $S_{jk}(x_3)$, we derive from (5.3.13) that

$$
(\Pr D_{jk} - \beta)u_{jk} = -\frac{j\alpha_1\pi}{\alpha_{jk}^2}(\Pr D_{jk} - \beta)Dw_{jk}, \tag{5.3.17}
$$

$$
(\Pr D_{jk} - \beta)v_{jk} = -\frac{k\alpha_2\pi}{\alpha_{jk}^2}(\Pr D_{jk} - \beta)Dw_{jk}, \tag{5.3.18}
$$

$$
(D_{jk} - \beta)T_{jk} = -w_{jk}, \tag{5.3.19}
$$

$$
(\text{Le}\, D_{jk} - \beta)S_{jk} = -\text{sign}(S_0 - S_1)w_{jk}. \tag{5.3.20}
$$

With the above calculations, all eigenvalues and eigenvectors of (5.3.11) can be derived, and they are given in the following three groups:

1. For $(j,k,l)=(j,k,0)$, we have

$$\beta_{jk0} = -\mathrm{Pr}\,\alpha_{jk}^2 = -\mathrm{Pr}\,(j^2\alpha_1^2+k^2\alpha_2^2),$$
$$\psi_{jk0} = (k\alpha_2 \sin j\alpha_1\pi x_1 \cos k\alpha_2\pi x_2, -j\alpha_1 \cos j\alpha_1\pi x_1 \sin k\alpha_2\pi x_2, 0,0,0).$$

2. For $(j,k,l)=(0,0,l)$ with $l\neq 0$, we have

$$\beta_{00l}^1 = -l^2\pi^2, \qquad \beta_{00l}^2 = -\mathrm{Le}\, l^2\pi^2,$$
$$\psi_{00l}^1 = (0,0,0,\sin l\pi x_3, 0), \qquad \psi_{00l}^2 = (0,0,0,0,\sin l\pi x_3).$$

3. For general (j,k,l) with $j^2+k^2\neq 0$ and $l\neq 0$, the solutions β of (5.3.16) are eigenvalues of (5.3.11). Let β_{jkl}^i $(1\leq i\leq 3)$ be three zeros of (5.3.16), with

$$\mathrm{Re}\beta_{jkl}^1 \geq Re\beta_{jkl}^2 \geq \mathrm{Re}\beta_{jkl}^3.$$

Then by (5.3.12), (5.3.15), and (5.3.17)–(5.3.20), the eigenvector ψ_{jkl}^i corresponding to β_{jkl}^i can be written as

$$\psi_{jkl}^i = \begin{cases} u_{jkl}^i \sin j\alpha_1\pi x_1 \cos k\alpha_2\pi x_2 \cos l\pi x_3 \\ v_{jkl}^i \cos j\alpha_1\pi x_1 \sin k\alpha_2\pi x_2 \cos l\pi x_3 \\ w_{jkl}^i \cos j\alpha_1\pi x_1 \cos k\alpha_2\pi x_2 \sin l\pi x_3 \\ T_{jkl}^i \cos j\alpha_1\pi x_1 \cos k\alpha_2\pi x_2 \sin l\pi x_3 \\ S_{jkl}^i \cos j\alpha_1\pi x_1 \cos k\alpha_2\pi x_2 \sin l\pi x_3, \end{cases} \tag{5.3.21}$$

where

$$u_{jkl}^i = -\frac{j\alpha_1 l\pi^2}{\alpha_{jk}^2} = -\frac{j\alpha_1 l}{j^2\alpha_1^2+k^2\alpha_2^2}, \qquad v_{jkl}^i = -\frac{k\alpha_2 l\pi^2}{\alpha_{jk}^2} = -\frac{k\alpha_2 l}{j^2\alpha_1^2+k^2\alpha_2^2},$$
$$w_{jkl}^i = 1, \qquad T_{jkl}^i = \frac{1}{\gamma_{jkl}^2+\beta_{jkl}^2}, \qquad S_{jkl}^i = \frac{\mathrm{sign}(S_0-S_1)}{\mathrm{Le}\,\gamma_{jkl}^2+\beta_{jkl}^2}.$$

The adjoint equations of (5.3.11) are given by

$$\begin{aligned} &\mathrm{Pr}\,(\Delta u^* - \nabla p^*) + (T^* + \mathrm{sign}(S_0-S_1)S^*)k = \overline{\beta} u^*, \\ &\Delta T^* + \mathrm{Pr}\, R u_3^* = \overline{\beta} T^*, \\ &\mathrm{Le}\,\Delta S^* - \mathrm{Pr}\,\widetilde{R}\mathrm{sign}(S_0-S_1)u_3^* = \overline{\beta} S^* \\ &\mathrm{div} u^* = 0. \end{aligned} \tag{5.3.22}$$

Thus, the conjugate eigenvectors ψ_{jkl}^{i*} of (5.3.22) corresponding to β_{jkl}^i are as follows:

$$\psi^{i*}_{jkl} = \begin{cases} u^{i*}_{jkl} \sin j\alpha_1 \pi x_1 \cos k\alpha_2 \pi x_2 \cos l\pi x_3, \\ v^{i*}_{jkl} \cos j\alpha_1 \pi x_1 \sin k\alpha_2 \pi x_2 \cos l\pi x_3, \\ w^{i*}_{jkl} \cos j\alpha_1 \pi x_1 \sin k\alpha_2 \pi x_2 \sin l\pi x_3, \\ T^{i*}_{jkl} \cos j\alpha_1 \pi x_1 \sin k\alpha_2 \pi x_2 \sin l\pi x_3, \\ S^{i*}_{jkl} \cos j\alpha_1 \pi x_1 \sin k\alpha_2 \pi x_3 \sin l\pi x_3, \end{cases} \tag{5.3.23}$$

where

$$u^{i*}_{jkl} = u^i_{jkl} = -\frac{j\alpha_1 l\pi^2}{\alpha^2_{jk}}, \qquad v^{i*}_{jkl} = v^i_{jkl} = -\frac{k\alpha_2 l\pi^2}{\alpha^2_{jk}}, \qquad w^{i*}_{jkl} = 1,$$

$$T^{i*}_{jkl} = \frac{\Pr R}{\gamma^2_{jkl} + \overline{\beta}^i_{jkl}}, \qquad S^{i*}_{jkl} = \frac{-\operatorname{sign}(S_0 - S_1)\Pr \widetilde{R}}{\operatorname{Re}\widetilde{\gamma}^2_{jkl} + \overline{\beta}^i_{jkl}}.$$

Thus, all eigenvectors of (5.3.11) consist of $\psi_{jk0}, \psi^1_{00l}, \psi^2_{00l}$, and ψ^i_{jkl} $(i = 1,2)$. All conjugate eigenvectors of (5.3.22) consist of $\psi^*_{jk0} = \psi_{jk0}, \psi^{1*}_{00l} = \psi^1_{00l}, \psi^{2*}_{00l} = \psi^2_{00l}$, and ψ^{i*}_{jkl} as in (5.3.23).

Principle of Exchange of Stability. We need to study the solutions β of (5.3.16), which is equivalent to the following form:

$$\begin{aligned} &\beta^3 + (\Pr + \mathrm{Le} + 1)\gamma^2_{jkl}\beta^2 + [(\Pr + \mathrm{Le} + \Pr \mathrm{Le})\gamma^4_{jkl} \\ &-\Pr \alpha^2_{jk}\gamma^{-2}_{jkl}(R - \widetilde{R})]\beta + \Pr \mathrm{Le}\, \gamma^6_{jkl} - \Pr \alpha^2_{jk}(\mathrm{Le}\, R - \widetilde{R}) = 0. \end{aligned} \tag{5.3.24}$$

First, we discuss the critical crossing of the real eigenvalues. To this end, we need to introduce a nondimensional parameter, called the R-Rayleigh number, defined by

$$\sigma = R - \mathrm{Le}^{-1}\widetilde{R} = \frac{gh^3}{\kappa_T \nu}(\alpha_T(T_0 - T_1) - \alpha_S \mathrm{Le}^{-1}(S_0 - S_1)). \tag{5.3.25}$$

It is clear that $\beta = 0$ is a zero of (5.3.24) if and only if

$$\mathrm{Le}\, \gamma^6_{jkl} - \alpha^2_{jk}(\mathrm{Le}\, R - \widetilde{R}) = 0.$$

In this case, we have

$$\sigma = \frac{\gamma^6_{jkl}}{\alpha^2_{jk}} \geq \sigma_c, \tag{5.3.26}$$

where σ_c, called the *critical R-Rayleigh number*, is defined by

$$\sigma_c = \min_{\substack{(j,k)\in\mathbb{Z}^2, j,k\geq 0, \\ j^2+k^2\neq 0, l\geq 1}} \frac{\pi^4(j^2L_1^{-2} + k^2L_2^{-2} + 1)^3}{j^2L_1^{-2} + k^2L_2^{-2}} = \frac{\pi^4(j_1^2L_1^{-2} + k_1^2L_2^{-2} + 1)^3}{j_1^2L_1^{-2} + k_1^2L_2^{-2}}, \tag{5.3.27}$$

for some integer pair (j_1,k_1) such that $j_1 \geq 0, k_1 \geq 0, j_1^2+k_1^2 \neq 0$. Here L_1 and L_2 are the nondimensional length scales in the zonal and meridional directions respectively (with the vertical length scaled to 1).

Next, we consider the critical crossing of the complex eigenvalues. For this case, we introduce another nondimensional parameter, called the C-Rayleigh number, defined by

$$\eta = R - \frac{\text{Pr} + \text{Le}}{\text{Pr} + 1}\widetilde{R} = \frac{gh^3}{\kappa_T \nu}\left[\alpha_T(T_0 - T_1) - \frac{(\text{Pr} + \text{Le})\alpha_S}{\text{Pr} + 1}(S_0 - S_1)\right]. \quad (5.3.28)$$

Let $i\rho_0$ $(\rho_0 \neq 0)$ be a zero of (5.3.24). Then we have

$$\rho_0^2 = (\text{Pr} + \text{Le} + \text{Pr Le})\gamma_{jkl}^4 - \text{Pr}\,\alpha_{jk}^2\gamma_{jkl}^{-2}(R - \widetilde{R}),$$

$$\rho_0^2 = \frac{\text{Pr Le}\,\gamma_{jkl}^6 - \text{Pr}\,\alpha_{jk}^2(\text{Le}\,R - \widetilde{R})}{\gamma_{jkl}^2(\text{Pr} + \text{Le} + 1)}.$$

Therefore, Eq. (5.3.24) has a pair of purely imaginary solutions $\pm i\rho_0$ if and only if the following condition holds:

$$\begin{aligned}&(\text{Pr} + \text{Le} + 1)(\text{Pr} + \text{Le} + \text{Pr Le})\gamma_{jkl}^6 - \text{Pr}\,\alpha_{jk}^2(\text{Pr} + \text{Le} + 1)(R - \widetilde{R})\\ &= \text{Pr Le}\,\gamma_{jkl}^6 - \text{Pr}\,\alpha_{jk}^2(\text{Le}\,R - \widetilde{R}) > 0.\end{aligned} \quad (5.3.29)$$

It follows from (5.3.29) that

$$\eta = R - \frac{\text{Pr} + \text{Le}}{\text{Pr} + 1}\widetilde{R} = \frac{(\text{Pr} + \text{Le})(1 + \text{Le})}{\text{Pr}}\frac{\gamma_{jkl}^6}{\alpha_{jk}^2}.$$

Hence we define the critical C-Rayleigh number by

$$\eta_c = \min_{(j,k,l)\in I}\frac{(\text{Pr} + \text{Le})(1 + \text{Le})}{\text{Pr}}\frac{\gamma_{jkl}^6}{\alpha_{jk}^2} = \frac{(\text{Pr} + \text{Le})(1 + \text{Le})}{\text{Pr}}\frac{\gamma_{j_1k_11}^6}{\alpha_{j_1k_1}^2}, \quad (5.3.30)$$

for the same integer pair (j_1,k_1) as in (5.3.27).

Definition 5.3.1 *Let (j_1,k_1) satisfy (5.3.27). We call σ_c (respectively η_c) the first critical Rayleigh number if for all eigenvalues β of (5.3.11), we have*

$$Re\beta(\sigma_c) \leq 0 \quad (\textit{respectively } Re\beta(\eta_c) \leq 0).$$

To determine the first critical Rayleigh number for σ_c and η_c, we introduce another nondimensional parameter, K:

$$K = \text{Le}^2\left(1 + \frac{1}{\text{Pr}}\right)\sigma_c - (1 - \text{Le})\widetilde{R}. \quad (5.3.31)$$

Theorem 5.3.2 *Let* Le $\neq 1$, *let* (j_1,k_1) *satisfy (5.3.27), and let* $\gamma^2 = \gamma^2_{j_1k_11}$, $\alpha^2 = \alpha^2_{j_1k_1}$.

(1) If $K > 0$, *where* K *is defined by (5.3.31), then the number* σ_c *is the first critical Rayleigh number, and*

$$\beta^1_{j_1k_11}\begin{cases} <0 & \text{if } \sigma < \sigma_c, \\ =0 & \text{if } \sigma = \sigma_c, \\ >0 & \text{if } \sigma > \sigma_c, \end{cases} \tag{5.3.32}$$

$$Re\beta^r_{jkl}(\sigma_c) < 0 \qquad \text{for all } (j,k,l) \text{ not satisfying } (5.3.27). \tag{5.3.33}$$

(2) If $K < 0$, *then the number* η_c *is the first critical Rayleigh number, and*

$$Re\beta^1_{j_1k_11} = Re\beta^2_{j_1k_11}\begin{cases} <0 & \text{if } \eta < \eta_c, \\ =0 & \text{if } \eta = \eta_c, \\ >0 & \text{if } \eta > \eta_c, \end{cases} \tag{5.3.34}$$

$$Re\beta^r_{jkl}(\eta_c) < 0 \qquad \text{for } (j,k,l) \text{ not satisfying } (5.3.27). \tag{5.3.35}$$

To prove Theorem 5.3.2, we begin with the following two lemmas. The first lemma ensures that the R-Rayleigh number σ defined by (5.3.25) is the parameter required to describe the critical crossing for the real eigenvalues of (5.3.11).

Lemma 5.3.3 *Assume that*

$$\widetilde{R} \neq \frac{-\text{Le}^2(1+\text{Pr}\,)\gamma^6_{j_1k_11}}{(1-\text{Le}\,)\text{Pr}\,\alpha^2_{j_1k_1}},$$

and that for the R-Rayleigh number σ *near* σ_c, *all real eigenvalues of (5.3.11) are real:*

$$\beta_1 \geq \beta_2 \geq \cdots \geq \beta_m \geq \beta_{m+1} \geq \cdots. \tag{5.3.36}$$

Then $\beta_1 = \beta^1_{j_1k_11}$, *and with* σ_c *defined by (5.3.27), we have*

$$\beta_{j_1k_11}(\sigma)\begin{cases} <0 & \text{if } \sigma < \sigma_c, \\ =0 & \text{if } \sigma = \sigma_c, \\ >0 & \text{if } \sigma > \sigma_c. \end{cases} \tag{5.3.37}$$

Proof. Let $\alpha = \alpha_{j_1k_1}, \gamma = \gamma_{j_1k_11}, R_0 \quad \text{Le}^{-1}\widetilde{R}_0 = \gamma^6/\alpha^2$, where (j_1,k_1) is as in (5.3.27). We shall show that

$$(\text{Pr}\, + \text{Le}\, + \text{Pr Le}\,)\gamma^4 - \text{Pr}\,\alpha^2\gamma^{-2}(R_0 - \widetilde{R}_0) > 0. \tag{5.3.38}$$

Assuming that (5.3.38) is false, we consider the case

$$(\text{Pr}\, + \text{Le}\, + \text{Pr Le}\,)\gamma^4 - \text{Pr}\,\alpha^2\gamma^{-2}(R_0 - \widetilde{R}_0) < 0. \tag{5.3.39}$$

Note that the solution $\beta(\sigma)=\beta^1_{j_1k_11}(\sigma)$ of (5.3.24) with $(j,k,l)=(j_1,k_1,1)$ is continuous on σ, and $\beta(\sigma_c)=0$. Hence

$$\beta(\sigma)\to 0 \quad \text{as } \sigma\to\sigma_c=R_0-\text{Le}^{-1}\widetilde{R}_0. \tag{5.3.40}$$

Thus, near $\sigma=\sigma_c$, Eq. (5.3.24) can be written as

$$\beta(\sigma)=\frac{-b_0(\sigma)}{b_1(\sigma)}+o(\beta(\sigma)), \tag{5.3.41}$$

where

$$\begin{aligned} b_0(\sigma) &= \text{Pr Le}\,\alpha^2\left(\frac{\gamma^6}{\alpha^2}-\sigma\right), \\ b_1(\sigma) &= (\text{Pr}+\text{Le}+\text{Pr Le})\gamma^4-\text{Pr}\,\alpha^2\gamma^{-2}(R-\widetilde{R}). \end{aligned} \tag{5.3.42}$$

It follows from (5.3.39)–(5.3.41) that for $\sigma=R-\text{Le}^{-1}\widetilde{R}$ near $\sigma_c=R_0-\text{Le}^{-1}\widetilde{R}_0$,

$$\beta(\sigma)\begin{cases} >0 & \text{if } \sigma<\sigma_c, \\ =0 & \text{if } \sigma=\sigma_c, \\ <0 & \text{if } \sigma>\sigma_c. \end{cases} \tag{5.3.43}$$

We write Eq. (5.3.24) in the following form:

$$\beta^3+b_2\beta^2+b_1(\sigma)\beta+b_0(\sigma)=0, \tag{5.3.44}$$

where $b_0(\sigma), b_1(\sigma)$ are as in (5.3.42), and

$$b_2=(\text{Pr}+\text{Le}+1)\gamma^2>0.$$

Meanwhile, it is easy to see that

$$b_0(\sigma)\to+\infty, b_1(\sigma)\to+\infty, \text{ as } \sigma\to-\infty.$$

This implies that when σ is sufficiently small, the real solutions of (5.3.44) must be negative. Thus, by (5.3.43), there exists a number $\sigma_0<\sigma_c$ such that the solution $\beta(\sigma)$ of (5.3.44) vanishes at $\sigma=\sigma_0$, which is a contradiction to $b_0(\sigma_0)\neq 0$. Thus, we derive $b_1(\sigma_c)\geq 0$. By the assumption in the lemma, $b_1(\sigma_c)\neq 0$. Hence, (5.3.38) holds, i.e., $b_1(\sigma_c)>0$. By (5.3.41), we can obtain (5.3.37).

In the following, we shall prove that $\beta^1_{j_1k_11}=\beta_1$ as in (5.3.36). We need to consider only the real eigenvalues β satisfying (5.3.24). Let $\beta_m=\beta_{jkl}(\sigma_c)$ be a solution of (5.3.24) at $\sigma=\sigma_c$. We consider the coefficients of (5.3.24) at $\sigma=\sigma_c$. Thanks to (5.3.27),

$$\begin{aligned} &(\text{Pr}+\text{Le}+1)\gamma^2_{jkl}>0, \\ &\text{Pr Le}\,\gamma^6_{jkl}-\text{Pr}\,\alpha^2_{jk}(\text{Le}\,R_0-\widetilde{R})=\text{Pr}\,\alpha^2_{jk}\text{Le}\left[\frac{\gamma^6_{jkl}}{\alpha^2_{jk}}-\sigma_c\right]\geq 0. \end{aligned}$$

Thanks to (5.3.38), we have

$$\frac{(\text{Pr} + \text{Le} + \text{Pr Le})\gamma^6}{\text{Pr}\,\alpha^2} > R_0 - \widetilde{R}_0.$$

Thus, we obtain

$$\begin{aligned}
&(\text{Pr} + \text{Le} + \text{Pr Le})\gamma_{jkl}^4 - \text{Pr}\,\alpha_{jk}^2\gamma_{jkl}^{-2}(R_0 - \widetilde{R}_0) \\
&= \text{Pr}\,\alpha_{jk}^2\gamma_{jkl}^{-2}\left[\frac{\text{Pr} + \text{Le} + \text{Pr Le}}{\text{Pr}}\frac{\gamma_{jkl}^6}{\alpha_{jk}^2} - (R_0 - \widetilde{R}_0)\right] \\
&> \alpha_{jk}^2\gamma_{jkl}^{-2}(\text{Pr} + \text{Le} + \text{Pr Le})\left(\frac{\gamma_{jkl}^6}{\alpha_{jkl}^2} - \frac{\gamma^6}{\alpha^2}\right) \\
&> 0 \quad (\text{by } (5.3.27)).
\end{aligned}$$

Hence, the coefficients of (5.3.24) at $\sigma = \sigma_c$ are nonnegative, and are strictly positive if $\gamma_{jkl}^6/\alpha_{jkl}^2 \neq \sigma_c$. It follows that

$$\beta_m = \beta_{jkl}(\sigma_c) < 0 \quad \forall \sigma_c \neq \gamma_{jkl}^6/\alpha_{jkl}^2.$$

Thus, we derive that $\beta_{j_1k_11} = \beta_1$ for σ near σ_c. The proof is complete. □

The following lemma shows that the C-Rayleigh number η defined by (5.3.28) characterizes the critical crossing at η_c for the complex eigenvalues of (5.3.11).

Lemma 5.3.4 *Let (j_1, k_1) satisfy (5.3.30), and suppose that the condition (5.3.29) holds for $(j,k,l) = (j_1, k_1, 1)$. Then the eigenvalues $\beta_{j_1k_11}^1(\eta) = \overline{\beta}_{j_1k_11}^2(\eta)$ constitute a critical crossing at $\eta = \eta_c$:*

$$Re\beta_{j_1k_11}^1(\eta)\begin{cases} < 0 & \text{if } \eta < \eta_c, \\ = 0 & \text{if } \eta = \eta_c, \\ > 0 & \text{if } \eta > \eta_c. \end{cases}$$

Proof. Near $\eta = \eta_c$, the solution $\beta_{j_1k_11}^1(\eta)$ of (5.3.44) takes the form

$$\begin{aligned}
&\beta_{j_1k_11}^1(\eta) = \lambda(\eta) + i\rho(\eta), \\
&\lambda(\eta) \to 0, \quad \rho(\eta) \to \rho_0 \quad \text{as } \eta \to \eta_c.
\end{aligned} \tag{5.3.45}$$

Inserting (5.3.45) into (5.3.44), we get

$$\begin{aligned}
&(-3\rho^2 + b_1)\lambda + b_0 - b_2\rho^2 + o(\lambda) = 0, \\
&-\rho^3 + \rho b_1 + 2\rho b_2\lambda + o(\lambda) = 0.
\end{aligned} \tag{5.3.46}$$

Since $\rho_0 \neq 0$, we derive from (5.3.46) that

$$\lambda(\eta)+o(\lambda)=\frac{b_0-b_2\rho^2}{3\rho^2-b_1}=\frac{b_0-b_1b_2-2b_2^2\lambda}{2b_1+6b_2\lambda}+o(\lambda),$$

which yields

$$(1+\frac{b_2^2}{b_1})\lambda(\eta)+o(\lambda)=\frac{b_0-b_1b_2}{2b_1+6b_2\lambda}. \tag{5.3.47}$$

Note that

$$\rho_0^2=b_1(\eta_c)>0, \quad \lambda(\eta_c)=0.$$

We derive from (5.3.47) that

$$\mathrm{Re}\beta^1_{j_1k_11}(\eta)=\lambda(\eta)\begin{cases} <0 & \text{if } b_0<b_1b_2,\\ =0 & \text{if } b_0=b_1b_2,\\ >0 & \text{if } b_0>b_1b_2,\end{cases} \tag{5.3.48}$$

for η near η_c. It is easy to check that

$$b_0\begin{cases} <b_1b_2 & \text{if } \eta<\eta_c,\\ =b_1b_2 & \text{if } \eta=\eta_c,\\ >b_1b_2 & \text{if } \eta>\eta_c.\end{cases}$$

Thus, the lemma follows from (5.3.48). □

Proof of Theorem 5.3.2. Let $\sigma=R-\mathrm{Le}^{-1}\widetilde{R}$ be at the critical state

$$R-\mathrm{Le}^{-1}\widetilde{R}=\sigma_c=\gamma^6/\alpha^2. \tag{5.3.49}$$

Assume that

$$\eta=R-\frac{\mathrm{Pr}+\mathrm{Le}}{\mathrm{Pr}+1}\widetilde{R}>\eta_c=\frac{(\mathrm{Pr}+\mathrm{Le})(1+\mathrm{Le})}{\mathrm{Pr}}\frac{\gamma^6}{\alpha^2}. \tag{5.3.50}$$

Then we deduce from (5.3.49) and (5.3.50) that

$$\widetilde{R}\begin{cases} >\dfrac{\mathrm{Le}^2(\mathrm{Pr}+1)}{(1-\mathrm{Le})\mathrm{Pr}}\dfrac{\gamma^6}{\alpha^2} & \text{if Le} <1,\\ <\dfrac{\mathrm{Le}^2(\mathrm{Pr}+1)}{(1-\mathrm{Le})\mathrm{Pr}}\dfrac{\gamma^6}{\alpha^2} & \text{if Le} >1,\end{cases}$$

which implies that $K<0$.

In addition, let $\eta=\eta_c$, namely

$$R-\frac{\mathrm{Pr}+\mathrm{Le}}{\mathrm{Pr}+1}\widetilde{R}=\frac{(\mathrm{Pr}+\mathrm{Le})(1+\mathrm{Le})}{\mathrm{Pr}}\frac{\gamma^6}{\alpha^2}. \tag{5.3.51}$$

Then noticing that $K < 0$, we can infer from (5.3.51) that

$$\rho_0^2 = (\mathrm{Pr} + \mathrm{Le} + \mathrm{Pr}\,\mathrm{Le})\gamma^4 - \mathrm{Pr}\,\alpha^2\gamma^{-2}(R - \widetilde{R}) > 0. \tag{5.3.52}$$

Thus by (5.3.52) and Lemma 5.3.4, the conditions (5.3.49) and (5.3.50) imply that η_c is the first critical Rayleigh number, and (5.3.34) and (5.3.35) hold.

Likewise, let

$$\eta = R - \frac{\mathrm{Pr} + \mathrm{Le}}{\mathrm{Pr} + 1}\widetilde{R} < \frac{(\mathrm{Pr} + \mathrm{Le})(1 + \mathrm{Le})}{\mathrm{Pr}} \frac{\gamma^6}{\alpha^2}. \tag{5.3.53}$$

Then it is clear that (5.3.49) and (5.3.53) imply that σ_c is the first critical Rayleigh number, and

$$\widetilde{R} \begin{cases} < \dfrac{\mathrm{Le}^2(\mathrm{Pr}+1)}{(1-\mathrm{Le})\,\mathrm{Pr}} \dfrac{\gamma^6}{\alpha^2} & \text{if Le} < 1, \\ > \dfrac{\mathrm{Le}^2(\mathrm{Pr}+1)}{(1-\mathrm{Le})\,\mathrm{Pr}} \dfrac{\gamma^6}{\alpha^2} & \text{if Le} > 1, \end{cases}$$

which implies $K > 0$. By Lemma 5.3.3, the conclusions (5.3.32) and (5.3.33) are valid. The theorem is proved. □

5.3.3 Nonlinear Dynamic Transitions

Transitions and Onset of the First Real Eigenvalues. By Theorem 5.3.2, we know that under the conditions $K > 0$ with K defined by (5.3.31), the Boussinesq equations (5.3.6) with (5.3.8) always undergo a dynamic transition at $\sigma = \sigma_c$.

In this section, we study the transition from a real simple eigenvalue. We know that generically, the first eigenvalues of (5.3.11) are simple. Hence we always assume that the first real eigenvalue of (5.3.11) near $\sigma = \sigma_c$ is simple.

The following main theorem shows that the transition is either a Type I (continuous) transition, or a Type II (jump) transition, depending on the sign of the following parameter:

$$b_1 = \sigma_c - \frac{1 - \mathrm{Le}^2}{\mathrm{Le}^3}\widetilde{R}. \tag{5.3.54}$$

A Type II transition leads also to the existence of metastable states, saddle-node bifurcations, and the associated hysteresis. In this case, as the R-Rayleigh number goes beyond the critical value σ_c, the physical reality is represented by local attractors away from the basic equilibrium state.

Theorem 5.3.5 *Assume that* Le $\neq 1$, $K > 0$, *and* $b_1 > 0$, *where* K *is defined by (5.3.31), and* b_1 *is defined by (5.3.54). Then the problem (5.3.6) with (5.3.8) undergoes a Type I (continuous) transition at the critical R-Rayleigh number* σ_c, *and the following assertions hold:*

(1) If the R-Rayleigh number σ crosses σ_c, the problem bifurcates to two steady-state solutions $\psi_i^\sigma = (u_i^\sigma, T_i^\sigma, S_i^\sigma)$, $i = 1, 2$.
(2) There is an open set $U \subset H$ with $\psi = 0 \in U$ such that $\overline{U} = \overline{U}_1 + \overline{U}_2$, $U_1 \bigcap U_2 = \emptyset$, $\psi = 0 \in \partial U_1 \bigcap \partial U_2$, and $\psi_i^\sigma \in U_i$ attracts U_i $(i = 1, 2)$.
(3) ψ_i^σ can be expressed as

$$\psi_i^\sigma = (-1)^i a\sqrt{\beta(\sigma)}\psi^1_{j_1k_11} + o(\beta^{1/2}(\sigma)) \qquad \text{for } i = 1, 2,$$

where $a > 0$ is a constant, $\beta(\sigma) = \beta^1_{j_1k_11}(\sigma)$ is the first eigenvalue satisfying (5.3.37), and $\psi^1_{j_1k_11}$ is the first eigenvector given by (5.3.21).
(4) If $\psi_0 \in U_i$ $(i = 1, 2)$, there exists $t_0 > 0$ such that when $t > t_0$, the velocity component of $\psi(t, \psi_0)$ is topologically equivalent to the structure as shown in Fig. 5.12, with either the same or the reversed orientation. Here $\psi(t, \psi_0)$ is the solution of (5.3.6) with (5.3.8) with initial data ψ_0.

Theorem 5.3.6 *Assume that* Le $\neq 1$, $K > 0$, *and* $b_1 < 0$. *Then the problem (5.3.6)–(5.3.8) undergoes a Type II (jump) transition at σ_c, and the following assertions hold:*

(1) The transition of this problem at σ_c is a subcritical bifurcation, i.e., there are steady-state solutions bifurcated on $\sigma < \sigma_c$ that are repellers, and there are no steady-state solutions bifurcated on $\sigma > \sigma_c$.
(2) There is a saddle-node bifurcation of steady-state solutions from (ψ_1^*, σ^*) and (ψ_2^*, σ^*) with $\sigma^* < \sigma_c$, and there are four branches of steady-state solutions $\psi_\sigma^i (1 \leq i \leq 4)$ in which ψ_σ^1 and ψ_σ^2 are bifurcated from (ψ_1^*, σ^*), and $\psi_\sigma^3, \psi_\sigma^4$ bifurcated from (ψ_2^*, σ^*) on $\sigma > \sigma^*$, such that

$$\lim_{\sigma \to \sigma_c} \psi_\sigma^1 = \lim_{\sigma \to \sigma_c} \psi_\sigma^2 = 0, \quad \psi_{\sigma_c}^3, \psi_{\sigma_c}^4 \neq 0,$$

and ψ_σ^3 and ψ_σ^4 are attractors for $\sigma^* < \sigma < \sigma_c + \varepsilon$ with some $\varepsilon > 0$.

Proof of Theorems 5.3.5 and 5.3.6. Let $J_1 = (j_1, k_1, 1)$, $\psi_{J_1} = \psi^1_{j_1k_11}$. The reduced equation of (5.3.6)–(5.3.8) in H reads

$$\frac{dx}{dt} = \beta^1_{J_1}(\sigma)x + \frac{1}{(\psi_{J_1}, \psi^*_{J_1})}(G(\psi, \psi), \psi^*_{J_1}), \tag{5.3.55}$$

where $\psi \in H$ is written as $\psi = x\psi_{J_1} + \Phi$, Φ is the center manifold function, and for $\psi_i = (u_i, T_i, S_i) \in H$ $(i = 1, 2, 3)$,

$$G(\psi_1, \psi_2) = -P((u_1 \cdot \nabla)u_2, (u_1 \cdot \nabla)T_2, (u_1 \cdot \nabla)S_2),$$
$$(G(\psi_1, \psi_2), \psi_3) = -\int_\Omega \left[\sum_{i,j=1}^3 u_{1i}\frac{\partial u_{2j}}{\partial x_i}u_{3j} + \sum_{i,j=1}^3 \left(u_{1i}\frac{\partial T_2}{\partial x_i}T_3 + u_{1i}\frac{\partial S_2}{\partial x_i}S_3\right)\right] dx.$$

By Theorem A.1.1, the center manifold function Φ satisfies

$$-L_\lambda \Phi = x^2 P_2 G(\psi_{J_1}, \psi_{J_1}) + \text{higher-order terms.}$$

Hence it is routine to calculate that

$$\begin{aligned}\Phi =& x^2\Big[\frac{(G(\psi_{J_1}, \psi_{J_1}), \psi_{2J}^*)}{-\beta_{2J}(\psi_{2J}, \psi_{2J}^*)}\psi_{2J} + \frac{(G(\psi_{J_1}, \psi_{J_1}), \psi_{002}^{1*})}{-\beta_{002}^1(\psi_{002}^1, \psi_{002}^{1*})}\psi_{002}^1 \\ &+ \frac{(G(\psi_{J_1}, \psi_{J_1}), \psi_{002}^{2*})}{-\beta_{002}^2(\psi_{002}^2, \psi_{002}^{2*})}\psi_{002}^2\Big] + o(2), \end{aligned} \tag{5.3.56}$$

where

$$\begin{aligned} &-\beta_{2J} = 4\text{Pr}\,\alpha^2, \quad -\beta_{002}^1 = 4\pi^2, \quad -\beta_{002}^2 = 4\text{Le}\,\pi^2, \\ &\psi_{2J} = \psi_{2J}^* = (k_1\alpha_2 \sin 2j_1\alpha_1\pi x_1 \cos 2k_1\alpha_2\pi x_2, \\ &\qquad\qquad - j_1\alpha_1 \cos 2j_1\alpha_1\pi x_1 \sin 2k_1\alpha_2\pi x_2, 0, 0, 0), \\ &\psi_{002}^1 = \psi_{002}^{1*} = (0, 0, 0, \sin 2\pi x_3, 0), \\ &\psi_{002}^2 = \psi_{002}^{2*} = (0, 0, 0, 0, \sin 2\pi x_3). \end{aligned}$$

Inserting $\psi = x\psi_{J_1} + \Phi$ into (5.3.55), we have

$$\frac{dx}{dt} = \beta_{J_1}^1(\sigma)x - \delta(\sigma)x^3 + o(3), \tag{5.3.57}$$

where

$$\begin{aligned}\delta(\sigma_c) =& \sum_{(J,i)\neq(J_1,1)} \frac{(G(\psi_{J_1}, \psi_{J_1}), \psi_J^{i*})\left[(G(\psi_{J_1}, \psi_{J_1}^*), \psi_J^i) + (G(\psi_J^i, \psi_{J_1}^*), \psi_{J_1})\right]}{-\beta_J^i(\psi_{J_1}, \psi_{J_1}^*)(\psi_J^i, \psi_J^{i*})} \\ =& \frac{1}{(\psi_{J_1}, \psi_{J_1}^*)}\Big[\frac{(G(\psi_{J_1}, \psi_{J_1}), \psi_{2J}^*)(G(\psi_{J_1}, \psi_{J_1}^*), \psi_{2J})}{-\beta_{2J}(\psi_{2J}, \psi_{2J}^*)} \\ &+ \frac{(G(\psi_{J_1}, \psi_{J_1}), \psi_{002}^{1*})(G(\psi_{J_1}, \psi_{J_1}^*), \psi_{002}^1)}{-\beta_{002}^1(\psi_{002}^1, \psi_{002}^{1*})} \\ &+ \frac{(G(\psi_{J_1}, \psi_{J_1}), \psi_{002}^{2*})(G(\psi_{J_1}, \psi_{J_1}^*), \psi_{002}^2)}{-\beta_{002}^2(\psi_{002}^2, \psi_{002}^{2*})}\Big] \\ =& \frac{\pi^2}{8(\psi_{J_1}, \psi_{J_1}^*)}\Big[\frac{(k_1\alpha_2 u_{J_1} - J_1\alpha_1 v_{J_1})(k_1\alpha_2 u_{J_1}^* - J_1\alpha_1 v_{J_1}^*)}{\text{Pr}\,\alpha^4} + \frac{T_{J_1}T_{J_1}^*}{\pi^4} + \frac{S_{J_1}S_{J_1}^*}{\pi^4\text{Le}}\Big]. \end{aligned}$$

Here

$$\begin{aligned} &u_{J_1} = u_{J_1}^* = -\frac{j_1\alpha_1\pi^2}{\alpha^2}, \quad v_{J_1} = v_{J_1}^* = -\frac{k_1\alpha_2\pi^2}{\alpha^2}, \quad T_{J_1} = \frac{1}{\gamma^2 + \beta_{J_1}^1}, \\ &T_{J_1}^* = \frac{\text{Pr}R}{\gamma^2 + \beta_{J_1}^1}, \quad S_{J_1} = \frac{\text{sign}(S_0 - S_1)}{\text{Le}\,\gamma^2 + \beta_{J_1}^1}, \quad S_{J_1}^* = \frac{\text{Pr}\widetilde{R}\,\text{sign}(S_0 - S_1)}{\text{Le}\,\gamma^2 + \beta_{J_1}^1}. \end{aligned}$$

From $\beta_{J_1}(\sigma_c)=0$, direct calculation yields that at $\sigma=\sigma_c$,

$$\delta(\sigma_c)=\frac{\Pr(\mathrm{Le}^3R-\widetilde{R})}{8\pi^2\gamma^4\mathrm{Le}^3(\psi_{J_1},\psi^*_{J_1})}=\frac{\mathrm{Le}^3R-\widetilde{R}}{8\pi^2\mathrm{Le}^3\left(\frac{\sigma_c}{\Pr}+R-\frac{\widetilde{R}}{\mathrm{Le}^2}\right)}.$$

Putting $R=\mathrm{Le}^{-1}\widetilde{R}+\sigma_c$ into $\delta(\sigma_c)$, we derive that

$$\delta=\delta(\sigma_c)=\frac{1}{8\pi^2\mathrm{Le}^3}\frac{\mathrm{Le}^3\sigma_c-(1-\mathrm{Le}^2)\widetilde{R}}{(1+\frac{1}{\Pr})\sigma_c-\frac{1-\mathrm{Le}}{\mathrm{Le}^2}\widetilde{R}}.$$

Hence using $K>0$, we see that the sign of δ is the same as the sign of b_1 defined by (5.3.54), and then the theorems follow from Theorems 2.3.1 and 2.5.3 and from (5.3.57). □

Transition to Spatiotemporal Oscillatory Patterns. If $K<0$, then by Theorem 5.3.2, the problem (5.3.6)–(5.3.8) undergoes a dynamic transition to the periodic solutions at the critical Rayleigh number η_c given by (5.3.30). It is easy to see that $L_1/L_2\gg 1,\eta_c$ is given by

$$\eta_c=\frac{(\Pr+\mathrm{Le})(1+\mathrm{Le})}{\Pr}\frac{\gamma^6_{j_101}}{\alpha^2_{j_10}},\tag{5.3.58}$$

for some $(j_1,k_1)=(j_1,0)$ with $j_1>0$. Here, we always assume that condition (5.3.58) holds.

We now introduce a parameter for determining the transition type:

$$\begin{aligned}b_2=\;&R_c(R_2B_1-I_2B_2)C_3+R_c(I_2B_1+R_2B_2)C_4\\&+\widetilde{R}(R_3B_1-I_3B_2)C_5+\widetilde{R}(I_3B_1+R_3B_2)C_6,\end{aligned}\tag{5.3.59}$$

where R_2,I_2,C_3,C_4 are as in (4.4.73), $\alpha^2=\alpha^2_{j_10}=j_1^2\alpha_1^2\pi^2$, and

$$R_3=\mathrm{Re}A_3=\frac{\mathrm{Le}\,(\alpha^2+\pi^2)}{\mathrm{Le}^2(\alpha^2+\pi^2)^2+\rho^2},\qquad I_3=\mathrm{Im}A_3=\frac{-\rho}{\mathrm{Le}^2(\alpha^2+\pi^2)^2+\rho^2},$$

$$B_1=\widetilde{R}I_3^2-R_cI_2^2,\qquad B_2=\widetilde{R}I_3R_3-RI_2R_2,$$

$$C_5=\frac{3R_3}{2\pi\mathrm{Le}}-\frac{\rho^2R_3}{2\pi\mathrm{Le}\,(4\pi^4\mathrm{Le}^2+\rho^2)}-\frac{\rho I_3}{4\pi^4\mathrm{Le}^2+\rho^2},$$

$$C_6=-\frac{I_3}{2\pi\mathrm{Le}}+\frac{\rho^2I_3}{2\pi\mathrm{Le}\,(4\pi^4\mathrm{Le}^2+\rho^2)}-\frac{\rho R_3}{4\pi^4\mathrm{Le}^2+\rho^2},$$

$$\rho^2=\frac{\Pr(1-\mathrm{Le})\widetilde{R}}{\Pr+1}\frac{\alpha^2}{(\alpha^2+\pi^2)}-\mathrm{Le}^2(\alpha^2+\pi^2)^2,$$

$$R_c=\frac{(\Pr+\mathrm{Le})(1+\mathrm{Le})}{\Pr}\frac{(\alpha^2+\pi^2)^3}{\alpha^2}+\frac{\Pr+\mathrm{Le}}{\Pr+1}\widetilde{R}.$$

The following theorem characterizes the spatiotemporal transition.

Theorem 5.3.7 *Let b_2 be given by (5.3.59), and $K < 0$ with K defined by (5.3.31). Then for problem (5.3.6)–(5.3.8), we have the following assertions:*

(1) If $b_2 < 0$, then the problem has a Type I transition at $\eta = \eta_c$ and bifurcates to a periodic solution on $\eta > \eta_c$ that is an attractor; the periodic solution can be expressed as

$$\psi = x(t)\mathrm{Re}\psi_{j_1 01} + y(t)\mathrm{Im}\psi_{j_1 01} + o(|x| + |y|),$$

$$(x(t), y(t)) = \left[\frac{2\pi \mathrm{Re}\beta(\eta)|\rho|}{|b_2|}\right]^{1/2} (\sin\rho t, \cos\rho t),$$

where $\beta(\eta) = \lambda(\eta) + i\rho(\eta)$ is the first eigenvalue with $\lambda(\eta_c) = Re\beta(\eta_c) = 0, \rho(\eta_c) = \rho$ as given in (5.3.59).

(2) If $b_2 > 0$, then the transition at $\eta = \eta_c$ is of Type II, and there is a singular separation of periodic solutions at some (Γ^*, η^*) for $\eta^* < \eta_c$, where Γ^* is a periodic solution. In particular, there is a branch Γ_η of periodic solutions separated from Γ^* that are repellers such that $\Gamma_\eta \to 0$ as $\eta \to \eta_c$.

Remark 5.3.8 The following example shows that both cases with $b_2 < 0$ and $b_2 > 0$ may appear in different physical regimes.

Let $\rho^2 \cong 0$. In this case, we have

$$R_c \cong \frac{\mathrm{Pr} + \mathrm{Le}}{\mathrm{Pr}\,(1 - \mathrm{Le})} \frac{\gamma^6}{\alpha^2}, \qquad \widetilde{R} \cong \frac{(\mathrm{Pr} + 1)\mathrm{Le}^2}{\mathrm{Pr}\,(1 - \mathrm{Le})} \frac{\gamma^6}{\alpha^2} \begin{cases} > 0 & \text{for } 1 > \mathrm{Le}, \\ < 0 & \text{for } 1 < \mathrm{Le}. \end{cases}$$

Thus, the number b_2 defined by (5.3.59) is given by

$$\begin{aligned} b_2 &\cong -\gamma^{-12}(R_c - \mathrm{Le}^{-4}\widetilde{R})(R_c - \mathrm{Le}^{-3}\widetilde{R}) \\ &= \frac{-1}{\mathrm{Pr}\,\alpha^2(1 - \mathrm{Le})\mathrm{Le}^3}(\mathrm{Le}\,(\mathrm{Pr} + \mathrm{Le}) - \mathrm{Pr} - 1)(\mathrm{Le}^2(\mathrm{Pr} + \mathrm{Le}) - \mathrm{Pr} - 1). \end{aligned}$$

It is clear that for Le $\simeq 1$ and for Pr small,

$$b_2 \begin{cases} < 0 & \text{if } \mathrm{Le} < 1, \\ > 0 & \text{if } \mathrm{Le} > 1. \end{cases}$$

Proof of Theorem 5.3.7. Let $J = (j_1, 0, 1)$ and let $\beta(\eta) = \beta_J^1(\eta) = \lambda(\eta) + i\rho(\eta)$ be the first eigenvalue of (5.3.45) near $\eta = \eta_c$ with

$$\lambda(\eta_c) = 0, \quad \rho(\eta_c) = \rho > 0. \tag{5.3.60}$$

By (5.3.60), the eigenvectors $\psi_J = \psi_J^1 + i\psi_J^2$ corresponding to $\beta(\eta)$ are given by

$$\begin{aligned} \psi_J^1 &= (-\frac{1}{j_1\alpha_1}\sin\phi\cos\pi x_3, 0, \cos\phi\sin\pi x_3, \\ &\qquad ReA_2(\beta)\cos\phi\sin\pi x_3, ReA_3(\beta)\cos\phi\sin\pi x_3), \\ \psi_J^2 &= (0, 0, 0, \mathrm{Im}A_2(\beta)\cos\phi\sin\pi x_3, \mathrm{Im}A_3\cos\phi\sin\pi x_3), \end{aligned}$$

where $\phi = j_1\alpha_1\pi x_1$. By (5.3.23), the conjugate eigenvectors $\psi_J^* = \psi_J^{1*} + i\psi_J^{2*}$ are given by

$$\begin{aligned}\psi_J^{1*} &= (-\frac{1}{j_1\alpha_1}\sin\phi\cos\pi x_3, 0, \cos\phi_1\sin\pi x_3,\\ &\quad \Pr R\mathrm{Re}A_2(\overline{\beta})\cos\phi\sin\pi x_3, -\Pr \widetilde{R}\mathrm{Re}A_3(\overline{\beta})\cos\phi\sin\pi x_3),\\ \psi_J^{2*} &= (0,0,0,\Pr R\mathrm{Im}A_2(\overline{\beta})\cos\phi\sin\pi x_3, -\Pr \widetilde{R}\mathrm{Im}A_3(\overline{\beta})\cos\phi\sin\pi x_3)\end{aligned}$$

and

$$A_2(\beta) = \frac{1}{\alpha^2+\pi^2+\beta}, \quad A_3(\beta) = \frac{1}{\mathrm{Le}\,(\alpha^2+\pi^2)+\beta}. \tag{5.3.61}$$

The conjugate eigenvectors Φ_J^{1*} and Φ_J^{2*}, satisfying

$$(\psi_J^1, \Phi_J^{1*}) = (\psi_J^2, \Phi_J^{2*}) \neq 0, \qquad (\psi_J^1, \Phi_J^{2*}) = (\psi_J^2, \Phi_J^{1*}) = 0,$$

are given by

$$\Phi_J^{1*} = \psi_J^{1*} + C\psi_J^{2*}, \qquad \Phi_J^{2*} = -C\psi_J^{1*} + \psi_J^{2*}, \tag{5.3.62}$$

where

$$C = \frac{(\psi_J^1, \psi_J^{2*})}{(\psi_J^1, \psi_J^{1*})} = -\frac{(\psi_J^2, \psi^{1*})}{(\psi_J^2, \psi_J^{2*})} = \frac{B_2}{B_1}. \tag{5.3.63}$$

The reduced equations of (5.3.6)–(5.3.8) read

$$\begin{aligned}\frac{dx}{dt} &= \lambda x + \rho y + \frac{1}{(\psi_J^1, \Phi_J^{1*})}(G(\psi,\psi), \Phi_J^{1*}),\\ \frac{dy}{dt} &= -\rho x + \lambda y + \frac{1}{(\psi_J^2, \Phi_J^{2*})}(G(\psi,\psi), \Phi_J^{2*}),\end{aligned} \tag{5.3.64}$$

where $\psi = x\psi_J^1 + y\psi_J^2 + \Phi$, and Φ is the center manifold function.

Note that for any gradient field $\nabla\varphi$, the Leray projection $P(\nabla\varphi)$ is equal to 0. Therefore, by (5.3.62)–(5.3.63), we have

$$\begin{aligned}G(\psi_J^1, \psi_J^1) &= \frac{\pi}{2}(0,0,0,\mathrm{Re}A_2\sin 2\pi x_3, \mathrm{Re}A_3\sin 2\pi x_3),\\ G(\psi_J^1, \psi_J^2) &= \frac{\pi}{2}(0,0,0,\mathrm{Im}A_2\sin 2\pi x_3, \mathrm{Im}A_3\sin 2\pi x_3),\\ G(\psi_J^1, \Phi_J^{1*}) &= \frac{\pi \Pr}{2B_1}(0,0,0,R(B_1\mathrm{Re}A_2 - B_2\mathrm{Im}A_2)\sin 2\pi x_3,\\ &\quad -\widetilde{R}(B_1\mathrm{Re}A_3 - B_2\mathrm{Im}A_3)\sin 2\pi x_3),\end{aligned}$$

$$\begin{aligned}G(\psi_J^1, \Phi_J^{2*}) &= \frac{\pi \Pr}{2B_1}(0,0,0,-R(B_2\mathrm{Re}A_2 + B_1\mathrm{Im}A_2)\sin 2\pi x_3,\\ &\quad \widetilde{R}(B_2\mathrm{Re}A_3 + B_1\mathrm{Im}A_3)\sin 2\pi x_3)\\ G(\psi_J^2, \psi) &= 0, \quad \forall \psi \in H.\end{aligned}$$

Hence by (A.1.15), the center manifold function Φ can be expressed as

$$\Phi = \Phi_1 + \Phi_2 + \Phi_3 + o(2), \tag{5.3.65}$$

where $G_{ij} = G(\psi_J^i, \psi_J^j)$ and

$$\begin{aligned}
\Phi_1 = & -\frac{\psi_{002}^1}{\beta_{002}^1 \|\psi_{002}^1\|^2}[x^2(G_{11}, \psi_{002}^1) + xy(G_{12}, \psi_{002}^1)] \\
& - \frac{\psi_{002}^2}{\beta_{002}^2 \|\psi_{002}^2\|^2}[x^2[(G_{11}, \psi_{002}^2) + xy(G_{12}, \psi_{002}^2)], \\
\Phi_2 = & \frac{2\rho^2 \psi_{002}^1}{\beta_{002}^1(\beta_{002}^{12} + 4\rho^2)\|\psi_{002}^1\|^2}[(x^2 - y^2)(G_{11}, \psi_{002}^1) + 2xy(G_{12}, \psi_{002}^1)] \\
& + \frac{2\rho^2 \psi_{002}^2}{\beta_{002}^2((\beta_{002}^2)^2 + 4\rho^2)\|\psi_{002}^2\|^2}[(x^2 - y^2)(G_{11}, \psi_{002}^2) + 2xy(G_{12}, \psi_{002}^2)], \\
\Phi_3 = & \frac{\rho \psi_{002}^1}{((\beta_{002}^1)^2 + 4\rho^2)\|\psi_{002}^1\|^2}[2xy(G_{11}, \psi_{002}^1) + (y^2 - x^2)(G_{12}, \psi_{002}^1)] \\
& + \frac{\rho \psi_{002}^2}{((\beta_{002}^2)^2 + 4\rho^2)\|\psi_{002}^2\|^2}[2xy(G_{11}, \psi_{002}^2) + (y^2 - x^2)(G_{11}, \psi_{002}^2)].
\end{aligned}$$

It is clear that

$$(G(\Phi, \Phi_J^{i*}), \psi_J^j) = o(2) \quad \forall i, j = 1, 2.$$

Then, inserting $\psi = x\psi_J^1 + y\psi_J^2 + \Phi$ into (5.3.64), one gets

$$\begin{aligned}
\frac{dx}{dt} &= \lambda x + \rho y - \frac{x(G(\psi_J^1, \Phi_J^{1*}), \Phi)}{(\psi_J^1, \Phi_J^{1*})} + o(3), \\
\frac{dy}{dt} &= -\rho x + \lambda y - \frac{x(G(\psi_J^1, \Phi_J^{2*}), \Phi)}{(\psi_J^2, \Phi_J^{2*})} + o(3).
\end{aligned} \tag{5.3.66}$$

From (5.3.65) and (5.3.66), it follows that

$$\begin{aligned}
\frac{dx}{dt} &= \lambda x + \rho y + \frac{\Pr \pi x[a_1 x^2 + a_2 xy + a_3 y^2]}{(\psi_J^1, \psi_J^{1*})^2 + (\psi_J^1, \psi_J^{2*})^2} + o(3), \\
\frac{dy}{dt} &= -\rho x + \lambda y + \frac{\Pr \pi x[b_1 x^2 + b_2 xy + b_3 y^2]}{(\psi_J^1, \psi_J^{1*})^2 + (\psi_J^1, \psi_J^{2*})^2} + o(3),
\end{aligned} \tag{5.3.67}$$

where

$$\begin{aligned}
a_1 &= R(-R_2 B_1 + I_2 B_2)D_1 + \widetilde{R}(R_3 B_1 - I_3 B_2)F_1, \\
a_3 &= R(-R_2 B_1 + I_2 B_2)D_3 + \widetilde{R}(R_3 B_1 - I_3 B_2)F_3, \\
b_2 &= R(I_2 B_1 + R_2 B_2)D_2 - \widetilde{R}(I_3 B_1 + R_3 B_2)F_2,
\end{aligned}$$

where B_1, B_2 are as in (5.3.63), and

$$
\begin{aligned}
D_1 &= \frac{R_2}{8\pi} - \frac{\rho^2 R_2}{16\pi(4\pi^4+\rho^2)} + \frac{\rho\pi I_2}{4(4\pi^4+\rho^2)},\\
D_2 &= \frac{I_2}{8\pi} - \frac{\rho^2 I_2}{8\pi(4\pi^4+\rho^2)} - \frac{\rho\pi R_2}{4(4\pi^4+\rho^2)},\\
D_3 &= \frac{\rho^2 R_2}{16\pi(4\pi^4+\rho^2)} - \frac{\rho\pi I_2}{8(4\pi^4+\rho^2)},\\
F_1 &= \frac{R_3}{8\pi\mathrm{Le}} - \frac{\rho^2 R_3}{16\pi\mathrm{Le}\,(4\pi^4\mathrm{Le}^2+\rho^2)} - \frac{\rho I_3}{8(4\pi^4\mathrm{Le}^2+\rho^2)},\\
F_2 &= \frac{I_3}{8\pi\mathrm{Le}} - \frac{\rho^2 I_3}{8\pi\mathrm{Le}\,(4\pi^4\mathrm{Le}^2+\rho^2)} + \frac{\rho R_3}{4(4\pi^4\mathrm{Le}^2+\rho^2)},\\
F_3 &= \frac{\rho^2 R_3}{16\pi\mathrm{Le}\,(4\pi^4\mathrm{Le}^2+\rho^2)} + \frac{\rho I_3}{8(4\pi^4\mathrm{Le}^2+\rho^2)}.
\end{aligned}
$$

Then we obtain

$$
\begin{aligned}
b &= 3a_1 + a_3 + b_2\\
&= \frac{1}{4}[R(R_2B_1 - I_2B_2)C_3 + R(I_2B_1 + R_2B_2)C_4\\
&\quad + \widetilde{R}(R_3B_1 - I_3B_2)C_5 + \widetilde{R}(I_3B_1 + R_3B_2)C_6]\\
&= \frac{1}{4}b_2,
\end{aligned}
\tag{5.3.68}
$$

where b_2 is the number defined by (5.3.59).

By (5.3.67)–(5.3.68), this theorem follows from Theorems 2.3.7 and 2.5.7; the expression of bifurcated periodic solution in assertion (1) can be derived using the formula (2.3.22). The proof is complete. □

5.3.4 Convection Scales and Dynamic Transition

Revised Boussinesq Equations and Critical Parameters. Consider (5.3.1) with a damping term δu. The revised nondimensional equations of (5.3.1) are given by

$$
\begin{aligned}
&\frac{\partial u}{\partial t} = \Pr\left[\Delta u - \delta u + RTk - \mathrm{sign}(S_0 - S_1)\widetilde{R}Sk - \nabla p\right] - (u\cdot\nabla)u,\\
&\frac{\partial T}{\partial t} = \Delta T + u_3 - (u.\nabla)T,\\
&\frac{\partial S}{\partial t} = \mathrm{Le}\,\Delta S + \mathrm{sign}(S_0 - S_1)u_3 - (u\cdot\nabla)S,\\
&\mathrm{div} u = 0,
\end{aligned}
\tag{5.3.69}
$$

supplemented with the boundary condition (5.3.8), where $\delta u = (\delta_0 u_1, \delta_0 u_2, \delta_1 u_3)$.

As in (5.3.16), the eigenvalues of the revised problem satisfy the following equation:

$$\begin{aligned}&\alpha_{jk}^2(\gamma_{jk}^2+\beta)(\text{Le }\gamma_{jk}^2+\beta)(\text{Pr }\gamma_{jk}^2+\text{Pr }\delta_1+\beta)\\&+\pi^2(\gamma_{jk}^2+\beta)(\text{Le }\gamma_{jk}^2+\beta)(\text{Pr }\gamma_{jk}^2+\text{Pr }\delta_0+\beta)\\&-\text{Pr }R\alpha_{jk}^2(\text{Le }\gamma_{jk}^2+\beta)+\text{Pr }\widetilde{R}\alpha_{jk}^2(\gamma_{jk}^2+\beta)=0,\end{aligned}\tag{5.3.70}$$

which is equivalent to

$$\begin{aligned}&\gamma^2\beta^3+[\gamma^4(1+\text{Pr }+\text{Le })+\text{Pr }(\alpha^2\delta_1+\pi^2\delta_0)]\beta^2\\&+[(\text{Pr }+\text{Le }+\text{Pr Le })\gamma^6+\text{Pr }(1+\text{Le })\gamma^2(\alpha^2\delta_1+\pi^2\delta_0)-\text{Pr }\alpha^2(R-\widetilde{R})]\beta\\&+\gamma^8\text{Pr Le }+\gamma^4\text{Pr Le }(\alpha^2\delta_1+\pi^2\delta_0)-\text{Pr }\gamma^2\alpha^2(\text{Le }R-\widetilde{R})=0.\end{aligned}\tag{5.3.71}$$

Let $\beta=0$. Then we derive from (5.3.71) the critical Rayleigh number for the first real eigenvalues as

$$R_{c_1}=\text{Le}^{-1}\widetilde{R}+\min_{\alpha^2}\left[\gamma^2\delta_1+\frac{\gamma^6+\pi^2\gamma^2\delta_0}{\alpha^2}\right],\tag{5.3.72}$$

where $\gamma^2=\alpha^2+\pi^2$. Let $\beta=i\rho_0$ $(\rho_0\neq 0)$ in (5.3.71). Then we get the critical Rayleigh number for the first complex eigenvalues as follows:

$$R_{c_2}=\min_{\alpha^2}(\Gamma_1+\Gamma_2+\Gamma_3),\tag{5.3.73}$$

where

$$\begin{aligned}\Gamma_1&=\frac{(\text{Pr }+\text{Le })\gamma^4+\text{Pr }(\alpha^2\delta_1+\pi^2\delta_0)}{(\text{Pr }+1)\gamma^4+\text{Pr }(\alpha^2\delta_1+\pi^2\delta_0)}\widetilde{R},\\ \Gamma_2&=\frac{(\text{Pr }+1)(\text{Pr }+\text{Le })(1+\text{Le })\gamma^{10}+\text{Pr }(1+\text{Le })(1+2\text{Pr }+\text{Le })\gamma^6(\alpha^2\delta_1+\pi^2\delta_0)}{\text{Pr }\alpha^2((\text{Pr }+1)\gamma^4+\text{Pr }(\alpha^2\delta_1+\pi^2\delta_0))},\\ \Gamma_3&=\frac{\text{Pr}^2(1+\text{Le })\gamma^2(\alpha^2\delta_1+\pi^2\delta_0)^2}{\text{Pr }\alpha^2((\text{Pr }+1)\gamma^4+\text{Pr }(\alpha^2\delta_1+\pi^2\delta_0))}.\end{aligned}$$

It is readily checked that the first eigenvectors of the revised problem are the same as in (5.3.21) and (5.3.23).

For (5.3.69), we need to modify the transition theorems, Theorems 5.3.5, 5.3.6, and 5.3.7, i.e., to modify the numbers b_1 and b_2 given by (5.3.54) and (5.3.59) associated with the new parameters (5.3.72) and (5.3.73). Thus, we can make a comparison between the new and old theories.

Physical Parameters in Oceanography. In oceanography, the main physical parameters are

$$\begin{aligned}&\text{Pr }=8, &&\nu=1.1\times 10^{-6}\,\text{m}^2\text{s}^{-1}, &&\alpha_T=2.1\times 10^{-4}\,K^{-1},\\&\text{Le }=10^{-2}, &&\kappa_T=1.4\times 10^{-7}\,\text{m}^2\text{s}^{-1}, &&h=4\times 10^3\,\text{m}.\end{aligned}\tag{5.3.74}$$

The haline contraction coefficient α_S satisfies

$$\alpha_S = \frac{1}{\rho}\frac{d\rho}{dS} \times 10^{-3}, \qquad \rho = S\rho_S + (1-S)\rho_W,$$

where S is the concentration of salt with unit $0/_{00}(psu)$, ρ_S and ρ_W are the densities of salt and water given by

$$\rho_W = 10^3\,\text{kg/m}^3, \quad \rho_S = 1.95 \times 10^3\,\text{kg}\,\text{m}^{-3}.$$

At the seawater density $\rho = 1.01 \sim 1.05 \times 10^3\,\text{kg}\,\text{m}^{-3}$, the haline concentration coefficient α_S is about

$$\alpha_S = \frac{1}{\rho}(\rho_S - \rho_W) \times 10^{-3} \cong 0.92 \times 10^{-3}(\text{psu})^{-1}.$$

Thus, for the oceanic motion we obtain the thermal and saline Rayleigh numbers as follows:

$$R = \frac{g\alpha_T(T_0 - T_1)}{\kappa_T \nu} h^3 = 0.86 \times 10^{21}(T_0 - T_1)[^\circ C^{-1}], \tag{5.3.75}$$

$$\widetilde{R} = \frac{g\alpha_S(S_0 - S_1)}{\kappa_T \nu} h^3 = 3.75 \times 10^{21}(S_0 - S_1)(\text{psu})^{-1}. \tag{5.3.76}$$

For the thermohaline circulation, we know that its scale is tens of thousands of kilometers, namely

$$L_c = O(10^4) \quad (h = 4\,\text{km as unit}). \tag{5.3.77}$$

For the damping coefficients δ_0 and δ_1, by (4.5.17), we have

$$\delta_i = C_i h^4/\nu = 2.33 \times 10^{20} C_i\,\text{m}^2 \cdot \text{s} \quad (i = 0, 1).$$

Phenomenologically, the constants C_i depend on the density ρ, or equivalently on the pressure p, and we assume that C_i is proportional to ρ. Note that the ratio of densities of water and air is about 10^3, i.e.,

$$\rho_W/\rho_a \cong 10^3.$$

Therefore, the constant C_1 of water is about 10^3 times that of air. Thus, by (4.5.17), for water we take

$$C_1 \approx 10^3(\text{m}^2 \cdot \text{s})^{-1}. \tag{5.3.78}$$

However, the constant C_0 of seawater is not larger than that of air. Thus, we take

$$C_0 \cong \frac{1}{2} \times 10^{-12}. \tag{5.3.79}$$

Then, we obtain the damping coefficients as

$$\delta_0 = 1.17 \times 10^8, \qquad \delta_1 = 2.33 \times 10^{23}. \tag{5.3.80}$$

We remark that (5.3.78) and (5.3.79) are theoretically estimated values. However, they provide a fairly reasonable explanation of the critical Rayleigh numbers and the convective scales in both the atmospheric and oceanic circulations.

Critical Rayleigh Numbers. Due to (5.3.80), the critical Rayleigh numbers (5.3.72) and (5.3.73) can be approximatively expressed as

$$R_{c_1} - \mathrm{Le}^{-1}\widetilde{R} \cong \min_{\alpha^2} \left[(\pi^2+\alpha^2)\delta_1 + \frac{\pi^4\delta_0}{\alpha^2} \right],$$
$$R_{c_2} - \widetilde{R} \cong (1+\mathrm{Le}\,) \min_{\alpha^2} \left[(\pi^2+\alpha^2)\delta_1 + \frac{\pi^4\delta_0}{\alpha^2} \right].$$

Thus, they have the same α_c^2, given by

$$\alpha_c^2 = \pi^2 \left[\frac{\delta_0}{\delta_1} \right]^{1/2} = 2.24 \times 10^{-7}. \tag{5.3.81}$$

The convective scale is

$$L_c = \frac{\pi}{\alpha_c} = \left[\frac{\delta_1}{\delta_0} \right]^{1/4} = 0.67 \times 10^4. \tag{5.3.82}$$

The critical Rayleigh numbers for real and complex eigenvalues are respectively given by

$$\sigma_c = (\pi^2+\alpha_c^2)\delta_1 + \frac{\pi^4\delta_0}{\alpha_c^2} = 2.33 \times 10^{24}, \tag{5.3.83}$$

$$\eta_c = (1+\mathrm{Le}\,) \left[(\pi^2+\alpha_c^2)\delta_1 + \frac{\pi^4\delta_0}{\alpha_c^2} \right] = (1+\mathrm{Le}\,)\sigma_c. \tag{5.3.84}$$

Here, as in (5.3.25) and (5.3.26), we define the R-Rayleigh number σ and C-Rayleigh number η as follows:

$$\sigma = R - \mathrm{Le}^{-1}\widetilde{R}, \qquad \eta = R - \widetilde{R}. \tag{5.3.85}$$

It is seen that the theoretical value (5.3.82) agrees with the realistic length scale given in (5.3.77).

Dynamic Transition of the Revised System. From (5.3.83)–(5.3.85), we can see that if

$$\sigma = R - \mathrm{Le}^{-1}\widetilde{R} = \sigma_c, \quad \text{and} \quad \eta = R - \widetilde{R} < \eta_c = (1+\mathrm{Le}\,)\sigma_c,$$

then $\widetilde{R} < \mathrm{Le}^2\sigma_c/(1-\mathrm{Le}\,)$. In this case, σ_c is the first critical Rayleigh number.

In addition, if

$$R - \mathrm{Le}^{-1}\widetilde{R} = \sigma_c, \quad R - \widetilde{R} > (1+\mathrm{Le}\,)\sigma_c,$$

then $\widetilde{R} > \mathrm{Le}^2\sigma_c/(1-\mathrm{Le}\,)$, and η_c is the first critical Rayleigh number. Thus by Theorem 5.3.2, we obtain the following physical conclusions.

Physical Conclusion 5.3.9 *For the thermohaline circulation, we have the following assertions:*

1. *If*

$$\widetilde{R} < \mathrm{Le}^2\sigma_c = 2.35 \times 10^{20},$$

the number σ_c given by (5.3.83) is the first critical Rayleigh number.

2. *If*

$$\widetilde{R} > \frac{\mathrm{Le}^2}{1-\mathrm{Le}}\sigma_c = 2.35 \times 10^{20},$$

the number η_c given by (5.3.84) is the first critical Rayleigh number.

Note that the first eigenvectors of the linearized equations of (5.3.69) are the same as those of (5.3.11). For the transition of (5.3.69) from real eigenvalues, the parameter b_1 should be as in (5.3.54), namely,

$$b_1 = \sigma_c - \frac{1-\mathrm{Le}^2}{\mathrm{Le}^3}\widetilde{R}.$$

Here $R = \sigma_c + \mathrm{Le}^{-1}\widetilde{R}$, $\mathrm{Le} \ = 10^{-2}$, $\sigma_c = 2.33 \times 10^{24}$, $\mathrm{Pr} = 8$, and $\alpha_c^2 = 2.24 \times 10^{-7}$. Thus we have

$$b_1 \cong 10^6(2.33 \times 10^{18} - \widetilde{R}), \tag{5.3.86}$$

where $\widetilde{R}$ is given by (5.3.76).

By Physical Conclusion 5.3.9, for $\widetilde{R} < 2.35 \times 10^{20}$, the transition of (5.3.69) is at $\sigma = \sigma_c$. The following physical conclusion is a revised version of Theorems 5.3.5 and 5.3.6.

Physical Conclusion 5.3.10 *Let $\widetilde{R} < 2.35 \times 10^{20}$, and let b_1 be as in (5.3.86). Then the problem (5.3.69) with (5.3.8) undergoes a dynamic transition from $\sigma = \sigma_c$, and the following assertions hold:*

(1) If $\widetilde{R} < 2.33 \times 10^{18}$, the transition is continuous, and the problem bifurcates to two steady-state solutions ψ_1^σ and ψ_2^σ for $\sigma > \sigma_c$, with σ as in (5.3.85); ψ_1^σ and ψ_2^σ have expressions as in assertion (3) of Theorem 5.3.5, with U_1 and U_2 their respective basins of attraction in H.

(2) If $2.33 \times 10^{18} < \widetilde{R} < 2.35 \times 10^{20}$, the transition is a jump, and there are two saddle-node bifurcations from (ψ_1^*, σ^*) and (ψ_2^*, σ^*) with $\sigma^* < \sigma_c$.

(3) When the transition is of Type I, then if the initial value $\widetilde{\psi}$ is in U_i, then there is a time t_0 such that for $t > t_0$, the velocity component u in the solution $\psi(t, \psi_0)$ with initial value ψ_0 is topologically equivalent to the structure shown in Fig. 5.12.

For the transition to periodic solutions, the revised parameters are as follows:

$$\begin{aligned}
&\alpha^2 = \alpha_c^2 = 2.24 \times 10^{-7} \ll 1,\\
&\gamma^2 = \alpha^2 + \pi^2 \cong \pi^2,\\
&R = R_{c_2} = \widetilde{R} + \eta_c \quad (\eta_c \text{ as in } (5.3.84)).
\end{aligned}$$

From (5.3.71), we can obtain the imaginary part $\beta = i\rho$ as

$$\rho^2 = (\text{Pr} + \text{Le} + \text{Pr Le})\gamma^4 + \text{Pr}(1+\text{Le})(\alpha^2\delta_1 + \pi^2\delta_0) + \frac{\text{Pr}\,\alpha^2\eta_c}{\gamma^2} \cong 7.3 \times 10^{16}.$$

Thus, we derive that

$$\begin{aligned}
&R_2 \cong \frac{\pi^2}{\rho^2}, && R_3 \cong \frac{\text{Le}\,\pi^2}{\rho^2}, && I_2 \cong I_3 \cong -\frac{1}{\rho},\\
&C_3 \cong -\frac{2\pi}{\rho^2}, && C_4 \cong \frac{\pi(\pi^2-2)}{\rho^3}, && C_5 \cong \frac{\pi+1}{\rho^2},\\
&C_6 \cong -\frac{\text{Le}\,\pi^2(2\pi+1)}{\rho^3}, && B_1 \cong \frac{1}{\rho^2}(\widetilde{R} - R_{c_2}), && B_2 \cong \frac{\pi^2(R_{c_2} - \text{Le}\,\widetilde{R})}{\rho^3}.
\end{aligned}$$

Then the parameter b_2 in (5.3.59) reads

$$\begin{aligned}
b_2 \cong{}& \frac{\pi}{\rho^6}[\text{Le}\,\pi(2\pi+1)\widetilde{R}^2 + (\pi^2-2)R_{c_2}^2\\
&- (2\pi^2 - \pi - 2 + \text{Le}\,\pi(\pi+2))\widetilde{R}R_{c_2}]\\
\cong{}& \frac{\pi}{\rho^6}[-(1-\text{Le})\pi(\pi-1)\widetilde{R}^2 + (\pi^2-2)\eta_c^2\\
&+ (\pi - 2 - \text{Le}\,\pi^2 - 2\text{Le}\,\pi)\widetilde{R}\eta_c]\\
\cong{}& \frac{\pi}{\rho^6}[-\pi(\pi-1)\widetilde{R}^2 + (\pi^2-2)\eta_c^2 + \widetilde{R}\eta_c].
\end{aligned} \tag{5.3.87}$$

Then Theorem 5.3.7 can be rewritten as follows.

Physical Conclusion 5.3.11 *Let $\widetilde{R} > 2.35 \times 10^{20}$, and let b_2 be a parameter as in (5.3.87). Then for the problem (5.3.69) with (5.3.8), assertions (1) and (2) of Theorem 5.3.7 hold.*

The above Physical Conclusions 5.3.9-5.3.11 provide the possible dynamical behaviors for the great ocean conveyer, depending on the saline Rayleigh number $\widetilde{R}$.

We note that the temperature and salinity gradients $T_0 - T_1$ and $S_0 - S_1$ between the oceanic bottom and upper surfaces are different from the observed data. In fact, the observed values should be as follows:

$$\begin{aligned}
&\Delta T = T_0 - T_1 + T(x_1,x_2,0) - T(x_1,x_2,1),\\
&\Delta S = S_0 - S_1 + S(x_1,x_2,0) - S(x_1,x_2,1),
\end{aligned}$$

where $T(x)$ and $S(x)$ are the transition solutions as described by assertion (3) of Theorem 5.3.5. Due to heat sources within the Earth's crust, the bottom temperature T_0 of the ocean remains essentially a constant, and the bottom salinity S_0 is also a constant, which equals the average value

$$S_0 = \frac{1}{|V|}\int_V S(x)dx, \quad |V| \text{ the volume of the ocean.}$$

It is easy to see that on the upper surface, the temperature and salinity vary in different regions. At the equator, T is about $20 \sim 40°C$, and near the poles, T is $2 \sim -1.8°C$. Here, T_1 takes an average value on the upper surface. Since there is an elevated rate of evaporation in the tropical areas and the freezing of polar sea ice, both of which leave the salt behind in the remaining seawater, the densities of tropical and polar water are quite high. Hence, on average, we determine that

$$T_0 - T_1 \leq 0, \quad S_0 - S_1 < 0. \tag{5.3.88}$$

By Physical Conclusion 5.3.9 and (5.3.88), the saline Rayleigh number $\widetilde{R}$ is negative. Therefore, the transition of (5.3.69) with (5.3.8) is from real eigenvalues. In addition, by $\widetilde{R} < 0$, we see that the number in (5.3.86) satisfies $b_1 > 0$. Thus, by Physical Conclusion 5.3.10, the dynamical behavior of the THC is a Type I transition to a pair of stable equilibrium states, provided

$$\sigma = R - \mathrm{Le}^{-1}\widetilde{R} > \sigma_c,$$

which, by (5.3.75), (5.3.76), and (5.3.83), is equivalent to

$$0.86 \times 10^{21}(T_0 - T_1) + 3.75 \times 10^{23}(S_1 - S_0) > 2.33 \times 10^{24}. \tag{5.3.89}$$

For the large-scale ocean circulation, $T_0 - T_1 < 100°C$. Hence, (5.3.89) shows that if $S_1 - S_0 = O(10)$, then the great ocean conveyer is driven by doubly diffusive convection.

In assertion (1) of Physical Conclusion 5.3.10, the steady-state solutions describing the great conveyer are approximatively expressed by

$$v^{\pm} = \begin{cases} \pm C\beta^{1/2}(\sigma)L_1 \sin\dfrac{\pi x_1}{L_1}\cos\pi x_3, \\ \mp C\beta^{1/2}(\sigma)\cos\dfrac{\pi x_1}{L_1}\sin\pi x_3, \end{cases} \tag{5.3.90}$$

$$T^{\pm} = T_0 + (T_1 - T_0)x_3 \mp \frac{C\beta^{1/2}(\sigma)}{\alpha_c^2 + \pi^2}\cos\frac{\pi x_1}{L_1}\sin\pi x_3, \tag{5.3.91}$$

$$S^{\pm} = S_0 + (S_1 - S_0)x_3 \mp \frac{\mathrm{sign}(S_0 - S_1)C\beta^{1/2}}{\mathrm{Le}\,(\alpha_c^2 + \pi^2)}\cos\frac{\pi x_1}{L_1}\sin\pi x_3, \tag{5.3.92}$$

where $C > 0$ is a constant, $\beta(\sigma)$ is the first real eigenvalue, and by Theorem 5.3.5, $\beta(\sigma)$ can be expressed as

$$\beta(\sigma) = k(\sigma - \sigma_c) + o(|\sigma - \sigma_c|) \tag{5.3.93}$$

for some constant $k > 0$.

In oceanic dynamics, the Rayleigh number $\sigma = R - \mathrm{Le}^{-1}\widetilde{R}$ is a main driving force for the great ocean conveyer; (5.3.90) and (5.3.93) show that the velocity v of the oceanic circulation is proportional to $\sqrt{\sigma - \sigma_c}$.

The velocity field v^+ given by (5.3.90) is topologically equivalent to the structure shown in Fig. 5.12, which is consistent with the real oceanic flow structure.

Assertion (3) in Physical Conclusion 5.3.10 shows also that the theoretical results are in agreement with the real thermohaline ocean circulation.

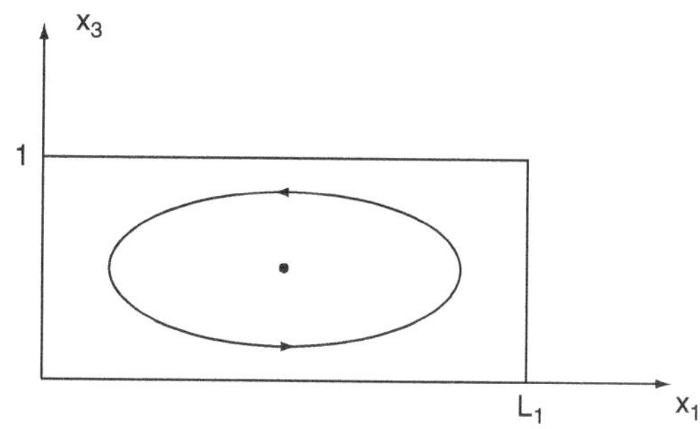

Fig. 5.12 The flow structure described by v^+ in (5.3.90)

At $x_3 = 1/2$, the temperature T^+ and salinity S^+ given by (5.3.91) and (5.3.92) are as follows:

$$T^+_{1/2}(x_1) = \frac{1}{2}(T_0 + T_1) - \frac{C\beta^{1/2}}{\alpha_c^2 + \pi^2}\cos\frac{\pi x_1}{L_1},$$

$$S^+_{1/2}(x_1) = \frac{1}{2}(S_0 + S_1) + \frac{C\beta^{1/2}}{\mathrm{Le}\,(\alpha_c^2 + \pi^2)}\cos\frac{\pi x_1}{L_1}.$$

The distributions $T^+_{1/2}(x_1)$ and $S^+_{1/2}(x_1)$ are illustrated in Figs. 5.13 and 5.14 respectively. In particular, if $x_1 = 0$ stands for the North Atlantic, $x_1 = L_1$ for the North Pacific, and $x_3 = 1/2$ for the deep basin of the ocean, then the profiles of $T^+_{1/2}(x_1)$ and $S^+_{1/2}(x_1)$ conform with observations in which the North Atlantic Deep Water is cold and salty water, and the North Pacific Deep Water is warmer and fresher water.

Physical Conclusions 5.3.9 and 5.3.10 show that although the THC is mainly due to a continuous transition to multiple equilibria, the analysis for other possible dynamical behaviors is still interesting.

We can see that when $\widetilde{R}$ satisfies

$$\frac{\mathrm{Le}^3}{1-\mathrm{Le}^2}\sigma_c < \widetilde{R} < \frac{\mathrm{Le}^2}{1-\mathrm{Le}}\sigma_c, \tag{5.3.94}$$

the number b_1 in (5.3.86) is less than zero, i.e., $b_1 < 0$. Then due to Physical Conclusions 5.3.9 and 5.3.10, the transition is a jump transition, leading to a saddle-node oscillation.

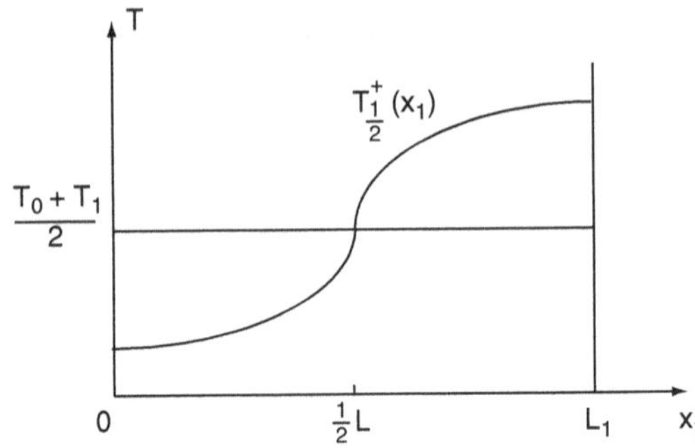

Fig. 5.13 The temperature profile in the deep water of the ocean

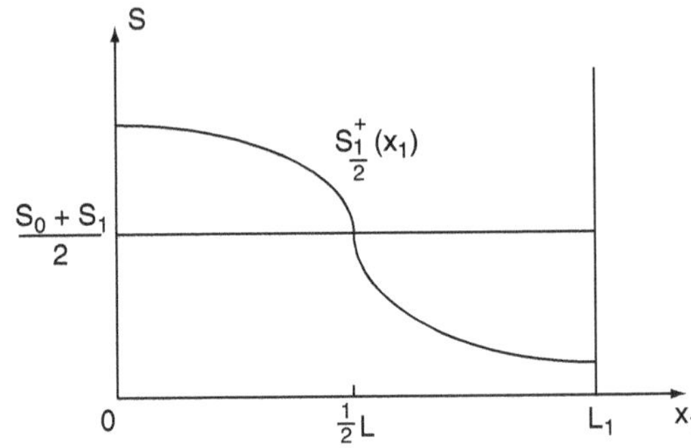

Fig. 5.14 The salinity profile in the deep water of the ocean

When $\widetilde{R} > \frac{\text{Le}^2}{1-\text{Le}}\sigma_c = 2.35 \times 10^{20}$, the problem undergoes a dynamic transition to a periodic solution for $\eta > \eta_c$. This transition can be either of Type I or of Type II, described as follows.

First, if $\widetilde{R} > 2.35 \times 10^{24}$, the number b_2 in (5.3.87) is less than zero, and by Physical Conclusion 5.3.11, this system undergoes a Type I transition to a stable periodic solution for $\eta = R - \widetilde{R} > \eta_c$. Its velocity field is written as

$$u(x,t) = \begin{cases} -\left(\dfrac{2\pi\lambda(\eta)}{|b_2|}\right)^{1/2} L_1 \sin\rho_0 t \sin\dfrac{\pi x_1}{L_1}\cos\pi x_3, \\ \left(\dfrac{2\pi\lambda(\eta)}{|b_2|}\right)^{1/2} \sin\rho_0 t \cos\dfrac{\pi x_1}{L_1}\sin\pi x_3, \end{cases} \tag{5.3.95}$$

where $\lambda(\eta) = \text{Re}\beta_1(\eta), \eta = R - \widetilde{R}$, and ρ_0 is as in (5.3.87).

Second, if $2.35 \times 10^{20} < \widetilde{R} < 2.35 \times 10^{24}$, the number b_2 is positive. Hence the transition is a jump transition to a stable periodic solution at $\eta = \eta_c$, leading to an oscillation between a time-periodic solution and the trivial equilibrium near η_c.

Finally, some remarks on the oceanic thermohaline circulation are in order.

First, the condition (5.3.88) provides a more realistic scenario, where $\widetilde{R}$ satisfies $\widetilde{R} < \frac{\text{Le}^3}{1-\text{Le}^2}\sigma_c$. In this case, the circulation is a continuous transition to multiple stable equilibria, with flow structure as illustrated in Fig. 5.12, and with the temperature and salinity profiles for the deep water of the ocean as shown in Figs. 5.13 and 5.14. This is consistent with observations.

Second, it is known that the velocity takes κ/h as its unit. Thus, the maximal value of v_1 in (5.3.90) is

$$v_{\max} = \delta^{1/2}\beta^{1/2}\kappa h^{-1}L_1,$$

where $\delta = \delta(\sigma_c)$ is as in (5.3.57), and $\beta = \beta(\sigma)$ is as in (5.3.93).

The ratio between the vertical and horizontal velocities is given by

$$\frac{v_3}{v_1} = \frac{1}{L_1} = 1.56 \times 10^{-4}.$$

In contrast, the oceanic circulation takes about 1,600 years for its a journey, i.e., the real velocity is also very small.

Third, condition (5.3.94) would lead to a saddle-node bifurcation to metastable states. However, such states have not been observed in oceanography.

Fourth, if $\widetilde{R} > 2.35 \times 10^{23}$, then the circulation is time-periodic, and we see from (5.3.95) that the period is

$$\tau = \frac{2\pi}{\rho_0} \cdot \frac{h^2}{\kappa} = 1.1 \times 10^6 s,$$

where ρ_0 is as in (5.3.87), and $\rho_0 \cong 6.5 \times 10^8$. This period is about four months. It is not realistic.

Fifth, when $2.35 \times 10^{20} < \widetilde{R} < 2.35 \times 10^{24}$, the oceanic system undergoes a time-periodic oscillation phenomenon. However, this behavior has not been observed in a realistic oceanic regime.

5.4 Arctic Ocean Circulations

The Arctic Ocean is almost entirely within the Arctic Circle (at $67°N$). It is surrounded mostly by land, and its outlets are the Bering Strait between Alaska and Russia, the Davis Strait between Greenland and Canada, and the Denmark Strait and the Norwegian Sea between Greenland and Europe. This region has an area of $14 \times 10^6\,\text{km}^2$ and an average depth of 3700 m off the continental shelf. Most of the Arctic Ocean is over 1000 m deep and is covered with sea ice, whose thickness varies between 1 and 10 m. About one-third of the Arctic Ocean is over the shallow continental shelf. Although it is found year-round over the deep ocean, sea ice is absent over the shelf regions for part of the year.

The Arctic Ocean is divided into two basins, the Eurasian basin and the North American basin, by the Lomonosov Ridge. There are also submarine ridges between the Arctic and Atlantic Oceans. Since the land and the submarine ridges block water from flowing out, the Arctic Ocean consists of an almost closed pool of cold water.

Unlike tropical oceans, which are temperature-stratified (i.e., there is a thermocline), the Arctic Ocean is salinity-stratified. Because of its salt, ocean water freezes only at $-1.8°C$. And salt water is densest at its freezing point, unlike fresh water, which is densest at $4°C$. The temperature gradient between the surface and bottom of the Arctic Ocean is about $2 \sim 3°C$, i.e., at the bottom, $T_0 \cong 1°C$, and at the surface, $T_1 \cong -1.8°C$.

The main objective of this section is to derive the onset and mechanism of the Arctic oceanic circulation. In particular, the analysis shows that the transition to Arctic oceanic circulation is a continuous (Type I) transition as multiple equilibria and with transients represented by the unstable manifolds of the multiple equilibria. In addition, the analysis indicates the possibility of the existence of a thermohaline circulation in the Arctic region. These results are presented in this section for the first time.

5.4.1 Model

We begin with the rotating Boussinesq equations:

$$\begin{aligned}
&\frac{\partial u}{\partial t} + (u\cdot\nabla)u = \nu\Delta u - 2\Omega e_z \times (u_1,u_2,0) - \frac{1}{\rho_0}\rho g k - \frac{1}{\rho_0}\nabla p,\\
&\frac{\partial T}{\partial t} + (u\cdot\nabla)T = \kappa_T \Delta T,\\
&\frac{\partial S}{\partial t} + (u\cdot\nabla)S = \kappa_S \Delta S,\\
&\text{div} u = 0,
\end{aligned} \tag{5.4.1}$$

where Ω is the angular velocity of the Earth, and the other values are the same as in (5.3.1). We take $D = (-l_1, l_1) \times (-l_2, l_2) \times (0, h)$ to be the spatial domain, where l_1 and $l_2 > 0$ are the horizontal length scales, and h is the height.

As for (5.3.6), the nondimensional equations of (5.4.1) are given by

$$\begin{aligned}
&\frac{\partial u_1}{\partial t} + (u\cdot\nabla)u_1 = \Pr\,\Delta u_1 + \omega u_2 - \Pr\,\frac{\partial p}{\partial x_1},\\
&\frac{\partial u_2}{\partial t} + (u\cdot\nabla)u_2 = \Pr\,\Delta u_2 - \omega u_1 - \Pr\,\frac{\partial p}{\partial x_2},\\
&\frac{\partial u_3}{\partial t} + (u\cdot\nabla)u_3 = \Pr\,\Delta u_3 + \Pr\,(RT - \text{sign}(S_0 - S_1)\widetilde{R}S) - \Pr\,\frac{\partial p}{\partial x_3},\\
&\frac{\partial T}{\partial t} + (u\cdot\nabla)T = \Delta T + u_3,\\
&\frac{\partial S}{\partial t} + (u\cdot\nabla)S = \text{Le}\,\Delta S + \text{sign}(S_0 - S_1)u_3,\\
&\text{div}\, u = 0,
\end{aligned} \tag{5.4.2}$$

for $x = (x_1, x_2, x_3)$ in the nondimensional domain $D = (-L_1, L_1) \times (-L_2, L_2) \times (0,1)$, where $l_i = L_i h$ $(i = 1,2)$, and the nondimensional parameters $R, \widetilde{R}$, Pr , Le are as in (5.2.29), and

$$\omega = \frac{2\Omega h^2}{\kappa_T}.$$

Equations (5.4.2) are supplemented with the following boundary conditions:

$$(u_3, T, S) = 0, \quad \frac{\partial u_1}{\partial x_3} = \frac{\partial u_2}{\partial x_3} = 0 \qquad \text{at} \quad x_3 = 0, 1, \tag{5.4.3}$$

$$(u, T, S)(x) = (u, T, S)(x_1 + 2jL_1, x_2 + 2kL_2, x_3) \qquad \text{for } j, k \in \mathbb{Z}. \tag{5.4.4}$$

We remark here that $x' = (x_1, x_2) = (0,0)$ represents the North Pole, and it is then natural to impose the following oddness condition on the horizontal velocity field:

$$(u_1, u_2)(-x', x_3) = -(u_1, u_2)(x', x_3). \tag{5.4.5}$$

The function spaces for (5.4.2)–(5.4.5) are taken as

$$H = \{(u, T, S) \in L^2(D)^5 \mid \text{div} u = 0, (u, T, S) \text{ satisfies } (5.4.5)\},$$
$$H_1 = \{(u, T, S) \in H^2(D)^5 \bigcap H \mid (u, T, S) \text{ satisfies } (5.4.3)\text{–}(5.4.5)\}.$$

5.4.2 Linear Theory

To study the dynamic transitions of the problem, we need to consider the following linear eigenvalue problem:

$$\begin{aligned}
&\Pr \Delta u_1 + \omega u_2 - \Pr \frac{\partial p}{\partial x_1} = \beta u_1, \\
&\Pr \Delta u_2 - \omega u_1 - \Pr \frac{\partial p}{\partial x_2} = \beta u_2, \\
&\Pr \Delta u_3 + \Pr (RT - \text{sign}(S_0 - S_1)\widetilde{R}S) - \Pr \frac{\partial p}{\partial x_3} = \beta u_3, \\
&\Delta T + u_3 = \beta T, \\
&\text{Le}\, \Delta S + \text{sign}(S_0 - S_1) u_3 = \beta S, \\
&\text{div} u = 0,
\end{aligned} \tag{5.4.6}$$

with the boundary conditions (5.4.3)–(5.4.5). Let the eigenvalues $\psi = (u, T, S)$ be expressed as

$$\begin{aligned}
&(u_1, u_2) = (u_{jk}(x_3), v_{jk}(x_3)) \sin \pi (j\alpha_1 x_1 + k\alpha_2 x_2), \\
&u_3 = w_{jk}(x_3) \cos \pi (j\alpha_1 x_1 + k\alpha_2 x_2), \\
&(T, S, p) = (T_{jk}(x_3), S_{jk}(x_3), p_{jk}(x_3)) \cos \pi (j\alpha_1 x_1 + k\alpha_2 x_2),
\end{aligned} \tag{5.4.7}$$

where $\alpha_i = L_i^{-1}$. Putting (5.4.7) into (5.4.6), we obtain

$$\begin{aligned}
&\Pr D_{jk}u_{jk} + \Pr j\alpha_1\pi p_{jk} + \omega v_{jk} = \beta u_{jk},\\
&\Pr D_{jk}v_{jk} + \Pr k\alpha_2\pi p_{jk} - \omega u_{jk} = \beta v_{jk},\\
&\Pr D_{jk}w_{jk} - \Pr Dp_{jk} + \Pr RT_{jk} - \Pr \widetilde{R}\mathrm{sign}(S_0 - S_1)S_{jk} = \beta w_{jk},\\
&D_{jk}T_{jk} + w_{jk} = \beta T_{jk},\\
&\mathrm{Le}\, D_{jk}S_{jk} + \mathrm{sign}(S_0 - S_1)w_{jk} = \beta S_{jk},\\
&j\alpha_1\pi u_{jk} + k\alpha_2\pi v_{jk} + Dw_{jk} = 0,\\
&Du_{jk} = Dv_{jk} = w_{jk} = T_{jk} = S_{jk} = 0 \qquad \text{at} \quad x_3 = 0,1,
\end{aligned} \tag{5.4.8}$$

where

$$D = \frac{d}{dz}, \quad D_{jk} = D^2 - \alpha_{jk}^2, \quad \alpha_{jk}^2 = \pi^2(j^2\alpha_1^2 + k^2\alpha_2^2).$$

Equations (5.4.8) can be reduced to a single equation for w_{jk}:

$$\begin{aligned}
&\Big[(D_{jk} - \beta)(\mathrm{Le}\, D_{jk} - \beta)(\Pr D_{jk} - \beta)^2 D_{jk}\\
&\quad + \omega^2(D_{jk} - \beta)(\mathrm{Le}\, D_{jk} - \beta)D^2\\
&\quad + \Pr R\alpha_{jk}^2(\Pr D_{jk} - \beta)(\mathrm{Le}\, D_{jk} - \beta)\\
&\quad - \Pr \widetilde{R}\alpha_{jk}^2(\Pr D_{jk} - \beta)(D_{jk} - \beta)\Big] w_{jk} = 0,
\end{aligned} \tag{5.4.9}$$

$$D^{2k}w_{jk} = 0, \quad 0 \le k \le 4 \qquad \text{at} \quad x_3 = 0,1. \tag{5.4.10}$$

The solutions of (5.4.9)–(5.4.10) are given by

$$w_{jk}(x_3) = \sin l\pi x_3 \qquad \text{for } l \in \mathbb{N}. \tag{5.4.11}$$

Putting (5.4.11) into (5.4.9), we derive that the eigenvalues β of (5.4.6) with (5.4.3)–(5.4.5) satisfy the following fourth-order equation:

$$\begin{aligned}
&\gamma_{jkl}^2(\gamma_{jkl}^2 + \beta)(\mathrm{Le}\, \gamma_{jkl}^2 + \beta)(\Pr \gamma_{jkl}^2 + \beta)^2\\
&+ l^2\pi^2\omega^2(\gamma_{jkl}^2 + \beta)(\mathrm{Le}\, \gamma_{jkl}^2 + \beta) - \Pr R\alpha_{jkl}^2 + \beta)(\mathrm{Le}\, \gamma_{jkl}^2 + \beta)\\
&+ \Pr \widetilde{R}\alpha_{jk}^2(\Pr \gamma_{jkl}^2 + \beta)(\gamma_{jkl}^2 + \beta) = 0,
\end{aligned} \tag{5.4.12}$$

where $\gamma_{jkl}^2 = \alpha_{jk}^2 + l^2\pi^2$.

For each given (j,k,l), there are four eigenvalues β_{jkl}^i $(1 \le i \le 4)$ satisfying (5.4.12). We infer from (5.4.8) that

$$(\Pr D_{jk} - \beta)u_{jk} = \frac{-j\alpha_1\pi}{\alpha_{jk}^2}(\Pr D_{jk} - \beta - \frac{k\alpha_2}{j\alpha_1}\omega)Dw_{jk}, \tag{5.4.13}$$

$$(\Pr D_{jk} - \beta)u_{jk} = \frac{-k\alpha_2\pi}{\alpha_{jk}^2}(\Pr D_{jk} - \beta + \frac{j\alpha_1}{k\alpha_2}\omega)Dw_{jk}, \tag{5.4.14}$$

$$(D_{jk}-\beta)T_{jk}=-w_{jk}, \tag{5.4.15}$$

$$(\text{Le}\,D_{jk}-\beta)S_{jk}=-\text{sign}(S_0-S_1)w_{jk}. \tag{5.4.16}$$

By (5.4.13)–(5.4.16), we obtain the following eigenvector corresponding to β^i_{jkl} $(1\le i\le 4)$:

$$\psi^i_{jkl}=\begin{cases} u^i_{jkl}\sin(j\alpha_1\pi x_1+k\alpha_2\pi x_2)\cos l\pi x_3,\\ v^i_{jkl}\sin(j\alpha_1\pi x_1+k\alpha_2\pi x_2)\cos l\pi x_3,\\ w^i_{jkl}\cos(j\alpha_1\pi x_1+k\alpha_2\pi x_2)\sin l\pi x_3,\\ T^i_{jkl}\cos(j\alpha_1\pi x_1+k\alpha_2\pi x_2)\sin l\pi x_3,\\ S^i_{jkl}\cos(j\alpha_1\pi x_1+k\alpha_2\pi x_2)\sin l\pi x_3,\end{cases} \tag{5.4.17}$$

where

$$u^i_{jkl}=-\frac{l\pi^2}{\alpha^2_{jk}}(j\alpha_1+\frac{k\alpha_2\omega}{\text{Pr}\,\gamma^2_{jkl}+\beta^i_{jkl}}),\qquad v^i_{jkl}=-\frac{l\pi^2}{\alpha^2_{jk}}\left(k\alpha_2-\frac{j\alpha_1\omega}{\text{Pr}\,\gamma^2_{jkl}+\beta^i_{jkl}}\right),$$

$$w^i_{jkl}=1,\qquad T^i_{jkl}=\frac{1}{\gamma^2_{jkl}+\beta^i_{jkl}},\qquad S^i_{jkl}=\frac{\text{sign}(S_0-S_1)}{\text{Le}\,\gamma^2_{jkl}+\beta^i_{jkl}}.$$

The conjugate equations of (5.4.6) are as follows:

$$\begin{aligned}
&\text{Pr}\,\Delta u_1^*-\omega u_2^*-\text{Pr}\,\frac{\partial p^*}{\partial x_1}=\overline{\beta}u_1^*,\\
&\text{Pr}\,\Delta u_2^*+\omega u_1^*-\text{Pr}\,\frac{\partial p^*}{\partial x_2}=\overline{\beta}u_2^*,\\
&\text{Pr}\,\Delta u_3^*+T^*+\text{sign}(S_0-S_1)S^*-\text{Pr}\,\frac{\partial p}{\partial x_3}=\overline{\beta}u_3^*,\\
&\Delta T^*+\text{Pr}\,Ru_3^*=\overline{\beta}T^*,\\
&\text{Le}\,\Delta S^*-\text{Pr}\,\widetilde{R}\text{sign}(S_0-S_i)u_i^*=\overline{\beta}S^*,\\
&\text{div}u^*=0.
\end{aligned}$$

Therefore, the conjugate eigenvector ψ^{i*}_{jkl} corresponding to β^i_{jkl} is given by

$$\psi^{i*}_{jkl}=\begin{cases} u^{i*}_{jkl}\sin(j\alpha_1\pi x_1+k\alpha_2\pi x_2)\cos l\pi x_3,\\ v^{i*}_{jkl}\sin(j\alpha_1\pi x_1+k\alpha_2\pi x_2)\cos l\pi x_3,\\ w^{i*}_{jkl}\cos(j\alpha_1\pi x_1+k\alpha_2\pi x_2)\sin l\pi x_3,\\ T^{i*}_{jkl}\cos(j\alpha_1\pi x_1+k\alpha_2\pi x_2)\sin l\pi x_3,\\ S^{i*}_{jkl}\cos(j\alpha_1\pi x_1+k\alpha_2\pi x_2)\sin l\pi x_3,\end{cases} \tag{5.4.18}$$

where

$$u^{i*}_{jkl} = -\frac{l\pi^2}{\alpha^2_{jk}}(j\alpha_1 - \frac{k\alpha_2\omega}{\text{Pr}\,\gamma^2_{jkl} + \overline{\beta}^i_{jkl}}), \qquad v^{i*}_{jkl} = -\frac{l\pi^2}{\alpha^2_{jk}}(k\alpha_2 + \frac{j\alpha_1\omega}{\text{Pr}\,\gamma^2_{jkl} + \overline{\beta}^i_{jkl}}),$$

$$w^{i*}_{jkl} = 1, \qquad T^{i*}_{jkl} = \frac{\text{Pr}\,R}{\gamma^2_{jkl} + \overline{\beta}^i_{jkl}}, \qquad S^{i*}_{jkl} = -\frac{\text{sign}(S_0 - S_1)\text{Pr}\,\widetilde{R}}{\text{Le}\,\gamma^2_{jkl} + \overline{\beta}^i_{jkl}}.$$

We now examine the PES. To this end, we write (5.4.12) as

$$\begin{aligned}
&\gamma^2_{jkl}\beta^4 + (2\text{Pr} + \text{Le} + 1)\gamma^4_{jkl}\beta^3 \\
&+ [(\text{Le} + 2\text{Pr}\,\text{Le} + \text{Pr}^2 + 2\text{Pr})\gamma^6_{jkl} + l^2\pi^2\omega^2 - \text{Pr}\,R\alpha^2_{jk} + \text{Pr}\,\widetilde{R}\alpha^2_{jk}]\beta^2 \\
&+ \gamma^2_{jkl}[2\text{Pr}\,\text{Le}\,\gamma^6_{jkl} + \text{Pr}^2\text{Le}\,\gamma^6_{jkl} + \text{Pr}^2\gamma^6_{jkl} + l^2\pi^2\omega^2(\text{Le} + 1) \\
&- \text{Pr}\,R\alpha^2_{jk}(\text{Pr} + \text{Le}) + \text{Pr}\,(\text{Pr} + 1)\widetilde{R}\alpha^2_{jk}]\beta + \gamma^4_{jkl}(\text{Le}\,\text{Pr}^2\gamma^6_{jkl} + l^2\pi^2\omega^2\text{Le} \\
&- \text{Pr}^2R\alpha^2_{jk}\text{Le} + \text{Pr}^2\widetilde{R}\alpha^2_{jk}) = 0. \qquad (5.4.19)
\end{aligned}$$

Obviously, $\beta = 0$ is a zero of (5.4.19) if and only if

$$R = \frac{\gamma^6_{jkl}}{\alpha^2_{jk}} + \frac{\pi^2\omega^2}{\text{Pr}^2\alpha^2_{jk}} + \text{Le}^{-1}\widetilde{R}. \qquad (5.4.20)$$

Hence, the critical Rayleigh number for the first real eigenvalues is given by

$$R_{c_1} = \min_{(j,k,l)\in I} \frac{\gamma^6_{jkl}}{\alpha^2_{jk}} + \frac{\pi^2\omega^2}{\text{Pr}^2\alpha^2_{jk}} + \text{Le}^{-1}\widetilde{R} = \frac{\gamma^6_{j_1k_1l}}{\alpha^2_{j_1k_1}} + \frac{\pi^2\omega^2}{\text{Pr}^2\alpha^2_{j_1k_1}} + \text{Le}^{-1}\widetilde{R}, \qquad (5.4.21)$$

for some integer pair (j_1, k_1).

If $\beta = i\rho$ ($\rho \neq 0$), a purely imaginary number, is a zero of (5.4.19), then

$$\begin{aligned}
&a_0\gamma^2_{jkl}\rho^2 = a_1 - \text{Pr}\,(\text{Pr} + \text{Le})\alpha^2_{jk}R, \\
&\gamma^2_{jkl}\rho^4 - (a_2 - \text{Pr}\,\alpha^2_{jk}R)\rho^2 + \gamma^4_{jkl}(a_3 - \text{Pr}^2\text{Le}\,\alpha^2_{jk}R) = 0,
\end{aligned} \qquad (5.4.22)$$

where

$$\begin{aligned}
a_0 &= 2\text{Pr} + \text{Le} + 1, \\
a_1 &= (\text{Pr}^2 + \text{Pr}^2\text{Le} + 2\text{Pr}\,\text{Le})\gamma^6_{jkl} + l^2\pi^2\omega^2(\text{Le} + 1) + \text{Pr}\,(\text{Pr} + 1)\alpha^2_{jk}\widetilde{R}, \\
a_2 &= (\text{Pr}^2 + 2\text{Pr} + 2\text{Pr}\,\text{Le} + \text{Le})\gamma^6_{jkl} + l^2\pi^2\omega^2 + \text{Pr}\,\alpha^2_{jk}\widetilde{R}, \\
a_3 &= \text{Pr}^2\text{Le}\,\gamma^6_{jkl} + l^2\pi^2\omega^2\text{Le} + \text{Pr}^2\alpha^2_{jk}\widetilde{R}.
\end{aligned}$$

We infer from (5.4.22) that

$$b_2r^2 - b_1R + b_0 = 0, \qquad (5.4.23)$$

where

$$b_0 = (a_0a_1a_2 - a_0^2a_3\gamma_{jkl}^6)\alpha_{jk}^{-2}$$
$$b_1 = a_0a_1\text{Pr} + a_0a_2\text{Pr}(\text{Pr} + \text{Le}) - 2a_1\text{Pr}(\text{Pr} + \text{Le}) - a_0^2\text{Pr}^2\text{Le}\,\gamma_{jkl}^6,$$
$$b_2 = \text{Pr}^2(\text{Pr} + \text{Le})(\text{Pr} + 1)\alpha_{jk}^2.$$

Thus, by (5.4.23), we obtain

$$R^{\pm} = \frac{b_1 \pm \sqrt{b_1^2 - 4b_0b_2}}{2b_2}. \tag{5.4.24}$$

When $R^{\pm}$ are real, $R^- \leq R^+$. Therefore, the critical Rayleigh number for the first complex eigenvalues is given by

$$R_{c_2} = \min_{(j,k,l)\in I} R^-, \tag{5.4.25}$$

provided that the following conditions hold:

$$b_1^2 - 4b_0b_2 \geq 0, \quad a_1 - \text{Pr}(\text{Pr} + \text{Le})\alpha_{jk}^2R^- > 0. \tag{5.4.26}$$

Then, as in the proofs of Lemmas 5.3.3 and 5.3.4, we obtain the PES.

Lemma 5.4.1

(1) Assume that

$$\widetilde{R} \neq \frac{(Pr + 1)Le^2\gamma_{jkl}^6}{(1 - \text{Le})\alpha_{jk}^2 Pr} + \frac{(Pr - 1)Le^2 l^2\pi^2\omega^2}{(1 - \text{Le})Pr^3\alpha_{jk}^2}.$$

Then the real eigenvalues $\beta_{jkl}(R)$ of (5.4.19) with $\beta_{jkl} = 0$ at $R = R_c$ satisfying (5.4.20) cross the imaginary axis, i.e.,

$$\beta_{jkl}(R)\begin{cases} < 0 & \text{if } R < R_{c_1}, \\ = 0 & \text{if } R = R_{c_1}, \\ > 0 & \text{if } R > R_{c_1}. \end{cases}$$

(2) If (5.4.26) is valid, then the complex eigenvalues β_{jkl} of (5.4.19) with $Re\beta_{jkl} = 0$ at R_{c_2} satisfying (5.4.24) cross the imaginary axis:

$$Re\beta_{jkl}(R)\begin{cases} < 0 & \text{if } R < R_{c_2}, \\ = 0 & \text{if } R = R_{c_2}, \\ > 0 & \text{if } R > R_{c_2}. \end{cases}$$

5.4.3 Transition Theorems

Transition from R_{c_1}. We first consider the case $R_{c_1} < R_{c_2}$. In this case, Problem (5.4.2)–(5.4.5) transitions to steady-state solutions of this problem at $R = R_{c_1}$. Here, we assume that the first real eigenvalue $\beta = \beta^1_{jkl}(R)$ is simple. Then we define a physical parameter as follows:

$$b_1 = \frac{\alpha^2 \mathrm{Pr}^4 \gamma^6 - \pi^2(\pi^2 - \mathrm{Pr}^2\alpha^2)\omega^2 - \mathrm{Le}^{-3}\mathrm{Pr}^4\alpha^4(1-\mathrm{Le}^2)\widetilde{R}}{\mathrm{Le}^2\mathrm{Pr}^2(\mathrm{Pr}+1)\gamma^6 + (\mathrm{Pr}-1)\pi^2\mathrm{Le}^2\omega^2 - \mathrm{Pr}^3\alpha^2(1-\mathrm{Le})\widetilde{R}}, \tag{5.4.27}$$

where $\alpha^2 = \alpha^2_{j_1k_1} = (j_1^2\alpha_1^2 + k_1^2\alpha_2^2)\pi^2, \gamma^2 = \gamma^2_{j_1k_1l} = \alpha^2_{j_1k_1} + \pi^2$, and (j_1, k_1) satisfies (5.4.21).

Theorem 5.4.2 *Let $R_{c_1} < R_{c_2}$, and let b_1 be the parameter defined by (5.4.27). Then the following assertions hold:*

(1) When $b_1 > 0$, the problem (5.4.2)–(5.4.5) undergoes a continuous transition at $R = R_{c_1}$ and bifurcates from $((u,T,S),R) = (0, R_{c_1})$ to two equilibrium solutions (u^1_R, T^1_R, S^1_R) and (u^2_R, T^2_R, S^2_R) in H_1 that are attractors, given by

$$(u^i_R, T^i_R, S^i_R) = (-1)^i a\sqrt{\beta(R)}\,\psi^1_{j_1k_1l} + o(|\beta|^{1/2}),$$

where $a > 0$ is a constant and $\psi^1_{j_1k_1l}$ the first eigenvector corresponding to the first eigenvalue $\beta = \beta^1_{j_1k_1l}$.

(2) When $b_1 < 0$, the problem (5.4.2)–(5.4.5) undergoes a jump transition at $R = R_{c_1}$ and has a saddle-node bifurcation from (ψ_1^*, R^*) and (ψ_2^*, R^*) with $R^* < R_{c_1}$.

Proof. Let $J_1 = (j_1, k_1, 1), \psi_{J_1} = \psi^1_{j_1k_1l}$. By (5.4.17) and (5.4.18), we have

$$\psi_{J_1} = \begin{cases} u_{J_1}\sin(j_1\alpha_1\pi x_1 + k_1\alpha_2\pi x_2)\cos\pi x_3, \\ v_{J_1}\sin(j_1\alpha_1\pi x_1 + k_1\alpha_2\pi x_2)\cos\pi x_3, \\ \cos(j_1\alpha_1\pi x_1 + k_1\alpha_2\pi x_2)\sin\pi x_3, \\ T_{J_1}\cos(j_1\alpha_1\pi x_1 + k_1\alpha_2\pi x_2)\sin\pi x_3, \\ S_{J_1}\cos(j_1\alpha_1\pi x_1 + k_1\alpha_2\pi x_2)\sin\pi x_3, \end{cases} \tag{5.4.28}$$

where

$$u_{J_1} = -\frac{\pi^2}{\alpha^2}\left(j_1\alpha_1 + \frac{k_1\alpha_2\omega}{\mathrm{Pr}\,\gamma^2 + \beta^1_{J_1}}\right), \qquad v_{J_1} = -\frac{\pi^2}{\alpha^2}\left(k_1\alpha_2 - \frac{j_1\alpha_1\omega}{\mathrm{Pr}\,\gamma^2 + \beta^1_{J_1}}\right),$$

$$T_{J_1} = \frac{1}{\gamma^2 + \beta^1_{J_1}}, \qquad S_{J_1} = \frac{\mathrm{sign}(S_0 - S_1)}{\mathrm{Le}\,\gamma^2 + \beta^1_{J_1}},$$

and the conjugate eigenvector $\psi^*_{J_1}$ is given by

$$\psi^*_{J_1} = \begin{cases} u^*_{J_1} \sin(j_1\alpha_1\pi x_1 + k_1\alpha_2\pi x_2)\cos\pi x_3, \\ v^*_{J_1} \sin(j_1\alpha_1\pi x_1 + k_1\alpha_2\pi x_2)\cos\pi x_3, \\ \cos(j_1\alpha_1\pi x_1 + k_1\alpha_2\pi x_2)\sin\pi x_3, \\ T^*_{J_1} \cos(j_1\alpha_1\pi x_1 + k_1\alpha_2\pi x_2)\sin\pi x_3, \\ S^*_{J_1} \cos(j_1\alpha_1\pi x_1 + k_1\alpha_2\pi x_2)\sin\pi x_3, \end{cases} \tag{5.4.29}$$

where

$$u^*_{J_1} = -\frac{\pi^2}{\alpha^2}\left(j_1\alpha_1 - \frac{k_1\alpha_2\omega}{\Pr\gamma^2+\beta^1_{J_1}}\right), \qquad v^*_{J_1} = -\frac{\pi^2}{\alpha^2}\left(k_1\alpha_2 + \frac{j_1\alpha_1\omega}{\Pr\gamma^2+\beta^1_{J_1}}\right),$$

$$T^*_{J_1} = \frac{\Pr R}{\gamma^2+\beta^1_{J_1}}, \qquad S^*_{J_1} = -\frac{\mathrm{sign}(S_0-S_1)}{\mathrm{Le}\,\gamma^2+\beta^1_{J_1}}\Pr\widetilde{R}.$$

The reduced equation of (5.4.2)–(5.4.5) reads

$$\frac{dx}{dt} = \beta^1_{J_1}(R)x + \frac{1}{(\psi_{J_1},\psi^*_{J_1})}(G(\psi,\psi),\psi^*_{J_1}), \tag{5.4.30}$$

where $\psi = x\psi_{J_1} + \Phi$, Φ is the center manifold function, and

$$(G(\psi_1,\psi_2),\psi_3) = -\int_D [((U_1\cdot\nabla)U_2)\cdot U_3 + ((U_1\cdot\nabla)T_2)\cdot T_3 + ((U_1\cdot\nabla)S_2)\cdot S_3]\,dx,$$

for $\psi_k = (U_k, T_k, S_k) \in H_1, U_k = (u_{k1}, u_{k2}, u_{k3})$, $1 \le k \le 3$.

As in (5.3.57), putting $\psi = x\psi_{J_1} + \Phi$ into (5.4.30), we obtain

$$\frac{dx}{dt} = \beta^1_{J_1}(R)x - \delta(R)x^3 + o(3),$$

where $\delta(R)$ can be calculated as

$$\begin{aligned}\delta = \frac{1}{(\psi_{J_1},\psi^*_{J_1})}\Bigg[&\frac{(G(\psi_{J_1},\psi_{J_1}),\psi_{2j})(G(\psi_{J_1},\psi^*_{J_1}),\psi_{2j})}{-\beta_{2j}(\psi_{2J},\psi_{2J})} \\ &+ \frac{(G(\psi_{J_1},\psi_{J_1}),\psi^1_{002})\cdot(G(\psi_{J_1},\psi^*_{J_1}),\psi^1_{002})}{-\beta^1_{002}(\psi^1_{002},\psi^1_{002})} \\ &+ \frac{(G(\psi_{J_1},\psi_{J_1}),\psi^2_{002})\cdot(G(\psi_{J_1},\psi^*_{J_1}),\psi^2_{002})}{-\beta^2_{002}(\psi^2_{002},\psi^2_{002})}\Bigg],\end{aligned}$$

where $-\beta_{2J} = 4\Pr\alpha^2, -\beta^1_{002} = 4\pi^2, -\beta^2_{002} = 4\mathrm{Le}\,\pi^2$, and

$$\begin{aligned}\psi_{2J} &= (k_1\alpha_2\sin 2\phi, -j_1\alpha_1\sin 2\phi, 0, 0, 0), \\ \psi^1_{002} &= (0,0,0,\sin 2\pi x_3, 0), \\ \psi^2_{002} &= (0,0,0,0,\sin 2\pi x_3).\end{aligned}$$

Direct calculation shows that

$$\delta(R) = \frac{\pi^2}{8(\psi_{J_1},\psi_{J_1}^*)}\left[\frac{(k_1\alpha_2 u_{J_1} - j_1\alpha_1 v_{J_1})(k_1\alpha_2 u_{J_1}^* - j_1\alpha_1 v_{J_1}^*)}{\mathrm{Pr}\,\alpha^4} + \frac{T_{J_1}T_{J_1}^*}{\pi^4} + \frac{S_{J_1}S_{J_1}^*}{\pi^4 L_e}\right]. \tag{5.4.31}$$

From $\beta_{J_1}^1(R_{c_1}) = 0$, we obtain that at $R = R_{c_1}$,

$$\delta(R_{c_1}) = \frac{L_e^2}{2\mathrm{Pr}\,\alpha^2} b_1,$$

where b_1 is as in (5.4.27). The proof is complete. □

Transition at R_{c_2}. Now we consider the case $R_{c_2} < R_{c_1}$. In this case, Problem (5.4.2-(5.4.5) undergoes a transition to periodic solutions at $R = R_{c_2}$. We assume that the critical Rayleigh number R_{c_2} defined by (5.4.25) is reached at some integer triple $(j,k,l) = J_1 = (j_1,k_1,l)$, and the first complex eigenvalues $\beta_{j_1k_1l}^1 = \overline{\beta}_{j_1k_1l}^2$ are complex and simple. Let $\beta_{J_1} = \beta_{j_1k_1l}^1(R) = \lambda(R) + i\rho(R)$ with $\lambda(R_{c_2}) = 0, \rho(R_{c_2}) > 0$. Then the first eigenvectors $\psi_{J_1}^1$ and $\psi_{J_1}^2$ are given by

$$\psi_{J_1}^1 = \mathrm{Re}\psi_{J_1}, \qquad \psi_{J_1}^2 = \mathrm{Im}\psi_{J_1},$$

and the conjugate eigenvectors $\psi_{J_1}^{1*}$ and $\psi_{J_1}^{2*}$ are given by

$$\psi_{J_1}^{1*} = \mathrm{Re}\psi_{J_1}^*, \qquad \psi_{J_1}^{2*} = \mathrm{Im}\psi_{J_1}^*,$$

where ψ_{J_1} is as in (5.4.28), and $\psi_{J_1}^*$ is defined by (5.4.29) with $\beta_{J_1}^1$ being replaced by $\overline{\beta}_{J_1}^1 = \lambda - i\rho$. The first conjugate eigenvectors $\Phi_{J_1}^{1*}$ and $\Phi_{J_1}^{2*}$ satisfying

$$(\psi_{J_1}^i, \Phi_{J_1}^{j*}) \begin{cases} = 0 \text{ if } i \neq j, \\ \neq 0 \text{ if } i = j, \end{cases}$$

are as in (5.3.62) and (5.3.63).

We now define a parameter as

$$\begin{aligned} b_2 =& \pi^3\omega^2[(R_1B_1 - I_1B_2)C_1 + (I_1B_1 + R_1B_2)C_2] \\ &+ \mathrm{Pr}\,R[(R_2B_1 - I_2B_2)C_3 + (I_2B_1 + R_2B_2)C_4] \\ &+ \mathrm{Pr}\,\widetilde{R}[(R_3B_1 - I_3B_2)C_5 + (I_3B_1 + R_3B_2)C_6], \end{aligned} \tag{5.4.32}$$

where R_i, I_i $(1 \leq i \leq 3)$ and C_k $(1 \leq k \leq 6)$ are as in (4.4.73) and (5.3.59) with $\alpha^2 = \alpha^2_{j_2 k_2}, R = R_{c_2}$ satisfying (5.4.25), and

$$\begin{aligned}
\rho^2 = &\frac{1}{(2\mathrm{Pr} + \mathrm{Le} + 1)(\alpha^2 + \pi^2)}[\pi^2\omega^2(1 + \mathrm{Le}) + \mathrm{Pr}\,(\mathrm{Pr} + 1)\alpha^2\widetilde{R} \\
&- \mathrm{Pr}\,(\mathrm{Pr} + \mathrm{Le})\alpha^2 R + (\mathrm{Pr}^2 + \mathrm{Pr}^2\mathrm{Le} + 2\mathrm{Pr}\,\mathrm{Le})(\alpha^2 + \pi^2)^3], \\
B_1 = &\pi^2\omega^2 I_1^2 - \mathrm{Pr}\, R\alpha^2 I_2^2 + \mathrm{Pr}\,\widetilde{R}\alpha^2 I_3^2, \\
B_2 = &\pi^2\omega^2 R_1 I_1 - \mathrm{Pr}\, R\alpha^2 R_2 I_2 + \mathrm{Pr}\,\widetilde{R}\alpha^2 R_3 I_3.
\end{aligned}$$

Then we have the following theorem.

Theorem 5.4.3 *Let $R_{c_2} < R_{c_1}$, and let b_2 be the number defined by (5.4.32). Then we have the following assertions:*

(1) When $b_2 < 0$, the problem (5.4.2)–(5.4.5) has a Type I transition at $R = R_{c_2}$ and bifurcates to only one periodic solution on $R > R_{c_2}$, which is an attractor.
(2) When $b_2 > 0$, this problem has a type-II transition at $R = R_{c_2}$ and bifurcates to a unique periodic solution on $R < R_{c_2}$, which is a repeller. Moreover, there is a singular separation of periodic solutions at some R^* with $0 < R^* < R_{c_2}$.

The proof is similar to the proof of Theorem 5.3.7, and we omit the details.

Remark 5.4.4 The two parameters b_1 and b_2 defined by (5.4.27) and (5.4.32) become the two numbers given by (4.4.54) and (4.4.73) when the salinity Rayleigh number $\widetilde{R}$ is equal to 0. In this case, the problem (5.4.2)–(5.4.5) is reduced to the rotating convection problem. These two numbers will be the same as the two numbers given by (5.3.54) and (5.3.59) when the rotating parameter ω is equal to 0, and in this case, the problem is reduced to a doubly diffusive system.

5.4.4 Revised Transition Theory

Consider the rotating double-diffusive equations with a turbulent friction term:

$$\begin{aligned}
&\frac{\partial u_1}{\partial t} + (u\cdot\nabla)u_1 = \mathrm{Pr}\,(\Delta u_1 - \delta_0 u_1 - \frac{\partial p}{\partial x_1}) + \omega u_2, \\
&\frac{\partial u_2}{\partial t} + (u\cdot\nabla)u_2 = \mathrm{Pr}\,(\Delta u_2 - \delta_0 u_2 - \frac{\partial p}{\partial x_2}) - \omega u_1 \\
&\frac{\partial u_3}{\partial t} + (u\cdot\nabla)u_3 = \mathrm{Pr}\,(\Delta u_3 - \delta_1 u_3 + RT - \mathrm{sign}(S_0 - S_1)\widetilde{R}S - \frac{\partial p}{\partial x_3}), \\
&\frac{\partial T}{\partial t} + (u\cdot\nabla)T = \Delta T + u_3, \\
&\frac{\partial S}{\partial t} + (u\cdot\nabla)S = \mathrm{Le}\,\Delta S + \mathrm{sign}(S_0 - S_1)u_3, \\
&\mathrm{div}u = 0,
\end{aligned} \tag{5.4.33}$$

with the boundary conditions (5.4.3)–(5.4.5).

As in (5.4.12), the eigenvalues β of the linearized equations of (5.4.33) satisfy

$$\begin{aligned}&\alpha^2(\gamma^2+\beta)(\mathrm{Le}\,\gamma^2+\beta)(\mathrm{Pr}\,\gamma^2+\mathrm{Pr}\,\delta_0+\beta)(\mathrm{Pr}\,\gamma^2+\mathrm{Pr}\,\delta_1+\beta)\\&+(\gamma^2+\beta)(\mathrm{Le}\,\gamma^2+\beta)(\mathrm{Pr}\,\gamma^2+\mathrm{Pr}\,\delta_0+\beta)^2\pi^2\\&+(\gamma^2+\beta)(\mathrm{Le}\,\gamma^2+\beta)\omega^2\pi^2-\mathrm{Pr}\,R\alpha^2(\mathrm{Le}\,\gamma^2+\beta)(\mathrm{Pr}\,\gamma^2+\mathrm{Pr}\,\delta_0+\beta)\\&+\mathrm{Pr}\,\widetilde{R}\alpha^2(\gamma^2+\beta)(\mathrm{Pr}\,\gamma^2+\mathrm{Pr}\,\delta_0+\beta)=0,\end{aligned} \tag{5.4.34}$$

where $\alpha^2=\alpha_{jk}^2, \gamma^2=\alpha^2+\pi^2$.

Let $\beta=0$ in (5.4.34); we derive the critical Rayleigh number for the first real eigenvalues:

$$R_c-\mathrm{Le}^{-1}\widetilde{R}=\min_{\alpha^2}\left[\gamma^2\delta_1+\frac{\gamma^6+\pi^2\gamma^2\delta_0}{\alpha^2}+\frac{\gamma^2\pi^2\omega^2}{\mathrm{Pr}^2(\gamma^2+\delta_0)\alpha^2}\right]. \tag{5.4.35}$$

Similar to (4.5.21)–(4.5.23), for the oceanic motion, from (5.4.35) we can obtain the critical Rayleigh number σ_c as

$$\sigma_c=R_c-\mathrm{Le}^{-1}\widetilde{R}\cong\pi^2\delta_1+\frac{\pi^4\omega^2}{\alpha_c^2\mathrm{Pr}\,(\pi^2+\delta_0)}+\frac{\pi^6+\pi^4\delta_0}{\alpha_c^2}, \tag{5.4.36}$$

and the convective scale L_c as

$$L_c^2=\frac{\pi^2}{\alpha_c^2}, \tag{5.4.37}$$

where

$$\alpha_c^2=\pi^2\left(\pi^2+\delta_0+\frac{\omega^2}{\mathrm{Pr}^2(\pi^2+\delta_0)}\right)^{1/2}\delta_1^{-1/2}.$$

Let $\beta=i\rho$ ($\rho\neq0$) in (5.4.34). The critical Rayleigh number R_{c_2} corresponds to the first complex eigenvalues. Here we do not give the value of R_{c_2} again.

For the transition of (5.4.33) from σ_c, the parameter b_1 can be derived by (5.4.31), where we replace $u_{J_1}, v_{J_1}, u_{J_1}^*$, and $v_{J_1}^*$ by

$$\begin{aligned}u_{J_1}&=-\frac{\pi^2}{\alpha^2}\left(j_1\alpha_1+\frac{k_1\alpha_2\omega}{\mathrm{Pr}\,\gamma^2+\delta_0}\right), & v_{J_1}&=-\frac{\pi^2}{\alpha^2}\left(k_1\alpha_2-\frac{j_1\alpha_1\omega}{\mathrm{Pr}\,\gamma^2+\delta_0}\right),\\u_{J_1}^*&=-\frac{\pi^2}{\alpha^2}\left(j_1\alpha_1-\frac{k_1\alpha_2\omega}{\mathrm{Pr}\,\gamma^2+\delta_0}\right), & v_{J_1}^*&=-\frac{\pi^2}{\alpha^2}\left(k_1\alpha_2+\frac{j_1\alpha_1\omega}{\mathrm{Pr}\,\gamma^2+\delta_0}\right).\end{aligned}$$

Then direction calculation implies that

$$b_1=\frac{(\mathrm{Pr}\,\gamma_c^2+\delta_0)^2\mathrm{Pr}^2\alpha_c^4(R_c-\mathrm{Le}^{-3}\widetilde{R})-\gamma_c^4\pi^4\omega^2}{(\mathrm{Pr}\,\gamma_c^2+\delta_0)^2(\gamma_c^6+\mathrm{Pr}\,\alpha_c^2R_c-\mathrm{Pr}\,\alpha_c^2\mathrm{Le}^{-2}\widetilde{R})-\gamma_c^4\pi^2\omega^2}, \tag{5.4.38}$$

where R_c is as in (5.4.36), $\gamma_c^2=\alpha_c^2+\pi^2$, and α_c^2 is as in (5.4.37). Then we have the following transition theorem for the problem (5.4.33) with (5.4.3)–(5.4.5).

Theorem 5.4.5 *Let the number σ_c given by (5.4.36) be the first critical Rayleigh number, and let b_1 be defined by (5.4.38). Then the assertions of Theorem 5.4.2 hold for the problem (5.4.33) with (5.4.3)–(5.4.5).*

5.4.5 Physical Conclusions

We proceed by examining some typical parameters. Using our scaling formula, we have $\delta_i = C_i h^4/\nu$ $(i = 0,1)$, and we take the effective height for the convective motion in the Arctic Ocean as $h = 2 \times 10^3$ m. We infer from (5.3.78), (5.3.79), and (5.3.74) that

$$\delta_0 = 0.73 \times 10^7, \qquad \delta_1 = 1.45 \times 10^{22}, \qquad \omega^2 = \frac{4\Omega h^4}{\kappa^2} = 1.72 \times 10^{19}. \quad (5.4.39)$$

The thermal and saline Rayleigh numbers are then given by

$$\begin{aligned} R &= \frac{ag(T_0 - T_1)}{\kappa\nu} h^3 = 1.07 \times 10^{20}(T_0 - T_1)/°C, \\ \widetilde{R} &= \frac{bg(S_0 - S_1)}{\kappa\nu} h^3 = 4.68 \times 10^{20}(S_0 - S_1)/\text{psu}. \end{aligned} \quad (5.4.40)$$

By (5.4.39), the convective length scale (5.4.37) and the critical Rayleigh number (5.4.36) are given by

$$L_c = 0.8 \times 10^3, \qquad \sigma_c = R_c - \text{Le}^{-1}\widetilde{R} = 1.45 \times 10^{23}. \quad (5.4.41)$$

The analysis in this section leads to the following conclusions for the Arctic oceanic circulation:

Physical Conclusion 5.4.6 *Necessary and sufficient conditions for the Arctic oceanic circulation are*

$$\widetilde{R} < 0 \quad or \quad \Delta S = S_0 - S_1 < \Delta S_c \cong -3^0/_{00}, \quad (5.4.42)$$

and

$$L \simeq 3000\,km. \quad (5.4.43)$$

Physical Conclusion 5.4.7 *The onset of the Arctic oceanic circulation occurs at the first real eigenvalue σ_c, and the system undergoes a Type I (continuous) transition leading to multiple equilibria.*

Note that it is clear that (5.4.43) is a valid condition for the Arctic Ocean, and the validity of (5.4.42) is based on the freezing and evaporating processes, causing larger salinity on the top.

Proof of Physical Conclusion 5.4.6. By (5.4.40) and (5.4.41), if $\widetilde{R} > 0$, then the critical temperature ΔT_c is given by $T_0 - T_1 \sim O(10^3)$, which is unrealistic. Also, with

(5.4.41), $h = 2\,\text{km}$, and the above critical salinity gradient, the Arctic oceanic circulation requires the horizontal length scale given by (5.4.43). □

Proof of Physical Conclusion 5.4.7. Condition (5.4.42) implies that σ_c is the first critical Rayleigh number. For simplicity, we use (5.4.21) and (5.4.25) to show this point. By (5.3.74), (5.4.39), and (5.4.40),

$$|\text{Le}^{-1}\widetilde{R}| = \frac{\pi^2\omega^2}{\text{Pr}^2\alpha^2}, \quad |\widetilde{R}| \ll \frac{\pi^2\omega^2}{\text{Pr}^2\alpha^2}, \quad |\widetilde{R}| \gg \gamma^6/\alpha^2. \tag{5.4.44}$$

Hence, we infer from (5.4.24) that

$$R_{c_2} = \frac{b_1}{2b_2}\left(1 - (1 - \frac{4b_0b_2}{b_1^2}\right)^{1/2}) \cong \frac{b_0}{b_1},$$

where

$$b_1 \cong 2\text{Pr}^3(\pi^2\omega^2 + \text{Pr}\,\alpha^2\widetilde{R}), \quad b_0 \cong 2\text{Pr}\,(\pi^2\omega^2 + \text{Pr}^2\alpha^2\widetilde{R})(\pi^2\omega^2 + \text{Pr}\,\alpha^2\widetilde{R})/\alpha^2.$$

Thus, by (5.4.21) and (5.4.44), we have

$$R_{c_2} \cong \frac{\pi^2\omega^2}{\text{Pr}^2\alpha^2} + \widetilde{R}, \qquad R_{c_1} \cong \frac{\pi^2\omega^2}{\text{Pr}^2\alpha^2} + \text{Le}^{-1}\widetilde{R} = \frac{\pi^2\omega^2}{\text{Pr}^2\alpha^2} + 100\widetilde{R}.$$

Under the condition (5.4.44), it follows that

$$R_{c_2} > R_{c_1} \quad \text{as} \quad \widetilde{R} < 0,$$
$$R_{c_2} < R_{c_1} \quad \text{as} \quad \widetilde{R} > 0.$$

Therefore, $\sigma_c = R_{c_1} - \text{Le}^{-1}\widetilde{R}$ is the first critical Rayleigh number if $\widetilde{R} < 0$.

By conditions (5.4.39)–(5.4.42), it is easy to check that the number b_1 in (5.4.38) is positive, i.e., $b_1 > 0$. By Theorem 5.4.5, under the conditions (5.4.42)–(5.4.43), the problem (5.4.33) with (5.4.3)–(5.4.5) has a Type I transition from $\sigma = \sigma_c$ to a stable equilibrium. □

5.5 Large-Scale Meridional Atmospheric Circulation

5.5.1 Polar, Ferrel, and Hadley Cells

There are three primary large-scale meridional atmospheric circulation cells: the polar, the Ferrel, and the Hadley cells. Note that the Hadley cell may contain several cells within the equatorial zone that may shift, merge, and decouple in a complicated process over time. However, the Hadley cell is referred to as the unique large-scale meridional cell.

The polar cell is a relatively simple system. Warm air rises at lower latitudes and moves poleward through the upper troposphere at both the North and South Poles. When the air reaches the polar areas, it has cooled considerably, and it descends as a cold, dry high-pressure area, moving away from the pole along the surface but twisting westward as a result of the Coriolis effect to generate the polar easterlies.

The Ferrel cell is located between the polar cell and the Hadley cell and comes about as a result of the eddy circulations of the middle latitudes. At its southern extent, it overrides the Hadley cell, and at its northern extent, it overrides the polar cell. Just as the trade wind can be found beneath the Hadley cell, the westerlies can be found below the Ferrel cell.

The Hadley cell provides an explanation for the trade winds and matches fairly well with observations. It is a closed-circulation loop, which begins at the equator with warm, moist air raised in equatorial low-pressure areas to the tropopause and carried poleward. At about $30°N/S$ latitude, it descends in a cooler high-pressure area. Part of the descending air travels toward the equator along the surface, closing the loop of the Hadley cell and creating the trade winds.

The meridional circulation is very different from the longitudinal circulation. In the equatorial motion, equations (5.2.3), we see that the Coriolis force $(-\omega u_2, \omega u_1)$ can be balanced by a potential term, playing the same role as the pressure gradient (5.2.16), so that the Coriolis effect has little influence on the equatorial circulation. However, in the latitudinal regions, the rotational effect of the Earth is crucial:

- The Coriolis force deflects the poleward-moving air to the east and the equatorward moving air to the west, which complicates the atmospheric motion.
- The rotating effect reduces the convective scale, leading to an increase in the cells, i.e., there are twice as many loops in the latitudinal circulation as in the equatorial circulation.

Also, the surface temperature gradient/deviation in the meridional direction is an order-one term, and cannot be treated as a small perturbation to the average temperature gradient $\Delta T = T_0 - T_1$. This large surface temperature gradient leads to a strong climate variance, and to substantial difficulties for mathematical treatments.

The main objective of this section is to study dynamic transitions and pattern formations associated with polar, the Ferrel, and the Hadley cells.

5.5.2 *β-Plane Assumption*

Consider the Boussinesq equations on a hemispheric region with the β-plane assumption (Pedlosky 1987; Salby 1996). Namely, let

$$(x_1, x_2, x_3) = \left(r_0 \varphi \cos\theta_0, r_0(\frac{\pi}{2} - \theta), z\right),$$

for some $0 < \theta_0 < \frac{\pi}{2}$, and the northern hemispheric region D becomes a rectangle:

$$D = (0, 2\pi r_0) \times (0, \pi r_0/2) \times (r_0, r_0 + h).$$

The velocity field $(u_\varphi, u_\theta, u_r)$ becomes

$$u_1 = u_\varphi, u_2 = -u_\theta, u_3 = u_z.$$

The Coriolis force takes the following form:

$$2\Omega k \times u = (-2\Omega \sin\theta_0 u_2, 2\Omega \sin\theta_0 u_1, 0).$$

With the above considerations, the Boussinesq model takes the following form: (Pedlosky 1987; Salby 1996):

$$\begin{aligned}
&\frac{\partial u_1}{\partial t} + (u\cdot\nabla)u_1 = \nu\Delta u_1 - \sigma_0 u_1 - f u_2 - \frac{1}{\rho_0}\frac{\partial p}{\partial x_1},\\
&\frac{\partial u_2}{\partial t} + (u\cdot\nabla)u_2 = \nu\Delta u_2 - \sigma_0 u_2 + f u_1 - \frac{1}{\rho_0}\frac{\partial p}{\partial x_2},\\
&\frac{\partial u_3}{\partial t} + (u\cdot\nabla)u_3 = \nu\Delta u_3 - \sigma_1 u_3 - (1 - a(T - T_0))g - \frac{1}{\rho_0}\frac{\partial p}{\partial x_3},\\
&\frac{\partial T}{\partial t} + (u\cdot\nabla)T = \kappa\Delta T,\\
&\frac{\partial u_1}{\partial x_1} + \frac{\partial u_2}{\partial x_2} + \frac{\partial u_3}{\partial x_3} = 0,
\end{aligned} \tag{5.5.1}$$

where $f = 2\Omega\sin\theta_0$, and $\Delta, (u\cdot\nabla)$ are the usual operators in $\mathbb{R}^3$.

Let T_1 be the temperature at the upper boundary $x_3 = r_0 + h$, and let T_0 be the average temperature at the lower boundary $x_3 = r_0$, where T_1 and T_0 are constants. Take the standard nondimensional form as used in (5.2.3). Then equations (5.5.1) for the deviation of the motionless state with linear temperature profile become

$$\begin{aligned}
&\frac{\partial u}{\partial t} = \mathrm{Pr}\,(\Delta u + RTk - \delta u - \nabla p) + \omega k \times u - (u\cdot\nabla)u,\\
&\frac{\partial T}{\partial t} = \Delta T + u_3 - (u\cdot\nabla)T,\\
&\mathrm{div} u = 0,
\end{aligned} \tag{5.5.2}$$

where $k = (0,0,1)$, $\delta u = (\delta_0 u_1, \delta_0 u_2, \delta_1 u_3)$, and

$$\omega = 2\Omega\sin\theta_0 h^2/\kappa, \qquad \delta_i = \frac{h^2}{\nu}\sigma_i, \quad (i = 0, 1).$$

The boundary conditions in the (x_1, x_2)-plane are periodic:

$$(u,T)(x_1,x_2) = (u,T)(x_1 + 2\pi j r_0, x_2 + kL), \text{ for } j,k \in \mathbb{Z}, L = 2L_0 = \pi r_0. \tag{5.5.3}$$

At the pole, $(x_1,x_2)=0$, no horizontal flows appear, and we have to impose an additional condition:

$$u_1(0,0,x_3)=u_2(0,0,x_3)=0. \tag{5.5.4}$$

For convenience, at the top and bottom boundaries, we consider only the free–free boundary condition

$$\begin{aligned} & u_3=0, \quad \frac{\partial u_1}{\partial x_3}=\frac{\partial u_2}{\partial x_3}=0 && \text{at } x_3=r_0,\ r_0+1, \\ & T=0 && \text{at } x_3=r_0+1, \\ & T=\varphi(x_1,x_2) && \text{at } x_3=r_0, \end{aligned} \tag{5.5.5}$$

where $\varphi(x_1,x_2)$ is the temperature deviation satisfying

$$\int_0^{2\pi r_0}\int_0^{\frac{\pi}{2}r_0}\varphi(x_1,x_2)dx_2dx_1=0.$$

Namely, the real temperature distribution on the surface of the Earth is $T=T_0+\varphi(x_1,x_2)$. In contrast to the situation on the equator, here $\varphi(x_1,x_2)$ is not small in comparison with $\Delta T=T_0-T_1$.

5.5.3 Meridional Circulation Under Idealized Conditions

Model and Physical Parameters. Consider the idealized case that the temperature distribution on the surface $x_3=r_0$ is a constant, i.e.,

$$\varphi(x_1,x_2)=0 \quad \forall(x_1,x_2)\in(0,2\pi r_0)\times(0,\frac{\pi}{2}r_0).$$

Based on Table 5.1, at $T_0\cong 0°C$, the main parameters of air are taken as follows:

$$\begin{aligned} & \nu=1.3\times10^{-5}\,\text{m}^2/\text{s}, && \kappa=1.8\times10^{-5}\,\text{m}^2/\text{s}, \\ & a=3.7\times10^{-3}/\text{k}, && \text{Pr}=0.72, \\ & \Omega=0.73\times10^{-4}, && h=8\times10^3\,\text{m}, \\ & \sin\theta_0=1/2, && \omega^2=4\Omega^2h^4/\kappa^2=6.325\times10^{16}. \end{aligned} \tag{5.5.6}$$

Then the Rayleigh number is expressed as

$$R=\frac{g\cdot a(T_0-T_1)}{\kappa\nu}h^3=7.9\times10^{19}(T_0-T_1). \tag{5.5.7}$$

We take δ_0 and δ_1 as in (5.2.31). Then by (4.5.21)–(4.5.23), we infer from (5.5.6) and (5.5.7) the following theoretical values of the critical length scale L_c of the circulation cells, the critical Rayleigh number R_c, and the critical temperature gradient ΔT_c:

$$L_c \cong \left[\frac{\delta_1}{\delta_0 + \frac{\omega^2}{\Pr^2 \delta_0}}\right]^{1/4} \cdot h = 5000\,\text{km}, \tag{5.5.8}$$

$$R_c \cong \pi^2 \delta_1 + \frac{\pi^4 \omega^2}{\alpha_c^2 \Pr^2 \delta_0} \cong 2.7 \times 10^{21}, \tag{5.5.9}$$

$$\Delta T_c = T_0 - T_1 \cong 34°C. \tag{5.5.10}$$

It is known that the scale of the polar, Ferrel, and Hadley cells is

$$L_0 = \frac{\pi}{6} r_0 \cong 3300\,\text{km}. \tag{5.5.11}$$

The average surface temperature is $T_0 = 0°C \sim 10°C$, and the tropopause temperature is $T_1 \cong -70°C$. Hence, the real temperature gradient is about

$$\Delta T_c = 70°C \sim 80°C. \tag{5.5.12}$$

Linearized Eigenvalue Problem. Consider the eigenvalue problem

$$\begin{aligned} &\Pr\,(\Delta u + RTk - \delta u - \nabla p) + \omega k \times u = \beta u, \\ &\Delta T + u_3 = \beta T, \\ &\text{div} u = 0, \end{aligned} \tag{5.5.13}$$

with boundary conditions (5.5.1)–(5.5.3).

The eigenvalues of (5.5.12) can be expressed in the form of (4.4.9) and (4.4.10), and the functions $(u_{jk}, v_{jk}, W_{jk}, T_{jk}, p_{jk})$ satisfy the ordinary differential equations

$$\Pr\,(D_{jk} u_{jk} - \delta_0 u_{jk} + j\alpha_1 p_{jk}) - \omega v_{jk} = \beta u_{jk}, \tag{5.5.14}$$

$$\Pr\,(D_{jk} v_{jk} - \delta_0 v_{jk} + k\alpha_2 p_{jk}) + \omega u_{jk} = \beta v_{jk}, \tag{5.5.15}$$

$$\Pr\,(D_{jk} W_{jk} - \delta_1 w_{jk} - D p_{jk} + RT_{jk}) = \beta W_{jk}, \tag{5.5.16}$$

$$D_{jk} T_{jk} + W_{jk} = \beta T_{jk}, \tag{5.5.17}$$

$$j\alpha_1 u_{jk} + k\alpha_2 v_{jk} + DW_{jk} = 0, \tag{5.5.18}$$

with the boundary condition

$$Du_{jk} = Dv_{jk} = W_{jk} = T_{jk} = 0, \quad \text{at} \quad x_3 = 0, 1, \tag{5.5.19}$$

where D_{jk}, D, α_{jk}^2 are the same as in (4.4.14).

From (5.5.14), (5.5.15) and (5.5.18) we obtain

$$\begin{aligned} &-(\Pr D_{jk} - \Pr \delta_0 - \beta) DW_{jk} + \Pr \alpha_{jk}^2 p_{jk} = \omega (j\alpha_1 v_{jk} - k\alpha_2 u_{jk}), \\ &-(\Pr D_{jk} - \Pr \delta_0 - \beta)(j\alpha_1 v_{jk} - k\alpha_2 u_{jk}) = -\omega DW_{jk}. \end{aligned}$$

It follows that

$$(\Pr D_{jk} - \Pr \delta_0 - \beta)^2 DW_{jk} - \Pr \alpha_{jk}^2 (\Pr D_{jk} - \Pr \delta_0 - \beta) p_{jk} + \omega^2 DW_{jk} = 0. \tag{5.5.20}$$

Then, from (5.5.16), (5.5.17), and (5.5.20), we infer that W_{jk} satisfies the ordinary differential equation

$$\begin{aligned}&[\alpha_{jk}^2(\Pr D_{jk} - \Pr \delta_0 - \beta)(\Pr D_{jk} - \Pr \delta_1 - \beta)(D_{jk} - \beta)\\&-(\Pr D_{jk} - \Pr \delta_0 - \beta)^2(D_{jk} - \beta)D^2 - \omega^2(D_{jk} - \beta)D^2\\&-\alpha_{jk}^2 \Pr R(\Pr D_{jk} + \Pr \delta_0 - \beta)]W_{jk} = 0.\end{aligned} \tag{5.5.21}$$

From (5.5.19) and (5.5.14)–(5.5.18), we deduce that

$$W_{jk} = D^2 W_{jk} = D^4 W_{jk} = D^6 W_{jk} = 0. \tag{5.5.22}$$

It is clear that the solutions of (5.5.21)–(5.5.22) are sine functions $W_{jk} = \sin l\pi x_3 (l \in Z)$. Inserting $W_{jk} = \sin \pi x_3$ into (5.5.21), we obtain (4.5.19), which the first eigenvalues β of (5.5.13) satisfy.

Similar to (4.4.18)–(4.4.20), u_{jk}, v_{jk}, and T_{jk} satisfy

$$\begin{aligned}&(\Pr D_{jk} - \Pr \delta_0 - \beta)u_{jk} = -\frac{j\alpha_1\pi}{\alpha_{jk}^2}(\Pr D_{jk} - \Pr \delta_0 - \beta + \frac{k\alpha_2}{j\alpha_1}\omega)DW_{jk},\\&(\Pr D_{jk} - \Pr \delta_0 - \beta)v_{jk} = -\frac{k\alpha_2\pi}{\alpha_{jk}^2}(\Pr D_{jk} - \Pr \delta_0 - \beta - \frac{j\alpha_1}{k\alpha_2}\omega)DW_{jk},\\&(D_{jk} - \beta)T = -W_{jk}.\end{aligned} \tag{5.5.23}$$

Let $W_{jk} = \sin \pi x_3$. Then from (5.5.23), we get the eigenvectors of (5.5.13) corresponding to the eigenvalue β as follows:

$$\psi_{jk}^{\beta} = \begin{cases} u_{jk}^{\beta} \sin 2\pi(j\alpha_1 x_1 + k\alpha_2 x_2)\cos \pi x_3,\\ v_{jk}^{\beta} \sin 2\pi(j\alpha_1 x_1 + k\alpha_2 x_2)\cos \pi x_3,\\ \cos 2\pi(j\alpha_1 x_1 + k\alpha_2 x_2)\sin \pi x_3,\\ T_{jk}^{\beta} \cos 2\pi(j\alpha_1 x_1 + k\alpha_2 x_2)\sin \pi x_3,\end{cases} \tag{5.5.24}$$

where $\alpha_1 = (2\pi r_0)^{-1}$, $\alpha_2 = L^{-1}$, and

$$(u_{jk}^{\beta}, v_{jk}^{\beta}, T_{jk}^{\beta}) = \left(-\frac{\pi^2}{\alpha_{jk}^2}(j\alpha_1 - k\alpha_2\omega A_1(\beta)), -\frac{\pi^2}{\alpha_{jk}^2}(k\alpha_2 + j\alpha_1\omega A_1(\beta)), A_2(\beta)\right),$$

$$A_1(\beta) = \frac{1}{\Pr r_{jk}^2 + \Pr \delta_0 + \beta}, \qquad A_2(\beta) = \frac{1}{r_{jk}^2 + \beta}, (r_{jk}^2 = \alpha_{jk}^2 + \pi^2).$$

Physically, (4.5.14) requires that $\alpha_{jk}^2 = k^2\pi^2/L_0^2$, i.e., $(j,k) = (0,k)$. In each period there are two cells, and the size of every cell in the meridional direction is about $\pi r_0/6$. Thus, when the length scale L in (5.5.3) takes $\pi r_0/3$, for the first eigenvectors, the integer pair (j,k) takes $(j,k) = (0,1)$ in (5.5.22). Then we obtain the first eigenvalue of (5.5.13) at R_c as

$$\psi_1 = \begin{cases} L\omega \mathrm{Pr}^{-1}(r^2+\delta_0)^{-1}\sin\frac{2\pi x_2}{L}\cos\pi x_3, \\ -L\sin\frac{2\pi x_2}{L}\cos\pi x_3, \\ \cos\frac{2\pi x_3}{L}\sin\pi x_3, \\ r^{-2}\cos\frac{2\pi x_3}{L}\sin\pi x_3, \end{cases} \tag{5.5.25}$$

where $L = 2L_0 = \frac{\pi}{3}r_0$.

Dynamic Transitions. It is readily checked that Theorems 4.4.9 and 4.4.10 are still valid for (5.5.2)–(5.5.5) if we replace b_1 by the following parameter:

$$b = \mathrm{Pr}^4(r^2+\delta_0)^2 R_{c_1} - \pi^4\omega^2\gamma^4/\alpha^4, \tag{5.5.26}$$

where $\alpha^2 = \pi^2 L^{-2}$, $\gamma^2 = \alpha^2 + \pi^2$, and R_{c_1} is as in (4.5.20).

Theorem 5.5.1 *Let R_{c_1} and R_{c_2} be given by (4.5.20) and (4.5.24), and let b be a parameter as in (5.5.26). If $R_{c_1} < R_{c_2}$, then the following assertions hold:*

1. *When $b > 0$, the system (5.5.2)–(5.5.5) undergoes a continuous dynamic transition at $R = R_{c_1}$ and bifurcates to exactly two steady-state solutions ψ_R^+ and ψ_R^- for $R > R_{c_1}$, which are attractors and can be expressed as*

$$\psi_R^{\pm} = \pm C\beta_1^{1/2}(R)\psi_1 + o(|\beta_1|^{1/2}), \tag{5.5.27}$$

where ψ_1 is as in (5.5.25), $C > 0$ is a constant, and $\beta_1(R)$ is the first eigenvalue of (5.5.13).

2. *When $b < 0$, the system undergoes a jump transition, and has exactly two saddle-node bifurcation points (ψ_1^*, R^*) and (ψ_2^*, R^*) with $R^* < R_{c_1}$ and $\psi_i^* \neq 0$ $(i = 1,2)$.*

Also, Theorem 4.4.12 holds for the revised problem (5.5.2)–(5.5.5). The parameter b_2 is still given by (4.4.73), with the following modifications:

$$R_1 = \frac{\mathrm{Pr}\,\gamma^2 + \mathrm{Pr}\,\delta_0}{\mathrm{Pr}^2(\gamma^2+\delta_0)^2 + a^2}, \qquad I_1 = \frac{-a}{\mathrm{Pr}^2(\gamma^2+\delta_0)^2 + a^2},$$

$$\begin{aligned} a^2 = \frac{1}{\gamma^2}[&\pi^2\omega^2 - \alpha^2\mathrm{Pr}\,R_{c_2} + \alpha^2\mathrm{Pr}^2(r^2+\delta_0)(r^2+\delta_1) + \alpha^2\mathrm{Pr}\,\gamma^2(\gamma^2+\delta_1) \\ &+ \alpha^2\mathrm{Pr}\,r^2(r^2+\delta_0) + \pi^2\mathrm{Pr}^2(r^2+\delta_0)^2 + 2\pi^2\mathrm{Pr}\,r^2(r^2+\delta_0)], \end{aligned}$$

where $r^2 = \alpha^2 + \pi^2$ and α^2 is as in (4.5.24).

5.5.4 Physical Implications

With an idealized model, we derive some physical conclusions. First, the derived theoretical critical length scale and the critical temperature gradient of the large-scale meridional atmospheric circulation are in gross agreement with observations. The slight differences of these values are due essentially to highly idealization of the model.

Second, we set typical parameters in the revised problem (5.5.2)–(5.5.5) as

$$\delta_0 = 3.5\times 10^8, \qquad \delta_1 = 2.7\times 10^{20}, \qquad \mathrm{Pr} = 0.72.$$

By (4.5.21), (4.5.20), (4.5.24), and (5.5.26), we obtain

$$\alpha_c^2 \cong \pi^2\delta_0^{1/2}/\delta_1^{1/2} = 1.16\times 10^{-5}, \quad R_{c_1} \cong \pi^2\delta_1 = 2.7\times 10^{21},$$
$$R_{c_2} \cong \mathrm{Pr}^2\alpha_c^2\delta_0\delta_1 = 5.5\times 10^{23}, \qquad b \cong \mathrm{Pr}^4\delta_0^2 R_{c_1} - \pi^8\omega^2/\alpha_c^4 = 8.26\times 10^{37}.$$

Hence Theorem 5.5.1 indicates that the system (5.5.2)–(5.5.5) undergoes a continuous dynamic transition at $R = R_{c_1}$, and the polar, Ferrel, and Hadley cells are essentially depicted by two local attractors given by (5.5.27). In particular, the leading-order approximation of the velocity field $u = (u_1, u_2, u_3)$ governing the roll structure of the meridional circulation is expressed by

$$u \approx \left(\frac{L\omega}{\mathrm{Pr}(\gamma^2+\delta_0)}\sin\frac{2\pi x_2}{L}\cos\pi x_3, -L\sin\frac{2\pi x_2}{L}\cos\pi x_3, \cos\frac{2\pi x_3}{L}\sin\pi x_3\right).$$

The easterlies and westerlies are characterized by u_1, and the ratio between the zonal and meridional components is

$$\frac{u_1}{u_2} = -\frac{\omega}{\mathrm{Pr}\,(\delta_0 + r^2)} \cong -2.$$

Hence along a straight line $x_2 = \frac{1}{2}x_1 + C$ (C is a constant), three closed loops of u in a hemisphere are formed in the interval $kL < x_2 < (k+1)L$ for $k = 0, 1, 2$ and $L = \frac{\pi}{6}r_0$. They stand for the polar cell for $0 < x_2 < L$, the Ferrel cell for $L < x_2 < 2L$, and the Hadley cell for $2L < x_2 < 3L$; see Fig. 5.15. We remark here that the type

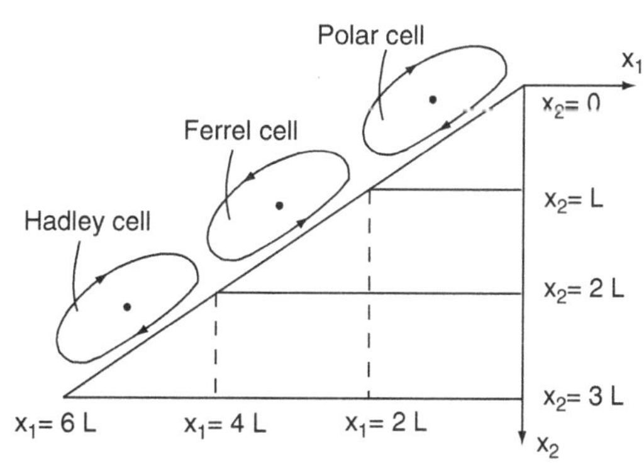

Fig. 5.15 The polar, Ferrel, and Hadley cells

of transitions derived in this section is independent of the added turbulent frictional terms, although these frictional terms are crucial for deriving proper critical convections of the large-scale circulation.

Finally, the study here can be used for the atmospheric circulation on planets other than Earth. In particular, it is easy to derive from the theorems in this section that if the radius r_0 of a planet is sufficiently large, then the system can undergo a dynamic transition, leading to spatiotemporal oscillations.

Chapter 6
Dynamical Transitions in Chemistry and Biology

Chemical reaction systems and biological models were among the first dissipative systems described by the Belgian school; see, e.g., (Prigogine and Lefever 1968; Glansdorff and Prigogine 1971). Among many other features, dissipative systems in nature demonstrate self-organized and self-assembled structures; see, for example, (Nicolis and Prigogine 1977; Kapral and Showalter 1995; Pismen 2006; Desai and Kapral 2009; Cross and Hohenberg 1993; Swinney et al. 1990; Murray 2002).

This chapter studies dynamic transitions of a few typical nonequilibrium chemical and biological models. The main focus is on Belousov–Zhabotinsky models (Oregonator and Brusselator), the chemotactic model, and a population model. As in the previous chapters, a crucial component of our study is based on the general principle, Principle 1, following Philosophy 2. Namely, we classify dynamic transitions as continuous, catastrophic, or random and determine their patterns by searching for a complete set of transition states, whereby the chemical and biological implications of our rigorous analysis become transparent.

In all these models, linear analysis provides the onset of the dynamic transition, and nonlinear analysis offers precise information on the types of transitions and the phase diagrams in terms of typical biological and chemical parameters as well as the spatial geometry of the domain. As we know, catastrophic (jump) and random (mixed) transitions are far-from-equilibrium states. Hence the study here provides a precise characterization of the conditions and mechanisms of such states. This has always been a difficult subject due to its complexity and global nature.

Pattern formation is an important topic in the study of nonequilibrium phase transitions. In this chapter, we focus mainly on dynamic transitions and touch only briefly on pattern formation. However, as mentioned earlier, our approach provides a new avenue for studying the formation and mechanisms of important patterns appearing in nature.

T. Ma and S. Wang, *Phase Transition Dynamics*, DOI 10.1007/978-1-4614-8963-4_6,

6.1 Modeling

6.1.1 Dynamical Equations of Chemical Reactions

Fick's Law of Diffusion. If u stands for the density or temperature, then by Fick's law, or Fourier's law, the flux of density or temperature is $J = -k\nabla u$, where k is the diffusion (or conduction) coefficient. The rate of change of density or temperature is then given by

$$\frac{\partial u}{\partial t} = -\text{div}\, J = k\Delta u. \tag{6.1.1}$$

Thus Δu represents the diffusion. The heterogeneous distribution of density and temperature in $\Omega \subset \mathbb{R}^3$ implies that there exists a potential energy in some generalized sense, and the Laplace operator ∇^2 plays the role of reducing potential energy by diffusion.

Reaction Process. We begin with an example. Consider the chemical reaction of synthetic ammonia as follows:

$$3\mathrm{H}_2 + \mathrm{N}_2 \underset{k_2}{\overset{k_1}{\rightleftharpoons}} 2\mathrm{NH}_3, \tag{6.1.2}$$

where H_2 is the hydrogen, N_2 is the nitrogen, NH_3 is the ammonia, and k_1 is the synthetic coefficient, k_2 the decomposition coefficient. For simplicity, we denote H_2 by X_1, N_2 by X_2, and NH_3 by X_3. Then the reaction equation (6.1.2) can be expressed in the abstract form

$$3\mathrm{X}_1 + \mathrm{X}_2 \xrightarrow{k_1} 2\mathrm{X}_3, \tag{6.1.3}$$

$$2\mathrm{X}_3 \xrightarrow{k_2} 3\mathrm{X}_1 + \mathrm{X}_2. \tag{6.1.4}$$

Let u_i be the concentrations of X_i $(1 \leq i \leq 3)$. For the synthetic reaction (6.1.3), the rate of change of u_1 is proportional to the probability of collision of one X_2 and three X_1, i.e.,

$$\frac{\delta_1 u_1(x)}{\delta t} = CP(3X_1 \bigcap X_2, x), \quad x \in \Omega \subset \mathbb{R}^3,$$

where $C \in \mathbb{R}^1$ is a constant, and $P(3X \bigcap X_2, x)$ represents the probability density of one X_1 and three X_2 at $x \in \Omega$. It is known that

$$P(3X_1 \bigcap X_2, x) = P^3(X_1, x) \cdot P(X_2, x),$$

where $P(X_i, x)$ is the probability density of X_i at x. It is clear that

$$P(X_i, x) = Cu_i(x), \quad C > 0 \text{ a constant}.$$

Hence we derive that

$$\frac{\delta_1 u_1(x)}{\delta t} = C u_1^3 u_2. \tag{6.1.5}$$

Since each synthetic reaction in (6.1.3) loses three X_1, the proportionality constant C in (6.1.5) is $C = -3k_1$. Thus, we get

$$\frac{\delta_1 u_1(x)}{\delta t} = -3k_1 u_1^3 u_2. \tag{6.1.6}$$

Equations (6.1.5) and (6.1.6) are also called the laws of mass action.

For the decomposition reaction (6.1.4), the rate of change of u_1 is

$$\frac{\delta_2 u_1(x)}{\delta t} = 3k_2 u_3^2. \tag{6.1.7}$$

Taking the diffusion effect into account, see (6.1.1), the total rate of change of u_1 in (6.1.3) and (6.1.4) is given by

$$\frac{\partial u_1}{\partial t} = \mu_1 \Delta u_1 + \frac{\delta_1 u_1}{\delta t} + \frac{\delta_2 u_1}{\delta t}.$$

Therefore, it follows from (6.1.6) and (6.1.7) that

$$\frac{\partial u_1}{\partial t} = \mu_1 \Delta u_1 - 3k_1 u_1^3 u_2 + 3k_2 u_3^2. \tag{6.1.8}$$

In the same fashion, we can derive the following rate equations of u_2 and u_3:

$$\frac{\partial u_2}{\partial t} = \mu_2 \Delta u_2 - k_1 u_1^3 u_2 + k_2 u_3^2, \tag{6.1.9}$$

$$\frac{\partial u_3}{\partial t} = \mu_3 \Delta u_3 + 2k_1 u_1^3 u_3 - 2k_2 u_3^2. \tag{6.1.10}$$

Equations (6.1.8)–(6.1.10) describe the dynamical properties of the synthetic reaction and decomposition of ammonia.

General Case. Let $X_1, \cdots, X_n$ represent n chemical reaction species, and suppose that there are r reactions:

$$\begin{aligned} &\alpha_{11}\mathrm{X}_1 + \cdots + \alpha_{1n}\mathrm{X}_n \xrightarrow{k_1} \beta_{11}\mathrm{X}_1 + \cdots + \beta_{1n}\mathrm{X}_n, \\ &\qquad \vdots \\ &\alpha_{r1}\mathrm{X}_1 + \cdots + \alpha_{rn}\mathrm{X}_n \xrightarrow{k_r} \beta_{r1}\mathrm{X}_1 + \cdots + \beta_{rn}\mathrm{X}_n, \end{aligned} \tag{6.1.11}$$

where $\alpha_{ij}, \beta_{ij} \geq 0$ are the stoichiometric factors, and k_i $(1 \leq i \leq r)$ are the reaction constants. Let u_j represent the concentration of X_j $(1 \leq j \leq n)$. Then we derive from (6.1.11) the following chemical reaction–diffusion model:

$$\begin{aligned}\frac{\partial u_1}{\partial t} &= \mu_1 \Delta u_1 + \sum_{i=1}^{r} (\beta_{i1} - \alpha_{i1}) k_i u_1^{\alpha_{i1}} \cdots u_n^{\alpha_{in}}, \\ &\vdots \\ \frac{\partial u_n}{\partial t} &= \mu_n \Delta u_n + \sum_{i=1}^{r} (\beta_{in} - \alpha_{in}) k_n u_1^{\alpha_{i1}} \cdots u_n^{\alpha_{in}},\end{aligned} \tag{6.1.12}$$

where the μ_j $(1 \leq j \leq n)$ are the diffusive coefficients.

Equations (6.1.12) constitute the model governing the chemical reaction (6.1.11); they are supplemented with the Neumann boundary condition if there is no exchange of materials on the boundary, or the Dirichlet boundary condition if the system permits the flux of X_j $(1 \leq j \leq n)$ to pass through the boundary $\partial\Omega$.

When the reaction system (6.1.11) is stirred in Ω, then (6.1.12) reduces to a system of ordinary differential equations,

$$\frac{du_j}{dt} = \sum_{i=1}^{r} (\beta_{ij} - \alpha_{ij}) k_i u_1^{\alpha_{i1}} \cdots u_n^{\alpha_{in}} \qquad \text{for } 1 \leq j \leq n. \tag{6.1.13}$$

We have to impose a natural condition for (6.1.12) and (6.1.13) as follows:

$$u \geq 0. \tag{6.1.14}$$

6.1.2 Population Models of Biological Species

Models with No Environmental Effect. We consider the population problem of m species. Let $X_1, \cdots, X_m$ represent m species that coexist in a spatial domain $\Omega \subset \mathbb{R}^2$, and let u_i denote the population density of X_i $(1 \leq i \leq m)$. When we ignore the effect of the environment, the unknown functions u_i satisfy the following ordinary differential equations:

$$\frac{du_i}{dt} = r_i u_i + \sum_{j=1}^{m} a_{ij} u_i u_j + g_i(u) \qquad \forall 1 \leq i \leq m, \tag{6.1.15}$$

where $r_i > 0$ is the natural growth rate of X_i, the constant matrix

$$A = \begin{pmatrix} a_{11} & \cdots & a_{1m} \\ \vdots & & \vdots \\ a_{m1} & \cdots & a_{mm} \end{pmatrix} \tag{6.1.16}$$

is called the interaction matrix of the species, and

$$g(u) = o(|u|^2). \tag{6.1.17}$$

Environmental Effect. The environment effect on the population density is given by $\mu\Delta u$. Equation (6.1.15) should be revised as follows:

$$\frac{\partial u_i}{\partial t}=\mu_i\Delta u_i+\sum_{j=1}^{m}b_{ij}(x)u_j+r_iu_i+\sum_{j=1}^{n}a_{ij}u_iu_j+g_i(u), \tag{6.1.18}$$

where the $\mu_i>0$ are the environment constants, and the function matrix

$$B(x)=\begin{pmatrix} b_{11}(x) & \cdots & b_{1m}(x) \\ \vdots & & \vdots \\ b_{m1}(x) & \cdots & b_{mm}(x) \end{pmatrix} \tag{6.1.19}$$

is called the environmental factor.

As usual, equations (6.1.18) are supplemented with either the Dirichlet or the Neumann boundary condition. The Dirichlet boundary condition implies that the environment conditions near the boundary $\partial\Omega$ are not suitable for the species to exist. For example, many animals cannot live in a desert or near the North and South Poles. The Neumann boundary condition describes the species living in an aqueous area.

For (6.1.15) and (6.1.18), the nonnegative condition (6.1.14) is also necessary. Hence, the environmental factors (6.1.19) are required to satisfy that the first eigenfunctions of the equations

$$\begin{aligned} &\mu\cdot\Delta e+B(x)e+re=\lambda e,\\ &e|_{\partial\Omega}=0 \quad (\text{or } \frac{\partial e}{\partial n}|_{\partial\Omega}=0), \end{aligned} \tag{6.1.20}$$

are nonnegative in Ω, i.e.,

$$e_i(x)\geq 0,\quad \forall x\in\Omega,\quad e=(e_1,\cdots,e_m). \tag{6.1.21}$$

The eigenfunctions $e_i(x)$ essentially determine the distribution of population density u_i. This is why we need the term Bu in (6.1.18), which characterizes the interaction between the species X_i $(1\leq i\leq m)$ and the environment.

The biological significance of the interaction matrix (6.1.16) is as follows:

$a_{ij}\cdot a_{ji}<0$	the predator–prey relation between X_i and X_j;
$a_{ij}>0, a_{ji}>0$	the mutually beneficial relation between X_i and X_j;
$a_{ij}<0, a_{ji}<0$	the competition relation between X_i and X_j;
$a_{ij}=a_{ji}=0$	X_i and X_j are independent of each other;
$a_{ij}=0, a_{ji}<0$	X_j is independent of X_i, but X_i is harmful to X_j; and
$a_{ij}=0, a_{ji}>0$	X_j is independent of X_i, but X_i is beneficial to X_j,

where $i \neq j$. Based on biological rules, a species with a large population density $u_i \gg 1$ is harmful to itself. Therefore we assume that

$$a_{ii} \geq 0 \quad \text{and} \quad g_i(u)u_i < 0 \qquad \forall 1 \leq i \leq m. \tag{6.1.22}$$

In general, the environmental factor B is a diagonal matrix. However, here we do not exclude the possibility that $b_{ij} \neq 0$ for $i \neq j$. The conditions (6.1.14), (6.1.21), and (6.1.22) are basic for the coexistence models (6.1.18) of m species.

6.2 Belousov–Zhabotinsky Chemical Reactions: Oregonator

In the 1950s, B.P. Belousov discovered experimentally a spatiotemporal oscillation phenomenon in the concentrations of intermediaries when citric acid was oxidized by acid bromate in the presence of a cerium ion catalyst (Belousov 1959). It was also observed by Zhabotinski 1964 that organic acids and metal ions could be used as well in the reaction, leading to spatiotemporal oscillations. It is considered nowadays that all chemical reactions give rise to oscillations, and the actions of a catalyst are called Belousov–Zhabotinsky (BZ) reactions. BZ reactions are now a classical example of nonequilibrium thermodynamics.

The Brusselator was introduced by Prigogine and Lefever 1968. Some numerical and theoretical analyses were made by Glansdorff and Prigogine 1971; Nicolis and Prigogine 1977, and a linear stability analysis was given by Reichl 1998. The existence of periodic solutions for the Oregonator was first proved by Hastings and Murray 1975. In this section, Theorem 6.2.1 was proved in Ma and Wang 2005b, 2007b, and the other results are from Ma and Wang 2011c,d.

6.2.1 The Field–Kőrös–Noyes Equations

The Model. In this section, we study the dynamic phase transitions associated with the spatiotemporal oscillations of the BZ reactions given by Field et al. 1972. This BZ reaction consists of the following five irreversible steps:

$$\begin{aligned}
&\mathrm{A}+\mathrm{Y} \xrightarrow{k_1} \mathrm{X},\\
&\mathrm{X}+\mathrm{Y} \xrightarrow{k_2} \mathrm{P},\\
&\mathrm{B}+\mathrm{X} \xrightarrow{k_3} 2\mathrm{X}+\mathrm{Z},\\
&2\mathrm{X} \xrightarrow{k_4} \mathrm{Q},\\
&\mathrm{Z} \xrightarrow{k_5} \gamma Y,
\end{aligned} \tag{6.2.1}$$

where γ is a stoichiometric factor, P and Q are products that do not join the reaction again, and $X = HBrO_2$, $Y = Br^-$, $Z = Ce^{4+}$, $A = B = BrO_3^-$. It is the temporal oscillation of the curium ion ratio Ce(IV)/Ce(III) that, with a suitable indicator, is displayed by a color change when the reagent is stirred; see among others (Field and Noyes 1974), where a system of ordinary differential equations for this reaction was derived.

Here we consider the general cases. Let u_1,u_2,u_3,a,b represent the concentrations of X,Y,Z,A, and B, with u_i treated as variable functions, and with a and b treated as constants. Then the equations describing the chemical reaction process (6.2.1) are given by

$$\begin{aligned}\frac{\partial u_1}{\partial t} &= \sigma_1 \Delta u_1 + k_1 a u_2 - k_2 u_1 u_2 + k_3 b u_1 - 2k_4 u_1^2,\\ \frac{\partial u_2}{\partial t} &= \sigma_2 \Delta u_2 - k_1 a u_2 - k_2 u_1 u_2 + \gamma k_5 u_3,\\ \frac{\partial u_3}{\partial t} &= \sigma_3 \Delta u_3 + k_3 b u_1 - k_5 u_3,\end{aligned} \tag{6.2.2}$$

where σ_i $(i = 1,2,3)$ are the diffusivities of u_i, and k_j $(1 \le j \le 5)$ are the reaction coefficients as in (6.2.1). The model (6.2.2) is also called the Oregonator.

The coefficients σ_i and k_j are functions of the temperature T. In fact, $k_j = k_j^0 e^{-E_i/RT}$, E_i is the activation energy, and R is the Boltzmann constant.

The dimensions of the relevant quantities are given by

$$\begin{aligned}&k_j : M^{-1}t^{-1} \text{ for } 1 \le j \le 4, && k_5 : t^{-1},\\ &a,b,u_i : M \text{ for } 1 \le i \le 3, && \sigma_i : l^2 t^{-1} \text{ for } 1 \le i \le 3,\end{aligned}$$

where t is the time, M is the molar density, and l is the length. Then we introduce the following nondimensional variables:

$$\begin{aligned}&u_1 = \frac{k_1 a}{k_2} u_1', && u_2 = \frac{k_3 b}{k_2} u_2', && u_3 = \frac{k_1 k_3}{k_2 k_5} a b u_3',\\ &t = (k_1 k_3 a b)^{-1/2} t', && x = L x', && \alpha = \left(\frac{k_3 b}{k_1 a}\right)^{1/2},\\ &\beta = \frac{2k_1 k_4 a}{k_2 k_3 b}, && \delta = k_5 (k_1 k_3 a b)^{1/2}, && \mu_i = \frac{\sigma_i}{l^2 (k_1 k_2 a b)^{1/2}}.\end{aligned}$$

Omitting the primes, we obtain the following nondimensional form of (6.2.2):

$$\begin{aligned}\frac{\partial u_1}{\partial t} &= \mu_1 \Delta u_1 + \alpha(u_1 + u_2 - u_1 u_2 - \beta u_1^2),\\ \frac{\partial u_2}{\partial t} &= \mu_2 \Delta u_2 + \frac{1}{\alpha}(\gamma u_3 - u_2 - u_1 u_2),\\ \frac{\partial u_3}{\partial t} &= \mu_3 \Delta u_3 + \delta(u_1 - u_3),\end{aligned} \tag{6.2.3}$$

where the unknown functions are $u_i \geq 0$ $(1 \leq i \leq 3)$, and the parameters are positive constants:

$$\mu_1,\ \mu_2,\ \mu_3,\ \alpha,\ \beta,\ \gamma, \delta > 0.$$

Hence the control parameter is given by

$$\lambda = (\mu_1, \mu_2, \mu_3, \alpha, \gamma, \delta) \in \mathbb{R}^6_+.$$

We consider the case in which the spatial domain is the bounded domain $\Omega \subset \mathbb{R}^n$. When there is an exchange of materials on the boundary $\partial\Omega$ to maintain the level of concentrations of X, Y, Z, then equations (6.2.3) are supplemented with the Dirichlet boundary condition

$$u = (u_1, u_2, u_3) = 0 \quad \text{on } \partial\Omega. \tag{6.2.4}$$

If there is no exchange of materials on the boundary, the Neumann boundary condition is used:

$$\frac{\partial u}{\partial n} = 0 \quad \text{on } \partial\Omega. \tag{6.2.5}$$

Invariant Regions and Mathematical Setup. It is shown in Smoller 1983 that the region

$$D = \{(u_1, u_2, u_3) \in L^2(\Omega)^3 \mid 0 < u_i < a_i,\ 1 \leq i \leq 3\} \tag{6.2.6}$$

is invariant for (6.2.3)–(6.2.4), where a_i $(1 \leq i \leq 3)$ satisfy

$$a_1 > \max\{1, \beta^{-1}\}, \quad a_2 > \gamma a_3, \quad a_3 > a_1.$$

Since the solution $u = (u_1, u_2, u_3)$ of (6.2.3) represents concentrations of chemical materials, only nonnegative functions $u_i \geq 0$ $(1 \leq i \leq 3)$ are chemically realistic.

We define the following function spaces:

$$\begin{aligned} &H = L^2(\Omega)^3, \\ &H_1 = \begin{cases} H^2(\Omega)^3 \bigcap H^1_0(\Omega)^3 & \text{for boundary condition (6.2.4)}, \\ \left\{ u \in H^2(\Omega)^3 \mid \dfrac{\partial u}{\partial n} = 0 \text{ on } \partial\Omega \right\} & \text{for boundary condition (6.2.5)}. \end{cases} \end{aligned}$$

Define the operators $L_\lambda = A_\lambda + B_\lambda$ and $G_\lambda : H_1 \to H$ by

$$\begin{aligned} A_\lambda u &= (\mu_1 \Delta u_1, \mu_2 \Delta u_2, \mu_3 \Delta u_3), \\ B_\lambda u &= \left(\alpha u_1 + \alpha u_2, -\frac{1}{\alpha} u_2 + \frac{\gamma}{\alpha} u_3, \delta u_1 - \delta u_3 \right), \\ G(u, \lambda) &= \left(-\alpha(u_1 u_2 + \beta u_1^2), -\frac{1}{\alpha} u_1 u_2, 0 \right). \end{aligned}$$

Thus the Field–Kőrös–Noyes equations (6.2.3), with either (6.2.4) or (6.2.5), take the following operator form:

$$\frac{du}{dt} = L_\lambda u + G(u,\lambda), \qquad u(0) = \varphi. \tag{6.2.7}$$

6.2.2 Transition Under the Dirichlet Boundary Condition

We consider the dynamic transitions of (6.2.3)–(6.2.4). Let ρ_k and e_k be the kth eigenvalue and eigenvector of the Laplacian $-\Delta$ with the Dirichlet boundary condition

$$-\Delta e_k = \rho_k e_k, \quad e_k|_{\partial\Omega} = 0, \qquad \text{for } k = 1,2,\cdots, \tag{6.2.8}$$

and the first eigenvector is simple and positive:

$$e_1(x) > 0 \quad \forall x \in \Omega. \tag{6.2.9}$$

Then eigenvalues of L_λ are associated with eigenvalues of $E_k(\lambda)u = \beta_k(\lambda)u$, where

$$E_k(\lambda) = \begin{pmatrix} -(\mu_1\rho_k - \alpha) & \alpha & 0 \\ 0 & -(\mu_2\rho_k + \alpha^{-1}) & \alpha^{-1}\gamma \\ \delta & 0 & -(\mu_3\rho_k + \delta) \end{pmatrix}. \tag{6.2.10}$$

Let $N(\lambda)$ be a function of $\lambda = (\mu_1, \mu_2, \mu_3, \alpha, \gamma, \delta) \in \mathbb{R}^6_+$ defined by

$$N(\lambda) = \alpha \det E_1(\lambda) = \alpha\gamma\delta - (\mu_1\rho_1 - \alpha)(\mu_2\alpha\rho_1 + 1)(\mu_3\rho_1 + \delta).$$

Let $\lambda_0 = (\mu_1^0, \mu_2^0, \mu_3^0, \alpha^0, \gamma^0, \delta^0) \in \mathbb{R}^6_+$ be a solution of

$$N(\lambda_0) = 0. \tag{6.2.11}$$

Then the set $\Gamma \subset \mathbb{R}^6_+$ defined by

$$\Gamma = \{\lambda_0 \in \mathbb{R}^6_+ \mid \lambda_0 \text{ satisfies } (6.2.11)\}$$

is a 5-dimensional manifold that separates $\mathbb{R}^6_+$ into two connected components U_1 and U_2 such that

$$N(\lambda) \begin{cases} < 0 & \text{if } \lambda \in U_1, \\ > 0 & \text{if } \lambda \in U_2. \end{cases}$$

Then we have the following dynamic transition theorem for (6.2.3)–(6.2.4), which characterizes the dynamic phase transition to a steady-state solution for the Belousov–Zhabotinsky reaction (6.2.1).

Theorem 6.2.1 *For any $\lambda_0 = (\mu_1^0, \mu_2^0, \mu_3^0, \alpha^0, \gamma^0, \delta^0) \in \Gamma$, the system (6.2.3)–(6.2.4) undergoes a Type I dynamic transition at $(u,\lambda) = (0,\lambda_0)$, and the following assertions hold:*

1. *If $\lambda \in \mathbb{R}_+^6$ with $\lambda \neq \lambda_0$ and $N(\lambda) < N(\lambda_0) = 0$, then $u = 0$ is a stable equilibrium.*
2. *If $N(\lambda) > N(\lambda_0) = 0$, then the problem (6.2.3)–(6.2.4) bifurcates from $(u,\lambda) = (0,\lambda_0)$ to a unique steady-state solution $u_\lambda \in H_1$ with $u_\lambda > 0$ in Ω, given by*

$$\begin{aligned} u_\lambda(x) &= \beta_1(\lambda) K_1 K_2^{-1} v_0(x) + o(|\beta_1(\lambda)|), \\ v_0(x) &= \left(\alpha^0(\delta^0 + \mu_3^0 \rho_1) e_1, \frac{\alpha^0 \gamma^0 \delta^0}{1 + \mu_2^0 \alpha^0 \rho_1} e_1, \alpha^0 \delta^0 e_1 \right), \end{aligned} \tag{6.2.12}$$

where $\beta_1(\lambda)$ is the positive eigenvalue of (6.2.10), and $K_1, K_2 > 0$ are defined by

$$\begin{aligned} K_1 &= \left[\alpha^0 \delta^0 (1 + \mu_2^0 \alpha^0 \rho_1)(\mu_3^0 \rho_1 + \delta^0) + \frac{(\alpha^0)^3 (\delta^0)^2 \gamma^0}{1 + \mu_2^0 \alpha^0 \rho_1} + \frac{(\alpha^0)^2 \gamma^0 (\delta^0)^2}{\delta^0 + \mu_3^0 \rho_1} \right] \cdot \|e_1\|^2, \\ K_2 &= \delta^0 (\alpha^0)^2 (\mu_3^0 \rho_1 + \delta^0) \Big[\beta(\mu_3^0 \rho_1 + \delta^0)(\mu_2^0 \alpha^0 \rho_1 + 1) \\ &\quad + \alpha^0 \delta^0 \gamma^0 + \frac{\alpha^0 \gamma^0 \delta^0}{1 + \mu_2^0 \alpha^0 \rho_1} \Big] \cdot \int_\Omega e_1^3 dx. \end{aligned}$$

In addition, u_λ is an attractor. Namely, there exists an open set $\mathcal{O} \subset H$ such that $0 \in \mathcal{O}$ and u_λ attracts $\mathcal{O} \bigcap D$, where D is the invariant set given by (6.2.6), i.e., for any $\varphi \in \mathcal{O} \bigcap D$, the solution $u(t,\varphi)$ of (6.2.7) satisfies

$$\lim_{t \to \infty} \|u(t,\varphi) - u_\lambda\| = 0.$$

Proof. We use Theorem 2.3.1 to prove this theorem, and proceed in two steps.

Step 1. Verification of the principle of exchange of stabilities (PES). It is well known that the operators L_λ and G in (6.2.7) satisfy (2.1.2) and (2.1.3). We now verify the PES: (1.2.4) and (1.2.5).

All eigenvalues $\beta_{jk}(\lambda)$ of L_λ $(1 \leq j \leq 3, k = 1, 2, \cdots)$ satisfy

$$\det(\beta_{jk} \mathrm{I} - E_k(\lambda)) = 0. \tag{6.2.13}$$

Observe that

$$\operatorname{grad} N(\lambda) = \left(\frac{\partial N}{\partial \mu_1}, \frac{\partial N}{\partial \mu_2}, \frac{\partial N}{\partial \mu_3}, \frac{\partial N}{\partial \alpha}, \frac{\partial N}{\partial \gamma}, \frac{\partial N}{\partial \delta} \right) \neq 0 \quad \forall \lambda \in \Gamma,$$

and Γ can be expressed as zeros of the following equation:

$$\gamma = \frac{(\mu_1 \rho_1 - \alpha)(\mu_2 \alpha \rho_1 + 1)(\mu_3 \rho_1 + \delta)}{\alpha \delta},$$

where $\alpha, \delta, \mu_1, \mu_2, \mu_3 > 0$. Hence, it follows that $\Gamma \subset \mathbb{R}^6_+$ is a 5-dimensional manifold that divides $\mathbb{R}^6_+$ into two disjoint open sets:

$$\mathbb{R}^6_+ = \overline{U}_1 + \overline{U}_2, \quad U_1 \bigcap U_2 = \emptyset, \quad \partial U_1 \bigcap \partial U_2 = \Gamma,$$

such that

$$\det E_1(\lambda) = \beta_{11}(\lambda)\beta_{21}(\lambda)\beta_{31}(\lambda) \begin{cases} < 0 & \forall \lambda \in U_1, \\ > 0 & \forall \lambda \in U_2. \end{cases} \tag{6.2.14}$$

On the other hand, the trace of $E_1(\lambda)$ is given by

$$\sum_{j=1}^{3} \operatorname{Re}\beta_{j1}(\lambda) = -\left[(\mu_1\rho_1 - \alpha) + (\mu_2\rho_1 + \alpha^{-1}) + (\mu_3\rho_1 + \delta)\right]. \tag{6.2.15}$$

By (6.2.14) we deduce that

$$\mu_1\rho_1 - \alpha > 0, \quad \forall \lambda = (\mu_1, \mu_2, \mu_3, \alpha, \gamma, \delta) \in \overline{U}_1. \tag{6.2.16}$$

It follows from (6.2.14)–(6.2.16) that

$$\sum_{j=1}^{3} \operatorname{Re}\beta_{j1}(\lambda) < 0, \quad \forall \lambda \in \overline{U}_1. \tag{6.2.17}$$

Then by (6.2.14) and (6.2.17), we obtain

$$\begin{aligned} &\beta_{11}(\lambda) = 0, \ \operatorname{Re}\beta_{21}(\lambda) < 0, \ \operatorname{Re}\beta_{31}(\lambda) < 0 && \text{for } \lambda \in \Gamma, \\ &\beta_{11}(\lambda) < 0, \ \operatorname{Re}\beta_{21}(\lambda) < 0, \ \operatorname{Re}\beta_{313}(\lambda) < 0 && \text{for } \lambda \in U_1, \\ &\beta_{11}(\lambda) > 0 && \text{for } \lambda \in U_2. \end{aligned} \tag{6.2.18}$$

Therefore, (1.2.4) follows from (6.2.18). Now we need to prove that

$$\operatorname{Re}\beta_{jk}(\lambda_0) < 0 \qquad \forall \lambda_0 \in \Gamma, \quad 1 \leq j \leq 3, \ k \geq 2. \tag{6.2.19}$$

By (6.2.16), we deduce that for $k \geq 2$,

$$\begin{aligned} \det E_k(\lambda_0) &= \frac{1}{\alpha^0}\left[\alpha^0\gamma^0\delta^0 - (\mu_1^0\rho_k - \alpha^0)(\mu_2^0\rho_k\alpha^0 + 1)(\mu_3^0\rho_k + \delta^0)\right] \\ &< \frac{1}{\alpha^0}\left[\alpha^0\gamma^0\delta^0 - (\mu_1^0\rho_1 - \alpha^0)(\mu_2^0\rho_1\alpha^0 + 1)(\mu_3^0\rho_1 + \delta^0)\right]. \end{aligned}$$

Namely,

$$\det E_k(\lambda_0) < \det E_1(\lambda_0) = 0 \qquad \forall k \geq 2, \ \lambda_0 \in \Gamma. \tag{6.2.20}$$

In addition, it is easy to see that

$$E_k(\lambda_0) = E_1(\widetilde{\lambda}), \quad \widetilde{\lambda} = (\mu_1^0\rho_k\rho_1^{-1}, \mu_2^0\rho_k\rho_1^{-1}, \mu_3^0\rho_k\rho_1^{-1}, \alpha^0, \gamma^0, \delta^0). \tag{6.2.21}$$

Hence, we have

$$\beta_{jk}(\lambda_0) = \beta_{j1}(\widetilde{\lambda}), \quad \forall k \geq 1,\ j = 1,2,3. \tag{6.2.22}$$

Thus from (6.2.20), (6.2.21), and (6.2.14) we obtain $\widetilde{\lambda}_1 \in U_1$. Then (6.2.19) follows from (6.2.18) and (6.2.22).

Step 2. By Theorem 2.3.1, we now verify

$$\frac{(G(v_0,\lambda_0),v_0^*)}{(v_0,v_0^*)} < 0, \tag{6.2.23}$$

where v_0 and v_0^* are the eigenvectors of L_{λ_0} and $L_{\lambda_0}^*$ corresponding to $\beta_{11}(\lambda_0) = 0$. Let

$$v_0 = (x_1e_1, x_2e_1, x_3e_1), \qquad v_0^* = (x_1^*e_1, x_2^*e_1, x_3^*e_1),$$

where e_1 satisfies (6.2.9). Then (x_1,x_2,x_3) is the eigenvector of the matrix $E_1(\lambda_0)$:

$$\begin{pmatrix} -(\mu_1^0\rho_1 - \alpha^0) & \alpha^0 & 0 \\ 0 & -(\mu_2^0\rho_1 + \frac{1}{\alpha^0}) & \gamma^0/\alpha^0 \\ \delta^0 & 0 & -(\mu_3^0\rho_1 + \delta^0) \end{pmatrix} \begin{pmatrix} x_1 \\ x_2 \\ x_3 \end{pmatrix} = 0,$$

and (x_1,x_2,x_3) is expressed as

$$(x_1,x_2,x_3) = \left(\alpha^0(\delta^0 + \mu_3^0\rho_1),\ \frac{\alpha^0\gamma^0\delta^0}{1+\mu_2^0\alpha^0\rho_1},\ \alpha^0\delta^0\right).$$

Hence, the eigenvector $v_0(x)$ is as given in (6.2.12).

The conjugate eigenvector (x_1^*,x_2^*,x_3^*) satisfies

$$\begin{pmatrix} -(\mu_1^0\rho_1 - \alpha^0) & 0 & \delta^0 \\ \alpha^0 & -\left(\mu_2^0\rho_1 + \frac{1}{\alpha^0}\right) & 0 \\ 0 & \gamma^0/\alpha^0 & -(\mu_3^0\rho_1 + \delta^0) \end{pmatrix} \begin{pmatrix} x_1^* \\ x_2^* \\ x_3^* \end{pmatrix} = 0.$$

Hence we have

$$v_0^*(x) = \left(\delta^0(1+\mu_2^0\alpha^0\rho_1)e_1,\ \delta^0(\alpha^0)^2e_1,\ \frac{\alpha^0\gamma^0\delta^0}{\delta^0+\mu_3^0\rho_1}e_1\right).$$

Then inserting v_0 and v_0^* into (6.2.23), we deduce that

$$K_2 = -(G(v_0,\lambda_0),v_0^*) = \int_\Omega [\alpha^0(x_1x_2 + \beta x_1^2)x_1^* + x_1x_2x_2^*/\alpha^0]e_1^3dx,$$

$$K_1 = (v_0,v_0^*) = (x_1x_1^* + x_2x_2^* + x_3x_3^*)\int_\Omega e_1^2dx.$$

Then with simple substitutions, the theorem follows. □

6.2.3 Transitions Under the Neumann Boundary Condition

Basic States. We now consider the Neumann boundary condition (6.2.5). Now the system (6.2.3) admits two constant steady states:

$$\begin{aligned} &U_0 = (0,0,0), \qquad U_1 = (u_1^0, u_2^0, u_3^0), \\ &u_1^0 = u_3^0 = \sigma, \qquad u_2^0 = \frac{\gamma\sigma}{1+\sigma} = \frac{1}{2}(1+\gamma-\beta\sigma), \\ &\sigma = \frac{1}{2\beta}\left[(1-\gamma-\beta) + \sqrt{(1-\gamma-\beta)^2 + 4\beta(1+\gamma)}\right]. \end{aligned} \tag{6.2.24}$$

It is easy to check that the steady-state solution $U_0 = 0$ is always unstable, because the linearized equations of (6.2.3) have three real eigenvalues λ_1, λ_2, and λ_3 satisfying

$$\lambda_1\lambda_2\lambda_3 = \det\begin{pmatrix} \alpha & \alpha & 0 \\ 0 & -\frac{1}{\alpha} & \frac{\gamma}{\alpha} \\ \delta & 0 & -\delta \end{pmatrix} = \delta(\gamma+1) > 0,$$

and there is at least one eigenvalue with positive real part. Therefore, we need to consider the transition of (6.2.3) with (6.2.5) only at the steady-state solution U_1.

Consider the deviation

$$u_i = u_i' + u_i^0 \quad (1 \le i \le 3). \tag{6.2.25}$$

Omitting the primes, the problem (6.2.3) with (6.2.5) becomes

$$\begin{aligned} &\frac{\partial u_1}{\partial t} = \mu_1\Delta u_1 + \alpha[(1-u_2^0-2\beta u_1^0)u_1 + (1-u_1^0)u_2 - u_1u_2 - \beta u_1^2], \\ &\frac{\partial u_2}{\partial t} = \mu_2\Delta u_2 + \frac{1}{\alpha}\left[-u_2^0u_1 - (1+u_1^0)u_2 + \gamma u_3 - u_1u_2\right], \\ &\frac{\partial u_3}{\partial t} = \mu_3\Delta u_3 + \delta(u_1-u_3), \\ &\frac{\partial u}{\partial n}|_{\partial\Omega} = 0. \end{aligned} \tag{6.2.26}$$

The invariant region for the Neumann boundary condition case is the same as for the Dirichlet boundary condition. The following lemma is well known (Smoller 1983; Temam 1997).

Lemma 6.2.2 *The region D given by (6.2.6) is also invariant for the problem (6.2.3) with (6.2.5). In particular, this problem possesses a global attractor in D.*

Linearized Problem. In view of (6.2.24), the eigenvalue equations of (6.2.26) read

$$\begin{aligned}
&\mu_1\Delta u_1-\alpha\left[\frac{1}{2}(\gamma+3\beta\sigma-1)u_1+(\sigma-1)u_2\right]=\lambda u_1,\\
&\mu_2\Delta u_2-\frac{1}{\alpha}\left[\frac{1}{2}(1+\gamma-\beta\sigma)u_1+(\sigma+1)u_2-\gamma u_3\right]=\lambda u_2,\\
&\mu_3\Delta u_3+\delta(u_1-u_3)=\lambda u_3,\\
&\frac{\partial u}{\partial n}|_{\partial\Omega}=0.
\end{aligned} \tag{6.2.27}$$

For brevity, we still use $\rho_k\geq 0$ and e_k to denote the kth eigenvalue and eigenvector of the Laplace operator Δ with the Neumann boundary condition:

$$\Delta e_k=-\rho_k e_k,\quad \frac{\partial e_k}{\partial n}|_{\partial\Omega}=0. \tag{6.2.28}$$

Let M_k be the following matrix:

$$M_k=\begin{pmatrix} -\mu_1\rho_k-\frac{\alpha}{2}(\gamma+3\beta\sigma-1) & -\alpha(\sigma-1) & 0\\ -\frac{1}{2\alpha}(1+\gamma-\beta\sigma) & -\mu_2\rho_k-\frac{1}{\alpha}(\sigma+1) & \frac{\gamma}{\alpha}\\ \delta & 0 & -\mu_3\rho_k-\delta \end{pmatrix}. \tag{6.2.29}$$

Thus all eigenvalues $\lambda=\beta_{kj}$ of (6.2.27) satisfy

$$M_k x_{kj}=\beta_{kj}x_{kj},\quad 1\leq j\leq 3,\ k=1,2,\cdots,$$

where $x_{kj}\in\mathbb{R}^3$ is the eigenvector of M_k corresponding to β_{kj}. Hence, the eigenvector u_{kj} of (6.2.27) corresponding to β_{kj} is

$$u_{kj}(x)=x_{kj}e_k(x), \tag{6.2.30}$$

where $e_k(x)$ is as in (6.2.28). In particular, $\rho_1=0$ and e_1 is a constant, and

$$M_1=\begin{pmatrix} -\frac{\alpha}{2}(\gamma+3\beta\sigma-1) & -\alpha(\sigma-1) & 0\\ -\frac{1}{2\alpha}(1+\gamma-\beta\sigma) & -\frac{1}{\alpha}(\sigma+1) & \frac{\gamma}{\alpha}\\ \delta & 0 & -\delta \end{pmatrix}. \tag{6.2.31}$$

The eigenvalues $\lambda=\beta_{1j}$ $(1\leq j\leq 3)$ of M_1 satisfy

$$\begin{aligned}
&\lambda^3+A\lambda^2+B\lambda+C=0,\\
&A=\delta+\left[\frac{\alpha}{2}(3\beta\sigma+\gamma-1)+\frac{1}{\alpha}(\sigma+1)\right],\\
&B=2\beta\sigma^2+\gamma-(1-\beta)\sigma+\delta\left[\frac{\alpha}{2}(3\beta\sigma+\gamma-1)+\frac{1}{\alpha}(\sigma+1)\right],\\
&C=\delta\sigma(2\beta\sigma+\beta+\gamma-1).
\end{aligned} \tag{6.2.32}$$

It is known that all solutions of (6.2.32) have negative real parts if and only if

$$A>0,\quad C>0,\quad AB-C>0. \tag{6.2.33}$$

Direct calculation shows that the two parameters A and C in (6.2.32) are positive:

$$A > 0, \quad C > 0; \tag{6.2.34}$$

see also (Hastings and Murray 1975). Note that

$$\beta_{11}\beta_{12}\beta_{13} = -C < 0.$$

Hence all real eigenvalues of (6.2.31) do not change their signs, and at least one of these real eigenvalues is negative.

In addition, we derive from $AB - C = 0$ a critical number

$$\begin{aligned} \delta_0 &= \frac{c + b - a^2 + \sqrt{(c+b-a^2)^2 + 4a^2 b}}{2a}, \\ a &= \frac{\alpha}{2}(3\beta\sigma + \gamma - 1) + \frac{1}{\alpha}(\sigma + 1), \\ b &= (1-\beta)\sigma - 2\beta\sigma^2 - \gamma, \\ c &= \sigma(2\beta\sigma + \beta + \gamma - 1), \end{aligned} \tag{6.2.35}$$

such that $\delta_0 > 0$ if and only if

$$b = (1-\beta)\sigma - 2\beta\sigma^2 - \gamma > 0, \tag{6.2.36}$$

and under the condition (6.2.36),

$$AB - C \begin{cases} > 0 & \text{if } \delta > \delta_0, \\ = 0 & \text{if } \delta = \delta_0, \\ < 0 & \text{if } \delta < \delta_0. \end{cases} \tag{6.2.37}$$

Now we need to check the other eigenvalues β_{kj} with $k \geq 2$. By (6.2.29), $\lambda_k = \beta_{kj}$ ($k \geq 2$) satisfy

$$\lambda_k^3 + A_k \lambda_k^2 + B_k \lambda_k + C_k = 0, \tag{6.2.38}$$

where

$$\begin{aligned} A_k =& A + (\mu_1 + \mu_2 + \mu_3)\rho_k, \\ B_k =& B + (\mu_1 + \mu_2)\rho_k + a\mu_3\rho_k + \left(\frac{\sigma+1}{\alpha}\mu_1 + \frac{\alpha}{2}(3\beta\sigma| + \gamma - 1)\mu_2\right)\rho_k \\ &+ (\mu_1\mu_2 + \mu_1\mu_3 + \mu_2\mu_3)\rho_k^2, \\ C_k =& C + \left[\left(\frac{\sigma+1}{\alpha}\mu_1 + \frac{\alpha}{2}(3\beta\sigma + \gamma - 1)\mu_2\right)\delta - b\mu_3\right]\rho_k \\ &+ \left[\mu_1\mu_2\delta + \left(\frac{1+\sigma}{\alpha}\mu_1 + \frac{\alpha}{2}(3\beta\sigma + \gamma - 1)\mu_2\right)\mu_3\right]\rho_k^2 + \mu_1\mu_2\mu_3\rho_k^3, \end{aligned}$$

where a, b are as in (6.2.35), and A, B, C are as in (6.2.32).

By (6.2.24), it is easy to check that

$$3\beta\sigma+\gamma-1>0.$$

With the condition (6.2.36) we introduce another critical number,

$$\delta_1=\max_{\rho_k\neq 0}\left[\frac{\mu_3\rho_k b-C}{\left(\frac{\sigma+1}{\alpha}\mu_1+\frac{\alpha}{2}(3\beta\sigma+\gamma-1)\mu_2\right)\rho_k+\mu_1\mu_2\rho_k^2}-\mu_3\rho_k\right], \tag{6.2.39}$$

where b is as in (6.2.35).

Then the following lemma establishes the principle of exchange of stability for the eigenvalues β_{kj} of (6.2.27) as functions of δ.

Lemma 6.2.3 *Let δ_0 and δ_1 be the numbers given by (6.2.35) and (6.2.39), and let b be given by (6.2.35). When $b<0$, all eigenvalues of (6.2.27) have negative real part, and when $b>0$, the following assertions hold:*

1. *Let $\delta_0<\delta_1$ and let $k_0\geq 2$ denote the integer for which δ_1 in (6.2.39) reaches its maximum at ρ_{k_0}. Then $\beta_{k_0 l}$ is a real eigenvalue of (6.2.27), and*

$$\beta_{k1}(\delta)\begin{cases}<0 & \text{if } \delta>\delta_1,\\ =0 & \text{if } \delta=\delta_1,\\ >0 & \text{if } \delta<\delta_1\end{cases} \qquad \text{for } \rho_k=\rho_{k_0},$$

$$Re\beta_{ij}(\delta_1)<0 \qquad \forall (i,j)\neq(k,1) \text{ with } \rho_k=\rho_{k_0}.$$

2. *Let $\delta_0>\delta_1$. Then $\beta_{11}(\delta)=\overline{\beta}_{12}(\delta)$ are a pair of complex eigenvalues of (6.2.27) near $\delta=\delta_0$, and*

$$Re\beta_{11}=Re\beta_{12}\begin{cases}<0 & \text{if } \delta>\delta_0,\\ =0 & \text{if } \delta=\delta_0,\\ >0 & \text{if } \delta<\delta_0,\end{cases}$$

$$Re\beta_{kj}(\delta_0)<0 \qquad \forall (k,j)\neq(1,1),(1,2).$$

Proof. In (6.2.32), we see that $A=a+\delta$, $B=a\delta-b$. By direct calculation, we can see that

$$A_k>0,\quad A_kB_k-C_k>0,\quad \forall k\geq 2. \tag{6.2.40}$$

For $b<0$, by (6.2.34)–(6.2.36), (6.2.38), and (6.2.40), for all physically sound parameters $\mu_1,\mu_2,\mu_3,\alpha,\beta,\delta>0$, the following relations hold:

$$A_i>0,\quad C_i>0,\quad A_iB_i-C_i>0,\quad \forall i\geq 1.$$

Hence it follows that all eigenvalues β_{kj} $(k\geq 1, 1\leq j\leq 3)$ of (6.2.31) have negative real parts.

For $b>0$ and $\delta_0<\delta_1$, we infer from (6.2.34) and (6.2.37) that

$$\mathrm{Re}\beta_{1j}(\delta_1)<0,\quad \forall 1\leq j\leq 3. \tag{6.2.41}$$

In addition, it is clear that

$$
\begin{aligned}
&C_{k_1}\begin{cases} >0 & \text{if } \delta>\delta_1,\\ =0 & \text{if } \delta=\delta_1,\\ <0 & \text{if } \delta<\delta_1,\end{cases}\\
&C_k>0 \qquad \text{at } \delta=\delta_1 \text{ for all } k\neq k_1.
\end{aligned}
\tag{6.2.42}
$$

Thus assertion (1) follows from (6.2.40)–(6.2.42).

For $\delta_0>\delta_1$, by (6.2.42) we know that $C_k>0$ at $\delta=\delta_0$ for all $k\geq 2$. Since the real eigenvalues of β_{1j} ($1\leq j\leq 3$) are negative, the condition (6.2.37) implies that there is a pair of complex eigenvalues $\beta_{11}=\overline{\beta}_{12}$ that cross the imaginary axis at $\delta=\delta_0$. Then assertion (2) follows. The lemma is proved. □

Spatiotemporal Oscillations. By Lemma 6.2.3, if $\delta_1<\delta_0$, the problem (6.2.26) undergoes a dynamic transition to a periodic solution from $\delta=\delta_0$. To determine the types of transition, we introduce a parameter as follows:

$$
\begin{aligned}
b_1=\frac{\rho^2}{D^2E}\Big[&\frac{3}{D_0}((2D-6-\alpha\gamma\rho^2D_3)F_3-(2\gamma D_3D_4+\alpha D_5D-6)F_1)\\
&-\frac{1}{D_0}((\alpha\gamma\rho^2D_3-2D_6)F_1+(2\gamma D_3D_4)+\alpha D_5D_6)F_3)\\
&+\frac{1}{D_0}(2\gamma\rho D_3+\alpha\rho D_6+\alpha\gamma\rho D_3D_5-2\rho^{-1}D_4D_6)F_2\\
&+\frac{\alpha}{2D^2E\rho}(2\gamma D_3D_8+\alpha D_6D_7)(2\rho^{-1}D_6D_8-\alpha\gamma\rho D_3D_7)\\
&+\frac{\alpha\rho^2}{2D^2E}(2\gamma D_3+\alpha D_6)(2\gamma D_3D_8+\alpha D_6D_7)\\
&-\frac{\alpha\rho^2}{2D^2E}(2D_6-\alpha\gamma\rho^2D_3)(2\rho^{-2}D_6D_8-\alpha\gamma D_3D_7)\Big],
\end{aligned}
\tag{6.2.43}
$$

where

$$
\begin{aligned}
&D^2=\frac{\gamma^2\delta_0^2\rho^2(\sigma-1)^2+\rho^2((\delta_0^2+\rho^2)^2+\gamma\delta_0^2(\sigma-1))^2}{N^2(\sigma-1)^2(\delta^2+\rho^2)^4}, && D_0=\frac{\gamma\delta_0}{\alpha a^2}+\frac{a+2\delta_0}{\alpha(\delta-1)},\\
&D_1=\frac{\beta}{\alpha}+\frac{(\alpha A-2)(\beta\sigma+2\beta+\gamma-1)}{2\alpha(\sigma-1)^2}, && D_2=\frac{2-\alpha A}{\alpha^2(\sigma-1)^2},\\
&D_4=A+\alpha\beta\sigma^2+\alpha-3\alpha\beta\sigma-\alpha\beta-\alpha\gamma, && D_3=\delta_0(\sigma-1),\\
&D_5=A+\alpha-\alpha\beta\sigma-2\alpha\beta-\alpha\gamma, && D_6=(\delta_0^2+\rho^2)^2+\gamma\delta_0^2(\sigma-1),\\
&D_7=1-\beta\sigma-\beta-\gamma, && D_8=\beta\sigma^2+\beta+1-3\beta\sigma-\gamma,\\
&F_1=\frac{D_1}{A}-\frac{2\rho^2D_1}{A(A^2+4\rho^2)}-\frac{\rho D_2}{A^2+4\rho^2}, && \rho=\sqrt{B}=\sqrt{\frac{C}{A}},
\end{aligned}
$$

$$F_2 = \frac{D_2}{A} - \frac{4\rho^2 D_2}{A(A^2+4\rho^2)} + \frac{2\rho D_1}{A^2+4\rho^2}, \qquad E = \alpha^3(\sigma-1)^3(\delta_0^2+\rho^2)^2,$$

$$F_3 = \frac{2\rho D_1}{A(A^2+4\rho^2)} + \frac{\rho D_2}{A^2+4\rho^2}.$$

Here A, B, C are as in (6.2.32), and a is as in (6.2.35).

Then we have the following dynamic transition theorem.

Theorem 6.2.4 *Let $\delta_1 < \delta_0$, and let b_1 be the number given by (6.2.43).*

1. *If $b_1 < 0$, the problem (6.2.26) undergoes a Type I dynamic transition at $\delta = \delta_0$ and bifurcates to a periodic solution on $\delta < \delta_0$ that is an attractor.*
2. *If $b_1 > 0$, the problem (6.2.26) undergoes a Type II dynamic transition at $\delta = \delta_0$ and bifurcates on $\delta > \delta_0$ to a periodic solution that is a repeller.*

Proof. By Lemma 6.2.3, at $\delta = \delta_0$ there is a pair of imaginary eigenvalues $\beta_{11} = \overline{\beta}_{12} = -i\rho$ of (6.2.27). Let $z = \xi + i\eta$ and $z^* = \xi^* + i\eta^*$ be the eigenvectors and conjugate eigenvectors of (6.2.27) corresponding to $-i\rho$, i.e., z and z^* satisfy

$$(M_1 + i\rho)z = 0, \qquad (M_1^* - i\rho)z^* = 0, \tag{6.2.44}$$

where M_1 is the matrix in (6.2.31), and M_1^* the transpose of M. Because $\pm i\rho$ are solutions of (6.2.32), and $AB = C$ at $\delta = \delta_0$, we deduce that

$$\rho^2 = B = C/A. \tag{6.2.45}$$

For $z = (z_1, z_2, z_3)$, from the first equation of (6.2.44) we get

$$\delta_0 z_1 = (\delta_0 - i\rho)z_3, \qquad \alpha(\sigma-1)z_2 = \left(-\frac{\alpha}{2}(\gamma + 3\beta\sigma - 1) + i\rho\right)z_1. \tag{6.2.46}$$

Thus, we derive from (6.2.46) the eigenvectors $z = \xi + i\eta$ as follows:

$$\begin{aligned} \xi &= (\xi_1, \xi_2, \xi_3) = \left(1, -\tfrac{\gamma+3\beta\sigma-1}{2(\sigma-1)}, \tfrac{\delta_0^2}{\delta_0^2+\rho^2}\right), \\ \eta &= (\eta_1, \eta_2, \eta_3) = \left(0, \tfrac{\rho}{\alpha(\sigma-1)}, \tfrac{\rho\delta_0}{\delta_0^2+\rho^2}\right). \end{aligned} \tag{6.2.47}$$

In the same fashion, from the second equation of (6.2.44) we derive the conjugate eigenvectors $z^* = \xi^* + i\eta^*$ as

$$\begin{aligned} \xi^* &= (\xi_1^*, \xi_2^*, \xi_3^*) = \left(-\tfrac{\sigma+1}{\alpha^2(\sigma-1)}, 1, \tfrac{\gamma\delta_0}{\alpha(\delta_0^2+\rho^2)}\right), \\ \eta^* &= (\eta_1^*, \eta_2^*, \eta_3^*) = \left(-\tfrac{\rho}{\alpha(\sigma-1)}, 0, -\tfrac{\gamma\rho}{\alpha(\delta_0^2+\rho^2)}\right). \end{aligned} \tag{6.2.48}$$

As in (4.4.74), we have

$$(\xi, \xi^*) = (\eta, \eta^*) = -\frac{\gamma\rho^2 D_3}{H_1}, \quad (\xi, \eta^*) = -(\eta, \xi^*) = -\frac{\rho D_6}{H_1}, \tag{6.2.49}$$

where $H_1 = \alpha(\sigma-1)(\delta_0^2+\rho^2)^2, D_3$ and D_6 are as in (6.2.43). It is known that functions $\Psi_1^* + i\Psi_2^*$ given by

$$\begin{aligned}\Psi_1^* &= \tfrac{1}{(\xi,\xi^*)}[(\xi,\xi^*)\xi^* + (\xi,\eta^*)\eta^*],\\ \Psi_2^* &= \tfrac{1}{(\eta,\eta^*)}[(\eta,\xi^*)\xi^* + (\eta,\eta^*)\eta^*],\end{aligned} \tag{6.2.50}$$

also satisfy the second equation of (6.2.44) with

$$(\xi,\Psi_1^*) = (\eta,\Psi_2^*) \neq 0, \qquad (\xi,\Psi_2^*) = (\eta,\Psi_1^*) = 0. \tag{6.2.51}$$

Let $u \in H$ be a solution of (6.2.26) expressed as

$$u = x\xi + y\eta + \Phi(x,y), \quad x,y \in \mathbb{R}^1,$$

where $\Phi(x,y)$ is the center manifold function of (6.2.26) at $\delta = \delta_0$. By (6.2.51), the reduced equations of (6.2.26) on the center manifold are given by

$$\begin{aligned}\frac{dx}{dt} &= -\rho y + \frac{1}{(\xi,\Psi_1^*)}(G(x\xi + y\eta + \Phi), \Psi_1^*),\\ \frac{dy}{dt} &= \rho x + \frac{1}{(\eta,\Psi_2^*)}(G(x\xi + y\eta + \Phi), \Psi_2^*),\end{aligned} \tag{6.2.52}$$

where $G(u) = G(u,u)$ is the bilinear operator defined by

$$G(u,v) = (-\alpha u_1 v_2 - \alpha\beta u_1 v_1, -\frac{1}{\alpha}u_1 v_2, 0) \tag{6.2.53}$$

for $u = (u_1,u_2,u_3)$, $v = (v_1,v_2,v_3) \in H_1$.

We are now in a position to find the center manifold function $\Phi(x,y)$. To this end, we need to determine the third eigenvalue β_{13} and eigenvector ζ of M_1 at $\delta = \delta_0$. We know that

$$\beta_{13}\cdot(i\rho)(-i\rho) = -\rho^2\beta_{13} = \det M_1 = C.$$

By (6.2.45), we obtain

$$\beta_{13} = -A = -(\delta_0 + a),$$

and a is as in (6.2.35). Then from the equation

$$(M_1 - \beta_{13})\zeta = 0,$$

we derive the eigenvector

$$\zeta = (\zeta_1,\zeta_2,\zeta_3) = \left(1, \frac{A - \frac{\alpha}{2}(\gamma + 3\beta\sigma + 1)}{\alpha(\sigma-1)}, -\frac{\delta_0}{a}\right). \tag{6.2.54}$$

In the same fashion, from $(M_1^* - \beta_{13})\zeta^* = 0$, we obtain the conjugate eigenvector

$$\zeta^* = (\zeta_1^*,\zeta_2^*,\zeta_3^*) = \left(\frac{A - \frac{1}{\alpha}(\sigma+1)}{\alpha(\sigma-1)}, 1, -\frac{\gamma}{\alpha a}\right). \tag{6.2.55}$$

On the other hand, from (6.2.47) and (6.2.53), it follows that

$$\begin{aligned} G_{11} &= G(\xi,\xi) = (-\alpha(\xi_2+\beta), -\frac{\xi_2}{\alpha}, 0), \quad & G_{22} &= G(\eta,\eta) = 0, \\ G_{12} &= G(\xi,\eta) = (-\alpha\eta_2, -\frac{\eta_2}{\alpha}, 0), & G_{21} &= G(\eta,\xi) = 0. \end{aligned} \tag{6.2.56}$$

Direct calculation shows that

$$(\zeta,\zeta^*) = D_0, \quad (G_{11},\zeta^*) = D_1, \quad (G_{12},\zeta^*) = D_2,$$

and D_0, D_1, D_2 are as in (6.2.43).

By (A.1.19), the center manifold function Φ satisfies

$$\Phi = \Phi_1 + \Phi_2 + \Phi_3 + o(2), \tag{6.2.57}$$

where

$$\begin{aligned} &\mathscr{L}\Phi_1 = -x^2 P_2 G_{11} - xy P_2 G_{12}, \\ &(\mathscr{L}^2 + 4\rho^2)\mathscr{L}\Phi_2 = 2\rho^2(x^2 - y^2)P_2 G_{11} + 4\rho^2 xy P_2 G_{12}, \\ &(\mathscr{L}^2 + 4\rho^2)\Phi_3 = \rho(y^2 - x^2)P_2 G_{12} + 2\rho xy P_2 G_{11}, \end{aligned}$$

$P_2 : H \to E_2$ is the canonical projection, E_2 the orthogonal complement of $E_1 = \text{span}\{\xi,\eta\}$, and $\mathscr{L}$ is the linearized operator of (6.2.26). By the spectral theorem, Theorem 3.4 in Ma and Wang 2005b, and (6.2.30), it is clear that

$$P_2 G_{11} = (G_{11},\zeta^*)\zeta = D_1\zeta, \quad P_2 G_{12} = (G_{12},\zeta^*)\zeta = D_2\zeta. \tag{6.2.58}$$

Hence $\Phi_1, \Phi_2, \Phi_3 \in \text{span}\{\zeta\}$, which implies that

$$\mathscr{L}\Phi_j = M_1\Phi_j = -A\Phi_j. \tag{6.2.59}$$

We derive from (6.2.57)–(6.2.59) the center manifold function as follows:

$$\begin{aligned} \Phi =& \frac{\zeta}{D_0}\left[\left(\frac{D_1}{A} - \frac{2\rho^2 D_1}{A(A^2+4\rho^2)} - \frac{\rho D_2}{A^2+4\rho^2}\right)x^2\right. \\ &+ \left(\frac{D_2}{A} - \frac{4\rho^2 D_2}{A(A^2+4\rho^2)} + \frac{2\rho D_1}{A^2+4\rho^2}\right)xy \\ &+ \left.\left(\frac{2\rho^2 D_1}{A(A^2+4\rho^2)} + \frac{\rho D_2}{A^2+4\rho^2}\right)y^2\right] + o(2) \\ =& \frac{1}{D_0}(F_1x^2 + F_2xy + F_3y^2)\zeta + o(2), \end{aligned} \tag{6.2.60}$$

where F_1, F_2, F_3 are as in (6.2.43). The remaining part of the proof is to use Theorem 2.3.7 with routine substitutions, and we omit the details; see also (Ma and Wang 2011c). □

Theorem 6.2.4 describes the temporal oscillation for the Belousov–Zhabotinsky reactions of (6.2.1) in a nonstirred condition. If the reagent is stirred, then equations (6.2.26) are reduced to a system of ordinary differential equations as follows:

$$\begin{aligned} \frac{du_1}{dt} &= -\alpha\left(\frac{1}{2}(\gamma+3\beta\sigma-1)u_1+(\sigma-1)u_2-u_1u_2-\beta u_1^2\right), \\ \frac{du_2}{dt} &= -\frac{1}{\alpha}\left(\frac{1}{2}(1+\gamma-\beta\sigma)u_1+(\sigma+1)u_2-\gamma u_3-u_1u_2\right), \\ \frac{du_3}{dt} &= \delta(u_1-u_3). \end{aligned} \tag{6.2.61}$$

In this case, only the transition to periodic solutions can take place.

Theorem 6.2.5 *Let $\delta_0 > 0$ be the number given by (6.2.35), and let b be as in (6.2.43). Then the system (6.2.61) undergoes a dynamic transition to periodic solutions at $\delta = \delta_0$. Furthermore, the system bifurcates to an unstable periodic orbit on $\delta > \delta_0$ when $b > 0$, and to a stable periodic orbit on $\delta < \delta_0$ when $b < 0$. In addition, the bifurcated periodic solution can be expressed in the following form:*

$$u_\delta = [-b^{-1}Re\beta_{11}(\delta)]^{1/2}(\xi\cos\rho t+\eta\sin\rho t)+o(|Re\beta_{11}|^{1/2}), \tag{6.2.62}$$

where ξ,η are as in (6.2.55) and (6.2.57), and $\beta_{11}(\delta)$ is the first complex eigenvalue as described in Lemma 6.2.3.

Also, by Lemma 6.2.2, for all physically sound parameters, each of the two systems (6.2.26) and (6.2.61) has a global attractor in the invariant region D in (6.2.6). Hence, when $b > 0$, the bifurcated periodic solution u_δ is a repeller, and its stable manifold divides D into two disjoint open sets D_1 and D_2, i.e., $\overline{D} = \overline{D}_1 + \overline{D}_2, D_1 \bigcap D_2 = \emptyset$, such that the equilibrium $U_1 = (u_1^0, u_2^0, u_3^0)$ in (6.2.24) attracts D_1, and there is another attractor $\mathscr{A}_2 \subset D_2$ that attracts D_2.

Transition to Steady-State Solutions. Due to Lemma 6.2.3, for $\delta_1 > \delta_0$, the transition of (6.2.26) occurs at $\delta = \delta_1$, which is from real eigenvalues. Let δ_1 achieve its maximum at ρ_{k_0} ($k_0 \geq 2$). Assume that $\beta_{k_0 l}(\delta)$ is simple near δ_1.

First, we consider the case

$$\int_\Omega e_{k_0}^3 dx \neq 0. \tag{6.2.63}$$

In general, the condition (6.2.63) holds for the case $\Omega \neq (0,L) \times D$, and $D \subset \mathbb{R}^{n-1}$ $(1 \leq n \leq 3)$ is a bounded open set.

Let

$$b_0 = [\alpha\beta(\sigma-1)(\mu_2\rho_{k_0}+\frac{1}{\alpha}(\sigma+1))-(\alpha\mu_2\rho_{k_0}+2\sigma)(\mu_1\rho_{k_0}+\frac{\alpha}{2}(\gamma+3\beta\sigma-1))].$$

Theorem 6.2.6 *Let the above number b_0 be nonzero, and suppose $\delta_1 > \delta_0$. Under the condition (6.2.63), the transition of (6.2.26) at $\delta = \delta_1$ is mixed (Type III), and the system bifurcates from $(u,\delta) = (0,\delta_1)$ to a unique steady-state solution u^δ such* *that u^δ is a saddle on $\delta > \delta_1$ and an attractor on $\delta < \delta_1$. The solution u_δ can be expressed as*

$$u^\delta = C\beta_{k_0 l}(\delta)\xi e_{k_0} + o(|\beta_{k_0 l}|), \tag{6.2.64}$$

where $\xi = (\xi_1, \xi_2, \xi_3)$ and constant C are given by

$$\begin{aligned}
&\xi_1 = \alpha(\sigma - 1)(\mu_3\rho_{k_0} + \delta_1),\\
&\xi_2 = -(\mu_1\rho_{k_0} + \frac{\alpha}{2}(\gamma + 3\beta\sigma - 1))(\mu_3\rho_{k_0} + \delta_1),\\
&\xi_3 = \alpha\delta_1(\sigma - 1),\\
&C = \frac{[(\mu_3\rho_{k_0} + \delta_1)^2(\mu_1\rho_{k_0} + \mu_2\rho_{k_0} + \frac{1}{\alpha}(\sigma+1) + \frac{\alpha}{2}(\gamma + 3\beta\sigma - 1)) - (\sigma - 1)\gamma\delta_1]\int_\Omega e_{k_0}dx}{b_0(\mu_3\rho_{k_0} + \delta_1)^3\int_\Omega e_{k_0}^3 dx}.
\end{aligned}$$

Proof. We apply Theorem 2.3.2 to prove this theorem. Let ξ and $\xi^* \in \mathbb{R}^3$ be the eigenvectors of M_{k_0} and $M_{k_0}^*$ corresponding to $\beta_{k_0 l}(\delta_1) = 0$, i.e.,

$$M_{k_0}\xi = 0, \quad M_{k_0}^*\xi^* = 0,$$

where M_{k_0} is the matrix (6.2.29) with $k = k_0$. Clearly, ξ is as in (6.2.64), and

$$\xi^* = (-(\mu_2\rho_{k_0} + \frac{1}{\alpha}(\sigma + 1))(\mu_3\rho_{k_0} + \delta_1), \alpha(\mu_3\rho_{k_0} + \delta_1)(\sigma - 1), \gamma(\sigma - 1)).$$

For the operator G in (6.2.7), we can derive that

$$\frac{1}{(\xi e_{k_0}, \xi^* e_{k_0})}(G(y\xi e_{k_0}), \xi^* e_{k_0}) = -\frac{1}{C}y^2 + o(2),$$

where C is as in (6.2.64). Therefore, the theorem follows from Theorem 2.3.2. □

In the following, we assume that

$$\int_\Omega e_{k_0}^3 dx = 0.$$

Define

$$\begin{aligned}
b_1 = &\frac{1}{(\xi, \xi^*)}\Big[-(\mu_1\rho_{k_0} + \frac{\alpha}{2}(\gamma + 3\beta\sigma - 1))(\alpha\mu_2\rho_{k_0} + 2)\int_\Omega \phi_1 e_{k_0}^2 dx \\
&+ 2\alpha^2\beta(\sigma - 1)(\mu_2\rho_{k_0} + \frac{\sigma + 1}{\alpha})\int_\Omega \phi_1 e_{k_0}^2 dx \\
&+ \alpha(\sigma - 1)(\alpha\mu_2\rho_{k_0} + 2)\int_\Omega \phi_2 e_{k_0}^2 dx\Big],
\end{aligned} \tag{6.2.65}$$

where

$$\begin{aligned}
(\xi, \xi^*) = \alpha(1 - \sigma)\big[&(\mu_1\rho_{k_0} + \mu_2\rho_{k_0} + \frac{\sigma + 1}{\alpha} + \frac{\alpha}{2}(\gamma + 3\beta\sigma - 1)) \\
&\times(\mu_3\rho_{k_0} + \delta_1)^2 - \gamma\delta_1(\sigma - 1)\big],
\end{aligned}$$

the function $\phi = (\phi_1, \phi_2, \phi_3)$ satisfies

$$L\phi = -G(\xi)e_{k_0}^2, \tag{6.2.66}$$

and the operators L and G are defined by

$$L\phi = \begin{cases} \mu_1 \Delta \phi_1 - \alpha \left[\frac{1}{2}(\gamma + 3\beta\sigma - 1)\phi_1 + (\sigma - 1)\phi_2\right], \\ \mu_2 \Delta \phi_2 - \frac{1}{\alpha}\left[\frac{1}{2}(1 + \gamma - \beta\sigma)\mu_1 + (\sigma + 1)\phi_2\right], \\ \mu_3 \Delta \phi_3 + \delta(\phi_1 - \phi_3), \end{cases}$$

$$G(\xi) = \begin{cases} \alpha^2(\sigma - 1)(\mu_3 \rho_{k_0} + \delta_1)^2 \left[\mu_1 \rho_{k_0} + \frac{\alpha}{2}(\gamma + 3\beta\sigma - 1) - \alpha\beta(\sigma - 1)\right], \\ (\sigma - 1)(\mu_3 \rho_{k_0} + \delta_1)^2(\mu_1 \rho_{k_0} + \frac{\alpha}{2}(\gamma + 3\beta\sigma - 1)), \\ 0. \end{cases}$$

Theorem 6.2.7 *Let $b_1 \neq 0$ be the parameter given by (6.2.65), and $\delta_0 < \delta_1$.*

1. *If $b_1 < 0$, then the system (6.2.26) undergoes a Type I transition at $\delta = \delta_1$ and bifurcates to two steady-state solutions u_+^δ and u_-^δ on $\delta < \delta_1$ that are attractors, and there is no bifurcation on $\delta > \delta_1$.*
2. *If $b_1 > 0$, then the system undergoes a Type II transition at $\delta = \delta_1$ and bifurcates from $(u, \delta) = (0, \delta_1)$ to two steady-state solutions u_+^δ and u_-^δ on $\delta > \delta_1$, which are saddles, and there is no bifurcation on $\delta < \delta_1$.*
3. *The bifurcated solutions $u_\pm^\delta$ can be expressed as*

$$u_\pm^\delta = \pm \left[-\frac{\int_\Omega e_{k_0}^2 dx}{(\mu_2 \rho_{k_0} + \delta_1)^2 b_1} \beta_{k_0 l}(\delta)\right]^{1/2} \xi e_{k_0} + o(|\beta_{k_0 l}|^{1/2}),$$

where $\xi \in \mathbb{R}^3$ is as in (6.2.64), and $\beta_{k_0 l}(\delta)$ is as in Lemma 6.2.3.

Proof. We use Theorem 2.3.1 to verify this theorem. Let

$$u = y\xi e_{k_0} + \Phi(y), \quad y \in \mathbb{R}^1,$$

and let $\Phi(y)$ be the center manifold function. Set $\Phi = y^2\phi = o(2)$. Then by Theorem A.1.1, ϕ satisfies

$$L\phi = -G(\xi e_{k_0}) = -G(\xi)e_{k_0}^2,$$

which is (6.2.66).

We see that

$$\begin{aligned}(G(y\xi e_{k_0} + \Phi), \xi^* e_{k_0}) = &-\int_\Omega [\alpha(y\xi_1 e_{k_0} + \Phi_1)(y\xi_2 e_{k_0} + \Phi_2)\xi_1^* e_{k_0} \\ &+ \alpha\beta(y\xi_1 e_{k_0} + \Phi_1)^2 \xi_1^* e_{k_0} + \frac{1}{\alpha}(y\xi_1 e_{k_0} + \Phi_1)(y\xi_2 e_{k_0} + \Phi_2)\xi_2^* e_{k_0}] dx\end{aligned}$$

$$
\begin{aligned}
&= y^3[-(\alpha\xi_2\xi_1^* + 2\alpha\beta\xi_1\xi_1^* + \frac{1}{\alpha}\xi_2\xi_2^*)\int_\Omega \phi_1 e_{k_0}^2 dx \\
&- (\alpha\xi_1\xi_1^* + \frac{1}{\alpha}\xi_1\xi_2^*)\int_\Omega \phi_2 e_{k_0}^2 dx] + o(3).
\end{aligned}
$$

Hence, we have

$$
\frac{1}{(\xi e_{k_0}, \xi^* e_{k_0})}(G(y\xi e_{k_0} + \Phi), \xi^* e_{k_0}) = \frac{1}{\int_\Omega e_{k_0}^2 dx}(\mu_2 \rho_{k_0} + \delta_1)^2 b_1 y^3 + o(y^3).
$$

Thus the theorem follows from Theorem 2.3.1. □

6.2.4 Phase Transition in the Realistic Oregonator

We begin with chemical parameters. In the chemical reaction (6.2.1), the constants given in Field and Noyes 1974 are as follows; see also (Hastings and Murray 1975):

$$
\begin{aligned}
&k_1 = 1.34 M^{-1}S^{-1}, && k_2 = 1.6 \times 10^9 M^{-1}S^{-1}, \\
&k_3 = 8 \times 10^3 M^{-1}S^{-1}, && k_4 = 4 \times 10^7 M^{-1}S^{-1},
\end{aligned}
$$

where

$$
a = b = 6 \times 10^{-2} M, \quad \gamma = O(1), \quad k_5 = O(1)\ S^{-1}.
$$

Here we take

$$
\gamma = 1, \quad k_5 = 1\ S^{-1},
$$

and $a = b$ as a control parameter. The nondimensional parameters α, β, δ and μ_i $(1 \leq i \leq 3)$ are given by

$$
\begin{aligned}
\alpha &= \left(\frac{k_3 b}{k_1 a}\right)^{1/2} = 7.727 \times 10, \\
\beta &= \frac{2k_1k_4 a}{k_2 k_3 b} = 8.375 \times 10^{-6}, \\
\delta &= k_5(k_1k_3ab)^{1/2} = 1.035 \times 10^2 a M^{-1}, \\
\mu_i &= \frac{1}{(k_1k_2ab)^{1/2}}\frac{\sigma_i}{l^2} = 4.664 \times 10^{-6}\frac{\sigma_i}{l^2 a} M^{-1}S^{-1}.
\end{aligned}
\tag{6.2.67}
$$

Dirichlet Boundary Condition Case. When the concentrations of $X = HBrO_2$, $Y = Br^-$, and Ce^{4+} are maintained so that they vanish at the boundary, then by Theorem 6.2.1, the reaction system (6.2.1) undergoes a dynamic transition to an equilibrium $u_\lambda \neq 0$ in Ω at the critical values $\lambda_0 = (\mu_1^0, \mu_2^0, \mu_3^0, \alpha^0, \gamma^0, \delta^0) \in \Gamma$. Furthermore, we see that the state of the system (6.2.1) depends on the sign of the parameter $N(\lambda)$ given by

$$N(\lambda) = \alpha\gamma\delta - (\mu_1\rho_1 - \alpha)(\mu_2\rho_1\alpha + 1)(\mu_3\rho_1 + \delta). \tag{6.2.68}$$

When $N(\lambda) < 0$, this system lies in the trivial state $u = 0$, and for $N(\lambda) > 0$ it is at the state $u_\lambda \neq 0$ given by (6.2.12).

Let

$$\alpha^0 = \frac{\mu_1\mu_2\rho_1^2 + \sqrt{\mu_1^2\mu_2^2\rho_1^4 + 4} - 2}{2\mu_2\rho_1},$$
$$\delta^0 = \frac{(\mu_1\rho_1 - \alpha)(\mu_2\rho_1\alpha + 1)}{\alpha(2 + \mu_2\rho_1\alpha) - \mu_1\mu_2\rho_1^2\alpha - \mu_1\rho_1}.$$

It is easy to derive from (6.2.68) the following statements:

- If $\mu_1\rho_1 \leq \alpha$, then $N(\lambda) > 0$.
- If $\alpha^0 < \alpha < \mu_1\rho_1$, then

$$N(\lambda)\begin{cases} > 0 & \text{for } \delta > \delta^0, \\ < 0 & \text{for } \delta < \delta^0. \end{cases}$$

- If $0 < \alpha \leq \alpha_0$, then $N(\lambda) < 0$.

Then in view of (6.2.67), we claim that when $\mu_1\rho_1 \leq 77.27$, the reaction system (6.2.1) is in a nontrivial stable state; when $\alpha^0 < 77.27 < \mu_1\rho_1$, at $\delta = \delta^0$ this system undergoes a dynamic transition from the trivial state $u = 0$ to the equilibrium u_λ of (6.2.12); and when $77.27 \leq \alpha^0$, the system lies at $u = 0$. The first eigenvalue ρ_1 of (6.2.8) can be expressed as

$$\rho_1 = \frac{C}{L^2},$$

where $C > 0$ is a constant and L is the length scale (or radius) of Ω. Then from statements (1)–(3) above, we derive the following conclusion for the Belousov–Zhabotinsky reactions.

Let Ω_0 be a given domain, and Ω the extension or contraction of Ω_0 from x_0 in the scale L $(0 < L < \infty)$ defined by

$$\Omega(L, x_0) = \{(x - x_0)L \mid x, x_0 \in \Omega_0, x_0 \text{ is fixed}\}. \tag{6.2.69}$$

Physical Conclusion 6.2.8 *Let $\Omega(L, x_0)$ be the domain defined by (6.2.69). Then for the reaction system (6.2.1) with vanishing boundary condition (6.2.4), there exist two critical scales L_0 and L_1 $(L_0 < L_1)$, depending on Ω_0 and x_0, such that (a) when $L_1 < L$, the system is at a chemical state far from $u = 0$; (b) when $L_0 < L < L_1$, this system bifurcates from $(u, \delta) = (0, \delta^0)$ to the stable equilibrium u_λ given by (6.2.12) on $\delta > \delta^0$; and (c) when $L < L_0$, the system will tend to the trivial state $u = 0$.*

Stirred Case. In view of (6.2.67), the numbers in (6.2.35) are as follows:

$$\sigma = 7 \times 10^2, \quad a = 9.74, \quad b = 690.79, \quad c = 8.21. \tag{6.2.70}$$

Hence

$$\delta_0 = 71.67. \tag{6.2.71}$$

Now we need to compute the parameter b_1 in (6.2.43). At the critical value $\delta_0 = 71.67$, the numbers in (6.2.32) are

$$A = a + \delta_0 = 81.41, \quad B = a\delta_0 - b = 7.27, \quad C = \delta_0 c = 588.41. \tag{6.2.72}$$

Then we have

$$\begin{aligned}
&\rho = \sqrt{B} \cong 2.7, && E \cong 4 \times 10^{21}, && D_0 \cong 1.3 \times 10^{-2},\\
&F_1 \cong 8.3 \times 10^{-9}, && F_2 \cong -2.6 \times 10^{-7}, && F_3 \cong -8.4 \times 10^{-10},\\
&D_3 \cong 5 \times 10^4, && D_4 \cong 4 \times 10^2, && D_5 \cong 80,\\
&D_6 \cong 3 \times 10^7, && D_7 \cong -6 \times 10^{-3}, && D_8 \cong 4.
\end{aligned} \tag{6.2.73}$$

Thus by (6.2.67) and (6.2.70)–(6.2.73), the number b_1 in (6.2.43) is given by

$$\begin{aligned}
b_1 \cong & \frac{\rho^2}{D^2 E D_0} \times \left[(\alpha\rho D_6 - 2\rho^{-1} D_4 D_6) F_2 - \alpha D_5 D_6 (3F_1 - F_3)\right]\\
& + \frac{\rho^2}{D^2 E}\left[\frac{\alpha^2 D_6^2 D_7 D_8}{E D^2 \rho^2} + \frac{\alpha^3 \rho^2 D_6^2 D_7}{2ED^2} - \frac{3.2 \times 10^7 D_6 D_8}{\rho^2 E D^2}\right]\\
\cong & -\frac{\rho^2}{E D_0 D^2} \times 4 \times 10^3 - \frac{3.2 D_6 D_8}{E^2 D^4} \times 10^7\\
& - \frac{6\alpha^2 D_6^2 D_8}{E^2 D^4} \times 10^{-3} - \frac{6\alpha^3 \rho^4 D_6^2}{2E^2 D^4} \times 10^{-3}.
\end{aligned} \tag{6.2.74}$$

Hence in the stirred case, by Theorem 6.2.5 and (6.2.71)–(6.2.74), when $\delta < \delta_0 = 71.67$, the system (6.2.61) bifurcates from $(u, \delta) = (0, \delta_0)$ to a stable periodic solution, i.e., the reaction system (6.2.1) undergoes a temporal oscillation in the concentrations of $X = HBrO_2$, $Y = Br^-$, and $Z = Ce^{4+}$.

Nonstirred Case. We consider the problem (6.2.26). In this case, there is another critical parameter δ_1 defined by (6.2.39). When $\delta_0 < \delta_1$, the system undergoes a transition to multiple equilibria at $\delta = \delta_1$, and when $\delta_0 > \delta_1$, the system has a transition to periodic solutions at $\delta = \delta_0$.

By (6.2.67) and (6.2.70), the number δ_1 is given by

$$\delta_1 = \max_{\rho_k}\left[\frac{690.8\mu_3}{9.1\mu_1 + 0.7\mu_2 + \mu_1\mu_2\rho_k} - \frac{588.4}{(9.1\mu_1 + 0.7\mu_2)\rho_k + \mu_1\mu_2\rho_k^2} - \mu_3\rho_k\right].$$

It is clear that if

$$\mu_3 \leq \mu_1, \mu_2, \tag{6.2.75}$$

then in view of (6.2.71), we have

$$\delta_1 \leq \frac{690.8\mu_3}{9.1\mu_1 + 0.7\mu_2} < \frac{690.8}{9.8} < \delta_0.$$

If (6.2.75) is not satisfied, we consider the domain $\Omega = \Omega(L, x_0)$ as defined by (6.2.69). In this case, the eigenvalues ρ_k $(k \geq 2)$ of (6.2.28) can be expressed by

$$\rho_k = \frac{C_k}{L^2} \quad \text{with } C_k > 0 \quad \text{and } C_k \to \infty \text{ as } k \to \infty.$$

It follows that there exists an $L_0 > 0$ such that $\delta_1 < \delta_0$ for all $L < L_0$. Thus by Theorem 6.2.4 and (6.2.74), we have the following conclusion.

Physical Conclusion 6.2.9 *Under the condition of (6.2.75) or $\Omega = \Omega(L, x_0)$ with $L < L_0$, the system (6.2.26) bifurcates from $(u, \delta) = (0, \delta_0)$ to a stable periodic solution on $\delta < \delta_0$, which implies that the reaction system (6.2.1) with the Neumann boundary condition undergoes a temporal oscillation for $\delta < \delta_0$. However, when $\delta > \delta_0$, this system is in the trivial state U_1 in (6.2.24).*

Phase Transition at $\delta = \delta_1$. By the above discussion, when $\mu_3 > \mu_i$ $(i = 1, 2)$ and $L > L_0$ is large enough, the condition $\delta_0 < \delta_1$ may hold. In the following, we assume that $L > L_0$ is sufficiently large, and

$$\mu_1 = \mu_2, \quad \mu_3 = z\mu_1 \quad (z > 1).$$

Then δ_1 becomes

$$\delta_1 \simeq \max_{x>0} \left[\frac{691z}{10+x} - \frac{589}{(10+x)x} - xz \right], \quad (x = \mu_1 \rho_k).$$

It follows that $3 < x < 4$ for $z > 1$. We take $x = \mu_1 \rho_{k_0} = 3.5$. Then

$$\delta_1 = 47.7z - 12.5.$$

Thereby, we deduce that

$$\delta_1 > \delta_0 = 71.8 \quad \text{as } \frac{\mu_3}{\mu_1} = z > 1.8.$$

Now we investigate the transition of (6.2.26) at δ_1. For simplicity, we consider the case

$$\Omega = (0, L) \subset \mathbb{R}^1.$$

In this case, the eigenvalues and eigenvectors of (6.2.28) are

$$\rho_k = (k-1)^2 \pi^2 / L^2; \quad e_k = \cos(k-1)\pi x / L.$$

It is clear that

$$e_{k_0}^2 = (e_1 + e_j)/2 \qquad \text{with } j = 2k_0 - 1,$$

$$\int_\Omega e_{k_0}^3 dx = 0 \qquad \text{by } k_0 \geq 2.$$

We need to compute the number b_1 in (6.2.65). By (6.2.67), (6.2.70)–(6.2.71), and

$$\mu_3\rho_{k_0} = z\mu_1\rho_{k_0} = z\mu_2\rho_{k_0} = 3.5z \quad (z > 1.8),$$

the vectors $G(\xi)$ and ϕ in (6.2.66) are given by

$$\begin{aligned}
&G(\xi) \approx (1.6\times 10^7(\mu_3\rho_{k_0}+\delta_1)^2, 3\times 10^3(\mu_3\rho_{k_0}+\delta_1)^2, 0),\\
&\phi = \varphi_0 e_1 + \varphi_j e_j \ (j = 2k_0 - 1),\\
&M_1\varphi_0 = -\frac{1}{2}G(\xi),\\
&M_j\varphi_j = -\frac{1}{2}G(\xi),
\end{aligned}$$

where

$$M_1 = \begin{pmatrix} -0.7 & -5.4\times 10^4 & 0\\ -1.25\times 10^{-2} & -12.5 & 1.25\times 10^{-2}\\ \delta_1 & 0 & -\delta_1 \end{pmatrix},$$

$$M_j = \begin{pmatrix} -14.7 & -5.4\times 10^4 & 0\\ -1.25\times 10^{-2} & -26.5 & 1.25\times 10^{-2}\\ \delta_1 & 0 & -(\mu_3\rho_j+\delta_1) \end{pmatrix}.$$

Direct calculation shows that

$$\begin{aligned}
&\det M_1 = -8.75\delta_1,\\
&\det M_j \simeq -1.5\times 10^4 z + 4.9\times 10^3,\\
&M_1^{-1} = \frac{1}{\det M_1}\begin{pmatrix} 12.5\delta_1 & -5.4\times 10^4\delta_1 & *\\ 0 & 0.7\delta_1 & *\\ * & * & * \end{pmatrix},\\
&M_j^{-1} = \frac{1}{\det M_j}\begin{pmatrix} 26.5(\mu_3\rho_j+\delta_1) & -5.4\times 10^4(\mu_3\rho_j+\delta_1) & *\\ -1.25\times 10^{-2}\mu_3\rho_j & 14.7(\mu_3\rho_j+\delta_1) & *\\ * & * & * \end{pmatrix}.
\end{aligned}$$

Then we get

$$\begin{aligned}
&\varphi_0 = (\varphi_{01},\varphi_{02},\varphi_{03}) = -\frac{1}{2}M_1^{-1}G(\xi), && \varphi_{01} = 2.5\times 10^6(\mu_3\rho_{k_0}+\delta_1)^2,\\
&\varphi_{02} = 1.1\times 10^2(\mu_3\rho_{k_0}+\delta_1)^2, && \varphi_j = (\varphi_{j1},\varphi_{j2},\varphi_{j3}) = -\frac{1}{2}M_j^{-1}G(\xi),\\
&\varphi_{j1} = \frac{13.4\times 10^3}{1.5z-0.5}(\mu_3\rho_{k_0}+\delta_1)^2, && \varphi_{j2} = \frac{2z+27}{1.5z-0.5}(\mu_3\rho_{k_0}+\delta_1)^2.
\end{aligned}$$

Namely,

$$\phi_1 = \varphi_{01}e_1 + \varphi_{j_1}e_j = (\mu_3\rho_{k_0} + \delta_1)^2\left[2.5\times 10^6 e_1 + \frac{13.4\times 10^3}{1.5z-0.5}e_j\right],$$
$$\phi_2 = \varphi_{02}e_1 + \varphi_{j2}e_j = (\mu_3\rho_{k_0} + \delta_1)^2\left[1.1\times 10^2 e_1 + \frac{2z+27}{1.5z-0.5}e_j\right].$$

Putting ϕ_1 and ϕ_2 into (6.2.65), we derive that

$$b_1 = \frac{(\mu_3\rho_{k_0}+\delta_1)^2 L}{(\xi,\xi^*)}\left(5\times 10^8 - \frac{5.7\times 10^6}{1.5z-0.5} - \frac{2z+27}{1.5z-0.5}\right),$$
$$(\xi,\xi^*) = -699\alpha\left[16.7\times(\delta_1+3.5z)^2 - 699\delta_1\right].$$

By $\delta_1 > \delta_0 = 71.8$ and $z > 1.8$, we see that $b_1 < 0$. Hence by Theorem 6.2.7, the system (6.2.1) undergoes a dynamic transition at $\delta = \delta_1$ to the following steady-state solutions:

$$u_\pm = U_1 \pm \left(-\frac{L}{2(\mu_2\rho_{k_0}+\delta_1)^2 b_1}\beta_{k_0 l}(\delta)\right)^{1/2}\xi\cos\frac{(k_0-1)\pi}{L}x + o\left(\beta_{k_0 l}^{1/2}\right),$$

which are attractors.

6.3 Belousov–Zhabotinsky Reactions: Brusselator

The main objective of this section is to study the dynamic phase transitions of the Belousov–Zhabotinsky reactions focusing on the Brusselator, first introduced by Prigogine and Lefever 1968. The Brusselator is one of the simplest models in nonlinear chemical systems. It has six components, four of which retain constants, and the other two permit their concentrations to vary in time and space. The chemical reaction consists of four irreversible steps, given by

$$\begin{aligned}
&\mathrm{A} \xrightarrow{k_1} \mathrm{X},\\
&\mathrm{B}+\mathrm{X} \xrightarrow{k_2} \mathrm{Y}+\mathrm{D},\\
&2\mathrm{X}+\mathrm{Y} \xrightarrow{k_3} 3\mathrm{X},\\
&\mathrm{X} \xrightarrow{k_4} \mathrm{E},
\end{aligned} \qquad (6.3.1)$$

where A and B are constant components, D and E are products, and X and Y are the two components variable in time and space. Over the years, there have been extensive studies of the Brusselator and related chemical reaction problems; see among many others (Tzou et al. 2009; Guo and Han 2009; Bao 2002; Kaper and Kaper 2002; Shi 2009) and the references therein.

Dynamic phase transition of the Brusselator model is briefly addressed in this section without proofs, and we refer the interested reader to Ma and Wang 2011d for details. We derive a complete characterization of the transition from the homogeneous state. There are two aspects of this characterization. First, our analysis shows that the transitions both to multiple equilibria and to time-periodic solutions (spatiotemporal oscillations) can occur for the Brusselator model, and they are precisely determined by the sign of an explicit nondimensional parameter $\delta_0 - \delta_1$ as defined by (6.3.12) and (6.3.13). Then in both transition cases, the dynamic behavior of the transition is classified, and the type of transitions is determined also by the signs of some nondimensional and computable parameters.

6.3.1 Prigogine–Lefever Model

Let u_1, u_2, a, and b stand for the concentrations of X, Y, A and B. Then the reaction equations of (6.3.1) are given by

$$\begin{aligned} \frac{\partial u_1}{\partial t} &= \sigma_1 \Delta u_1 + k_1 a - (k_2 b + k_4) u_1 + k_3 u_1^2 u_2, \\ \frac{\partial u_2}{\partial t} &= \sigma_2 \Delta u_2 + k_2 b u_1 - k_3 u_1^2 u_2. \end{aligned} \tag{6.3.2}$$

To get the nondimensional form of (6.3.2), let

$$\begin{aligned} &t = k_4^{-1} t', && x = l x', && u_i = \left(\frac{k_4}{k_3}\right)^{1/2} u_i', \\ &a = (k_4^3 / k_1^2 k_3)^{1/2} \alpha, && b = (k_4/k_2)\lambda, && \sigma_i = l^2 k_4 \mu_i. \end{aligned}$$

Omitting the primes, equations (6.3.2) become

$$\begin{aligned} \frac{\partial u_1}{\partial t} &= \mu_1 \Delta u_1 + \alpha - (\lambda + 1) u_1 + u_1^2 u_2, \\ \frac{\partial u_2}{\partial t} &= \mu_2 \Delta u_2 + \lambda u_1 - u_1^2 u_2, \end{aligned} \tag{6.3.3}$$

where $u_1, u_2 \geq 0$, $\Omega \subset \mathbb{R}^n$ $(1 \leq n \leq 3)$ is a bounded domain, and

$$\mu_1, \mu_2, \alpha, \lambda > 0.$$

The equations (6.3.3) have a constant steady-state solution

$$u_0 = (\alpha, \lambda/\alpha).$$

Make the translation

$$u_1 = \alpha + v_1, \quad u_2 = \frac{\lambda}{\alpha} + v_2.$$

Then (6.3.3) can be written as

$$\begin{aligned}\frac{\partial v_1}{\partial t} &= \mu_1 \Delta v_1 + (\lambda - 1)v_1 + \alpha^2 v_2 + \frac{2\lambda}{\alpha} v_1^2 + 2\alpha v_1 v_2 + v_1^2 v_2, \\ \frac{\partial v_2}{\partial t} &= \mu_2 \Delta v_2 - \lambda v_1 - \alpha^2 v_2 - \frac{2\lambda}{\alpha} v_1^2 - 2\alpha v_1 v_2 - v_1^2 v_2.\end{aligned} \tag{6.3.4}$$

There are two types of physically sound boundary conditions: the Dirichlet boundary condition

$$v = (v_1, v_2) = 0 \quad \text{on } \partial\Omega, \tag{6.3.5}$$

and the Neumann boundary condition

$$\frac{\partial v}{\partial n} = 0 \quad \text{on } \partial\Omega. \tag{6.3.6}$$

Define two function spaces

$$H = L^2(\Omega, \mathbb{R}^2), \qquad H_1 = \begin{cases} H^2(\Omega, \mathbb{R}^2) \bigcap H_0^1(\Omega, \mathbb{R}^2) & \text{for (6.3.5)}, \\ \{v \in H^2(\Omega, \mathbb{R}^2) | \frac{\partial v}{\partial n}|_{\partial\Omega} = 0\} & \text{for (6.3.6)}. \end{cases}$$

Define the operators $L_\lambda = A + B_\lambda$ and $G : H_1 \to H$ by

$$\begin{aligned} Av &= (\mu_1 \Delta v_1, \mu_2 \Delta v_2), \\ B_\lambda v &= ((\lambda - 1)v_1 + \alpha^2 v_2, -\lambda v_1 - \alpha^2 v_2), \\ G(v, \lambda) &= \left(\frac{2\lambda}{\alpha} v_1^2 + 2\alpha v_1 v_2 + v_1^2 v_2, -\frac{2\lambda}{\alpha} v_1^2 - 2\alpha v_1 v_2 - v_1^2 v_2 \right). \end{aligned} \tag{6.3.7}$$

Then (6.3.4) with (6.3.5) or with (6.3.6) can be written in the following abstract form:

$$\frac{dv}{dt} = L_\lambda v + G(v, \lambda). \tag{6.3.8}$$

6.3.2 Linearized Problem

Consider the linearized eigenvalue problem of (6.3.4),

$$\begin{aligned} \mu_1 \Delta v_1 + (\lambda - 1)v_1 + \alpha^2 v_2 &= \beta v_1, \\ \mu_2 \Delta v_2 - \lambda v_1 - \alpha^2 v_2 &= \beta v_2, \end{aligned} \tag{6.3.9}$$

with the boundary condition (6.3.5) or (6.3.6).

Let ρ_k and e_k be the kth eigenvalue and eigenvector of the Laplacian with either the Dirichlet or the Neumann condition. All eigenvalues $\beta_k^{\pm}$ and eigenvectors $\phi_k^{\pm}$ of (6.3.9) satisfy

$$\phi_k^{\pm} = \xi_k^{\pm} e_k, \qquad M_k \xi_k^{\pm} = \beta_k^{\pm} \xi_k^{\pm}, \tag{6.3.10}$$

where $\xi_k^{\pm} \in \mathbb{R}^2$ are the eigenvectors of M_k, and $\beta_k^{\pm}$ are the eigenvalues of M_k:

$$\begin{aligned} \beta_k^{\pm}(\lambda) =& \frac{1}{2}[\lambda - (\mu_1\rho_k + \mu_2\rho_k + \alpha^2 + 1)] \\ & \pm \frac{1}{2}\big[(\lambda - \mu_1\rho_k - \mu_2\rho_k - \alpha^2 - 1)^2 \\ & + 4(\lambda\mu_2\rho_k - (\mu_1\rho_k + 1)(\mu_2\rho_k + \alpha^2))\big]^{1/2}. \end{aligned} \tag{6.3.11}$$

It is clear that $\beta_k^-(\lambda) < \beta_k^+(\lambda) = 0$ if and only if

$$\lambda = \frac{1}{\mu_2\rho_k}(\mu_1\rho_k + 1)(\mu_2\rho_k + \alpha^2) < \mu_1\rho_k + \mu_2\rho_k + \alpha^2 + 1,$$

and $\beta_k^{\pm}(\lambda) = \pm\sigma_k(\lambda)i$ with $\sigma_k \neq 0$ if and only if

$$\lambda = \mu_1\rho_k + \mu_2\rho_k + \alpha^2 + 1 < \frac{1}{\mu_2\rho_k}(\mu_1\rho_k + 1)(\mu_2\rho_k + \alpha^2).$$

Thus we introduce two critical parameters

$$\lambda_0 = \min_{\rho_k} \frac{1}{\mu_2\rho_k}(\mu_1\rho_k + 1)(\mu_2\rho_k + \alpha^2), \tag{6.3.12}$$

$$\lambda_1 = \mu_1\rho_1 + \mu_2\rho_1 + \alpha^2 + 1. \tag{6.3.13}$$

Lemma 6.3.1 *Let λ_0 and λ_1 be the two numbers given by (6.3.12) and (6.3.13). Then we have the following assertions:*

(1) Let $\lambda_0 < \lambda_1$, and let $k_0 \geq 1$ be the integer such that the minimum is achieved at ρ_{k_0} in the definition of λ_0. Then $\beta_{k_0}^+(\lambda)$ is the first real eigenvalue of (6.3.9) near $\lambda = \lambda_0$ satisfying

$$\begin{aligned} &\beta_k^+(\lambda)\begin{cases} <0 & \text{if } \lambda < \lambda_0 \\ =0 & \text{if } \lambda = \lambda_0 \\ >0 & \text{if } \lambda > \lambda_0 \end{cases} && \forall k \in \mathbb{N} \text{ with } \rho_k = \rho_{k_0}, \\ &Re\beta_j^{\pm}(\lambda_0) < 0 && \forall \beta_j^{\pm} \neq \beta_k^+ \text{ with } \rho_k = \rho_{k_0}. \end{aligned} \tag{6.3.14}$$

(2) Let $\lambda_1 < \lambda_0$. Then $\beta_1^+(\lambda) = \overline{\beta}_1^-(\lambda)$ are a pair of first complex eigenvalues of (6.3.9) near $\lambda = \lambda_1$, and

$$\begin{aligned} &Re\beta_1^+(\lambda) = Re\beta_1^-(\lambda)\begin{cases} <0 & \text{if } \lambda < \lambda_1, \\ =0 & \text{if } \lambda = \lambda_1, \\ >0 & \text{if } \lambda > \lambda_1, \end{cases} \\ &Re\beta_k^{\pm}(\lambda_1) < 0 \qquad \forall k > 1. \end{aligned} \tag{6.3.15}$$

6.3.3 Transition from Real Eigenvalues

Consider the case $\lambda_0 < \lambda_1$. Based on Lemma 6.3.1, the system (6.3.8) always undergoes a dynamic transition at the real eigenvalue $\lambda = \lambda_0$. Hereinafter, we always assume that the eigenvalue $\beta_{k_0}^+$ in (6.3.14) is simple. Let ρ_{k_0} be as in Lemma 6.3.1, and let e_{k_0} be an eigenvector of the Laplacian with either the Dirichlet or the Neumann boundary condition, corresponding to ρ_{k_0}.

Theorem 6.3.2 *Let $\lambda_0 < \lambda_1$ and assume*

$$\int_\Omega e_{k_0}^3 dx \neq 0. \tag{6.3.16}$$

Then the system (6.3.4) with (6.3.5) or with (6.3.6) undergoes a Type III (mixed) dynamic transition at $\lambda = \lambda_0$ and bifurcates on each side of $\lambda = \lambda_0$ to a unique branch v^λ of steady-state solutions such that the following assertions hold:

1. *On $\lambda < \lambda_0$, the bifurcated solution v^λ is a saddle, and the stable manifold Γ_λ^1 of v^λ separates the space H into two disjoint open sets U_1^λ and U_2^λ such that $v = 0 \in U_1^\lambda$ is an attractor, and the orbits of (6.3.8) in U_2^λ are far from $v = 0$.*
2. *On $\lambda > \lambda_0$, the stable manifold Γ_λ^0 of $v = 0$ separates the neighborhood $\mathcal{O}$ of $u = 0$ into two disjoint open sets $\mathcal{O}_1^\lambda$ and $\mathcal{O}_2^\lambda$ such that the transition is a jump in $\mathcal{O}_1^\lambda$ and is continuous in O_2^λ. The bifurcated solution $v^\lambda \in \mathcal{O}_2^\lambda$ is an attractor such that for all $\varphi \in \mathcal{O}_2^\lambda$,*

$$\lim_{t\to\infty} \|v(t,\varphi) - v^\lambda\| = 0,$$

   where $v(t,\varphi)$ is the solution of (6.3.8) with $v(0,\varphi) = \varphi$.
3. *The bifurcated solution v^λ can be expressed as*

$$\begin{aligned} v^\lambda &= C\beta_{k_0}^+(\lambda)\xi_{k_0}^+ e_{k_0} + o(|\beta_{k_0}^\pm|),\\ \xi_{k_0}^+ &= (-\mu_2\rho_{k_0}, \mu_1\rho_{k_0}+1),\\ C &= \frac{(\alpha\mu_2\rho_{k_0}(\mu_2\rho_{k_0}+\alpha^2) - \alpha^3(\mu_1\rho_{k_0}+1))\int_\Omega e_{k_0}^2 dx}{2\mu_2^3\rho_{k_0}^3(\mu_1\rho_{k_0}+1)\int_\Omega e_{k_0}^3 dx}. \end{aligned} \tag{6.3.17}$$

For the case

$$\int_\Omega e_{k_0}^3 dx = 0, \tag{6.3.18}$$

we introduce the following parameter:

$$\begin{aligned} b_1 = &\Big[\alpha\mu_2^2\rho_{k_0}^2(\mu_1\rho_{k_0}+1)\int_\Omega e_{k_0}^4 dx - 2\alpha^2\mu_2\rho_{k_0}\int_\Omega \psi_2 e_{k_0}^2 dx\\ &-2(\mu_1\rho_{k_0}+1)(2\mu_2\rho_{k_0}+\alpha^2)\int_\Omega \psi_1 e_{k_0}^2 dx\Big]\\ &\times[\alpha^2(\mu_1\rho_{k_0}+1) - \mu_2\rho_{k_0}(\mu_2\rho_{k_0}+\alpha^2)]^{-1}, \end{aligned} \tag{6.3.19}$$

where $\psi = (\psi_1, \psi_2)$ satisfies

$$\begin{aligned}
&\mu_1 \Delta \psi_1 + (\lambda_0 - 1)\psi_1 + \alpha^2 \psi_2 = -\frac{2\mu_2^2 \rho_{k_0}^2}{\alpha}(\mu_1 \rho_{k_0} + 1)e_{k_0}^2,\\
&\mu_2 \Delta \psi_2 - \lambda_0 \psi_1 - \alpha^2 \psi_2 = \frac{2\mu_2^2 \rho_{k_0}^2}{\alpha}(\mu_1 \rho_{k_0} + 1)e_{k_0}^2,\\
&\psi|_{\partial\Omega} = 0 \quad (\text{or } \frac{\partial \psi}{\partial n}|_{\partial\Omega} = 0).
\end{aligned} \tag{6.3.20}$$

By the Fredholm alternative, under the condition (6.3.18), equation (6.3.20) has a unique solution.

Theorem 6.3.3 *Let $\lambda_0 < \lambda_1$, let b_1 be the number given by (6.3.19), and assume (6.3.18).*

1. *If $b_1 > 0$, then (6.3.8) undergoes a Type II transition at $\lambda = \lambda_0$. Moreover, (6.3.8) has no bifurcation on $\lambda > \lambda_0$ and bifurcates on $\lambda < \lambda_0$ to exactly two solutions v_+^λ and v_-^λ, which are saddles.*
   Moreover, the stable manifolds Γ_+^λ and Γ_-^λ of the two bifurcated solutions divide H into three disjoint open sets U_+^λ, U_0^λ, U_-^λ such that $v = 0 \in U_0^\lambda$ is an attractor, and the orbits in $U_\pm^\lambda$ are far from $v = 0$.
2. *If $b_1 < 0$, then (6.3.8) undergoes a Type I transition at $\lambda = \lambda_0$, has no bifurcation on $\lambda < \lambda_0$, and bifurcates on $\lambda > \lambda_0$ to exactly two solutions v_+^λ and v_-^λ, which are attractors.*
   In addition, there is a neighborhood $\mathcal{O} \subset H$ of $v = 0$ such that the stable manifold Γ of $v = 0$ divides $\mathcal{O}$ into two disjoint open sets $\mathcal{O}_+^\lambda$ and $\mathcal{O}_-^\lambda$ such that $v_+^\lambda \in \mathcal{O}_+^\lambda$, $v_-^\lambda \in \mathcal{O}_-^\lambda$, and $v_\pm^\lambda$ attracts $\mathcal{O}_\pm^\lambda$.
3. *The bifurcated solutions $v_\pm^\lambda$ can be expressed as*

$$\begin{aligned}
&v_\pm^\lambda = \pm C(\beta_{k_0}^+(\lambda))^{1/2} \xi_{k_0}^+ e_{k_0} + o(|\beta_{k_0}^+|^{1/2}),\\
&\xi_{k_0}^+ = (-\mu_2 \rho_{k_0}, \mu_1 \rho_{k_0} + 1),\\
&C = \left[\frac{-\alpha}{\mu_2 \rho_{k_0} b_1} \int_\Omega e_{k_0}^2 dx\right]^{1/2},
\end{aligned} \tag{6.3.21}$$

   where b_1 is as in (6.3.19).

6.3.4 Transition from Complex Eigenvalues

For the case $\lambda_1 < \lambda_0$, (6.3.8) undergoes a dynamic transition at $\lambda = \lambda_1$, leading to spatiotemporal oscillations. For the Neumann boundary condition, $\lambda_1 = \alpha^2 + 1$, and we have the following theorem.

Theorem 6.3.4 *Let $\lambda_1 < \lambda_0$. The problem (6.3.4) with (6.3.6) undergoes a Type I dynamic transition at $\lambda = \lambda_1$ and bifurcates on $\lambda > \lambda_1$ to a periodic solution that is an attractor. Moreover, the bifurcated periodic solution $v^\lambda = (v_1^\lambda, v_2^\lambda)$ can be expressed as*

$$\begin{aligned} v_1^\lambda &= 2[C\beta_0(\lambda)]^{1/2}\alpha^2 \sin(\alpha t + \frac{\pi}{4}) + o(|\beta_0|^{1/2}), \\ v_2^\lambda &= [C\beta_0(\lambda)]^{1/2}\alpha((\alpha-1)\cos\alpha t - (\alpha+1)\sin\alpha t) + o(|\beta_0|^{1/2}), \end{aligned} \tag{6.3.22}$$

where $\beta_0 = \frac{1}{2}(\lambda - \lambda_1)$ and $C = (2\pi\alpha^2 + \frac{3}{2}\pi\alpha^4)^{-1}$.

For the Dirichlet boundary condition case, $\rho_1 > 0$ and $\lambda_1 = (\mu_1 + \mu_2)\rho_1 + \alpha^2 + 1$. We define the following parameter:

$$\begin{aligned} b_1 =& \frac{2\pi\alpha^2[\int_\Omega e_1^3 dx]^2}{\sigma_0^2[\int_\Omega e_1^2 dx]^2}(\mu_1\rho_1 + 1)(\mu_2^2\rho_1^2 + 2\mu_2\rho_1(\mu_1\rho_1 + 1) - \sigma_0^2) \\ &- \frac{\pi\alpha^2\int_\Omega e_1^4 dx}{2\int_\Omega e_1^2 dx}(2\mu_2\rho_1 + 3\alpha^2) \\ &+ \frac{2\pi\alpha^2}{\int_\Omega e_1^2 dx}[(3\mu_1\rho_1 + \mu_2\rho_1 + 3)A_1 + (\mu_1\mu_2\rho_1^2 + \mu_2\rho_1 + \delta_0)B_1] \\ &- \frac{8\pi\alpha^2\sigma_0^2}{\int_\Omega e_1^2 dx}[(\mu_1\rho_1 + \mu_2\rho_1 + 1)A_2 - (\mu_1\mu_2\rho_1^2 + \mu_2\rho_1 - \sigma_0)B_2] \\ &- \frac{4\pi\alpha^2}{\int_\Omega e_1^2 dx}[(\mu_1\mu_2\rho_1^2 + \mu_2\rho_1 - \sigma_0^2)A_3 + \sigma_0^2(\mu_1\rho_1 + \mu_2\rho_1 + 1)B_3], \end{aligned} \tag{6.3.23}$$

where

$$\begin{aligned} \sigma_0 &= [\alpha^2(\mu_1\rho_1 + 1) - \mu_2\rho_1(\mu_2\rho_1 + \alpha^2)]^{1/2}, \\ A_i &= (2\lambda_1 - \mu_2\rho_1 - \alpha^2)B_i + \alpha^2 C_i \quad \text{for } i = 1,2,3, \\ B_1 &= \sum_{k>1}^\infty \frac{\mu_2\rho_k[\int_\Omega e_1^2 e_k dx]^2}{\det M_k \int_\Omega e_k^2 dx}, \\ B_2 &= \sum_{k>1}^\infty \frac{[\int_\Omega e_1^2 e_k dx]^2}{\det M_k \det(M_k^2 + 4\sigma_0^2)\int_\Omega e_k^2 dx}[(\mu_2\rho_k + \alpha^2)D_k + \alpha^2 Q_k], \\ B_3 &= \sum_{k>1}^\infty \frac{[\int_\Omega e_1^2 e_k dx]^2}{\det(M_k^2 + 4\sigma_0^2)\int_\Omega e_k^2 dx} D_k, \\ C_1 &= -\sum_{k>1}^\infty \frac{(\mu_1\rho_1 + 1)[\int_\Omega e_1^2 e_k dx]^2}{\det M_k \int_\Omega e_k^2 dx}, \\ C_2 &= \sum_{k>1}^\infty \frac{[\int_\Omega e_1^2 e_k dx]^2}{\det M_k \det(M_k^2 + 4\sigma_0^2)\int_\Omega e_k^2 dx}[(\mu_1\rho_k + 1 - \lambda_1)Q_k - \lambda_1 D_k], \end{aligned}$$

$$
\begin{aligned}
C_3 &= \sum_{k>1}^{\infty} \frac{[\int_\Omega e_1^2 e_k dx]^2}{\det(M_k^2 + 4\sigma_0^2) \int_\Omega e_k^2 dx} Q_k, \\
D_k &= (\mu_2\rho_k + \alpha^2)^2 + 4\sigma_0^2 - \alpha^2(\mu_1 + \mu_2)\rho_k - \alpha^2(\alpha^2 + 1), \\
Q_k &= \lambda_1\alpha^2 - \lambda_1(\mu_1 + \mu_2)(\rho_k - \rho_1) - (\mu_1\rho_k + 1 - \lambda_1)^2 - 4\sigma_0^2.
\end{aligned}
$$

Here M_k is the matrix defined by

$$
M_k = \begin{pmatrix} -\mu_1\rho_k + \lambda_1 - 1 & \alpha^2 \\ -\lambda_1 & -\mu_2\rho_k - \alpha^2 \end{pmatrix}. \tag{6.3.24}
$$

Theorem 6.3.5 *Let b_1 be the parameter given by (6.3.23) and let $\lambda_1 < \lambda_0$.*

1. *If $b_1 < 0$, then the problem (6.3.4)–(6.3.5) undergoes a Type I transition at $\lambda = \lambda_1$ and bifurcates to a stable periodic solution on $\lambda > \lambda_1$.*
2. *If $b_1 > 0$, then the problem undergoes a Type II transition at $\lambda = \lambda_1$ and bifurcates to an unstable periodic solution on $\lambda < \lambda_1$.*
3. *The bifurcated periodic solution $v^\lambda = (v_1^\lambda, v_2^\lambda)$ can be expressed as*

$$
\begin{aligned}
v_1^\lambda &= 2\alpha^2[-\gamma(\lambda)/b_1]^{1/2} e_1 \sin(\sigma_0 t + \frac{\pi}{4}) + o(|\gamma|^{1/2}), \\
v_2^\lambda &= 2(\sigma_0^2 + (\mu_2\rho_1 + \alpha^2)^2) \left[-\frac{\gamma(\lambda)}{b_1}\right]^{12} e_1 \cos(\sigma_0 t + \theta) + o(|\gamma|^{1/2}), \\
\theta &= \tan^{-1} \frac{\sigma_0 + \mu_2\rho_1 + \alpha^2}{\sigma_0 - \mu_2\rho_1 - \alpha^2},
\end{aligned} \tag{6.3.25}
$$

where $\gamma = (\lambda - \lambda_1)/2$.

6.4 Bacterial Chemotaxis

Chemotaxis is an important phenomenon occurring in many organisms. It involves mobility and aggregation of the species in two aspects: random walk and chemically directed movement. For example, in the slime mold *Dictyostelium discoideum*, the single-cell amoebae move toward regions of relatively high concentration of a chemical called cyclic-AAMP, which is secreted by the amoebae themselves.

Many experiments demonstrate that under proper conditions, a bacterial colony can form a rather regular pattern that is relatively stable on a certain time scale. For example, a series of experimental results on the patterns formed by the bacterium *Escherichia coli* (*E. coli*) and *Salmonella typhimurium* (*S. Typhimurium*) were derived in Budrene and Berg 1991, 1995, where two types of experiments were conducted: one is in semisolid medium, and the other is in liquid medium. Both showed that when the bacteria are exposed to intermediates of the TCA cycle, they can form various regular patterns, typically as ringlike and sunflowerlike formations. In all these experiments, the bacteria are known to secrete aspartate, a potent chemoattractant; also see (Murray 2002; Brenner et al. 1998).

In their pioneering work (Keller and Segel 1970), E.F. Keller and L.A. Segel proposed a model, called the Keller–Segel equations, to describe the chemotactic behavior of slime mold amoebae. In their equations, the growth rate of amoeba cells was ignored, i.e., the model can depict the chemotaxis process on only a small time scale, as exhibited in the liquid-medium experiments with *E. Coli* and *S. Typhimurium* by Budrene and Berg 1991, 1995. However, in the semisolid-medium experiments, the time scale of the pattern-formation process is long enough to accommodate many generations of bacteria. Therefore, various revised models were presented by many authors, taking into consideration the effects of the stimulant (i.e., food source) and the growth rate of the population; see among others (Murray 2002) and the references therein. Also, there is a vast literature on mathematical studies for the Keller–Segel model; see among others (Perthame et al. 2011; Guo and Hwang 2010; Perthame and Dalibard 2009; Nadin et al. 2008).

The main objective of this section is to study dynamic transition and pattern formation of the Keller–Segel model, and the presentation is based mainly on Ma and Wang 2011a.

6.4.1 Keller–Segel Models

The general form of the revised Keller–Segel model is given by

$$\begin{aligned} \frac{\partial u_1}{\partial t} &= k_1 \Delta u_1 - \nabla(\chi(u)\nabla u_2) + f(u), \\ \frac{\partial u_2}{\partial t} &= k_2 \Delta u_2 + g(u), \\ \frac{\partial u_3}{\partial t} &= k_3 \Delta u_3 + h(u) + q(x), \end{aligned} \tag{6.4.1}$$

where $u = (u_1, u_2, u_3)$, u_1 is the population density of individual organisms, u_2 is the chemoattractant concentration, u_3 is the stimulant concentration, $q(x)$ is the nutrient source, and χ, f, g, h are functions of u, which will be determined based on a few biological and mathematical rules as follows.

Basic Modeling Rules. The chemotaxis equations (6.4.1) obey a few biological and mathematical rules as follows:

(R1) As stated in Sect. 1.1, all models describing natural phenomena obey basic physical laws. The first step to model chemotaxis is to seek basic laws describing the underlying phenomena.

(R2) It is clear that a medium consisting of only nutrients without organisms and chemoattractants can be observed in a realistic world. In particular, if some bacteria are inoculated in a nutrient medium, the cells will proliferate, and if inoculated in a nutrient-free medium the cells will vanish.

Equivalently, there exists a steady-state solution $U_0 = (0,0,u_3^0)$ for (6.4.1) such that $u_3^0 \geq 0$ satisfies the equation

$$\begin{aligned} &-k_3\Delta u_3 = q(x), \qquad q \geq 0 \text{ in } \Omega, \\ &\frac{\partial u_3}{\partial n} = 0 \text{ on } \partial\Omega, \end{aligned}$$

and U_0 is stable if $q(x) = 0$, and unstable if $q \neq 0$.

(R3) It is clear that with constant stimulant supply, the bacteria and the chemicals produced by them will exist permanently, but they cannot increase indefinitely even with an abundant food supply.
Equivalently, when $q \neq 0$, the system possesses a nonnegative steady-state solution $u^* = (u_1^*, u_2^*, u_3^*)$ such that

$$u_1^*, u_2^* \leq M, \qquad u^* \to 0 \text{ as } q \to 0, \qquad \|u_3^*\| \to \infty \text{ as } \|q\| \to \infty,$$

where M is a constant independent of q.

(R4) There are no stable steady states of the following types for the system:

$$(\widetilde{u}_1, \widetilde{u}_2, 0), \quad (\widetilde{u}_1, 0, \widetilde{u}_3), \quad (0, \widetilde{u}_2, \widetilde{u}_3), \quad (\widetilde{u}_1, 0, 0), \quad (0, \widetilde{u}_2, 0),$$

for $\widetilde{u}_j \geq 0\ (\not\equiv 0)$, $1 \leq j \leq 3$.

(R5) The space of solutions of (6.4.1) is the set of nonnegative functions, denoted by H^+, that are positively invariant, i.e., for any initial value $u_0 \in H^+$, the solution $u(t,u_0)$ of (6.4.1) with $u(0,u_0) = u_0$ satisfies

$$u(t,u_0) \in H^+, \quad \forall t > 0.$$

(R6) For any initial data in H^+, there exists a global in time weak solution in H^+, and there exists a global attractor in H^+.

(R7) SIMPLICITY PRINCIPLE: A model usually takes its simplest form.

First Equation of (6.4.1). The first basic physical law for chemotaxis is the conservation of bacterial population u_1, which is written as

$$\frac{\partial u_1}{\partial t} = -\nabla J + f(u), \tag{6.4.2}$$

where J is the cell flux, and f the growth rate of u_1.

The mobility of organisms is characterized by a random walk and a directed movement toward a higher concentration of the chemoattractant:

$$J = J_d + J_c. \tag{6.4.3}$$

Physically, the random walk is similar to Brownian motion. Therefore, J_d follows Fick's diffusion law:

$$J_d = -k_1 \nabla u_1, \tag{6.4.4}$$

where $k_1 > 0$ is the diffusion coefficient. It is reasonable to consider chemotaxis as a motion of the attractants in the direction of a concentration gradient:

$$J_c = \chi(u)\nabla u_2, \tag{6.4.5}$$

where $\chi(u)$ is the chemotaxis response function, depending on u_1, u_2, and the bacteria types. Based on items (R5) and (R7), we take $\chi(u)$ in the form

$$\chi(u) = \chi u_1, \tag{6.4.6}$$

with a coefficient $\chi > 0$.

The growth rate f of u_1 is defined by

$$f = \text{birth rate } f_1 - \text{death rate } f_2.$$

By (R2) and the fact that f is independent of the chemoattractant u_2, we assume

$$f_1 = \alpha(u_3)u_1, \quad f_2 = \alpha_1 u_1^n,$$

with a coefficient $\alpha_1 > 0$, an integer $n \geq 2$, and a function $\alpha_3(u_3) \geq 0$. Then (R6) suggests that $n \geq 3$, and by (R7), we take $n = 3$. According to (R3) and (R7), we employ

$$\alpha(u_3) = \frac{ku_3}{\alpha_0 + u_3},$$

where $k > 0$ is a constant. Thus we obtain that

$$f(u) = \alpha_1 u_1 \left(\frac{\alpha_2 u_3}{\alpha_0 + u_3} - u_1^2 \right), \quad (\alpha_1 \alpha_2 = k). \tag{6.4.7}$$

By (6.4.2)–(6.4.7), the first equation of (6.4.1) reads

$$\frac{\partial u_1}{\partial t} = k_1 \Delta u_1 - \chi \nabla (u_1 \nabla u_2) + \alpha_1 u_1 \left(\frac{\alpha_2 u_3}{\alpha_0 + u_3} - u_1^2 \right). \tag{6.4.8}$$

Remark 6.4.1 There has been much discussion as to how to find a precise expression for the chemotaxis function $\chi(u)$. (Lapidus and Schiller 1976) suggest the following form:

$$\chi(u) = \frac{\chi_1 u_1}{(\beta + u_2)^2}, \tag{6.4.9}$$

with two parameters $\chi_1, \beta > 0$, which is regarded as optimal in comparison with experimental data; see (Murray 2002). In (6.4.6), for simplicity we use an average $\overline{u}_2$ to replace u_2 in (6.4.9), i.e., the parameter χ in (6.4.6) is considered to be

$$\chi = \frac{\chi_1}{(\beta + \overline{u}_2)^2}, \tag{6.4.10}$$

where $\overline{u}_2 = \frac{1}{|\Omega|} \int_\Omega u_2^* dx$ for some steady state u^*. □

Remark 6.4.2 Mathematical theories show that for a parabolic system with the Dirichlet or Neumann boundary condition given by

$$\begin{aligned}
&\frac{\partial u_j}{\partial t} = k_j \Delta u_j + F_j(x,u,\nabla u), \quad 1 \le j \le m, \\
&u|_{\partial\Omega} = 0 \quad \text{or} \quad \left.\frac{\partial u}{\partial n}\right|_{\partial\Omega} = 0, \\
&u(0) = u_0,
\end{aligned} \tag{6.4.11}$$

if F_j are of the form

$$\begin{aligned}
&F_j = f_j(x,u,\nabla u)u_j + \sum_{i=1}^{n} g_i(x,u,\nabla u)\frac{\partial u_j}{\partial x_i} = h_j(x,u,\nabla u), \\
&h_j(x,\xi,\eta) \ge 0, \quad \forall (x,\xi,\eta) \in \Omega \times \mathbb{R}^n \times \mathbb{R}^{n^2},
\end{aligned} \tag{6.4.12}$$

then the nonnegative function space H^+ is an invariant set of (6.4.11). Namely, under the condition (6.4.12), if $u_0 \in H^+$, then the solution $u(t,u_0)$ of (6.4.11) is nonnegative, i.e.,

$$u_j(t,u_0) \ge 0 \text{ in } \Omega \qquad \forall t > 0 \quad \text{for } 1 \le i \le m;$$

see (Smoller 1983). Hence due to (R5) and (6.4.12), the expression of $\chi(u)$ must be taken in the following form:

$$\chi(u) = \beta(u)u_1,$$

with some function $\beta(u)$.

The Second and Third Equations of (6.4.1). The chemoattractant u_2 and stimulant u_3 are two types of chemicals governed by the conservation law of mass and Fick's law of diffusion:

$$\begin{aligned}
\frac{\partial u_2}{\partial t} &= k_2 \Delta u_2 - \text{decay of } u_2 + \text{production of } u_2 \text{ by } u_1, \\
\frac{\partial u_3}{\partial t} &= k_3 \Delta u_3 - \text{consumption by } u_1 + \text{nutrient source}.
\end{aligned}$$

Based on (R2)–(R5) and (R7), we take

$$\begin{aligned}
&\text{decay of } u_2 = r_2 u_2, \\
&\text{production of } u_2 \text{ by } u_1 = r_1 u_1, \\
&\text{consumption of } u_3 \text{ by } u_1 = r_3 u_3 u_1,
\end{aligned}$$

with coefficients $r_i > 0$ $(1 \le i \le 3)$. Finally, the food supplies of u_1 and u_3 are given by external sources:

$$\text{nutrient source} = q(x) \ge 0.$$

Explicit Formula of (6.4.1). We are now in a position to write down a more explicit form of (6.4.1) for the Keller–Segel equations:

$$\begin{aligned}
\frac{\partial u_1}{\partial t} &= k_1 \Delta u_1 - \chi \nabla(u_1 \nabla u_2) + \alpha_1 u_1 \left(\frac{\alpha_2 u_3}{\alpha_0 + u_3} - u_1^2 \right), \\
\frac{\partial u_2}{\partial t} &= k_2 \Delta u_2 + r_1 u_1 - r_2 u_2, \\
\frac{\partial u_3}{\partial t} &= k_3 \Delta u_3 - r_3 u_1 u_3 + q(x),
\end{aligned} \tag{6.4.13}$$

which are supplemented with the Neumann boundary condition

$$\frac{\partial u}{\partial n} = 0 \text{ on } \partial \Omega, \tag{6.4.14}$$

and with the initial condition

$$u(0) = u_0. \tag{6.4.15}$$

We remark that equations (6.4.13) are the simplest model satisfying the basic modeling rules (R1)–(R7). From the pattern formation and dynamic transition points of view, we shall see that there is no essential difference between the two terms of the simplest formula (6.4.6) and the Lapidus–Schiller formula (6.4.9) for $\chi(u)$.

Two-Component Systems. For the case in which the stimulant u_3 is ample, i.e., $q, u_3 = \infty$ in (6.4.13), the third equation of (6.4.13) can be ignored, and the birth rate in the first equation reads

$$\frac{\alpha_1 \alpha_2 u_1 u_3}{\alpha_0 + u_3} = \alpha_1 \alpha_2 u_1.$$

Then equations (6.4.13) can be rewritten as

$$\begin{aligned}
\frac{\partial u_1}{\partial t} &= k_1 \Delta u_1 - \chi \nabla(u_1 \nabla u_2) + \alpha_1 u_1 (\alpha_2 - u_1^2), \\
\frac{\partial u_2}{\partial t} &= k_2 \Delta u_2 + r_1 u_1 - r_2 u_2.
\end{aligned} \tag{6.4.16}$$

This model depicts the chemotactic aggregating exhibited by the semisolid-medium experiments with sufficient nutrient supply as mentioned earlier.

In liquid-medium experiments with *E. coli* and *S. typhimurium*, the pattern formation process of the bacteria is on a time scale of minutes. This implies that the organisms neither are born nor die during the aggregating period, and the nutrient u_3 is also considered as being sufficiently supplied. Thus, equations (6.4.16) become

$$\begin{aligned}
\frac{\partial u_1}{\partial t} &= k_1 \Delta u_1 - \chi \nabla(u_1 \nabla u_2), \\
\frac{\partial u_2}{\partial t} &= k_2 \Delta u_2 + r_1 u_1 - r_2 u_2, \\
\int_\Omega & u_1(x,t) dx = \text{constant}.
\end{aligned} \tag{6.4.17}$$

In fact, the third equation (conservation of bacterial population density) in (6.4.17) is compatible with the first equation. To see this, with the boundary condition (6.4.14) we deduce that

$$\frac{d}{dt}\int_{\Omega} u_1 dx = \int_{\partial\Omega}\left[k_1\frac{\partial u_1}{\partial n} - \chi u_1\frac{\partial u_2}{\partial n}\right] ds = 0.$$

Equations (6.4.17) are the original Keller–Segel model with an addition of the conservation of population density equation, which appears to be crucial for eliminating the existence of an infinite number of steady-state solutions.

Chemoattractant Balance Systems. Often, the following form of the Keller–Segel equations is discussed in the literature:

$$\begin{aligned}
&\frac{\partial u_1}{\partial t} = k_1\Delta u_1 - \chi\nabla(u_1\nabla u_2) + \alpha_1 u_1\left(\frac{\alpha_2 u_3}{\alpha_0 + u_3} - u_1^2\right),\\
&\frac{\partial u_3}{\partial t} = k_3\Delta u_3 - r_3 u_1 u_3 + q,\\
&-k_2\Delta u_2 + r_2 u_2 = r_1 u_1.
\end{aligned} \tag{6.4.18}$$

The biological significance of (6.4.18) is that the diffusion and degradation of the chemoattractant secreted by the bacteria are essentially balanced by their production. The advantage of (6.4.18) lies in its mathematical simplicity, preserving the main characteristics of pattern formation and phase transition.

Nondimensional Form. The main objective of this section is to explicate the pattern formation and dynamic transition of the Keller–Segel models (6.4.13)–(6.4.15). For simplicity, the domain $\Omega = (0,l_1)\times(0,l_2)$ is a rectangle with $l_1 \neq l_2$. Also, let

$$(t,x) = (t'/r_2, \sqrt{k_2/r_2}x'), \quad (u_1,u_2,u_3) = (\sqrt{\alpha_2}u_1', k_2u_2'/\chi, \alpha_0 u_3'), \tag{6.4.19}$$

and let us introduce some nondimensional parameters

$$\begin{aligned}
&\lambda = r_1\sqrt{\alpha_2}\chi/r_2k_2, \quad && \alpha = \alpha_1\alpha_2/r_2, \quad && \mu = k_1/k_2,\\
&r = k_3/k_2, && \delta = r_3\sqrt{\alpha_2}/r_2, && \delta_0 = q/r_2\alpha_0.
\end{aligned} \tag{6.4.20}$$

Then the nondimensional form of (6.4.13)–(6.4.15) reads (dropping the primes)

$$\begin{aligned}
&\frac{\partial u_1}{\partial t} = \mu\Delta u_1 - \nabla(u_1\nabla u_2) + \alpha u_1\left(\frac{u_3}{1+u_3} - u_1^2\right),\\
&\frac{\partial u_2}{\partial t} = \Delta u_2 - u_2 + \lambda u_1,\\
&\frac{\partial u_3}{\partial t} = r\Delta u_3 - \delta u_1 u_3 + \delta_0,\\
&\left.\frac{\partial u}{\partial n}\right|_{\partial\Omega} = 0,\\
&u(0) = u_0 \text{ in } \Omega.
\end{aligned} \tag{6.4.21}$$

The nondimensional form of Ω is written as

$$\Omega = (0,L_1)\times(0,L_2), \quad \text{with } L_1 \neq L_2.$$

The nondimensional form of the balanced chemoattractant system is given by

$$\begin{aligned}
&\frac{\partial u_1}{\partial t} = \mu\Delta u_1 - \nabla(u_1\nabla u_2) + \alpha u_1\left(\frac{u_3}{1+u_3} - u_1^2\right),\\
&\frac{\partial u_3}{\partial t} = r\Delta u_3 - \delta u_1 u_3 + \delta_0,\\
&-\Delta u_2 + u_2 = \lambda u_1.\\
&\left.\frac{\partial u}{\partial n}\right|_{\partial\Omega} = 0,\\
&u(0) = u_0.
\end{aligned} \tag{6.4.22}$$

6.4.2 Dynamic Transitions for a Rich Stimulant System

Rich Stimulant System. We know that when nutrient u_3 is richly supplied, the Keller–Segel model (6.4.13) is reduced to a two-component system:

$$\begin{aligned}
&\frac{\partial u_1}{\partial t} = \mu\Delta u_1 - \nabla(u_1\nabla u_2) + \alpha u_1(1-u_1^2),\\
&\frac{\partial u_2}{\partial t} = \Delta u_2 - u_2 + \lambda u_1,\\
&\left.\frac{\partial(u_1,u_2)}{\partial n}\right|_{\partial\Omega} = 0,\\
&u(0) = u_0.
\end{aligned} \tag{6.4.23}$$

It is easy to see that $u^* = (1,\lambda)$ is a steady state of (6.4.23). Consider the deviation from u^*:

$$u = u^* + u'.$$

Suppressing the primes, we see that the system (6.4.23) is then transformed into

$$\begin{aligned}
&\frac{\partial u_1}{\partial t} = \mu\Delta u_1 - 2\alpha u_1 - \Delta u_2 - \nabla(u_1\nabla u_2) - 3\alpha u_1^2 - \alpha u_1^3,\\
&\frac{\partial u_2}{\partial t} = \Delta u_2 - u_2 + \lambda u_1,\\
&\left.\frac{\partial(u_1,u_2)}{\partial n}\right|_{\partial\Omega} = 0,\\
&u(0) = u_0.
\end{aligned} \tag{6.4.24}$$

Dynamic Transition and Pattern Formation for the Diffusion and Degradation Balanced Case. We start with an important case in which the diffusion and degradation of the chemoattractant secreted by the bacteria themselves are almost balanced by their production. In this case, the second equation in (6.4.24) is given by

$$0=\triangle u_2-u_2+\lambda u_1.$$

With the Neumann boundary condition for u_2, we have $u_2=[-\triangle+1]^{-1}u_1$, and the functional form of the resulting equation is given by

$$\frac{\partial u_1}{\partial t}=\mathscr{L}_\lambda u_1+G(u_1,\lambda), \tag{6.4.25}$$

where the operators $\mathscr{L}_\lambda : H_1\to H$ and $G: H_1\times\mathbb{R}\to\mathbb{R}$ are defined by

$$\begin{aligned}&\mathscr{L}_\lambda u_1=\mu\Delta u_1-2\alpha u_1-\lambda\Delta[-\Delta+I]^{-1}u_1,\\&G(u_1,\lambda)=-\lambda\nabla(u_1\nabla[-\Delta+I]^{-1}u_1)-3\alpha u_1^2-\alpha u_1^3.\end{aligned} \tag{6.4.26}$$

Here the two Hilbert spaces H and H_1 are defined by

$$H=L^2(\Omega),\qquad H_1=\left\{u_1\in H^2(\Omega)\mid\frac{\partial u_1}{\partial n}=0\text{ on }\Omega\right\}.$$

To study the dynamic transition of this problem, we need to consider the linearized eigenvalue problem of (6.4.25):

$$\mathscr{L}_\lambda e=\beta(\lambda)e. \tag{6.4.27}$$

Let ρ_k and e_k be the eigenvalues and eigenfunctions of $-\Delta$ with the Neumann boundary condition given by

$$e_k=\cos\frac{k_1\pi x_1}{L_1}\cos\frac{k_2\pi x_2}{L_2},\qquad \rho_k=\pi^2\left(\frac{k_1^2}{L_1^2}+\frac{k_2^2}{L_2^2}\right), \tag{6.4.28}$$

for $k=(k_1,k_2)\in\mathbb{N}_+^2$. Here $\mathbb{N}_+$ is the set of all nonnegative integers. In particular, $e_0=1$ and $\rho_0=0$.

Obviously, the functions in (6.4.28) are also eigenvectors of (6.4.27), and the corresponding eigenvalues β_k are

$$\beta_k(\lambda)=-\mu\rho_k-2\alpha+\frac{\lambda\rho_k}{1+\rho_k}. \tag{6.4.29}$$

Define a critical parameter by

$$\lambda_c=\min_{\rho_k}\frac{(\rho_k+1)(\mu\rho_k+2\alpha)}{\rho_k}. \tag{6.4.30}$$

Let

$$\mathscr{S} = \left\{K = (K_1, K_2) \in \mathbb{N}_+^2 \text{ achieves its minimum in (6.4.30)}\right\}.$$

It follows from (6.4.29) and (6.4.30) that

$$\beta_K(\lambda) \begin{cases} < 0 & \text{if } \lambda < \lambda_c \\ = 0 & \text{if } \lambda = \lambda_c \\ > 0 & \text{if } \lambda > \lambda_c \end{cases} \qquad \forall K = (K_1, K_2) \in \mathscr{S}, \tag{6.4.31}$$

$$\beta_k(\lambda_c) < 0 \qquad \forall k \in \mathbb{Z}^2 \text{ with } k \notin \mathscr{S}. \tag{6.4.32}$$

Notice that for all $K = (K_1, K_2) \in \mathscr{S}$, $K \neq 0$ and

$$\lambda_c = \frac{(\rho_K + 1)(\mu\rho_K + 2\alpha)}{\rho_K}. \tag{6.4.33}$$

We note that for a properly chosen spatial geometry, we have

$$\rho_K = \pi^2\left(\frac{K_1^2}{L_1^2} + \frac{K_2^2}{L_2^2}\right) = \sqrt{\frac{2\alpha}{\mu}} \qquad \forall K = (K_1, K_2) \in \mathscr{S}, \tag{6.4.34}$$

$$\lambda_c = 2\alpha + \mu + 2\sqrt{2\alpha\mu}. \tag{6.4.35}$$

Conditions (6.4.31) and (6.4.32) give rise to a dynamic transition of (6.4.25) from $(u, \lambda) = (0, \lambda_c)$. For simplicity, we define

$$K_1 \triangleq (K_1, 0), \quad K_2 \triangleq (0, K_2),$$

and $K = (K_1, K_2) \in \mathscr{S}$. Also, we introduce a parameter

$$\begin{aligned} b = &- 3\mu\rho_K + \left[12 - \frac{24 - 15\text{sign}(K_1K_2)}{4 - 2\text{sign}(K_1K_2)}\right]\alpha \\ &- \frac{(2\mu\rho_K + \alpha)(2\mu\rho_K^2 + 28\alpha\rho_K + 4\alpha - \mu\rho_K)}{[1 + \text{sign}(K_1K_2)] \cdot [(\mu\rho_{2K} + 2\alpha)(1 + \rho_{2K}) - \lambda_c\rho_{2K}]} \\ &- \frac{2(2\mu\rho_K\rho_{K_1} + 4\alpha\rho_{K_1} - 3\alpha\rho_K)}{(1 + \text{sign}K_1)\rho_K^2[(\mu\rho_{2K_1} + 2\alpha)(1 + \rho_{2K_1}) - \lambda_c\rho_{2K_1}]} \\ &\times [(\mu\rho_K + 2\alpha)(2\rho_{K_1}^2 - 6\rho_{K_1}\rho_{K_2} - \rho_K) + 6\alpha\rho_K(4\rho_{K_1} + 1)] \\ &- \frac{2(2\mu\rho_K\rho_{K_2} + 4\alpha\rho_{K_2} - 3\alpha\rho_K)}{(1 + \text{sign}K_2)\rho_K^2((\mu\rho_{2K_2} + 2\alpha)(1 + \rho_{2K_2}) - \lambda_c\rho_{2K_2})} \\ &\times [(\mu\rho_K + 2\alpha)(2\rho_{K_2}^2 - 6\rho_{K_1}\rho_{K_2} - \rho_K) + 6\alpha\rho_K(4\rho_{K_2} + 1)]. \end{aligned} \tag{6.4.36}$$

The following is the main dynamic transition theorem, providing a precise criterion for the transition type and the pattern-formation mechanism of the system.

Theorem 6.4.3 *Let b be the parameter defined by (6.4.36). Assume that the eigenvalue β_k satisfying (6.4.31) is simple. Then for the system (6.4.25), we have the following assertions:*

(1) The system always undergoes a dynamic transition at $(u,\lambda)=(0,\lambda_c)$. Namely, the basic state $u=0$ is asymptotically stable for $\lambda<\lambda_c$, and is unstable for $\lambda>\lambda_c$.

(2) For the case $b<0$, this transition is continuous (Type I). In particular, the system bifurcates from $(0,\lambda_c)$ to two steady-state solutions on $\lambda>\lambda_c$, which can be expressed as

$$u_1^{\pm}(x,\lambda)=\pm\frac{1}{2}\sqrt{\frac{\beta_K(\lambda)}{2|b|}}\cos\frac{K_1\pi x_1}{L_1}\cos\frac{K_2\pi x_2}{L_2}+o\left(\beta_K^{1/2}\right), \tag{6.4.37}$$

and $u_1^{\pm}(x,\lambda)$ are attractors.

(3) For the case $b>0$, this transition is a jump (Type II), and the system has two saddle-node bifurcation solutions at some $\lambda^*(0<\lambda^*<\lambda_c)$ such that there are two branches v_1^λ and v_2^λ of steady states bifurcated from (v^*,λ^*), and there are two other branches v_3^λ and v_4^λ bifurcated from (u^*,λ^*). In addition, v_1^λ and v_3^λ are saddles, v_2^λ and v_4^λ are attractors, with $v_1^\lambda, v_3^\lambda\to 0$ as $\lambda\to\lambda_c$.

Two remarks are now in order.

Remark 6.4.4 From the pattern-formation point of view, for the Type I transition, the patterns described by the transition solutions given in (6.4.37) are either lamellar or rectangular:

lamellar pattern	for $K_1K_2=0$,
rectangular pattern	for $K_1K_2\neq 0$.

When $b>0$, the system undergoes a more drastic change. For $\lambda^*<\lambda<\lambda_c$, the homogeneous state, the new patterns v_2^λ and v_4^λ are metastable. For $\lambda>\lambda_c$, the system undergoes transitions to more complex patterns away from the basic homogeneous state form.

Remark 6.4.5 If we take the growth term $f(u)$ as $f=\alpha u_1(1-u_1)$ instead of $f=\alpha u_1(1-u_1^2)$ in (6.4.23), (6.4.24), and (6.4.25), then Theorem 6.4.3 still holds except the assertion on the existence of the two saddle-node bifurcation solutions, and the parameter should be replaced by

$$\begin{aligned}
b=&-\mu\rho_K+\alpha-\frac{(2\mu\rho_K+\alpha)(2\mu\lambda_K^2+10\alpha\rho_K+\alpha-\mu\rho_K)}{2(1+\mathrm{sign}K_1K_2)[(\mu\rho_{2K}+\alpha)(1+\rho_{2K})-\lambda_c\rho_{2K}]}\\
&-\frac{(2\mu\rho_K\rho_{K_1}+2\alpha\rho_{K_1}-\alpha\rho_K)}{(1+\mathrm{sign}K_1)\lambda_K^2[(\mu\rho_{2K_1}+\alpha)(1+\rho_{2K_1})-\lambda_c\rho_{2K_1}]}\\
&\times[(\mu\rho_K+\alpha)(2\lambda_{K_1}^2-6\rho_{K_1}\rho_{K_2}-\rho_K)+2\alpha\rho_K(4\rho_{K_1}+1)]\\
&-\frac{(2\mu\rho_K\rho_{K_2}+2\alpha\rho_{K_2}-\alpha\rho_K)}{(1+\mathrm{sign}K_2)\lambda_K^2[(\mu\rho_{2K_2}+\alpha)(1+\rho_{2K_1})-\lambda_c\rho_{2K_2}]}\\
&\times[(\mu\rho_K+\alpha)(2\lambda_{K_1}^2-6\rho_{K_1}\rho_{K_2}-\rho_K)+2\alpha\rho_K(4\rho_{K_2}+1)].
\end{aligned}$$

Pattern Formation and Dynamic Transition for the General Case. Consider the general case (6.4.23). In this case, the unknown variable becomes $u = (u_1, u_2)$, and the basic function spaces are then defined by

$$H = L^2(\Omega, \mathbb{R}^2), \qquad H_1 = \{u \in H^2(\Omega, \mathbb{R}^2) \mid \frac{\partial u}{\partial n} = 0 \text{ on } \Omega\}.$$

Let $L_\lambda : H_1 \to H$ and $G : H_1 \to H$ be defined by

$$\begin{aligned} L_\lambda u &= \begin{pmatrix} \mu\Delta - 2\alpha & -\Delta \\ \lambda & \Delta - 1 \end{pmatrix} u, \\ G(u) &= \begin{pmatrix} -\nabla(u_1 \nabla u_2) - 3\alpha u_1^2 - \alpha u_1^3 \\ 0 \end{pmatrix}. \end{aligned} \tag{6.4.38}$$

The linearized eigenvalue problem of (6.4.24) is $L_\lambda \varphi = \beta\varphi$, where $L_\lambda : H_1 \to H$ is defined by (6.4.38). Let B_k^λ be the matrices given by

$$B_k^\lambda = \begin{pmatrix} -(\mu\rho_k + 2\alpha) & \rho_k \\ \lambda & -(\rho_k + 1) \end{pmatrix}, \tag{6.4.39}$$

where ρ_k are the eigenvalues in (6.4.28). It is easy to see that all eigenvectors φ_k and eigenvalues β_k of L_λ can be expressed as follows:

$$\varphi_k = \begin{pmatrix} \xi_{k1} e_k \\ \xi_{k2} e_k \end{pmatrix}, \qquad B_k^\lambda \begin{pmatrix} \xi_{k1} \\ \xi_{k2} \end{pmatrix} = \beta_k \begin{pmatrix} \xi_{k1} \\ \xi_{k2} \end{pmatrix}, \tag{6.4.40}$$

where e_k are as in (6.4.28), and β_k are also the eigenvalues of B_k^λ. By (6.4.39), β_k can be expressed by

$$\begin{aligned} \beta_k^\pm(\lambda) &= \frac{1}{2}\left[-B \pm \sqrt{B^2 - 4((\rho_k + 1)(\mu\rho_k + 2\alpha) - \lambda\rho_k)}\right], \\ B &= (\mu + 1)\rho_k + 2\alpha + 1. \end{aligned} \tag{6.4.41}$$

Let λ_c be the parameter defined by (6.4.30). It follows from (6.4.41) and (6.4.30) that

$$\beta_K^+(\lambda) \begin{cases} < 0 & \text{if } \lambda < \lambda_c, \\ = 0 & \text{if } \lambda = \lambda_c, \\ > 0 & \text{if } \lambda > \lambda_c, \end{cases} \tag{6.4.42}$$

$$\begin{cases} \text{Re}\beta_k^-(\lambda_c) < 0 \ \forall k \in \mathbb{Z}^2, \\ \text{Re}\beta_k^+(\lambda_c) < 0 \ \forall k \in \mathbb{Z}^2 \text{ with } \rho_k \neq \rho_K, \end{cases} \tag{6.4.43}$$

with $K = (K_1, K_2)$ as in (6.4.33).

Then we have the following dynamic transition theorem.

Theorem 6.4.6 *Let b be the parameter defined by (6.4.36). Assume that the eigenvalue β_K^+ satisfying (6.4.42) is simple. Then the assertions of Theorem 6.4.3 hold for (6.4.24), with the expression (6.4.37) replaced by*

$$\begin{aligned} u_\lambda^\pm &= \pm\sqrt{a\beta_K^+(\lambda)}\left(\frac{\rho_K+1}{\lambda_c}\right)\cos\frac{K_1\pi x_1}{L_1}\cos\frac{K_2\pi x_2}{L_2}+o(|\beta_K^+|^{1/2}),\\ a &= \frac{8(\mu\rho_K+\rho_K+2\alpha+1)}{(\rho_K+1)^3|b|}. \end{aligned}$$

Proof of Theorem 6.4.3. Assertion (1) follows directly from the general dynamic transition theorem in Chap. 2. To prove assertions (2) and (3), we need to reduce (6.4.25) to the center manifold near $\lambda=\lambda_c$. We note that although the underlying system is now quasilinear in this general case, the center manifold reduction holds as well; see (Liu et al. 2012) for details.

To this end, let $u = xe_k+\Phi$, where $\Phi(x)$ the center manifold function of (6.4.25). Since $L_\lambda : H_1 \to H$ is symmetric, the reduced equation is given by

$$\frac{dx}{dt} = \beta_K(\lambda)x+\frac{1}{(e_K,e_K)}(G(xe_K+\Phi,\lambda),e_K), \tag{6.4.44}$$

where $G: H_1\to H$ is defined by (6.4.26), and

$$(e_K,e_K) = \int_\Omega e_K^2dx = \frac{2-\text{sign}(K_1K_2)}{4}|\Omega|. \tag{6.4.45}$$

It is known that the center manifold function satisfies $\Phi(x)=O(x^2)$. A direct computation shows that

$$\begin{aligned} (G(xe_K+\Phi,\lambda_c),e_K) =& -\alpha x^3\int_\Omega e_K^4dx - 6\alpha x\int_\Omega e_K^2\Phi dx\\ &+\lambda_c x\int_\Omega\Big[e_K\nabla e_k\cdot\nabla(-\Delta+I)^{-1}\Phi\\ &+\Phi\nabla e_k\cdot\nabla(-\Delta+I)^{-1}e_K\Big]dx+o(x^3). \end{aligned}\tag{6.4.46}$$

It is clear that

$$(-\Delta+I)^{-1}e_K = \frac{1}{\rho_K+1}e_K,\quad \Delta e_K = -\rho_K e_K.$$

We infer from (6.4.46) that

$$\begin{aligned} (G(xe_K+\Phi,\lambda_c),e_K) =& -\alpha x^3\int_\Omega e_K^4dx - 6\alpha x\int_\Omega e_K^2\Phi dx\\ &+\lambda_c x\int_\Omega\Big[\frac{1}{\rho_K+1}|\nabla e_K|^2\Phi - |\nabla e_K|^2(-\Delta_I)^{-1}\Phi\\ &+\rho_K e_K^2(-\Delta+I)^{-1}\Phi\Big]dx+o(x^3). \end{aligned}\tag{6.4.47}$$

By (A.1.14), Φ satisfies the equation

$$\begin{aligned} -L_{\lambda_c}\Phi &= G_2(xe_K,\lambda_c)+o(x^2) \\ &= x^2\left[\left(\frac{\rho_K\lambda_c}{\rho_K+1}-3\alpha\right)e_K^2-\frac{\lambda_c}{\rho_K+1}|\nabla e_K|^2\right]+o(x^2). \end{aligned} \tag{6.4.48}$$

In view of (6.4.28), we obtain

$$\begin{aligned} e_K^2 &= \frac{1}{4}[e_0+e_{2K_1}+e_{2K_2}+e_{2K}], \\ |\nabla e_K|^2 &= \frac{1}{4}[\rho_K e_0+(\rho_{K_2}-\rho_{K_1})e_{2K_1}+(\rho_{K_1}-\rho_{K_2})e_{2K_2}-\rho_K e_{2K}]. \end{aligned} \tag{6.4.49}$$

Thus (6.4.48) can be written as

$$\begin{aligned} -L_{\lambda_c}\Phi =& \frac{x^2}{4}\Big[-3\alpha e_0+\left(\frac{2\rho_{K_1}\lambda_c}{\rho_K+1}-3\alpha\right)e_{2K_1}+\left(\frac{2\rho_{K_2}\lambda_c}{\rho_K+1}-3\alpha\right)e_{2K_2} \\ &+\left(\frac{2\rho_K\lambda_c}{\rho_K+1}e_{2K}-3\alpha\right)e_{2K}\Big]+o(x^2). \end{aligned} \tag{6.4.50}$$

Let

$$\Phi=\Phi_0 e_0+\Phi_{2K_1}e_{2K_1}+\Phi_{2K_2}e_{2K_2}+\Phi_{2K}e_{2K}. \tag{6.4.51}$$

Note that

$$-L_{\lambda_c}e_{2K}=\frac{1}{1+\rho_{2K}}[(1+\rho_{2K})(\mu\rho_K+2\alpha)-\lambda_c\rho_{2K}]e_{2K}. \tag{6.4.52}$$

Then by (6.4.33) and (6.4.50)–(6.4.52), we obtain

$$\begin{aligned} \Phi_0 &= -\frac{3}{8}, \\ \Phi_{2K_1} &= \frac{(1+\rho_{2K_1})(2\mu\rho_K\rho_{K_1}+4\alpha\rho_{K_1}-3\alpha\rho_K)}{4\rho_K[(1+\rho_{2K_1})(\mu\rho_{2K_1}+2\alpha)-\rho_{2K_1}\lambda_c]}, \\ \Phi_{2K_2} &= \frac{(1+\rho_{2K_2})(2\mu\rho_K\rho_{K_2}+4\alpha\rho_{K_2}-3\alpha\rho_K)}{4\rho_K[(1+\rho_{2K_2})(\mu\rho_{2K_2}+2\alpha)-\rho_{2K_2}\lambda_c]}, \\ \Phi_{2K} &= \frac{(1+\rho_{2K})(2\mu\rho_K+\alpha)}{4[(1+\rho_{2K})(\mu\rho_{2K}+2\alpha)-\rho_{2K}\lambda_c]}. \end{aligned} \tag{6.4.53}$$

Inserting (6.4.51) and (6.4.28) into (6.4.47), we get

$$\begin{aligned} &(G(xe_K+\Phi,\lambda_c),e_K) = -\alpha x^3\int_\Omega e_K^4 dx \\ &-\frac{6\alpha x(2-\text{sign}(K_1K_2))}{4}\int_\Omega\left[\Phi_0 e_0^2+\Phi_{2K_1}e_{2K_1}^2+\Phi_{2K_2}e_{2K_2}^2+\Phi_{2K}e_{2K}^2\right]dx \\ &+\frac{\lambda_c x(2-\text{sign}(K_1K_2))}{4(\rho_K+1)}\int_\Omega[\rho_K\Phi_0 e_0^2+(\rho_{K_2}-\rho_{K_1})\Phi_{2K_1}e_{2K_1}^2 \end{aligned}$$

$$
\begin{aligned}
&+(\rho_{K_1}-\rho_{K_2})\Phi_{2K_2}e_{2K_2}^2-\rho_K\Phi_{2K}e_{2K}^2\big]dx\\
&-\frac{\lambda_c x(2-\mathrm{sign}(K_1K_2))}{4}\int_\Omega\Big[\rho_K\Phi_0e_0^2+\frac{\rho_{K_2}-\rho_{K_1}}{1+\rho_{2K_1}}\Phi_{2K_1}e_{2K_1}^2\\
&+\frac{\rho_{K_1}-\rho_{K_2}}{1+\rho_{2K_2}}\Phi_{2K_2}e_{2K_2}^2-\frac{\rho_K}{1+\rho_{2K}}\Phi_{2K}e_{2K}^2\Big]dx\\
&+\frac{\lambda_c\rho_K x(2-\mathrm{sign}(K_1K_2))}{4}\int_\Omega\Big[\Phi_0e_0^2+\frac{\Phi_{2K_1}}{1+\rho_{2K_1}}e_{2K_1}^2+\frac{\Phi_{2K_2}}{1+\rho_{2K_2}}e_{2K_2}^2\\
&+\frac{\Phi_{2K}}{1+\rho_{2K}}e_{2K}^2\Big]dx+o(x^3)\\
=&-\alpha x^3\int_\Omega e_K^4dx+\frac{|\Omega|x(2-\mathrm{sign}(K_1K_2))}{4}\\
&\times\Big[(\mu\rho_K-4\alpha)\Phi_0+\frac{1}{1+\mathrm{sign}K_1}\left(\frac{\lambda_c(\rho_{K_2}-\rho_{K_1})}{1+\rho_K}+\frac{2\lambda_c\rho_{K_1}}{1+\rho_{2K_1}}-6\alpha\right)\Phi_{2K_1}\\
&+\frac{1}{1+\mathrm{sign}K_2}\left(\frac{\lambda_c(\rho_{K_1}-\rho_{K_2})}{1+\rho_K}+\frac{2\lambda_c\rho_{K_2}}{1+\rho_{2K_1}}-6\alpha\right)\Phi_{2K_2}\\
&+\frac{1}{2(1+\mathrm{sign}(K_1K_2))}\left(-\frac{\lambda_c\rho_K}{1+\rho_K}+\frac{2\lambda_c\rho_K}{1+\rho_{2K}}-6\alpha\right)\Phi_{2K}\Big]dx+o(x^3).
\end{aligned}
\tag{6.4.54}
$$

Also, we note that

$$\int_\Omega e_K^4=\int_0^{L_1}e_{K_1}^4dx_1\int_0^{L_2}e_{K_2}^4dx_2=\frac{24-15\mathrm{sign}(K_1K_2)}{64}.$$

Then, putting (6.4.53) into (6.4.54), we arrive at

$$(G(xe_K+\Phi,\lambda_c),e_K)=\frac{(2-\mathrm{sign}(K_1K_2))|\Omega|x^3}{32}b+o(x^3), \tag{6.4.55}$$

where b is the parameter given by (6.4.36).

By (6.4.44) and (6.4.55), we derive the reduced equation on the center manifold as follows:

$$\frac{dx}{dt}=\beta_K(\lambda)x+\frac{b}{8}x^3+o(x^3). \tag{6.4.56}$$

Based on the dynamic transition theory developed in Chap. 2, we obtain assertions (2) and (3), except that two saddle-node bifurcations occur at the same point $\lambda=\lambda^*$. To prove this conclusion, we note that if $u^*(x)$ is a steady-state solution of (6.4.25), then

$$v^*(x)=u^*(x+\pi)=u^*(x-\pi)$$

is also a steady-state solution of (6.4.25). This is because the eigenvectors (6.4.28) form an orthogonal base of H_1. Hence, two saddle-node bifurcations on $\lambda<\lambda_c$ imply that they must occur at the same point $\lambda=\lambda^*$. Thus the proof of the theorem is complete. □

Proof of Theorem 6.4.6. Assertion (1) follows from (6.4.42) and (6.4.43). To prove assertions (2) and (3), we need to get the reduced equation of (6.4.24) to the center manifold near $\lambda = \lambda_c$.

Let $u = x \cdot \varphi_K + \Phi$, where φ_K is the eigenvector of L_λ corresponding to β_K at $\lambda = \lambda_c$, and $\Phi(x)$ the center manifold function of (6.4.24). Then the reduced equation of (6.4.24) reads

$$\frac{dx}{dt} = \beta_K^+(\lambda)x + \frac{1}{(\varphi_K, \varphi_K^*)}(G(x \cdot \varphi_K + \Phi), \varphi_K^*). \tag{6.4.57}$$

Here φ_K^* is the conjugate eigenvector of φ_K.

By (6.4.40), φ_K can be written as

$$\varphi_K = \begin{pmatrix} \xi_1 e_K \\ \xi_2 e_K \end{pmatrix}, \qquad \begin{pmatrix} -(\mu\rho_K + 2\alpha) & \rho_K \\ \lambda_c & -(\rho_K + 1) \end{pmatrix} \begin{pmatrix} \xi_1 \\ \xi_2 \end{pmatrix} = 0, \tag{6.4.58}$$

which implies that

$$(\xi_1, \xi_2) = (\rho_K + 1, \ \lambda_c). \tag{6.4.59}$$

In the same manner, we derive

$$\varphi_K^* = (\xi_1^* e_K, \ \xi_2^* e_K)^T = (\rho_K + 1, \rho_K)^T e_K. \tag{6.4.60}$$

By (6.4.38), the nonlinear operator G is

$$\begin{aligned} G(u_1, u_2) &= G_2(u_1, u_2) + G_3(u_1, u_2), \\ G_2(u_1, u_2) &= -(\nabla u_1 \nabla u_2 + u_1 \Delta u_2 + 3\alpha u_1^2) \begin{pmatrix} 1 \\ 0 \end{pmatrix}, \\ G_3(u_1, u_2) &= -\alpha u^3 \begin{pmatrix} 1 \\ 0 \end{pmatrix}. \end{aligned}$$

It is known that the center manifold function is given by $\Phi(x) = (\Phi_1(x), \Phi_2(x)) = O(x^2)$. Then in view of (6.4.58) and (6.4.59), by direct computation we derive

$$\begin{aligned} &(G(x\xi_1 e_K + \Phi_1, \ x\xi_2 e_K + \Phi_2), \ \varphi_K^*) \\ &= (xG_2(\xi_1 e_K, \Phi_2) + xG_2(\Phi_1, \xi_2 e_K) + x^3 G_3(\xi_1 e_K, \xi_2 e_K), \varphi_K^*) + o(x^3) \\ &= x\xi_1^* \int_\Omega [\xi_2 \Phi_1 |\nabla e_K|^2 - \frac{1}{2}\xi_1 \Delta \Phi_2 e_K^2 - 6\alpha \xi_1 \Phi_1 e_K^2] dx \\ &\quad \alpha \xi_1^* \xi_1^3 x^3 \int_\Omega e_K^4 dx + o(x^3). \end{aligned} \tag{6.4.61}$$

Using the approximation formula for center manifold functions given in (A.11) in Ma and Wang 2010b, we see that $\Phi = (\Phi_1, \Phi_2)$ satisfies

$$\begin{aligned} -L_{\lambda_c}\Phi &= -x^2 G_2(\xi_1 e_k, \xi_2 e_K) + o(x^2) \\ &= -x^2(\xi_1 \xi_2 |\nabla e_K|^2 + (3\alpha\xi_1^2 - \xi_1 \xi_2 \rho_K) e_K^2) \begin{pmatrix} 1 \\ 0 \end{pmatrix} + o(x^2). \end{aligned} \tag{6.4.62}$$

From (6.4.28), we see that

$$
\begin{aligned}
e_K^2 =& \frac{1}{4}(1+e_{2K_1})(1+e_{2K_2}) = \frac{1}{4}(e_0+e_{2K_1}+e_{2K_2}+e_{2K}),\\
|\nabla e_K|^2 =& \frac{\rho_{K_1}}{4}(1-e_{2K_1})(1+e_{2K_2})+\frac{\rho_{K_2}}{4}(1+e_{2K_1})(1-e_{2K_2})\\
=& \frac{\rho_K}{4}e_0+\frac{\rho_{K_2}-\rho_{K_1}}{4}e_{2K_1}+\frac{\rho_{K_1}-\rho_{K_2}}{4}e_{2K_2}-\frac{\rho_K}{4}e_{2K}.
\end{aligned}
$$

Thus, (6.4.62) can be written as

$$
\begin{aligned}
L_{\lambda_c}\Phi =& -\frac{\xi_1 x^2}{4}(3\alpha\xi_1 e_0+(3\alpha\xi_1-2\xi_2\rho_{K_1})e_{2K_1}\\
&+(3\alpha\xi_1-2\xi_2\rho_{K_2})e_{2K_2}+(3\alpha\xi_1-2\xi_2\rho_K)e_{2K})\begin{pmatrix}1\\0\end{pmatrix}+o(x^3). \qquad (6.4.63)
\end{aligned}
$$

Let

$$
\begin{pmatrix}\Phi_1\\ \Phi_2\end{pmatrix}=\begin{pmatrix}\Phi_1^0\\ \Phi_2^0\end{pmatrix}e_0+\begin{pmatrix}\Phi_1^{2K_1}\\ \Phi_2^{2K_1}\end{pmatrix}e_{2K_1}+\begin{pmatrix}\Phi_1^{2K_2}\\ \Phi_2^{2K_2}\end{pmatrix}e_{2K_2}+\begin{pmatrix}\Phi_1^{2K}\\ \Phi_2^{2K}\end{pmatrix}e_{2K}. \qquad (6.4.64)
$$

It is clear that

$$
L_\lambda\begin{pmatrix}\Phi_1^k\\ \Phi_2^k\end{pmatrix}e_k=B_k^\lambda\begin{pmatrix}\Phi_1^k\\ \Phi_2^k\end{pmatrix}e_k,
$$

where B_k^λ is the matrix given by (6.4.39). Then by (6.4.63) and (6.4.64), we have

$$
\begin{pmatrix}\Phi_1^{2k}\\ \Phi_2^{2k}\end{pmatrix}=-\frac{(3\alpha\xi_1^2-2\xi_1\xi_2\rho_k)x^2}{4}B_{2k}^{-1}\begin{pmatrix}1\\0\end{pmatrix}, \qquad B_{2k}=B_{2k}^{\lambda_c} \qquad \text{for } k=K,K_1,K_2.
$$

Direct computation shows that

$$
\begin{pmatrix}\Phi_1^0\\ \Phi_2^0\end{pmatrix}=\frac{3\xi_1^2x^2}{8}\begin{pmatrix}1\\ \lambda_c\end{pmatrix}, \qquad (6.4.65)
$$

$$
\begin{pmatrix}\Phi_1^{2K_1}\\ \Phi_2^{2K_2}\end{pmatrix}=\frac{\xi_1(3\alpha\xi_1-2\xi_2\rho_{K_1})}{4\det B_{2K_1}}\begin{pmatrix}1+\rho_{2K_1}\\ \lambda_c\end{pmatrix}, \qquad (6.4.66)
$$

$$
\begin{pmatrix}\Phi_1^{2K_2}\\ \Phi_2^{2K_2}\end{pmatrix}=\frac{\xi_1(3\alpha\xi_1-2\xi_2\rho_{K_2})}{4\det B_{2K_2}}\begin{pmatrix}1+\rho_{2K_2}\\ \lambda_c\end{pmatrix}, \qquad (6.4.67)
$$

$$
\begin{pmatrix}\Phi_1^{2K}\\ \Phi_2^{2K}\end{pmatrix}=\frac{\xi_1(3\alpha\xi_1-2\xi_2\rho_K)}{4\det B_{2K}}\begin{pmatrix}1+\rho_{2K}\\ \lambda_c\end{pmatrix}. \qquad (6.4.68)
$$

Inserting (6.4.65) into (6.4.61), by (6.4.59) and (6.4.60) we get

$$\begin{aligned}(G(x\varphi_K+\Phi),\varphi_K^*)=&\frac{(2-\text{sign}K_1K_2)(\rho_K+1)|\Omega|}{8}\cdot\Big[-\frac{8\alpha(\rho_K+1)^3x^3\int_\Omega e_K^4dx}{(2-\text{sign}K_1K_2)|\Omega|}\\&+2(\xi_2\rho_K-6\alpha\xi_1)\Phi_1^0x+\frac{2(\xi_2(\rho_{K_2}-\rho_{K_1})-6\alpha\xi_1)\Phi_1^{2K_1}x}{1+\text{sign}K_1}\\&+\frac{2(\xi_2(\rho_{K_1}-\rho_{K_2})-6\alpha\xi_1)\Phi_2^{2K_2}x}{1+\text{sign}K_2}\\&-\frac{2-\text{sign}K_1K_2}{2}(\xi_2\rho_K+6\alpha\xi_1)\Phi_1^{2K}x\\&+\frac{(\rho_K+1)\rho_{2K_1}}{1+\text{sign}K_1}\Phi_2^{2K_1}x+\frac{(\rho_K+1)\rho_{K_2}}{1+\text{sign}K_2}\Phi_2^{2K_2}x\\&+\frac{(\rho_K+1)\rho_{2K}(2-\text{sign}K_1K_2)}{4}\Phi_2^{2K}x\Big]+o(x^3).\end{aligned}$$

By definition, we have

$$\begin{aligned}&\rho_{K_1}+\rho_{K_2}=\rho_K,\quad \rho_{2K}=4\rho_K\qquad\qquad \forall K=(K_1,K_2),\\&(\varphi,\varphi^*)=\left[(\rho_K+1)^2+\rho_K\lambda_c\right]\int_\Omega e_K^2dx\\&\qquad=\frac{2-\text{sign}(K_1K_2)}{4}(\rho_K+1)(\mu\rho_K+\rho_K+2\alpha+1)|\Omega|.\end{aligned}$$

In view of (6.4.65)–(6.4.68), the reduced equation (6.4.57) is given by

$$\frac{dx}{dt}=\beta_K^+(\lambda)x+\frac{(\rho_K+1)^3bx^3}{8(\mu\rho_K+\rho_K+2\alpha+1)}+o(x^3), \tag{6.4.69}$$

where b is the parameter in (6.4.36). Then the theorem follows readily from (6.4.69).
□

6.4.3 *Transition of Three-Component Systems*

The Model. Hereinafter we always assume that $\delta_0\geq 0$ is a constant. Then (6.4.22) possesses a positive constant steady state u^* given by

$$(u_1^*,u_2^*,u_3^*)\quad\text{with } u_1^*=\left(\frac{u_3^*}{1+u_3^*}\right)^{1/2},\quad u_2^*=\lambda u_1^*,\quad u_3^*u_1^*=\frac{\delta_0}{\delta}. \tag{6.4.70}$$

It is easy to see that u_3^* is the unique positive real root of the cubic equation

$$x^3-\left(\frac{\delta_0}{\delta}\right)^2x-\left(\frac{\delta_0}{\delta}\right)^2=0.$$

Consider the translation

$$(u_1,u_2,u_3) \to (u_1^*+u_1,u_2^*+u_2,u_1^*+u_1). \tag{6.4.71}$$

Then equations (6.4.22) are equivalent to

$$\begin{aligned}
&\frac{\partial u_1}{\partial t} = \mu\Delta u_1 - 2\alpha u_1^{*2}u_1 - u_1^*\Delta u_2 + \frac{\alpha u_1^*}{(1+u_3^*)^2}u_3 + g(u),\\
&\frac{\partial u_3}{\partial t} = r\Delta u_3 - \delta u_1^* u_3 - \delta u_3^* u_1 - \delta u_1 u_3,\\
&-\Delta u_2 + u_2 = \lambda u_1,\\
&\left.\frac{\partial (u_1,u_2,u_3)}{\partial n}\right|_{\partial\Omega} = 0,\\
&u(0) = u_0,
\end{aligned} \tag{6.4.72}$$

where $u=(u_1,u_3)$, $u_2=\lambda[-\triangle+1]^{-1}u_1$, and

$$\begin{aligned}
g(u) = {} & -\nabla(u_1\nabla u_2) - 3\alpha u_1^* u_1^2 - \alpha u_1^3 + \frac{\alpha(u_1+u_1^*)(u_3+u_3^*)}{1+u_3^*+u_3}\\
& - \frac{\alpha u_1^* u_3^*}{1+u_3^*} - \frac{\alpha u_1^* u_3}{(1+u_3^*)^2} - \frac{\alpha u_3^* u_1}{1+u_3^*}.
\end{aligned} \tag{6.4.73}$$

The Taylor expansion of g at $u=0$ is expressed as

$$\begin{aligned}
g(u) = {} & -\nabla(u_1\nabla u_2) - 3\alpha u_1^* u_1^2 + \frac{\alpha u_1 u_3}{(1+u_3^*)^2} - \frac{\alpha u_1^* u_3^2}{(1+u_3^*)^3}\\
& -\alpha u_1^3 - \frac{\alpha u_1 u_3^2}{(1+u_3^*)^3} + \frac{\alpha u_1^* u_3^3}{(1+u_3^*)^4} + o(3).
\end{aligned}$$

Let

$$H = L^2(\Omega,\mathbb{R}^2), \qquad H_1 = \left\{ u \in H^2(\Omega,\mathbb{R}^2) \;\middle|\; \frac{\partial u}{\partial n} = 0 \text{ on } \partial\Omega \right\}.$$

Define the operators $L_\lambda : H_1 \to H$ and $G_\lambda : H_1 \to H$ by

$$\begin{aligned}
&L_\lambda u = \begin{pmatrix} \mu\Delta - 2\alpha u_1^{*2} - \lambda u_1^*\Delta[-\Delta+I]^{-1} & \frac{\alpha u_1^*}{(1+u_3^*)^2} \\ -\delta u_3^* & r\Delta - \delta u_1^* \end{pmatrix} \begin{pmatrix} u_1 \\ u_3 \end{pmatrix},\\
&G(u,\lambda) = \begin{pmatrix} g(u) \\ -\delta u_1 u_3 \end{pmatrix}.
\end{aligned} \tag{6.4.74}$$

Then the problem (6.4.72) takes the following abstract form:

$$\frac{du}{dt} = L_\lambda u + G(u,\lambda), \quad u(0) = u_0. \tag{6.4.75}$$

It is known that the inverse mapping $[-\Delta+I]^{-1}: H \to H_1$ is a bounded linear operator. Therefore, we have that $L_\lambda : H_1 \to H$ is a sectorial operator, and $G_\lambda : H_\theta \to H$ is a C^∞ bounded operator for $\theta \geq 1/2$.

Theorems 6.4.3 and 6.4.6 show that a two-component system undergoes a dynamic transition only at real eigenvalues. However, the three-component system (6.4.21) can undergo dynamic transitions at both real and complex eigenvalues, leading to both a steady state and spatiotemporal oscillations.

Linearized Problem. The eigenvalue equations of (6.4.22) at the steady state (u_1^*, u_2^*, u_3^*) given by (6.4.70) in their abstract form are given by

$$L_\lambda \varphi = \beta \varphi, \tag{6.4.76}$$

where $L_\lambda : H_1 \to H$ as defined in (6.4.74). The explicit form of (6.4.76) is given by

$$\begin{pmatrix} \mu\Delta - 2\alpha u_1^{*2} - \lambda u_1^*\Delta[-\Delta+I]^{-1} & \frac{\alpha u_1^*}{(1+u_3^*)^2} \\ -\delta u_3^* & r\Delta - \delta u_1^* \end{pmatrix} \begin{pmatrix} \psi_1 \\ \psi_3 \end{pmatrix} = \beta \begin{pmatrix} \psi_1 \\ \psi_3 \end{pmatrix}.$$

As before, let ρ_k and e_k be the eigenvalue and eigenvector of $-\Delta$ with Neumann boundary condition given by (6.4.28), and let

$$\psi_k = (\psi_1^k, \psi_3^k) = (\xi_{k1} e_k, \xi_{k3} e_k).$$

Then it is easy to see that ψ_k is an eigenvector of (6.4.76), provided that $(\xi_{k1}, \xi_{k3}) \in \mathbb{R}^2$ is an eigenvector of the matrix A_k^λ:

$$A_k^\lambda \begin{pmatrix} \xi_{k1} \\ \xi_{k3} \end{pmatrix} = \beta_k \begin{pmatrix} \xi_{k1} \\ \xi_{k3} \end{pmatrix}, \quad A_k^\lambda = \begin{pmatrix} \frac{\lambda\rho_k u_1^*}{1+\rho_k} - \mu\rho_k - 2\alpha u_1^{*2} & \frac{\alpha u_1^*}{(1+u_3^*)^2} \\ -\delta u_3^* & -r\rho_k - \delta u_1^* \end{pmatrix}. \tag{6.4.77}$$

The eigenvalues β_k of A_k^λ, which are also eigenvalues of (6.4.76), are expressed by

$$\begin{aligned} \beta_k^\pm(\lambda) &= \frac{1}{2}\left[a \pm \sqrt{a^2 - 4\det A_k^\lambda}\right], \\ a = \mathrm{tr} A_k^\lambda &= \frac{\lambda\rho_k u_1^*}{1+\rho_k} - \mu\rho_k - 2\alpha u_1^{*2} - r\rho_k - \delta u_1^*. \end{aligned} \tag{6.4.78}$$

To derive the PES, we introduce two parameters as follows:

$$\Lambda_c = \min_{\rho_K} \frac{(\rho_K+1)}{\rho_K u_1^*}\left[\mu\rho_K + 2\alpha u_1^{*2} + r\rho_K + \delta u_1^*\right], \tag{6.4.79}$$

$$\lambda_c = \min_{\rho_K} \frac{(\rho_K+1)}{\rho_K u_1^*}\left[\mu\rho_K + 2\alpha u_1^{*2} + \frac{\alpha\delta}{(1+u_3^*)^2(r\rho_K + \delta u_1^*)}\right]. \tag{6.4.80}$$

Let $K = (K_1, K_2)$ and $K^* = (K_1^*, K_2^*)$ be the integer pairs such that ρ_K and ρ_{K^*} satisfy (6.4.79) and (6.4.80) respectively.

Theorem 6.4.7 *Let Λ_c and λ_c be the parameters defined by (6.4.79) and (6.4.80) respectively. Then we have the following assertions:*

1 *If $\Lambda_c < \lambda_c$, then the eigenvalues $\beta_K^{\pm}(\lambda)$ of (6.4.78) are a pair of conjugate complex numbers near $\lambda = \Lambda_c$, and all eigenvalues of (6.4.78) satisfy*

$$Re\ \beta_K^{\pm}(\lambda)\begin{cases} <0 & if\ \lambda < \Lambda_c, \\ =0 & if\ \lambda = \Lambda_c, \\ >0 & if\ \lambda > \Lambda_c, \end{cases} \tag{6.4.81}$$

$$Re\ \beta_k^{\pm}(\Lambda_c) < 0 \qquad \forall k \in \mathbb{Z}^2 \text{ with } \rho_k \neq \rho_K. \tag{6.4.82}$$

2 *If $\lambda_c < \Lambda_c$, then the eigenvalue $\beta_{K^*}^{+}(\lambda)$ is real near $\lambda = \lambda_c$, and all of (6.4.78) satisfy*

$$\beta_{K^*}^{+}(\lambda)\begin{cases} <0 & if\ \lambda < \lambda_c, \\ =0 & if\ \lambda = \lambda_c, \\ >0 & if\ \lambda > \lambda_c, \end{cases} \tag{6.4.83}$$

$$\begin{cases} Re\beta_k^{+}(\lambda_c) < 0 & \forall k \in \mathbb{Z}^2 \text{ with } \rho_k \neq \rho_{K^*}, \\ Re\beta_k^{-}(\lambda_c) < 0 & \forall |k| \geq 0. \end{cases} \tag{6.4.84}$$

Proof. By (6.4.78), we see that $\beta_k^{\pm}(\lambda)$ are a pair of complex eigenvalues of (6.4.76) near some $\lambda = \lambda^*$ satisfying

$$\text{Re}\beta_k^{\pm}(\lambda)\begin{cases} <0 & \text{if } \lambda < \lambda^*, \\ =0 & \text{if } \lambda = \lambda^*, \\ >0 & \text{if } \lambda > \lambda^*, \end{cases}$$

if and only if

$$\text{tr}A_k^{\lambda^*} = 0, \quad \det A_k^{\lambda^*} > 0.$$

Likewise, $\beta_k^{+}(\lambda)$ is real near $\lambda = \lambda^*$ and satisfies

$$\beta_k^{+}(\lambda)\begin{cases} <0 & \text{if } \lambda < \lambda^*, \\ =0 & \text{if } \lambda = \lambda^*, \\ >0 & \text{if } \lambda > \lambda^*, \end{cases}$$

if and only if

$$\text{tr}A_k^{\lambda^*} < 0, \quad \det A_k^{\lambda^*} = 0.$$

By the definition of λ_c and Λ_c, when $\Lambda_c < \lambda_c$, we have

$$\begin{aligned} &\text{tr}\ A_K^{\Lambda_c} = 0, \\ &\text{tr}\ A_k^{\Lambda_c} < 0 \qquad \forall k \in \mathbb{Z}^2 \text{ with } \rho_k \neq \rho_K, \\ &\det A_k^{\Lambda_c} > 0 \quad \forall |k| \geq 0, \end{aligned} \tag{6.4.85}$$

and when $\lambda_c < \Lambda_c$,

$$\begin{aligned}
&\det A_{K^*}^{\lambda_c} = 0, \\
&\det A_k^{\lambda_c} > 0 \quad \forall k \in \mathbb{Z}^2 \text{ with } \rho_k \neq \rho_K, \\
&\operatorname{tr} A_k^{\lambda_c} < 0 \quad \forall |k| \geq 0.
\end{aligned} \tag{6.4.86}$$

It is known that the real parts of $\beta_k^{\pm}$ are negative at λ if and only if

$$\det A_k^{\lambda} > 0, \quad \operatorname{tr} A_k^{\lambda} < 0.$$

Then assertions (1) and (2) follow from (6.4.85) and (6.4.86) respectively. □

Dynamic Transition Theorem for (6.4.22). Based on Theorem 6.4.7, we immediately get the following transition theorem for (6.4.72).

Theorem 6.4.8 *Let Λ_c and λ_c be given by (6.4.79) and (6.4.80) respectively. Then the following assertions hold for (6.4.72):*

(1) When $\Lambda_c < \lambda_c$, the system undergoes a dynamic transition to periodic solutions at $(u,\lambda) = (0,\Lambda_c)$. In particular, if the eigenvalues $\beta_K^{\pm}$ satisfying (6.4.81) are complex and simple, then there is a parameter b_0 such that the dynamic transition is continuous (Type I) for $b_0 < 0$, and is a jump (Type II) for $b_0 > 0$ with a singularity separation of periodic solutions at some $\lambda^* < \Lambda_c$.

(2) When $\lambda_c < \Lambda_c$, the system undergoes a dynamic transition to steady states at $(u,\lambda) = (0,\lambda_c)$. If $\beta_{K^*}^{+}(\lambda)$ satisfying (6.4.83) is simple, then there exists a parameter b_1 such that the transition is continuous for $b_1 < 0$ and a jump for $b_1 > 0$ with two saddle-node bifurcation points at some $\widetilde{\lambda} < \lambda_c$ from $(u^+,\widetilde{\lambda})$ and $(u^-,\widetilde{\lambda})$.

Remark 6.4.9 By applying the standard procedure used in the preceding sections, we can derive explicit formulas for the two parameters b_0 and b_1 in Theorem 6.4.8. However, due to their complexity, we omit the details.

Transition for the System (6.4.21). We are now in a position to discuss the transition of (6.4.21). With the translation (6.4.71), the system (6.4.21) can be rewritten in the following form:

$$\begin{aligned}
&\frac{\partial u_1}{\partial t} = \mu\Delta u_1 - 2\alpha u_1^{*2} u_1 - u_1^*\Delta u_2 + \frac{\alpha u_1^* u+3}{(1+u_3^*)^2} + g(u), \\
&\frac{\partial u_2}{\partial t} = \Delta u_2 - u_2 + \lambda u_1, \\
&\frac{\partial u_3}{\partial t} = r\Delta u_3 - \delta u_1^* u_3 - \delta u_3^* u_1 - \delta u_1 u_3, \\
&\left.\frac{\partial u}{\partial n}\right|_{\partial\Omega} = 0, \\
&u(0) = u_0,
\end{aligned} \tag{6.4.87}$$

where $g(u)$ is as in (6.4.73). Here the notation u stands for the three-component unknown $u=(u_1,u_2,u_3)$.

Let

$$L_\lambda u=\begin{pmatrix} \mu\Delta-2\alpha u_1^{*2} & -u_1^*\Delta & \frac{\alpha u_1^*}{(1+u_3^*)^2} \\ \lambda & \Delta-1 & 0 \\ -\delta u_3^* & 0 & r\Delta-\delta u_1^* \end{pmatrix}\begin{pmatrix} u_1 \\ u_2 \\ u_3 \end{pmatrix}.$$

Then all eigenvalues $\beta_k^j(\lambda)$ and eigenvectors ψ_k^j of L_λ satisfy

$$D_k^\lambda\begin{pmatrix} \xi_{k1}^j \\ \xi_{k2}^j \\ \xi_{k3}^j \end{pmatrix}=\beta_k^j(\lambda)\begin{pmatrix} \xi_{k1}^j \\ \xi_{k2}^j \\ \xi_{k3}^j \end{pmatrix}, \quad 1\le j\le 3,\ k\in\mathbb{Z}^2,$$

with

$$\psi_k^j=(\xi_{k1}^j e_k,\xi_{k2}^j e_k,\xi_{k3}^j e_k),$$

and e_k as in (6.4.28), with D_k^λ a 3×3 matrix given by

$$D_k^\lambda=\begin{pmatrix} -(\mu\rho_k+2\alpha u_1^{*2}) & u_1^*\rho_k & \frac{\alpha u_1^*}{(1+u_3^*)^2} \\ \lambda & -(\rho_k+1) & 0 \\ -\delta u_3^* & 0 & -(r\rho_k+\delta u_1^*) \end{pmatrix}.$$

We introduce the following three parameters:

$$\begin{aligned}
A_k^\lambda &= -\mathrm{tr}D_k^\lambda=\mu\rho_k+2\alpha u_1^{*2}+\rho_k+1+r\rho_k+\delta u_1^*,\\
B_k^\lambda &= \det\begin{pmatrix} -(\mu\rho_k+2\alpha u_1^{*2}) & u_1^*\rho_k \\ \lambda & -(\rho_k+1) \end{pmatrix}+\det\begin{pmatrix} -(\mu\rho_k+2\alpha u_1^{*2}) & \frac{\alpha u_1^*}{(1+u_3^*)^2} \\ -\delta u_3^* & -(r\rho_k+\delta u_1^*) \end{pmatrix}\\
&\quad +(\rho_k+1)(r\rho_k+\delta u_1^*),\\
C_k^\lambda &= -\mathrm{det}D_k^\lambda=(\mu\rho_k+2\alpha u_1^{*2})(\rho_k+1)(r\rho_k+\delta u_1^*)\\
&\quad -u_1^*\rho_k\lambda(r\rho_k+\delta u_1^*)+\frac{\alpha u_1^*}{(1+u_3^*)^2}\delta u_3^*(\rho_k+1).
\end{aligned}$$

By the Routh–Hurwitz theorem, we know that all eigenvalues β_k^j of D_k^j have negative real parts if and only if

$$A_k^\lambda>0,\quad A_k^\lambda B_k^\lambda-C_k^\lambda>0,\quad C_k^\lambda>0. \tag{6.4.88}$$

Let Λ^c and $K=(K_1,K_2)$ satisfy

$$\begin{aligned}
&A_K^{\Lambda_c}>0,\quad A_K^{\Lambda_c}B_K^{\Lambda_c}-C_K^{\Lambda_c}=0,\quad C_K^{\Lambda_c}>0,\\
&A_k^{\Lambda_c}>0,\quad A_k^{\Lambda_c}B_k^{\Lambda_c}-C_k^{\Lambda_c}>0,\quad C_k^{\Lambda_c}>0 \quad \forall k \text{ with } \rho_k\neq\rho_K.
\end{aligned} \tag{6.4.89}$$

Then Λ_c satisfies

$$\Lambda_c = \inf_{\rho_k} \frac{1}{\rho_k u_1^*}[(\mu+r)\rho_k + 2\alpha u_1^{*2} + \delta u_1^*] \times \left[(r+1)\rho_k + \delta u_1^* + 1 + \frac{\alpha\delta}{(\mu\rho_k + 2\alpha u_1^{*2} + \rho_k + 1)(1+u_3^*)^2}\right], \quad (6.4.90)$$

and ρ_K satisfies (6.4.90). In particular, under the condition (6.4.89), there is a pair of complex eigenvalues $\beta_K^1(\lambda)$ and $\beta_K^2(\lambda)$ of D_K^λ such that

$$\text{Re}\,\beta_K^{1,2}(\lambda) \begin{cases} <0 & \text{if } \lambda < \Lambda_c, \\ =0 & \text{if } \lambda = \Lambda_c, \\ >0 & \text{if } \lambda > \Lambda_c, \end{cases} \quad (6.4.91)$$

and the other eigenvalues $\beta_k^j(\lambda)$ of L_λ satisfy

$$\begin{aligned} &\text{Re}\,\beta_k^j(\Lambda_c) < 0 \quad \forall k \text{ with } \rho_k \neq \rho_K \text{ and } 1 \leq j \leq 3, \\ &\beta_K^3(\Lambda_c) < 0. \end{aligned} \quad (6.4.92)$$

Let λ_c and $K^* = (K_1^*, K_2^*)$ satisfy

$$\begin{aligned} &A_{K^*}^{\lambda_c} > 0, \quad A_{K^*}^{\lambda_c} B_{K^*}^{\lambda_c} - C_{K^*}^{\lambda_c} > 0, \quad C_{K^*}^{\lambda_c} = 0, \\ &A_k^{\lambda_c} > 0, \quad A_k^{\lambda_c} B_k^{\lambda_c} - C_k^{\lambda_c} > 0, \quad C_k^{\lambda_c} > 0 \quad \forall k \text{ with } \rho_k \neq \rho_{K^*}. \end{aligned} \quad (6.4.93)$$

Then λ_c is given by

$$\lambda_c = \inf_{\rho_k} \frac{(\rho_k + 1)}{\rho_k u_1^*} \left[\mu\rho_k + 2\alpha u_1^{*2} + \frac{\alpha\delta}{(1+u_3^*)^2(r\rho_k + \delta u_1^*)}\right], \quad (6.4.94)$$

and λ_c achieves its minimum at ρ_{K^*}. From the Routh–Hurwitz criterion (6.4.88), we deduce that with (6.4.93), there is a real eigenvalue $\beta_{K^*}^1(\lambda)$ of $D_{K^*}^{\lambda_c}$ such that

$$\beta_{K^*}(\lambda) \begin{cases} <0 & \text{if } \lambda < \lambda_c, \\ =0 & \text{if } \lambda = \lambda_c, \\ >0 & \text{if } \lambda > \lambda_c, \end{cases} \quad (6.4.95)$$

$$\begin{cases} \text{Re}\beta_{K^*}^j(\lambda_c) < 0 & j = 2,3, \\ \text{Re}\beta_k^j(\lambda_c) < 0 & \forall k \in \mathbb{Z}^2 \text{ with } \rho_k \neq \rho_{K^*} \text{ and } 1 \leq j \leq 3. \end{cases} \quad (6.4.96)$$

It is clear that (6.4.91) and (6.4.92) hold when $\Lambda_c < \lambda_c$, and (6.4.94)–(6.4.95) hold for $\lambda_c < \Lambda_c$. Hence, we have the following transition theorem for (6.4.87).

Theorem 6.4.10 *Let Λ_c and λ_c be given by (6.4.90) and (6.4.94) respectively. Then assertions (1) and (2) of Theorem 6.4.8 hold for the system (6.4.87).*

6.4.4 Biological Conclusions

Biological Significance of Transition Theorems. Pattern formation is one of the main characteristics of bacteria chemotaxis, and it is fully characterized by dynamic transitions. Theorems 6.4.3–6.4.10 tell us that the nondimensional parameter λ, given by

$$\lambda = \frac{\sqrt{\alpha_2} r_1 \chi}{r_2 k_2}, \tag{6.4.97}$$

plays a crucial role in determining the dynamic transition and pattern formation. In fact, the key factor in (6.4.97) is the product of the chemotactic coefficient χ and the production rate $r_1 : \chi r_1$, which depends on the type of bacteria. When λ is less than some critical value λ_c, the uniform distribution of individual organisms is a stable state. When λ exceeds λ_c, the bacteria cells aggregate to form more complex and stable patterns.

As seen in (6.4.33), (6.4.79), (6.4.80), and (6.4.90), under different biological conditions, the critical parameter λ_c takes different forms and values. But a general formula for λ_c is given by

$$\lambda_c = a_0 + \inf_{\rho_k} \left(a_1 \rho_k + \frac{a_2}{\rho_k} + \frac{a_3}{b_1 \rho_k + b_0} + \frac{a_4}{\rho_k (b_1 \rho_k + b_0)} \right), \tag{6.4.98}$$

where the ρ_k are taken as the eigenvalues of $-\Delta$ with the Neumann boundary condition. When Ω is a rectangular region, the ρ_k are given by (6.4.28), and the coefficients a_j $(1 \le j \le 4)$, $b_0, b_1 \ge 0$ depend on the parameters in (6.4.20), with

$$a_0, a_1, a_2, b_0, b_1 > 0, \quad a_3, a_4 \ge 0.$$

In particular, for the system with rich nutrient supplies, (6.4.98) becomes

$$\lambda_c = a_0 + \inf_{\rho_k} \left[a_1 \rho_k + \frac{a_2}{\rho_k} \right].$$

The eigenvalues ρ_k, depending on the geometry of Ω, satisfy

$$0 = \rho_0 < \rho_1 \le \cdots \le \rho_k \le \cdots, \qquad \lim_{k \to \infty} \rho_k = \infty, \qquad \rho_1 \propto \frac{1}{L^2}, \tag{6.4.99}$$

where L is the length scale of Ω. We infer from (6.4.98) and (6.4.99) that

$$\lambda_c \to \infty \quad \text{as} \quad |\Omega| \to 0 \quad (L \to 0).$$

This implies that when the container Ω is small, the homogenous state is stable and there is no pattern formation of bacteria under any biological conditions.

Spatiotemporal Oscillation. Theorems 6.4.8 and 6.4.10 show that there are two critical parameters λ_c and Λ_c such that if $\lambda_c < \Lambda_c$, the patterns formed by organisms are steady, as exhibited by many experimental results, and if $\Lambda_c < \lambda_c$, a spatiotemporal oscillatory behavior takes place.

For the case with rich nutrient, $u_1^* = 1, \quad u_3^* = \infty$. Also, λ_c in (6.4.80) is reduced to (6.4.30), and obviously we have

$$\lambda_c < \Lambda_c \quad \text{for both (6.4.79) and (6.4.90),}$$

and the dynamic transition and pattern formation are determined by Theorems 6.4.3 and 6.4.6. Hence there is no spatiotemporal oscillation for the rich nutrient case, and the time-periodic oscillation of chemotaxis occurs only for the case in which the nutrient is moderately supplied.

On the other hand, if $\mu, r \cong 0$ and

$$\delta^2 u_1^{*2}(1+u_3^*)^2 < \alpha\delta_0,$$

then for Λ_c defined by (6.4.79) and (6.4.90), we have $\Lambda_c < \lambda_c$. In this case, a spatiotemporal oscillation pattern is expected for $\lambda > \Lambda_c$.

Transition types. The main theorems provide precise information on the transition types. In all cases, types are precisely determined by the sign of some nondimensional parameters; see b, b_0, and b_1 respectively in the main theorems. Hence a global phase diagram can be obtained easily by setting the related parameter to zero.

For example, when $\Omega = (0, L_1)$ is one-dimensional or when $K = (K_1, 0)$ (respectively $K = (0, K_2)$), the parameter b in (6.4.36) can be simplified to the following form:

$$b = 2\left[-3\mu\rho_K + 9\alpha - \frac{(2\mu\rho_K+\alpha)(2\mu\lambda_K^2 + 28\alpha\rho_K + 4\alpha - \mu\rho_K)}{(\mu\rho_{2K}+2\alpha)(\rho_{2K}+1) - \rho_{2K}\lambda_c}\right]. \tag{6.4.100}$$

For a nongrowth system, we have $\alpha = 0$, $K = (1,0)$, $\lambda_c = \mu(\rho_K + 1)$. Then (6.4.100) becomes

$$b = \frac{\mu}{3}(1 - 20\lambda_1), \quad \lambda_1 = \frac{\pi^2}{L_1^2}, \tag{6.4.101}$$

and $\lambda = \frac{ar_1 x}{r_2 k_2}$, with $a = \frac{1}{|\Omega|}\int_\Omega u_1 dx$. It follows from (6.4.101) that

$$b \begin{cases} < 0 & \text{if } L_1 < 2\sqrt{5}\pi, \\ > 0 & \text{if } L_1 > 2\sqrt{5}\pi. \end{cases} \tag{6.4.102}$$

By Theorems 6.4.3 and 6.4.6, the phase transition of (6.4.25) and (6.4.23) from $(u, \lambda) = (u^*, \lambda_c)$ is continuous if the length scale L_1 of Ω is less than $2\sqrt{5}\pi$, and it is a jump if L_1 is greater than $2\sqrt{5}\pi$.

In addition, when we take $\chi(u) = \chi_1 u_1/(\beta + u_2)^2$ as the chemotaxis function, by Remark 6.4.9, the parameter b of (6.4.101) is replaced by

$$b_1 = \frac{\mu}{3}\left(1 - \frac{20\pi^2}{L_1^2}\right) + \frac{4\kappa\mu^2\pi^2}{L_1^2}, \qquad \kappa = \frac{k_2}{\beta\chi + k_2\lambda_c}, \qquad \lambda_c = \mu\left(\frac{\pi^2}{L_1^2} + 1\right).$$

The above conclusion shows that for a nongrowth system, the parameter

$$\lambda = \frac{r_1 \chi}{r_2 k_2} a, \quad \text{with } a = \frac{1}{|\Omega|} \int_\Omega u_1 dx,$$

is proportional to the average density of the initial condition of u_1. Note that u_1 is conservative. Hence, the individual organism is in a homogeneous state if

$$\frac{1}{|\Omega|} \int_\Omega \varphi dx < \frac{r_2 k_2}{r_1 \chi} \mu \left(\frac{\pi^2}{L_1^2} + 1 \right), \quad \varphi = u_1(0),$$

and the bacteria will aggregate to form high-density regions if that

$$\frac{1}{|\Omega|} \int_\Omega \varphi dx > \frac{r_2 k_2}{r_1 \chi} \mu \left(\frac{\pi^2}{L_1^2} + 1 \right). \tag{6.4.103}$$

Moreover, under the condition (6.4.103), if the scale L_1 of Ω is smaller than some critical value L_c (in (6.4.102) $L_c = 2\sqrt{5}\pi$), i.e., $L_1 < L_c$, the continuous transition implies that only one high-density region of bacteria forms, and if $L_1 > L_c$, then in the jump transition, one expects a large number of high-density regions to appear.

Pattern Formation. As mentioned before, the pattern formation behavior is dictated by the dynamic transition of the system. We analyzed the formation of two types of patterns, the lamellar and the rectangular, and the method and ideas can be generalized to study the formation of more complex patterns.

For a growth system, the critical parameter λ_c takes its value at some eigenvalue ρ_K of $-\Delta$ for $K = (K_1, K_2)$, as shown by (6.4.33) and (6.4.80). From the pattern formation point of view, for a Type I transition, the patterns described by the transition solutions in the main theorems are either lamellar or rectangular:

lamellar pattern	for $K_1 K_2 = 0$,
rectangular pattern	for $K_1 K_2 \neq 0$.

In the case $b > 0$, the system undergoes a more drastic change. For $\lambda^* < \lambda < \lambda_c$, the homogeneous state, the new patterns v_2^λ and v_4^λ are metastable. For $\lambda > \lambda_c$, the system undergoes transitions to more complex patterns away from the basic homogeneous state form.

6.5 Biological Species

6.5.1 Modeling

An important aspect of ecological balance is the interaction between biological species. The initial work in this field was on the Lotka–Volterra model for predator–prey systems (Lotka 1925; Volterra 1926), and a great many studies have appeared in the literature. In Sect. 6.1.2, we introduced a general model for population dynamics, taking into account both environmental and random-walk effects.

In this section, we consider the following n-species system, ignoring the environmental and random-walk effects:

$$\frac{dx_k}{dt} = \lambda_k x_k + \sum_{j=1}^{n} a_{kj} x_k x_j - \beta_k x_k^3 \qquad \text{for } 1 \leq k \leq n, \tag{6.5.1}$$

where $x = (x_1, \cdots, x_n)$ represents the population vector of n individual species $X_1, \cdots X_n$, and $\lambda_k, \beta_k, a_{kj}$ are constants.

The model (6.5.1) is derived based on the following basic modeling rules:

(B1) The nonnegative vector space $\mathbb{R}^n_+$ is positively invariant for (6.5.1).
(B2) The system (6.5.1) is a dissipative dynamical system. Namely, the system possesses a global attractor in $\mathbb{R}^n_+$.
(B3) A model usually takes its simplest form.

The rule (B2) requires that for $1 \leq k \leq n$,

$$\begin{aligned} &\beta_k \geq 0, \\ &\beta_k > 0 \quad \text{if } a_{kk} \geq 0. \end{aligned} \tag{6.5.2}$$

From the biological point of view, the case

$$\beta_k > 0 \text{ and } a_{kk} > 0 \text{ in (6.5.1)} \tag{6.5.3}$$

amounts to saying that the species X_k has an optimal population given by $x_k^* = a_{kk}/\beta_k$ such that X_k is mutually beneficial if $x_k > x_k^*$ and is mutually conflictive if $x_k < x_k^*$. Namely, the condition (6.5.3) displays a competition mechanism.

We note that many species obey the duality of mutual benefit and conflict. When their populations are in an inadequate state, they cooperate, and when they are in an excessive state, they compete with each other for limited resources.

Also, biologically speaking, a basic condition for (6.5.1) is the existence of a nontrivial steady-state solution in $\mathbb{R}^n_+$. Namely, the equations

$$\beta_k z_k^2 - \sum_{j=1}^{n} a_{kj} z_j - \lambda_k = 0 \quad (1 \leq k \leq n) \tag{6.5.4}$$

possess at least a nontrivial solution $z = (z_1, \cdots, z_n) \in \mathbb{R}^n_+$. Let $z \in \mathbb{R}^n_+$ be such a solution of (6.5.4). To investigate the stability and transition of (6.5.1) at z, we need to consider the deviation from this basic state:

$$x = x' + z.$$

Then (6.5.1) becomes (suppressing the primes)

$$\frac{dx_k}{dt} = \sum_{j=1}^{n} b_{kj} x_j + \sum_{j=1}^{n} c_{kj} x_k x_j - \beta_k x_k^3, \tag{6.5.5}$$

where b_{kj} and c_{kj} are given by

$$b_{kj} = z_k(a_{kj} - 2\delta_{kj}\beta_k z_k), \qquad c_{kj} = a_{kj} - 3\delta_{kj}\beta_k z_k, \tag{6.5.6}$$

and δ_{kj} is the Kronecker symbol.

6.5.2 Predator–Prey Systems

For simplicity, we begin with a predator–prey model as follows:

$$\begin{aligned} \frac{dx_1}{dt} &= \lambda_1 x_1 + a_{11}x_1^2 - a_{12}x_1x_2 - \beta_1 x_1^3, \\ \frac{dx_2}{dt} &= -\lambda_2 x_2 + a_{21}x_1x_2 - a_{22}x_2^2, \end{aligned} \tag{6.5.7}$$

where we always assume that

$$\beta_1 > 0, \quad \lambda_i > 0, \quad a_{ij} > 0 \quad \text{for } 1 \le i,j \le 2. \tag{6.5.8}$$

Namely, the prey x_1 in system (6.5.7) obeys the duality of mutual benefit and conflict.

Remark 6.5.1 For the case $a_{11} = a_{22} = \beta_1 = 0$, the system (6.5.7) is the classical Lotka–Volterra model. In this case, it is well known that (6.5.7) has a unique fixed point $z^* = (\lambda_2/a_{21}, \lambda_1/a_{12})$, and for any initial value $x_0 \in \mathbb{R}^2_+$ with $x \neq 0$ and $x \neq z^*$, the solution $x(t,x_0)$ of (6.5.7) with $x(0,x_0) = x_0$ is a periodic orbit.

Remark 6.5.2 For the case in which $\beta_1 = 0$ and $a_{11} < 0$, the following results are essentially known:

- If $\lambda = \lambda_1 a_{21} - \lambda_2|a_{11}| < 0$, the system (6.5.7) admits a unique singular point $z^* = (\lambda_1/|a_{11}|, 0)$, which is globally asymptotically stable in $\mathbb{R}^2_+$.
- If $\lambda = \lambda_1 a_{21} - \lambda_2|a_{11}| > 0$, the system bifurcates from $(x,\lambda) = (z^*,0)$ to a positive singular point $z_\lambda \in \mathbb{R}^2_+$:

$$z_\lambda = (x_1^\lambda, x_2^\lambda) = \left(\frac{\lambda_1 a_{22} + \lambda_2 a_{12}}{a_{12}a_{21} + |a_{11}|a_{22}}, \frac{\lambda}{a_{12}a_{21} + |a_{11}|a_{22}} \right),$$

which is globally asymptotically stable in $\mathbb{R}^2_+$. □

To introduce the nondimensional form of the system, let

$$\begin{aligned} &x_1 = \frac{\lambda_2}{a_{21}}x_1', \quad x_2 = \frac{\lambda_2}{a_{22}}x_2', \qquad t = \frac{1}{\lambda_2}t', \\ &\alpha_1 = \lambda_1/\lambda_2, \quad \alpha_2 = a_{11}/a_{21}, \quad \alpha_3 = a_{12}/a_{22}, \quad \alpha_4 = \beta_1\lambda_2/a_{21}^2. \end{aligned} \tag{6.5.9}$$

Then equations (6.5.7) can be rewritten as (suppressing the primes)

$$\begin{aligned}\frac{dx_1}{dt} &= \alpha_1 x_1 + \alpha_2 x_1^2 - \alpha_3 x_1 x_2 - \alpha_4 x_1^3,\\ \frac{dx_2}{dt} &= -x_2 + x_1 x_2 - x_2^2.\end{aligned} \tag{6.5.10}$$

In addition, we introduce two parameters λ and Λ:

$$\lambda = \alpha_1 + \alpha_2 - \alpha_4, \quad \Lambda = (A + \alpha_2 - \alpha_3)(A + 1 - \alpha_3) - 2\alpha_4, \tag{6.5.11}$$

where

$$A = \sqrt{(\alpha_2 - \alpha_3)^2 + 4\alpha_4(\alpha_1 + \alpha_3)}.$$

The following two theorems completely characterize the dynamics of the predator–prey system (6.5.7):

- When $\lambda < 0$, the single-species of prey is unique and stable state.
- When $\lambda > 0$ with $\Lambda > 0$, the coexistence of predator and prey is a unique stable state.
- When $\lambda > 0$ with $\Lambda < 0$, the system undergoes a dynamic transition to a stable periodic solution from the coexistence state.

Theorem 6.5.3 *Let λ and Λ be the parameters defined by (6.5.11), and assume (6.5.8).*

(1) If $\lambda < 0$, the system (6.5.10) possesses a unique singular point $z_0^* = (x_1^0, 0)$ in $\mathbb{R}_+^2$, which is globally asymptotically stable in $\mathbb{R}_+^2$. Here $x_1^0 = \frac{1}{2\alpha_4}(\alpha_2 + \sqrt{\alpha_2^2 + 4\alpha_1\alpha_4})$.

(2) The system (6.5.10) undergoes a continuous dynamic transition from $(x, \lambda) = (z_0^*, 0)$ to a unique singular point z_λ^* in $\mathbb{R}_+^2$ on $\lambda > 0$:

$$z_\lambda^* = (x_1^*, x_2^*) = \left(\frac{1}{2\alpha_4}(A + \alpha_2 - \alpha_3), \quad \frac{1}{2\alpha_4}(A + \alpha_2 - \alpha_3)\right), \tag{6.5.12}$$

where A is as defined in (6.5.11).

(3) If $\lambda = 0$, then the parameter Λ is positive, and the bifurcated solution z_λ^* is an attractor if and only if $\Lambda > 0$. Consequently, z_λ^* is a global attractor in the interior of $\mathbb{R}_+^2$ for $\lambda > 0$ sufficiently small.

If $\Lambda < 0$, the following theorem shows that (6.5.10) undergoes a dynamic transition to periodic solutions. We need to introduce the following parameter:

$$\begin{aligned} b =& x_1^* A \left(\frac{2x_1^* - 1}{2\alpha_3} - \frac{x_1^* - 1}{\alpha_3^2} - \alpha_4 x_1^{*2}\right) + \alpha_4^2 x_1^{*4} \\ &+ \frac{1}{2}\left(\alpha_4 x_1^{*2} + x_1^* - \frac{x_1^* - 1}{\alpha_3}\right) - \left(x_1^* - \frac{x_1^* - 1}{\alpha_3}\right)^2, \end{aligned} \tag{6.5.13}$$

where A is as in (6.5.11), x_1^* is given by (6.5.12), and

$$(A+\alpha_2-\alpha_3)(A+1-\alpha_3)=2\alpha_4. \tag{6.5.14}$$

By (6.5.11), it is easy to see that (6.5.14) is equivalent to $\Lambda=0$.

Theorem 6.5.4 *Let λ,Λ be given by (6.5.11), and let b be the parameter defined by (6.5.13). When $\lambda>0$, the following assertions hold:*

(1) If $b<0$, the system (6.5.10) undergoes a continuous dynamic transition from $(x,\Lambda)=(x^*,0)$ to a stable periodic solution on $\Lambda<0$ (Hopf bifurcation).
(2) If $b>0$, the system (6.5.10) undergoes a jump transition from $(x,\Lambda)=(x^*,0)$, and there exists a singular separation of periodic solutions at some $\Lambda_0>0$. Moreover, (6.5.13) has no periodic solutions in $\mathbb{R}^2_+$ for $\Lambda>\Lambda_0$, while it has exactly one periodic solution Γ_0 for $\Lambda=\Lambda_0$, and two periodic solutions Γ_1 and Γ_2 bifurcated from (Γ_0,Λ_0) when $\Lambda<\Lambda_0$ such that Γ_1 is a repeller and Γ_2 is an attractor in $\mathbb{R}^2_+$, and

$$\lim_{\Lambda\to 0}\Gamma_1=z_\lambda^*,\quad \lim_{\Lambda\to\Lambda_0}\Gamma_i=\Gamma_0\quad (i=1,2),$$

where $z_\lambda^*=(x_1^*,x_2^*)$ is as in (6.5.12) with $\alpha_j(1\le j\le 4)$ satisfying (6.5.14), and Γ_2 does not disappear on $\Lambda\le 0$. See Fig. 6.1.

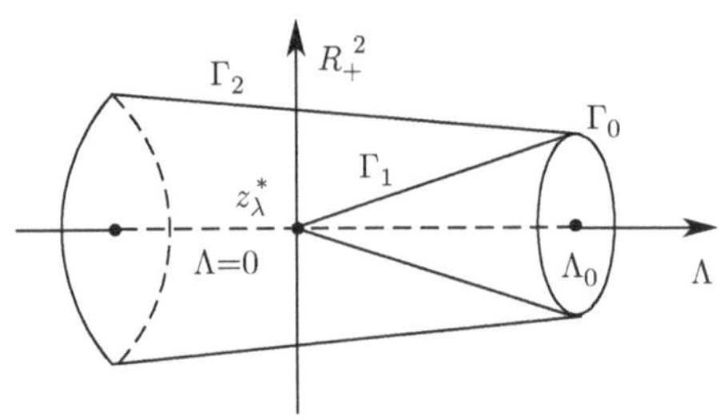

Fig. 6.1 Singular separation of periodic solution at Λ_0: for $0<\Lambda<\Lambda_0$, there are two limit cycles Γ_1 and Γ_2, and for $\Lambda\le 0$, there is only one limit cycle Γ_2

Remark 6.5.5 If $b\neq 0$, the bifurcated periodic solution of (6.5.10) from $(z_\lambda^*,0)$ on $\Lambda<0$ in Theorem 6.5.4 can be expressed as

$$x_1=x_1^*+\left(\frac{1}{|b|}\rho(\Lambda)\right)^{1/2}(\xi_1\sin\sigma t+\eta_1\cos\sigma t)+o(\rho^{1/2}(\Lambda)),$$

$$x_2=x_2^*+\left(\frac{1}{|b|}\rho(\Lambda)\right)^{1/2}(\xi_2\sin\sigma t+\eta_2\cos\sigma t)+o(\rho^{1/2}(\Lambda)),$$

where $\rho(\Lambda) > 0$ ($\Lambda < 0$) and σ satisfy

$$\lim_{\Lambda \to 0} \rho(\Lambda) = 0, \qquad \sigma = \sqrt{x_1^* x_2^* A} > 0,$$
$$(\xi_1, \xi_2) = \left(1, \frac{x_2^*}{\alpha_3 x_1^*}\right), \qquad (\eta_1, \eta_2) = \left(0, \frac{\sigma}{\alpha_3 x_1^*}\right).$$

Remark 6.5.6 For the parameter b defined by (6.5.13), the following analysis shows that both cases $b < 0$ and $b > 0$ may occur under the condition (6.5.14).

In fact, assume that $\alpha_1, \alpha_2, \alpha_4 \approx 0$. Then (6.5.14) implies that

$$x_1^* \approx 1 + \alpha_2, \qquad \alpha_3 \approx 1 + \frac{\alpha_1}{\alpha_2}.$$

Hence $A \approx \alpha_3$ and

$$\begin{aligned} b &\approx \alpha_3 \left(\frac{1}{2\alpha_3} - \frac{\alpha_2}{\alpha_3^2} - \alpha_4\right) + \frac{1}{2}(1 + \alpha_4) - (1 + \alpha_1)^2 \\ &\approx -2\alpha_1 - \alpha_3\alpha_4 + \frac{1}{2}\alpha_4. \end{aligned}$$

Namely, $b < 0$.

For the case $\alpha_2 > \alpha_3 \gg 1, \alpha_4\alpha_3 \approx 0$, and $\alpha_4\alpha_1 \approx 0$, (6.5.14) yields

$$\alpha_2 \approx 2\alpha_3, \qquad x_3^* \approx \frac{\alpha_3}{\alpha_4}, \qquad A \approx \alpha_3.$$

Thus we deduce

$$b \approx \left(\frac{1}{\alpha_3} + \frac{\alpha_4}{2}\right) x_1^{*2} + \frac{1}{2} x_1^* > 0.$$

Proof of Theorem 6.5.3. It is clear that the number of singular points of (6.5.10) is six, and they are given by

$$(0,0),\ (0,-1),\ (x_1^+,0),\ (x_1^-,0),\ (\widetilde{x}_1^+,\widetilde{x}_2^+),\ (\widetilde{x}_1^-,\widetilde{x}_2^-),$$

where

$$\begin{aligned} x_1^{\pm} &= \frac{1}{2\alpha_4}\left(\alpha_2 \pm \sqrt{\alpha_2^2 + 4\alpha_1\alpha_4}\right), \\ \widetilde{x}_1^{\pm} &= \frac{1}{2\alpha_4}\left[(\alpha_2 - \alpha_3) \pm \sqrt{(\alpha_2 - \alpha_3)^2 + 4\alpha_4(\alpha_1 + \alpha_3)}\right], \qquad \widetilde{x}_2^{\pm} = \widetilde{x}_1^{\pm} - 1. \end{aligned}$$

Obviously, the solutions $(0,-1), (x_1^-,0)$, and $(\widetilde{x}_1^-,\widetilde{x}_2^-)$ are not in $\mathbb{R}_+^2$, and $(0,0)$ is unstable. We also see that

$$\widetilde{x}_2^+ \begin{cases} < 0 & \text{if } \lambda < 0, \\ = 0 & \text{if } \lambda = 0, \\ > 0 & \text{if } \lambda > 0, \end{cases} \tag{6.5.15}$$

where λ is as in (6.5.11).

It is easy to check that under the condition (6.5.8), the system (6.5.10) has a global attractor in $\mathbb{R}^2_+$. Hence, assertion (1) follows from (6.5.15) and the Poincaré–Bendixon theorem.

In addition, it is readily proved that $z_0 = (x_1^+, 0)$ is globally asymptotically stable in $\mathbb{R}^2_+$ at $\lambda = 0$, and the eigenvalues of the Jacobian of (6.5.10) at z_0 satisfy

$$\beta_1(\lambda) = (x_1^+ - 1)\begin{cases} <0 & \text{if } \lambda < 0, \\ =0 & \text{if } \lambda = 0, \\ >0 & \text{if } \lambda > 0, \end{cases}$$
$$\beta_2(\lambda) = -x_1^+\sqrt{\alpha_2^2 + 4\alpha_1\alpha_4} < 0 \qquad \text{at } \lambda = 0.$$

Thus, by Theorem 2.2.3 and the existence of global attractors in $\mathbb{R}^2_+$, we derive assertion (2) and that the bifurcated singular point $z_\lambda = (\widetilde{x}_1^+, \widetilde{x}_2^+)$ is a global attractor in the interior of $\mathbb{R}^2_+$ for $\lambda > 0$ small.

We now show that z_λ is an attractor for $\Lambda > 0$. The Jacobian of (6.5.10) at z_λ^* is given by

$$B = \begin{pmatrix} (\alpha_2 - 2\alpha_4\widetilde{x}_1^+)\widetilde{x}_1^+ & -\alpha_3\widetilde{x}_1^+ \\ \widetilde{x}_2^+ & -\widetilde{x}_2^+ \end{pmatrix}. \tag{6.5.16}$$

Direct computation indicates that

$$\det B = \widetilde{x}_1^+\widetilde{x}_2^+ A > 0 \qquad \forall \alpha_i > 0\ (1 \le i \le 4). \tag{6.5.17}$$

Hence, the eigenvalues β_1 and β_2 of (6.5.16) satisfy

$$\mathrm{Re}\beta_{1,2}\begin{cases} <0 & \text{if } \Lambda > 0, \\ =0 & \text{if } \Lambda = 0, \\ >0 & \text{if } \Lambda < 0, \end{cases} \tag{6.5.18}$$

where Λ is as defined in (6.5.11), and

$$\Lambda = -2\alpha_4 \mathrm{tr} B = 2\alpha_4(\widetilde{x}_2^+ + \widetilde{x}_1^+(2\alpha_4\widetilde{x}_1^+ - \alpha_2)).$$

Finally, we prove that $\Lambda > 0$ when $\lambda = 0$. Noting that $(\widetilde{x}_1^+, \widetilde{x}_2^+)$ is bifurcated from $(x_1^+, 0)$ at $\lambda = 0$, we deduce that $\widetilde{x}_1^+ = x_1^+$ and $\widetilde{x}_2^+ = 0$ at $\lambda = 0$. It follows that

$$\Lambda = 2\alpha_4 x_1^+(2\alpha_4 x_1^+ - \alpha_2) = 2\alpha_4 x_1^+\sqrt{\alpha_2^2 + 4\alpha_1\alpha_4} > 0 \text{ at } \lambda = 0.$$

Thus, assertion (3) is proved. The proof is complete. □

Proof of Theorem 6.5.4. It follows from (6.5.17) that the eigenvalues β_1 and β_2 of (6.5.16) satisfying (6.5.18) are a pair of complex numbers. Thus the system always undergoes a dynamic transition to a spatiotemporal oscillatory solution.

To determine the types of transitions, consider

$$x \to x + z_\lambda^*,$$

with $z_\lambda^* = (x_1^*, x_2^*)$ as in (6.5.12). Then (6.5.10) can be rewritten as

$$\frac{dx}{dt} = Bx + G(x), \tag{6.5.19}$$

where B is the matrix defined by (6.5.16), and

$$G(x) = \begin{pmatrix} (\alpha_2 - 3\alpha_4 x_1^*)x_1^2 - \alpha_3 x_1 x_2 - \alpha_4 x_1^3 \\ x_1 x_2 - x_2^2 \end{pmatrix}.$$

At $\Lambda = 0$, we have

$$x_2^* = x_1^*(\alpha_2 - 2\alpha_4 x_1^*). \tag{6.5.20}$$

Hence the matrix B in (6.5.16) becomes

$$B = \begin{pmatrix} a & -c \\ a & -a \end{pmatrix} \qquad \text{at } \Lambda = 0, \tag{6.5.21}$$

where

$$a = x_2^* = x_1^* - 1, \quad c = \alpha_3 x_1^*,$$

and the eigenvalues of (6.5.21) are

$$\beta_\pm = \pm i\sigma \qquad \text{with } \sigma = \sqrt{ac - a^2}.$$

From $\Lambda = 0$, we obtain $\sigma = \sqrt{x_1^* x_2^* A}$.

The eigenvectors of (6.5.21) satisfy the following equations:

$$\begin{pmatrix} a & -c \\ a & -a \end{pmatrix}\begin{pmatrix} \xi_1 \\ \xi_2 \end{pmatrix} = \sigma \begin{pmatrix} \eta_1 \\ \eta_2 \end{pmatrix}, \qquad \begin{pmatrix} a & -c \\ a & -a \end{pmatrix}\begin{pmatrix} \eta_1 \\ \eta_2 \end{pmatrix} = -\sigma \begin{pmatrix} \xi_1 \\ \xi_2 \end{pmatrix}.$$

Solving these equations, we derive

$$(\xi_1, \xi_2) = \left(1, \frac{a}{c}\right), \quad (\eta_1, \eta_2) = \left(0, \frac{\sigma}{c}\right).$$

Furthermore, from the equations

$$\begin{pmatrix} a & a \\ -c & -a \end{pmatrix}\begin{pmatrix} \xi_1^* \\ \xi_2^* \end{pmatrix} = -\sigma \begin{pmatrix} \eta_1^* \\ \eta_2^* \end{pmatrix}, \qquad \begin{pmatrix} a & a \\ -c & -a \end{pmatrix}\begin{pmatrix} \eta_1^* \\ \eta_2^* \end{pmatrix} = \sigma \begin{pmatrix} \xi_1^* \\ \xi_2^* \end{pmatrix},$$

we derive the conjugate eigenvectors as

$$(\xi_1^*, \xi_2^*) = (1, 0), \quad (\eta_1^*, \eta_2^*) = \left(-\frac{a}{\sigma}, \frac{c}{\sigma}\right).$$

Take the transformation

$$x = Ez, \tag{6.5.22}$$

where E is the matrix

$$E = \begin{pmatrix} \xi_1 & \eta_1 \\ \xi_2 & \eta_2 \end{pmatrix} = \begin{pmatrix} 1 & 0 \\ \frac{a}{c} & \frac{\sigma}{c} \end{pmatrix}, \qquad E^{-1} = \begin{pmatrix} \xi_1^* & \xi_2^* \\ \eta_1^* & \eta_2^* \end{pmatrix} = \begin{pmatrix} 1 & 0 \\ -\frac{a}{\sigma} & \frac{c}{\sigma} \end{pmatrix}.$$

With the transformation (6.5.22), the system (6.5.19) becomes

$$\frac{dz}{dt} = (E^{-1}BE)z + E^{-1}G(Ez),$$

which can be explicitly expressed as

$$\begin{aligned} \frac{dz_1}{dt} &= -\sigma z_2 + a_{20}^1 z_1^2 + a_{11}^1 z_1 z_2 + a_{30}^1 z_1^3, \\ \frac{dz_2}{dt} &= \sigma z_1 + a_{20}^2 z_1^2 + a_{11}^2 z_1 z_2 + a_{02}^2 z_2^2 + a_{30}^2 z_1^3, \end{aligned} \tag{6.5.23}$$

with

$$\begin{aligned} a_{20}^1 &= \alpha_2 - 3\alpha_4 x_1^* - \frac{a\alpha_3}{c}, & a_{11}^1 &= -\frac{\alpha_3\sigma}{c}, \\ a_{20}^2 &= \frac{a}{\sigma c}(a\alpha_3 + c + 3\alpha_4 c x_1^* - \alpha_2 c - a), & a_{30}^1 &= -\alpha_4, \\ a_{11}^2 &= \frac{1}{c}(a\alpha_3 + c - 2a), & a_{02}^2 &= -\frac{\sigma}{c}, \\ a_{30}^2 &= \frac{a\alpha_4}{\sigma}, \end{aligned} \tag{6.5.24}$$

where $\sigma = \sqrt{x_1^* x_2^* A}, a = x_2^* = x_1^* - 1, c = \alpha_3 x_1^*$. By (6.5.20), we get

$$a_{20}^1 = -\alpha_4 x_1^*, \qquad a_{20}^2 = \frac{a}{\sigma c}\left(a\alpha_3 + \alpha_4 c x_1^* + \frac{c}{x_1^*} - a\right). \tag{6.5.25}$$

By Theorem 2.3.7, we know that the criterion for transitions of type (6.5.23) is given by

$$b = 3a_{30}^1 - \frac{2}{\sigma}a_{20}^1 a_{20}^2 + \frac{1}{\sigma}\left(a_{11}^1 a_{20}^1 - a_{11}^2 a_{20}^2 - a_{11}^2 a_{02}^2\right). \tag{6.5.26}$$

Inserting (6.5.24) and (6.5.25) into (6.5.26), we obtain

$$\begin{aligned} b = \frac{2(x_1^*-1)}{\sigma^2 x_1^{*2}} \Bigg[& x_1^* A \left(\frac{2x_1^*-1}{2\alpha_3} - \frac{x_1^*-1}{\alpha_3^2} - \alpha_4 x_1^{*2}\right) \\ & + \left(\alpha_4 x_1^{*2} + x_1^* - \frac{x_1^*-1}{\alpha_3}\right)\left(\alpha_4 x_1^{*2} - x_1^* + \frac{x_1^*-1}{\alpha_3} + \frac{1}{2}\right)\Bigg], \end{aligned}$$

which is equivalent to the parameter defined by (6.5.13). The proof is complete. □

6.5.3 Three-Species Systems

General theory. We consider a general three-species system governed by

$$\begin{aligned}\frac{dx_1}{dt} &= \lambda_1 x_1 + a_{11}x_1^2 + a_{12}x_1x_2 + a_{13}x_1x_3 - \beta_1 x_1^3,\\ \frac{dx_2}{dt} &= \lambda_2 x_2 + a_{21}x_2x_1 + a_{22}x_2^2 + a_{23}x_2x_3 - \beta_2 x_2^3,\\ \frac{dx_3}{dt} &= \lambda_3 x_3 + a_{31}x_3x_1 + a_{32}x_3x_2 + a_{33}x_3^2 - \beta_3 x_3^2.\end{aligned} \tag{6.5.27}$$

Let $z = (z_1, z_2, z_3) \in \mathbb{R}^3_+$ be a steady-state solution of (6.5.27) with $z_i > 0$ $(1 \leq i \leq 3)$. In view of (6.5.5) and (6.5.6), under the translation $x \to x + z$, (6.5.27) is transformed into

$$\begin{aligned}\frac{dx_1}{dt} &= z_1\alpha_{11}x_1 + z_1a_{12}x_2 + z_1a_{13}x_3 + \sum_{k=1}^{3} c_{1k}x_1x_k - \beta_1 x_1^3,\\ \frac{dx_2}{dt} &= z_2a_{21}x_1 + z_2\alpha_{22}x_2 + z_2a_{23}x_3 + \sum_{k=1}^{3} c_{2k}x_2x_k - \beta_2 x_2^3,\\ \frac{dx_3}{dt} &= z_3a_{31}x_1 + z_3a_{32}x_2 + z_3\alpha_{33}x_3 + \sum_{k=1}^{3} c_{3k}x_3x_k - \beta_3 x_3^3,\end{aligned} \tag{6.5.28}$$

where the c_{ij} are as in (6.5.6), and

$$\alpha_{kk} = a_{kk} - 2\beta_k z_k, \quad 1 \leq k \leq 3.$$

We set

$$\begin{aligned}\delta_2 &= -(z_1\alpha_{11} + z_2\alpha_{22} + z_3\alpha_{33}),\\ \delta_1 &= z_1z_2\det\begin{pmatrix}\alpha_{11} & \alpha_{12}\\ a_{21} & a_{22}\end{pmatrix} + z_1z_3\det\begin{pmatrix}\alpha_{11} & a_{13}\\ a_{31} & \alpha_{33}\end{pmatrix} + z_2z_3\det\begin{pmatrix}\alpha_{22} & a_{23}\\ a_{32} & \alpha_{33}\end{pmatrix},\\ \delta_0 &= -z_1z_2z_3\det\begin{pmatrix}\alpha_{11} & a_{12} & a_{13}\\ a_{21} & \alpha_{22} & a_{23}\\ a_{31} & a_{32} & \alpha_{33}\end{pmatrix}.\end{aligned} \tag{6.5.29}$$

Theorem 6.5.7 *Let $\delta_0, \delta_1, \delta_2$ be the parameters defined by (6.5.29), and $\lambda = \delta_1\delta_2 - \delta_0$. Assume that $\delta_2 > 0$. Then for the system (6.5.28), the following assertions hold:*

(1) When $\delta_0, \lambda > 0$, the singular point $x = 0$ of (6.5.28) is asymptotically stable, i.e., $x = z$ is an attractor of (6.5.27).
(2) When $\lambda > 0$, this system undergoes a dynamic transition from $(x, \delta_0) = (0,0)$ to some singular points as δ_0 crosses zero, leading to a real eigenvalue of the linearized problem crossing the imaginary axis.

(3) *When $\delta_0 > 0$, the system undergoes a dynamic transition from $(x,\lambda) = (0,0)$ to a time-periodic solution when λ crosses zero, leading to a pair of complex eigenvalues crossing the imaginary axis.*

Proof. Let A be the matrix given by

$$A = \begin{pmatrix} z_1\alpha_{11} & z_1a_{12} & z_1a_{13} \\ z_2a_{21} & z_2\alpha_{22} & z_2a_{23} \\ z_3a_{31} & z_3a_{32} & z_3\alpha_{33} \end{pmatrix}.$$

It is well known that if all eigenvalues of A have negative real part, then $x = 0$ is asymptotically stable for (6.5.28). The characteristic equation of A is in the form

$$\beta^3 + \delta_2\beta^2 + \delta_1\beta + \delta_0 = 0. \tag{6.5.30}$$

By the Routh–Hurwitz criterion, if

$$\delta_0 > 0,\ \delta_2 > 0,\ \lambda = \delta_1\delta_2 - \delta_0 > 0,$$

then all zeros of (6.5.30) have negative real part. Thus assertion (1) is derived.

Let β_j $(1 \leq j \leq 3)$ be the eigenvalues of A. Then we have

$$\begin{aligned} \delta_0 &= -\beta_1\beta_2\beta_3, \\ \delta_1 &= \beta_1\beta_2 + \beta_1\beta_3 + \beta_2\beta_3, \\ \delta_2 &= -(\beta_1 + \beta_2 + \beta_3). \end{aligned} \tag{6.5.31}$$

We infer from (6.5.31) that if $\delta_1, \delta_2 > 0$, and $\lambda = \delta_1\delta_2 - \delta_0 > 0$, then for δ_0 near zero,

$$\beta_1 \begin{cases} < 0 & \text{for } \delta_0 > 0, \\ = 0 & \text{for } \delta_0 = 0, \\ > 0 & \text{for } \delta_0 < 0, \end{cases} \tag{6.5.32}$$

$$\mathrm{Re}\beta_{2,3}(\delta_0) < 0. \tag{6.5.33}$$

When $\delta_2, \delta_0 > 0$, for λ near $\lambda = 0$ we have

$$\mathrm{Re}\beta_{1,2} \begin{cases} < 0 & \text{for } \lambda > 0, \\ = 0 & \text{for } \lambda = 0, \\ > 0 & \text{for } \lambda < 0, \end{cases} \tag{6.5.34}$$

$$\beta_3 < 0 \qquad \text{at } \lambda = 0, \tag{6.5.35}$$

where β_1 and β_2 are a pair of complex eigenvalues of A. Therefore, assertions (2) and (3) follow from (6.5.32)–(6.5.33) and (6.5.34)–(6.5.35) respectively. □

Predator–Prey Mutualism Chains. We consider the following equations, which describe an ecological system such that the species X_3 and X_1 are the predator and prey, X_1 and X_2 are beneficial to each other, and X_2 and X_3 are mutually independent. The model is given by

$$\begin{aligned}\frac{dx_1}{dt} &= \lambda_1 x_1 + a_{12}x_1x_2 - a_{13}x_1x_3 - \beta_1 x_1^3,\\ \frac{dx_2}{dt} &= \lambda_2 x_2 + a_{21}x_2x_1 - a_{22}x_2^2,\\ \frac{dx_3}{dt} &= -\lambda_3 x_3 + a_{31}x_3x_1 - a_{33}x_3^2,\end{aligned} \tag{6.5.36}$$

where all coefficients are positive.

We take the nondimensional scheme

$$\begin{aligned}
&t = \frac{1}{a_{31}}t', && x_1 = \sqrt{\frac{a_{31}}{\beta_1}}x_1', && x_2 = \frac{a_{31}}{a_{22}}x_2', && x_3 = \frac{a_{31}}{a_{33}}x_3',\\
&\mu_1 = \frac{\lambda_1}{a_{31}}, && \mu_2 = \frac{\lambda_2}{a_{31}}, && \mu_3 = \frac{\lambda_3}{a_{31}}, && \alpha_1 = \frac{a_{12}}{a_{22}},\\
&\alpha_2 = \frac{a_{13}}{a_{33}}, && \alpha_3 = \frac{a_{21}}{\sqrt{a_{31}\beta_1}}.
\end{aligned}$$

Then equation (6.5.36) becomes (dropping the primes)

$$\begin{aligned}\frac{dx_1}{dt} &= \mu_1 x_1 + \alpha_1 x_1x_2 - \alpha_2 x_1x_3 - x_1^3,\\ \frac{dx_2}{dt} &= \mu_2 x_2 + \alpha_3 x_1x_2 - x_2^2,\\ \frac{dx_3}{dt} &= -\mu_3 x_3 + x_1x_3 - x_3^2.\end{aligned} \tag{6.5.37}$$

Let

$$\lambda = \mu_1 + \alpha_1\mu_2 + \alpha_1\alpha_3\mu_3 - \mu_3^2. \tag{6.5.38}$$

Clearly, when $\lambda > 0$, (6.5.37) admits a singular point $z^* = (z_1, z_2, z_3) \in \mathbb{R}_+^3$:

$$\begin{aligned}
z_1 &= \frac{1}{2}\left[\alpha_1\alpha_3 - \alpha_2 + \sqrt{(\alpha_1\alpha_3 - \alpha_2)^2 + 4(\mu_1 + \alpha_1\mu_2 + \alpha_2\mu_3)}\right],\\
z_2 &= \alpha_3 z_1 + \mu_2,\\
z_3 &= z_1 - \mu_3, \quad (z_3 > 0 \text{ for } \lambda > 0).
\end{aligned} \tag{6.5.39}$$

Also, we introduce a parameter Λ as

$$\begin{aligned}\Lambda = {}& (z_2 + z_3)(4z_1^4 + 2z_1^2 z_2 + 2a_1^2 z_3 + z_2z_3)\\ &+ \alpha_2 z_1 z_3(2z_1^2 + z_3) - \alpha_1\alpha_3 z_1 z_2(2z_1^2 + z_2),\end{aligned} \tag{6.5.40}$$

where z_1, z_2, z_3 are as in (6.5.39).

Theorem 6.5.8 *Let λ and Λ be the two parameters defined by (6.5.38) and (6.5.40) respectively.*

(1) When $\lambda < 0$, the system (6.5.37) admits only one singular point in $\mathbb{R}^3_+$:

$$z_0 = (z_1^0, z_2^0, 0) = \left(\frac{1}{2}\left(\alpha_1\alpha_3 + \sqrt{\alpha_1^2\alpha_3^2 + 4(\mu_1 + \alpha_1\mu_2)}\right), \alpha_3 z_1^\circ + \mu_2, 0\right), \tag{6.5.41}$$

which is globally asymptotically stable in $\mathbb{R}^3_+$.

(2) The system (6.5.37) bifurcates from $(x,\lambda) = (z_0, 0)$ to a unique positive singular point z^* of (6.5.39) on $\lambda > 0$, and z^* is an attractor if and only if $\Lambda > 0$. In particular, $\Lambda > 0$ if $\lambda = 0$.

(3) When $\lambda > 0$, the system has a Hopf bifurcation at $(x,\Lambda) = (z^*, 0)$. Moreover, there exists a parameter b depending on μ_j and $\alpha_j (1 \le j \le 3)$ at $\Lambda = 0$ such that (6.5.37) bifurcates to a stable periodic solution from z^* on $\Lambda < 0$ if $b < 0$, and bifurcates to an unstable periodic solution on $\Lambda > 0$ if $b > 0$. Furthermore, if $b > 0$, there is a singular separation of periodic solutions at some $\Lambda_0 > 0$.

Proof. We divide the proof into several steps.

Step 1. Equations (6.5.37) have a global attractor in $\mathbb{R}^3_+$. It is clear that $\mathbb{R}^3_+$ is invariant for (6.5.37), and

$$\begin{aligned}\frac{1}{2}\frac{d}{dt}(x_1^2 + x_2^2 + x_3^2) &= x_1(\mu_1 x_1 + \alpha_1 x_1 x_2 - \alpha_2 x_1 x_3 - x_1^3)\\ &\quad + x_2(\mu_2 x_2 + \alpha_3 x_1 x_2 - x_2^2) + x_3(-\mu_3 x_3 + x_1 x_3 - x_3^2),\end{aligned}$$

which implies that

$$\begin{aligned}\frac{1}{2}\frac{d}{dt}|x|^2 &\le -x_1^4 - x_2^3 - x_3^3 + \mu_1 x_1^2 + \alpha_1 x_1^2 x_2 + \mu_2 x_2^2\\ &\quad + \alpha_3 x_1 x_2^2 + x_1 x_3^2,\end{aligned} \tag{6.5.42}$$

for all $x \in \mathbb{R}^3_+$. By Young's inequality, we have

$$\begin{aligned} &x_i^2 \le \varepsilon x_i^p + \varepsilon^{-\frac{1}{p-2}}, \quad \forall p > 2,\\ &x_1^2 x_2 \le \delta^{-1} x_2^2 + \delta x_1^4 \le \delta^{-1}\varepsilon^{-1} + \varepsilon\delta^{-1} x_2^3 + \delta x_1^4,\\ &x_1 x_j^2 \le \delta x_j^3 + \delta^{-2} x_1^3 \le \delta^{-2}\varepsilon^{-3} + \delta x_j^3 + \varepsilon\delta^{-2} x_1^4,\end{aligned} \tag{6.5.43}$$

for every $\varepsilon, \delta > 0$. Properly taking ε and δ small, from (6.5.42) and (6.5.43) we deduce that

$$\frac{d}{dt}|x|^2 \le -C_1(x_1^4 + x_2^3 + x_3^3) + C_2 \tag{6.5.44}$$

for some constants $C_1, C_2 > 0$. Then the existence of the global attractor of (6.5.37) follows from (6.5.44).

Step 2. Direct calculation indicates that when $\lambda \le 0$, (6.5.37) has no singular points in the interior of $\mathbb{R}^3_+$, and it possesses a unique singular point $z_0 = (z_1^0, z_2^0, 0)$

as given by (6.5.41), which is globally asymptotically stable in $\mathbb{R}^2_+ = \{(x_1, x_2, 0) \in \mathbb{R}^3 |\ x_1, x_2 \geq 0\}$, and asymptotically stable in $\mathbb{R}^3_+$. Hence, assertion (1) follows. In addition, we can prove assertion (2) in the same fashion as in the proof of Theorem 6.5.3.

Step 3. Let $\lambda > 0$. Then (6.5.37) has a singular point $z^* = (z_1, z_2, z_3) \in \mathbb{R}^3_+$ as defined by (6.5.39). Under the translation $x \to z^* + x$, (6.5.37) becomes

$$\frac{dx}{dt} = Ax + \begin{pmatrix} \alpha_1 x_1 x_2 - \alpha_2 x_1 x_3 - 3z_1 x_1^2 - x_1^3 \\ \alpha_3 x_1 x_2 - x_2^2 \\ x_1 x_3 - x_3^2 \end{pmatrix}, \tag{6.5.45}$$

where A is the matrix given by

$$A = \begin{pmatrix} -2z_1^2 & z_1\alpha_1 & -z_1\alpha_2 \\ z_2\alpha_3 & -z_2 & 0 \\ z_3 & 0 & -z_3 \end{pmatrix}. \tag{6.5.46}$$

As in (6.5.29), for (6.5.46) we introduce

$$\begin{aligned} \delta_0 &= \det A = z_1 z_2 z_3 (2z_1 + \alpha_2 - \alpha_1\alpha_3), \\ \delta_1 &= z_1 z_2 (2z_1 - \alpha_1\alpha_3) + z_1 z_3 (2z_1 + \alpha_2) + z_2 z_3, \\ \delta_2 &= -\mathrm{tr}A = 2z_1^2 + z_2 + z_3. \end{aligned}$$

By (6.5.39), we see that

$$\begin{aligned} &\delta_0 = z_1 z_2 z_3 \sqrt{(\alpha_1\alpha_3 - \alpha_2)^2 + 4(\mu_1 + \alpha_1\mu_2 + \alpha_2\mu_3)} > 0, \\ &\delta_2 > 0. \end{aligned} \tag{6.5.47}$$

It is easy to check that the parameter Λ defined by (6.5.40) is

$$\Lambda = \delta_1\delta_2 - \delta_0.$$

Therefore, by (6.5.47) and assertion (3) of Theorem 6.5.7, we can obtain assertion (3). The proof is complete. □

Remark 6.5.9 We show that there are $\alpha_j, \mu_j > 0$ $(1 \leq j \leq 3)$ such that $\lambda > 0$ and $\Lambda \leq 0$. Let

$$\mu_1, \mu_2 \approx 0, \quad \alpha_1\alpha_3 \approx \alpha_2, \quad \text{and } \alpha_1\alpha_3 > \mu_3. \tag{6.5.48}$$

Then by (6.5.39) and (6.5.48), we have

$$z_1 \approx \sqrt{\alpha_2\mu_3}, \quad z_2 \approx \alpha_3 z_1, \quad z_3 \approx z_1, \quad \lambda > 0. \tag{6.5.49}$$

By (6.5.48) and (6.5.49), we have

$$\begin{aligned} \Lambda \approx\ & -\alpha_1\alpha_3^2(\alpha_3 + 2z_1)z_1^3 + \alpha_2(2z_1 + 1)z_1^3 \\ & + (\alpha_3 + 1)(4\alpha_2\mu_3 + 2\alpha_3\sqrt{\alpha_2\mu_3} + 2\sqrt{\alpha_2\mu_3} + \alpha_3)z_1^3. \end{aligned}$$

This implies that

$$\Lambda < 0 \text{ if } \alpha_3 > 0 \text{ sufficiently large.}$$

Mutualism. Here we investigate the mutual benefit systems modeled by (6.5.27) with

$$\lambda_j, \beta_j > 0, \quad 1 \leq j \leq 3 \quad \text{and } a_{ij} > 0 \, \forall i \neq j. \tag{6.5.50}$$

We shall use Theorem 6.5.7 to prove the following theorem.

Theorem 6.5.10 *Under the conditions (6.5.50), the system (6.5.27) possesses a unique positive singular point z in the interior of $\mathbb{R}^3_+$, which is an attractor attracting the interior of $\mathbb{R}^3_+$.*

Proof. Let $F : \mathbb{R}^3 \to \mathbb{R}^3$ be the vector field of (6.5.27). Then by (6.5.50), we can deduce that there is a real number $R > 0$ such that

$$(F(x), x) < 0 \qquad \forall x \in \mathbb{R}^3 \text{ with } |x| \geq R. \tag{6.5.51}$$

For each $1 \leq i \leq 3$, F has a unique singular point z_i^* in the set

$$\Gamma_i = \left\{ x \in \mathbb{R}^3 \mid x_i = 0, \quad x_j > 0 \quad \forall j \neq i \right\}.$$

This point z_i^* attracts Γ_i, and z_i^* is a saddle point in $\mathbb{R}^3$. Note that $x = 0$ is an unstable node. It follows from (6.5.51) that the set

$$A_R^+(\varepsilon) = \left\{ x \in \mathbb{R}^3_+ \mid |x| \leq R, \operatorname{dist}(x, \partial \mathbb{R}^3_+) \geq \varepsilon \right\}$$

is an absorbing set of (6.5.27) in $\mathbb{R}^3_+$, provided that $\varepsilon > 0$ is sufficiently small. Hence, by the Brouwer degree theory, we have

$$\deg(F, A_R^+(\varepsilon), 0) = \deg(-(id - z^*), A_R^+(\varepsilon), 0) = (-1)^3. \tag{6.5.52}$$

We infer from (6.5.50) that (6.5.27) has at least a singular point in the interior of $\mathbb{R}^3_+$. Let $z^* = (z_1, z_2, z_3)$ be such a singular point of (6.5.27) in the interior of $\mathbb{R}^3_+$. We need to show that

$$\operatorname{ind}(F, z^*) = -1. \tag{6.5.53}$$

To this end, it suffices to prove that the eigenvalues of the Jacobian matrix $DF(z^*)$ have negative real part. By (6.5.28), $DF(z^*)$ can be written as

$$DF(z^*) = \begin{pmatrix} z_1\alpha_{11} & z_1 a_{12} & z_1 a_{13} \\ z_2 a_{21} & z_2 \alpha_{22} & z_2 a_{23} \\ z_3 a_{31} & z_3 a_{32} & z_3 a_{33} \end{pmatrix} \tag{6.5.54}$$

with $\alpha_{kk} = a_{kk} - 2\beta_k z_k$.

Since $z^* = (z_1, z_2, z_3)$ is a zero of F, we have

$$\begin{aligned}\alpha_{11}z_1 &= (a_{11} - 2\beta_1 z_1)z_1 = -(a_{12}z_2 + a_{13}z_3 + \lambda_1 + \beta_1 z_1^2),\\ \alpha_{22}z_2 &= (a_{22} - 2\beta_2 z_2)z_2 = -(a_{21}z_1 + a_{23}z_3 + \lambda_2 + \beta_2 z_2^2),\\ \alpha_{33}z_3 &= (a_3 - 2\beta_3 z_3)z_3 = -(a_{31}z_1 + a_{32}z_2 + \lambda_3 + \beta_3 z_3^2).\end{aligned}$$

Let

$$\begin{aligned}A_1 &= \lambda_1 + \beta_1 z_1^2 + a_{12}z_2 + a_{13}z_3,\\ A_2 &= \lambda_2 + \beta_2 z_2^2 + a_{21}z_1 + a_{23}z_3,\\ A_3 &= \lambda_3 + \beta_3 z_3^2 + a_{31}z_1 + a_{32}z_2.\end{aligned}$$

Then $\alpha_{kk}z_k = -A_k$, and these parameters of (6.5.29) read

$$\begin{aligned}\delta_2 =& A_1 + A_2 + A_3,\\ \delta_1 =& A_1A_2 + A_1A_3 + A_2A_3 - a_{12}a_{21}z_1z_2 - a_{13}a_{31}z_1z_3 - a_{23}a_{32}z_2z_3,\\ \delta_0 =& A_1A_2A_3 - A_1a_{23}a_{32}z_2z_3 - A_2a_{13}a_{31}z_1z_3\\ &- A_3a_{12}a_{21}z_1z_2 - a_{12}a_{23}a_{31}z_1z_2z_3 - a_{13}a_{21}a_{32}z_1z_2z_3.\end{aligned}$$

It is clear that

$$\begin{aligned}\delta_2 >& \ 0,\\ \delta_0 =& \ (\lambda_1 + \beta_1 z_1^2)(\lambda_2 + \beta_2 z_2^2)(\lambda_3 + \beta_3 z_3^2) + (\lambda_1 + \beta_1 z_1^2)(\lambda_3 + \beta_3 z_3^2)(a_{21}z_1 + a_{23}z_3)\\ &+(\lambda_2+\beta_2 z_2^2)(\lambda_3+\beta_3 z_3^2)(a_{12}z_2+a_{13}z_3)+(\lambda_1+\beta_1 z_1^2)(\lambda_2+\beta_2 z_2^2)(a_{31}z_1+a_{32}z_2)\\ &+(\lambda_1 + \beta_1 z_1^2)(a_{21}a_{31}z_1^2 + a_{21}a_{32}z_1z_2 + a_{23}a_{31}z_1z_3)\\ &+(\lambda_2 + \beta_2 z_2^2)(a_{12}a_{31}z_1z_2 + a_{12}a_{32}z_2^2 + a_{13}a_{32}z_2z_3)\\ &+(\lambda_3 + \beta_3 z_3^2)(a_{12}a_{23}z_2z_3 + a_{13}a_{21}z_1z_3 + a_{13}a_{23}z_3^2)\\ &> 0.\end{aligned}$$

Direct calculation shows that

$$\begin{aligned}\delta_1\delta_2 - \delta_0 = \ & 2A_1A_2A_3 + A_1^2A_2 + A_1^2A_3 + A_1A_2^2 + A_2^2A_3\\ &+A_1A_3^2 + A_2A_3^2 + a_{12}a_{23}a_{31}z_1z_2z_3 + a_{13}a_{21}a_{32}z_1z_2z_3\\ &-A_1a_{12}a_{21}z_1z_2 - A_1a_{13}a_{31}z_1z_3 - A_2a_{12}a_{21}z_1z_2\\ &-A_2a_{23}a_{32}z_2z_3 - A_3a_{13}a_{31}z_1z_3 - A_3a_{23}a_{32}z_2z_3.\end{aligned}$$

Obviously

$$\begin{aligned}&A_1^2A_2 > A_1(a_{12}z_2)(a_{21}z_1),\ A_1^2A_3 > A_1(a_{13}z_3)(a_{31}z_1),\\ &A_2^2A_1 > A_2(a_{12}z_2)(a_{21}z_1),\ A_2^2A_3 > A_2(a_{23}z_3)(a_{32}z_2),\\ &A_3^2A_1 > A_3(a_{13}z_3)(a_{31}z_1),\ A_3^2A_2 > A_3(a_{23}z_3)(a_{32}z_2).\end{aligned}$$

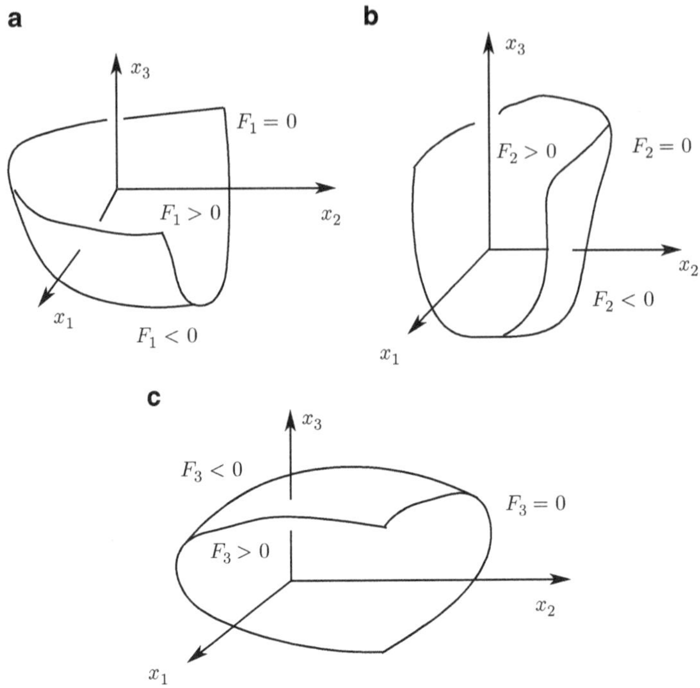

Fig. 6.2 Graphs of $F_i = 0$ $(i = 1,2,3)$

Hence we obtain

$$\delta_1 \delta_2 - \delta_0 > 0,$$

for all parameters satisfying (6.5.50). Thus, by assertion (1) of Theorem 6.5.7, we obtain that all eigenvalues of (6.5.54) have negative real part. It follows that z^* is locally asymptotically stable and (6.5.53) holds. We then infer from (6.5.52) and (6.5.53) that z^* is the unique singular point of (6.5.27) in the interior of $\mathbb{R}^3_+$.

Finally, we prove that z^* is globally asymptotically stable in the interior of $\mathbb{R}^3_+$. Let $F = (x_1F_1, x_2F_2, x_3F_3)$. Then for any $1 \leq i \leq 3$,

$$F_i = 0 \text{ if and only if } \sum_{j \neq i}^{3} a_{ij}x_j = \beta_i x_i^2 + a_{ii}x_i - \lambda_j.$$

The graphs of $F_i = 0$ $(1 \leq i \leq 3)$ are as shown in Fig. 6.2.

From Fig. 6.3a, we see that the orbits of F in the interior of $\mathbb{R}^3_+$ point to the intersection Σ_1 of the surfaces $F_2 = 0$ and $F_3 = 0$. Likewise, in Fig. 6.3b, c, the orbits of F in the interior of $\mathbb{R}^3_+$ point to the intersection Σ_2 of $F_1 = 0$ and $F_3 = 0$, and to the intersection Σ_3 of $F_1 = 0$ with $F_2 = 0$ respectively. Hence for any x_0 in the interior of $\mathbb{R}^3_+$, the solution $x(t,x_0)$ of (6.5.27) with $x(0,x_0) = x_0$ satisfies

$$\lim_{t \to \infty} x(t,x_0) = z^*,$$

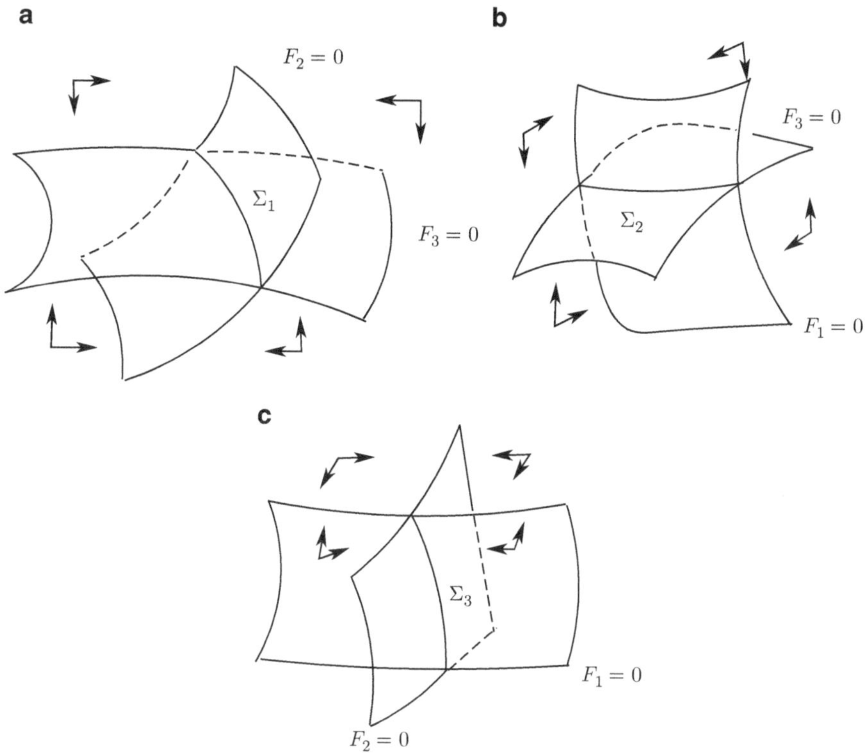

Fig. 6.3 (**a**) The interaction of graphs of $F_2 = 0$ and $F_3 = 0$, (**b**) the interaction of graphs of $F_1 = 0$ and $F_3 = 0$, and (**c**) the interaction of graphs of $F_1 = 0$ and $F_2 = 0$

where z^* is the intersection of $F_1 = 0$, $F_2 = 0$, and $F_3 = 0$. Namely, z_1^* is a singular point of (6.5.27) in the interior of $\mathbb{R}_+^3$. Therefore, the positive singular point z^* of (6.5.27) is globally asymptotically stable in the interior of $\mathbb{R}_+^3$. The proof is complete. □

Remark 6.5.11 Theorem 6.5.10 also holds for any n-species mutualism system. Namely, if λ_j, $k_j > 0$, and $a_{ij} > 0$ for every $i \neq j$, then the system (6.5.1) has a unique positive singular point that is globally asymptotically stable in the interior of $\mathbb{R}_+^n$.

Appendix A

A.1 Formulas for Center Manifold Functions

The stable and center manifold theorems are classical (Henry 1981; Ma and Wang 2005b). The main objective of this appendix is to give a brief account of the center manifold reduction, and to prove a few important formulas in Theorem A.1.1. Part of Theorem A.1.1 was first derived and proved in Ma and Wang (2005b, 2007b), and part of it is introduced for the first time in this appendix. These formulas play a crucial role in many applications addressed in this book.

Let X_1 and X be two Banach spaces, and $X_1 \subset X$ a dense and compact inclusion. Consider the following nonlinear evolution equation:

$$\begin{aligned} &\frac{du}{dt} = L_\lambda u + G(u,\lambda), \qquad \lambda \in \mathbb{R}^1, \\ &u(0) = u_0, \end{aligned} \tag{A.1.1}$$

where $L_\lambda = -A + B_\lambda : X_1 \to X$ are parameterized linear completely continuous fields depending continuously on $\lambda \in \mathbb{R}^1$, i.e.,

$$\begin{aligned} &A : X_1 \to X \quad \text{a linear homeomorphism}, \\ &B_\lambda : X_1 \to X \quad \text{parameterized linear compact operators}. \end{aligned} \tag{A.1.2}$$

Consider the case in which $L_\lambda : X_1 \to X$ is a sectorial operator. Then one can define fractional power operators L_λ^α for $\alpha \in \mathbb{R}^1$ with domains $X_\alpha = D(L_\lambda^\alpha)$. Assume that the nonlinear mappings $G(\cdot,\lambda) : X_\theta \to X$ for some $\theta < 1$ are parameterized $C^r (r \geq 1)$ bounded operators continuously depending on the parameter $\lambda \in \mathbb{R}^1$, such that

$$G(u,\lambda) = o(\|u\|_{X_\theta}) \qquad \forall \lambda \in \mathbb{R}^1 \qquad (\theta < 1). \tag{A.1.3}$$

In addition, let $G(u,\lambda)$ have a Taylor expansion at $u = 0$ as follows:

$$G(u,\lambda) = \sum_{m=k}^{r} G_m(u,\lambda) + o(\|u\|_{X_1}^r) \qquad \text{for some } 2 \leq k \leq r, \tag{A.1.4}$$

T. Ma and S. Wang, *Phase Transition Dynamics*, DOI 10.1007/978-1-4614-8963-4,

where $u \in X_1$, $G_m : X_1 \times \cdots \times X_1 \to X$ is an m-multiple linear operator, and $G_m(u,\lambda) = G_m(u,\cdots,u,\lambda)$.

Let $\{\beta_i(\lambda) \in \mathbb{C} \mid i = 1,2,\cdots\}$ be all eigenvalues of L_λ (counting multiplicities), and suppose that they satisfy

$$\text{Re } \beta_i(\lambda) \begin{cases} <0 & \text{if } \lambda < \lambda_0, \\ =0 & \text{if } \lambda = \lambda_0, \\ >0 & \text{if } \lambda > \lambda_0, \end{cases} \qquad \forall 1 \le i \le m, \tag{A.1.5}$$

$$\text{Re } \beta_j(\lambda_0) \neq 0 \qquad \forall j \ge m+1, \tag{A.1.6}$$

for some $\lambda_0 \in \mathbb{R}^1$. Let $\{e_1(\lambda),\cdots,e_m(\lambda)\}$ and $\{e_1^*(\lambda),\cdots,e_m^*(\lambda)\}$ be the eigenvectors of L_λ and the conjugate operator L_λ^* corresponding to the eigenvalues $\beta_i(\lambda)$ $(1 \le i \le m)$ respectively. By the spectral theorem, Theorem 3.4 in Ma and Wang (2005b), the spaces X_1 and X can be decomposed into the direct sum

$$X_1 = E_1^\lambda \bigoplus E_2^\lambda, \qquad X = E_1^\lambda \bigoplus \overline{E}_2^\lambda,$$

where

$$\begin{aligned} E_1^\lambda &= \text{span } \{e_1(\lambda),\cdots,e_m(\lambda)\}, \\ E_2^\lambda &= \{u \in X_1 \mid \ (u, e_i^*(\lambda)) = 0 \quad \forall 1 \le i \le m\}, \\ \overline{E}_2^\lambda &= \text{closure of } E_2^\lambda \text{ in } X. \end{aligned}$$

The linear operator L_λ can be decomposed into

$$L_\lambda = J_\lambda \bigoplus \mathscr{L}_\lambda, \tag{A.1.7}$$

where $J_\lambda : E_1^\lambda \to E_1^\lambda$ is the Jordan matrix of L_λ at $\beta_i(\lambda)$ $(1 \le i \le m)$, and $\mathscr{L}_\lambda = L_\lambda|_{E_2^\lambda} : E_2^\lambda \to \overline{E}_2^\lambda$ possesses eigenvalues $\beta_j(\lambda)$ $(j \ge m+1)$. In this case, (A.1.1) can be written as

$$\frac{dx}{dt} = J_\lambda x + P_1 G(x+y,\lambda), \tag{A.1.8}$$

$$\frac{dy}{dt} = \mathscr{L}_\lambda y + P_2 G(x+y,\lambda), \tag{A.1.9}$$

where $P_i : X \to E_i^\lambda$ $(i = 1,2)$ are the canonical projections.

The existence of center manifold functions is classical; see among others (Henry 1981; Ma and Wang 2005b). Since the center manifold function is defined implicitly, one ingredient in the theory and application of phase transition dynamics in this book is that an approximation of the center manifold function to a certain order will lead to a complete understanding of the transitions of a dynamical system. We now present some approximation formulas of the center manifold functions.

First, we show how the center manifold functions are constructed. Let $\rho_\varepsilon : E_1^\lambda \to [0,1]$ be a C^∞ cutoff function defined by

$$\rho_\varepsilon(x) = \begin{cases} 1 & \text{if } \|x\| < \varepsilon, \\ 0 & \text{if } \|x\| > 2\varepsilon, \end{cases}$$

for some $\varepsilon > 0$, and let

$$C^{0,1}(E_1^\lambda, E_2^\lambda(\theta)) = \{h : E_1^\lambda \to E_2^\lambda(\theta) |\ h(0) = 0, h \text{ is Lipschitz}\}.$$

As in Henry (1981), we need to find a function $h \in C^{0,1}(E_1^\lambda, E_2^\lambda(\theta))$ satisfying

$$h(\cdot) = \int_{-\infty}^{0} e^{-\mathscr{L}_\lambda \tau} \rho_\varepsilon(x(\tau,\cdot))G_2(x(\tau,\cdot) + h(x(\tau,\cdot)),\lambda)d\tau, \tag{A.1.10}$$

where $x(t,x_0)$ is a solution of the ordinary differential equation

$$\frac{dx}{dt} = J_\lambda x + \rho_\varepsilon(x)G_1(x + h(x),\lambda), \quad x(0) = x_0. \tag{A.1.11}$$

Then the function $y(t,h(x_0)) = h(x(t,x_0))$ is a solution of the equation

$$\frac{dy}{dt} = \mathscr{L}_\lambda y + \rho_\varepsilon(x(t,x_0))G_2(x(t,x_0),y), \quad y(0) = h(x_0).$$

Thus, $x(t,x_0) + h(x(t,x_0))$ is a local solution of (A.1.8) and (A.1.9), and the manifold M_λ is locally invariant for (A.1.1).

Hence, the existence of the center manifold of (A.1.1) is referred to as the existence of the fixed point of the mapping $F : C^{0,1}(E_1^\lambda, E_2^\lambda(\theta)) \to C^{0,1}(E_1^\lambda, E_2^\lambda(\theta))$ defined by

$$F(h) = \int_{-\infty}^{0} e^{-\mathscr{L}_\lambda \tau} \rho_\varepsilon(x(\tau))G_2(x(\tau),h(x(\tau)))d\tau. \tag{A.1.12}$$

Using some well-known properties of sectorial operatora (Pazy 1983; Henry 1981), one can prove that under the condition (A.1.3), the mapping F defined by (A.1.12) is a contraction operator for $\varepsilon > 0$ small. Therefore, F has a fixed point in $C^{0,1}(E_1^\lambda, E_2^\lambda(\theta))$. For the details of the proof, see (Temam 1997; Henry 1981).

The following theorem gives a first-order approximation formula for the center manifold function of (A.1.1) near $\lambda = \lambda_0$.

Theorem A.1.1 *Assume that the nonlinear operator $G(u,\lambda)$ has a Taylor expansion as in (A.1.4) at $u = 0$ and that the conditions (A.1.5) and (A.1.6) hold. Then the center manifold function $\Phi : E_1 \to E_2$ of (A.1.1) near $\lambda = \lambda_0$ can be expressed as*

$$\Phi(x,\lambda) = \int_{-\infty}^{0} e^{-\tau\mathscr{L}_\lambda} \rho_\varepsilon P_2 G_k(e^{\tau J_\lambda}x,\lambda)d\tau + o(\|x\|^k), \tag{A.1.13}$$

where J_λ *and* $\mathscr{L}_\lambda$ *are the linear operators given by (A.1.7),* $G_k(u,\lambda)$ *is the lowest-order k-multiple linear operator as in (A.1.4), and* $x = \sum_{i=1}^{m} x_i e_i \in E_1$*. In particular, we have the following assertions:*

(1)If J_λ *is diagonal near* $\lambda = \lambda_0$*, then (A.1.13) can be written as*

$$-\mathscr{L}_\lambda \Phi(x,\lambda) = P_2 G_k(x,\lambda) + o(\|x\|^k) + O(|\beta|\|x\|^k), \tag{A.1.14}$$

where $\beta(\lambda) = (\beta_1(\lambda),\cdots,\beta_m(\lambda))$ *is as in (A.1.5).*

(2)Let $m = 2$ *and* $\beta_1(\lambda) = \overline{\beta_2(\lambda)} = \alpha(\lambda) + i\rho(\lambda)$ *with* $\rho(\lambda_0) \neq 0$*. If* $G_k(u,\lambda) = G_2(u,\lambda)$ *is bilinear, then* $\Phi(x,\lambda)$ *can be expressed as*

$$\begin{aligned}
[(-\mathscr{L}_\lambda)^2 + 4\rho^2(\lambda)](-\mathscr{L}_\lambda)\Phi(x,\lambda) &= [(-\mathscr{L}_\lambda)^2 + 4\rho^2(\lambda)]P_2G_2(x,\lambda)\\
&- 2\rho^2(\lambda)P_2G_2(x,\lambda) + 2\rho^2 P_2 G_2(x_1e_2 - x_2e_1)\\
&+ \rho(-\mathscr{L}_\lambda)P_2[G_2(x_1e_1 + x_2e_2, x_2e_1 - x_1e_2)\\
&+ G_2(x_2e_1 - x_1e_2, x_1e_1 + x_2e_2)] + o(k),
\end{aligned} \tag{A.1.15}$$

where we have used $o(k) = o(\|x\|^k) + O(|Re\beta(\lambda)|\|x\|^k)$*.*

(3)Let $\beta_1(\lambda) = \cdots = \beta_m(\lambda)$ *have algebraic and geometric multiplicities* $m \geq 2$ *and* $r = 1$ *near* $\lambda = \lambda_0$*, i.e.,* J_λ *has the Jordan form*

$$J_\lambda = \begin{pmatrix} \beta(\lambda) & \delta & \cdots & 0 & 0\\ 0 & \beta(\lambda) & \cdots & 0 & 0\\ \vdots & \vdots & \vdots & \vdots & \vdots\\ 0 & 0 & \cdots & \beta(\lambda) & \delta\\ 0 & 0 & \cdots & 0 & \beta(\lambda) \end{pmatrix} \quad \text{for some } \delta \neq 0. \tag{A.1.16}$$

Let

$$z = \sum_{j=1}^{m} \xi_j e_j \in E_1 \quad \text{with} \quad \xi_j = \sum_{r=0}^{m-j} \frac{\delta^r t^r x_{j+r}}{r!},$$

where $x = (x_1,\cdots x_m) \in \mathbb{R}^m$*,* $\delta \neq 0$ *is as in* J_λ*, and* $t \geq 0$*. Then the k-linear term* $G_k(z,\lambda)$ *can be expressed as*

$$G_k(z,\lambda) = F_1(x) + tF_2(x) + \cdots + t^{m-1}F_m(x),$$

and the center manifold function Φ *can be expressed as*

$$\Phi = \sum_{j=1}^{m} \Phi_j + o(k), \quad -\mathscr{L}_\lambda^j \Phi_j = (j-1)! P_2 F_j(x) \quad \forall 1 \leq j \leq m. \tag{A.1.17}$$

Remark A.1.2 The formula (A.1.14) for center manifold functions can be equivalently expressed as

$$-\mathscr{L}_\lambda \Phi(x,\lambda) = \sum_{1\leq j_1,\cdots,j_k\leq m} g_{j_1\cdots j_k} x_{j_1}\cdots x_{j_k} + o(k), \tag{A.1.18}$$

where $g_{j_1\cdots j_k} = P_2G_k(e_{j_1},\cdots,e_{j_k},\lambda)$. Formula (A.1.15) is equivalent to the form

$$\Phi = \Phi_1 + \Phi_2 + \Phi_3 + o(3), \tag{A.1.19}$$

where

$$\begin{aligned}
&-\mathscr{L}_\lambda\Phi_1 = x_1^2G_{11} + x_2^2G_{22} + x_1x_2(G_{12}+G_{21}),\\
&[(-\mathscr{L}_\lambda)^2 + 4\rho^2](-\mathscr{L}_\lambda)\Phi_2 = 2\rho^2[(x_1^2 - x_2^2)(G_{22}-G_{11}) - 2x_1x_2(G_{12}+G_{21})],\\
&[(-\mathscr{L}_\lambda)^2 + 4\rho^2]\Phi_3 = \rho(-\mathscr{L}_\lambda)[(x_2^2 - x_1^2)(G_{12}+G_{21}) + 2x_1x_2(G_{11}-G_{22})],
\end{aligned}$$

and $G_{ij} = P_2G_2(e_i,e_j,\lambda), 1\leq i,j\leq 2$.

Proof of Theorem A.1.1. **Step 1.** By (A.1.10) and (A.1.11), the center manifold function $y=\Phi(x,\lambda)$ of (A.1.8) and (A.1.9) satisfies

$$\Phi(x,\lambda) = \int_{-\infty}^{0} e^{-\tau\mathscr{L}_\lambda}\rho_\varepsilon P_2G(z(\tau,x)+\Phi,\lambda)d\tau, \tag{A.1.20}$$

where $z(t,x)$ is a solution of the equations

$$\frac{dz}{dt} = J_\lambda z + \rho_\varepsilon(z)P_1G(z+\Phi(z),\lambda), \quad z(0)=x. \tag{A.1.21}$$

By (A.1.4) and $\Phi(x,\lambda) = o(\|x\|)$, the solution $z(t,x)$ of (A.1.21) is given by

$$z = e^{tJ_\lambda}x + o(\|x\|).$$

Inserting $z(t,x)$ into (A.1.20), we get (A.1.13).

Since J_λ is diagonal, $z=\sum_{i=1}^{m} z_ie_i$ can be expressed as

$$z_i(t,x) = x_ie^{\beta_it} + o(\|x\|).$$

Then from (A.1.20), we obtain

$$\begin{aligned}
\Phi(x,\lambda) &= \int_{-\infty}^{0} e^{-\tau\mathscr{L}_\lambda}\rho_\varepsilon P_2G_k(xe^{\beta\tau},\lambda)d\tau + o(\|x\|^k)\\
&= (-\mathscr{L}_\lambda)^{-1}\int_{-\infty}^{0} e^{-\tau\mathscr{L}_\lambda}\rho_\varepsilon P_2G_kd(-\tau\mathscr{L}_\lambda) + o(\|x\|^k)\\
&= (-\mathscr{L}_\lambda)^{-1}P_2G_k(x,\lambda) + o(\|x\|^k) + O(|\beta|\|x\|^k).
\end{aligned}$$

Hence assertion (1) is proved.

Step 2. When the eigenvalues satisfying (A.1.5) are a pair of complex eigenvalues $\beta_1 = \overline{\beta}_2 = \alpha + i\rho$, and $G_k = G_2$ is bilinear, the solution $z(t,x)$ of (A.1.21) reads

$$\begin{aligned} z &= z_1 e_1 + z_2 e_2 + o(\|x\|), \\ z_1 &= e^{\alpha t}(x_1 \cos\rho t - x_2 \sin\rho t), \\ z_2 &= e^{\alpha t}(x_1 \sin\rho t + x_2 \cos\rho t), \quad (\alpha(\lambda_0) = 0), \end{aligned} \tag{A.1.22}$$

and $G_2(e^{\tau J_\lambda} x, \lambda)$ is given by

$$\begin{aligned} P_2 G_2(e^{\tau J_\lambda} x, \lambda) &= P_2 G_2(z_1 e_1 + z_2 e_2, z_1 e_1 + z_2 e_2, \lambda) \\ &= z_1^2 G_{11} + z_1 z_2 (G_{12} + G_{21}) + z_2^2 G_{22}, \end{aligned}$$

where $G_{ij} = P_2 G_2(e_i, e_j, \lambda)$. It follows from (A.1.22) that

$$\begin{aligned} P_2 G_2(e^{tJ_{\lambda_0}} x, \lambda_0) =& [\frac{1}{2}(x_1^2 + x_2^2) + \frac{1}{2}(x_1^2 - x_2^2)\cos 2\rho t - x_1 x_2 \sin 2\rho t] G_{11} \\ &+ [\frac{1}{2}(x_1^2 + x_2^2) - \frac{1}{2}(x_1^2 - x_2^2)\cos 2\rho t + x_1 x_2 \sin 2\rho t] G_{22} \\ &+ [\frac{1}{2}(x_1^2 - x_2^2)\sin 2\rho t + x_1 x_2 \cos 2\rho t][G_{12} + G_{21}]. \end{aligned}$$

By the formulas

$$\int_{-\infty}^{0} e^{-\tau \mathscr{L}_\lambda} \cos 2\rho\tau d\tau = (-\mathscr{L}_\lambda)((-\mathscr{L}_\lambda)^2 + 4\rho^2)^{-1},$$

$$\int_{-\infty}^{0} e^{-\tau \mathscr{L}_\lambda} \sin 2\rho\tau d\tau = -2\rho((-\mathscr{L}_\lambda)^2 + 4\rho^2)^{-1},$$

we deduce that

$$\begin{aligned} &\int_{-\infty}^{0} e^{-\tau \mathscr{L}_\lambda} \rho_\varepsilon P_2 G_2(e^{\tau J_\lambda} x, \lambda) d\tau = \int_{-\infty}^{0} e^{-\tau \mathscr{L}_\lambda} P_2 G_2(e^{\tau J_{\lambda_0}} x, \lambda_0) d\tau + O(|\alpha| \|x\|^3) \\ &= (-\mathscr{L}_\lambda)^{-1} P_2 G_2(x, \lambda) - 2\rho^2((-\mathscr{L}_\lambda)^2 + 4\rho^2)^{-1} P_2 G_2(x, \lambda) \\ &+ 2\rho^2((-\mathscr{L}_\lambda)^2 + 4\rho^2)^{-1}(x_1^2 G_{22} + x_2^2 G_{11} - x_1 x_2 (G_{12} + G_{21})) \\ &+ \rho(-\mathscr{L}_\lambda)((-\mathscr{L}_\lambda)^2 + 4\rho^2)^{-1}[(x_2^2 - x_1^2)(G_{12} + G_{21}) \\ &+ 2x_1 x_2 (G_{11} - G_{22})] + O(|\alpha| \|x\|^3). \end{aligned}$$

Thus, from (A.1.20) we derive (A.1.15), and assertion (2) is proved.

Step 3. When J_λ has the Jordan form given by (A.1.16), the solution z of (A.1.21) can be written as

$$z = \sum_{j=1}^{m} z_j e_j + o(\|x\|), \qquad z_j = \xi_j e^{\beta(\lambda)t}, \qquad \xi_j = \sum_{r=0}^{m-j} \frac{\delta^r t^r x_{j+r}}{r!},$$

where $\beta = \beta_j$. Thus, (A.1.20) reads

$$\begin{aligned}\Phi(x,\lambda) &= \int_{-\infty}^{0} e^{-\tau\mathscr{L}_\lambda} P_2 G(\sum_{j=1}^{m} \xi_j e^{\beta\tau} e_j, \lambda) d\tau + o(k) \\ &= \sum_{j=1}^{m} \left[\int_{-\infty}^{0} \tau^{j-1} e^{-\tau\mathscr{L}_\lambda} d\tau \right] P_2 F_j(x) + o(k).\end{aligned}$$

Direct computation shows that

$$\int_{-\infty}^{0} \tau^{j-1} e^{-\tau\mathscr{L}_\lambda} d\tau = -(j-1)!\mathscr{L}_\lambda^{-j},$$

and assertion (3) follows. The proof is complete. □

A.2 Dynamics of Gradient-Type Systems

Equilibrium phase transitions in statistical physics are governed by gradient-type flows, where the Gibbs free energies are the corresponding functionals. We study in this appendix dynamical systems generated by gradient-type operators.

Let H_1 and H be two Hilbert spaces, $L : H_1 \to H$ a sectorial operator, and $G : H_\theta \to H$ a C^r $(r \geq 1)$ mapping for some $0 < \theta < 1$. Let $Y \subset H_\alpha$ $(0 < \alpha < 1)$ be a Banach space. Throughout this section, we always use $\|\cdot\|$ and $(\cdot,\cdot)$ to denote the norm and inner product in H. Consider the following dynamical system:

$$\frac{du}{dt} = Lu + G(u), \quad u(0) = u_0. \tag{A.2.1}$$

Definition A.2.1 *A mapping $L+G : H_1 \to H$ is called a gradient-type operator if there exist a C^1 function $F : Y \to \mathbb{R}^1$ and a constant $C > 0$ such that $DF(u) \in H$ for every $u \in H_\beta$ for some $\beta \leq 1$, and*

$$-(DF(u), Lu + Gu) \geq C\|DF(u)\|^2 \qquad \forall u \in H_1 \tag{A.2.2}$$

$$DF(u_0) = 0 \qquad \Leftrightarrow \qquad Lu_0 + Gu_0 = 0. \tag{A.2.3}$$

In this case, F is called the energy functional of $L+G$.

Obviously, a gradient operator $L+G = -DF$ is of gradient-type structure. Conversely, if $L+G : H_1 \to H$ is a gradient-type operator, then in general, $L+G$ is of the form

$$L+G = ADF + B : H_1 \to H, \tag{A.2.4}$$

where $A : H_1 \to H$ is a negative definite linear operator, and

$$\begin{aligned}&(Au,u) \leq -C\|u\|^2 && \text{for some } C > 0, \\ &(Bu, DF(u)) \leq 0 && \forall u \in H_1, \\ &DF(u_0) = 0 \Rightarrow Bu_0 = 0.\end{aligned} \tag{A.2.5}$$

In this section, we always assume that $L+G: H_1 \to H$ is a gradient-type operator with an energy functional $F: Y \to \mathbb{R}^1$. Let $u(t,\varphi) \in H_1$ be a solution of (A.2.1). Then

$$\frac{d}{dt}F(u(t,\varphi)) \le -C\|DF(u(t,\varphi))\|^2 \qquad \forall t \ge 0, \tag{A.2.6}$$

which implies that the function $t \to F(u(t,\varphi))$ is strictly decreasing except at critical points of DF:

$$F(u(t,\varphi)) < F(u(s,\varphi)) \qquad \forall 0 \le s < t \qquad \text{if } DF(\varphi) \ne 0. \tag{A.2.7}$$

The following theorem shows that all solutions of gradient-type equations converge to steady-state solutions, and it is a direct consequence of (A.2.6) and (A.2.7).

Theorem A.2.2 *Assume that the energy functional $F: Y \to \mathbb{R}^1$ is weakly lower semicontinuous, i.e., for every $u_n \rightharpoonup v$ in Y,*

$$\liminf_{n\to\infty} F(u_n) \ge F(v),$$

and $F(u) \to +\infty \Leftrightarrow \|u\|_Y \to \infty$. Then for every $\varphi \in H$, there exists $u_0 \in H$ such that

$$\lim_{t\to\infty} u(t,\varphi) = u_0 \text{ in } H, \quad Lu_0 + G(u_0) = 0, \tag{A.2.8}$$

where $u(t,\varphi) \in L^\infty(0,\infty;Y)$ is the weak solution of (A.2.1).

Now we consider a gradient operator

$$L+G = -DF, \tag{A.2.9}$$

where the function F maps Y to $\mathbb{R}^1$, and $Y \subseteq H_{1/2}$ is a Banach space. We also assume that

$$F(u) \to +\infty \qquad \Leftrightarrow \qquad \|u\|_Y \to \infty, \tag{A.2.10}$$

$$(Lu+Gu,u) \le -C_1\|u\|_Y^2 + C_2, \tag{A.2.11}$$

where $C_1, C_2 > 0$ are constants.

The following theorem can be derived using the existence theorem of global attractors in Ma et al. (2002).

Theorem A.2.3 *Under the assumptions (A.2.9)–(A.2.11), (A.2.1) generates an operator semigroup $S(t): H \to H$, which has a global attractor $\mathscr{A}$ in H attracting every bounded set in H-norm.*

The dynamical systems associated with gradient-type equations have some nice properties in their global attractors and asymptotic behaviors. In particular, for the systems generated by the Morse-type operators, the attractors possess a simpler topological structure. A gradient-type operator $L+G: H_1 \to H$ is called a Morse field if all singular points of $L+G$ are regular, i.e., the derivative operators of $L+G$ at their singular points are invertible,

$$L+DG(u_0): H_1 \to H.$$

It is known that a Morse field has a finite number of singular points, and the unstable manifolds at the singular points are homeomorphic to a finite-dimensional open ball.

The following theorem can be derived directly from the classical invariant manifold theorems and Lemma A.2.7 below.

Theorem A.2.4 *Assume that (A.2.1) possesses a global attractor $\mathscr{A}$ in H. Then we have the following assertions:*

(1)The attractor $\mathscr{A}$ consists of all singular points of $L+G$ and their unstable sets, and

$$\dim \mathscr{A} = \text{maximal dimension of unstable sets}.$$

(2)$\mathscr{A}$ contains a set S of singular points such that for all $\varphi \in H$, the solutions $u(t,\varphi)$ of (A.2.1) converge to S,

$$\lim_{t\to\infty} dist\,(u(t,\varphi),S) = 0, \quad \forall \varphi \in H.$$

(3)If $L+G$ is a Morse field, then $\mathscr{A}$ consists of a finite number of finite-dimensional closed balls and singular points.

In order to understand the global dynamics of gradient flows, we need to study the local topological structure of singular points. In particular, we are interested in isolated singular points because they have nice properties, and they are generic.

Let G_C be a continuous group that acts on H:

$$g\circ u \in H, \quad (g\circ u = 0 \Leftrightarrow u = 0), \quad \forall g \in G_C, \quad u \in H.$$

We can see that if under the action of the group G_C, the operator $L+G$ is invariant, i.e.,

$$g\circ (Lu+Gu) = L(g\circ u)+G(g\circ u), \quad \forall g \in G_C, \quad u \in H,$$

then for every singular point u_0 of $L+G$, the set

$$S = \{g\circ u_0 |\ g \in G_C\}$$

consists of singular points of $L+G$. Obviously, in this case $L+G$ has no isolated singular points in the usual sense. However, we can define (generalized) isolated singular points as follows.

Definition A.2.5 *Let G_C be a continuous group, and let $L+G$ be invariant under the action of G_C. A singular point u_0 of $L+G$ is called isolated if there is a neighborhood U of $S = \{g\circ u_0 | g \in G_C\}$ such that there are no other singular points $\tilde{u} \notin S$ of $L+G$ in U.*

Hereinafter, isolated singular points are always understood in the sense of Definition A.2.5.

Definition A.2.6 *Let $u_0 \in H$ be a singular point of $L+G$. The sets W^s and $W^u \subset H$ are called the stable and unstable sets of u_0, respectively, if*

$$W^s = \{\varphi \in H|\ u(t,\varphi) \to u_0 \text{ in } H \text{ as } t \to \infty\},$$
$$W^u = \{\varphi \in H|\ u(t,\varphi) \to u_0 \text{ in } H \text{ as } t \to -\infty\},$$

where $u(t,\varphi)$ is the solution of (A.2.1).

For any two singular points u_1 and u_2 of (A.2.1), we have

$$W_1^u \bigcap W_2^s = \emptyset \quad \text{if } F(u_2) \geq F(u_1).$$

Namely, the unstable manifold of the singular point with lower energy does not intersect the stable manifold of the singular point with higher energy.

Lemma A.2.7 *Let $u_0 \in H$ be an isolated singular point of (A.2.1), and let $U \subset H$ be a neighborhood of u_0. Then*

$$\begin{aligned} &U = W_0^s + W_0^u + D, \quad D = \{\varphi \in U|\ \lim_{t\to\pm\infty} u(t,\varphi) \neq u_0\}, \\ &W_0^s = W^s \bigcap U, \qquad W_0^u = W^u \bigcap U. \end{aligned} \tag{A.2.12}$$

Moreover, D is an open set, called the hyperbolic set of u_0, and

$$\dim W^u < \infty, \quad \dim W^s = \infty, \quad W^u \bigcap W^s = \{u_0\}. \tag{A.2.13}$$

Proof. Since $G: H_\theta \to H$ $(0<\theta<1)$ is a $C^r (r \geq 1)$ mapping, $L+DG(u_0): H_1 \to H$ is a linear completely continuous field and sectorial operator. Hence $L+DG(u_0)$ has an eigenvalue sequence $\{\lambda_k\} \subset \mathbb{C}$ (counting multiplicities) such that

$$\begin{aligned} &\text{Re } \lambda_i \geq 0 && \text{if } 1 \leq i \leq m, \\ &\text{Re } \lambda_j < 0 && \text{if } j \geq m+1. \end{aligned}$$

The corresponding eigenspaces are

$$\begin{aligned} &E_1 = \{\sum_{i=1}^m x_i e_i|\ (L+DG(u_0)-\lambda_j)^{k_j} e_j = 0, \quad 1 \leq j \leq m, \quad k_j \in \mathbb{N}\}, \\ &E_2 = \{u \in H|\ (u, e_i^*) = 0, \quad \forall 1 \leq i \leq m\}, \end{aligned}$$

where the e_j^* are the conjugate eigenvectors of e_j $(1 \leq j \leq m)$.

Equation (A.2.1) has an invariant manifold M^I and a stable manifold M^s at u_0 such that M^I and M^s are tangent to E_1 and E_2 respectively:

$$T_{u_0} M^I = E_1, \quad T_{u_0} M^s = E_2,$$

$$\dim M^I = \dim E_1 = m, \quad \dim M^s = \dim E_2 = \infty.$$

It is clear that

$$W_0^u \subset M^I, \quad M^s \bigcap U \subset W_0^s.$$

Hence

$$\dim W_0^u < \infty, \quad \dim W_0^s = \infty.$$

By (A.2.7), we get

$$W_0^u \bigcap W_0^s = \{u_0\}.$$

Thus, we derive (A.2.13).

From (A.2.7) it follows that only three cases can occur:

1. $\lim_{t\to\infty} u(t,\varphi) = u_0, \quad \lim_{t\to-\infty} u(t,\varphi) \neq u_0;$
2. $\lim_{t\to-\infty} u(t,\varphi) = u_0, \quad \lim_{t\to\infty} u(t,\varphi) \neq u_0$; and
3. $\lim_{t\to\pm\infty} u(t,\varphi) \neq u_0.$

Hence, we have

$$U = W_0^u + W_0^s + D.$$

Since $\dim M^I < \infty$ and $W_0^u \subset M^I$, it is easy to see that W_0^u is relatively closed in U. We need to prove that W_0^s is also relatively closed in U.

By assumption, $G : H_\theta \to H$ is $C^r (r \geq 1)$ for some $0 < \theta < 1$. It is known that (A.2.1) generates a local operator semigroup $S(t) : U \to H$ for $0 \leq t < \infty$. Since we have $S(t)\varphi \to u_0, \forall \varphi \in W_0^s$, and $S(t)$ is continuous for $0 \leq t < \infty$, it follows that if $\varphi_n \in W_0^s$ and $\varphi_n \to \varphi_0$ in H, then $S(t)\varphi_0 \in U, \forall t \geq 0$. Thus by (A.2.7), it follows that

$$\lim_{t\to\infty} S(t)\varphi_0 = u_0,$$

i.e., $\varphi_0 \in W_0^s$. Hence W_0^s is a relatively closed set in U. Therefore, $D = U/(W_0^u + W_0^s)$ is an open set. The proof is complete. □

Let $u_0 \in H$ be an isolated singular point of (A.2.1). The eigenvalues of $L + DG(u_0)$ satisfy

$$\begin{aligned} &\text{Re } \lambda_i = 0 && \text{for } 1 \leq i \leq n, \\ &\text{Re } \lambda_j > 0 && \text{for } n+1 \leq j \leq n+m, \\ &\text{Re } \lambda_k < 0 && \text{for } k \geq n+m+1. \end{aligned} \tag{A.2.14}$$

Equation (A.2.1) has three invariant manifolds at u_0: the stable manifold M^s, the unstable manifold M^u, and the center manifold M^c such that

$$\dim M^c = n, \quad \dim M^u = m, \quad \dim M^s = \infty. \tag{A.2.15}$$

If $n = 0$ in (A.2.15), then the local topological structure of u_0 in its neighborhood U is simple, which is characterized by

$$W^u = M^u, \quad W^s = M^s, \quad D = U/(M^u + M^s).$$

Moreover, M^u is homeomorphic to an m-dimensional open ball, and

$$\text{ind}\,(-(L+G),u_0)=(-1)^m.$$

Especially if $n=m=0$, then u_0 is a local attractor.

when $n\neq 0$, the local topological structure of u_0 in the center manifold is complex. By the index reduction theorem (Nirenberg 2001; Chen 1981), we derive that

$$\text{ind}\,(-(L+G),u_0)=(-1)^m\text{ind}\,(g,u_0), \tag{A.2.16}$$

where $g=-(L+G)|_{M^c}$ is the restriction of $-(L+G)$ on M^c.

We remark that it is useful to study the locally topological structure of singular points in the center manifolds M^c for the bifurcation and transition theory developed in this book.

Let $B_r^c\subset M^c$ be the open ball with center u_0 and radius r, and

$$W_r^u=W^u\bigcap B_r^c,\quad W_r^s=W^s\bigcap B_r^c,\quad D_r=D\bigcap B_r^c.$$

Lemma A.2.8 *If $D_r=\emptyset$, then $B_r^c=W_r^u$ or $B_r^c=W_r^s$. If $D_r\neq\emptyset$, then*

$$D_r=\sum_{j=1}^k D_j \qquad (1\leq k<\infty), \tag{A.2.17}$$

where each D_j is a connected component, and $\partial D_j\setminus\partial B_r^c$ consists of only two connected components Γ_j^+ and Γ_j^-:

$$\partial D_j\setminus\partial B_r^c=\Gamma_j^+ +\Gamma_j^-,$$

and $\Gamma_j^+\subset W_r^u,\Gamma_j^-\subset W_r^s$. Moreover, Γ_j^+ (respectively Γ_j^-) is a deformation retract of $D_j+\Gamma_j^+$ (respectively $D_j+\Gamma_j^-$).

Proof. Obviously, if the hyperbolic set D_r is empty, then the ball B_r^c in the center manifold is either a stable manifold or an unstable manifold, i.e., either $B_r^c=W_r^s$ or $B_r^c=W_r^u$.

Let $D_r\neq\emptyset$. We consider the intersections of $\overline{W}_r^s,\overline{W}_r^u$ and $\overline{D}_r$ with the sphere ∂B_r^c. Set

$$I=\overline{W}_r^s\bigcap\partial B_r^c,\quad O=\overline{W}_r^u\bigcap\partial B_r^c,\quad U=\overline{D}_r\bigcap\partial B_r^c.$$

Let $S(t)$ be the operator semigroup generated by (A.2.1). It is clear that

$$\begin{aligned}
&F(\varphi)>F(\psi) && \forall\varphi\in I \text{ and } \psi\in O\\
&S(t)\varphi\in B_r^c && \forall t>0,\quad \text{as } \varphi\in I,\\
&S(-t)\varphi\in B_r^c,\quad S(t)\varphi\notin B_r^c && \forall t>0,\quad \text{as } \varphi\in O;
\end{aligned} \tag{A.2.18}$$

namely, the flows passing through the points $\varphi\in I$ are into B_r^c, and the flows passing through $\varphi\in O$ are out of B_r^c forever.

To complete this proof, it suffices to consider the case in which D_r is connected. For the other cases, the proof is the same.

Let D_r be connected. Then $U = \overline{D}_r \bigcap \partial B_r^c$ is also connected in ∂B_r^c. It is clear that $D_r \neq B_r^c$, i.e., $I \neq \emptyset, O \neq \emptyset$, for otherwise, from (A.2.7) one could deduce that there are no singular points in B_r^c. Set

$$\begin{aligned} U^- &= \{\varphi \in U |\, S(t)\varphi \in B_r^c,\ \forall 0 < t < t_\varphi \quad \text{and } S(t_\varphi)\varphi \in \partial B_r^c\}, \\ U^+ &= \{\varphi \in U |\, S(-t)\varphi \in B_r^c, \quad \forall 0 < t < t_\varphi, \quad \text{and } S(-t_\varphi)\varphi \in \partial B_r^c\}. \end{aligned}$$

Namely, the flow $S(t)U^-$ enters B_r^c from U^- and leaves B_r^c from U^+. It is clear that U^- is homeomorphic to U^+ by the mapping $h : U^- \to U^+$ defined by

$$h(\varphi) = \psi, \quad \varphi \in U^-, \quad \text{and } \psi = S(t_\varphi)\varphi \in U^+. \tag{A.2.19}$$

Let $I_0 \subset I$ be a connected component. We set

$$T_\rho(I_0) = \{\varphi \in U |\, \text{dist}\,(\varphi, I_0) < \rho\}.$$

Then $T_\rho(I_0) + I_0$ is a collar neighborhood of I_0 in $U \subset \partial B_r^c$. It is known that I_0 is a deformation retract of $T_\rho(I_0) + I_0$. Obviously, for $\rho > 0$ sufficiently small, we have

$$T_\rho(I_0) \subset U^-.$$

On the other hand, by (A.2.18), we have

$$\begin{aligned} &\lim_{t\to\infty} S(t)\varphi = u_0 \qquad \forall \varphi \in I, \\ &\lim_{t\to\infty} S(-t)\varphi = u_0 \quad \forall \varphi \in O. \end{aligned} \tag{A.2.20}$$

By the stability of extended orbits, Theorem 2.1.15 in Ma and Wang (2005d), it follows from (A.2.19) and (A.2.20) that there is a connected component O_0 of O such that

$$\begin{aligned} &h(T_\rho(I_0)) + O_0 \text{ is a collar neighborhood of } O_0, \text{ and} \\ &\text{dist}\,(h(T_\rho(I)), O_0) \to 0 \text{ as } \rho \to 0. \end{aligned} \tag{A.2.21}$$

Hence, O_0 is a deformation retract of $h(T_\rho(I^-)) + O_o$, and

$$h(T_\rho(I_0)) \subset U^+.$$

Let U_0^- and U_0^+ be the connected components of U^- and U^+ containing $T_\rho(I_0)$ and $h(T_\rho(I_0))$ respectively. Then by the stability of extended orbits, Theorem 2.1.15 in Ma and Wang (2005d), we can deduce that $h : U_0^- \to U_0^+$ is a homeomorphism, where h is defined by (A.2.19), which implies that

$$\partial U_0^- / I_0 = \partial U_0^+ / O_0 = \partial U_0^- \bigcap \partial U_0^+,$$

and I_0 and O_0 are the deformation retracts of $U_0^- + I_0$ and $U_0^+ + O_0$ respectively. Hence $\partial U_0^-/I_0$ and $\partial U_0^+/O_0$ are connected. Thus, we deduce that

$$U = U_0^- + U_0^+ + \partial U_0^- \bigcap \partial U_0^+,$$

which means that

$$\partial U = A^- + A^+, \quad A^- \subset I_0 = I, \quad A^+ \subset O_0 = 0. \tag{A.2.22}$$

We note that D_r is homeomorphic to $U \times (0,r)$. Then by (A.2.22), we have

$$\partial D_r = \Gamma^- + \Gamma^+, \qquad \Gamma^- \subset W_r^s = I \times (0,r), \qquad \Gamma^+ \subset W_r^u = O \times (0,r).$$

Moreover, $\Gamma^- = I \times (0,r)$ (respectively $\Gamma^+ = O \times (0,r)$) is a deformation retract of $D_r + \Gamma^-$ (respectively $D_r + \Gamma^+$).

By (A.2.20), we can deduce that the number of connected components of I and O must be finite, since otherwise, the singular point u_0 would not be isolated. Hence, (A.2.19) holds.

The lemma is proved. □

References

Aftalion, A. and Q. Du (2002). The bifurcation diagrams for the Ginzburg-Landau system of superconductivity. *Phys. D 163*(1-2), 94–105.

Agmon, S., A. Douglis, and L. Nirenberg (1959). Estimates near the boundary for solutions of elliptic partial differential equations satisfying general boundary conditions. I. *Comm. Pure Appl. Math. 12*, 623–727.

Agmon, S., A. Douglis, and L. Nirenberg (1964). Estimates near the boundary for solutions of elliptic partial differential equations satisfying general boundary conditions. II. *Comm. Pure Appl. Math. 17*, 35–92.

Albert, J. P., J. L. Bona, and D. B. Henry (1987). Sufficient conditions for stability of solitary-wave solutions of model equations for long waves. *Phys. D 24*(1-3), 343–366.

Alikakos, N. D. and G. Fusco (1993). The spectrum of the Cahn-Hilliard operator for generic interface in higher space dimensions. *Indiana Univ. Math. J. 42*(2), 637–674.

Andronov, A. A., E. A. Leontovich, I. I. Gordon, and A. G. Maĭer (1973). *Theory of bifurcations of dynamic systems on a plane*. Halsted Press [A division of John Wiley & Sons], New York-Toronto, Ont. Translated from the Russian.

Angulo, J., J. L. Bona, F. Linares, and M. Scialom (2002). Scaling, stability and singularities for nonlinear, dispersive wave equations: the critical case. *Nonlinearity 15*(3), 759–786.

Assad, S. M. and C. C. Lim (2006). Statistical equilibrium distributions of baroclinic vortices in a rotating two-layer model at low Froude numbers. *Geophys. Astrophys. Fluid Dyn. 100*(6), 503–524.

Babin, A. V. and M. I. Vishik (1983). Regular attractors of semigroups and evolution equations. *J. Math. Pures Appl. (9) 62*(4), 441–491 (1984).

Bao, W. (2002). The random projection method for a model problem of combustion with stiff chemical reactions. *Appl. Math. Comput. 130*(2-3), 561–571.

Bao, W. (2004). Ground states and dynamics of multicomponent Bose-Einstein condensates. *Multiscale Model. Simul. 2*(2), 210–236 (electronic).

Bao, W. (2007). Ground states and dynamics of rotating Bose-Einstein condensates. In *Transport phenomena and kinetic theory*, Model. Simul. Sci. Eng. Technol., pp. 215–255. Boston, MA: Birkhäuser Boston.

Bao, W., D. Jaksch, and P. A. Markowich (2003). Numerical solution of the Gross-Pitaevskii equation for Bose-Einstein condensation. *J. Comput. Phys. 187*(1), 318–342.

Bates, P. W. and P. C. Fife (1993). The dynamics of nucleation for the Cahn-Hilliard equation. *SIAM J. Appl. Math. 53*(4), 990–1008.

Battisti, D. S. and A. C. Hirst (1989). Interannual variability in a tropical atmosphere-ocean model. influence of the basic state, ocean geometry and nonlinearity. *J. Atmos. Sci. 46*, 1687–1712.

Belousov, B. P. (1959). An oscillating reaction and its mechanism. *Sborn. referat. radiat. med.*, 145.

T. Ma and S. Wang, *Phase Transition Dynamics*, DOI 10.1007/978-1-4614-8963-4,

Bennett, D. P., R. W. Brown, S. E. Stansfield, J. D. Stroughair, and J. L. Bona (1983). The stability of internal solitary waves. *Math. Proc. Cambridge Philos. Soc. 94*(2), 351–379.

Bethuel, F., H. Brezis, and F. Hélein (1994). *Ginzburg-Landau vortices*. Progress in Nonlinear Differential Equations and their Applications, 13. Boston, MA: Birkhäuser Boston Inc.

Bjerknes, V. (1904). Das problem von der wettervorhersage, betrachtet vom standpunkt der. mechanik un der physik. *Meteor. Z. 21*, 1–7.

Bona, J. L., C.-H. Hsia, T. Ma, and S. Wang (2011). Hopf bifurcation for two-dimensional doubly diffusive convection. *Appl. Anal. 90*(1), 5–30.

Bona, J. L. and R. L. Sachs (1988). Global existence of smooth solutions and stability of solitary waves for a generalized Boussinesq equation. *Comm. Math. Phys. 118*(1), 15–29.

Bona, J. L., P. E. Souganidis, and W. A. Strauss (1987). Stability and instability of solitary waves of Korteweg-de Vries type. *Proc. Roy. Soc. London Ser. A 411*(1841), 395–412.

Bona, J. L. and J. Wu (2002). The zero-viscosity limit of the 2D Navier-Stokes equations. *Stud. Appl. Math. 109*(4), 265–278.

Branstator, G. W. (1987). A striking example of the atmosphere's leading traveling pattern. *J. Atmos. Sci. 44*, 2310–2323.

Brenner, M. P., L. S. Levitov, and E. O. Budrene (1998). Physical mechanisms for chemotactic pattern formation by bacteria. *Biophysical Journal 74*, 1677–1693.

Brézis, H. and L. Nirenberg (1978). Characterizations of the ranges of some nonlinear operators and applications to boundary value problems. *Ann. Scuola Norm. Sup. Pisa Cl. Sci. (4) 5*(2), 225–326.

Budrene, E. O. and H. C. Berg (1991). Complex patterns formed by motile cells of Escherichia coli. *Nature 349*, 630–633.

Budrene, E. O. and H. C. Berg (1995). Dynamics of formation of symmetric patterns of chemotactic bacteria. *Nature 376*, 49–53.

Busse, F. (1978). Non-linear properties of thermal convection. *Reports on Progress in Physics 41*, 1929–1967.

Caffarelli, L., R. Kohn, and L. Nirenberg (1982a). On the regularity of the solutions of navier-stokes equations. *Comm. Pure Appl. Math. 35*, 771–831.

Caffarelli, L., R. Kohn, and L. Nirenberg (1982b). Partial regularity of suitable weak solutions of the Navier-Stokes equations. *Comm. Pure Appl. Math. 35*(6), 771–831.

Cahn, J. and J. E. Hillard (1957). Free energy of a nonuniform system i. interfacial energy. *J. Chemical Physics 28*, 258–267.

Calvez, V. and B. Perthame (2006). A Lyapunov function for a two-chemical species version of the chemotaxis model. *BIT 46*(suppl.), S85–S97.

Cessi, P. and G. R. Ierley (1995). Symmetry-breaking multiple equilibria in quasi-geostrophic, wind-driven flows. *J. Phys. Oceanogr. 25*, 1196–1202.

Chalkin, P. M. and T. C. Lubensky (2000). *Principles of Condensed Matter Physics*. Cambridge University Press; Reprint edition.

Chandrasekhar, S. (1981). *Hydrodynamic and Hydromagnetic Stability*. Dover Publications, Inc.

Charney, J. (1947). The dynamics of long waves in a baroclinic westerly current. *J. Meteorol. 4*, 135–163.

Charney, J. (1948). On the scale of atmospheric motion. *Geofys. Publ. 17(2)*, 1–17.

Charney, J. and J. DeVore (1979). Multiple flow equilibria in the atmosphere and blocking. *J. Atmos. Sci. 36*, 1205–1216.

Chen, W. (1981). *Nonlinear functional analysis*. P. R. China: Gansu Renmin Press.

Cherfils, L., S. Gatti, and A. Miranville (2008). Existence of global solutions to the Caginalp phase-field system with dynamic boundary conditions and singular potentials. *J. Math. Anal. Appl. 343*(1), 557–566.

Chow, S. N. and J. K. Hale (1982). *Methods of bifurcation theory*, Volume 251 of *Grundlehren der Mathematischen Wissenschaften [Fundamental Principles of Mathematical Science]*. New York: Springer-Verlag.

Crandall, M. G. and P. H. Rabinowitz (1977). The Hopf bifurcation theorem in infinite dimensions. *Arch. Rational Mech. Anal. 67*(1), 53–72.

Cross, M. and P. Hohenberg (1993). Pattern formation outside of equilibrium. *Reviews of Modern Physics 65*(3), 851–1112.

de Gennes, P. (1966). *Superconductivity of Metals and Alloys*. W. A. Benjamin.

del Pino, M., P. L. Felmer, and P. Sternberg (2000). Boundary concentration for eigenvalue problems related to the onset of superconductivity. *Comm. Math. Phys. 210*(2), 413–446.

Desai, R. C. and R. Kapral (2009). *Self-Organized and Self-Assembled Structures*. Cambridge University Press.

Dijkstra, H., T. Sengul, and S. Wang (2013). Dynamic transitions of surface tension driven convection. *Physica D. Nonlinear Phenomena*, 247, 7–17. http://dx.doi.org/10.1016/j.physd.2012.12.008. doi: 10.1016/j.physd.2012.12.008.

Dijkstra, H. A. (2000a). *Nonlinear Physical Oceanography: A Dynamical Systems Approach to the Large Scale Ocean Circulation and El Niño*. Dordrecht, the Netherlands: Kluwer Academic Publishers.

Dijkstra, H. A. (2000b). *Nonlinear Physical Oceanography: A Dynamical Systems Approach to the Large-Scale Ocean Circulation and El Niño*. Kluwer Acad. Publishers, Dordrecht/ Norwell, Mass.

Dijkstra, H. A. and M. Ghil (2005). Low-frequency variability of the large-scale ocean circulations: a dynamical systems approach. *Review of Geophysics 43*, 1–38.

Dijkstra, H. A. and M. J. Molemaker (1997). Symmetry breaking and overturning oscillations in thermohaline-driven flows. *J. Fluid Mech. 331*, 195–232.

Dijkstra, H. A. and M. J. Molemaker (1999). Imperfections of the north-atlantic wind-driven ocean circulation: Continental geometry and wind stress shape. *J. Mar. Res. 57*, 1–28.

Dijkstra, H. A. and J. D. Neelin (1999). Imperfections of the thermohaline circulation: Multiple equilibria and flux-correction. *J. Clim. 12*, 1382–1392.

Dijkstra, H. A. and J. D. Neelin (2000). Imperfections of the thermohaline circulation: Latitudinal asymmetry versus asymmetric freshwater flux. *J. Clim. 13*, 366–382.

Ding, X. and C. C. Lim (2007). Phase transitions of the energy-relative enstrophy theory for a coupled barotropic fluid-rotating sphere system. *Phys. A 374*(1), 152–164.

Drazin, P. and W. Reid (1981). *Hydrodynamic Stability*. Cambridge University Press.

Du, Q., M. D. Gunzburger, and J. S. Peterson (1992). Analysis and approximation of the Ginzburg-Landau model of superconductivity. *SIAM Rev. 34*(1), 54–81.

Eckhardt, B., T. M. Schneider, B. Hof, and J. Westerweel (2007). Turbulence transition in pipe flow. *Annu. Rev. Fluid Mech. 39*, 447–468.

Eckmann, J. P. and D. Ruelle (1985). Ergodic theory of chaos and strange attractors. *Rev. Mod. Phys.*, 617–656.

Elliott, C. M. and Z. Songmu (1986). On the Cahn-Hilliard equation. *Arch. Rational Mech. Anal. 96*(4), 339–357.

Field, M. (1996). *Lectures on bifurcations, dynamics and symmetry*, Volume 356 of *Pitman Research Notes in Mathematics Series*. Harlow: Longman.

Field, R. J., E. Körös, and R. M. Noyes (1972). Oscillations in chemical systems, Part 2. thorough analysis of temporal oscillations in the bromate-cerium-malonic acid system. *J. Am Chem. Soc. 94*, 8649–8664.

Field, R. J. and R. M. Noyes (1974). Oscillations in chemical systems, IV. limit cycle behavior in a model of a real chemical reaction. *J. Chem. Physics. 60*, 1877–1884.

Fisher, M. (1964). Specific heat of a gas near the critical point. *Physical Review 136:6A*, A1599–A1604.

Foias, C., O. Manley, R. Rosa, and R. Temam (2001). *Navier-Stokes equations and turbulence*, Volume 83 of *Encyclopedia of Mathematics and its Applications*. Cambridge: Cambridge University Press.

Foiaş, C. and R. Temam (1979). Some analytic and geometric properties of the solutions of the evolution Navier-Stokes equations. *J. Math. Pures Appl. (9) 58*(3), 339–368.

Friedlander, S., M. Vishik, and V. Yudovich (2000). Unstable eigenvalues associated with inviscid fluid flows. *J. Math. Fluid Mech. 2*(4), 365–380.

Friedlander, S. and M. M. Vishik (1992). Instability criteria for steady flows of a perfect fluid. *Chaos 2*(3), 455–460.

Friedlander, S. and M. M. Vishik (1995). On stability and instability criteria for magnetohydrodynamics. *Chaos 5*(2), 416–423.

Getling, A. V. (1997). *Rayleigh-Benard convection: Structures and dynamics*. Advanced Series in Nonlinear Dynamics. World Scientific.

Ghil, M. (1976). Climate stability for a sellers-type model. *J. Atmos. Sci. 33*, 3–20.

Ghil, M. (2000). Is our climate stable? bifurcations, transitions and oscillations in climate dynamics, in science for survival and sustainable development, v. i. keilis-borok and m. sánchez sorondo (eds.), pontifical academy of sciences, vatican city. pp. 163–184.

Ghil, M. and S. Childress (1987). *Topics in Geophysical Fluid Dynamics: Atmospheric Dynamics, Dynamo Theory, and Climate Dynamics*. Springer-Verlag, New York.

Ghil, M., T. Ma, and S. Wang (2001). Structural bifurcation of 2-D incompressible flows. *Indiana Univ. Math. J. 50*(Special Issue), 159–180. Dedicated to Professors Ciprian Foias and Roger Temam (Bloomington, IN, 2000).

Ghil, M., T. Ma, and S. Wang (2005). Structural bifurcation of 2-D nondivergent flows with Dirichlet boundary conditions: applications to boundary-layer separation. *SIAM J. Appl. Math. 65*(5), 1576–1596 (electronic).

Ghil, M. and K.-C. Mo (1991a). Intraseasonal oscillations in the global atmosphere. part i: Northern hemisphere and tropics. *J. Atmos. Sci. 48*, 752–779.

Ghil, M. and K.-C. Mo (1991b). Intraseasonal oscillations in the global atmosphere. part ii: Southern hemisphere. *J. Atmos. Sci. 48*, 780–790.

Ghil, M. and A. W. Robertson (2002). “waves” vs. “particles” in the atmosphere’s phase space: A pathway ro long-range forecasting? *Proc. Natl. Acad. Sci. 99*, 2493–2500.

Gibbon, J. D. and M. J. McGuinness (1980). A derivation of the Lorenz equations for some unstable dispersive physical systems. *Phys. Lett. A 77*(5), 295–299.

Ginzburg, V. L. (2004). On superconductivity and superfluidity (what i have and have not managed to do), as well as on the ’physical minimum’ at the beginning of the xxi century. *Phys.-Usp. 47*, 1155–1170.

Glansdorff, P. and I. Prigogine (1971). Structure, stability, and fluctuations. *Wiley-Interscience, New York.*

Goldstein, S. (1937). The stability of viscous fluid flow between rotating cylinders. *Proc. Camb. Phil. Soc. 33*, 41–61.

Golubitsky, M. and D. G. Schaeffer (1985). *Singularities and groups in bifurcation theory. Vol. I*, Volume 51 of *Applied Mathematical Sciences*. New York: Springer-Verlag.

Golubitsky, M., I. Stewart, and D. G. Schaeffer (1988). *Singularities and groups in bifurcation theory. Vol. II*, Volume 69 of *Applied Mathematical Sciences*. New York: Springer-Verlag.

Gor’kov, L. (1968). Generalization of the Ginzburg-Landau equations for non-stationary problems in the case of alloys with paramagnetic impurities. *Sov. Phys. JETP 27*, 328–334.

Guckenheimer, J. and P. Holmes (1990). *Nonlinear oscillations, dynamical systems, and bifurcations of vector fields*, Volume 42 of *Applied Mathematical Sciences*. New York: Springer-Verlag. Revised and corrected reprint of the 1983 original.

Guo, B. and Y. Han (2009). Attractor and spatial chaos for the Brusselator in $\mathbf{R}^N$. *Nonlinear Anal. 70*(11), 3917–3931.

Guo, Y. and H. J. Hwang (2010). Pattern formation (I): the Keller-Segel model. *J. Differential Equations 249*(7), 1519–1530.

Hale., J. (1988). Asymptotic behaviour of dissipative systems. *AMS Providence RI*.

Hastings, S. P. and J. D. Murray (1975). The existence of oscillatory solutions in the Field-Noyes model for the Belousov-Zhabotinskii reaction. *SIAM J. Appl. Math. 28*, 678–688.

Held, I. M. and M. J. Suarez (1974). Simple albedo feedback models of the ice caps. *Tellus 26*, 613–629.

Henry, D. (1981). *Geometric theory of semilinear parabolic equations*, Volume 840 of *Lecture Notes in Mathematics*. Berlin: Springer-Verlag.

Holton, J. R. (2004). *An Introduction to Dynamic Meteorology (4th Edition)*. Elsevier Academic Press.

Hopf, E. (1942). Abzweigung einer periodischen lbsung von einer stationaren lösung eines differentialsystems. *Ber. Math.-Phys. K. Sachs. Akad. Wiss. Leipzig 94*, 1–22.

Hopf, E. (1951). Über die Anfangswertaufgabe für die hydrodynamischen Grundgleichungen. *Math. Nachr. 4*, 213–231.

Hsia, C.-H., C.-S. Lin, T. Ma, and S. Wang (2011). Tropical atmospheric circulations with humidity effects.

Hsia, C.-H., T. Ma, and S. Wang (2007). Stratified rotating Boussinesq equations in geophysical fluid dynamics: dynamic bifurcation and periodic solutions. *J. Math. Phys. 48*(6), 065602, 20.

Hsia, C.-H., T. Ma, and S. Wang (2008a). Attractor bifurcation of three dimensional double-diffusive convection. *ZAA 27*, 233–252.

Hsia, C.-H., T. Ma, and S. Wang (2008b). Bifurcation and stability of two-dimensional double-diffusive convection. *Commun. Pure Appl. Anal. 7*(1), 23–48.

Hsia, C.-H., T. Ma, and S. Wang (2010). Rotating Boussinesq equations: dynamic stability and transitions. *Discrete Contin. Dyn. Syst. 28*(1), 99–130.

Ierley, G. and V. A. Sheremet (1995). Multiple solutions and advection-dominated flows in the wind-driven circulation. i: Slip. *J. Marine Res. 53*, 703–737.

Iooss, G. and M. Adelmeyer (1998). *Topics in bifurcation theory and applications* (Second ed.), Volume 3 of *Advanced Series in Nonlinear Dynamics*. River Edge, NJ: World Scientific Publishing Co. Inc.

Jin, F. F. (1996). Tropical ocean-atmosphere interaction, the pacific cold tongue, and the el niño southern oscillation. *Science 274*, 76–78.

Jin, F. F. and M. Ghil (1990). Intraseasonal oscillations in the extratropics: Hopf bifurcation and topographic instabilities. *J. Atmos. Sci. 47*, 3007–3022.

Jin, F. F., D. Neelin, and M. Ghil (1996). El niño/southern oscillation and the annual cycle: subharmonic frequency locking and aperiodicity. *Physica D 98*, 442–465.

K. I. Chang, M. Ghil, K. I. and C.-C. A. Lai (2001). Transition to aperiodic variability in a wind-driven double-gyre circulation model. *J. Phys. Oceanogr. 31*, 1260–1286.

Kalnay, E. (2003). *Atmospheric Modeling, Data Assimilation and Predictability*. Cambridge Univ. Press.

Kaper, H., S. Wang, and M. Yari (2009). Dynamic transitions of turing patterns. *Nonlinearity 22*, 601–626.

Kaper, H. G. and T. J. Kaper (2002). Asymptotic analysis of two reduction methods for systems of chemical reactions. *Phys. D 165*(1-2), 66–93.

Kapral, R. and K. Showalter (1995). *Chemical Waves and Patterns*. Kluwer, Dordrecht.

Kato, T. (1995). *Perturbation theory for linear operators*. Classics in Mathematics. Berlin: Springer-Verlag. Reprint of the 1980 edition.

Keller, E. F. and L. A. Segel (1970). Initiation of slime mold aggregation viewed as an instability. *J. Theor. Biol. 26*, 399–415.

Kirchgässner, K. (1975). Bifurcation in nonlinear hydrodynamic stability. *SIAM Rev. 17*(4), 652–683.

Kiselev, A. A. and O. A. Ladyženskaya (1957). On the existence and uniqueness of the solution of the nonstationary problem for a viscous, incompressible fluid. *Izv. Akad. Nauk SSSR. Ser. Mat. 21*, 655–680.

Kleman, M. and O. D. Lavrentovich (2003). *Soft Matter Physics, An Introduction*. Springer.

Koschmieder, E. L. (1993). *Bénard cells and Taylor vortices*. Cambridge Monographs on Mechanics. Cambridge University Press.

Krasnosel'skii, M. A. (1956). *Topologicheskie metody v teorii nelineinykh integralnykh uravnenii*. Gosudarstv. Izdat. Tehn.-Teor. Lit., Moscow.

Kukavica, I. and M. Ziane (2007). On the regularity of the primitive equations of the ocean. *Nonlinearity 20*(12), 2739–2753.

Kushnir, Y. (1987). Retrograding wintertime low-frequency disturbances over the north pacific ocean. *J. Atmos. Sci. 44*, 2727–2742.

Ladyženskaya, O. A. (1958). On the nonstationary Navier-Stokes equations. *Vestnik Leningrad. Univ. 13*(19), 9–18.

Ladyzhenskaya, O. A. (1982). The finite-dimensionality of bounded invariant sets for the Navier-Stokes system and other dissipative systems. *Zap. Nauchn. Sem. Leningrad. Otdel. Mat. Inst. Steklov. (LOMI) 115*, 137–155, 308. Boundary value problems of mathematical physics and related questions in the theory of functions, 14.

Langer, J. (1971). Theory of spinodal decomposition in allays. *Ann. of Physics 65*, 53–86.

Lapidus, I. and R. Schiller (1976). Model for the chemotactic response of a bacterial population. *Biophys J. 16(7)*, 779–789.

Lappa, M. (2009). *Thermal convection: Patterns, evolution and stability*. Wiley.

Legras, B. and M. Ghil (1985). Persistent anomalies, blocking and variations in atmospheric predictability. *J. Atmos. Sci. 42*, 433–471.

Leray, J. (1933). Étude de diverses équations intégrales non linéaires et de quelques problèmes que pose l'dydrodynamique. *J. Math. Pures Appl 12*, 1–82.

Li, J. and S. Wang (2008). Some mathematical and numerical issues in geophysical fluid dynamics and climate dynamics. *Communications in Computational Physics 3:4*, 759–793.

Lim, C. C. (2007). Extremal free energy in a simple mean field theory for a coupled barotropic fluid-rotating sphere system. *Discrete Contin. Dyn. Syst. 19*(2), 361–386.

Lin, F. H. (1996). Some dynamical properties of Ginzburg-Landau vortices. *Comm. Pure Appl. Math. 49*(4), 323–359.

Lions, J.-L., R. Temam, and S. Wang (1992a). New formulations of the primitive equations of atmosphere and applications. *Nonlinearity 5*(2), 237–288.

Lions, J.-L., R. Temam, and S. Wang (1992b). On the equations of the large-scale ocean. *Nonlinearity 5*(5), 1007–1053.

Lions, J.-L., R. Temam, and S. Wang (1993). Models for the coupled atmosphere and ocean. (CAO I,II). *Comput. Mech. Adv. 1*(1), 120.

Liu, C. and H. Liu (2008). Boundary conditions for the microscopic FENE models. *SIAM J. Appl. Math. 68*(5), 1304–1315.

Liu, C. and J. Shen (2001). On liquid crystal flows with free-slip boundary conditions. *Discrete Contin. Dynam. Systems 7*(2), 307–318.

Liu, H., T. Sengul, and S. Wang (2012). Dynamic transitions for quasilinear systems and cahn-hilliard equation with onsager mobility. *Journal of Mathematical Physics 53, 023518, doi: 10.1063/1.3687414*, 1–31.

Liu, H., T. Sengul, S. Wang, and P. Zhang (2013). Dynamic transitions and pattern formations for cahn-hilliard model with long-range repulsive interactions. *Comm. Math Sci., revised.*

Lorenz, E. N. (1963a). Deterministic nonperiodic flow. *J. Atmos. Sci. 20*, 130–141.

Lorenz, E. N. (1963b). The mechanics of vacillation. *J. Atmos. Sci. 20*, 448–464.

Lorenz, E. N. (1967). *The Nature and Theory of the General Circulation of the Atmosphere*. World Meteorological Organization, Geneva, Switzerland.

Lorenz, E. N. (1969). Atmospheric predictability as revealed by naturally occurring analogues. *J. Atmos. Sci. 26*, 636–646.

Lotka, A. J. (1925). *Elements of physical biology*. Baltimore: Williams & Wilkins Co.

M. Ghil, Y. F. and L. Sushama (2002). Baroclinic and barotropic aspects of the wind-driven ocean circulation. *Physica D 167*, 1–35.

Ma, Q., S. Wang, and C. Zhong (2002). Necessary and sufficient conditions for the existence of global attractors for semigroups and applications. *Indiana Univ. Math. J. 51*(6), 1541–1559.

Ma, T. and S. Wang (2000). Structural evolution of the Taylor vortices. *M2AN Math. Model. Numer. Anal. 34*(2), 419–437. Special issue for R. Temam's 60th birthday.

Ma, T. and S. Wang (2001). Structure of 2D incompressible flows with the Dirichlet boundary conditions. *Discrete Contin. Dyn. Syst. Ser. B 1*(1), 29–41.

Ma, T. and S. Wang (2004a). Boundary layer separation and structural bifurcation for 2-D incompressible fluid flows. *Discrete Contin. Dyn. Syst. 10*(1-2), 459–472.

Ma, T. and S. Wang (2004b). Dynamic bifurcation and stability in the Rayleigh-Bénard convection. *Commun. Math. Sci. 2*(2), 159–183.

Ma, T. and S. Wang (2004c). Interior structural bifurcation and separation of 2D incompressible flows. *J. Math. Phys. 45*(5), 1762–1776.

Ma, T. and S. Wang (2005a). Bifurcation and stability of superconductivity. *J. Math. Phys. 46*(9), 095112, 31.

Ma, T. and S. Wang (2005b). *Bifurcation theory and applications*, Volume 53 of *World Scientific Series on Nonlinear Science. Series A: Monographs and Treatises*. World Scientific Publishing Co. Pte. Ltd., Hackensack, NJ.

Ma, T. and S. Wang (2005c). Dynamic bifurcation of nonlinear evolution equations. *Chinese Ann. Math. Ser. B 26*(2), 185–206.

Ma, T. and S. Wang (2005d). *Geometric theory of incompressible flows with applications to fluid dynamics*, Volume 119 of *Mathematical Surveys and Monographs*. Providence, RI: American Mathematical Society.

Ma, T. and S. Wang (2006). Stability and bifurcation of the Taylor problem. *Arch. Ration. Mech. Anal. 181*(1), 149–176.

Ma, T. and S. Wang (2007a). Rayleigh-Bénard convection: dynamics and structure in the physical space. *Commun. Math. Sci. 5*(3), 553–574.

Ma, T. and S. Wang (2007b). *Stability and Bifurcation of Nonlinear Evolutions Equations*. Science Press, Beijing.

Ma, T. and S. Wang (2008a). Dynamic model and phase transitions for liquid helium. *Journal of Mathematical Physics 49:073304*, 1–18.

Ma, T. and S. Wang (2008b). Dynamic phase transition theory in PVT systems. *Indiana University Mathematics Journal 57:6*, 2861–2889.

Ma, T. and S. Wang (2008c). Dynamic transitions for ferromagnetism. *Journal of Mathematical Physics 49:053506*, 1–18.

Ma, T. and S. Wang (2008d). Exchange of stabilities and dynamic transitions. *Georgian Mathematics Journal 15:3*, 581–590.

Ma, T. and S. Wang (2008e). Superfluidity of helium-3. *Physica A: Statistical Mechanics and its Applications 387:24*, 6013–6031.

Ma, T. and S. Wang (2009a). Boundary-layer and interior separations in the Taylor-Couette-Poiseuille flow. *J. Math. Phys. 50*(3), 033101, 29.

Ma, T. and S. Wang (2009b). Cahn-hilliard equations and phase transition dynamics for binary systems. *Dist. Cont. Dyn. Systs., Ser. B 11:3*, 741–784.

Ma, T. and S. Wang (2009c). Phase separation of binary systems. *Physica A: Statistical Physics and its Applications 388:23*, 4811–4817.

Ma, T. and S. Wang (2009d). Phase transition and separation for mixture of liquid he-3 and he-4, in *lev davidovich landau and his impact on contemporary theoretical physics*, horizons in world physics, edited by a. sakaji and i. licata. *264*, 107–119.

Ma, T. and S. Wang (2010a). Dynamic transition and pattern formation in Taylor problem. *Chin. Ann. Math. Ser. B 31*(6), 953–974.

Ma, T. and S. Wang (2010b). Dynamic transition theory for thermohaline circulation. *Phys. D 239*(3-4), 167–189.

Ma, T. and S. Wang (2010c). Tropical atmospheric circulations: dynamic stability and transitions. *Discrete Contin. Dyn. Syst. 26*(4), 1399–1417.

Ma, T. and S. Wang (2011a). Dynamic transition and pattern formation for chemotactic systems.

Ma, T. and S. Wang (2011b). El Niño southern oscillation as sporadic oscillations between metastable states. *Advances in Atmospheric Sciences 28:3*, 612–622.

Ma, T. and S. Wang (2011c). Phase transitions for Belousov-Zhabotinsky reactions. *Math. Methods Appl. Sci. 34*(11), 1381–1397.

Ma, T. and S. Wang (2011d). Phase transitions for the Brusselator model. *J. Math. Phys. 52*(3), 033501, 23.

Ma, T. and S. Wang (2011e). Third-order gas-liquid phase transition and the nature of andrews critical point. *AIP Advances 1*, 042101.

Madden, R. A. and P. R. Julian (1971). Detection of a 40-50 day oscillation in the zonal wind in the tropical pacific. *J. Atmos. Sci. 28*, 702–708.

Madden, R. A. and P. R. Julian (1972). Description of global-scale circulation cells in the tropics with a 40-50 day period. *J. Atmos. Sci. 29*, 1109–1123.

Majda, A. J. and X. Wang (2006). *Non-linear dynamics and statistical theories for basic geophysical flows*. Cambridge: Cambridge University Press.

Malkus, W. V. R. and G. Veronis (1958). Finite amplitude cellular convection. *J. Fluid Mech. 4*, 225–260.

Marsden, J. E. and M. McCracken (1976). *The Hopf bifurcation and its applications*. New York: Springer-Verlag. With contributions by P. Chernoff, G. Childs, S. Chow, J. R. Dorroh, J. Guckenheimer, L. Howard, N. Kopell, O. Lanford, J. Mallet-Paret, G. Oster, O. Ruiz, S. Schecter, D. Schmidt and S. Smale, Applied Mathematical Sciences, Vol. 19.

Meacham, S. and P. Berloff (1997). Instability of a steady, barotropic, wind driven circulation. *J. Mar. Res. 55*, 885–913.

Meacham, S. and P. Berloff (1998). On the stability of the wind-driven circulation. *J. Mar. Res. 56*, 937–993.

Mezić, I. (2001). Break-up of invariant surfaces in action-angle-angle maps and flows. *Phys. D 154*(1-2), 51–67.

Miranville, A. and A. Piétrus (2006). A new formulation of the Cahn-Hilliard equation. *Nonlinear Anal. Real World Appl. 7*(2), 285–307.

Miranville, A. and A. Rougirel (2006). Local and asymptotic analysis of the flow generated by the Cahn-Hilliard-Gurtin equations. *Z. Angew. Math. Phys. 57*(2), 244–268.

Murray, J. (2002). *Mathematical Biology, II*. 3rd Ed. Springer-Verlag.

Nadin, G., B. Perthame, and L. Ryzhik (2008). Traveling waves for the Keller-Segel system with Fisher birth terms. *Interfaces Free Bound. 10*(4), 517–538.

Neelin, J. D. (1990a). A hybrid coupled general circulation model for el niño studies. *J. Atmos. Sci. 47*, 674–693.

Neelin, J. D. (1990b). The slow sea surface temperature mode and the fast-wave limit: Analytic theory for tropical interannual oscillations and experiments in a hybrid coupled model. *J. Atmos. Sci. 48*, 584–606.

Neelin, J. D., D. S. Battisti, A. C. Hirst, F.-F. Jin, Y. Wakata, T. Yamagata, and S. E. Zebiak (1998). Enso theory. *J. Geophys. Res. 103*, 14261–14290.

Newton, P. K. (2001). *The N-vortex problem*, Volume 145 of *Applied Mathematical Sciences*. New York: Springer-Verlag. Analytical techniques.

Nicolis, G. and I. Prigogine (1977). Self-organization in nonequilibrium systems. *Wiley-Interscience, New York*.

Nirenberg, L. (1975). Lectures on linear partial differential equations. *Uspehi Mat. Nauk 30*(4(184)), 147–204. Translated from the English by Ju. V. Egorov (Amer. Math. Soc., Providence, R.I., 1973).

Nirenberg, L. (1981). Variational and topological methods in nonlinear problems. *Bull. Amer. Math. Soc. (N.S.) 4*(3), 267–302.

Nirenberg, L. (2001). *Topics in nonlinear functional analysis*, Volume 6 of *Courant Lecture Notes in Mathematics*. New York: New York University Courant Institute of Mathematical Sciences. Chapter 6 by E. Zehnder, Notes by R. A. Artino, Revised reprint of the 1974 original.

Nishikawa, K. and T. Morita (1998). Fluid behavior at supercritical states studied by small-angle X-ray scattering. *Journal of Supercritical Fluid 13*, 143–148.

Novick-Cohen, A. and L. A. Segel (1984). Nonlinear aspects of the Cahn-Hilliard equation. *Phys. D 10*(3), 277–298.

Onuki, O. (2002). Phase transition dynamics. *Combridge Univ. Press.*.

Palm, E. (1975). Nonlinear thermal convection. *Annual Review of Fluid Mechanics 7*(1), 39–61.

Pazy, A. (1983). *Semigroups of linear operators and applications to partial differential equations*, Volume 44 of *Applied Mathematical Sciences*. New York: Springer-Verlag.

Pedlosky, J. (1981). Resonant topographic waves in barotropic and baroclinic flows. *J. Atmos. Sci. 38*, 2626–2641.

Pedlosky, J. (1987). *Geophysical Fluid Dynamics* (second ed.). New-York: Springer-Verlag.

Perko, L. (1991). *Differential equations and dynamical systems*, Volume 7 of *Texts in Applied Mathematics*. New York: Springer-Verlag.

Perthame, B. (2004). PDE models for chemotactic movements: parabolic, hyperbolic and kinetic. *Appl. Math. 49*(6), 539–564.

Perthame, B. (2005). The general relative entropy principle—applications in Perron-Frobenius and Floquet theories and a parabolic system for biomotors. *Rend. Accad. Naz. Sci. XL Mem. Mat. Appl. (5) 29*(1), 307–325.

Perthame, B. (2007). *Transport equations in biology*. Frontiers in Mathematics. Basel: Birkhäuser Verlag.

Perthame, B. and A.-L. Dalibard (2009). Existence of solutions of the hyperbolic Keller-Segel model. *Trans. Amer. Math. Soc. 361*(5), 2319–2335.

Perthame, B., C. Schmeiser, M. Tang, and N. Vauchelet (2011). Travelling plateaus for a hyperbolic Keller-Segel system with attraction and repulsion: existence and branching instabilities. *Nonlinearity 24*(4), 1253–1270.

Philander, S. G. and A. Fedorov (2003). Is el niño sporadic or cyclic? *Annu. Rev. Earth Planet. Sci. 31*, 579–594.

Phillips, N. A. (1956). The general circulation of the atmosphere: A numerical experiment. *Quart J Roy Meteorol Soc 82*, 123–164.

Pismen, L. M. (2006). *Patterns and Interfaces in Dissipative Dynamics*. Springer, Berlin.

Prandtl, L. (1904). In *Verhandlungen des dritten internationalen Mathematiker-Kongresses*. Heidelberg, Leipeizig, pp. 484-491.

Prigogine, I. and R. Lefever (1968). Symmetry breaking instabilities in dissipative systems. *II. J. Chem. Phys. 48*, 1695.

Quon, C. and M. Ghil (1992). Multiple equilibria in thermosolutal convection due to salt-flux boundary conditions. *J. Fluid Mech. 245*, 449–484.

Quon, C. and M. Ghil (1995). Multiple equilibria and stable oscillations in thermosolutal convection at small aspect ratio. *J. Fluid Mech. 291*, 33–56.

Rabinowitz, P. H. (1968). Existence and nonuniqueness of rectangular solutions of the Bénard problem. *Arch. Rational Mech. Anal. 29*, 32–57.

Rabinowitz, P. H. (1971). Some global results for nonlinear eigenvalue problems. *J. Functional Analysis 7*, 487–513.

Rabinowitz, P. H. (1977). A bifurcation theorem for potential operators. *J. Functional Analysis 25*(4), 412–424.

Raguin, L. G. and J. G. Georgiadis (2004). Kinematics of the stationary helical vortex mode in taylor-couette-poiseuille flow. *J. Fluid Mech. 516*, 125–154.

Rayleigh, L. (1916). On convection currents in a horizontal layer of fluid, when the higher temperature is on the under side. *Phil. Mag. 32*(6), 529–46.

Reichl, L. E. (1998). *A modern course in statistical physics* (Second ed.). A Wiley-Interscience Publication. New York: John Wiley & Sons Inc.

Richardson, L. F. (1922). *Weather Prediction by Numerical Process*. Cambridge University Press.

Rogerson, A. M., P. D. Miller, L. J. Pratt, and C. K. R. T. Jones (1999). Lagrangian motion and fluid exchange in a barotropic meandering jet. *J. Phys. Oceanogr. 29*(10), 2635–2655.

Rooth, C. (1982). Hydrology and ocean circulation. *Prog. Oceanogr. 11*, 131–149.

Rossby, C.-G. (1926). On the solution of problems of atmospheric motion by means of model experiment. *Mon. Wea. Rev. 54*, 237–240.

Rubinstein, J. (2004). Topological methods in superconductivity. *Not. S. Afr. Math. Soc. 35*(1), 47–62. Invited papers from the 46th Annual SAMS Congress.

S. Jiang, F. F. J. and M. Ghil (1995). Multiple equilibria, periodic, and aperiodic solutions in a wind-driven, double-gyre, shallow-water model. *J. Phys. Oceanogr. 25*, 764–786.

Salby, M. L. (1996). *Fundamentals of Atmospheric Physics*. Academic Press.

Samelson, R. M. (2008). Time-periodic flows in geophysical and classical fluid dynamics. in: Handbook of numerical analysis, special volume on computational methods for the ocean and the atmosphere. r. temam and j. tribbia, eds. elsevier, new york. to appear.

Samelson, R. M. and S. Wiggins (2006). *Lagrangian transport in geophysical jets and waves*, Volume 31 of *Interdisciplinary Applied Mathematics*. New York: Springer. The dynamical systems approach.

Sardeshmukh, P. D., G. P. Compo, and C. Penland (2000). Changes of probability associated with el niño. *Journal of Climate*, 4268–4286.

Sattinger, D. H. (1978). Group representation theory, bifurcation theory and pattern formation. *J. Funct. Anal. 28*(1), 58–101.

Sattinger, D. H. (1979). *Group-theoretic methods in bifurcation theory*, Volume 762 of *Lecture Notes in Mathematics*. Berlin: Springer. With an appendix entitled "How to find the symmetry group of a differential equation" by Peter Olver.

Sattinger, D. H. (1980). Bifurcation and symmetry breaking in applied mathematics. *Bull. Amer. Math. Soc. (N.S.) 3*(2), 779–819.

Sattinger, D. H. (1983). *Branching in the presence of symmetry*, Volume 40 of *CBMS-NSF Regional Conference Series in Applied Mathematics*. Philadelphia, PA: Society for Industrial and Applied Mathematics (SIAM).

Schmid, A. (1966). A time dependent ginzburg-landau equation and its application to the problem of resistivity in the mixed state. *Phys. Kondens. Mater. 5*, 302–317.

Schneider, E. K., B. P. Kirtman, D. G. DeWitt, A. Rosati, L. Ji, and J. J. Tribbia (2003). Retrospective enso forecasts: Sensitivity to atmospheric model and ocean resolution. *Monthly Weather Review 131:12*, 3038–3060.

Schopf, P. S. and M. J. Suarez (1987). Vacillations in a coupled ocean-atmosphere model. *J. Atmos. Sci. 45*, 549–566.

Sengul, T. and S. Wang (2011). Pattern formation and dynamic transition for magnetohydrodynamic convection.

Sengul, T. and S. Wang (2012). Pattern formation in rayleigh–bénard convection. *Communication of Mathematical Sciences*.

Serfaty, S. (2006). Vortices in the Ginzburg-Landau model of superconductivity. In *International Congress of Mathematicians. Vol. III*, pp. 267–290. Eur. Math. Soc., Zürich.

Serre, D. (1991). Transitions de phase et oscillations de grande amplitude. In *Qualitative aspects and applications of nonlinear evolution equations (Trieste, 1990)*, pp. 172–184. World Sci. Publ., River Edge, NJ.

Shen, J. and X. Yang (2010). Numerical approximations of Allen-Cahn and Cahn-Hilliard equations. *Discrete Contin. Dyn. Syst. 28*(4), 1669–1691.

Shi, J. (2009). Bifurcation in infinite dimensional spaces and applications in spatiotemporal biological and chemical models. *Front. Math. China 4*(3), 407–424.

Smoller, J. (1983). *Shock waves and reaction-diffusion equations*, Volume 258 of *Grundlehren der Mathematischen Wissenschaften [Fundamental Principles of Mathematical Science]*. New York: Springer-Verlag.

Speich, S., H. Dijkstra, and M. Ghil (1995). Successive bifurcations in a shallow-water model, applied to the wind-driven ocean circulation. *Nonlin. Proc. Geophys. 2*, 241–268.

Speich, S. and M. Ghil (1994). Interannual variability of the mid-latitude oceans: a new source of climate variability? *Sistema Terra 3(3)*, 33–35.

Stanley, H. E. (1971). *Introduction to Phase Transitions and Critical Phenomena*. Oxford University Press, New York and Oxford.

Stern, M. E. (1960). The "salt fountain" and thermohaline convection. *Tellus 12*, 172–175.

Stommel, H. (1961). Thermohaline convection with two stable regimes of flow. *Tellus 13*, 224–230.

Surana, A., O. Grunberg, and G. Haller (2006). Exact theory of three-dimensional flow separation. I. Steady separation. *J. Fluid Mech. 564*, 57–103.

Swinney, H. L. (1988). Instabilities and chaos in rotating fluids. In *Nonlinear evolution and chaotic phenomena (Noto, 1987)*, Volume 176 of *NATO Adv. Sci. Inst. Ser. B Phys.*, pp. 319–326. New York: Plenum.

Swinney, H. L., N. Kreisberg, W. D. McCormick, Z. Noszticzius, and G. Skinner (1990). Spatiotemporal patterns in reaction-diffusion systems. In *Chaos (Woods Hole, MA, 1989)*, pp. 197–204. New York: Amer. Inst. Phys.

Tang, Q. and S. Wang (1995). Time dependent Ginzburg-Landau equations of superconductivity. *Phys. D 88*(3-4), 139–166.

Taylor, G. I. (1923). Stability of a viscous liquid contained between two rotating cylinders. *Philos. Trans. Royl London Ser. A 223*, 289–243.

Temam, R. (1997). *Infinite-dimensional dynamical systems in mechanics and physics* (Second ed.), Volume 68 of *Applied Mathematical Sciences*. New York: Springer-Verlag.

Temam, R. (2001). *Navier-Stokes equations*. AMS Chelsea Publishing, Providence, RI. Theory and numerical analysis, Reprint of the 1984 edition.

Thual, O. and J. C. McWilliams (1992). The catastrophe structure of thermohaline convection in a two-dimensional fluid model and a comparison with low-order box models. *Geophys. Astrophys. Fluid Dyn. 64*, 67–95.

Tinkham, M. (1996). *Introduction to Superconductivity*. McGraw-Hill, Inc.

Tziperman, E. (1997). Inherently unstable climate behavior due to weak thermohaline ocean circulation. *Nature 386*, 592–595.

Tziperman, E., J. R. Toggweiler, Y. Feliks, and K. Bryan (1994). Instability of the thermohaline circulation with respect to mixed boundary conditions: Is it really a problem for realistic models? *J. Phys. Oceanogr. 24*, 217–232.

Tzou, J. C., B. J. Matkowsky, and V. A. Volpert (2009). Interaction of Turing and Hopf modes in the superdiffusive Brusselator model. *Appl. Math. Lett. 22*(9), 1432–1437.

Velte, W. (1966). Stabilität and verzweigung stationärer lösungen der davier-stokeschen gleichungen. *Arch. Rat. Mech. Anal. 22*, 1–14.

Veronis, G. (1963). An analysis of wind-driven ocean circulation with a limited fourier components. *J. Atmos. Sci. 20*, 577–593.

Veronis, G. (1965). On finite amplitude instability in the thermohaline convection. *J. Marine Res. 23*, 1–17.

Veronis, G. (1966). Wind-driven ocean circulation, part ii: Numerical solution of the nonlinear problem. *Deep-Sea Res. 13*, 31–55.

Villani, C. (2008). Entropy production and convergence to equilibrium. In *Entropy methods for the Boltzmann equation*, Volume 1916 of *Lecture Notes in Math.*, pp. 1–70. Berlin: Springer.

Vishik, M. I. (1992). *Asymptotic behaviour of solutions of evolutionary equations*. Lezioni Lincee. [Lincei Lectures]. Cambridge: Cambridge University Press.

Volterra, V. (1926). Variazioni e fluttuazioni del numero d'individui in specie animali conviventi. *Mem. R. Accad. Naz. dei Lincei. Ser. VI 2*, 31–113.

von Neumann, J. (1960). Some remarks on the problem of forecasting climatic fluctuations. In R. L. Pfeffer (Ed.), *Dynamics of climate*, pp. 9–12. Pergamon Press.

Wiggins, S. (1990). *Introduction to applied nonlinear dynamical systems and chaos*, Volume 2 of *Texts in Applied Mathematics*. New York: Springer-Verlag.

Wolansky, G. (1988). Existence, uniqueness, and stability of stationary barotropic flow with forcing and dissipation. *Comm. Pure Appl. Math. 41*, 19–46.

Wolansky, G. (1989). The barotropic vorticity equation under forcing and dissipation: bifurcations of nonsymmetric responses and multiplicity of solutions. *SIAM J. Appl. Math. 49*, 1585–1607.

Yudovich, V. I. (1966). Secondary flows and fluid instability between rotating cylinders. *Appl. Math. Mech. 30*, 822–833.

Yudovich, V. I. (1967a). Free convection and bifurcation. *J. Appl. Math. Mech. 31*, 103–114.

Yudovich, V. I. (1967b). Stability of convection flows. *J. Appl. Math. Mech. 31*, 272–281.

Zebiak, S. E. and M. A. Cane (1987). A model el niño—southern oscillation. *Mon. Wea. Rev. 115*, 2262–2278.

Zhabotinski, A. (1964). Periodic process of the oxidation of malonic acid in solution (study of the kinetics of belousov's reaction). *Biofizika 9*, 306–311.

Index

T. Ma and S. Wang, *Phase Transition Dynamics*, DOI 10.1007/978-1-4614-8963-4,

FSC
www.fsc.org
MIX
Papier | Fördert gute Waldnutzung
FSC® C083411

Kolibri 360 GmbH
Ferdinand-Jühlke-Straße 7
99095 Erfurt, Deutschland
produktsicherheit@kolibri360.de